Water Science & Technology

Wetland Systems for Water Pollution Control 2000

Wetland Systems for Water Pollution Control 2000

Selected Proceedings of the 7th International Conference on Wetland Systems for Water Pollution Control 2000, held in Lake Buena Vista, Florida, USA, 11–16 November 2000

Issue Editors: KR Reddy* and RH Kadlec*

*Wetland Biogeochemistry Laboratory, Soil and Water Science Department, Institute of Food and Agricultural Sciences, University of Florida, Gainesville, FL 32611–0510, USA

**Wetland Management Services, Chelsea, MI 48118, USA

International Programme Committee
K. Ramesh Reddy (*Chair*) USA
R. H. Kadlec (*Co-chair*) USA
J. Bavor Australia
S. K. Billore India
Peter Breen Australia
Hans Brix Denmark
Paul Cooper UK
Michal Green Israel
Raimund Haberl Austria
D. A. Hammer USA
Alain Lienard France
Trond Maehlum Norway
R. Perfler Austria
Eneas Salati Filho Brazil
Brian Shutes UK
Karen Sundblad-Tonderski Sweden
Chris Tanner New Zealand
Jan Vymazal Czech Republic

Steering Committee
K. Ramesh Reddy (*Chair*) University of Florida
R. H. Kadlec (*Co-chair*) Wetland Management Services
Robert Bastian U.S. Environmental Protection Agency
Jim Bays CH2M Hill
Ronnie Best U.S. Geological Survey, Biological Resources Division
Tom Ferraro Ecology and Environment, Inc.
Gary Goforth South Florida Water Management District
Patrick Hunt U.S. Department of Agriculture
Robert L. Knight Environmental Consultant
Eric Livingston Florida Department of Environmental Protection
Ed Lowe St. Johns River Water Management District

Organised by
IWA Specialist Group on Use of Macrophytes in Water Pollution Control
University of Florida Institute of Food and Agricultural Sciences

Sponsored by
University of Florida Institute of Food and Agricultural Sciences
International Water Association
Florida Center for Environmental Studies
Florida Department of Environmental Protection
South Florida Water Management District
St Johns River Water Management District
US Department of Agriculture
US Environmental Protection Agency
US Geological Survey – Biological Resources Division
CH2M Hill
Ecology & Environment Inc.

British Library Cataloguing in Publication Data
A CIP catalogue record for this book is available from the British Library
ISBN 1 84339 407 3

Contents

Keynote Papers

1   **Fundamental processes within natural and constructed wetland ecosystems: short-term versus long-term objectives**   R.G. Wetzel

9   **Plants as ecosystem engineers in subsurface-flow treatment wetlands**
    C.C. Tanner

19 **State of the art for animal wastewater treatment in constructed wetlands**
    P.G. Hunt and M.E. Poach,

27 **Surface flow (SF) treatment wetlands as a habitat for wildlife and humans**
    R.L. Knight, R.A. Clarke, Jr. and R.K. Bastian

Phosphorus removal and transformations

39 **The use of macrophyte-based systems for phosphorus removal: an overview of 25 years of research and operational results in Florida**   T.A. DeBusk, F.E. Dierberg and K.R. Reddy

47 **Media selection for sustainable phosphorus removal in subsurface flow constructed wetlands**   H. Brix, C.A. Arias and M. del Bubba

55 **Phosphorus removal from trout farm effluents by constructed wetlands**
    Y. Comeau, J. Brisson, J.-P. Réville, C. Forget and A. Drizo

61 **The Impact of biomass harvesting on phosphorus uptake by wetland plants**
    S.-Y. Kim and P.M. Geary

69 **Phosphorus retention capacity of iron-ore and blast furnace slag in subsurface flow constructed wetlands**   B. Grüneberg and J. Kern

77 **The removal of nutrients from plant nursery irrigation runoff in subsurface horizontal-flow wetlands**   T.R. Headley, D.O. Huett and L. Davison

85 **Removing filterable reactive phosphorus from highly coloured stormwater using constructed wetlands**   M.A. Lund, P.S. Lavery and R.F. Froend

93 **Environmental impacts to the Everglades ecosystem: a historical perspective and restoration strategies**   M.J. Chimney and G. Goforth

101 **Phosphorus removal from Everglades agricultural area runoff by submerged aquatic vegetation/limerock treatment technology: an overview of research**   B. Gu, T.A. Debusk, F.E. Dierberg, M.J. Chimney, K.C. Pietro and T. Aziz

109 **Evaluation of phosphorus retention in a South Florida treatment wetland**
    M.K. Nungesser and M.J. Chimney

117 **The Everglades Nutrient Removal Project Test Cells: STA optimization – status of the research at the north site**   J. Majer Newman and T. Lynch

123 **Progress in the research and demonstration of Everglades periphyton-based stormwater treatment areas**   J.S. Bays, R.L. Knight, L. Wenkert, R. Clarke and S. Gong

131 **Performance of a recirculating wetland filter designed to remove particulate phosphorus for restoration of Lake Apopka (Florida, USA)**   M.F. Coveney, E.F. Lowe and L.E. Battoe

Nitrogen removal and transformations

137 **Nitrogen removal in a combined system: vertical vegetated bed over horizontal flow sand bed**   S. Kantawanichkul, P. Neamkam and R.B.E. Shutes

143 **Nitrogen and phosphorus budget in rewetted fens**   A. Lenz and U. Wild

149 **Nutrient removal in subsurface flow constructed wetlands for application in sensitive regions** H. Rustige and Chr. Platzer

157 **Distribution of ammonium-N in the water-soil interface of a surface-flow constructed wetland for swine wastewater treatment** A.A. Szögi and P.G. Hunt

163 **Denitrification in free water surface wetlands receiving carbon supplements** P.S. Burgoon

171 **Removal of nutrients from combined sewer overflows and lake water in a vertical-flow constructed wetland system** L. Gervin and H. Brix

Pathogen removal

177 **Microbial indicator removal in on-site constructed wetlands for wastewater treatment in the southeastern US** E.C. Barrett, M.D. Sobsey, C H. House and K. D. White

183 **Removal of bacterial indicators and pathogens from dairy wastewater by a multi-component treatment system** M.M. Karpiscak, L.R. Sanchez, R.J. Freitas and C.P. Gerba

191 **Protozoan predation as a mechanism for the removal of *Cryptosporidium* oocysts from wastewaters in constructed wetlands** R. Stott, E. May, E. Matsushita and A. Warren

199 **Distribution and retention of faecal coliforms in the Nakivubo Wetland in Kampala, Uganda** F. Kansiime and J.J.A. van Bruggen

207 **Bacterial dynamics in the sub-surface constructed wetland** J. Vymazal, J. Balcarov and H. Doušová

211 **Biota participating in wastewater treatment in a horizontal flow constructed wetland** J. Vymazal, V. Sládeček and J. Stach

215 **Removal of *Salmonella* and microbial indicators in constructed wetlands treating swine wastewater** V.R. Hill and M. D. Sobsey

223 **Occurrence and die-off of indicator organisms in the sediment in two constructed wetlands** T.A. Stenström and A. Carlander

Optimization and modeling

231 **Performance modeling of subsurface-flow constructed wetland systems** M.F. Dahab, R.Y. Surampalli and W. Liu

237 **Modelling nitrogen transformations in surface flow wastewater treatment wetlands in Sweden** S. Kallner and H.B. Wittgren

245 **The influence of water table fluctuations on nutrient dynamics in the rhizosphere of Common Reed (*Phragmites australis*)** O. Urbanc-Berčič and A. Gaberščik

251 **Thermal environments of subsurface treatment wetlands** R.H. Kadlec

259 **Cold climate wetlands: design and performance** S. Wallace, G. Parkin and C. Cross

267 **Behavior of organic carbon during subsurface wetland treatment in the Sonoran Desert** D.M. Quanrud, M.M. Karpiscak, K.E. Lansey and R.G. Arnold

273 **Development of a conceptual model for vertical flow wetland metabolism** E. Giraldo and E. Zárate

281 **Accumulation of organic matter fractions in a gravel-bed constructed wetland** L. Nguyen

Ecological considerations

289 **Determining ecologically acceptable nutrient loads to natural wetlands for water quality improvement** L.W. Keenan and E.F. Lowe

295 **Surmounting the engineering challenges of Everglades restoration** G.F. Goforth

303 **Changes in plant biomass and nutrient removal over 3 years in a constructed wetland in Cairns, Australia** M. Greenway and A. Woolley

311 **Performance of two macrophyte species in experimental wetlands receiving variable loads of anaerobically treated municipal wastewater** D.M.L. da Motta Marques, G.R. Leite and S.G.T. Giovannini

317 **Ecological characteristics of a natural wetland receiving secondary effluent** J.R. Martin, R.A. Clarke, Jr. and R.L. Knight

325 **Protection of surface water against contamination by wetland systems in Poland** H. Obarska-Pempkowiak, T. Ozimek and W. Chmiel

331 **Natural wastewater treatment in Hungary** A. Szabó, A. Osztoics and F. Szilágyi

339 **Constructed wetlands for wastewater treatment in central Italy** G. Conte, N. Martinuzzi, L. Giovannelli, B. Pucci and F. Masi

Subsurface flow wetland systems

345 **Residential subsurface flow treatment wetlands in northern Minnesota** R. Axler, J. Henneck and B. McCarthy

353 **On-site domestic wastewater treatment by reed bed in the moist subtropics** L. Davison, T. Headley and M. Edmonds

361 **Removal of hydrogen sulphide BOD from brackish water using vertical flow wetlands in a Caribbean environment** E. Giraldo and E. Zárate

369 **Constructed wetlands for wastewater treatment in the Czech Republic** J. Vymazal

375 **Subsurface-flow constructed wetlands treatment in the plains: five years of experience** M.F. Dahab and R.Y. Surampalli

381 **Application of constructed wetlands for wastewater treatment in Nepal** R.R. Shrestha, R. Haberl, J. Laber, R. Manandhar and J. Mader

387 **Experimental results on constructed wetland pilot system** J.M. González, G. Ansola and E. Luis

393 **Reed beds: constructed wetlands for municipal wastewater treatment plant sludge dewatering** J.S. Begg, R.L. Lavigne and P.L.M. Veneman

399 **Reciprocating constructed wetlands for treating industrial, municipal and agricultural wastewater** L. Behrends, L. Houke, E. Bailey, P. Jansen and D. Brown

407 **Zero-discharge of nutrients and water in a willow dominated constructed wetland** P. Gregersen and H. Brix

Surface flow wetland systems

413 **Long-term performance summary for the Boot Wetland Treatment System** J.R. Martin, C.H. Keller, R.A. Clarke, Jr. and R.L. Knight

421 **Diel changes in iron concentrations in surface-flow constructed wetlands** R.R. Goulet and F.R. Pick

427 **Wastewater treatment by algal turf scrubbing** R. Craggs

435 **Rerating capacity of a constructed wetland treatment system** J.A. Jackson and M. Sees

Industrial wastewaters

441 **Treatment of a molasses based distillery effluent in a constructed wetland in central India** S.K. Billore, N. Singh, H.K. Ram, J.K. Sharma, V.P. Singh, R.M. Nelson and P. Dass

449 **Use of constructed wetlands for acid mine drainage abatement and stream restoration** F.J. Brenner

455 **Nutrient and heavy metal uptake and storage in constructed wetland systems in Arizona** M.M. Karpiscak, L.R. Whiteaker, J.F. Artiola and K.E. Foster

463 **Retention of selected heavy metals: Cd, Cu, Pb in a hybrid wetland system** H. Obarska-Pempkowiak

469 **The integration of constructed wetlands into a treatment system for airport runoff**
D.M. Revitt, P. Worrall and D. Brewer

477 **Investigation of copper adsorption to peat using the simple metal sorption model**
S. Dierks

485 **Design and performance of experimental constructed wetlands treating coke plant effluents**   N. Jardinier, G. Blake, A. Mauchamp and G. Merlin

493 **The effect of refinery effluent on the aquatic macrophytes *Scirpus californicus*, *Typha subulata* and *Zizaniopsis bonariensis*** A.R. Campagna and D. da Motta Marques

499 **Treatment of laboratory wastewater in a tropical constructed wetlands comparing surface and subsurface flow**   A.A. Meutia

507 **Heavy metal contents and mobility of artificially inundated grassland along River Weser, Germany**   C. Erber and P. Felix-Henningsen

515 **Tolerance toward explosives, and explosives removal from groundwater in treatment wetland mesocosms**   E.P.H. Best, J.L. Miller and S.L. Larson

Agricultural wastewaters

523 **A constructed surface flow wetland for treating agricultural waste waters**   M. Borin, G. Bonaiti, G. Santamaria and L. Giardini

531 **Treatment of stormwater runoff from row crop farming in Ruskin, Florida**
B.T. Rushton and B.M. Bahk

539 **Evaluation of atrazine removal processes in a wetland**   C.M. Kao, J.Y. Wang and M.J. Wu

545 **Treatment of swine wastewater in marsh-pond-marsh constructed wetlands**
G.B. Reddy, P.G. Hunt, R. Phillips, K. Stone and A. Grubbs

551 **Reed bed dewatering of agricultural sludges and slurries**   J.K. Edwards, K.R. Gray, D.J. Cooper, A.J. Biddlestone and N. Willoughby

559 **Using wetlands for water quality improvement in agricultural watersheds; the importance of a watershed scale approach**   W.G. Crumpton

Stormwater

565 **Stormwater treatment: do constructed wetlands yield improved pollutant management performance over a detention pond system?**   H.J. Bavor, C.M. Davies and K. Sakadevan

571 **An experimental constructed wetland system for the treatment of highway runoff in the UK**   R.B E Shutes, D.M Revitt, L.N.L Scholes, M. Forshaw and B. Winter

579 **Performance of constructed wetland system for public water supply**   J.M. Elias, E. Salati Filho and E. Salati

585 **Application of a constructed wetland for non-point source pollution control**
C.M. Kao, J.Y. Wang, H.Y. Lee and C.K. Wen

591 **Buffer zones promoting oligotrophication in golf course runoffs: fiddler crabs as estuarine health indicators**   R.Y. George, G. Bodnar, S.L. Gerlach and R.M. Nelson

599 **River water quality improvement by natural and constructed wetland systems in the tropical semi-arid region of Northeastern Brazil**   B.S.O. de Ceballos, H. Oliveira, C.M.B.S. Meira, A. Konig, A.O. Guimarães and J.T. de Souza

607 **Metals in combined conventional and vegetated road runoff control systems**
H. Pontier, J.B. Williams and E. May

615 **Nitrogen and phosphorus variation in shallow groundwater and assimilation in plants in complex riparian buffer zones**   V. Kuusemets, Ü. Mander, K. Lõhmus and M. Ivask

Fundamental processes within natural and constructed wetland ecosystems: short-term versus long-term objectives

R.G. Wetzel*

* Department of Environmental Sciences and Engineering, The University of North Carolina, Chapel Hill, North Carolina 27599-7431, USA

Abstract Use of wetland ecosystems for water pollution control consists essentially of sustained induced disturbances as pollutants are loaded to complex biological communities. Objectives are to maximize pollutant loading, incorporation, and retention while maintaining highest levels of community metabolism and minimal alteration of community structure. Several basic processes are emphasized: (a) macrophyte productivity in relation to shoot:root ratios, and nutrient availability; (b) macrophyte life history strategies, succession, and biodiversity under constant pollutant stress; (c) importance of standing dead and particulate detritus; (d) functions and controlling mechanisms of heterotrophic and autotrophic periphyton in pollutant retention and recycling; (e) coupling of microbial metabolism to macrophyte retention of pollutants; (f) gaseous losses to the atmosphere; (g) losses of dissolved organic matter and its utilization; and (h) water losses by evapotranspiration and effects on wetland efficacy. Short-term wetland removal efficiencies are confounded by massive variations in retentive capacities diurnally, seasonally, and spatially, in exceeding physiological tolerance levels, and in species succession. Problems of channelization, altered microhydrology, and assimilation/retention are major in natural and non-engineered ecosystems. Wetlands are highly ephemeral and variable in their capabilities for sequestering and retention of nutrients and other pollutants.

Keywords Detritus; macrophyte productivity; microbial metabolism; nutrient recycling; organic matter; periphyton; wetlands

Human expansions and exploitation of resources have led to progressive removal of wetlands, largely for agricultural use of the land. As our understanding of the metabolically active components of land-water interface regions increases, wetlands in particular have been exploited to reduce loadings of sediment, nutrients, and pollutants as water moves to downgradient recipient lakes, rivers, and ground waters. It is apparent now that natural wetland ecosystems often do not function efficiently for purposes of permanent storage or controlled downgradient discharge of nutrients or pollutants. Moreover, various conservation efforts to maximize biodiversity have further inhibited use of natural wetlands for many applied purposes. As a result of these and many other factors, a rapid development of constructed wetlands has occurred in an attempt to simulate and hopefully enhance optimal properties of natural wetlands in performing the desired functions (e.g., Kadlec and Knight, 1996; Hammer, 1997).

A time for maturation – application of scientific understanding to management

Natural wetlands have often been assumed to have a specific capacity to absorb and retain particulate matter, nutrients, or other pollutants. Little recognition is given to either the highly dynamic growth or the metabolic relationships within these ecosystems on many different spatial and time scales. As will be noted briefly here, plant and microbial metabolism involved in nutrient/pollutant uptake, sequestering, and retention is highly dynamic on daily, seasonal, and long-term annual scales. Additionally, poor recognition is given to the effects of the loadings upon the metabolic and growth properties of the organisms within these ecosystems. Growth on both short-term and long-term bases is modified significantly along the concentration gradients from input sources. Furthermore, these loadings usually have marked effects upon natural or accelerated succession of the ecosystem components and, as a consequence, alter retentive capacities.

In particular, a wetland is commonly impacted with altered loadings, or plant communities of an artificially constructed wetland, after a brief initial colonization, are loaded with pollutants. The wetlands are expected to function in some retentive capacity in perpetuity with little or no management. Initial retentive capacities may be high for certain pollutants in the short-term, but they change markedly over the long-term. The changing efficacy of wetlands can be largely ignored, with little or no maintenance, with the objectives for inexpensive, "quick fix" solutions to disposal problems. The resulting wetlands often do not work efficiently or at all, and excessive loading can destroy both natural and constructed wetlands. Short-term wetland removal efficiencies are confounded by large variations in retentive capacities diurnally, seasonally, and spatially, by high loading that exceeds physiological tolerance levels of key microbial and plant species, and by succession of species within the communities. The metabolism of the macrophytes and microbes attached to the plants and their massive particulate detritus are physiologically coupled. Retentive capacities of the wetland ecosystem are highly dynamic with rates of photosynthesis of both the plants and the microbes. Such temporal variability in retention, particularly diurnal, is rarely measured or known and cannot be effectively evaluated by simple input-output models.

A primary problem is simply not understanding or ignoring biological constructs and operation of wetland ecosystems. Wetlands are highly complex and dynamic biological systems. Wetlands are by nature enormously heterogeneous spatially and biologically. For example, rates of water movement and access to microbes are critical to effective retention and storage capacities. Problems of channelization, altered microhydrology at the spatial scales of the microbes (e.g., Mann and Wetzel, 2000a, b) and assimilation vs. physical adsorptive retention are usually not incorporated into either design or management, particularly of natural wetlands.

Wetlands are highly ephemeral in capabilities. Despite this great variation, however, there are operational, functional similarities at the physiological levels. These rates of metabolism are dynamic in predictable ways as they are regulated by an array of basic physical, chemical, and biological properties. These properties should be exploited in more effective management and exploitation to achieve multiple use objectives.

There is a wealth of understanding of the physiology, life history characteristics, and competitive and mutualistic relationships of higher aquatic plants and associated microbiota. In addition, a remarkably detailed understanding exists of the functioning and coupled interrelationships of the biota emergent wetland and littoral zones, their growth dynamics, and regulatory mechanisms, and effects of these communities on the loadings to lakes and rivers. Yet, much of this important information is not effectively translated to the use of natural wetlands, the design of constructed wetlands, and the management of both natural and artificial wetland ecosystems. As a result, the retentive capabilities of wetlands can be misrepresented and exaggerated by misunderstanding of their efficacy. The ensuing remarks highlight a few examples of the importance of understanding natural ecosystems in relation to the effective design and utilization of land-water ecosystems for water quality management.

Nutrient/pollutant availability, uptake, and retention
Nutrients or pollutants are removed in large part from transporting water by biologically mediated uptake and retention. That retention occurs either by direct assimilation or by

physiologically mediated alterations of adsorptive sites on sediment and organic detrital particles. The essential property of simple physical contact between the dissolved substances and absorption/adsorption sites is frequently underestimated in wetlands: (a) Initial uptake is largely by microbiota, not the higher aquatic plants (see following discussion). Although some nutrients are assimilated by the mycorrhizae and roots of higher plants, most nutrients from influent sources, particularly in surface inflows, are assimilated directly by attached microbiota (bacteria, algae, fungi, protists) and intensively recycled amongst these microbial communities (e.g., Wetzel, 1990, 1993, 2001a, b, c). Even among submersed macrophytes, uptake from the water is predominantly by epiphytic microflora. Most nutrient acquisition for growth of the submersed angiosperm aquatic plant *per se* is from interstitial waters of sediments to rooting tissue with translocation to the foliage.

(b) Adsorption of ions and other solutes to sediment particles is largely mediated by redox changes induced by microbiota associated with the particle surfaces. Hence, the reducing conditions induced by the microbiota, generally under anaerobic conditions, are critical to retention capacities of the particles.

(c) Wetlands are highly channelized which accelerates rates of surface flows though the wetland and decreases both the spatial exposure of solutes physically with surfaces and the time needed for microbial assimilation (e.g., Mann and Wetzel, 2000a, b).

(d) Because of commonly low hydraulic conductivities in wetland hydrosoils, much of the transport of water to roots of higher plants is from surface water transported downward, mediated by evapotranspiration which is dynamic on a diurnal basis (e.g., Mann and Wetzel, 1999, 2000a).

Natural wetlands and most constructed wetlands are not efficient conduits for exposure of water to microbiota. Even where porous hydrosoils are used in constructed wetlands, flows are circuitous about root/rhizome masses with very slow exchanges within the rooting tissues. Flows to rooting tissues are mediated to a considerable extent by evapotranspiration processes that "draw" water downward to roots. The nutrients and/or pollutants sequestered in the microbiota are recycled and enter the detrital phases within the hydrosoils. Some of the substances are resolublized and move to the rhizosphere for plant assimilation. Once within the macrophytes, the nutrients tend to be stored for relatively long periods in comparison to those retained in the sessile microbiota.

The macrophytes are constantly recycling. Most aquatic plants are herbaceous perennial plants. Growth patterns exhibit multiple cohorts with high rates of biomass recycling over an annual period. For example, temperate cattails have approximately three cohorts per year, subtemperate rushes as many as eight, more or less continuous, cohorts, and water lilies as many a 10 complete cohorts per year (Dickerman and Wetzel, 1985; Wetzel and Howe, 1999; Wetzel, 2001a). Leaching of dissolved organic compounds and nutrients from the tissues accelerates as senescence progresses, usually over long periods of weeks or longer. Some of the released dissolved organic matter is readily sequestered and retained within the attached microflora of the detritus (Mann and Wetzel, 1996, 2000c). Additionally, some of these leached substances move downgradient out of the wetlands, dependent upon the highly variable hydrological flows. Time is essential for biotic retention and biotically mediated adsorptive retention. Channelization and variability in flow velocities are among the greatest limitations to maximizing retention capacities in wetlands.

In the final analysis, appreciable amounts of nutrients and certain pollutants can be stored in the macrophyte tissues. However, it is important to note that much of that storage occurs in the below-ground tissues. Particularly in emergent macrophytes, as much as two-thirds or more (up to 90%) of the plant productivity is in below-ground tissues. The nutrient content of the tissue fluctuates widely. As the plants enter maturity and senesce, much of the critical nutrient content is translocated to rooting tissues for storage and subsequent use in subsequent or simultaneous new cohort development.

During growth seasonally, shoot:root ratios decrease markedly, even though the above-ground tissues are increasing markedly (Ervin and Wetzel, 1997). Thus, much of the organic photosynthetic productivity is below-ground in a highly reducing anaerobic environment. That environment is conducive to long-term storage because of the relatively slow rates of decomposition and low hydraulic conductance in the deeper hydrosoils. Conversely, however, removal of the above-ground foliage of wetland plants, particularly of emergent plants, is not a practical method for removal of nutrients from the system because the amounts contained in the tissues are relatively low (e.g., ca. 5% of what is loaded artificially to the wetland). Most of the nutrients are sequestered into other compartments of the ecosystem.

Nutrient cycling and recycling within periphyton (biofilm) communities

Attached microbial communities colonize all wetted surfaces in aquatic ecosystems. In spite of differences in composition and other attributes, all attached communities share certain constraints to diffusion of nutrients, pollutants, and gases from overlying water into the community. Diffusion can be impeded by physical barriers associated with the surfaces of attached communities, such as velocity gradients (Vogel, 1994) as well as accumulations of extracellular mucilage around microorganisms (Riber and Wetzel, 1987; Characklis and Wilderer, 1989; Burkholder et al., 1990).

Algal communities attached to many surfaces of wetlands constitute a major or dominant primary producer and source of autochthonous organic matter. Much of that organic matter is utilized in coupled metabolism with microbial heterotrophs of the particulate detrital and sediment materials. These very high rates of photosynthesis are only possible because efficiencies of nutrient retention and recycling are much greater among closely aggregated attached algal-microbial-substratum communities than is the case among the plankton (reviewed in detail in Wetzel, 1996, 2001a). These attached communities function as the main mediator of nutrient and pollutant retention from water passing from land through the interface regions of wetlands to receiving lakes or rivers. All management strategies must maximize physical contact and duration of contact between water and biofilm communities.

Parameters regulating microbial metabolism within the complex attached communities indicate that periphytic microbial cells are efficient metabolically because the microbiota are in close juxtaposition, sometimes in direct contact to each other and to the living or dead supporting substrata. Diffusion distances are short and concentration gradients can be kept steep by constant metabolic utilization (sinks) within the attached communities.

Within these wetland and littoral zones of high primary productivity, microbial communities often experience low water flows that result in very large boundary layer thicknesses (several mm) (Losee and Wetzel, 1993) with reduced diffusion of solutes and gases from the surrounding environment (Riber and Wetzel, 1987; Stevenson and Glover, 1993). Diffusion rates are further reduced as periphyton biomass increases. Although these low rates would suggest limitations to productivity, very intensive rates of internal recycling of nutrients and essential gases are clearly necessary to maintain the high observed productivity (Wetzel, 1990, 1993, 1996). By means of selective metabolic inhibitors, close dependency of bacterial metabolism with photosynthesis of attached algae was demonstrated (Neely and Wetzel, 1995).

These couplings suggest microbial mutualisms that are important to the success of these critical uptake communities of wetland ecosystems. Recycling of both nutrients and gases clearly circumvents problems of slow diffusion from the environment across the boundary layers and internal diffusion limitations within the community sufficiently to permit the extremely high levels of productivity and retention to occur. As external nutrient loading and availability increase in surrounding waters, rates of recycling and retention are altered, and often decreased. In natural wetland ecosystems, algal photosynthesis is irrevocably coupled to heterotrophic attached microbial metabolism. As a result, mixed autotrophic-heterotrophic assemblages of periphyton form much more complex, metabolically dynamic communities with much greater retentive capacities than those that are only heterotrophic (i.e., bacteria-only biofilms). Any process that reduces photosynthesis of these attached microbial communities, such as shading by dense emergent macrophytes (cf. Grimshaw et al., 1997) or inorganic turbidity, will reduce the efficacy of the retentive capacities of these wetland communities.

Importance of ecosystem biodiversity

Maximum biodiversity of freshwater ecosystems occurs where wetland and littoral heterogeneity interfaces with pelagic regions (Wetzel, 1999a). Regulation of biotic regeneration and recycling of nutrients, and therefore the retention of nutrients or pollutants, is governed markedly by the sources of organic matter. Dissolved organic carbon emanating from decomposing higher plant tissues buffers the rates of nutrient recycling and increases the extent of microbial recycling and retention.

The most effective retentive wetland and littoral ecosystems are those that possess maximum biodiversity of higher aquatic plants and periphyton associated with the living and dead plant tissues and detritus. Natural wetlands do not consist of monocultures of a single species. Even apparent natural monocultures usually contain many other species, particularly in the understory, where associated microbial metabolism is most intense amongst a multitude of senescing and dead particulate organic matter from both foliar and rooting tissues. In addition to slowing and spreading hydrological flows, high biodiversity of higher vegetation maximizes the spatial structure that allows heterogeneous substrata structure and surfaces for maximum colonization of microorganisms. In addition, macrophyte diversity allows for light penetration to the understory to support the essential coupled mutualism of attached algal-bacterial-fungal-protistan microbial communities. That mutualism is essential to maximize retention, recycling, and storage in detritus and living macrophyte tissues.

Decomposition and retention of organic matter

Macrophytes provide a large quantity of organic matter that is critical to the operation of wetlands as a functional, retentive ecosystem. The very high productivity of wetland macrophytes provides large quantities of labile, readily decomposed dissolved organic matter which serves as a major energy and carbon source for the microbial consortia of the detritus and sediments. The more recalcitrant organic matter of much more complex chemical structures associated with lignocellulose of structural tissues is decomposed and utilized more slowly. That slow decomposition base is essential for ecosystem stability (Wetzel, 1995, 2001a).

Many humic and fulvic organic compounds are released during the complex sequences of degradation of polymers of structural plant tissues. These solubilized compounds function in many ways. Despite the relative chemical recalcitrance of these organic compounds, they are metabolized gradually, particularly when exposed to ultraviolet irradiance of natural sunlight. Partial photolysis releases organic fragments, such as volatile fatty acids, that serve as ideal substrates for bacterial metabolism and degradation (e.g., Wetzel et al., 1995). A portion of the dissolved organic matter from macrophyte tissues is photolytically degraded completely to CO_2 by natural irradiance (Vähätalo et al., 2000; Wetzel, 2001a, 2001b).

These organic compounds of higher plant origin function in many ways to suppress or regulate ecosystem metabolism (Wetzel, 1995, 2001a, 2001b). Examples are manifold. These compounds complex metals and micronutrients and regulate availability, and toxicity in some cases, to biota. Important extracellular and membrane-bound enzymes that function in recycling of nutrients can be inactivated by non-competitive inhibition and stored for long periods, even permanently. Phosphatases in particular can be inactivated by humic and fulvic acids of higher plant tissue origin. Some of these complexes can be subsequently cleaved by mild UV photolysis of natural sunlight and the enzymes reactivated (e.g., Boavida and Wetzel, 1998). Additionally, these compounds, often predominantly of wetland origins, can influence selective absorption of light, heating and thermal stratification cycles, nutrient distributions, and other important environmental parameters regulating biological metabolism.

Conduits of gases and water

The macrophytes of wetlands and littoral areas also function in regulation of gases to and from the sediments and with the atmosphere. Plant evapotranspiration by nearshore littoral and wetland vegetation can cause highly dynamic, bidirectional seepage conditions over various periods of time. These loss processes alter nearshore recharge from shallow groundwater sources as well as hydrosoil water storage and fluctuations in water levels (Wetzel, 1999b). The high productivity of emergent littoral and wetland vegetation can rapidly increase rates of sedimentation of partially decomposed organic matter. The resulting decreases in basin volume, expanding habitat and emergent vegetation can result in gradual transitions to conditions where evapotranspiration losses exceed collective water inputs. As a result of biologically mediated water losses, eventual greater losses than inputs set forth a progressively accelerating transition of some shallow lakes to terrestrial ecosystems (Wetzel, 1979).

Wetlands are sites of high organic sedimentation with high rates of decomposition. Diffusion of gases within sediments is slow, and microbes consume oxygen entering the sediments, essentially instantly. The gas-conducting lacunal systems of all higher plants, often constituting as much as 70% of the plant volume, are extraordinary morphological adaptations to service the high respiratory demands of rooting tissues growing in anaerobic hydrosoils. Although the mechanisms driving gas exchange with the atmosphere are physical, based on thermal-pressure differences (Brix *et al*., 1992; Armstrong *et al*., 1991, 1996), other gases generated by the metabolism of the tissues, as well as diffusing into the tissues from the sediments, are also transported to the atmosphere by this ventilation system. In particular, CO_2, CH_4, and N_2 can be ventilated to the air via the plants.

Wetlands form major sources of gaseous endproducts of fermentative metabolism of organic carbon. In particular wetlands release massive amounts of CO_2 and CH_4, the latter a particularly important addition to other sources of gases that influence the thermal retention in the stratosphere (Segers, 1998). These products, as well as many humic substances, must be evaluated in relation the other functions, such as denitrification and the desirable conversion of nitrogen compounds to N_2 and evasion of nitrogen to the atmosphere. Positive values of wetland metabolism, such as denitrification, are countered by negative products, such as methane release.

Aquatic plants senesce rapidly and die, within a few hours, if deprived of oxygen generated from photosynthesis or transported from the atmosphere to internal tissues. Rooting tissues in particular enter into fermentative metabolism and rapidly form toxic metabolic endproducts as oxygen is depleted. A small amount of oxygen is released into the microrhizosphere surrounding roots. This oxidized microzone counteracts toxic fermentative endproducts such as hydrogen sulfide and certain volatile fatty acids, and can support microaerophilic bacterial communities. However, the function of this gaseous transport is to supply the metabolic needs of the rooting tissues, not to oxidize the sediments. Even though the plants are very efficient in this regard, they cannot be expected to function to convert sediments to oxidized substrates. Expectations that macrophytes can be adequately efficient to aerate saturated organic-rich sediments are not realistic.

Conclusions

Wetlands are the foremost biological systems that experience distinct physiological and growth periodicities at diurnal, seasonal, and long-term annual scales. Spatial variations in physical and biological structures of the wetland environment and communities result in further wide excursions in the interactive contact with flow-though water masses and exchange processes. As a result wetlands are highly ephemeral in capabilities and efficiencies for uptake and especially biologically-mediated retention of nutrients and other pollutants. Both natural and constructed wetlands require fundamental hydrology with extensive contact and time for biological reactions and retention to occur. Clearly retention functions of wetlands as desired for human exploitation are nearly always optimal when loads are applied at very low and slow rates. Artificial acceleration of loading and flow-through are counter-intuitive functionally and management provides very limited options for improvement of retention. Use of natural wetlands for pollution regulation at any level of loading greater than final wastewater polishing will inevitably lead to wetland alteration and possible loss of many other beneficial functions and values of these ecosystems.

References

Armstrong, W., Armstrong, J., Becket, P.M. and Justin, S.H.F.W. (1991). Convective gas-flows in wetland plant aeration. In: Jackson, M.B., Davies, D.D. and Lambers, H., Editors. *Plant Life under Oxygen Deprivation*. SPB Academic Publishers, B.V., The Hague. Pp. 283–302.

Armstrong, W., Armstrong, J. and Beckett, P.M. (1996). Pressurised ventilation in emergent macrophytes: The mechanism and mathematical modelling of humidity-induced convection. *Aquat. Bot.* **54**(2), 121–135.

Boavida, M.-J. and Wetzel, R.G. (1998). Inhibition of phosphatase activity by dissolved humic substances and hydrolytic reactivation by natural UV. *Freshwat. Biol.* **40**(3), 285–293.

Brix, H., Sorrell, B.K. and Orr, P.T. (1992). Internal pressurization and convective gas flow in some emergent freshwater macrophytes. *Limnol. Oceanogr.* **37**(6), 1420–1433.

Burkholder, J.A., Wetzel, R.G. and Klomparens, K.L. (1990). Direct comparison of phosphate uptake by adnate and loosely attached microalgae within an intact biofilm matrix. *Appl. Environ. Microbiol.* **56**(6), 2882–2890.

Characklis, W.G. and Wilderer, P.A., Editors. (1989). *Structure and Function of Biofilms*. Wiley, Chichester.

Dickerman, J.A. and Wetzel, R.G. (1985). Clonal growth in *Typha latifolia*: Population dynamics and demography of the ramets. *J. Ecol.* **73**(3), 535–552.

Ervin, G.N. and Wetzel, R.G. (1997). Shoot:root dynamics during growth stages of the rush *Juncus effusus* L. *Aquat. Bot.* **59**(1), 63–73.

Grimshaw, H.J., Wetzel, R.G., Brandenburg, M., Segerblom, K., Wenkert, L.J., Marsh, G.A., Sharnetky, W., Haky, J.E. and Carraher, C. (1997). Shading of periphyton communities by wetland emergent macrophytes: Decoupling of algal photosynthesis from microbial nutrient retention. *Arch. Hydrobiol.* **139**(1), 17–27.

Hammer, D.A. (1997). *Creating Freshwater Wetlands*. 2nd Edition. CRC Press, Boca Raton, FL. 406 pp.

Kadlec, R.H. and Knight, R.L. (1996). *Treatment Wetlands*. Lewis Publishers, Boca Raton, FL 893 pp.

Losee, R.F. and Wetzel, R.G. (1993). Littoral flow rates within and around submersed macrophyte communities. *Freshwat. Biol.* **29**(1), 7–17.

Mann, C.J. and Wetzel, R.G. (1996). Loading and bacterial utilization of dissolved organic carbon from emergent macrophytes. *Aquat. Bot.* **53**(1), 61–72.

Mann, C.J. and Wetzel, R.G. (1999). Photosynthesis and stomatal conductance of *Juncus effusus* L. in a temperate wetland ecosystem. *Aquat. Bot.* **63**(2), 127–177.

Mann, C.J. and Wetzel, R.G. (2000a). Hydrology of an impounded lotic wetland – wetland sediment characteristics. *Wetlands* **20**(1), 23–32.

Mann, C.J. and Wetzel, R.G. (2000b). Hydrology of an impounded lotic wetland – subsurface hydrology. *Wetlands* **20**(1), 33–47.

Mann, C.J. and Wetzel, R.G. (2000c). Effects of the emergent macrophyte *Juncus effusus* L. on the chemical composition of interstitial water and bacterial productivity. *Biogeochemistry* **48**(3), 307–322.

Neely, R.K. and Wetzel, R.G. (1995). Simultaneous use of ^{14}C and ^{3}H to determine autotrophic production and bacterial protein production in periphyton. *Microb. Ecol.* **30**(3), 227–237.

Riber, H.H. and Wetzel, R.G. (1987). Boundary-layer and internal diffusion effects on phosphorus fluxes in lake periphyton. *Limnol. Oceanogr.* **32**(5), 1181–1194.

Segers, R. (1998). Methane production and methane consumption: A review of processes underlying wetland methane fluxes. *Biogeochemistry* **41**(1), 23–51.

Stevenson, R.J. and Glover, R. (1993). Effects of algal density and current on ion transport through periphyton communities. *Limnol. Oceanogr.* **38**(6), 1276–1281.

Vähätalo, A.V., Salkinoja-Salonen, M., Taalas, P. and Salonen, K. (2000). Spectrum of the quantum yield for photochemical mineralization of dissolved organic carbon in a humic lake. *Limnol. Oceanogr.* **45**(3), 664–676.

Vogel, S. (1994). *Life in Moving Fluids: The Physical Biology of Flow*. 2nd Ed., Princeton Univ. Press, Princeton, NJ. 467 pp.

Wetzel, R.G. (1979). The role of the littoral zone and detritus in lake metabolism. In: Likens, G.E., Rodhe, W. and Serruya, C., Editors. *Symposium on Lake Metabolism and Lake Management. Arch. Hydrobiol. Beih. Ergebnisse Limnol.* **13**, 145–161.

Wetzel, R.G. (1990). Land-water interfaces: Metabolic and limnological regulators. *Verhand. Internat. Verein. Limnol.* **24**(1), 6–24.

Wetzel, R.G. (1993). Microcommunities and microgradients: Linking nutrient regeneration, microbial mutualism, and high sustained aquatic primary production. *Netherlands J. Aquat. Ecol.* **27**(1), 3–9.

Wetzel, R.G. (1996). Benthic algae and nutrient cycling in lentic freshwater ecosystems. In: Stevenson, R.J., Bothwell, M.L. and Lowe, R.L., Editors. *Algal Ecology: Freshwater Benthic Ecosystems*. Academic Press, New York. Pp. 641–667.

Wetzel, R.G. (1999a). Biodiversity and shifting energetic stability within freshwater ecosystems. *Arch. Hydrobiol. Spec. Issues Advanc. Limnol.* **54**, 19–32.

Wetzel, R.G. (1999b). Plants and water in and adjacent to lakes. In: Baird, A.J. and Wilby, R.L., Editors. *Eco-hydrology: Plants and Water in Terrestrial and Aquatic Environments*. Routledge, London. Pp. 269–299.

Wetzel, R.G. (2000). Natural photodegradation by UV-B of dissolved organic matter of different decomposing plant sources to readily degradable fatty acids. *Verhand. Internat. Verein. Limnol.* **28**(4), 3007–3116.

Wetzel, R.G. (2001a). *Limnology: Lake and River Ecosystems*. 3rd Edition. Academic Press, San Diego. 1006 pp.

Wetzel, R.G. (2001b). Origins, fates, and ramifications of natural organic compounds of wetlands. In: Clark, A., Editor. *Chemical Ecology and Natural Organic Compounds*. Academic Press, San Diego. (In press).

Wetzel, R.G. (2001c). Dissolved organic carbon: Detrital energetics, metabolic regulators, and drivers of ecosystem stability of aquatic ecosystems. In: Findley, S.E.G. and Sinsabaugh, R.L., Editors. *Integrating Approaches to Microbial-dissolved Organic Carbon Trophic Linkages*. Academic Press, San Diego. (In press).

Wetzel, R.G., Hatcher, P.G. and Bianchi, T.S. (1995). Natural photolysis by ultraviolet irradiance of recalcitrant dissolved organic matter to simple substrates for rapid bacterial metabolism. *Limnol. Oceanogr.* **40**(5), 1369–1380.

Wetzel, R.G. and Howe, M.J. (1999). High production in a herbaceous perennial plant achieved by continuous growth and synchronized population dynamics. *Aquat. Bot.* **64**(2), 111–129.

Plants as ecosystem engineers in subsurface-flow treatment wetlands

C.C. Tanner

National Institute of Water and Atmospheric Research (NIWA), PO Box 11-115, Hamilton, New Zealand
(E-mail: c.tanner@niwa.cri.nz)

Abstract Mass balance performance data from side by side studies of planted and unplanted gravel-bed treatment wetlands with horizontal subsurface-flow are compared. Planted systems showed enhanced nitrogen and initial phosphorus removal, but only small improvements in disinfection, BOD, COD and suspended solids removal. Direct nutrient uptake by plants was insufficient to account for more than a fraction of the improved removal shown by planted systems. Roles of plants as ecosystem engineers are summarised, with organic matter production and root-zone oxygen release identified as key factors influencing nutrient transformation and sequestration.
Keywords Aeration; constructed wetlands; natural systems; nutrient removal; phytoremediation; wetland plants

Introduction

The success of human civilisation is largely due to our skills as physical ecosystem engineers. Although these engineering activities are primarily directed towards achieving some specific purpose (e.g. agricultural production), many have major indirect and unintended effects on ecosystems (e.g. water pollution, global warming). We are not, however, the only organisms that directly or indirectly control or modulate the availability of resources to other organisms by modifying the physical state of biotic or abiotic materials (Jones *et al.*, 1997). Beaver dams, termite mounds, coral reefs and forests are obvious examples of non-human engineering that come to mind, but all manner of organisms including microbes, play a role in the creation, modification and maintenance of habitats at a range of spatial scales.

The ecosystem engineering role of plants in treatment wetlands is probably most obvious in surface-flow (SF) systems. Here emergent plant shoots and litter form the main physical structure in the water column, moderating water flow, stabilising sediments, shading and sheltering the water column, providing surfaces for biofilm growth, and providing refuge and habitat for other biota. Here, however, I wish to focus primarily on the role of plants in horizontal subsurface-flow (SSF) wetlands, where wastewaters pass laterally through gravel media in which emergent plants are rooted. In this case, the media provides the main physical structure and the plants' role, apart from nutrient uptake, is more indirect.

Do plants influence treatment performance?

Dense beds of emergent wetland plants are the most obvious visual feature of SSF wetlands. They undoubtedly play a major role in enhancing their wildlife habitat values, aesthetics and perceived naturalness, but do they actually make much difference to treatment performance?

Because evapotranspiration by plants can significantly affect the hydrological balance of SSF treatment wetlands, comparative assessments ideally need to be made on the basis of mass balances (Howard-Williams, 1985). Many early studies (e.g Wolverton *et al.*,

1983) reported greater reductions in the concentration of particular contaminants in planted than unplanted gravel-beds, but most either did not measure flows or did not present data in a way which enabled mass balances to be readily calculated. A range of studies have also reported enhanced transformation of other contaminants in the rhizosphere of wetland plants, such as metals, pesticides and organic compounds.

Published data from side-by-side studies of planted and unplanted SSF treatment wetlands where mass balances could be calculated for biochemical or chemical oxygen demand (BOD, COD), nitrogen or faecal coliform bacteria removal was compiled and compared (see Figures 1 and 2). The studies included hydraulic loading rates of 25–182 mm d^{-1}, domestic and agricultural wastewaters with varying levels of preceding treatment, and eight different emergent plant species. Sizes of pilot-scale systems ranged from 18–400 m^2 and experimental mesocosm studies from 0.08–6 m^2. Plant effects may have been exaggerated in the smaller-scale experimental studies, where high edge to volume ratios result in ramification of rhizomes and roots at the edges of the container and elevated shoot densities and plant biomass (e.g. Tanner, 1994b). Although all of these studies monitored inflow rates, not all reported outflows (Bavor *et al.*, 1987; Gersberg *et al.*, 1983; 1986; 1987) and here mass removals have been calculated assuming overall hydrological balance.

Oxygen demand reduction

Comparison of oxygen demand removal (relative to mass loading) for planted and unplanted SSF wetlands shows obvious monotonic relationships between BOD (and COD) mass loading and removal rates (Figure 1), with little difference between planted and unplanted beds. Comparison of paired systems does, however, commonly show reduced (by 2–5 g m^{-3}) effluent BOD concentrations for planted beds. Removal of suspended solids, which is primarily a physical process of settling and retention, is also very similar for planted and unplanted beds.

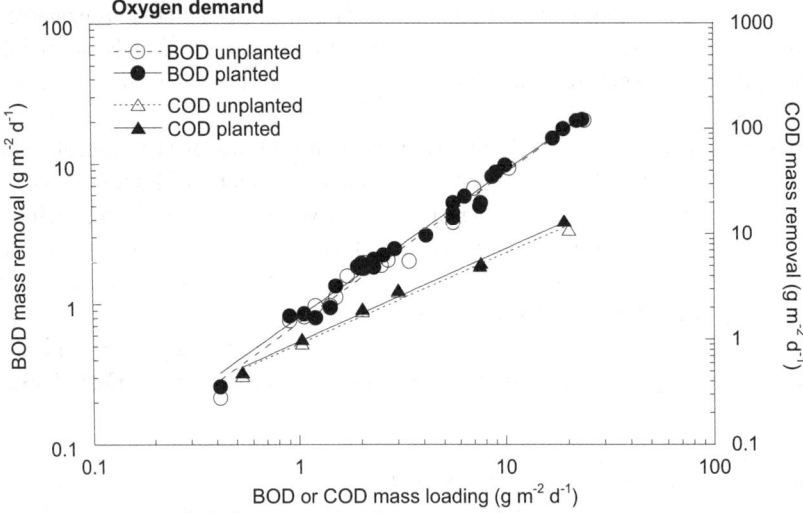

Figure 1 Comparison of mass loading and removal rates of BOD and COD for planted and unplanted wetlands. Log scales are used to improve visibility of clumped data; note different vertical scales. Lines shown are power fits ($R^2 \geq 0.99$); unplanted n = 16 and 7, planted n = 30 and 7, respectively. Data sources: pilot-scale studies at Santee, California, Gersberg *et al.* (1983, 1986); Sydney; Bavor *et al.* (1987) and Melbourne, Australia, Cottingham *et al.* (1999); Hamilton, NZ, Van Oostrom and Cooper (1990) and Tanner *et al.* (1995a); and experimental mesocosm studies reported by DeBusk *et al.* (1990)), Burgoon *et al.* (1991a), Tanner *et al.* (1999) and Soto *et al.* (1999)

Disinfection

Although comparative data for faecal bacterial indicator removal in planted and unplanted beds is more limited, mean areal removal rates show a very consistent linear relationship to areal loading over a broad range of application rates (10^7–10^{13} m^{-2} d^{-1}; data from comparative studies listed for Figures 1 and Gersberg et al., 1987; n = 6, includes Santee CA data for total, not faecal, coliforms). When comparative data for planted and unplanted systems are plotted together the data points and linear regression lines (R^2 >0.999) almost completely obscure each other. However, in side by side comparisons the planted beds commonly show small but consistent improvements in inactivation rates of faecal coliform and a range of other bacterial indicators (e.g. Soto et al., 1999). Planted SSF systems have also shown lower effluent concentrations of viruses than unplanted systems (e.g. Gersberg et al., 1987).

The water column in SSF systems is not exposed to sunlight and does not undergo significant diurnal variations in pH and dissolved oxygen, which together are predominant means of disinfection in natural treatment systems such as waste stabilisation ponds. Other mechanisms including settling, adsorption, protozoan grazing, and possibly release of anti-microbial compounds, are believed to account for the pathogen attenuation observed in SSF treatment wetlands. Decamp et al. (1999) found different ciliate community composition in planted and unplanted SSF wetlands, with greater abundance and grazing rates of fluorescently labelled E. coli in planted systems. They considered that the more oxidised conditions in the plant rhizosphere provided a more favourable habitat for ciliate protozoa.

Nutrient reduction

Comparison of total nitrogen (TN) removal performance for planted and unplanted systems is shown in Figure 2. Here, despite considerably more data scatter, the planted wetlands show a clear trend of improved TN removal. Some of the systems showed markedly poorer overall performance than others. These included systems receiving highly nitrified, low BOD wastewaters (presumably organic C limited, Gersberg et al., 1983) and others receiving ammonia-rich meat-processing wastewaters containing high levels of COD and sulphur (presumably oxygen limited, Van Oostrom and Cooper, 1990).

Planted wetlands have also shown enhanced P and metal mass removal compared to unplanted controls (e.g. DeBusk et al., 1990; Dunbabin et al., 1988; Soto et al., 1999; Tanner et al., 1995b). Unless specialised P sorbing media are employed, the primary long-term mechanism for wetland P removal is accumulation in accreting sediments (Kadlec and Knight, 1996). The studies noted above involved relatively immature systems where plant uptake and sediment adsorption pools were still actively filling, and it is unlikely that reported P removal rates would be sustainable (Richardson and Craft, 1993; Tanner et al., 1998c).

Can plant uptake explain differences in nutrient removal?

The quantity of nutrients able to be taken up and accumulated in live plant biomass per unit of wetland surface area is finite for a given plant species, nutrient regime and set of environmental conditions. Once live plant storage pools approach this limit, little further net annual uptake is possible (Howard-Williams, 1985). In pilot-scale trials where plant storage pools were still actively filling, Gersberg et al. (1986) estimated potential plant uptake could only account for 12–16% of the N removal recorded in systems planted with bulrushes. This was 5–7 times less than the additional removal recorded for the planted systems (over that of an unplanted system). In higher loaded systems achieving relatively low N removal (see above), Van Oostrom and Cooper (1990) estimated net N uptake by bulrush over an annual period accounted for 25% of wetland TN removal, representing 66% of the additional removal recorded for the planted systems.

Detailed measurements of seasonal uptake by bulrush during the second growth season in four equivalent SSF systems operated over a range of loading rates (Tanner, 2001) showed mean rates of 0.2–0.3 g N m^{-2} d^{-1} and 0.05–0.1 g P m^{-2} d^{-1} during late spring. Maximum autumn accumulations (including below-ground and standing dead tissues) of 26–47 g N m^{-2} and 5.8–12.2 g P m^{-2} were recorded in late summer of the second growth season, rising to 48–69 g N m^{-2} and 8.8–13.4 g P m^{-2} in the third growth season. Seasonal senescence after the second growth season resulted in the net release of 0.1–0.25 g N m^{-2} d^{-1} and 0.02–0.06 g P m^{-2} d^{-1} from live plant tissues. Over an annual period net storage in live plant tissues thus only accounted for 2–8% of TN removal and 1.9–5.3% of TP removal (Tanner et al., 1995b). Comparative measurements in planted and unplanted systems showed net annual plant uptake was responsible for only 3–19% of the additional TN removal and 3–60% of the additional TP removal recorded for the planted systems. This shows that, even in immature systems where plant nutrient pools are actively building, uptake and storage of N and P in live plant biomass can usually only account for a fraction of the improved performance of planted systems. This suggests plants primarily facilitate improved nutrient removal indirectly through their effects on other removal processes.

Ecosystem engineering by plants
Nutrient spiralling

Although, on an annual basis, little net accumulation of nutrients in plant tissues may occur in mature wetlands, this does not mean nothing has happened. Depending on the phenology of the species and the climate, senescence of aboveground tissues may occur on a seasonal basis or be spread more gradually over the year. Nutrients taken up by plants in dissolved inorganic forms are thus returned sometime later in complex organic forms. This means uptake and return of nutrients is often separated in time and occurs on different temporal scales. In temperate climates, this generally provides a period of rapid plant uptake in early spring when rising temperatures stimulate mineralisation of organic matter accumulated in the wetland over the previous winter. Although some of the nutrients assimilated in active

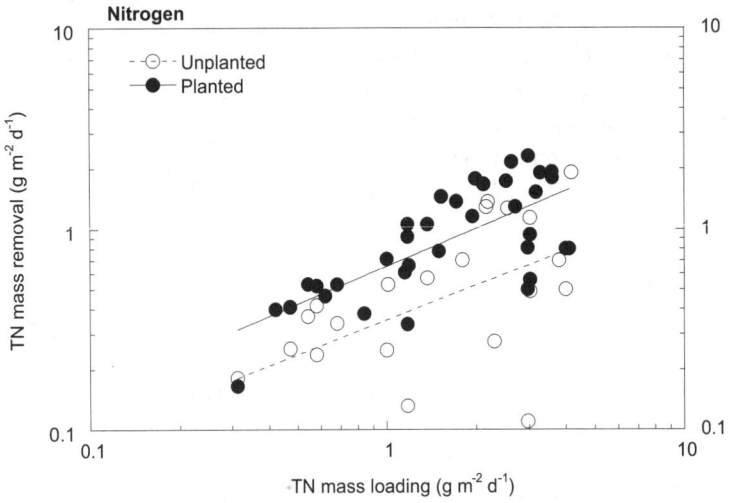

Figure 2 Comparison of mass loading and removal rates of TN for planted and unplanted wetlands. Lines shown are power fits (R^2 = 0.34 and 0.39). Data sources listed in Table 1; unplanted n = 24, planted n = 35. Data sources as for Figure 1, plus Burgoon et al. (1991b), Rogers et al. (1991), Tanner et al. (1995b)

growth phases are internally recycled during senescence by translocation to below-ground storage tissues (rhizomes, corms and bulbs), the magnitude of this is likely to be reduced in the eutrophic conditions of wastewater treatment wetlands (Shaver and Melillo, 1984).

Senescing aerial shoots of tall emergent species such as *Phragmites*, *Schoenoplectus* and *Typha* are not generally abscised. Dead shoots thus remain standing for a period of months during which considerable leaching of mobile nutrients (e.g. K) and in situ decomposition occurs. By the time the dead shoots fall much of the readily available carbon compounds will have already been consumed, so that mainly slowly degradable and recalcitrant litter fractions are returned to the wetland surface. Interactions with invertebrate shredders, grazers and burrowing worms are likely to be important in the decomposition of this fallen litter and its incorporation in the substratum.

After 5 years operation, Tanner *et al.* (1998a) found substantial accumulation of organic matter (6.8–15 kg m^{-2}) in SSF wetlands treating farm dairy wastewaters. Organic matter accumulated particularly on the surface and in the top 100 mm of the gravel media. Comparison with unplanted controls after ~2 y operation showed 1.6 to 6-fold higher organic matter accumulation in the presence of plants (Tanner and Sukias, 1995). Molar C:N ratios in the accumulated organic matter were also higher in the planted (mean 15.6) than unplanted (mean 11.0) wetlands. Organic matter accumulating in the sediments and as fallen litter in these systems represented a significant pool of both N (~ twice that present in live and standing dead plant tissues at maximum seasonal biomass, Tanner, 2001) and P (~ 6-fold higher than present in live and standing dead plant tissues at maximum seasonal biomass, Tanner *et al.*, 1998c). Organic matter accumulated from wastewaters and plant turnover provides additional sorption sites, sources of complexing and biochemically active substances (humic and fulvic acids), and substrates for microbial processes (e.g. denitrification). This intensifies nutrient spiralling (repeated cycling) along treatment wetlands, markedly elevating the residence time of nutrients relative to that of the wastewaters passing through them (Howard-Williams, 1985; Tanner *et al.*, 1998b).

Root-zone aeration

Diffusion, and in some cases convective flows, of oxygen down through the internal spaces (aerenchyma) of wetland plants enables root growth and survival in flooded sediments (Armstrong *et al.*, 1990). However, there has been controversy about how much of this oxygen is actually released into the root-zone (Sorrell and Armstrong, 1994). Increased rates of BOD removal and ammonia oxidation from wastewaters and elevated dissolved oxygen concentrations have been recorded in the root-zone of wetland plants (Dunbabin *et al.*, 1988; Reddy *et al.*, 1989a). Higher interstitial redox potentials (indicating more oxidised conditions) have also been reported for planted than unplanted SSF systems in comparative studies (Figure 3, Burgoon *et al.*, 1995; Dunbabin *et al.*, 1988; Tanner *et al.*, 1999). Root oxygen release has been postulated to account for improved rates of ammoniacal N removal in SSF wetland by stimulating nitrification, the rate limiting step (Gersberg *et al.*, 1986; Reddy *et al.*, 1989b), and higher densities and activity of nitrifiers have been recorded in biofilms associated with wetland plant roots and rhizomes than in the gravel media (Williams *et al.*, 1994).

The predicted depth of plant root penetration, and thus potential for oxygen release, has been proposed as a rational basis for determining the appropriate depth of SSF treatment wetlands (e.g. Reed *et al.*, 1995). In common with many other studies, Tanner (1996; 2001) found considerably shallower root penetration (mostly <300 mm) in gravel-bed systems than those reported by Gersberg *et al.* (1986; 300–760 mm depending on species) and commonly cited in guidelines. In mesocosm studies comparing bulrush root growth over a range of wastewater dilutions in the presence of excess nutrients, Tanner (1994a) found

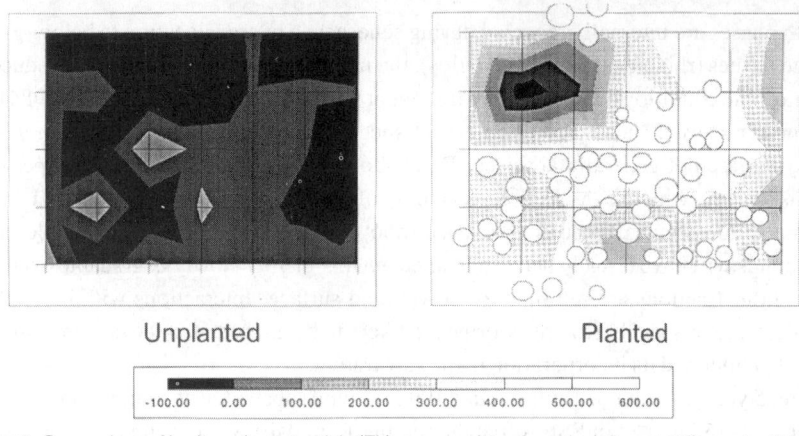

Figure 3 Comparison of in situ redox potentials (Eh) measured 3 m from the inflow at 100 mm depth in the gravel media of unplanted (left) and planted (right) SSF wetlands treating farm dairy wastewaters. Plant shoots are shown as open circles. Measurements made at 50mm spacings using platinum disc electrodes (Tanner et al., 1997)

that increasing concentration of BOD rather than nutrients was the primary environmental factor influencing the depth of root penetration. This is consistent with current theories and models of plant aeration (Armstrong et al., 1990; Sorrell et al., 2000) which predict that for a given root morphology, length will be controlled by cumulative radial and longitudinal consumption due to root respiration and leakage from the root. The extent of leakage is dependant both on anatomical properties of the root (e.g. diameter, degree of endodermal and lateral development, Sorrell et al., 2000) and the intensity of rhizospheric oxygen consumption (Sorrell and Armstrong, 1994). Obviously, it will normally be in the plants' interest to restrict radial oxygen losses from the root as much as possible; generally to areas of active root growth and nutrient assimilation. Further studies are needed to quantify and compare the oxygen release characteristics of different emergent species and their response to environmental conditions in the root-zone of SSF treatment wetlands.

As well as acting as a conduit for gas transport into (O_2) and out (e.g. CH_4, CO_2, N_2O, H_2S) of the substratum (Sebacher et al., 1985), the internal ventilation systems of emergent wetland plants may regulate the balance between gas production and consumption processes in the sediments. Tanner et al. (1997) found lower methane emissions for planted than unplanted SSF treatment wetlands, postulating that plant oxygen release was suppressing methanogenesis (a strictly anaerobic process) in the gravel media and/or enhancing root-zone methane oxidation.

Synthesis

Table 1 uses the ecosystem engineering paradigm of Jones et al. (1997), to synthesise the concepts discussed above with a range of other roles identified for plants in treatment systems (e.g. Brix, 1994). Some of the ecosystem engineering roles identified probably only have tenuous links to treatment performance. However, this approach does emphasise the importance of the indirect effects of plants on carbon and nutrient cycling, and ultimately the structure and functioning of treatment wetland ecosystems. The fundamental importance of factors such as the decomposition dynamics of plant litter and the efficacy of root-zone aeration, which ultimately determine the rate and sustainability of carbon and nutrient transformation and sequestration, are highlighted.

Conclusions

Data from comparative studies of planted and unplanted SSF systems suggests that, in

Table 1 Examples of ecosystem engineering by wetland plants in SSF treatment wetlands

Action	Examples of effects on resource availability to other organisms		
	Creation of resource	Control over resources	Modulation of forces
• Growth of plant canopy	• Shaded, moist & concealed habitat • Aerial shoot habitat • Shoot habitat for burrowing invertebrates	• Nutrient uptake & accumulation • Light attenuation • Evapotranspirative water loss • Conduit for oxygen & respiratory gas transport	• Temperature buffering • Shelter from wind & rain impact • Increased humidity & evapotranspirative cooling
• Senescence of above ground growth • Litter fall	• Standing dead shoot habitat for microbes & burrowing invertebrates • Nest & case building materials • Aerobic/anoxic/an aerobic surface litter habitats	• Nutrient retention, immobilisation & release • Rainfall interception & evaporation • Nutrient retention, immobilisation & release • Supply of humus-modifying sediment texture, cation exchange capacity etc. • Supply of humic substances – buffering of pH, binding/chelation of nutrients, heavy metals & toxins • Substrate for production of toxic fermentation products e.g. fatty acids	• Shelter from wind & rain • Shelter from wind & rain impact • Surface insulation • Clogging of media – reduced hydraulic conductivity
• Root & rhizome growth	• Additional surface area & attachment sites • Aerobic rhizosphere habitat • Root exudates, enzymes etc. • Aerenchymous shoot base rhizome and root habitat for burrowing invertebrates • Dead root & rhizome habitat	• Nutrient uptake & accumulation • Nutrient & metal retention in iron plaques • Oxygenation controls microbial pathways, decomposition & mineralisation rates, nutrient and toxin transformations (including ammonia), & methane oxidation • Transpirational draw of water & dissolved constituents to root-zone	• Binding/armouring of media surface • Occupation of interstitial spaces causing reduced hydraulic conductivity & increased short-circuiting • Filtering of suspended solids

general, wetland plants provide only small improvements in BOD and COD removal, and disinfection performance. However, wetland plants provide measurable enhancement of nutrient removal, mainly by promoting transformations to gaseous forms and sequestration in accumulating organic matter. Unless nutrient loadings are very low, net removal by direct plant uptake is generally a relatively small proportion of total removal. Plants primarily affect treatment performance through ecosystem engineering, enhancing key nutrient transformation processes (e.g. nitrification and denitrification) by root-zone oxygen release and supply of organic matter. Cycling and accumulation of plant-derived organic matter provides a sustained supply of organic C for microbes, sequesters organically bound nutrients, and buffers nutrient release.

Acknowledgements

This study was funded by the New Zealand Foundation for Research, Science and Technology. I am grateful to John Clayton and James Sukias (NIWA) for reviewing the manuscript, and Joan Garcia (Technical University of Catalonia) for constructive discussion.

References

Armstrong, W., Armstrong, J. and Beckett, P.M. (1990). Measurement and modelling of oxygen release from roots of *Phragmites australis*. In, *Constructed Wetlands in Water Pollution Control*. P.F. Cooper and B.C. Findlater (eds.). Pergamon Press, Oxford. pp. 41–52.

Bavor, H.J., Roser, D.J. and McKersie, S.A. (1987). Treatment of wastewater using artificial wetlands, large-scale fixed-film bioreactors. *Aust. J. Biotech.* **1**, 67–73.

Burgoon, P.S., DeBusk, T.A., Reddy, K.R. and Koopman, B. (1991a). Vegetated submerged beds with artificial substrates. I. BOD removal. *J. Environ. Eng.* **117**, 394–407.

Burgoon, P.S., DeBusk, T.A., Reddy, K.R. and Koopman, B. (1991b). Vegetated submerged beds with artifical substrates. II, N and P removal. *J. Environ. Eng.* **117**, 408–424.

Burgoon, P.S., Reddy, K.R. and DeBusk, T.A. (1995). Performance of subsurface flow wetlands with batch-load and continuous-flow conditions. *Water Environ. Res.* **67**, 855–862.

Brix, H. (1994). Functions of macrophytes in constructed wetlands. *Water Sci. Tech.* **29**(4), 71–78.

Cottingham, P.D., Davies, T.H. and Hart, B.T. (1999). Aeration to promote nitrification in constructed wetlands. *Environ. Tech.* **20**, 69–75.

DeBusk, T.A., Langston, M.A., Burgoon, P.S. and Reddy, K.R. (1990). A performance comparison of vegetated submerged beds and floating macrophytes for domestic wastewater treatment. In: *Constructed Wetlands in Water Pollution Control*, P.F. Cooper and B.C. Findlater (eds.). Pergamon Press, Oxford, UK. pp. 301–308.

Decamp, O., Warren, A. and Sanchez, R. (1999). The role of ciliated protozoa in subsurface flow wetlands and their potential as bioindicators. *Water Sci. Tech.* **40**(3), 91–98.

Dunbabin, J.S., Pokorny, J. and Bowmer, K.H. (1988). Rhizosphere oxygenation by Typha domingensis in minature artificial wetland filters used for metal removal from wastewaters. *Aquat. Bot.* **29**, 303–317.

Gersberg, R.M., Elkins, B.V. and Goldman, C.R. (1983). Nitrogen removal in artificial wetlands. *Water Res.* **20**, 363–367.

Gersberg, R.M., Elkins, B.V., Lyon, S.R. and Goldman, C.R. (1986). Role of aquatic plants in wastewater treatment by artificial wetlands. *Water Res.* **20**(3), 363–368.

Gersberg, R.M., Brenner, R., Lyon, S.R. and Elkins, B.V. (1987). Survival of bacteria and viruses in municipal wastewaters applied to artificial wetlands. In: *Aquatic Plants for Water Treatment and Resource Recovery*, K.R. Reddy and W.H. Smith (eds.). Magnolia Pubishing, Orlando FL. pp. 237–246.

Howard-Williams, C. (1985). Cycling and retention of nitrogen and phosphorus in wetlands, a theoretical and applied perspective. *Freshwat. Biol.* **15**, 391–431.

Jones, C.G., Lawton, J.H. and Shachak, M. (1997). Positive and negative effects of organisms as physical ecosystem engineers. *Ecology* **78**, 1946–1957.

Kadlec, R.H. and Knight, R.L. (1996). *Treatment Wetlands*, CRC Press, Boca Raton, FL.

Reddy, K.R., D'Angelo, E.M. and DeBusk, T.A. (1989a). Oxygen transport through aquatic macrophytes, the role in wastewater treatment. *J. Environ. Qual.* **19**, 261–267.

Reddy, K.R., Patrick, W.H. and Lindau, C.W. (1989b). Nitrification-denitrification at the plant root-sediment interface in wetlands. *Limnol. Oceanogr.* **34**, 1004–1013.

Reed, S.C., Middlebrooks, E.J. and Crites, R.W., eds. (1995). *Natural Systems for Waste Management and Treatment*. 2nd ed. McGraw-Hill, New York.

Rogers, K.H., Breen, P.F. and Chick, A.J. (1991). Nitrogen removal in experimental wetland treatment systems, evidence for the role of aquatic plants. *J. Water Poll. Contr. Fed.* **63**, 934–941.

Richardson, C.J. and Craft, C.B. (1993). Effective phosphorus retention in wetlands: fact or fiction? In: *Constructed Wetlands for Water Quality Improvement*, G.A. Moshiri (ed.). Lewis Publishers, Boca Raton FL. pp. 271–282.

Sebacher, D.I., Harriss, R.C. and Bartlett, K.B. (1985). Methane emissions to the atmosphere through aquatic plants. *J. Environ. Qual.* **14**, 40–46.

Shaver, G.R. and Melillo, J.M. (1984). Nutrient budgets of marsh plants, efficiency concepts and relation to availability. *Ecology* **65**, 1491–1510.

Sorrell, B.K. and Armstrong, W. (1994). On the difficulties of measuring oxygen release by root systems of wetland plants. *J. Ecol.* **82**, 177–183.

Sorrell, B.K., Mendelssohn, I.A., McKee, K.L. and Woods, R.A. (2000). Ecophysiology of wetland plant roots, a modelling comparison of aeration in relation to species distribution. *Ann. Bot.* **86**, 675–685.

Soto, F., Garcia, M., de Luis, E. and Becares, E. (1999). Role of *Scirpus lacustris* in bacterial and nutrient removal from wastewater. *Water Sci. Tech.* **40**(3), 241–247.

Tanner, C.C. (1994a). Growth and nutrition of *Schoenoplectus validus* in agricultural wastewaters. *Aquat. Bot.* **47**, 131–153.

Tanner, C.C. (1994b). Treatment of dairy farm wastewaters in horizontal and up-flow gravel-bed constructed wetlands. *Water Sci. Tech.* **29**(4), 85–94.

Tanner, C.C. (1996). Plants for constructed wetland treatment systems – A comparison of the growth and nutrient uptake characteristics of eight emergent species. *Ecol. Eng.* **7**, 59–83.

Tanner, C.C. (2001). Growth and nutrient dynamics of soft-stem bulrush in constructed wetlands treating nutrient-rich wastewaters. *Wetl. Ecol. Manag.* **9**, 49–73.

Tanner, C.C., Adams, D.D. and Downes, M.T. (1997). Methane emissions from constructed wetlands treating agricultural wastewaters. *J. Environ. Qual.* **26**, 1056–1062.

Tanner, C.C., Clayton, J.S. and Upsdell, M.P. (1995a). Effect of loading rate and planting on treatment of dairy farm wastewaters in constructed wetlands I. Removal of oxygen demand, suspended solids and faecal coliforms. *Water Res.* **29**, 17–26.

Tanner, C.C., Clayton, J.S. and Upsdell, M.P. (1995b). Effect of loading rate and planting on treatment of dairy farm wastewaters in constructed wetlands II. Removal of nitrogen and phosphorus. *Water Res.* **29**, 27–34.

Tanner, C.C., D'Eugenio, J., McBride, G.B., Sukias, J.P.S. and Thompson, K. (1999). Effect of water level fluctuation on nitrogen removal from constructed wetland mesocosms. *Ecol. Eng.* **12**, 67–92.

Tanner, C.C. and Sukias, J.P. (1995). Accumulation of organic solids in gravel-bed constructed wetlands. *Water Sci. Tech.* **32**(3), 229–240.

Tanner, C.C., Sukias, J.P.S. and Upsdell, M.P. (1998a). Organic matter accumulation during maturation of gravel-bed constructed wetlands treating farm dairy wastewaters. *Water Res.* **32**, 3046–3054.

Tanner, C.C., Sukias, J.P.S. and Upsdell, M.P. (1998b). Relationships between loading rates and pollutant removal during maturation of gravel-bed constructed wetlands. *J. Environ. Qual.* **27**, 448–458.

Tanner, C.C., Sukias, J.P.S. and Upsdell, M.P. (1998c). Substratum phosphorus accumulation during maturation of gravel-bed constructed wetlands. *Water Sci. Tech.* **40**(3), 647–659.

Van Oostrom, A.J. and Cooper, R.N. (1990). Meat processing effluent treatment in surface-flow and gravel-bed constructed wastewater wetlands. In: *Constructed Wetlands in Water Pollution Control*, P.F. Cooper and B.C. Findlater (eds.). Pergamon Press, Oxford. pp. 321–332.

Williams, J.B., May, E., Ford, M.G. and Butler, J.E. (1994). Nitrogen transformations in gravel bed hydroponic beds used as a tertiary treatment stage for sewage effluents. *Water Sci. Tech.* **29**(4), 29–36.

Wolverton, B.C., McDonald, R.C. and Duffer, W.R. (1983). Microorganisms and higher plants for waste water treatment. *J. Environ. Qual.* **12**, 236–242.

State of the art for animal wastewater treatment in constructed wetlands

P.G. Hunt and M.E. Poach

USDA-ARS, 2611 W. Lucas St., Florence, SC 29501, USA

Abstract Although confined animal production generates enormous per-unit-area quantities of waste, wastewater from dairy and swine operations has been successfully treated in constructed wetlands. However, solids removal prior to wetland treatment is essential for long-term functionality. Plants are an integral part of wetlands; cattails and bulrushes are commonly used in constructed wetlands for nutrient uptake, surface area, and oxygen transport to sediment. Improved oxidation and nitrification may also be obtained by the use of the open water of marsh-pond-marsh designed wetlands. Wetlands normally have sufficient denitrifying population to produce enzymes, carbon to provide microbial energy, and anaerobic conditions to promote denitrification. However, the anaerobic conditions of wetland sediments limit the rate of nitrification. Thus, denitrification of animal wastewaters in wetlands is generally nitrate-limited. Wetlands are also helpful in reducing pathogen microorganisms. On the other hand, phosphorus removal is somewhat limited by the anaerobic conditions of wetlands. Therefore, when very high mass removals of nitrogen and phosphorus are required, pre- or in-wetland procedures that promote oxidation are needed to increase treatment efficiency. Such procedures offer potential for enhanced constructed wetland treatment of animal wastewater.

Keywords Dairy; denitrification; nitrification; phosphorus; redox; swine

Introduction

Animal production is a major component of agriculture in the USA. It is vital for both food stability and economic health. However, there are increasingly more environmental problems associated with the present-day scale of animal production. These problems include nuisance odors, pathogens, concentrated wastewater, inadequate land treatment sites, residential encroachment, and new regulations. Increasingly, large-scale animal production occurs in confinement where enormous per-unit-area quantities of waste are generated. Additionally, industry expansion and relocation have introduced these significant waste treatment issues to non-traditional areas. For example, much of the Florida dairy industry moved from Lake Okeechobee to the Suwanee River Region. New large dairies have moved into Texas and New Mexico during the 1990s. Similarly, swine production grew from 2 to 12 million pigs in North Carolina during the 1990s.

Currently, most enterprises apply both solid and liquid waste to land for terminal treatment. This traditional method of waste management was not only used for centuries; it was essential for food production since it was the primary source of cropland fertilization. Nonetheless, application of liquid animal waste to land has unique problems, such as, high solids content, high nutrient concentrations, and limited pumping distances. Regulators and the public are demanding improved alternatives. One of these alternatives is constructed wetlands, which are generally perceived to be a technology that is relatively affordable and operationally simple. Wetlands have been used successfully for advanced treatment of municipal and residential wastewaters in the USA and around the world for over three decades (Kadlec and Knight, 1996). Compared to conventional systems, they have less construction, operation, and energy costs plus more flexibility in pollutant loading. They are also flexible in soil specificity; constructed wetlands can be built on aerated upland soils, and hydric soil conditions will develop when the soils are flooded. These hydric conditions will then support aquatic plant life and wetland processes. Currently, there are livestock producers in at least 26 states across the USA using constructed wetlands to treat animal wastewaters. However, there are limited data for the treatment of animal wastewater in constructed wetlands.

Two types of wetlands (subsurface and free-water-surface systems) are typically used (Hammer, 1989). Gumbricht (1993) categorized free-water-surface systems into free-floating-plant ponds, submersed-plant ponds, and constructed wetlands with emergent plants. Payne and Knight (1997) as well as others considered wetlands with surface-flow emergent plants to be the only likely candidate for wide scale adoption. Subsurface systems are subject to clogging and limited oxygen (O_2) diffusion, and floating aquatic systems are more affected by pests and cold temperatures. Additionally, these wetlands are more expensive to construct and operate than surface-flow wetlands. This paper will focus on free-water-surface-flow systems – their components and performance.

Wetland components
Plants
Extensive work on plant material selection has been done by the Soil Conservation Service (Marburger, 1992). The selection of appropriate plants for constructed wetlands depends on the functional requirements of the system. The most commonly used plant genera in constructed wetlands for animal waste treatment are *Scirpus*, *Typha*, and *Juncus*. Generally, a polyculture of submersed, floating, and emergent plants occupies the wetlands. Additionally, the plant community is not static. It changes with conditions of plant community health, wetland operation, and weather.

Oxygen transport
Wetland plants transport O_2 from leaves and stems to roots (Armstrong, 1964), providing an oxidized microenvironment in the anaerobic root zone. The juxtaposition of aerobic and anaerobic zones at the root-water-soil interface is critical to the treatment of wastewater (Good and Patrick, 1987). The efficient use of wetlands for wastewater treatment depends on the O_2 transport capacity of the plant-root system and O_2 diffusion across the free soil-water interface. Diffusion rates of O_2 can be lower than 0.12 g m^{-2} h^{-1} in anaerobic soil (Stolzy and Flühler, 1978), while O_2 transport and diffusion through wetland plants range from 0.02 to 1.2 g m^{-2} h^{-1} (Kadlec and Knight, 1996). The higher values of O_2 transport could be very important in the nitrification of ammonia to nitrate, a process that requires 4.33 g O_2 per gram of ammonia-N (NH_3-N) to nitrate-N. Different plant species have different capacities for O_2 transport. For instance, bulrushes have higher rates of O_2 transport and more oxidized sediments than cattails (Reddy *et al.*, 1989; Szögi *et al.*, 1994). Oxygen availability is also affected by the O_2 demand of the wetland. This of course could be a significant problem for concentrated animal wastewaters. Oxygen concentration in wetland waters will vary with the season of the year and the time of day. With cooler water temperature, O_2 saturation is greater and the O_2 demand is smaller. Oxygen also varies diurnally with photosynthesis during the light and dark periods. This is particularly true in constructed wetlands in open water areas that produce O_2 via submersed macrophytes and phytoplankton. These open water areas can be used for animal wastewater treatment, and they can be designed into the system. For instance, the marsh-pond-marsh system as described by Hammer (1989) takes advantage of this O_2 production from submersed macrophytes and phytoplankton. Such systems have been used for swine wastewater treatment in Mississippi by Cathcart *et al.* (1994) and at NCA&T State University, Greensboro, NC, by Reddy *et al.* (2000). Additionally, the depth of water in the wetland

can affect denitrification. Hunt et al. (2000) reported the wetlands were generally nitrate limited for denitrification and that denitrification enzyme potential decreased as water depth increased in wetlands used for swine wastewater treatment.

Carbon removal

Constructed wetlands will not completely remove carbon (C) because plant litter and plant/root exudate continually adds C to the system (Hunt et al., 1994). Yet, low levels of soluble C are not a problem because it is necessary for anaerobic respiration and denitrification. Furthermore, carbon removal is generally not a land-limited aspect of animal waste treatment. In fact, removal of large quantities of C from wastewater is a strongpoint of land treatment systems; and dewatered waste materials, particularly if composted, can be transported and spread to available land. The main consideration for wetlands is avoidance of large application of suspended solids, which will fill the wetland and degrade its treatment functionality. Hunt and Vanotti (1999) discussed both passive and high tech methods of solids removal from animal waste.

Phosphorus removal

Phosphorus (P) removal is via sedimentation, plant uptake, organic matter accumulation, immobilization, and soil sorption. Phosphorus is present in soils and sediments in organic and inorganic forms. The relative proportions of organic and inorganic P vary widely. Organic P may comprise a substantial reservoir because the litter-sediment processes control the long-term P removal capability of wetland ecosystems (Faulkner and Richardson, 1989). Inorganic P is retained by calcite, clay minerals, organometallic complexes, and Fe and Al oxides and hydroxides (Parfitt, 1978; Gale et al., 1994). Numerous investigators have found that oxalate-extractable iron is associated with P adsorption (Syers et al., 1973). This fraction comprises the poorly crystalline iron oxi-hydroxides that become highly soluble under prevalent-reduced conditions. This increased P solubility explains why wetlands treated with wastewater can become P-saturated and export excessive quantities of phosphate in a few years (Richardson, 1985). A rapid decline of P-removal efficiency (from 99 to 78% in one year) for constructed wetlands that treated swine wastewater was observed by Szögi et al. (1994). This rapid decline was probably related to the high content of poorly crystalline iron oxi-hydroxides of the wetland soil, strong soil reduction, and high load of P.

Nitrogen removal

Nitrogen (N) is removed from wastewater through processes including filtration, sedimentation, uptake by plants and microorganisms, adsorption, nitrification-denitrification, and volatilization. Organic N can be initially removed via filtration and sedimentation, but it will be mineralized and released over time as NH_3-N. Ammonia-N in the form of ammonium (NH_4-N) ion can be absorbed either by wetland plants through roots or by anaerobic microorganisms and converted back to organic-N or immobilized as an exchangeable ion in soil. Szögi et al. (2000) reported that NH_4-N pore water profiles peaked just below the sediment-water interface (0 to 5-cm) and decreased with depth at all sites in surface flow wetlands used for swine treatment. It was postulated that the NH_4-N peak levels in the 0-5 cm layer were related to plant uptake, soil adsorption, microbial assimilation and mineralization of sediment organic matter.

Gaseous losses of N through nitrification-denitrification can be very large; they are generally the most significant N-removal mechanisms for natural and constructed wetlands (Bowden, 1987; Faulkner and Richardson, 1989). Under anaerobic conditions, NH_4-N would normally build up to excessive levels. However, O_2 diffusion from the atmosphere to the overlying floodwater and O_2 transport by plants to the rooting zone can form oxidized microsites. The gradient between high concentrations of NH_4-N in the reduced soil and low concentrations in the oxidized microsites and layers causes diffusion of NH_4-N into these oxidized microsites and layers where nitrification can occur (Patrick and Reddy, 1976; Reddy and Graetz, 1988). Nitrification requires pH values above 5, aerobic conditions, and autotrophic nitrifying bacteria. Rapid nitrification can occur at oxidized interfaces of the root or liquid surface or when the wetland is periodically dry. Nitrate-N can then diffuse into the anoxic zone where it is denitrified to dinitrogen gas (N_2).

Additionally, ammonia can be lost through volatilization under alkaline pH conditions. This mechanism was initially thought to be of little consequence since constructed wetlands for animal wastewater treatment are generally < 8 in pH. However, the high ammonia concentrations in the wastewater have caused concern that volatilization may be a significant factor even at neutral pH. The very high rates of ammonia loss and low dissolved O_2 values add to this concern because there may not be enough oxygen to account for the complete nitrification-denitrification of the lost ammonia. Large volatilization is, of course, undesirable because NH_3-N can be absorbed by the surrounding ecosystems (i.e., cropland, pastureland, and wooded zones) and cause ecosystems shifts. Pre-wetland nitrification may be necessary to fully exploit the denitrification potential of constructed wetlands, which could be over 50 kg ha^{-1} day^{-1} (Hunt et al., 1999). Ammonia is generally a large portion of the total nitrogen content, and concentration > 200 mg L^{-1} may cause significant plant growth problems. Thus, there are plant health aspects that also encourage pre-wetland nitrification. However, our initial measurements indicate NH_3 volatilization is not a major factor.

Design

Constructed wetlands should be considered only as a component of a total animal wastewater treatment system. At a minimum, wastewater treatment for solids removal is needed ahead of the wetlands. Payne and Knight (1997) discussed the various design approaches in animal wastewater wetlands. The main methods are the NRCS guidelines, Reed et al. (1995), and Kadlec and Knight (1996). Stone et al. (2000) evaluated the design approaches in relation to the performance of swine wastewater wetlands. They found the design procedures reasonably accurate for nitrogen but phosphorus removal was overestimated.

Case studies of dairy and swine wastewater treatment

The results of several studies on constructed wetland treatment of dairy and swine wastewater are presented in Table 1 (Cathcart et al., 1994; Skarda et al., 1994; Hunt et al., 1995; Cooper and Testa, 1997; Hermans and Pries, 1997; McCaskey and Hannah, 1997; Moore and Niswander, 1997; Reaves and DuBowy, 1997; Reddy et al., 2000). All parameters, except fecal coliform, exhibited large variabilities between sites. For example, BOD$_5$ ranged from 53 to 93%, TKN ranged from 37 to 86%, and TP ranged from 42 to 83% removal. The treatment systems in Kosciusko Co., IN, and in Alabama showed the best performances with > 90% removal of total solids and BOD, > 80% removal of N, and > 75% removal of P.

The focus of our research in Duplin Co., NC, was to determine wetland treatment efficiency of swine wastewater and define redox conditions, denitrification potentials, and agronomic cropping potentials of constructed wetlands used for swine wastewater treatment (Hunt et al., 1999; Szögi et al., 2000). Three sets of two, 3.6 by 36 m wetland cells were constructed in Duplin Co., NC, in 1992; they contained either natural wetland plants or water-tolerant agronomic plants. Nitrogen loading rates of 3 to 25 kg ha^{-1} day^{-1} were used (Table 2). Mass N removal ranged from 81 to 94%. Phosphorus removal was much lower as the rate of application increased, < 35%. Redox conditions were highly anaerobic in the soils of all wetland cells in summer. Hunt *et al.* (1994) reported that rush (*Juncus effusus*) and bulrushes (*Scirpus americanus*, *Scirpus cyperinus*, and *Scirpus validus*) were not greatly different from bur-reed (*Sparganium americanum*) and cattails *(Typha angustifolia* and *Typha latifolia)* in effective treatment of swine wastewater. However, the bulrushes had more oxidized sediment than did the cattails. The higher O_2 transport rates of the bulrush allowed mildly oxidized soil conditions in the winter. Denitrification enzyme assay indicated that the wetland soils were nitrate-limited for denitrification. Szögi *et al.* (2000) also reported that saturation culture soybean and flooded rice satisfactorily treated swine wastewater in a constructed wetland, and the seed harvest removed significant amounts of nutrients as grain. The agronomic plants treated swine effluent very similarly to natural wetland plants when loading rates were < 10 kg ha^{-1} d^{-1} of N.

Ammonia volatilization was also measured at the Duplin Co. site with an open chamber device. These tests indicated that ammonia volatilization was occurring. From average hourly rates, it was estimated that 7 to 17% of the nitrogen load to the wetlands was removed through NH_3 volatilization. Because tests were conducted at only one wetland site, we need additional data before we can make a definitive conclusion on ammonia volatilization, but these results indicated that NH_3 volatilization was not responsible for removing the majority of nitrogen from the swine wastewater. This suggests that either oxygen diffusion and hence nitrification-denitrification is underestimated or there is another mechanism for ammonia loss that is being overlooked. Two novel nitrogen pathways may account for discrepancies in the data. The first has been labeled anaerobic ammonia oxidation. It is described by the equation:

Table 1 Operational reductions in dairy and swine wastewater parameters using constructed wetlands (Cathcart et al., 1994; Skarda et al., 1994; Hunt et al., 1995; Cooper and Testa, 1997; Hermans and Pries, 1997; McCaskey and Hannah, 1997; Moore and Niswander, 1997; Reaves and DuBowy, 1997; Reddy et al., 2000)

Site	BOD$_5$	COD	TSS	NH$_3$-N	TKN	TP	Coliform
			% reduction in concentration				
Dairy							
Kosciusko Co., In	93	–	94	89	86	83	–
Oregon State University	61	47	73	54	57	66	94
DeSoto Co., MS (11°C)	68	50	59	68	–	42	89
DeSoto Co., MS (22°C)	84	77	70	81	–	63	97
Mercer Co., KY	66	–	88	87	37	59	–
Essex, Ontario	66	–	66	80	–	69	99
Swine							
MS (Marsh-Pond-Marsh)	54	–	69	71	–	44	–
NCA&TSU, NC (Marsh-Pond-Marsh)	53	–	68	60	51	44	–
AL (continuous wetland)	90	–	89	85	83	76	–

Table 2 Mass removal of N in constructed wetlands, Duplin Co., NC (June '93–Nov. '97)

Nitrogen Load	System	
	Rush/bulrush	Cattails/bur-reed
kg ha^{-1} day^{-1}	Mass Removal, %	
3	94	94
8	88	86
15	85	81
25	90	84

% Mass Removal = % mass reduction of N (NH$_3$-N + NO$_3$-N) in the effluent with respect to the nutrient mass inflow

$$NH_4^+ + NO_2^- \rightarrow N_2 + 2H_2O$$

Recent research indicates that this process is performed by ammonia oxidizing bacteria and a newly discovered bacterium (Jetten *et al.*, 1999). To convert ammonia to N_2 through this pathway requires only half of the oxygen needed to convert ammonia to N_2 by conventional nitrification-denitrification. It also may be possible by some as yet unidentified process for ammonia to be converted to N_2 under completely anoxic conditions. This possibility is supported by recent data on N_2 production from animal waste lagoons (Lowry Harper, USDA-ARS, Athens, GA, personal communication).

The combined results of these studies on wastewater treatment suggest that constructed wetlands are excellent for mass removal of N. However, at the high loading rates necessary for substantive mass removal, constructed wetlands do not produce an effluent acceptable for discharge. Thus, subsequent land application is necessary. Croplands, vegetative strips, and woodlands are viable options for final treatments. Because terminal land application does not require a polished effluent, it is an approach that fits well with the capacities of constructed wetlands. Furthermore, pre- and post-wetland treatments will allow wetland adaptations to fit unique and changing water quality requirements.

References

Armstrong, W. (1964). Oxygen diffusion from the roots of some British bog plants. *Nature*, **204**, 801–802.

Bowden, W.B. (1987). The biogeochemistry of nitrogen in freshwater wetlands. *Biogeochemistry*, **4**, 313–348.

Cathcart, T.P., Hammer, D.A. and Triyono, S. (1994). Performance of a constructed wetland-vegetated strip system used for swine waste treatment. In: *Constructed Wetlands for Animal Waste Management*, P.J. DuBowy and R.P. Reaves (eds.), Purdue Research Foundation, West Lafayette, IN, pp. 9–22.

Cooper, C.M. and Testa, S., III. (1997). A constructed bulrush wetland for treatment of cattle waste. In: *Constructed Wetlands for Animal Waste*, E.P.A. Special Publication, Gulf of Mexico Program – Nutrient Enrichment Committee, Payne Engineering and CH2M Hill, pp. II.14–II.24.

Faulkner, S.P. and Richardson, C.J. (1989). Physical and chemical characteristics of freshwater wetland soils. In: *Constructed Wetlands for Wastewater Treatment – Municipal, Industrial and Agricultural*, D.A. Hammer (ed.), Lewis Publishers, Chelsea, MI, pp. 41–72.

Gale, P.M., Reddy, K.R. and Graetz, D.A. (1994). Phosphorus retention by wetland soils used for treated wastewater disposal. *J. Environ. Qual.*, **23**, 370–377.

Good, B.J. and Patrick, W.H., Jr. (1987). Root-water-sediment interface processes. In: *Aquatic Plants for Water Treatment and Resource Recovery*, K.R. Reddy and W.H. Smith (eds.), Magnolia Publishing, Orlando, FL, pp. 359–370.

Gumbricht, T. (1993). Nutrient removal processes in freshwater submersed macrophyte systems. *Ecol. Eng.*, **2**, 1–30.

Hammer, D.A. (ed.) (1989). *Constructed Wetlands for Wastewater Treatment – Municipal, Industrial, and Agricultural*. Lewis Publishers, Chelsea, MI.

Hermans, P. and Pries, J. (1997). Essex treatment wetland, Essex, Ontario, Canada. In: *Constructed Wetlands for Animal Waste Treatment*, E.P.A. Special Publication, Gulf of Mexico Program – Nutrient Enrichment Committee, Payne Engineering and CH2M Hill, pp. II.25–II.29.

Hunt, P.G. and Vanotti, M.B. (1999). Animal residual treatment and soil and water resource management. *Animal Residuals Management Conf. Proc.* (CD-ROM), 10 pp.

Hunt, P.G., Szögi, A.A., Humenik, F.J., Rice, J.M. and Stone, K.C. (1994). Swine wastewater treatment by constructed wetlands in the southeastern U.S. In: *Constructed Wetlands for Animal Waste Management*, P.J. DuBowy and R.P. Reaves (eds.), Purdue Research Foundation, West Lafayette, IN, pp. 144–154.

Hunt, P.G., Thom, W.O., Szogi, A.A. and Humenik, F.J. (1995). State of the art for animal wastewater treatment in constructed wetlands. *Proc. 7th Intl. Symp. on Agricultural and Food Processing Wastes*, pp. 53–65.

Hunt, P.G., Szogi, A.A., Humenik, F.J. and Rice, J.M. (1999). Treatment of animal wastewater in construction wetlands. *8th Intl. Conf. on the FAO ESCORENA Network on Recycling of Agricultural, Municipal and Industrial Residues in Agriculture Proc.* (Rennes, France), pp. 305–313.

Hunt, P.G., Szogi, A.A., Humenik, F.J., Reddy, G.B., Poach, M.E., Sadler, E.J. and Stone, K.C. (2000). Treatment of swine wastewater in wetlands with natural and agronomic plants. *9th Workshop of FAO ESCORENA Network on Recycling of Agricultural, Municipal and Industrial Residues in Agriculture Proc.* (Gargnano, Italy). (accepted for publication June 20, 2000).

Jetten, M.S.M, Strous, M., Schoonen, K.T., Schalk, J., van Dongen, L.G.J.M., van de Graaf, A.A., Logemann, S., Muyzer, G., van Loosdrecht, M.C.M. and Kuenen, J.G. (1999). The anaerobic oxidation of ammonium [Review]. *FEMS Microbiology Reviews*, **22**(5), 421–437.

Kadlec, R.H. and Knight, R.L. (1996). *Treatment Wetlands*. Lewis Publishers, Boca Raton, FL.

Marburger, J.E. (1992). Wetland plants. Plant materials technology needs and development for wetland enhancement, restoration, and creation in cool temperate regions of the United States, Terrene Institute, Washington, DC.

McCaskey, T.A. and Hannah, T.C. (1997). Performance of a full scale constructed wetland treating swine lagoon effluent in northern Alabama. In: *Constructed Wetlands for Animal Waste Treatment*, E.P.A. Special Publication, Gulf of Mexico Program – Nutrient Enrichment Committee, Payne Engineering and CH2M Hill, pp. II.5–II.8.

Moore, J.A. and Niswander, S.F. (1997). Oregon State University dairy wetland In: *Constructed Wetlands for Animal Waste Treatment*, E.P.A. Special Publication, Gulf of Mexico Program – Nutrient Enrichment Committee, Payne Engineering and CH2M Hill, pp. II.30–II.33.

Parfitt, R.L. (1978). Anion adsorption by soils and soil materials. *Adv. Agron.*, **30**, 1–50.

Patrick, W.H., Jr. and Reddy, K.R. (1976). Nitrification-denitrification reactions in flooded soils and sediments: Dependence on oxygen supply and ammonium diffusion. *J. Environ. Qual.*, **5**, 469–472.

Payne, V.W.E. and Knight, R.L. (1997) Constructed wetlands for treating animal wastes – Section I: Performance, design, and operation. In: *Constructed Wetlands for Animal Waste Treatment*, E.P.A. Special Publication, Gulf of Mexico Program – Nutrient Enrichment Committee, Payne Engineering and CH2M Hill.

Reaves, R.P. and DuBowy, P.J. (1997). Tom Brothers' dairy constructed wetland. In: *Constructed Wetlands for Animal Waste Treatment*, E.P.A. Special Publication, Gulf of Mexico Program – Nutrient Enrichment Committee, Payne Engineering and CH2M Hill, pp. II.9–II.13.

Reddy, K.R. and Graetz, D.A. (1988). Carbon and nitrogen dynamics in wetland soils. In: *The Ecology and Management of Wetlands*, D.D. Hook et al. (eds.), Vol. I, Timber Press Portland, OR, pp. 307–318.

Reddy, K.R., D'Angelo, E.M. and DeBusk, T.A. (1989). Oxygen transport through aquatic macrophytes: The role in wastewater treatment. *J. Environ. Qual.*, **19**, 261–267.

Reddy, G.B., Hunt, P.G., Phillips, R., Stone, K.C. and Grubbs, A. (2000). Treatment of swine wastewater in marsh-pond-marsh constructed wetlands. *7th Intl. Conf. on Wetland Systems for Water Pollution Control*, 8 pp.

Reed, S.C., Crites, R.W. and Middlebrooks, E.J. (1995). *Natural Systems for Waste Management and Treatment*. 2nd edn, McGraw-Hill, NY.

Richardson, C.J. (1985). Mechanisms controlling phosphorus retention capacity in freshwater wetlands. *Science*, **228**, 1424–1427.

Skarda, S.M., Moore, J.A., Niswander, S.F. and Gamroth, M.J. (1994). Preliminary results of wetland for treatment of dairy farm wastewater. In: *Constructed Wetlands for Animal Waste Management*, Purdue Research Foundation, West Lafayette, IN, .J. DuBowy and R.P. Reaves (eds.), pp. 34–42.

Stolzy, L.H. and Flühler, H. (1978). Measurement and prediction of anaerobiosis in soils. In: *Nitrogen in the Environment*, D.R. Nielsen and J.G. McDonald (eds.), Vol. 1, Academic Press, New York, pp. 363–426.

Stone, K.C., Hunt, P.G., Szogi, A.A., Humenik, F.J. and Rice, J.M. (2000). Constructed wetland design and performance for swine lagoon waste water treatment. ASAE Paper #00-4148, 11 pp.

Syers, J.K., Harris, R.F. and Armstrong, D.E. (1973). Phosphate chemistry in lake sediments. *J. Environ. Qual.*, **2**, 1–14.

Szögi, A.A., Hunt, P.G., Humenik, F.J., Stone, K.C., Rice, J.M. and Sadler, E.J. (1994). Seasonal dynamics of nutrients and physico-chemical conditions in a constructed wetland for swine wastewater treatment. ASAE Paper #94-2602.

Szögi, A.A., Hunt, P.G. and Humenik, F.J. (2000). Treatment of swine wastewater using a saturated-soil-culture soybean and flooded rice system. *Trans. ASAE*, **43**(2), 327–335.

Surface flow (SF) treatment wetlands as a habitat for wildlife and humans

R.L. Knight*, R.A. Clarke, Jr.* and R.K. Bastian**

* Environmental Consultant, Wetland Solutions, Inc., 2809 N.W. 161 Court, Gainesville, FL 32609, USA
** Senior Scientist, USEPA, OWM (4204M), 1200 Penn. Ave. NW, Washington, DC 20460, USA

Abstract Water quality improvement is generally the primary objective of treatment wetlands. Creation of wildlife habitat is an inevitable outcome of these projects. However, an increasing number of treatment wetland projects have been purposely designed and operated to enhance their beneficial utility to wildlife and humans. This trend to multi-purpose treatment wetlands has broadened the basis for assessing the advantages of this natural treatment alternative. There are at least 21 treatment wetlands in the U.S. that were implemented with wildlife habitat creation and/or human use as principal goals. A number of treatment wetlands outside the U.S. also share these priorities. Hundreds of other wetlands have collected and reported quantitative data on wildlife and/or human uses. The North American Treatment Wetland Database (NADB) has been expanded to include critical wildlife habitat and human use data. This paper provides a preliminary inventory of these habitat and human use treatment wetlands, summarizes lessons learned, and identifies additional data needs.

Keywords Avifauna; constructed wetlands; nature study; recreational use; wastewater treatment; wildlife habitat

Introduction

Wilhelm et al. (1989) describe the planning and design of the City of Show Low, Arizona treatment wetlands, one of the earliest intentional multi-use constructed treatment wetlands in the U.S. This system was observed to have high waterfowl and other wildlife usage. This wetland was also designed to be user-friendly for humans interested in nature study and hunting. A number of authors have described the ancillary wildlife and human use benefits resulting from treatment wetlands (Sather 1989; Freierabend 1989; Knight 1992, 1997). Researchers have also pointed out potential problems that might result from the use of wetlands for receiving wastewaters (Guntenspergen and Stearns, 1985; Freierabend 1989; Bastian et al., 1989; Knight, 1992; Wren et al., 1997). Bioaccumulation of toxics that are present in some wastewaters, as well as transmission of disease, create potential hazards that might outweigh the benefits of some treatment wetland projects.

Treatment wetlands that avoid the creation of environmental hazards can be considered as a beneficial environmental reuse option for wastewater treatment as contrasted to the more common notion that reuse only includes benefits directly to humans. Knight (1992, 1997) and Kadlec and Knight (1996) describe the engineering considerations important for optimizing beneficial wildlife and human use of treatment wetlands while minimizing the potential for nuisance conditions.

The U.S. EPA conducted a pilot study of wildlife usage and habitat functions of constructed treatment wetlands during the summer of 1992. A consistent rapid-assessment protocol was utilized at six constructed surface flow treatment wetlands to evaluate their habitat structure and function and the possibility of environmental hazards (McAllister 1992, 1993a, 1993b). That study represents the only known attempt to critically compare habitat and wildlife usage between treatment wetland sites. Similar studies at other treatment wetland sites are sorely needed.

A report prepared by the Canadian Wildlife Service (Wren *et al.*, 1997) summarizes information on wildlife usage of stormwater treatment wetlands, identifies areas of potential concern related to accumulation of hazardous pollutants, and concludes that insufficient data are available to document detrimental effects. That report recommends the need to require detailed monitoring of potential wildlife hazards in treatment wetlands. However, increased monitoring raises project costs that might discourage beneficial projects and should not be required without a sound purpose.

A recent inventory of habitat values of SF treatment wetlands was funded by the U.S. Environmental Protection Agency's Environmental Technology Intitiative (ETI) program (CH2M HILL, 1998). This project created an electronic database of habitat, wildife, human use, and ecological risk data from North American treatment wetlands. This database was appended to the existing NADB (Knight *et al.*, 1993) to form the NADB v. 2.0. This paper summarizes the structure and content of the NADB v 2.0 and describes key findings. It is hoped that this review of existing data might direct limited monitoring resources to improving the understanding of the most significant benefits and risks associated with treatment wetlands.

Inventory of U.S. habitat and human use treatment wetlands

Little effort has been made to collect or organize published and unpublished information concerning the habitat functions of treatment wetlands. New treatment wetland systems are being designed with very little guidance on how to optimize plant diversity and attract wildlife, or whether such habitat creation is even compatible with the goal of protecting wetland biota. While it is generally conceded that treatment wetlands provide habitat for wildlife, the amount and quality of that habitat has not been widely recorded. Moreover, the potential for this habitat to threaten the health of wildlife attracted to treatment wetlands has been described, but the documented occurrence of undesirable side effects has not been thoroughly reviewed.

The ETI Treatment Wetland Habitat Project represents the first effort to summarize the wide-ranging information concerning the habitat and wildlife use data from SF treatment wetlands. These data are assembled in an electronic data base format that allows researchers to take a critical look at the actual benefits and hazards that have been documented in treatment wetlands. This project involved a focused search of project reports and researcher files for qualitative and quantitative information concerning surface flow treatment wetland plant communities, animal populations, concentrations of trace metals and organics, and human use. However, this effort was not comprehensive and some habitat data sets were likely missed.

The primary purpose of the ETI Treatment Wetland Habitat Project was to gather existing wildlife and habitat use data from diverse sources into a consistent format and to make these data available to regulators, designers, owners, and researchers. This paper and the companion report (CH2M HILL, 1998) provide a preliminary summary of these data to begin identifying any apparent benefits or hazards. It is anticipated that other interested researchers will conduct more detailed analyses of these data.

Database structure

Five new database files with data pertinent to habitat quality and wildlife use of treatment wetlands were added to the existing 7 files in NADB v. 1.0 to create the NADB v. 2.0. The structure of the five new database files follows the hierarchical structure of the NADB v. 1.0. Each record identifies the treatment wetland site, system, and cell, allowing linkages between the 12 individual database files. The NADB v. 2.0 also has been updated by adding treatment wetland site, design, and operational performance data from confined animal feeding operations (CAFOs) summarized by a separate project completed for the Gulf of Mexico Program with U.S. EPA funding (CH2M HILL and Payne Engineering, 1997).

The *Vegetation* database file contains qualitative and quantitative plant community data for treatment wetland sites. It provides vegetation data of cells within each system for a specified period. Data include species lists, percent cover, biomass, density, basal area, and importance values.

The *Wildlife* database file contains qualitative and quantitative population data for benthic macroinvertebrates (benthos), fish, amphibians, reptiles, avifauna (birds), and mammals. Data include species lists, species density, species diversity, and reproductive success for a given period.

The *Metals/Organics* database file contains data on water, sediment, plant, and wildlife tissue concentrations for trace metals and organics. Data are identified by the sample matrix type (water, sediment, or tissue) and sample parameter. Water sampling data are recorded as influent and effluent concentrations for each system at a given site. Sediment data are identified by the station location. Plant and wildlife tissue data are identified by species and type of tissue.

The *Biomonitoring* database file includes information on acute and chronic toxicity tests, reproduction, and mortality tests on various test organisms. Each record identifies the sampling location (influent or effluent), dilution for each test, and the organism used. These data are not described further in this paper.

The *Human Use* database file contains information on how the public uses wetland treatment sites for recreation, research, hunting, and other activities. Data include use density, number of use days, and harvest totals per site.

Summary of NADB v. 2.0 contents

Sites and systems. The NADB v. 2.0 has information for a total of 257 sites, 367 systems, and 831 cells from treatment wetlands in North America. These numbers reflect the fact that some sites have multiple systems and some individual systems have multiple cells. Of these 257 sites, 160 sites treat municipal wastewater, 12 receive industrial effluents, 68 receive livestock wastewaters, and 17 receive other wastewater types including stormwaters and river waters. Of the systems described in NADB v. 2.0, 305 are surface flow, 54 are subsurface flow, and 8 are hybrids of these two designs.

The five new files in NADB v. 2.0 contain habitat and related data for 109 SF treatment wetland sites with 168 separate systems and 386 individual cells located in 31 states or provinces. Eighty-five percent of the sites within the five new database files are constructed treatment wetlands; the rest are natural treatment wetlands. Of the 29,960 new records in these five new database files, 65 percent come from constructed SF treatment wetland sites.

SF treatment wetlands as habitat

The word ***habitat*** refers to a place or environment that provides support for the needs of a plant or animal. Many plant species are typically found in wetland environments, including vascular plants, algae, mosses, ferns, and thousands of animal species. To be beneficial, habitats must provide one or more life history requirements that contribute to a species' sustainable population size. If the amount or quality of available habitat is limiting a given species' overall population size, then the addition of more habitat or the enhancement of existing habitat, will lead to a higher sustainable population of that organism.

Vegetation in treatment wetlands

The wetland environment is generally characterized by a high diversity and abundance of plants. In many cases, wetland plant communities include multiple vertical strata ranging from groundcover species to shrubs and sub-canopy trees to canopy tree species. Wetland plant diversity is important in determining wildlife diversity because of the creation of niches associated with differing vegetative structure, reproduction strategies, flowering and seeding phenologies, gross productivity, and rates of decomposition (Mitsch and Gosselink, 1993). In addition to their diversity of species and growth habitats, wetland plants are important for treatment wetland pollutant removal performance because the physical and chemical structure they provide supports microbial populations (Kadlec and Knight, 1996; Vymazal et al., 1997).

Of the more than 800 species of macrophytic plants that have been reported in natural and constructed SF treatment wetlands in North America, 693 species are emergent herbaceous macrophytes, 36 are floating aquatic species, 12 are submerged aquatics, 57 are shrubs, 55 are trees, and 18 are vines. A total of 593 macrophytic plant species has been reported from constructed SF treatment wetlands, and 427 species from natural treatment wetlands. Emergent herbaceous macrophytes account for 501 species in constructed treatment wetlands and 290 species in natural treatment wetlands. A significant variety of tree and shrub species occur in some constructed wetlands. Tree and shrub species are well represented in natural treatment wetlands with 88 different species recorded.

SF constructed treatment wetlands are typically dominated by emergent marsh, floating aquatic plant, or submerged aquatic plant communities. In some cases these constructed treatment wetlands are dominated by populations of filamentous algae because marsh plant species have had difficulty becoming established. Emergent marsh species are frequently intermingled and co-dominant with populations of small floating aquatic plants such as duckweed (*Lemna* spp.). Many constructed treatment marshes in the United States are dominated by cattails (*Typha* spp.) or bulrush (*Scirpus* = *Schenoplectus* spp.); however, some treatment marshes are dominated by other plant species or by complex admixtures that include cattails and bulrush. Natural wetlands used for water quality treatment may be dominated by emergent marsh plant species, by tree species, or by shrub species. Dominant species in natural wetlands used for water quality treatment vary regionally, depending upon the types of wetlands that are locally available.

While most SF constructed treatment wetlands are marshes, a few constructed treatment systems are developing shrub and swamp characteristics over time, either intentionally or through volunteer plant colonization and succession. On the other hand, natural forested wetlands receiving secondary treated municipal wastewaters have been partially converted to marshes in several areas of the United States. Other forested wetlands receiving higher quality municipal wastewaters (advanced secondary with nitrification or tertiary with phosphorus removal) have maintained their canopy dominance over significant periods of time. Water quality effects on plant diversity are greatest near the point of inflow, while water regime effects may occur over the entire area of a natural treatment wetland. Over the scale of the entire wetland, plant diversity may be increased by the addition of new plant species associated with the discharge.

Wetland plant diversity is a poorly understood subject, both in unaffected natural wetlands and in treatment wetlands. Non-treatment natural wetlands are frequently dominated by only a few plant species (for example, cypress swamps, cattail, sedge, or sawgrass marshes, etc.) that are adapted to stressful environmental conditions such as low nutrient levels, low soil oxygen levels, or fluctuating water levels. Other unaffected natural wetlands have higher plant diversity and greater evenness between multiple dominant plant species. Both constructed and natural treatment wetlands cover the same range of plant dominance and diversity of unaffected natural wetlands. Even when SF treatment wetlands are dominated by cattails or bulrush, dozens of other herbaceous and woody plant species are typically present as subdominants.

Total plant cover and dominance data do not indicate any observable difference between treatment and non-treatment wetlands for these indices. However, biomass data indicate that discharge of secondary municipal wastewater to natural, low-nutrient wetlands will greatly increase plant biomass. This enrichment effect is typical of wetlands receiving treated municipal discharges and is most observable in the discharge area.

Data requirements. Very few data have been collected that measure the ecological function of treatment wetland plant communities. High plant growth rates are apparent based on standing crop; however, clip plots, gas metabolism studies, litterfall and decomposition studies, or other methods for estimating net primary production have been conducted at only a few locations. More long-term, ecosystem-level studies are needed for both constructed and natural SF treatment wetlands under a variety of geographical and pollutant loading conditions to fully describe the parameters most predictive of plant community development in treatment wetlands. Also, more studies of the basic quantitative ecology of natural wetlands would be helpful for comparison to treatment wetland structure and function.

Wildlife in treatment wetlands

Over 1,400 species of wildlife have been reported for constructed and natural SF treatment wetlands in the NADB v. 2.0. These include more than 700 species of invertebrates, 78 species of fish, 21 species of amphibians, 31 species of reptiles, 412 species of birds, and 40 species of mammals. Over 800 animal species have been reported in constructed SF treatment wetlands alone. Because species lists have been determined for only a small fraction of the treatment wetland sites listed in NADB v. 2.0, and because of the widely disparate methods and seasons of measurement, these species totals underestimate the diversity that exists in treatment wetlands in North America.

Invertebrates. A total of 824 species of aquatic invertebrates have been recorded from treatment wetlands in NADB v. 2.0. These include 15 species of aschelminthes, 81 species of crustaceans, 12 species of arachnids, 29 species of molluscs, and 589 species of insects. Twenty-three treatment wetland systems listed in NADB v. 2.0 have invertebrate data. In most cases, only species lists are available. A few systems reported quantitative data, although sampling techniques varied. Although a total of 342 species of benthic macroinvertebrates have been reported for constructed treatment wetland sites, the average diversity (H') is low at 1.36 units. The average benthic macroinvertebrate diversity for natural treatment wetlands is 2.29 units with a total of 349 species reported for all sites. These low diversities are typical of unaltered wetland environments due to low ambient dissolved oxygen levels and fluctuating water availability.

Fish. Seventy-eight fish species are reported from 13 treatment wetland sites in the NADB v. 2.0 (64 species from constructed SF treatment wetlands and 24 species from natural treatment wetlands). Mosquitofish (*Gambusia affinis*) were reported from 5 constructed and 4 natural treatment wetlands. This species, found in 69 percent of the treatment wetlands where fish were sampled, is the only species known to be intentionally introduced into these treatment wetlands; all others are apparently present as a result of volunteer colonization.

Amphibians. Twenty-one amphibian species are reported from 6 constructed and 3 natural treatment wetlands in the NADB v. 2.0. Ten species are reported from constructed treatment wetlands and 14 species from natural treatment wetlands. No quantitative data on amphibian populations are included in the database.

Reptiles. Thirty-one reptile species are reported from 5 constructed and 4 natural treatment wetlands in NADB v. 2.0. These species include snakes, alligators, lizards, and turtles. Seven species are reported from constructed treatment wetlands and 28 species from natural sites. No quantitative data on reptile populations are included in the database.

Birds. Bird data are reported for 21 constructed treatment wetland sites and 7 natural treatment wetland sites in the NADB v. 2.0. The majority of these data are species lists and population densities. Very few data on breeding success, nesting, brood production, and mortality rates were found for this review.

A total of 412 bird species are reported from these treatment wetlands. Constructed treatment wetlands are represented by 361 bird species and natural treatment wetlands by 170 bird species. Of the bird species listed, 51 are waterfowl, 23 are wading birds, 24 are terns or gulls, 45 are shorebirds, 29 are raptors or scavengers, 7 are fowl-like, and 235 are passerine or non-passerine land birds. Only 45 percent of the total of 412 species reported from treatment wetlands are commonly considered to be wetland-dependent for some portion of their life history. This finding indicates that a majority of the bird species recorded at these treatment wetland sites are facultative wetland inhabitants.

Migratory waterfowl concentrated in wetlands are particularly susceptible to avian cholera, a highly infectious disease caused by the bacterium *Pasteurella multocida* (Friend, 1987). Avian cholera has been reported at one treatment wetland, the Hayward Marsh on the east shore of San Francisco Bay, south of Oakland, California. Annual episodes of avian cholera have been noted at this site. The average number of infected waterfowl collected during a 6 year period was 127 per year (15 to 340 birds per year). This wetland supports very high waterfowl populations during the fall months, with peak numbers above 30,000 birds per day. Avian cholera is encountered in nearly all wetlands in and around San Francisco Bay and there appears to be no relationship between the avian cholera observed at this location and the source or quality of the water treated at this system.

Mammals. Forty mammal species are recorded in NADB v. 2.0. A total of 22 species are reported from 6 constructed treatment wetland sites and 27 species from 4 natural treatment wetland sites. Quantitative data on mammal populations are limited in the database to small mammal surveys at the constructed treatment wetland in Iron Bridge, Florida (from the downstream Seminole Ranch wetlands that receive the discharge from Iron Bridge) and at the natural treatment marsh in Houghton Lake, Michigan.

Summary and Data Requirements. Qualitative and quantitative studies of animals inhabiting constructed and natural SF treatment wetlands have revealed that these ecosystems provide attractive and productive habitats. All trophic levels are represented, from microscopic invertebrates to macroinvertebrates, fish, herptiles, birds, and mammals. Numbers of species appear to be generally similar between constructed and natural wetland sites. However, insufficient quantitative faunal data currently exist to correlate population diversity or density with treatment wetland design criteria such as pretreatment water quality, mass loading for key pollutants and nutrients, water depth, vegetation types, etc. Essentially all conclusions concerning relationships between wildlife populations and wetland design must be based on non-treatment wetland studies or are currently anecdotal. This lack of correlative power emphasizes the need for well designed, quantitative studies of wildlife populations conducted in the context of controlled treatment wetland research projects.

Toxic metals and trace organics in treatment wetlands

A variety of data for metals and trace organic compounds have been collected from 26 wetland treatment wetland sites. Many data records for metal and trace organic compound concentrations are below detection limits (BDL) in the raw data in the NADB v. 2.0.

Available data for 25 metals and related elements measured in surface waters, sediments, and biological tissues from treatment wetlands are summarized in the NADB v. 2.0. These data confirm numerous published reports that treatment wetlands reduce surface water concentrations of metals. They also provide a basis for comparing treatment wetland sediment and tissue metals data to published criteria that are considered to be protective of environmental health. Mean treatment wetland metal concentrations in the NADB v. 2.0 for arsenic, cadmium, chromium, copper, lead, mercury, nickel, selenium, silver, and zinc are less than ambient water quality criteria; however, maximum recorded values are above criteria.

Data for more than 120 trace organic compounds are reported from treatment wetlands. Detectable levels for some of these trace organics were found in treatment wetland surface waters, sediments, and in biological tissues. A total of 29 compounds were detected in constructed treatment wetland sediments.

An important issue needing to be scrutinized is the extent to which these potentially toxic chemicals bioaccumulate and whether they are present in amounts that are in fact toxic to the biota that normally inhabit these wetland environments. Most criteria are based on laboratory tests using sensitive species. Comparisons of treatment wetland trace metal and organics concentrations to published criteria should be tempered by research that demonstrates that criteria levels actually create effects in wetland environments. The data summarized in the NADB v. 2.0 provide a basis from which to begin finding answers to these questions. However, additional data from controlled, realistic-scale treatment wetland research will need to be collected and analyzed to fully evaluate treatment performance and the potential for detrimental effects from each metal or organic compound of interest.

Human use of treatment wetlands

Recognized human uses of treatment wetlands in addition to water purification include: nature study, exercise activities, recreational harvest, and public education.

Summaries of human use data exist for only a few treatment wetland systems. The Arcata, California, constructed wetland is used by an estimated 100,000 visitors per year (Benjamin, 1993). This level of activity is sustained because the system is located in a progressive, coastal California community near a trail system and park-like setting. Data from Arcata summarized in NADB v. 2.0 indicate that from 27,000 to 64,000 human use-days per year (HUD/y) are devoted to general picnicking and relaxing. These data may also be expressed on a unit area basis as a total of about 1,600 HUD per hectare per year (HUD/ha/y) for the entire Arcata Marsh and Wildlife Sanctuary. At the Show Low, Arizona, constructed treatment wetland, human use data are lumped for all categories and averaged about 370 HUD/y or about 7 HUD/ha/yr. The Iron Bridge, Florida, constructed wetland has an overall estimated human use of about 4,800 HUD/y or about 10 HUD/ha/y.

Nature study includes a variety of activities that may be associated with treatment wetland projects: bird study, plant observation and identification, observation and identification of other wildlife groups, plant and wildlife photography, and plant and wildlife art. Few data are available that specifically describe any of these activities. Arcata, California, has reported data indicating about 10,000 HUD/y or 165 HUD/ha/y for bird watching. Photography and art account for about 360 to 900 HUD/y at Arcata. Anecdotal information is available that indicates that bird-watching groups regularly use treatment wetlands at West Jackson County, Mississippi; Hillsboro, Oregon; Show Low, Arizona; Pinetop-Lakeside, Arizona; Lakeland, Florida; and Iron Bridge, Florida. Some of these sites are visited by organized groups on a regular basis (once a week or month), while others are visited by individuals or groups on a less regular schedule.

When treatment wetlands are open to the public, they are frequently used for activities that provide exercise. Forms of exercise known to occur in treatment wetlands include hiking, jogging, and off-road bicycling. Treatment wetland sites that are open to the general public for these activities include Show Low, Arizona; Pinetop-Lakeside, Arizona; Tres Rios, Arizona; Arcata, California; Sea Pines, South Carolina; Iron Bridge, Florida; Cannon Beach, Oregon; Hillsboro, Oregon; and Mountain View, California. Hiking and jogging at the Arcata, California, constructed wetland is estimated as about 18,000 HUD/y.

A small number of treatment wetlands are open to the public or to private individuals for hunting and/or fishing. A borrow pit at the Arcata Marsh and Wildlife Sanctuary in California is open for fishing, but use is reported to be light and seasonal. At Incline Village, Nevada, duck blinds are available on a lottery basis. Typical hunter use days are about 877 HUD/y or 5.6 HUD/ha/y. About 817 ducks and 60 geese are harvested per year. The Iron Bridge, Florida, constructed wetland is closed to the public from September through March of each year and is available to former land-owners for waterfowl hunting and fishing during this period. The Houghton Lake, Michigan natural treatment wetland; the Show Low, Arizona, constructed wetland; and the area downstream of the Columbia, Missouri, treatment wetland are open to hunters as state-controlled wildlife management areas. About 836 HUD/y or 1.6 HUD/ha/y are available for duck hunting at the Columbia, Missouri, site.

Treatment wetlands have been used for a variety of educational opportunities. Some sites are open for controlled access of grade school and high school students and for various college classes and individual undergraduate and graduate research. Miscellaneous activities that have been observed include: school projects to name constructed wetlands, community service outings to help plant new constructed wetlands, clear trash, and install bird and bat houses, Boy Scout projects to build public use facilities, and citizen groups and government officials meeting to review wastewater management options. These activities are known to exist but have been difficult to quantify. Additional data on human use in treatment wetlands are needed to determine the significance of these activities and to provide information to designers on how to provide the best opportunities for cost-effective use.

Summary and conclusions

The ETI Treatment Wetland Habitat Project is the first attempt to provide a comprehensive summary of the state of our knowledge concerning the relationship between treatment wetlands and their interaction with wildlife and human use. While this summary indicates significant areas of incomplete understanding, it also provides a clearer view of those areas where conclusions are warranted.

The information summarized in the NADB v. 2.0 indicates that treatment wetlands typically have the following properties.
- Their biological structure is substantial and is dominated by relatively diverse assemblages of wetland plant species, typically including a few dominants and many less common species that have specific adaptations to grow in saturated soils.
- All major animal groups and trophic levels that occur in natural wetlands are represented in treatment wetlands; population size and diversity in treatment wetlands are generally as high or higher as in other wetlands; no documented occurrences of detrimental effects to wildlife caused by the pollutant-cleansing function of treatment wetlands were noted.
- Contaminant data from treatment wetlands for heavy metals and trace organics are available for sediments and biological tissues; treatment wetlands are effective at

reducing concentrations of these pollutants; these data do not generally indicate a threat to flora and fauna based on the existing range of contaminant loadings.
- Humans are using treatment wetlands for a variety of purposes in addition to water quality enhancement.

Contaminants in wastewaters are known to affect the wetland environment. These effects are highly variable depending on the specific constituents and the biological components of the wetland in question. Research efforts should be designed to correlate these water quality conditions with treatment wetland environmental conditions. The most basic comparisons have not been made between treatment wetlands with varying dissolved oxygen and nutrient conditions and their ability to support diverse plant and animal populations. Although pH requirements for some individual plant and animal species are known, there are no studies of the effect of varying pH in treatment wetlands. Although the toxicity of many trace metals and organics are known in laboratory studies with one or a few plant or animal species, there is very little information on the ecosystem-level effects of these substances in treatment wetlands. The information collected for this ETI Habitat Project only provides a starting point for the studies needed to develop empirically based treatment/habitat wetland design criteria.

During the review of new and existing discharge permits to treatment wetlands, environmental agency personnel are frequently faced with the difficulty of assessing the potential for harmful environmental effects. The potential receptors of most interest are typically the vertebrate inhabitants of the wetlands including fish, amphibians, reptiles, birds, and to a lesser extent, mammals. These organisms tend to be more highly visible to people than their invertebrate neighbors, and concern for their fate is highest in the public's priorities. While it is recognized that the livelihood of invertebrates is also of importance, their protection is generally justified based on their place in the food chain supporting the vertebrate forms.

While the use of wetlands to improve the quality of wastewaters is considered an important goal, it is also important to balance the benefits of meeting that goal with the avoidance of harm to those organisms that will ultimately reside in the living treatment system. The information gathered for the ETI Treatment Wetland Habitat Project indicates that biological changes can occur in response to discharges of treated effluents. These changes cover the spectrum from obvious to subtle. Many of the changes that have been noted favor one group of species over another. The most common changes result in an increase of wetland structure and function at an ecosystem level.

There is currently no evidence that treated wastewater effluents cause increased risks for vertebrates in treatment wetlands. This lack of evidence does not prove that there are no effects, but it indicates that most treatment wetland projects can be permitted without special requirements other than reasonable caution. Greater caution should be exercised when project wastewaters are known or suspected to contain unusually elevated concentrations of heavy metals, trace organics, un-ionized ammonia, or other chemicals that are likely to be acutely or chronically toxic to aquatic and wetland biota. These potentially toxic chemicals are only of special interest when they are at concentrations above the range typical of normal wastewaters from the same general source.

Very little information is available about how to best integrate human use with treatment wetlands. Benjamin (1993) represents a highly useful summary of the issues related to public perception and use of the most-visited treatment wetland in the United States, the Arcata Marsh and Wildlife Sanctuary in California. That study concluded that the Arcata Marsh is a great success in its role as a community open space and as a recreational, ecological, and educational resource. Interviews identified birds and wildlife viewing as the most popular public uses of the marsh. The second most common response to questions about the benefits of the marsh focused on its aesthetic qualities, including scenery, beauty, and open space. The most common response to the survey question concerning what the public disliked about the Arcata Marsh was "nothing." These obvious benefits are being accomplished even as the Arcata Marsh meets its primary goal of water quality protection. Studies similar to the one conducted by Benjamin (1993) should be conducted at a number of treatment wetlands that are open to the public to develop wider guidance for how humans interact with wetlands.

This project represents a first step at collecting and summarizing the information on habitat and wildlife use of treatment wetlands. The number of new treatment wetland projects gathering and reporting these types of data is increasing yearly. Because of the schedule of this project, these various works-in-progress could not be fully documented. While sufficient data about the diversity of plants and animals inhabiting treatment wetlands are available, it is hoped that these new and ongoing studies will shed greater light on the ecological functions of these systems and their full potential for environmental benefit or harm.

As data concerning each of these items continue to become more available, the next step is to apply this information to the design and operation of new and existing treatment wetlands. New projects are certain to benefit from an expanding information base relating wetland design to ecological function.

References

Bastian, R.K., Shanaghan, P.E., and Thompson, B.P. (1989). Use of Wetlands for Municipal Wastewater Treatment and Disposal-Regulatory Issues and EPA Policies. Chapter 22, pp. 265–278. In D.A. Hammer (ed.), *Constructed Wetlands for Wastewater Treatment: Municipal, Industrial, and Agricultural*. Lewis Publishers, Chelsea, Michigan.

Benjamin, T.S. (1993). *Alternative Wastewater Treatment Methods as Community Resources: The Arcata Marsh and Beyond*. Master of Landscape Architecture Thesis, University of California at Berkeley.

CH2M HILL (1998). Treatment Wetland Habitat and Wildlife Use Assessment Project. Report prepared for the U.S. Environmental Protection Agency, U.S. Bureau of Reclamation, and City of Phoenix with funding from the Environmental Technology Initiative Program. Available on CD-rom from Ron Clarke at CH2M HILL (352)-335-7991 (USA).

CH2M HILL and Payne Engineering (1997). *Constructed Wetlands for Livestock Wastewater Management. Literature Review, Database, and Research Synthesis*. Report Prepared for the Gulf of Mexico Program, Nutrient Enrichment Committee, Stennis Space Center, MS.

Freierabend, J.S. (1989). Wetlands: The Lifeblood of Wildlife Chapter 7, pp. 107–118. In D.A. Hammer (ed.), *Constructed Wetlands for Wastewater Treatment: Municipal, Industrial, and Agricultural*. Chelsea, Michigan: Lewis Publishers.

Friend, M. (1987). (ed.) *Field Guide to Wildlife Diseases. Volume 1. General Field Procedures and Diseases of Migratory birds*. U.S. Department of the Interior, Fish and Wildlife Service. National Wildlife Health Center, Madison, Wisconsin.

Guntenspergen, G.R. and Stearns, F. (1985). Ecological Perspectives on Wetland Systems. Chapter 5, pp. 69–97 in: P.J. Godfrey, E.R. Kaynor, S. Pelczanski and J. Benforado (eds.). *Ecological Considerations in Wetlands Treatment of Municipal Wastewaters*. Van Nostrand Reinhold, New York.

Kadlec, R.H. and Knight, R.L. (1996). *Treatment Wetlands*. Lewis Publishers, Boca Raton, FL, 893 pp.

Knight, R.L. (1992). Ancillary Benefits and Potential Problems With the Use of Wetlands for Nonpoint Source Pollution Control. *Ecological Engineering* 1, 97–113.

Knight, R.L. (1997). Wildlife Habitat and Public Use Benefits of Treatment Wetlands. *Water Science and Technology* 35(5), 35–43.

Knight, R.L., Ruble R.W., Kadlec, R.H. and Reed, S. (1993). Wetlands for Wastewater Treatment: Performance Database. Chapter 4, pp. 35–58 in: G.A. Moshiri (ed.). *Constructed Wetlands for Water Quality Improvement*. Lewis Publishers, Boca Raton, FL.

McAllister, L.S. (1992). *Habitat Quality Assessment of Two Wetland Treatment Systems in Mississippi: A Pilot Study*. U.S. Environmental Protection Agency. Environmental Research Laboratory, Corvallis, Oregon. November 1992. EPA/600/R-92/229.

McAllister, L.S. (1993a). *Habitat Quality Assessment of Two Wetland Treatment Systems in the Arid West:*

A Pilot Study. U.S. Environmental Protection Agency. Environmental Research Laboratory, Corvallis, Oregon. July 1993. EPA/600/R-93/117.

McAllister, L.S. (1993b). *Habitat Quality Assessment of Two Wetland Treatment Systems in Florida: A Pilot Study.* U.S. Environmental Protection Agency. Environmental Research Laboratory, Corvallis, Oregon. November 1993. EPA/600/R-93/222.

Mitsch, W.J. and Gosselink, J.G. (1993). *Wetlands.* Second Edition. Van Nostrand Reinhold, New York. 722 pp.

Sather, J.H. (1989). Ancillary Benefits of Wetlands Constructed Primarily for Wastewater Treatment. Chapter 28a, pp. 353–358 in: D.A. Hammer (ed.) *Constructed Wetlands for Wastewater Treatment, Municipal, Industrial, and Agricultural.* Lewis Publishers, Chelsea, MI.

U.S. Environmental Protection Agency (U.S. EPA) (1993). *Constructed Wetlands for Wastewater Treatment and Wildlife Habitat – 17 Case Studies.* EPA832-R-93-005, 174 pp.

Wengrzynek, R.J. and Terrell, C.R. (1990). Using Constructed Wetlands to Control Agricultural Nonpoint Source Pollution. In: P.F. Cooper and B.C. Findlater (eds.). Preprints of the International Conference on Use of Constructed Wetlands in Water Pollution Control. Cambridge, United Kingdom.

Wilhelm, M., Lawry, S.R. and Hardy, D.D. (1989). Creation and Management of Wetlands Using Municipal Wastewater in Northern Arizona: A Status Report. Chapter 13a, pp. 179–185 in D.A. Hammer (ed.), *Constructed Wetlands for Wastewater Treatment: Municipal, Industrial, and Agricultural.* Lewis Publishers, Chelsea, MI.

Wren, C.D., Bishop, C.A., Stewart, D.L. and Barrett, G.C. (1997). *Wildlife and Contaminants in Constructed Wetlands and Stormwater Ponds: Current State of Knowledge and Protocols for Monitoring Contaminant Levels and Effects in Wildlife.* Canadian Wildlife Service Ontario Region, Technical Report Series Number 269.

The use of macrophyte-based systems for phosphorus removal: an overview of 25 years of research and operational results in Florida

T.A. DeBusk*, F.E. Dierberg* and K.R. Reddy**

* DB Environmental Laboratories, Inc., 414 Richard Rd. Suite 1, Rockledge, FL 32955, USA
** Department of Soil and Water Sciences, University of Florida, Gainesville, FL 32611, USA

Abstract Phosphorus (P) removal from wastewaters and surface runoff using macrophyte-based systems (MBS) has been a topic of great interest in Florida for over 25 years. During this period, P removal by both treatment wetlands and floating aquatic macrophyte systems has been evaluated from both a research and operational standpoint. Several factors have contributed to the increased focus on the use of MBS for P removal. First, there exist no conventional technologies that can cost-effectively achieve the low outflow P concentrations required to protect the integrity of Florida's relatively pristine surface waters. Second, because MBSs typically provide some water storage, they can accommodate the wide ranges of flows typical for runoff sources such as agricultural drainage waters. Finally, many regions in Florida have sufficient area for deployment of the relatively land-intensive MBS technologies.

The first P removal work in Florida was initiated in the mid-1970s, and involved pilot-scale research on domestic wastewater treatment by natural wetlands. Parallel studies were performed with managed (periodically harvested) floating plant systems (i.e., *Eichhornia crassipes*) for tertiary treatment. Since that time, the range of operational systems that have been deployed include emergent macrophyte-based and forested wetlands, managed floating plant systems, and submerged macrophyte-based systems. Waters treated by MBS include domestic effluents, agricultural runoff and eutrophic lake waters. Phosphorus removal targets for MBS in Florida have been as low as 10 μg/L. In this paper, we summarize research and operational results for MBS in Florida over the past 25 years.

Keywords Aquatic macrophytes; phosphorus removal; treatment wetlands

Introduction

The state of Florida (USA) has numerous surface waters, including streams, estuaries and lakes, that have been degraded in the past five decades by phosphorus (P) loading from treated wastewater and urban and agricultural runoff. Since the mid-1970s, interest in Florida has increased concerning the use of aquatic plants and treatment wetlands for controlling P discharges. Interest in such "macrophyte-based system" (MBS) treatment technologies has been based in part on the state's warm climate, high availability of land, and preponderance of natural wetlands. As a consequence, Florida has a long and extensive history on the research and operation of MBS.

During the three decades in which MBS have been studied and utilized in Florida, the characteristics and treatment goals of these systems have evolved in response to three factors: increasingly stringent point source discharge requirements; improvements in the ability of conventional wastewater treatment technologies to achieve low effluent P levels; and, an increasing interest in treating non-point discharges, such as agricultural runoff, as well as eutrophic lake waters. In this document, we provide an overview of the key MBS research and operational systems for P removal in Florida.

Phosphorus removal from domestic wastewaters

Environmental regulations governing the discharge of treated domestic wastewaters have changed markedly in Florida during the past three decades. Prior to the 1970s, it was an accepted practice for many small communities to discharge secondary-treated, and at times, primary-treated effluent, into adjacent natural wetlands (Boyt et al., 1977). Since that time, effluent quality requirements have become much more stringent, and in many regions, surface discharge of treated domestic effluent has been completely eliminated.

Some of the wetlands originally used by small communities for wastewater effluent disposal have provided useful sites for investigating the fate of wastewater-borne P. In the City of Waldo, primary treated septic tank effluent was discharged for over 45 years into a cypress strand. Studies on a 2.6 ha portion of this strand demonstrated average surface water total P reductions from 3.9 to 1.3 mg/L during passage through the strand (Table 1) (Nessel and Bayley, 1984). Surface outflow volumes were 74% of surface water inflows, indicating a loss of water to evapotranspiration and seepage. The soil and belowground plant tissues were found to be the dominant P sinks in the system, with mean surface water concentrations being reduced from 1.53 mgP/L to 0.35 mgP/L during infiltration through the soil profile. Surface P export was calculated to be 2.1 g P/m^2-yr.

During the late 1970s, University of Florida scientists performed an exhaustive study on the feasibility of using cypress domes to treat wastewater effluent. Unlike the cypress strand at Waldo, cypress domes are hydraulically isolated, and at low hydraulic loading rates (HLRs) provide little surface outflow. In a cypress dome that received treated effluent (ca. 8.50 mgP/L) at hydraulic and P loading rates of 0.4 cm/day and 11.23 gP/m^2-yr, the bulk (91%) of the P was found to be sequestered in the soils (Table 1). Soil P removal (10.33 gP/m^2-yr) was attributed to storage in roots and adsorption by organic matter and clays (Dierberg and Brezonik, 1983a). Relative to control cypress domes that did not receive treated effluent, the water quality interior to the "treatment" domes was degraded, but during the five-year study the underlying organic soils and clay sands effectively protected the shallow aquifer from contamination (Dierberg and Brezonik, 1983b).

An herbaceous wetland in central Florida also was studied during the late 1970s for its P assimilation capability. Four 0.2 ha plots were established within a 32 ha freshwater marsh adjacent to the City of Clermont's wastewater treatment plant (Table 1) (Dolan et al., 1981). The marsh was dominated by the emergent macrophytes *Sagittaria lancifolia, Pontederia cordata, Panicum* spp. and *Hibiscus* spp. The marsh plots were fed secondary effluent (8.88 mgP/L) at HLRs of 0.18, 0.54 and 1.5 cm/day, which provided P loadings ranging from 5.9 to 45.9 gP/m^2-yr. Phosphorus budgets calculated from the highest loading rate treatment demonstrated that 69.2% of the loaded P was stored in the soil complex, 23.2% was stored in below-ground biomass, and 5.2% in aboveground biomass. Similar to the cypress dome investigations, most of the outflow from these herbaceous wetlands during the 11-month experimental period was through infiltration. These investigators speculated that the P adsorption capability of the soils, coupled with the peat production rate, would ultimately control the ability of the wetlands to treat P.

Table 1 Operational and performance characteristics of selected natural and constructed wetlands for treating domestic wastewaters in Florida

	Wetland size (ha)	HLR[a] (cm/day)	Influent [TP] (mg/L)	Effluent [TP] (mg/L)	P loading (gP/m^2-yr)
Waldo	2.6	0.6	3.9	1.3	4.3
Cypress domes	1.1	0.4	8.5	NA	11.2
Clermont	0.2	1.5	8.88	NA	45.9
ESA[b]	134	0.4	0.27	0.13	0.65
Easterly[c]	494	1.0	0.646	0.086	2.2

a HLR = Hydraulic loading rate
b Eastern Service Area (ESA) performance data represent first three years of operation
c Easterly data represent first two years of operation

Not all wetlands in Florida used for wastewater treatment have provided consistent P removal. A mixed cypress – hardwood forest was operated from 1978 – 1989 by the Reedy Creek Improvement District in Orange Co., FL, for nutrient removal from secondary treated effluent. This 34 ha system received relatively high hydraulic (1.6–4.9 cm/day) and mass P loadings (ca. 17 gP/m^2-yr). Despite excellent N removal performance (average reduction from 8.6 to 1.9 mgN/L), only occasional reductions from the mean inflow P levels of 1.4 mg/L were observed. Prominent internal hydraulic short-circuits, coupled with poor adsorption capabilities of the organic soils likely contributed to the poor P removal in this wetland (DeBusk, Merrick and Reddy, unpublished data).

During the 1970s, the proliferation of aquatic weeds, such as water hyacinth (*Eichhornia crassipes*), in Florida's lakes and rivers, captured the interest of many scientists and entrepreneurs, who realized that the periodic harvest of productive macrophytes could provide high levels of wastewater P removal. Extensive research on the productivity and nutrient removal performance of aquatic macrophytes also was performed at this time because of an interest in using productive aquatic plants as a feedstock for methane production (Reddy *et al.*, 1983). Most studies demonstrated that the floating water hyacinth was the most productive aquatic species (Reddy *et al.*, 1983; Reddy and DeBusk, 1985), capable of sustained biomass production rates of 42 g dry wt./m^2-day and P assimilation rates of 135 gP/m^2-yr under optimum conditions. The macrophyte pennywort (*Hydrocotyle umbellata*) was found to be an effective counterpart to water hyacinth for winter season P removal (Clough *et al.*, 1987). Much of the MBS research performed in Florida during this period was summarized in Reddy and Smith (1987).

From 1978 through 1987, several operational water hyacinth-based treatment systems were deployed in central and south Florida. These systems ranged in size from 0.5 ha to 12.2 ha, and in most cases, were established for purposes other than effluent P control, such as suspended solids and nitrogen removal. Nevertheless, P removal performance of most water hyacinth systems was documented, and P mass removal rates ranging from 11.3 to 51.9 gP/m^2-yr were reported (Table 2) (DeBusk and Reddy, 1989; Stewart *et al.*, 1987). Outflow P concentrations as low as 0.1 mgP/L were achieved, with systems that received either low inflow TP concentrations or a low TP mass loading providing lowest outflow TP concentrations (Table 2).

At the Coral Springs water hyacinth-based wastewater treatment facility, an experiment was conducted to assess the role of plant harvest in wastewater P removal. During a five month period, P removal calculated from inflow – outflow concentrations for ponds (0.32 ha total area) from which water hyacinths were routinely harvested averaged 29.9 gP/m^2-yr. An adjacent 0.18 ha pond from which plants were not harvested during this period

Table 2 Operational and performance characteristics of water hyacinth systems used for treating domestic wastewaters in Florida

System	Size (ha)	HLR (cm/day)	Influent [TP] (mg/L)	Effluent [TP] (mg/L)	P loading (g/m^2-yr)	P removal (g/m^2-yr)
Coral Springs	0.5	6.8	4.68	4.23	116	11.3
Melbourne	4.9	22.4	4.33	3.70	354	51.5
Kissimmee	1.5	3.9	1.46	0.27	20.8	16.9
Loxahatchee	3.4	27.9	1.06	0.55	108	51.9
Iron Bridge[1]	12.2	20.8	0.74	0.35	56.2	29.6
Iron Bridge[2]	12.2	24.6	0.30	0.10	26.9	18.0

Operational periods for water hyacinth systems were as follows:
Coral Springs, March–August 1981; Melbourne, November 1985–July 1986; Kissimmee, March 1985–March 1986; Loxahatchee, April 1985–April 1986; Iron Bridge[1], July 1985–July 1986; Iron Bridge[2], April 1987–March 1988

provided a lower P removal rate of 6.6 gP/m^2-yr (DeBusk et al., 1983). In the late 1980s, water hyacinth-based systems fell out of favor as a MBS technology for wastewater treatment. Large monocultures of water hyacinths proved susceptible to pest damage, which reduced treatment performance (Stewart et al., 1987). In addition, water hyacinth harvesting often required specialized equipment, and the harvested biomass had little product value. Finally, the ability of free water surface (FWS) constructed wetlands to provide effective nutrient removal (in particular, N removal via denitrification) without plant harvest made these wetlands more attractive nutrient control candidates than the harvested water hyacinth systems.

With the exception of a few small, on-site systems, subsurface flow (SSF) wetlands have not been used for domestic wastewater treatment in Florida, primarily because the subtropical climate is amenable to good year-round performance by free water surface (FWS) wetlands. Some small-scale investigations on P removal by SSF wetlands have, however, been performed. In a summer season mesocosm study, the P removal performance of several SSF wetlands was compared to that of a floating plant (pennywort) system (DeBusk et al., 1990). These outdoor mesocosms received a HLR of 10 cm/day of secondary domestic effluent, which provided a P mass loading of 347 gP/m^2-yr. From April–August 1989, non-vegetated, horizontal flow SSF wetlands reduced influent P concentrations of 9.48 mg/L to 9.02 mg/L. Vegetated SSF wetlands containing *Sagittaria latifolia* were more effective, providing outflow TP concentrations of 8.11 mg/L. This system was comparable in performance to a floating plant system containing *Hydrocotyle umbellata*, which provided an outflow TP concentration of 8.30 mg/L (DeBusk et al., 1990).

During the mid-1980s, more stringent nutrient concentration discharge requirements for domestic wastewaters, coupled with the deployment of "biological nutrient removal" wastewater treatment plants (WWTP) that could produce an advanced quality effluent, led to the construction of several extremely large FWS wetlands designed for effluent "polishing". During the first three years of operation, the 134 ha Eastern Service Area (ESA) wetland in Orlando received advanced treated domestic effluent averaging 0.27 mgP/L, and produced an outflow of 0.13 mgP/L (Table 1), thereby meeting the permitted discharge concentration of 0.20 mgP/L (Schwartz et al., 1994). Phosphorus loading to this system from March 1988 through January 1991, which consisted of a sequence of FWS wetlands constructed on mineral soils and natural, forested wetlands on organic soils, averaged 0.65 gP/m^2-yr. The mass P removal rate during this period was 0.43 gP/m^2-yr (Kadlec and Newman, 1992). Field and laboratory studies revealed that the mineral wetland soils provided slightly greater P retention than did the organic soils, and that highest P retention was achieved under oxic conditions. Soil analyses suggested that P chemistry in the natural and constructed wetlands is controlled by the organic P pool and the iron/aluminium bound fraction (Gale et al., 1994).

Operation of the 494 ha Easterly wetland, which receives effluent from the Orlando Iron Bridge WWTP, was initiated at about the same time as the ESA system. The FWS Easterly wetland is compartmentalized into several cells, with a wetland sequence of deep marsh, shallow mixed marsh, hardwood swamp, and lake (Swindell and Jackson, 1990). During the first two years of operation, this system received advanced treated effluent (0.646 mgP/L) at a HLR of 1 cm/day, and P loading of 2.2 gP/m^2-yr (Table 1). Outflow concentrations averaged 0.086 mgP/L, below the discharge permit limit of 0.20 mgP/L. An analysis of internal water quality samples during 1988 and 1989 showed that 98% of the observed P removal occurred within the first 11% of the system (Swindell and Jackson, 1990). This rapid P concentration reduction in the inflow region suggests that the wetland's design was somewhat conservative, and also that there is a lower limit, probably in the range of 0.050

to 0.100 mgP/L, to water column P levels for the ecological communities of this wetland. From 1990 through 1999, inflow and outflow TP concentrations to the Easterly wetland averaged 0.190 and 0.050 mg/L, respectively.

At present, there are approximately 20 permitted FWS wetlands in Florida that provide nutrient removal from domestic wastewater, with the ESA, Easterly and City of Lakeland systems (566 ha) being by far the largest (DeBusk and Krottje, 1996).

Phosphorus removal from eutrophic lake waters

Florida has numerous lakes that have been adversely impacted by point and non-point source discharges. Water quality in Lake Apopka, one of Florida's largest lakes at 125 km^2, has gradually deteriorated since the 1950s due to nutrient loadings from agricultural runoff and domestic wastewater. The ten year (1987–1997) average water column chlorophyll a, total P and total N concentrations for this lake were 0.092, 0.204 and 5.14 mg/L, respectively (Battoe et al., 1999).

In the early 1980s, studies on using MBS to directly remove P from Lake Apopka waters were initiated in an agricultural area adjacent to the lake (Fisher and Reddy, 1987). Three 372 m^2 raceways were each fed 37 cm/day of lake water, with average total P concentration of 0.29 mg/L, most of which was in a particulate organic form (phytoplankton cells). Soluble reactive P concentrations therefore comprised only a small fraction of the inflow P, averaging 0.020 mg/L. One raceway contained water hyacinths that were periodically harvested, a second contained non-harvested water hyacinths, and the third was a non-macrophyte control. Over 18 months (June 1984 – December 1985), the "harvested" water hyacinth raceway provided 63% P removal (24.6 gP/m^2-yr), the non-harvested raceway provided 57% P removal (22.2 gP/m^2-yr), and the control raceway removed 42% of the influent P (16.4 gP/m^2-yr). Plant assimilation comprised only 22% (non-harvested system) to 40% (harvested system) of the observed P removal, so sediment accumulation of P (primarily as a settled phytoplankton floc) was a prominent P sink (Fisher and Reddy, 1987).

Water managers in the late 1980s constructed a 200 ha "demonstration" marsh designed to filter the P-laden phytoplankton from Lake Apopka waters. This wetland was created from previously farmed fields that contain Histosol soils. Lake water is fed by gravity into the wetland inflow region, and pumped back into the lake from the wetland outflow region. Operation of a wetland that receives such a high particulate load presents unique challenges. As expected, a floc sediment of low bulk density (0.024 g/cm^3) accumulated in the wetland, particularly near the inflow region. The organic soils also released SRP upon flooding, although this release rate declined from 2.4 mgP/m^2-day three months after flooding, to 0.8 mgP/m^2-day ten months later (D'Angelo and Reddy, 1994).

The Lake Apopka demonstration marsh has been subjected to a range of operational regimes, including drydown. This varied operational protocol, coupled with the dynamic nature (and cycling) of P within this wetland, has made it difficult to define "steady-state" performance. However, Coveney et al. (2000) provide an analysis of marsh performance data from December 1991 through October 1992. During this period, the wetland received a hydraulic loading of 1.8 to 11.5 cm/day, which provided a wetland hydraulic retention time of between 5.6 and 27 days. The inflow (lake water) contained 0.12 – 0.23 mgP/L, with particulate organic P comprising 86 to 98% of TP. Mass P removal, based both on measurements and modeling results, was projected to be about 4 gP/m^2-yr (Coveney et al., 2000).

Phosphorus removal from agricultural drainage waters

Experimentation and applied work on the use of MBS for P removal from agricultural drainage waters (ADW) has been performed in three regions in Florida: the vegetable farms north of Lakes Apopka and the agricultural areas both north and south of Lake Okeechobee. The earliest work was performed in the late 1970s using a series of reservoirs and flooded fields to remove P from vegetable farm ADWs (Reddy et al., 1982; Reddy, 1983). In one experiment, two 0.12 ha sequential reservoirs, the first containing water hyacinths, and the second, the submerged macrophyte elodea (*Egeria densa*), were fed agricultural drainage waters at a HLR of 11.7 cm/day. The water hyacinth system removed 9.8% of the P load, reducing inflow total P concentrations from 0.66 to 0.60 mg/L. Mass loading to this system was 56.6 gP/m^2-yr, and the mass removal rate by the water hyacinths was 5.5 gP/m^2-yr. By contrast, the downstream elodea-dominated reservoir reduced total P concentrations from 0.595 to 0.366 mg/L. The P loading to this submerged macrophyte-dominated reservoir was 51.5 gP/m^2-yr, and the mass removal rate was 19.6 gP/m^2-yr. The alkalinity of the ADW was high, at 280 mg/L as $CaCO_3$. Because 70 to 80% of the influent P in the ADW was soluble reactive P, the investigators proposed that the high P removal rate by the submerged macrophyte-dominated reservoir was related to the co-precipitation of P with $CaCO_3$ (Reddy et al., 1982). An adjacent, shallow (0.2m) flooded field (0.37 ha) stocked with cattail (*Typha latifolia*) received the same drainage water at a HLR of 3.6 cm/day (P loading of 8.6 gP/m^2-yr). The inflow P concentration of 0.66 mg/L was reduced to 0.56 mg/L, representing a P mass removal rate of 1.35 gP/m^2-yr.

At a dairy north of Lake Okeechobee, mesocosm studies were performed to evaluate the effectiveness of floating and emergent macrophytes for removing P from dairy lagoon waste waters (DeBusk et al., 1996). One mesocosm experiment compared the P removal performance of water hyacinth with common duckweed (*Lemna obscura*) during both June and February. In batch mesocosms operated at a 7 day HRT, water hyacinth reduced inflow total P levels from 7.3 mg/L to 0.2 mg/L, whereas duckweed reduced P concentrations to only 2.4 mg/L. Based on plant assimilation of P during both June and February experiments, mass removal by water hyacinth and duckweed averaged 47 gP/m^2-yr and 7.3 gP/m^2-yr, respectively (DeBusk et al., 1996).

Additional experiments were conducted at this site to determine the feasibility of culturing and harvesting a commercially valuable emergent macrophyte as a sustainable means of P removal (DeBusk et al., 1995). Pickerelweed (*Pontederia cordata*), an aquatic macrophyte used for aquascaping, was cultured from small seedlings to a commercially useful size. Dairy lagoon wastewater (1.7 mgP/L) was provided to small mesocosms at a HLR of 42 cm/day. Pickerelweed grew poorly during the winter, so nine months was determined to be the maximum growing season attainable in S. Florida. Mesocosm data projected a mass P removal rate of 7 gP/m^2-yr for an operational pickerelweed "nursery" facility (DeBusk et al., 1995).

The largest treatment wetlands in Florida, the Stormwater Treatment Areas (STAs), are currently nearing completion in S. Florida. These six FWS wetlands, ranging in size from 1,420 to 4,700 ha, are designed to intercept ADW from the 285,000 ha Everglades Agricultural Area prior to discharge into the Everglades. These wetlands originally were designed to treat P levels in ADW (typically 0.100–0.200 mg/L) to 0.050 mg/L (Walker, 1995). The STAs are sized to receive an average flow of 2.7 cm/day, and an average P loading of 1.43 gP/m^2-yr. Since the 1994 startup of the Everglades Nutrient Removal (ENR) Project, a 1540 ha "demonstration-scale" STA wetland (Guardo et al., 1995), data collected by the South Florida Water Management District and contractors has demonstrated that water column concentrations substantially lower than 0.050 mg/L can be achieved by certain plant communities. Portions of the ENR that have been dominated by submerged vegetation (*Najas guadalupensis* and *Ceratophyllum demersum*) have been particularly effective for P removal. The influent ADW typically exhibits high levels of hardness and alkalinity, so the superior P removal performance of the submerged species may be due to P co-precipitation with $CaCO_3$, a removal mechanism previously proposed in small-scale studies in Florida (Reddy et al., 1987). Discussions are under way among state and federal officials regarding a lowering of the desired STA outflow TP concentration, perhaps to as low as 0.010 to 0.020 mg/L, so the submerged macrophyte communities may play a prominent role in the STA wetlands.

Conclusions

Studies documenting the P removal performance of MBS in Florida began in the mid-1970s, using natural wetlands that received secondary treated domestic effluent. Effective P removal was observed in systems that exhibited little surface discharge, with subsurface soils and associated plant tissues comprising the prominent P sink. Floating macrophyte systems also were used to treat secondary effluents, with routine plant harvest providing the principal P removal mechanism. Increasingly stringent discharge permits in the late 1980s led to the construction of several large FWS wetlands for nutrient removal. These systems received advanced treated effluents at low mass P loadings, and produced outflow P concentrations in the range of 0.05 to 0.15 mg/L. Research and demonstration-scale studies have shown that FWS and floating macrophyte systems can be used to remove particulate P (phytoplankton) from eutrophic lake waters. In the 1990s, research and demonstration work focused on the use of large FWS systems for treating agricultural drainage waters in S. Florida. Under hard water conditions, the wetlands dominated by submerged macrophytes appear most effective for treating the drainage waters to the desired low outflow concentrations (0.010–0.050 mgP/L).

References

Battoe, L.E., Coveney, M.F., Lowe, E.F. and Stites, D.S. (1999). The role of phosphorus reduction and export in the restoration of Lake Apopka, FL. In: *Phosphorus Biogeochemistry in Subtropical Ecosystems*, K.R. Reddy, G.A.O'Conner and C.L. Schelske, (eds), Lewis Publishers, Boca Raton, FL, pp. 511–526.

Boyt, F.L., Bayley, S.E. and Zoltek, J., Jr. (1977). Removal of nutrients from treated municipal wastewater by wetland vegetation. *J. Water Pollut. Control Fed.*, **49**, 789–799.

Clough, K.S., DeBusk, T.A. and Reddy, K.R. (1987). Model water hyacinth and pennywort systems for the secondary treatment of domestic wastewater. In: *Aquatic Plants for Water Treatment and Resource Recovery*. K.R. Reddy and W.H. Smith (eds.), Magnolia Pub. Inc., Orlando, FL.

Coveney, M.F., Lowe, E.F. and Battoe, L.E. (2000). Performance of a recirculating wetland filter designed to remove particulate phosphorus for restoration of Lake Apopka (Florida, USA). *Preprint of 7th International Conference on Wetland Systems for Winter Pollution Control, Orlando, Florida, USA, November 2000*.

D'Angelo, E.M. and Reddy, K.R. (1994). Diagenesis of organic matter in a wetland receiving hypereutrophic lake water: I. Distribution of dissolved nutrients in the soil and water column. *J. Environ. Qual.* **23**, 928–936.

DeBusk, T.A. and Krottje, P. (1996). The use of wetlands for wastewater treatment: A Florida overview. In: *Proc., 71st Florida Water Resources Conf.*, Florida Water Environment Assoc., Gainesville, FL, pp.189–194.

DeBusk, T.A. and Reddy, K.R. (1989). Wastewater nutrient removal in Florida using aquatic macrophytes. In: *Proceedings, Biological Nitrogen and Phosphorus Removal: The Florida Experience II*. University of Florida TREEO Center, Gainesville, FL, USA.

DeBusk, T.A., Williams, L.D. and Ryther, J.H. (1983). Removal of nitrogen and phosphorus from waste water in a water hyacinth-based treatment system. *J. Environ. Qual.*, **12**, 257–262.

DeBusk, T.A., Langston, M.A., Burgoon, P.S. and Reddy, K.R. (1990). A performance comparison of vegetated submerged beds and floating macrophytes for domestic wastewater treatment. In: *Constructed Wetlands in Water Pollution Control*, P.F. Cooper and B.C. Findlater, (eds), Pergamon Press, Oxford, pp. 301–308.

DeBusk, T.A., Peterson, J.E. and Jensen, K.R. (1995). Phosphorus removal from agricultural runoff: An assessment of macrophyte and periphyton-based treatment systems. In: K.L. Campbell, Ed.,

Proceedings, Versatility of Wetlands in the Agricultural Landscape Conference, Am. Society Agricultural Engineers, pp. 619–626.

DeBusk, T.A., Peterson, J.E. and Reddy, K.R. (1996). Use of aquatic and terrestrial plants for removing phosphorus from dairy wastewaters. *Ecol. Eng.* **5**, 371–390.

Dierberg, F.E. and Brezonik, P.L. (1983a). Nitrogen and phosphorus mass balances in natural and sewage-enriched cypress domes. *J. Appl. Ecol.* **20**, 323–337.

Dierberg, F.E. and Brezonik, P.L.. (1983b). Tertiary treatment of municipal wastewater by cypress domes. *Water Res.* **17**, 1027–1040.

Dolan, T.J., Bayley, S.E., Zoltek, J., Jr. and Hermann, A.J. (1981). Phosphorus dynamics of a Florida freshwater marsh receiving treated wastewater. *J. App. Ecol.*, **18**, 205–219.

Fisher, M.M. and Reddy, K.R. (1987). Water hyacinth (*Eichhornia crassipes* [Mart] Solms) for improving eutrophic lake water: water quality and mass balance. In: *Aquatic Plants for Water Treatment and Resource Recovery*, K.R. Reddy and W.H. Smith (eds.), Magnolia Pub. Inc., Orlando, FL, pp. 969–976.

Gale, P.M., Reddy, K.R. and Graetz, D.A. (1994). Phosphorus retention by wetland soils used for treated wastewater disposal. *J. Environ. Qual.*, **23**, 370–377.

Guardo, M., Fink, L., Fontaine, T.D., Newman, S., Chimney, M.J., Bearzotti, R. and Goforth, G. (1995). Large-scale constructed wetlands for nutrient removal from stormwater runoff: An Everglades restoration project. *Environ. Mgmt.* **19**, 879–889.

Kadlec, R.H. and Newman, S. (1992). Phosphorus Removal in Wetland Treatment Areas: Principles and Data. Report #321 to South Florida Water Management District, West Palm Beach, FL.

Nessel, J.K. and Bayley, S.E. (1984). Distribution and dynamics of organic matter and phosphorus in a sewage-enriched cypress swamp. In: *Cypress Swamps*, K.C. Ewel and H.T. Odum (eds), University Presses of Florida, Gainesville, FL, pp. 472.

Reddy, K.R. (1983). Fate of nitrogen and phosphorus in a waste water retention reservoir containing aquatic macrophytes. *J. Environ. Qual.* **12**, 137–141.

Reddy, K.R. and DeBusk, W.F. (1985). Nutrient removal potential of selected aquatic macrophytes. *J. Environ. Qual.* **14**, 459–462.

Reddy, K.R. and Smith, W.H. (eds.) (1987). *Aquatic Plants for Water Treatment and Resource Recovery*, Magnolia Publishing, Inc., Orlando, FL, p. 1030.

Reddy, K.R., Sacco, P.D., Graetz, D.A., Campbell, K.L. and Sinclair, L.R. (1982). Water treatment by aquatic ecosystem: Nutrient removal by reservoirs and flooded fields. *Environ. Mgmt.*, **6**(3), 261–271.

Reddy, K.R., Sutton, D.L. and Bowes, G. (1983). Freshwater aquatic plant biomass production in Florida. *Proc. Soil and Crop Sci. Soc.* Florida, **42**, 28–40.

Reddy, K.R., Tucker, J.C. and DeBusk, W.F. (1987). The role of egeria in removing nitrogen and phosphorus from nutrient-enriched waters. *J. Aquat. Plant Manage.* **25**, 14–19.

Schwartz, L.N., Wallace, P.M., Gale, P.M., Smith, W.F., Wittig, J.T. and McCarty, S.L. (1994). Orange County Florida Eastern Service Area Reclaimed Water Wetlands Reuse System. *Wat. Sci. Tech.* **29**(4), 273–281.

Stewart, E.A., Haselow, D.L. and Wyse, N.M. (1987). Review of operations and performance data on five water hyacinth-based treatment systems in Florida. In: *Aquatic Plants for Water Treatment and Resource Recovery*, K.R. Reddy and W.H. Smith (eds.), Magnolia Publishing, Orlando, FL, pp. 279–288.

Swindell, C.E. and Jackson, J.A. (1990). Constructed wetlands design and operation to maximize nutrient removal capabilities. In: *Constructed Wetlands in Water Pollution Control*, P.F. Cooper and B.C. Findlater (eds.), Pergamon Press, Oxford, England, pp. 107–114.

Walker, W.W., Jr. (1995). Design basis for Everglades Stormwater Treatment Areas. *Water Resources Bulletin*, **31**(4), 671–685.

Media selection for sustainable phosphorus removal in subsurface flow constructed wetlands

H. Brix*, C.A. Arias** and M. del Bubba***

* Department of Plant Ecology, University of Aarhus, Nordlandsvej 68, 8240 Risskov, Denmark
** Polytechnic University of Catalunya, Department of Hydraulics, 08034 Barcelona, Spain
*** University of Florence, Department of Analytical Chemistry, 50121 Florence, Italy

Abstract Sorption of phosphorus (P) to the bed sand medium is a major removal mechanism for P in subsurface flow constructed wetlands. Selecting a sand medium with a high P-sorption capacity is therefore important to obtain a sustained P-removal. The P-removal capacities of 13 Danish sands were evaluated and related to their physico-chemical characteristics. The P-removal properties of sands of different geographical origin varied considerably and the suitability of the sands for use as media in constructed reed beds thus differs. The P-sorption capacity of some sands would be used up after only a few months in full-scale systems, whereas that of others would subsist for a much longer time. The most important characteristic of the sands determining their P-sorption capacity was their Ca-content. Also the P-binding capacities of various artificial media were tested (light-expanded-clay-aggregates (LECA), crushed marble, diatomaceous earth, vermiculite and calcite). Particularly calcite and crushed marble were found to have high P-binding capacities. It is suggested that mixing one of these materials into the sand or gravel medium can significantly enhance the P-sorption capacity of the bed medium in a subsurface-flow constructed wetland system. It is also possible to construct a separate unit containing one of these artificial media. The media may then be replaced when the P-binding capacity is used up.
Keywords Constructed wetland; filter media; phosphorus; phosphorus sorption; reed beds; vertical-flow system; water treatment

Introduction

In concert with the provision of efficient nutrient removal at large wastewater treatment facilities, the treatment of sewage discharged from single houses and small rural communities in Denmark and other European countries are becoming still more important for improving the environmental quality in streams and lakes (Harremoës, 1998). There is therefore an urgent need to find efficient and at the same time economically reasonable solutions for these small wastewater producers. Different on-site solutions have been launched, such as different compact bio-film systems, sand filters and constructed reed beds. The degree of treatment required in the systems is determined by regulations, which may contain standards for suspended solids (SS), biochemical oxygen demand (BOD), nitrification and total phosphorus, among others. Most systems are able to fulfil the requirements for SS and BOD removal, and nitrification can also frequently be obtained. However, it is a paramount problem to remove phosphorus (P) in these small-scale, on-site systems.

Constructed reed beds may be an appropriate and economically attractive on-site treatment solution for single houses and small communities in rural areas (Vymazal *et al.*, 1998). Traditionally the reed beds in Denmark have been constructed with soil as the growth medium for the plants. However, this has caused problems with overland flow and short-circuiting of the wastewater between inlet and outlet because of the low hydraulic conductivity of soils. Therefore present adopted design guidelines are based on sand or gravel and in some cases intermittently loaded vertical-flow beds instead of horizontal-flow beds (Brix and Johansen, 1999; EPA, 1999).

Subsurface flow constructed reed beds generally have a greater potential to remove nitrogen than phosphorus (Vymazal et al., 1998). The only sustainable removal mechanism for phosphorus is plant uptake and subsequent harvesting (Lantzke et al., 1998). However, the amount of phosphorus that can be removed by harvesting the plant biomass usually constitutes only an insignificant fraction of the amount of phosphorus loaded into the system with the sewage (Brix, 1997). Phosphorus may also be bound in the media of the reed bed mainly as a consequence of adsorption and precipitation reactions with calcium (Ca), aluminium (Al) and iron (Fe) in the sand or gravel substrate. The capacity of a reed bed to remove P may therefore be dependent on the contents of these minerals in the substrate. This hypothesis is supported by the observation that P-removal has been found to be particularly efficient in constructed reed beds containing ferruginous sand (Netter, 1992). The P removal efficiency is often high initially and then decreases after some time as the P-sorption capacity of the sand is being used up (Ciupa, 1996). It is therefore of importance to select a sand medium with a high P-sorption capacity in order to obtain a sustained P-removal in the long term in constructed reed bed systems.

Various artificial media have been tested in order to improve the P-removal in subsurface-flow constructed wetlands: factory made light-weight expanded clay aggregates (LECA) (e.g. Zhu et al., 1997), granulated laterite (Wood and McAtamney, 1996), shale (Drizo et al., 1997), and crushed marble (Gervin and Brix, 2001), among others. It is generally found that several of these materials have the potential to enhance the P-removal in constructed wetland systems.

The objective of this study was to evaluate the P-removal capacities of locally available sands in Denmark for use as media in constructed reed beds. Furthermore, we tested the P-sorption capacity of different "artificial" medias that might be used to enhance P-removal in subsurface flow constructed wetlands.

Materials and methods
Sands and media tested
The composition of sands that are commercially available for use in constructed reed beds differs depending on the location of origin. The sands used in this study were obtained from thirteen gravel pits located in different regions of Denmark. See Arias et al. (2000) for details. In addition, the following artificial media were selected: (i) a *calcite* product (supplied by the company Damolin A/S) which is readily commercially available as cat litter; (ii) crushed Norwegian *marble* (supplied by Copenhagen Water) which has been used as the bed medium in a vertical-flow wetland system for removal of phosphorus from combined sewer overflows and lake water (Gervin and Brix, 2001); (iii) granulated and calcined (burnt) *diatomaceous earth* (supplied by the company Damolin A/S) which is commercially available as cat litter; (iv) Light Expanded Clay Aggregates (*LECA*) produced from Danish clay and delivered by the company Dansk Leca A/S; and (v) a *vermiculite* granulate supplied by the company Skamol A/S. A number of other potential artificial media were also tested initially, but were disregarded in further tests largely because of their physical properties, which were not suitable for use as media in subsurface flow constructed wetland systems.

Characterisation of media
The particle-size distribution on a weight basis was analysed in triplicate by conventional dry-sieving technique (Day, 1965). The grain-size distribution plots were used to estimate d_{10} (the effective grain size) and d_{60}, and the uniformity of the particle size distribution (the uniformity coefficient) was calculated as the ratio between d_{60} and d_{10}. Porosities were determined from the amount of water needed to saturate a known volume of medium

($n = 3$), and the bulk density (g cm^{-3}) was based on the ratio between the dry weight and the bulk volume of the media ($n = 3$). Saturated hydraulic conductivity was determined using the constant-head method ($n = 5$) in the laboratory (Klute, 1965). The concentrations of P, Fe, Ca, Al and Mg in the <2 mm size fractions of the media were analysed by plasma emission spectrometry (Perkin Elmer Plasma II Emission Spectrometer, USA) after extraction in boiling HNO_3–H_2O_2 ($n = 2$).

Sorption isotherm experiments
Equilibrium isotherm experiments were performed on all media ($n=2$ or 3). Approximately 5 g of material (only 1 g for calcite) were placed in 200 ml polyethylene bottles. Hundred-millilitre aliquots of tap water spiked with KH_2PO_4 to give one of 9 levels of phosphorus (0, 2.5, 5, 10, 20, 40, 80, 160 and 320 mg P l^{-1}) were then added. Tap water was used in order to mimic the mineral composition of wastewater. The bottles were sealed with screw-type lids and were continuously agitated in a rotating wheel at laboratory temperature for 20 hours. Blinds containing no media were always included in the experiments. After settling, an aliquot of the supernatant was filtered through Whatman GF/C filters and after adequate dilution analysed for phosphorus using the molybdenum blue-ascorbic acid method. A second unfiltered aliquot was used for measurement of pH and conductivity. Amount of phosphorus sorbed by the media was calculated from the decrease in the solution P concentration.

Column experiments
A total of 44 columns were constructed from inverted one-litre polyethylene bottles (diameter 95 mm). The columns were packed with c. 700 cm^3 of sand or artificial media ($n = 2$). Columns containing Quartz sand and 5 and 10% of calcite or marble by weight were also tested. Water spiked with KH_2PO_4 to a P concentration of 10 mg l^{-1} (pH adjusted to c. 6.9) was supplied continuously at a rate of approx. 240 ml day^{-1} per column (equivalent to a nominal retention time in the columns of approximately 12–14 hours) using a constant-head feeding tank. The water level in the feeding tank was kept constant by continuous pumping from a stirred storage tank kept at ground level. Water in the storage tank was renewed every third day. The columns were kept water saturated, as the effluent levels of the columns were set just above the surface of the sands in the columns. The effluent from each column was collected daily and the volume measured to estimate actual loading. Furthermore, pH and conductivity was measured and an aliquot filtered through Whatman GF/C filters and analysed for P as described above. After 12 weeks of continuous loading the experiment was terminated.

Results
Characteristics of media
The media all had textures and hydraulic conductivities that would make them suitable for use as substrates in constructed reed beds systems (Table 1). The effective grain size (d_{10}) varied between 0.2 mm for the finest sand to 1.4 mm for the coarsest sand. The hydraulic conductivities were generally highest for the coarsest textures, and particularly the artificial media had high hydraulic conductivities. The bulk densities of all the artificial media, except marble, were significantly lower than that of the sands. The mineral content of the sands differed markedly with the origin of the sand (Table 2). The concentrations of Ca were particularly low in the Vestergård sand reflecting the leached out conditions in the area of origin of the sand. In most other sands the Ca content was in the range of 20–70 mg g^{-1} dry weight. Vestergård sand also had low concentrations of Al and Mg, compared to the other sands, but contained some Fe. The quartz sand had low contents of all the analysed

minerals. Calcite and marble had – as expected – high contents of Ca; diatomaceous earth had high contents of Al; LECA had high contents of Fe and Al; and vermiculite had high contents of Mg, Fe and Al. The differences in equilibrium pH and conductivity were generally small for the sands and largely reflect the pH and conductivity of the tap water used in the extractions. Calcite increased the pH and the conductivity, and LECA decreased the pH slightly.

Table 1 Physical characteristics of sands and artificial media tested for P-sorption properties. Values for porosity, d_{10}, d_{60} and the uniformity coefficient (d_{60}/d_{10}) are means of triplicate analyses. Values for hydraulic conductivity (K_s) are means ± 1 SD ($n = 5$).

Material	Porosity (%)	Bulk density (g cm³)	d_{10} (mm)	d_{60} (mm)	d_{60}/d_{10}	K_s (m day⁻¹)
Natural sands:						
Birkesig	31	1.83	0.32	1.4	4.4	67 ± 10
Vestergård	36	1.70	0.45	1.2	2.7	202 ± 18
Almind	43	1.46	1.40	3.7	2.6	770 ± 180
Bedsted 1	32	1.79	0.28	1.1	3.9	143 ± 6
Bedsted 2	36	1.72	0.22	0.7	3.1	74 ± 7
Bedsted 3	38	1.66	0.25	0.9	3.7	77 ± 4
Nymølle	44	1.40	0.80	3.2	4.0	1130 ± 64
Aunsøgård	30	1.86	0.24	1.0	4.2	22 ± 1
Løgtved	36	1.67	0.21	0.6	3.0	69 ± 10
Sorø	35	1.66	0.23	0.7	3.1	40 ± 4
Darup	36	1.71	0.61	3.4	5.6	360 ± 40
Farum	33	1.70	0.20	0.7	3.5	30 ± 6
Quartz sand	32	1.74	0.29	0.9	3.1	111 ± 12
Artificial media:						
Calcite	42	0.83	0.80	3.3	4.1	349 ± 29
Marble	39	1.53	0.70	1.7	2.4	1760 ± 210
Diatomaceous earth	32	0.54	0.74	2.4	2.4	1460 ± 280
LECA	56	0.47	0.66	2.6	3.9	1310 ± 60
Vermiculite	42	0.14	0.88	3.3	4.1	1540 ± 40

Table 2 Equilibrium pH and conductivity in tap water extractions, and concentrations (mg g⁻¹ dry weight) of phosphorus (P), iron (Fe), calcium (Ca), aluminium (Al) and magnesium (Mg) in the natural sands and artificial media tested for P-sorption properties. All values are means of duplicate analyses

Material	pH	Conductivity (µS cm⁻¹)	P (mg g⁻¹ dw)	Fe (mg g⁻¹ dw)	Ca (mg g⁻¹ dw)	Al (mg g⁻¹ dw)	Mg (mg g⁻¹ dw)
Natural sands:							
Birkesig	8.25	563	0.21	3.54	24.7	1.73	0.83
Vestergård	8.20	492	0.14	3.33	0.2	0.61	0.18
Almind	8.30	510	0.30	8.13	22.8	1.70	1.21
Bedsted 1	8.65	481	0.35	4.89	31.1	2.40	1.45
Bedsted 2	8.65	496	0.19	3.75	23.7	1.28	0.92
Bedsted 3	8.68	453	0.18	2.79	27.7	1.63	0.98
Nymølle	8.30	580	0.28	4.77	69.9	2.53	2.08
Aunsøgård	8.26	557	0.45	8.47	23.6	4.18	2.23
Løgtved	8.32	525	0.34	3.65	3.5	2.36	1.05
Sorø	8.20	496	0.30	4.46	39.2	2.35	1.26
Darup	8.32	540	0.30	3.59	62.3	1.90	1.28
Farum	8.39	533	0.21	2.91	40.3	1.47	0.75
Quartz sand	8.53	496	0.04	1.21	0.6	0.32	0.08
Artificial media:							
Calcite	11.43	2350	0.21	1.25	240	3.49	3.59
Marble	8.13	514	0.11	0.39	389	0.94	2.57
Diatomaceous earth	8.06	612	0.77	2.89	2.3	13.5	3.15
LECA	7.69	602	0.36	14.1	8.6	10.2	1.74
Vermiculite	8.44	540	0.49	43.6	2.2	43.1	125

Phosphorus sorption isotherms

When plotting the amount of sorbed phosphorus as a function of the P concentration in the final equilibrium solution it was obvious that the media tested behaved differently. In Figure 1 the isotherms for the sorption of P on the five artificial media tested and the two extreme natural sands are shown. Sorption isotherms for the other natural sands were in between Quartz sand and Darup sand (Arias *et al.*, 2000). Sands with a high content of Ca generally had higher P-sorption capacity than the inert quartz sand and other sands with a low Ca content. Of the artificial media especially calcite and marble had high P-sorption capacities, whereas vermiculite, diatomaceous earth and LECA had low P-sorption capacities.

Column experiments

In Figure 2 are the removal efficiency of the columns containing artificial media and quartz sand shown as a function of loading rate. The removal of phosphorus in the columns generally decreased over time and most for the columns with low P-sorption capacity (quartz sand, vermiculite, LECA and 5 and 10% marble). The columns containing calcite showed the best performance of all the materials tested. Among the natural sands the best performance was obtained with Darup sand, which maintained a removal efficiency >80% throughout the incubation period, and showed only a slight tendency of decreased performance towards the end of the incubation period (data not shown). Also Nymølle and Sorø sands maintained fairly stable removal efficiencies throughout the incubation period, although at a lower level (approx. 70% removal). The removal efficiency of the other sands all decreased significantly during the incubation period to levels <50% removal after 12 weeks. Data for the natural sands are presented by Arias *et al.* (2000). The removal efficiency of diatomaceous earth and marble decreased slowly but was still c. 60 and 70%, respectively, at the termination of the experiment. Unfortunately, because of time and resource constraints, the column experiments were terminated before the columns were completely P-saturated, and it is thus not possible to estimate with more accuracy the final P-removal capacity of the materials.

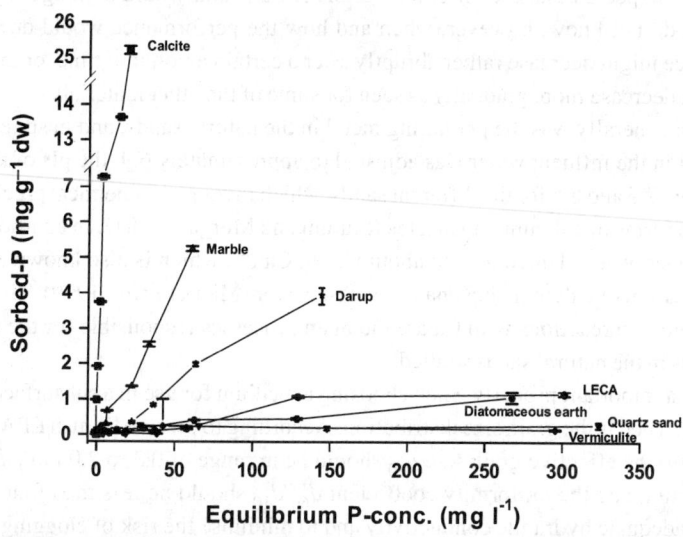

Figure 1 Isotherms for the sorption of phosphorus on the five artificial media and the two extreme natural sands tested related to the equilibrium concentration of P remaining in the solution. Error bars represent range of measured values (n = 2 or 3)

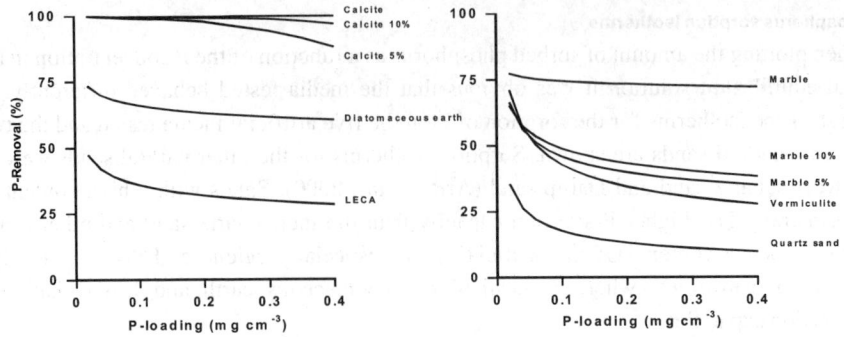

Figure 2 Relation between the amount of phosphorus removed (% of inlet concentration) and P loading on a media volume basis. The columns were continuously loaded with a 10 mg l^{-1} P-solution at a rate equivalent to a 12 to 14 hour nominal retention time. The curves represent the average of replicate columns

Discussion

The results of this study clearly demonstrate that the phosphorus removal properties of sands of different geographical origin in Denmark vary considerably. The suitability of the sands for use as media in constructed reed beds established for P-removal thus differs. The amount of phosphorus loaded onto the columns during the 12-week experimental period corresponds approximately to the loading during one year on a full-scale system using normally adopted dimensioning criteria (i.e. 5 m^2 bed area per person equivalent and a bed depth of 0.6 m). From this it can be deduced that the P-sorption capacity of sands like the quartz sand would be used up during the initial months, and that the P-removal therefore would be poor thereafter. However, if the Darup sand was used, the P-removal would be excellent at least during the initial year, and thereafter a good P-removal could probably be sustained for a number of years. Unfortunately, the study did not provide direct data to document for how long a good P-removal could be expected. However, as the P-removal obtained in the columns was strongly correlated with the amount of P sorbed in the isotherm experiments, these data might give some indication. The amount of P sorbed by the Darup sand at 320 mg l^{-1} is roughly 14 times higher than the amount sorbed by quartz sand. It may therefore be expected that the "lifetime" of the Darup sand would be longer by a similar factor. We do not know, however, when and how the performance would decrease. The performance might decrease rather abruptly after a certain amount of time, or the performance might decrease more gradually as seen for some of the other materials.

Calcium generally was the prevailing metal in the natural sands, and in spite of the fact that the pH in the influent water was adjusted to approximately 6.9, the pH of the effluent was between 7.5 and 8.5 for the different sands, which favours the chemical precipitation of the various forms of calcium phosphates (Stumm and Morgan, 1981). In addition, we used rather hard tap water that contained about 90 mg Ca l^{-1}, which is also known to facilitate the precipitation of calcium phosphates in wastewater (Maurer *et al.*, 1999). This indicates that precipitation reactions with Ca are the main processes responsible for the removal of phosphorus in the natural sands studied.

The most important property when choosing a medium for use in a subsurface flow constructed reed bed is the grain size distribution. According to present Danish EPA guidelines (EPA, 1999) the effective grain size d_{10} should be in range of 0.3 to 2.0 mm, d_{60} between 0.5 and 8 mm, and the uniformity coefficient d_{60}/d_{10} should be less than four in order to secure an adequate hydraulic conductivity and to minimise the risk of clogging. However, if P-removal is important, additional properties of the medium are important. Our data show that particularly the content of Ca is important, whereas Fe and Al are of less importance. A high content of Ca in the medium will, because of the relatively high pH of domestic sewage, favour precipitation reactions with Ca in the medium. In situations where the wastewater to be treated is more acid, the contents of Fe and Al may be more important as the precipitation reactions with these ions are favoured at low pH (Stumm and Morgan, 1981).

Among the artificial media tested, the Ca-based materials (calcite and marble) showed the highest capacity to remove P. Diatomaceous earth also sorbed P, but the 12-h retention time in the columns was probably too short. LECA and vermiculite did not bind P very well. The studies indicated that particularly calcite may be used to increase the P-sorption capacity of the bed medium in subsurface-flow constructed wetlands. If sands are used, in order to enhance the sorption capacity of it, then calcite can be mixed into the sand medium during construction. Another option would be to construct a separate modular unit containing a bed medium with calcite that could be replaced when the P-sorption capacity is used up. These options need to be studied in longer-term field studies. Further studies conducted with calcite showed a P removal capacity of approximately 25 kg P m^{-3}. Considering this capacity, around 30 kg PE^{-1} year^{-1} of calcite would be needed to comply with the discharge standards. The pH in the calcite effluent was moderately high at the beginning of the test but dropped to normal levels shortly after. This was particularly so for columns with pure calcite. Therefore, a filter unit containing pure calcite should not be placed at the effluent end of the constructed wetland system, but rather in-between two beds so that the final bed would neutralise pH. Another option would be to mix calcite with the sand in a section of the bed. It should also be remembered that the P-removal in full-scale systems occurs not only by P-sorption to the bed medium, but also through incorporation into organisms (biofilms and plants) and subsequent accumulation of organic matter in the systems. Therefore, even when the P-sorption capacity of the medium is used up, some P-removal in the system will occur.

Acknowledgements
We thank the various suppliers of sand material for providing the sands. The study was supported by the Danish Environmental Protection Agency through the programme "Action plan for Promotion of Ecological Urban Renewal and Wastewater Treatment", project no. M 226-0014.

References
Arias, C.A., Bubba, M. del and Brix, H. (2000). Phosphorus removal by sands for use as media in subsurface flow constructed reed beds. *Wat. Res.* **35**, 1159–1168.
Brix, H. (1997). Do macrophytes play a role in constructed treatment wetlands? *Wat. Sci. Tech.* **35**(5), 11–17.
Brix, H. and Johansen, N.H. (1999). Treatment of domestic sewage in a two-stage constructed wetland – design principles. In: Vymazal, J. (Ed.), *Nutrient Cycling and Retention in Natural and Constructed Wetlands*. Backhuys Publishers, Leiden, The Netherlands, 155–163.
Ciupa, R. (1996). The experience in the operation of constructed wetlands in North-Eastern Poland. Proc. 5th Int. Conf. on Wetland Systems for Water Pollution Control, Vienna, Austria 2, IX6.1–IX6.8.
Day, P.R. (1965). Particle fractionation and particle-size analysis. In: Black, C.A., Evans, D.D., Ensminger, L.E., White, J.L. and Clark, F.E. (Eds.), *Methods of Soil Analysis*. American Society of Agronomy. Inc., Publishers, Madison, Wisconsin, USA, 545–567.
Drizo, A., Frost, C.A., Smith, K.A. and Grace, J. (1997). Phosphate and ammonium removal by constructed wetlands with horizontal subsurface flow, using shale as a substrate. *Wat. Sci. Tech.* **35**(5), 95–102.
EPA (1999). Rodzoneanlæg op til 30 PE (in Danish). Environmental Guidelines 1, 1–46.
Gervin, L. and Brix, H. (2001). Removal of nutrients from combined sewer overflows and lake water in a vertical-flow constructed wetland system. *Wat. Sci. Tech.* **44**(11–12) 171–176 (this issue).
Harremoes, P. (1998). The challenge of managing water and material balances in relation to eutrophication. *Wat. Sci. Tech.* **37**(3), 9–17.

Klute, A. (1965). Laboratory measurement of hydraulic conductivity of saturated soil. In: Black, C.A., Evans, D.D., Ensminger, L.E., White, J.L. and Clark, F.E. (Eds.), *Methods of Soil Analysis*. American Society of Agronomy. Inc., Publishers, Madison, Wisconsin, USA, 210–221.

Lantzke, I.R., Heritage, A.D., Pistillo, G. and Mitchell, D.S. (1998). Phosphorus removal rates in bucket size planted wetlands with a vertical hydraulic flow. *Wat. Res.* **32**, 1280–1286.

Maurer, M., Abramomovich, D., Siegrist, H. and Gujer, W. (1999). Kinetics of biologically induced phosphorus precipitation in waste-water treatment. *Wat. Res.* **33**, 484–493.

Netter, R. (1992). The purification efficiency of planted soil filters for wastewater treatment. *Wat. Sci. Tech.* **26**(10–12), 2317–2320.

Stumm, W. and Morgan, J.J. (1981). *Aquatic Chemistry*. 2nd edition, John Wiley & Sons, Inc.

Vymazal, J., Brix, H., Cooper, P.F., Green, M.B. and Haberl, R.(eds.) (1998). *Constructed Wetlands for Wastewater Treatment in Europe*. Backhuys Publishers, Leiden, The Netherlands.

Wood, R.B. and McAtamney, C.F. (1996). Constructed wetlands for waste water treatment: the use of laterite in the bed medium in phosphorus and heavy metal removal. *Hydrobiol.* **340**, 323–331.

Zhu, T., Jenssen, P.D., Mæhlum, T. and Krogstad, T. (1997). Phosphorus sorption and chemical characteristics of lightweight aggregates (LWA) – potential filter media in treatment wetlands. *Wat. Sci. Tech.* **35**(5), 103–108.

Phosphorus removal from trout farm effluents by constructed wetlands

Y. Comeau*[1], J. Brisson**, J.-P. Réville***, C. Forget* and A. Drizo*

* Department of Civil, Geological and Mining Engineering, Ecole Polytechnique, Montreal (Quebec), Canada H3C 3A7
** Institut de recherche en biologie végétale, 4101 Sherbrooke St. East, Montreal (Quebec), Canada, H1X 2B2
*** Pisciculture du Lac William, 2423 R.R. 165, St-Ferdinand (Quebec), Canada G0N 1N0
[1] Author to whom all correspondence should be addressed: E-mail: *yves.comeau@polymtl.ca*

Abstract Freshwater trout farms need a high and continuous clean water flow to keep fish exposed to a non-toxic ammonium concentration. As a result, the concentration of effluents from these farms are even below standard effluent criteria for municipal wastewater effluent for solids, nitrogen and phosphorus. Nevertheless, the mass of pollutants discharged, originating mostly from excreta and undigested fish food, must be reduced by simple and economical treatment processes. We designed and operated a three-stage system aimed at retaining solids by a 60 μm nylon rotating microscreen followed by treatment with a phosphorus-retaining constructed wetland system. Washwater from the microscreen was pumped to a series of two horizontal flow beds of 100 m^3 each (0.6 m deep). Coarse (2 mm) and finer (< 2 mm) crushed limestone were used in each bed, respectively, with the first one being planted with reeds (*Phragmites australis*) and the second one designed to remove even more phosphorus by adsorption and precipitation. Preliminary results indicated that the microscreen captured about 60% of the suspended solids and that greater than 95% of the suspended solids and greater than 80% of the total phosphorus mass loads were retained by the beds. The potential of constructed wetlands as an ecologically attractive and economical method for treating fish farm effluents to reduce solids and phosphorus discharge appears promising.
Keywords Constructed wetland; fish farming; phosphorus removal; sludge treatment; trout; wastewater

Introduction

Freshwater fish farming has increased considerably over the last 25 years and is expected to continue to supply an ever increasing demand for seafood notably because of depleting natural fish stocks. In the Province of Quebec, for example, trout farm production increased from 40 tons/yr in 1976 to 1900 tons/yr in 1996 (MENV, 1999). With increasing examination from public and government groups, the sustainability of this industry requires minimizing environmental impacts from the extraction of ground and surface water, and from the release of effluents into receiving water bodies. Environmental certification of fish farms for fish health certification will soon become a reality. A typical medium size trout farm producing 100 tons/yr needs about 10,000 m^3/d of fresh water with the free ammonia concentration being the dilution controlling component (limit of 12.5 μg N/L of NH_3 for mature trout growth at pH 7.5 and 15°C; MAPAQ, 1997). Such a flowrate corresponds to that used to supply a community of about 40,000 people.

Effluents from trout farms are typically 20 to 25 times more diluted than medium strength municipal wastewaters and even below municipal secondary treatment criteria (Table 1). With respect to receiving water quality objectives, the most constraining element to remove from freshwater fish farms is phosphorus.

Phosphorus reduction in effluents is best achieved by source reduction using a low phosphorus diet with a high digestibility by the fish. Effluent treatment for trout farms is partly achieved with settling ponds. Infrequent solids removal from these ponds, however, results

Table 1 Relative strength of a typical trout farm effluent

Parameter	Units	Municipal WW – medium strength[a]	Municipal WW 2ry effluent criteria[a]	Trout farm effluent[b]	Water quality objectives[b]
BOD_5	mg/L	220	30	15	3
SS	mg/L	220	30	12	10[c]
NH_4	mg N/L	25	n/applic.	1.3	function of pH
TP	mg P/L	8	1	0.29	0.03

Notes: [a] Metcalf & Eddy (1991); [b] MENV (1999); [c] the water quality objective for suspended solids is for no more than a 10 mg/L increase in the receiving water body

in much mineralization and release of soluble ammonia and phosphorus in the effluent. Improved solids recovery from rearing and settling basins, or the use of microsreens can assist in reducing solids mineralization.

Physical, chemical or biological treatment processes to remove phosphorus are difficult to apply or expensive due to the low effluent phosphorus concentration. Constructed wetlands have been shown to be promising as an ecological and relatively passive treatment method with a low operating cost.

Phosphorus removal in constructed wetland systems (CWS) is mostly due to substrate adsorption and precipitation, biomass harvesting having not yet been proven feasible (Richardson, 1985; Kadlec and Knight, 1996; Laouali et al., 1998). In a horizontal flow system, the hydraulic loading rate, expressed in terms of vertical flow (vHLR), of 22 m/yr was used by Drizo et al. (1997) who obtained a greater than 98% phosphorus removal efficiency from a synthetic "municipal" wastewater with shale. For a 10,000 m³/d trout farm, such a vHLR would require an eight football-field size wetland totaling 400 m by 400 m. Treating the whole fish farm effluent appears prohibitive at such an HLR but treating solids to mineralize sludge while retaining releasable phosphorus is an interesting approach.

In this work we designed and operated a three-stage system to treat solids from a trout farm effluent. The functions of each stage were to : 1) capture of solids from the effluent with a microscreen; 2) treat the microscreen washwater in a subsurface flow bed planted with reeds (*Phragmites australis*) and 3) polish the effluent in a second bed containing a finer phosphorus-retaining substrate.

Materials and methods

The pilot plant was set up at the Pisciculture du Lac William near St-Ferdinand, southeast of Quebec city. Each of the two 100 m³ horizontal subsurface flow beds were designed as shown in Figure 1. The limestone contained 55% CaO, 1.5% MgO and some silica. Iron and aluminium contents were both less than 0.5%. The relative density of the grains was 2.7 g/mL with an in-place stamped void ratio of 46% and 39% for beds 1 and 2, respectively. The specific surfaces of the materials were 6 and 1 m²/kg for beds 1 and 2, respectively (ASTM D854-92, 1995), and the hydraulic conductivities were 1.5 to 54 cm/s for bed 1 and 1.9 to 70 cm/s for bed 2 (method of Aubertin et al., 1996).

The calendar of activities is summarized in Table 2. The effluent added to each bed, the presence of reeds and the vHLR are presented. Period A corresponded to a period during which some leakage from bed 2 (at the point where the outflow pipes goes through the clay layer) prevented its normal use. During periods B1 to B4, both beds were in operation. The water temperature in the beds was 5°C during winter and 18°C during summer with intermediate values during spring and fall.

The treatment system and the operating flowrates for period B4 are shown in Figure 2. A

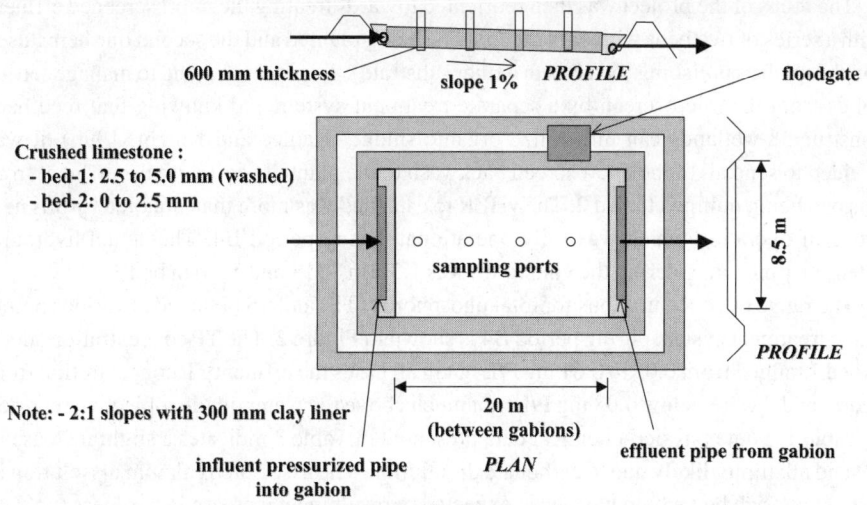

Figure 1 Design of a subsurface horizontal flow bed

60 μm nylon, continuously washed, rotating microscreen was used to capture solids. During the first 13 months, the microscreen was backwashed with its own effluent but clogging problems resulted in clean groundwater being used for this purpose afterwards. At the end of the 13th month, clumps (about 1 kg) of reeds *(Phragmites australis)* were planted in staggered rows at a density of 1 clump per m². An infiltration test was conducted by closing the outflow from the beds and by monitoring the evolution of the water level over time. Analytical methods used for chemical analyses are described in APHA *et al.* (1998).

Results and discussion

The solids removal efficiency by the microscreen was about 60% (57 ± 17%) of the suspended solids (SS) using about 11% of the wash water for backwash. Intermittent backwashing could have reduced this ratio to less than 2% (Bergheim *et al.*, 1993).

During periods A, B1 and B2, the microscreened settled effluent was added to bed-1 and bed-2 in parallel to compare the relative efficiency of fine versus coarse calcareous substrate. With a TP concentration ranging between 0.2 and 0.5 mg P/L being added to bed-1, its effluent remained consistently below 0.1 mg P/L giving a removal efficiency of 78 ± 15%. Operating difficulties with a leaking bed-2 and winter conditions making difficult its repair, this bed was put back in operation during month 13.

Table 2 Calendar of activities

Month	1	2	3	4	5	6	7	8	9	10	11	12	13	14	15	16	17	18
	1998								1999									
Date	May	June	July	Aug	Sep	Oct	Nov	Dec	Jan	Feb	Mar	Apr	May	June	July	Aug	Sep	Oct
Period	startup		A										B1	B2	B3	B4		
Bed-1																		
Effluent	microscreened, settled															microscreened		
Reeds	no												yes					
vHLR (m/yr)	25		30										80					
Bed-2																		
Effluent	microscreened, settled												bed-1 effluent					
vHLR (m/yr)	25	0	25	0	25				0				30	70				

Note: vHLR: vertical hydraulic loading rate

The focus of the project was then reoriented towards treating the microscreened effluent with a series of two beds with only the first one being planted and the second one being used for phosphorus polishing (bed 2 with a finer substrate). To prevent having to manage settled solids from the microscreen by a separate treatment system and knowing that reed beds constructed wetlands can mineralize organic sludge (Kadlec and Knight, 1996), it was decided to send all of the microscreen backwash to the planted bed 1 with the effluent from this bed being pumped to bed 2. The vHLR to each bed was more than doubled by this new mode of operation which was fully operational during period B4. The actual hydraulic retention time (considering the void ratio) was 1.3 d in bed 1 and 1.2 d in bed 2.

The range of concentrations for total phosphorus (TP) and SS obtained at various points of the treatment system during period B4 is shown in Figure 2. The TP concentration added to bed 1 ranged from 0.03 to 0.61 mg P/L but in all cases the effluent TP concentration from beds 1 and 2 were below 0.08 mg P/L, not much above the water quality objective reported in Table 1. Some physico-chemical data presented in Table 3 indicates a slight increase in pH and alkalinity likely due to carbonate dissolution. The associated calcium dissolution in this calcite rich bed would have been expected to result in an increase in hardness (no data available), favoring phosphorus precipitation.

Mass balance results for TP and SS during period B4 are more meaningful than concentration results and are shown in Figure 3. A significant amount of water was lost from both beds amounting to 11% of the incoming flow in bed 1 and 25% in bed 2. The installation of the effluent pipe through the clay liner is probably what impaired its imperviousness.

The overall removal efficiency for TP was 92% of which 87% was provided by bed 1 and the overall efficiency for SS was 96% of which 94% was contributed by bed 1. Due to the short period of testing, these results can only be considered as preliminary but, nevertheless, quite promising for the phosphorus removal of the treatment process considered. Limited reed growth took place during 5 months after planting. A fully grown reed bed SWF-CWS, however, should have a high potential to mineralize organic matter.

Solids capture can be achieved with a microscreen, as used in this study, or with settling zones or settling basins (Figure 4). Settling zones can be implemented at the end of raceways or in central solids evacuating circular tanks. Alternatively, washwaters can be recovered and settled in basins used intermittently for this purpose.

Solids recovered can then be sent to a storage basin for disposal by landspreading or composting. Care should be taken not to directly recirculate the supernatant from such basins as they would contain a very high concentration of soluble phosphorus due to sludge hydrolysis. The potential of reed bed CWS for organic matter mineralization can be exploited for sludge treatment as tested in the study. The substrate should have P retaining capacity normally associated with having a high content in calcium, as tested here, or in

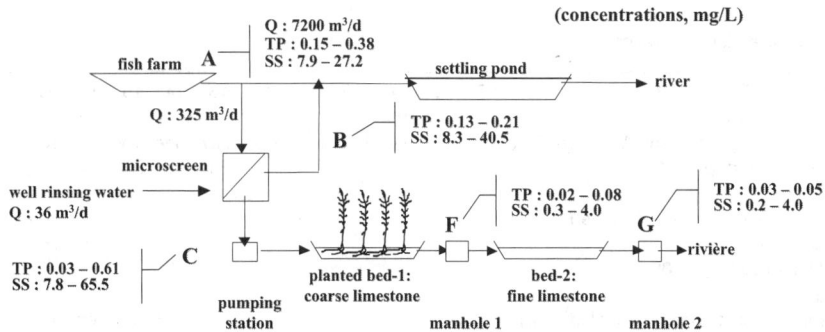

Figure 2 Range of concentrations for total P and suspended solids observed during the B4 period (August 11 to October 6, 1999)

Table 3 Physico-chemical characteristics of the influent and effluent of the two CWS beds during the B4 period

Parameter	Units	Sampling point		
		C-CWS1$_{in}$	F-CWS2$_{in}$	G-CWS2$_{out}$
Temp	°C	14.7	15.9	16.1
O_2	mg/L	6.4	5.8	5.4
pH		7.1	7.3	na
Alkalinity	mg $CaCO_3$/L	70	110	120
Hardness	mg $CaCO_3$/L	80	na	na

Na: not available

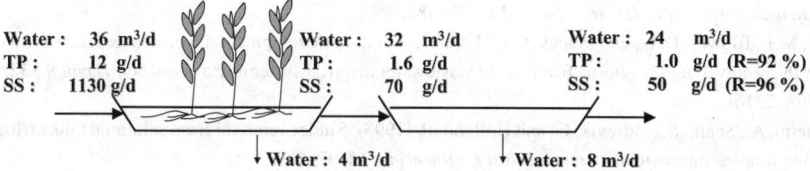

Figure 3 Hydraulic and mass balances for total P and suspended solids in the two constructed wetlands during the B4 period

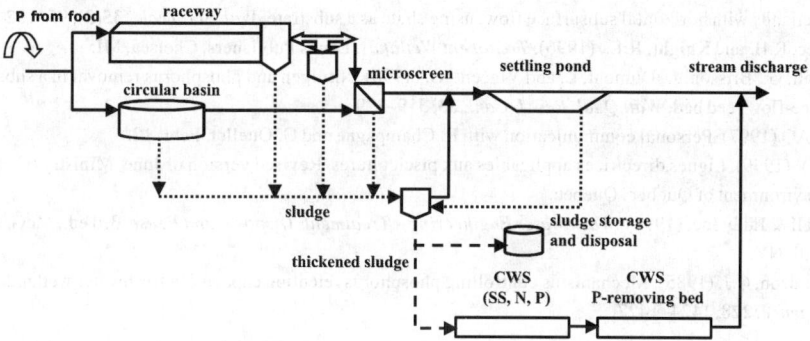

Figure 4 Alternative modes of solids retention and disposal from salmonid fish farm waste effluents

iron or aluminium (Baker et al., 1998). The overall size of the CWS should be calculated on the basis of the organic matter loading rate per unit volume, in addition to the vHLR.

Conclusions

A two-stage subsurface water flow constructed wetland system using a calcium rich calcareous substrate was found to be very efficient in removing suspended solids and phosphorus from the microscreened solids of a trout farm effluent. The first stage SWF-CWS was planted by reeds and served for sludge mineralization and phosphorus removal by substrate adsorption and precipitation. An unplanted second stage CWS with finer medium was used to improve the phosphorus removal efficiency. A proper substrate (rich in calcium, iron or aluminium) should be selected to maximize phosphorus capture by the bed. The potential of constructed wetlands as an ecologically attractive and economical method for treating fish farm effluents to reduce solids and phosphorus discharge appears promising. A summary of methods to reduce phosphorus discharge from fish farm effluents was presented that included source reduction from better feed compositions, solids management and effluent treatment.

Acknowledgements

The financial assistance of the Society for Research and Development of Continental Aquaculture (SORDAC) is acknowledged. We thank Denis Bouchard, Marcel Dugal and Antonio Gatien of Ecole Polytechnique, and Bernard Joseph of the PLW for technical assistance.

References

APHA, AWWA and WEF (1998). *Standard Methods for the Determination of Water and Wastewater.* 20th ed., American Public Health Association, Washington, D.C.

ASTM (1995). D854-92 – Standard test method for specific gravity of soils. American Society for Testing and Materials, Philadelphia, PA.

Aubertin, M., Bussière, B. and Chapuis, R.P. (1996). Hydraulic conductivity of homogenized tailings from hard rock mines. *Can. J. Geotechnol.*, **33**, 470–482.

Baker, M.J., Blowes, D.W. and Ptacek, C.J. (1998). Laboratory development of permeable reactive mixtures for the removal of phosphorus from onsite wastewater disposal systems. *Environ. Sci. Technol.*, **32**, 2308–2316.

Bergheim, A., Sanni, S., Indrevik, G. and Holland, P. (1993). Sludge removal from salmonid tank effluent using rotating microsieves. *Aquacultural Engineering*, **12**, 97–109.

Chapuis, R.P. and Legare, P.P. (1992). Simple method for determining the surface area of fine aggregates and fillers in bituminous mixtures. *Proc. of a Symposium on Effects of Aggregates and Mineral Fillers on Asphalt Mixture Performance, Dec. 10 1991*, San Diego, CA, p 177–186. ASTM, Philadelphia, PA.

Drizo, A., Frost, C.A., Smith, K.A. and Grace, J. (1997). Phosphate and ammonium removal by constructed wetlands with horizontal subsurface flow, using shale as a substrate. *Wat. Sci. Tech.*, **35**(5), 95–102.

Kadlec, R.H. and Knight, R.L.. (1996). *Treatment Wetlands*. Lewis Publishers, Chelsea, MI.

Laouali, G., Brisson, J., Dumont, L. and Vincent, G. (1998). Nitrogen and phosphorus removal in a subsurface-flow reed bed. *Wat. Qual. Res. J. Can.*, **33**, 319–329.

MAPAQ (1997). Personal communication with R. Champagne and G. Ouellet. Febr. 12.

MENV (1999). Lignes directrices applicables aux piscicultures. Revised version of June, Ministry of Environment of Quebec, Quebec.

Metcalf & Eddy Inc. (1991). *Wastewater Engineering – Treatment, Disposal and Reuse.* 3rd ed., McGraw-Hill, NY.

Richardson, C.J. (1985). Mechanisms controlling phosphorus retention capacity in freshwater wetlands. *Science*, **228**, 1424–1427.

The impact of biomass harvesting on phosphorus uptake by wetland plants

S-Y. Kim* and P.M. Geary**

* Department of Environmental Science and Engineering, Kwangju Institute of Science and Technology (K-JIST), Korea
** School of Geosciences, The University of Newcastle, N.S.W. Australia

Abstract Two species of macrophytes, *Baumea articulata* and *Schoenoplectus mucronatus*, were examined for their capacity to remove phosphorus under nutrient-rich conditions. Forty large bucket systems with the two different species growing in two types of substrate received artificial wastewaters for nine months, simulating a constructed wetland (CW) under high loading conditions. Half of the plants growing in the topsoil and gravel substrates were periodically harvested whereas the other half remained intact. Plant tissue and substrate samples were regularly analysed to determine their phosphorus concentrations.

With respect to phosphorus uptake and removal, the *Schoenoplectus* in the topsoil medium performed better than the *Baumea*. Biomass harvesting enhanced P uptake in the *Schoenoplectus*, however the effect was not significant enough to make an improvement on the overall P removal, due to the slow recovery of plants and regrowth of biomass after harvesting. From P partitioning, it was found that the topsoil medium was the major P pool, storing most of total P present in the system. Plant parts contributed only minor storage with approximately half of that P stored below ground in the plant roots. The overall net effect of harvesting plant biomass was to only remove less than 5% of total phosphorus present in the system.

Keywords Biomass harvesting; constructed wetland; macrophytes; phosphorus removal

Introduction

Pollutants in wetlands are removed by a combination of physical, chemical and biological processes occurring between the sediments, vegetation, and water. These activities achieve high removal efficiencies for BOD, SS and some nutrients from waste sources, as well as the natural die-off of pathogens under certain conditions. As one of the main constituents of agricultural, municipal and industrial wastewaters, phosphorus is removed in CW systems by plant uptake, soil adsorption and precipitation processes. While soil substrates are known as the ultimate sinks of phosphorus, P uptake and storage by wetland vegetation are considered to be minor at high loading rates which are present in these treatment situations.

For a CW system, the selection of substrates is important as their nutrient storage capacities significantly differ from one another. Soil type has little impact on the removal of suspended solids and organics, but it influences the removal of important contaminants such as ammonia, phosphorus and metals. Phosphorus and nitrogen removal are regulated by soil type through ion exchange and adsorption onto humic and fulvic substances and clay particles. Organic soils and clay minerals have higher exchange capacities than coarse mineral substances such as gravel.

Wetland plants also have different P uptake and storage capacities, depending on the species of macrophyte. While the ability of wetlands to retain added phosphorus depends on their P sorption and physico-chemical properties of the substrate, P removal is highly sensitive to loading rate and declines after an initial equilibration period (Tanner *et al.*, 1999). While some studies of aquatic plant systems, such as Brix (1997), have found the uptake capacity of emergent macrophytes to be quite high (30 to 50 kgP/ha/yr), other studies have typically found concentrations of P in plant tissue restricted to a narrow range. For example, an experiment by Asher and Loneragan (1967) used a 625 fold range of P concentrations continuously maintained at the root surfaces but only found a 10 fold range in P concentrations in the shoots of plants growing in these solutions. Plants appear to be able to tune the rate at which they utilise nutrients from their rooting medium to the demand established by their growth rate, however, they are only a short-term sink for phosphorus unless the biomass is harvested (Richardson, 1985).

As an option to enhance the capacity of wetland plants to uptake P, biomass harvesting is therefore often considered. Reed *et al.* (1995) found a direct correlation between the frequency of harvesting and P removal. Where routine harvesting and removal are practiced and the retention time is maintained to be more than seven days, the macrophyte uptake may become significant accounting for 20–30 per cent of the applied P. It has been suggested by Richardson and Craft (1993) that wetlands can fully assimilate phosphorus at low loadings up to 1 $g/m^2/yr$, whereas under high loading rates, the principal mechanism for P removal is usually storage in substrate rather than the biomass (Geary and Moore, 1999). It follows that the removal of the substrate may assist the longer-term operation of the wetland system, but it is not practical in most wastewater treatment conditions where continuous treatment is required. The only option then to sustain high loading rates may be regular plant harvesting and removal, yet research is scant about the impact of biomass harvesting on nutrient removal. A study by Sharma *et al.* (1998) examined the impacts of seasonal harvesting of plant biomass on sewage inflows and found that the P outflow concentrations increased temporarily after cutting for approximately 15 days but decreased thereafter as biomass uptake of phosphorus increased.

The aims of this study were:
1. To investigate the impact of biomass harvesting on phosphorus removal in bucket plant systems which simulated conditions in small CW systems,
2. To examine the partitioning of phosphorus in the bucket plant wetland systems and the uptake of phosphorus in biomass,
3. To suggest whether these macrophyte species and medium types would be suitable for wastewater treatment in CW systems.

Materials and methods

Experimental design

Table 1 illustrates the design of the experimental units. Two species of macrophytes were used with two different substrates. Each unit had five replicates which meant that 20 bucket systems were utilised and samples of the plant tissue and substrate were regularly analysed following weekly additions of phosphorus in solution. A parallel set of 20 bucket systems

Table 1 Design of experimental units

Macrophyte	Medium	Species	Replicates
With harvesting	Topsoil	*Baumea*	×5
		Schoenoplectus	×5
	Gravel	*Baumea*	×5
		Schoenoplectus	×5
Without harvesting	Topsoil	*Baumea*	×5
		Schoenoplectus	×5
	Gravel	*Baumea*	×5
		Schoenoplectus	×5

Figure 1 Bucket systems in the shade house

was also prepared and these were subjected to periodic harvesting of their biomass. The regrowth tissue and substrate were also regularly analysed for their phosphorus concentrations. Figure 1 shows the forty-bucket plant systems which were grown in the shade house area at The University of Newcastle, N.S.W.

P sorption curve

Prior to undertaking the experimental work, a phosphorus sorption curve was developed for the alluvial topsoil to be used as substrate in the experimental units. Soil samples were mixed with graded amounts of P solutions and shaken overnight (17hrs) in a standard test (Rayment and Higginson, 1992). The changes in P concentration in solution were then used to compute P sorbed into the soil sample and to predict the P sorption capacities of the bucket systems.

P analyses of soil and plant tissue samples

The P concentrations in the samples of plant tissue and the soil substrate medium were analysed regularly in the laboratory. The P concentrations were monitored at eight-week-intervals following the weekly addition of 1L of an artificial nutrient solution from January 1998 to September 1998. The nutrient solution contained $NH_4(H_2PO_4)$, KNO_3 and $Ca(NO_3)_2 4H_2O$, resulting in the final concentration of 50mgP/L, 350mgN/L, 385mgK/L and 161mgNH_3/L. The P concentration in plant tissue was determined by digestion with nitric acid on a heating block followed by a colorimetric procedure to measure the absorbance of clear extract (John, 1970). The Bray 1-P level in topsoil and gravel substrate was measured by mixing the samples with ammonium fluoride solution containing hydrochloric acid. The extracts were then analyzed by measuring absorbance (Rayment and Higginson, 1992). When the experiment was completed, individual parts of plant samples were collected and their P concentrations were analyzed for P partitioning. Statistical analysis was carried out with the data from P monitoring during the experiment. Table 2 lists the characteristics of the bucket plant systems.

Results and discussion

P sorption curve

P sorption curves were developed for the topsoil samples in the bucket plant systems, before and after the nutrient additions. The P sorption curve for the pre-treated soil samples showed the topsoil medium had a moderate P sorption capacity (358 mgP/kg). Using the mass and bulk density of the soil substrate, and the measured P sorption capacity of the soil, the estimated "life-time" of the bucket systems (based on P additions) was approximately 45 weeks. Figure 2 indicates that P sorption capacity decreased significantly after nine months of the experiment, however at the completion of the experiment, the topsoil had still not been fully saturated with P and sorption capacity still existed.

P analysis of the substrate and plant tissue samples

The changes in the P concentration of the substrate medium and tissue samples during the experiment are shown in Figures 3a and b. Results represent the mean of five replicates.

Table 2 Bucket plant system characteristics

Hydraulic loading rate:	1 L/week
P concentration used:	50 mg/L
P loading rate:	1.1 g/m^2/week
Bulk density of topsoil medium:	1.74 g/cm^3
Mass of the buckets with topsoil medium:	6.27 kg
Estimated life time before leaching P	45 weeks

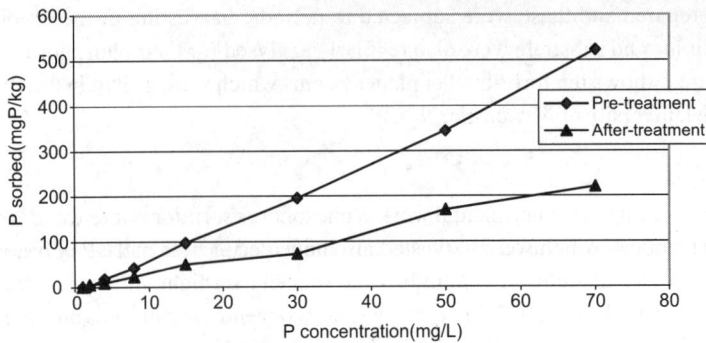

Figure 2 P sorption curve for topsoil substrate

The available topsoil substrate capacity for P adsorption is clearly shown in the results as soil P concentrations continued to increase during the nine months of monitoring. In all cases, the increases between the first and second sampling were greatest. While work was not undertaken to assess the P sorption of the gravel substrate used, it was expected that the topsoil medium had a greater capacity to sorb P compared to the gravel medium. P adsorption capacity varies considerably between soil types but is dependent on particle size, pH,

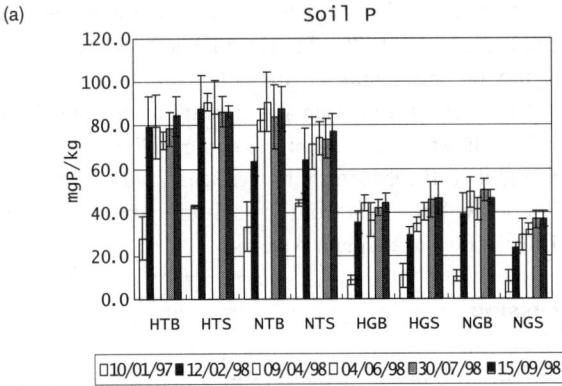

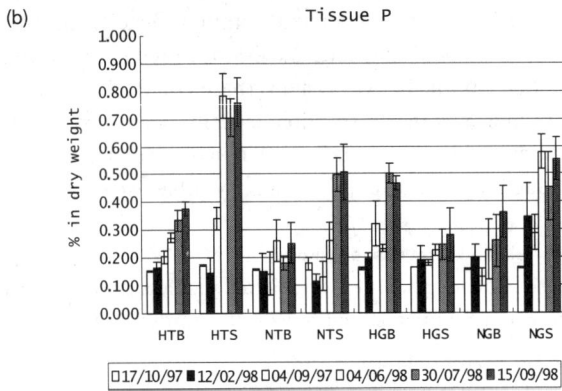

Figure 3 Changes of P concentration in medium and plant tissue samples during the experiment. (a) Bray 1 P changes in substrate over time. (b) Total P changes in plant tissue over time. HTB – *Baumea* planted in topsoil harvested; HTS – *Schoenoplectus* planted in topsoil harvested; NTB – *Baumea* planted in topsoil not harvested; NTS – *Schoenoplectus* planted in topsoil not harvested; HGB – *Baumea* planted in gravel harvested; HGS – *Schoenoplectus* planted in gravel harvested; NGB – *Baumea* planted in gravel not harvested; NGS – *Schoenoplectus* planted in gravel not harvested

redox and the chemical composition of the mineral matter. As the alluvial topsoil samples used had fine particle size with good hydraulic conductivity, there was more opportunity for contact between the soil particles and the nutrient solution and therefore more opportunity for soil adsorption.

In the same way the Bray1-P levels in the soil medium increased, the total P levels in plant tissue also increased during the experimental period. The increases however were more gradual between samplings as biomass increased, particularly in the non-harvested systems. The plants which were harvested also appeared to hold more P in biomass, however, the regrowth of biomass in harvested systems was usually much less than the non-harvested systems. While the harvesting of systems did encourage more P uptake in new biomass, the regrowth was often poor and the plants took time to recover from the effects of cutting. In terms of the macrophytes species, *Shoenoplectus* appeared to uptake and store more P compared to *Baumea*.

Statistical analysis

Table 3 is the summary of three factor-ANOVA results for two variables; (a) Bray1-P levels in substrate medium, and (b) plant tissue P levels. In the case of Bray1-P concentrations, there were significant differences ($P < 0.0001$) between the substrate mediums (topsoil and gravel) throughout the whole experimental period. The soil substrate removed more P than the gravel substrate. Any differences in P concentrations in plant tissues did not appear to be influenced by the different substrate mediums. The treatment by harvesting did not significantly explain P concentration differences in plant tissue. While harvesting may initially appear to enhance P uptake because of the new growth of biomass, the statistical analysis did not show significant differences between the harvested and non-harvested systems. This is apparently due to the physical impact to the plant due to cutting and the time taken for regrowth to occur. The only factor which appeared to be significant in terms of plant tissue P was that *Shoenoplectus* utilised more P than *Baumea* throughout the experiment.

Table 3 Summary of 3 factor-ANOVA results for two variables. Analysis limited to 2-way interactions for ease of interpretation

(a) Bray1-P (available phosphorus in substrate mediums)

Source	t = 1 F-ratio	P	t = 2 F-ratio	P	t = 3 F-ratio	P	t = 4 F-ratio	P	t = 5 F-ratio	P	t = 6 F-ratio	P
Harvest	0.980	NS	3.422	NS	0.579	NS	0.495	NS	0.390	NS	0.256	NS
Medium	196.73	<0.0001	81.66	<0.0001	102.56	<0.0001	102.81	<0.0001	66.84	<0.0001	129.74	<0.0001
Species	12.218	<0.005	0.041	NS	1.609	NS	0.177	NS	0.112	NS	0.431	NS
H × M	0.439	NS	7.411	<0.05	4.273	<0.05	0.017	NS	0.286	NS	0.430	NS
H × S	0.564	NS	0.344	NS	1.283	NS	4.583	<0.05	0.669	NS	0.304	NS
M × S	9.325	<0.005	0.927	NS	0.369	NS	0.004	NS	0.001	NS	1.383	NS

(b) Plant tissue P

Source	t = 1 F-ratio	P	t = 2 F-ratio	P	t = 3 F-ratio	P	t = 4 F-ratio	P	t = 5 F-ratio	P	t = 6 F-ratio	P
Harvest	0.426	NS	2.544	NS	10.981	<0.005	0.042	NS	4.387	NS	1.212	NS
Medium	1.328	NS	14.799	<0.001	0.134	NS	2.015	NS	0.712	NS	2.306	NS
Species	26.893	<0.0001	1.353	NS	2.505	NS	6.449	<0.05	3.596	NS	10.584	<0.01
H × M	2.988	NS	6.496	<0.05	3.757	NS	2.781	NS	1.929	NS	6.215	<0.05
H × S	0.213	NS	1.481	NS	0.638	NS	0.005	NS	1.034	NS	1.459	NS
M × S	8.749	<0.01	6.019	<0.05	0.187	NS	0.181	NS	3.941	NS	8.132	<0.05

t – sampling time (1 October, 1997; 2–12/02/98; 3–09/04/98; 4–04/06/98; 5–30/07/98; 6–15/09/98)
NS – not significant at 0.05 level

Overall, the analyses of P concentrations in the substrate and plant tissue samples provided some evidence that harvesting may marginally enhance P uptake by macrophytes growing in high nutrient load. This effect is however countered by the time taken for the new biomass to regrow. Among the selected species and the substrate medium, the *Schoenoplectus* grown in the topsoil medium was more effective in removing P than the systems with *Baumea* and a gravel medium.

Phosphorus partitioning in topsoil systems
At the completion of the bucket plant experiment, an analysis of the storage of P in the specific components of each topsoil system was undertaken. In all cases it appeared that substrate was the main component of P storage as more than 90% of total P present in the system was in the bucket substrate. Of the remaining P, the systems that were harvested had less P in plant parts than the systems that were not harvested. This was primarily due to the smaller biomass in the systems that had been harvested. Most of these systems recovered only slowly after the periodic harvesting and consequently ended up with very low biomass compared to those that were not harvested. Whether they were harvested or not, the overall contribution of macrophytes to the P removal in treatment systems appeared to be relatively small. The bucket plant experiments revealed that only a small proportion of nutrient (P) is stored in the plant parts (less than 10% by total P), with more than 90% total P remaining in the topsoil substrate at the conclusion of the experiment. Because around 50% of the P in plant parts is stored below ground, harvesting above ground biomass would have removed less than 5% of total P in the system. Figure 5 illustrates estimated P partitioning for the bucket plant systems with topsoil medium.

Conclusions
The results from this study suggest that species differ in their ability to utilise phosphorus and that the substrate is the major storage mechanism with respect to P removal in constructed wetland systems. The major findings from this study are as follow.
- Biomass harvesting appeared to enhance the P concentration in plant tissues for a short period, and the effect was stronger when the plants were grown in the topsoil medium than the gravel medium. *Shoenoplectus* responded more effectively to the harvesting than *Baumea*.
- In the longer term, the harvesting of biomass to assist with additional P removal may only assist in lightly loaded systems but would not be practicable in highly loaded systems given that the major storage pool of P is in the substrate and not in the plant tissue.
- The productivity of macrophytes was affected significantly by the harvesting, and this resulted in lower P removal (partitioning) by the wetland vegetation compared to systems that were not harvested.

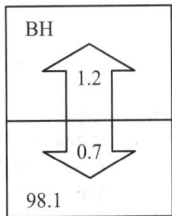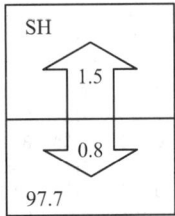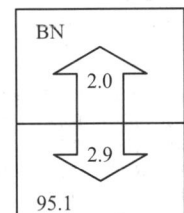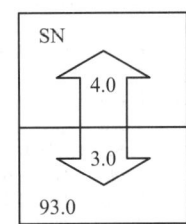

Figure 5 Estimated phosphorus partitioning (% by total P) in the bucket systems with topsoil at September 1998 (BH-*Baumea* harvested; SH-*Schoenoplectus* harvested; BN-*Baumea* not harvested; SN-*Schoenoplectus* not harvested)

- *Schoenoplectus* appeared to perform better than *Baumea* in P removal in the bucket systems.
- The topsoil medium had a significant capacity for the sorption of P compared to the gravel. It was found that more than 95% of P present in the bucket system was stored in the topsoil medium.
- Although the P partitioning in the wetland vegetation pool is small, the presence of macrophytes in CW systems is still important because of the role they play in assisting other reactions relevant to nutrient removal processes.
- Biomass harvesting is not a realistic management option for most CW systems because the harvesting of macrophytes may disturb substrate and potentially release additional P to the water column.

References

Asher, C.J. and Loneragan, J.F. (1967). Response of plants to phosphate concentration in solution culture. I Growth and phosphorus content, *Soil. Sci.* **103**, 225–233.

Brix, H. (1997). Do macrophytes play a role in constructed treatment wetlands? *Wat. Sci. Tech.* **35**(5), 11–17.

Geary, P.M. and Moore, J.A. (1999). Suitability of a treatment wetland for dairy wastewater, *Wat. Sci. Tech.* **40**(3), 179–186.

John, M.K. (1970). Colorimetric determination of phosphorus in soil and plant materials with ascorbic acid. *Soil Sci.* **109**(4), 214–220.

Lanzke, I.R., Heritage, A.D., Pistillo, G. and Mitchell, D.S. (1998). Phosphorus removal rates in bucket size planted wetlands with a vertical hydraulic flow, *Wat. Res.* **32**(4), 1280–1286.

Rayment, G.E. and Higginson, F.R. (1992). *Australian laboratory handbook of soil and water chemical methods*. Inkata Press, Melbourne.

Reed, S.C., Crites, R.W. and Middlebrooks, E.J. (1995). *Natural systems for waste management and treatment*. 2nd Ed. McGraw Hill Inc.

Richardson, C.J. (1985). Mechanisms controlling phosphorus retention capacity in freshwater wetlands, *Science*, **228**, 1424–27.

Richardson, C.J. and Craft, C.B. (1993). Effective phosphorus retention in wetlands: fact or fiction? In: *Constructed Wetlands for Water Quality Improvement*, G.A.Moshiri (ed), Lewis Publishers, Boca Raton, pp 271–282.

Tanner, C.C., Sukias, J.P. and Upsdell, M.P. (1999). Substratum phosphorus accumulation during maturation of gravel-bed constructed wetlands, *Wat. Sci. Tech.* **40**(3), 147–154.

Phosphorus retention capacity of iron-ore and blast furnace slag in subsurface flow constructed wetlands

B. Grüneberg and J. Kern

Institute of Agricultural Engineering, Bornim, Max-Eyth-Allee 100, 14469 Potsdam, Germany

Abstract The suitability of iron-ore and blast furnace slag for subsurface flow (SSF) constructed wetlands was studied over a period of four months. Dairy farm wastewater (TP 45 mg l^{-1}) was percolated through buckets planted with reed (volume 9.1 l; hydraulic load 15 l $m^{-2}d^{-1}$). One group of buckets was kept under aerobic conditions and the other group under anaerobic conditions, monitored by continuous redox potential measurements. Even at high mass loading rates of 0.65 g P $m^{-1}d^{-1}$ the slag provided 98% removal efficiency and showed no decrease in performance with time. However, phosphorus fractionation data indicate that the high phosphorus retention capacity under aerobic conditions is to a great extent attributable to unstable sorption onto calcium compounds (NH_4Cl-P). Phosphorus sorption of both the slag (200 μg P g^{-1}) and the iron-ore (140 μg P g^{-1}) was promoted by predominantly anaerobic conditions due to continuous formation of amorphous ferrous hydroxides. None of the substrates had adverse affects on reed growth.

Keywords Phosphorus retention; constructed wetlands; iron-ore; blast furnace slag; dairy farm wastewater

Introduction

Constructed wetlands (CWs) have become one of the preferred systems for wastewater treatment in rural areas. For most subsurface flow (SSF) constructed wetlands, local gravel or sand is used as filter medium. In contrast to other major pollutants (C, N), phosphorus (P) has no relevant gaseous phase in its biochemical cycle. Consequently, the P sorption capacity of the filter media is the main factor limiting the life expectancy of SSF constructed wetlands.

P retention mechanisms in wetlands were described by Faulkner and Richardson (1989), Kadlec and Knight (1996), Richardson *et al.* (1997), and have recently been reviewed by Reddy *et al.* (1999). P can be retained in the substrate via adsorption and precipitation, with both mechanisms being controlled by properties of the substrate (Fe-, Al-, Ca-minerals, porosity) and the physico-chemical environment (pH, Eh, dissolved ions). Therefore, the study focused on (1) filter media with high P-binding capacity and (2) physicochemical parameters that promote sorption of P.

Apart from the gravel mainly used in SSF systems, the suitability of various media with potentially high P sorption capacity was examined. Laboratory-scale experiments (Yamada *et al.*, 1986; Mann, 1997; Sakadevan and Bavor, 1998; Johansson, 1999) have proved that large amounts of P are sorbed onto the surface of blast furnace slag, predominantly associated with light elements such as calcium, aluminium, magnesium and silicon. Calcium hydroxide was found to adsorb the largest amounts of phosphate at an optimum pH of around 8. Only a few researchers have investigated the potentially high P-binding capacity of materials containing iron via specific adsorption of P onto iron hydroxides. Among the media studied are iron-scrap (Hirth and Schönborn, 1994), laterite (Wood and McAtamney, 1996), iron-ore (Malessa *et al.*, 1999) and used sand from groundwater iron elimination (Rustige and Platzer, 2000).

Methods

Properties and preparation of substrates

Blast furnace slag and iron-ore were selected for investigation. Additionally, pure quartz sand was chosen as a control to relate differences in treatment performances to substrate properties. The iron-ore used has a high content of Fe(III) in the form of hematite (Table 1). Only the fraction 0.3 to 1.0 mm was used for the experiment to establish equivalent conditions among the substrates regarding particle size and hydraulic conductivity. For the experiment, one part of the selected particle size of iron-ore was mixed with five parts of quartz sand to produce a substrate containing 16.5% iron-ore (Table 2). Granulated blast furnace slag is formed when molten slag is cooled rapidly by high velocity water jets. The resulting material consists of an amorphous matrix and is characterised by high porosity and low bulk density. The slag was mixed with quartz sand to produce a substrate containing 63% slag.

Experiment design and operation

For the experiment 16 buckets (surface area: 0.0314 m^2; total volume: 9.1 dm^3) were used, each of them representing an individual wetland system. As shown in Table 2, six buckets contained the slag substrate, another six contained the iron-ore substrate, and four buckets were filled with pure quartz sand as control. The buckets were planted with reed (*Phragmites australis*), filled with a nutrient solution and allowed to establish for 6 weeks prior to the trial period, which was divided into two phases of 6 weeks each. All buckets were kept under waterlogged conditions, although an occasional drying out of the buckets could not be prevented on a few occasions. For one group of the buckets aerobic conditions were maintained by bubbling air (flow approx. 100 l h^{-1}) through the saturated

Table 1 Properties of the original materials (source: ECO Stahl GmbH Eisenhüttenstadt)

	Quartz sand	Blast furnace slag	Iron-ore
Density	2.65 g cm^{-3}	2.4–2.5 g cm^{-3}	~5.3 g cm^{-3}
Bulk density	1.5 g cm^{-3}	1.1 g cm^{-3}	2.3 g cm^{-3}
SiO$_2$	97.2%	35.0%	1.00%
FeO		0.24%	1.30%
Fe$_2$O$_3$	0.03%		95.9%
TiO$_2$	0.08%	1.00%	0.05%
K$_2$O	0.86%	0.63%	0.01%
Na$_2$O	0.03%	0.37%	0.01%
CaO	0.02%	38.6%	0.05%
MgO	0.01%	10.3%	0.05%
Al$_2$O$_3$	1.45%	11.5%	0.50%
P		1.25%	0.04%
Mn		0.42%	0.40%
S		1.50%	0.002%
C	0.25%		0.01%

Table 2 Properties of substrate mixtures used for experiments

Applied substrate mixture	Pure quartz sand		63% slag in quartz sand		16.5% iron-ore in quartz sand	
Particle size distribution	0.3–0.8 mm		0.3–1.0 mm		0.2–2.0 mm (90% <1.0 mm)	
Hydraulic conductivity	1.4 × 10^{-3} m s^{-1}		5 × 10^{-3} m s^{-1}		2.5 × 10^{-3} s^{-1}	
Number of replicates	with aeration	without aeration	with aeration	without aeration	with aeration	without aeration
	2	2	3	3	3	3

Table 3 Mean influent concentrations, mass loading rates on a mass per area and mass per volume basis of major wastewater compounds. C = concentration (mg l^{-1}); A = mass loading rate per area (g m^{-2} d^{-1}); V = mass loading rate per volume (g m^{-3} d^{-1}); # = mean influent parameters of 300 European CWs compiled by Börner (1992).

	TP			PO$_4$-P			COD			TN		
	C	A	V	C	A	V	C	A	V	C	A	V
Phase 1	46	0.52	2.4	39	0.45	2.0	300	2.8	13	43	0.55	2.4
Phase 2	42	0.65	3.0	18	0.28	1.3	1199	19	86	213	3.3	15
#	13	0.40	0.6				350	14	19	60	1.6	2.3

substrate. Buckets without aeration became anaerobic after a few days of operation. The influent was dairy farm wastewater (Table 3), which is produced in large amounts during the milking process; it comprises cattle slurry, milk, washing detergents, disinfectants and water (Kern, 1999). The buckets were batch loaded three times a week with a hydraulic loading rate of 15 l m^{-2} d^{-1}. The wetland systems were operated at a constant water level. The effluent volume for each bucket was recorded and an aliquot part was stored at –18°C for chemical analysis. As a result of this sampling procedure, a mass balance could be calculated for each wastewater constituent.

Chemical analysis

Conductivity, pH and redox potential were measured in fresh samples every week using electrodes and pH/mV meters (WTW). Dissolved ions (Cl, NO$_2$ NO$_3$, PO$_4$, SO$_4$, NH$_4$, Na, K, Mg, Ca) were determined by ion chromatography. Chemical oxygen demand (COD) was measured photometrically after digestion in potassium dichromate. Total nitrogen (TN) was measured after digestion in a mixture of concentrated sulphuric acid and selenium (Kjeldahl digest) at 400°C for at least one hour. Total phosphorus (TP) was measured in the same digest by the vanadate-molybdate method.

At the end of the experiment substrate samples were taken from three different depths. Immediately after sampling, a sequential extraction (0.5 M NH$_4$Cl, 1 M NaOH, 0.5 M HCl) was started using a modified method adopted from Hieltjes and Lijklema (1980) and Psenner *et al.* (1984) to determine different fractions of P. TP of substrate samples was determined with the molybdenum blue method after ignition at 550°C and extraction in sulphuric acid. Iron and calcium concentrations were measured in the extracts by the ICP-OES technique. Oxalate extractable Fe was measured by AAS after extraction in a mixture of ammonium oxalate and oxalic acid (Schwertmann, 1964). The amount of total exchangeable iron was measured after extraction with sodium dithionite and sodium citrate (CDB). The reed plants were harvested and the above-ground biomass and the mass of rhizomes and roots were determined after drying at 60°C for 48 hours. Finally, N and P content were measured after Kjeldahl digestion.

Results and discussion
Effluent water quality

During the whole study period aerobic conditions could be maintained in the buckets with aeration system. Redox potential (Eh) did not fall below 300 mV. On the other hand, predominantly anaerobic conditions prevailed in the buckets without aeration with mean redox potentials of –50 mV (measured in fresh effluent). It is important to note that low redox potentials are characteristic for horizontal flow SSF systems, despite the vertical down flow regime within the buckets.

The systems with aeration generally showed higher removal performances and lower effluent concentrations for PO$_4$ and TP (Figure 1). The supply of oxygen via aeration

Table 4 Effluent concentrations (mg l^{-1}) of major wastewater constituents; removal (%) in parentheses is based on mass balance calculations

		Without aeration (anaerobic)			With aeration (aerobic)		
		quartz sand	iron-ore	slag	quartz sand	iron-ore	slag
TP	phase 1	17±9.4 (83)	7.0±2.8 (93)	0.8±0.5 (99)	6.7±2.2 (93)	6.3±2.5 (94)	1.2±0.5 (99)
	phase 2	48±7.1 (62)	24±2.9 (82)	4.5±0.9 (97)	8.2±2.6 (94)	7.6±0.6 (95)	2.5±0.7 (98)
PO$_4$-P	phase 1	14±6.9 (83)	4.0±1.9 (96)	0.5±0.5 (99)	4.4±1.8 (95)	4.2±1.6 (95)	0.9±0.5 (99)
	phase 2	21±6.6 (55)	7.2±2.1 (86)	0.6±0.7 (99)	4.8±3.2 (90)	4.4±0.5 (92)	1.0±0.7 (98)

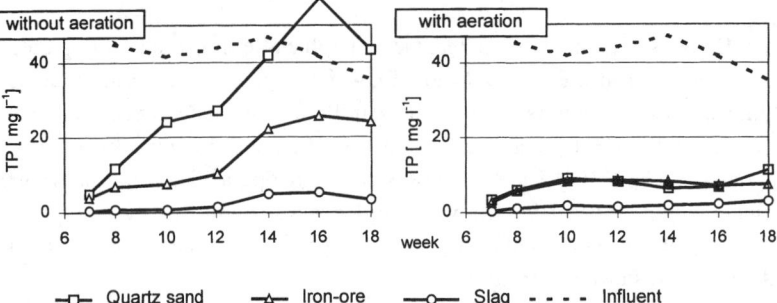

Figure 1 Influent and effluent concentrations of total phosphorus

should have led to a more rapid degradation of organic matter which might have enhanced the release and subsequent substrate sorption of formerly incorporated P. Within the anaerobic systems the slag media showed the best treatment performance, followed by iron-ore and the quartz sand control. Despite high loading rates particularly in phase 2 (Table 3), effluent concentrations from buckets with slag did not exceed 5 mg l^{-1} and 1 mg l^{-1} for TP and PO$_4$-P, respectively (Table 4). There was no evidence of saturation up to the end of the study period.

Phosphorus mass balance and sorption characteristics

During the whole study period 1.83 g P entered each bucket. Mass balance calculations at the end of the study period indicate that the substrates generally stored the largest amount of P, followed by total reed biomass (Table 5). Within the anaerobic systems the quartz sand released about one quarter of total input P via the effluent, while the slag media almost completely retained P. Between 4.7 and 6.3 g P m^{-2} was stored in above-ground biomass and up to 4.7 g P m^{-2} in roots and rhizomes. These numbers are quite high compared to data from other studies where P removal potential of *Phragmites* usually ranges from 2.3 g P m^{-2} to

Table 5 Phosphorus mass balance; data in percentage of total phosphorus input for each bucket

	without aeration (anaerobic)			with aeration (aerobic)		
	quartz sand	iron-ore	slag	quartz sand	iron-ore	slag
effluent	26±1.8	11±0.4	2±0.3	6±0.1	5±0.4	1±0.1
total reed biomass	31±1.5	36±4.7	27±0.8	24±0.3	28±1.8	27±2.8
above-ground reed biomass	11	10	10	9	8	8
below-ground reed biomass	20	26	17	15	19	18
substrate	43±0.3	53±4.3	71±0.5	70±0.2	67±1.7	72±2.8

☐ effluent
▨ total reed biomass
■ substrate

4.3 g P m^{-2} (Breen, 1990; Börner, 1992; Kadlec and Knight, 1996; Drizo et al., 1997). No significant differences regarding potential influence of substrate properties on plant growth could be detected. Despite high P biomass storage, these data indicate that differences in P treatment performance have to be almost entirely attributed to differences in sorption characteristics of the filter substrates.

The P fractions derived from the sequential extraction are shown in Figure 2. TP (83–283 µg P g^{-1}) was relatively low for all substrates due to the short duration of the experiment. We suggest that the accumulation of organic P (17–35% of substrate P) did not depend on the filter substrate or the redox status. This fraction may only increase with the time of operation. The same applies to the HCl-P fraction (Ca- and Mg-bound P) which is generally the smallest storage compartment with only about 10 µg P g^{-1}. This means that the high P sorption capacity of the slag can not be explained by precipitation of P as stable calcium phosphates. Consequently, only two fractions are responsible for major differences in P retention. NH$_4$Cl-P was the major fraction for the systems with aeration, whereas without aeration NaOH-P was the major P fraction.

High amounts of P (up to 148 µg P g^{-1}) were stored as NH$_4$Cl-P in the slag substrate (Figure 2). Similar results were reported by Johansson (1999) who found 51% of total P sorbed as NH$_4$Cl-P after 56 weeks of P loading during a column experiment. This fraction is defined as readily available or loosely bound P (Hieltjes and Lijklema, 1980; Psenner and Pucsko, 1988) or P associated with free Ca or Mg carbonates (van Eck, 1982). Especially during the start of the experiment, high concentrations of Ca^{2+} (>200 mg l^{-1}), high pH (>7) and constant supply of PO$_4$ may have favoured the precipitation of calcium phosphates. High pH values prevailed in the systems with aeration, so that about 50% of P occurred as NH$_4$Cl-P in these substrates. Johansson and Gustafsson (2000) presume that hydroxyapatite is precipitated when there is supersaturation for Ca^{2+} and PO$_4^{3-}$ and high pH. Blast furnace slag is known to release large amounts of Ca^{2+}, Mg^{2+} and OH$^-$ ions within the first weeks of contact with water (Larn, 1998), resulting in intensive sorption of P. From laboratory-scale sorption experiments and subsequent observations with scanning electron microscope and X-ray microanalyzer (Yamada et al., 1986), it was found that sorption sites of P were coincident with the sites of mainly calcium, magnesium and other light elements. Regardless which P sorption mechanism prevailed during the experiment, sequential extraction data indicate that the quantity of P sorption onto slag is highly pH-dependent. Decreasing pH (and probably Ca^{2+}, Mg^{2+} concentrations) in the NH$_4$Cl solution of the first extraction step will lead to a rapid release of sorbed P from the slag. Consequently, the high P sorption capacity of blast furnace slag always found in laboratory scale adsorption experiments does not allow any prediction of long-term performance of these media in construct-

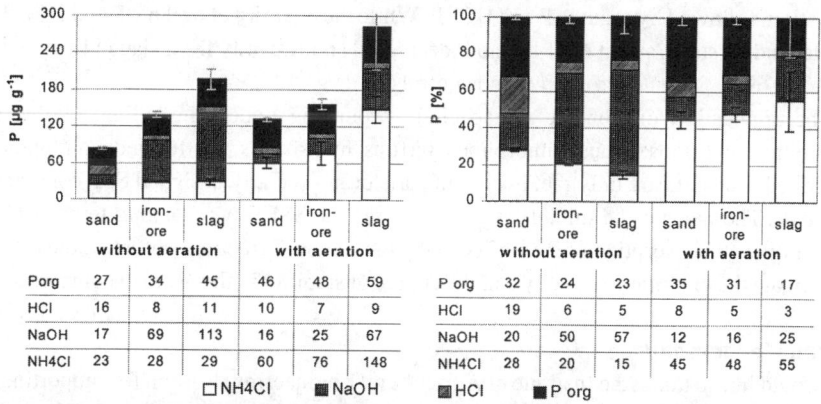

Figure 2 Fraction of phosphorus in the substrates (dry weight) according to the sequential extraction.

ed wetlands. Under anaerobic conditions, characteristic for horizontal flow SSF systems, the slag did not sorb more NH_4Cl-P than the sand or the iron-ore (Figure 2).

The following extraction step using sodium hydroxide (NaOH) dissolves iron- and aluminium-bound P. Due to the high Fe(III) content it was expected that the iron-ore would be an efficient sorption medium, but the slag clearly stored more NaOH-P than the iron-ore and the quartz sand control. Under continuous aerobic conditions (Eh > 300 mV), less P was bound than under conditions of lower redox status. The iron-ore in its original form mainly consists of crystalline iron oxides. Under aerobic conditions the material remains unchanged and the P sorption capacity is low. Anaerobic conditions have caused reduction of ferric iron and subsequent formation of amorphous ferrous hydroxides. The newly formed Fe compounds are very efficient in binding P by specific adsorption (Faulkner and Richardson, 1989). The amount of Fe_{ox} in the iron-ore had significantly increased from 1.2 to 14% during the study period. Oxygen may have entered the systems without aeration due to occasional drying out of the buckets and via root zone aeration by the reed plant. Root zone aeration is likely to gain importance with increasing plant biomass and decreasing substrate depth and hence high root density. As reported by Wang and Peverly (1999) large amounts of re-oxidised iron were found attached to the roots of reed. In the slag all iron was in the amorphous form and increased as well. The molar ratio Fe:P in the NaOH-extract of the slag was below 0.5. Therefore, other elements too, such as Al or Mn, must have acted as important sorption partners for P. This is also indicated by high amounts of NaOH-P in the slag under both aerobic and anaerobic conditions, suggesting that this storage compartment is to a great extent not redox-sensitive. The future applicability of the slag for P elimination will depend on the maximum quantity of NaOH-P. This fraction comprises the only significant redox-independent long-term P storage compartment in contrast to the highly unstable NH_4Cl-P.

Conclusions

The blast furnace slag showed the highest P retention capacity, followed by iron-ore and the quartz sand control. Effluent concentrations from buckets with slag did not exceed 5 mg l^{-1} for TP and 1 mg l^{-1} for PO_4-P, respectively. Substratum sorption was the major retention mechanism, but large amounts of P were also stored in above-ground (4.7–6.3 g P m^{-2}) and below-ground (4.7 P m^{-2}) biomass. A major part of the short-term P-binding capacity of the slag is attributed to highly unstable sorption onto calcium compounds (NH_4Cl-P). After a few weeks of slag application, the leaching of Ca^{2+} and OH^- from the slag will decrease and weakly bound P might be released. In addition to widely applied laboratory-scale adsorption experiments, long-term studies under authentic conditions are essential for further evaluation of the P sorption capacity of blast furnace slag. NaOH-P was the most important P fraction, followed by organic P and HCl-P. While the quantity of organic P and HCl-P did not depend on substrates, Fe/Al-bound P decreased significantly from slag (113 µg g^{-1}) to iron-ore (69 µg g^{-1}) and the quartz sand control (17 µg g^{-1}).

The application of iron-ore is especially promising under alternating aerobic and anaerobic conditions, where amorphous ferrous hydroxides are formed continuously. These requirements are only met in certain parts or in specially designed SSF constructed wetlands. The slag may be suitable as filter substrate for SSF CWs on a long-term basis due to the rather stable sorption of P as reflected by the NaOH-P fraction. Future studies should investigate the maximum capacity and the degree of stability in this storage compartment.

Acknowledgements

We would like to thank the Institute of Agricultural Engineering, Bornim for supporting the study, Dr. Blumenstein for encouragement and discussions, Dr. Bukowsky and Mrs.

Limpach for chemical analyses and EKO Stahl GmbH Eisenhüttenstadt for supplying the substrate materials.

References

Börner, T. (1992). *Einflußfaktoren für die Leistungsfähigkeit von Pfanzenkläranlagen.* PhD thesis, Schriftenreihe WAR Heft 58, Institut für Wasserversogung, Abwasserbeseitgung und Raumplanug der TH Darmstadt, Germany.

Breen, P.F. (1990). A mass balance method for assessing the potential of artificial wetlands for wastewater treatment. *Wat. Res.*, **24**(6), 689–697.

Drizo, A., Frost, C.A., Smith, K.A. and Grace, J. (1997). Phosphate and ammonium removal by constructed wetlands with horizontal subsurface flow, using shale as a substrate. *Wat. Sci. Tech.*, **35**(5), 95–102.

Faulkner, S.P. and Richardson, C.J. (1989). Physical and chemical characteristics of freshwater wetlands soils. In: *Constructed Wetlands for Wastewater Treatment*, D.A. Hammer (ed.), Lewis Publishers, Chelsea, Michigan, USA, pp. 41–72.

Hieltjes, A.H.M. and Lijklema, L. (1980). Fractionation of inorganic phosphates in calcareous sediments. *J. Environ. Qual.*, **9**(3), 405–407.

Hirth, A. and Schönborn, A. (1994). *Verteilung und Bindungsformen von Phosphur und Eisen in Pflanzenfilter Schwattweig*. Centrum für angewandte Ökologie Schattweid; CH-6114 Steinhuserberg, Germany, 1–31.

Johansson, L. (1999). Blast furnace slag as phosphorus sorbents – column studies. *Sci. Tot. Envir.*, **229**, 89–97.

Hohansson, L. and Gustafsson, J.P. (2000). Phosphate removal using blast furnace slags and opoka-mechanisms. *Wat. Res.*, **34**(1), 259–265.

Kadlec, R.H. and Knight, R.L. (1996). *Treatment Wetlands*. Lewis Publishers, Boca Raton, Florida, USA.

Kern, J. (1999). Seasonal efficiency of a constructed wetland for treating dairy farm wastewater. *Proceedings of the 4th International Conference on: Managing the wastewater resource*, Às, Norway.

Larn, A. (1998). *Entwicklung des LAGA Richtlinien-Entwurfs für die Verwendung von Eisenhüttenschlacken basierund auf Originalsubstanzgehalten und Eluatkonzentrationen*. Bildungszentrum für die Entsorgung und Wasserwirtschaft, Essen, Germany.

Malessa, R., Kampf, c. and Niehus, C. (1999). *Schwermetallentfernung aus Klärschlamm*. Abschlußbericht, Fachhoschule Brandenburg, Germany.

Mann, R.A. (1997). Phosphorus adsorption and desorption characteristics of constructed wetland gravels and steelwork by-products. *Aust. J. Soil Res.*, **35**, 375–384.

Psenner, R. and Pucsko, R. (1998). Phosphorus fractionation: advantages and limits of the method for the study of sediment P origins and interactions. *Arch. Hydriobiol. Beih.*, **30**, 43–59.

Psenner, R., Puscko R. and Sager, M. (1984). Die Fraktioneierung orcanischer und anorganischer Phosphoverbindungen von Sedimenten – Versuch einer Definition ökologisch wichtiger Faktoren. *Arch. Hydrobiol.Beih,* **70**, 111–155.

Reddy, K.R., Kadlec, R.H., Flaig, E.G. and Gale, P.M. (1999). Phosphorus retention in streams and wetlands: a review. *Critical Reviews in Environmental Science and Technology*, **29**(1), 83–146.

Rustige, H. and Platzer, C. (2001). Nutrient removal in substrate flow constructed wetlands for application in sensitive regions. *Wat. Sci. Tech.*, **44**(11–12), 149–155 (this issue).

Richardson, C.J., Qian, S.S., Craft, C.B. and Qualls, R.G. (1997). Predictive models for phosphorus retention in wetlands. *Wetlands Ecology and Management*, **4**(3), 159–175.

Sakadevan, K. and Bavor, H.J. (1998). Phosphate adsorption characteristics of soils, slags and zeolite to be used as substrates in constructed wetland systems. *Wat. Res.*, **32**, 393–399.

Schwertmann, U. (1964). Differenzierung der Eisenoxide des Bodens durch Extraktion mit Ammoniumoxalat-Lösung. *Z. Planfz. Bodenk.*, **105**(3), 194–202.

van Eck, G.T.M. (1982). Forms of phosphorus in particulate matter from the Hollands Diep/Haringvliet, The Netherlands. *Hydrobiologia* **92**, 665–681.

Wang, T. and Peverly, J.H. (1999). Iron oxidation status of root surfaces of a wetland plant (Phragmites australis). *Soil. Sci. Soc. Amer. J.*, **63**, 247–252.

Wood, R.B. and McAtamney, C.F. (1996). Constructed wetlands for waste water treatment: The use of laterite in the bed medium in phosphorus and heavy metal removal. *Hydrobiologia*, **340**, 323–331.

Yamada, H., Kayama, M., Saito, K. and Hara, M. (1986). A fundamental research on phosphate removal by using slag. *Wat. Res.*, **20**(5), 547–557.

The removal of nutrients from plant nursery irrigation runoff in subsurface horizontal-flow wetlands

T.R. Headley*, D.O. Huett** and L. Davison*

* School of Resource Science and Management, Southern Cross University, PO Box 157, Lismore, NSW 2480, Australia. (E-mail: thead10@scu.edu.au)
** Tropical Fruit Research Station, NSW Agriculture, Bruxner Hwy, Alstonville, NSW 2477, Australia

Abstract In New South Wales (NSW) Australia, the recent introduction of legislation to control runoff and charge for water used in agricultural production has encouraged commercial plant nurseries to collect and recycle their irrigation drainage. Runoff from a nursery typically contains around 6 mg/L TN (> 70% as NO_3), 0.5 mg/L TP (> 50% as PO_4), and virtually no organic matter (BOD < 5 mg/L; DOC < 20 mg/L). As a result, algal blooms frequently occur in storage dams. This paper describes a study evaluating the effectiveness of subsurface flow wetlands in the removal of nutrients from nursery runoff on the sub-tropical northern coast of NSW, Australia. Four experimental subsurface flow wetlands (1 m × 4 m × 0.5 m water depth) were planted with *Phragmites australis* in April 1999. TN and TP load removals were > 84% and > 65% respectively at HRTs of between 5 and 2 days, with the majority of out-flowing TN and TP being organic in form. Internal generation of organic N and P resulted in persistent background levels of 0.45 mg/L TN and 0.15 mg/L TP in the reed bed effluent. TN, NH_4 and TP removal was affected by HRT ($P < 0.05$). Greater than 90% load removal of NH_4, NO_2, NO_3 and Ortho-P was achieved at all HRTs, with outlet concentrations generally < 0.01 mg/L for all. For TN, a strong relationship existed between removal rate (g/m^2/day) and loading rate ($r^2 = 0.995$), while a weaker relationship existed for TP ($r^2 = 0.47$). It is estimated that a 1 ha nursery would require a reed bed area of 200 m^2 for a 2 day HRT.

Keywords Background levels; hydraulic residence time; nitrogen; nursery runoff; phosphorus; subsurface horizontal-flow wetlands

Introduction

The nursery industry in Australia has a farm gate value of $A540 million, and in the most populous state, New South Wales (NSW), a value of $A167 million (Atkinson, 1998). Production is largely outdoors, with some under shade cloth. Irrigation is typically by fixed sprinklers. To date, few nurseries have collected and recycled runoff. However, recent changes to legislation governing water use have resulted in restrictions being placed on nutrient levels in water leaving nurseries. Agricultural industries are now required to pay for the water they use, providing further incentive for nurseries to recycle irrigation water. A recent study monitored nutrient concentrations in runoff water from a nursery at Alstonville (29°S, 153°E) on the northern coast of NSW and the mean annual nitrate-N (NO_3) and phosphate-P (PO_4) concentrations were 9 and 2 mg/L respectively (Huett, 1999). Runoff from this nursery is collected in dams for recycling and despite a drop in NO_3 and PO_4 concentrations to 0.6 and 0.8 mg/L respectively, algal blooms frequently occur. The pH of this stored water rises and it must be filtered, acidified and chlorinated-brominated before it is suitable for re-use in the nursery. Thus, there is a need to reduce the nutrient content of nursery runoff and prevent algal blooms from occurring in storage dams.

Subsurface flow wetlands (reed beds) are a treatment device with the advantages of being relatively inexpensive, simple to operate and requiring little maintenance. Over the last 25 years reed beds have been used to treat a range of nutrient enriched waters with favourable nutrient removal performance, particularly for nitrate. However, reported nutrient removal efficiencies vary from 20% to 95% (Bastian and Hammer, 1993; Cooper,

1993; Juwarkar et al., 1995; Reed et al., 1995; Vymazal, 1995; Ottova et al., 1997). The majority of studies have focused on wastewaters with a relatively high nutrient content, such as sewage and dairy effluent.

The present study was undertaken to evaluate the effectiveness of subsurface, horizontal-flow reed beds for the removal of nitrogen and phosphorus from nursery irrigation runoff on the sub-tropical northern coast of NSW, Australia.

Materials and methods
Composition of nursery runoff
Nursery irrigation runoff from four nurseries was monitored between December 1998 and April 1999 and analysed for macro- and micro-elements. The nurseries varied in size and plant species. The nutrient concentration of runoff varied somewhat depending on irrigation efficiency (TN 1.35–18 mg/L, TP 0.05–1.5 mg/L). The distinguishing features of this runoff were a) the majority of nitrogen (> 70%) occured as nitrate, b) over 50% of the phosphorus was in the form of orthophosphate (Ortho-P), and c) there were negligible suspended solids and BOD. For the purpose of this study, a "test solution" of simulated nursery runoff containing 6 mg/L TN (> 70% as NO_3), 0.5 mg/L TP (> 40% as PO_4), and representative concentrations of other elements was prepared.

Site description
Four intermediate scale horizontal flow reed beds were set up at the Tropical Fruit Research Station (NSW Agriculture) in Alstonville on the northern coast of New South Wales, Australia. Figure 1 shows a plan view of the study site. All four reed beds are identical in design, thus allowing replication in experiments. The reed beds consist of a fibreglass shell, 4 m long × 1 m wide, filled with 10 mm basaltic gravel to give a water depth of 0.5 m. A longitudinal section is shown in Figure 2. The beds were planted with rhizome cuttings of *Phragmites australis* in April 1999. To promote reed growth and establishment, the reed beds were regularly topped-up with a soluble fertiliser until monitoring commenced in September 1999 (5 months later).

The test solution was mixed in two header tanks, with each tank supplying runoff to two reed beds. Timer activated pumps intermittently dosed test solution to each reed bed (6 times per day). Water meters were used to record the volume entering the reed beds on a weekly basis. The treated test solution from each reed bed was then held in a collection tank. The volume of reed bed outflow was recorded weekly via a level tube on the side of each collection tank.

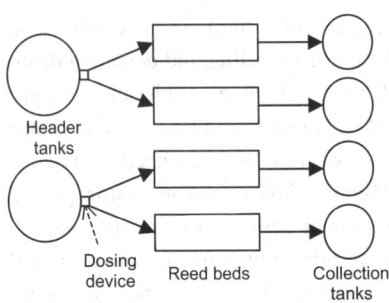

Figure 1 Plan view of system arrangement at Alstonville

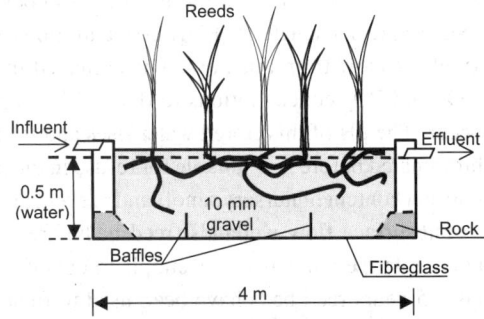

Figure 2 Simplified longitudinal section through reed bed

Water quality monitoring

Weekly inlet and outlet samples were collected from September 1999. An additional sample from halfway along each reed bed was also collected from November 1999. This paper presents results of monitoring up until April 2000 (7 months). Halfway samples were obtained from a vertical length of 50 mm perforated pipe running from the gravel surface to the bottom of each reed bed. Weekly samples were analysed for TN, NH_4, NO_2, NO_3, TP, Ortho-P, and pH, while water temperature was logged daily. Dissolved organic carbon (DOC) was determined less frequently. An on-site weather station records meteorological data.

Nutrient samples were analysed using Flow Injection Analysis on a Lachat QuickChem 8000 Automated Ion Analyser, and DOC analysed on a Shimadzu TOC 5000 analyser. pH was measured in the field using a hand held probe. All methods were in accordance with those described in the *Standard Methods for the Examination of Water and Wastewater* (1995).

Experimental approach

The primary objective of the monitoring program at this stage of the project is to examine the nutrient removal performance of the reed beds under a number of Hydraulic Residence Times (HRTs), and subsequently define an "optimal HRT" – the shortest residence time at which the desired level of treatment can be maintained. While monitoring is on-going, only results from 5, 4 and 2 day HRTs are presented here. All four reed beds were subjected to a 5 day HRT from September 1999 until late December 1999. The HRT of the reed beds was decreased to four days from January 2000 to April 2000. The results from the halfway sampling points during the 4 day HRT period are presented as being representative of a 2 day HRT. Using the weekly data from the four reed beds, a one-way analysis of variance (ANOVA) was conducted to examine the effect of 5, 4 and 2 day HRTs on nutrient load reduction for the various forms of N and P. Where a significant difference was observed, the Tukey HSD test was used to ascertain between which HRTs the difference existed.

Results and discussion

Nutrient removal

At HRTs of 5 and 4 days, greater than 86% load removal of all N and P species was achieved, with the exception of TP at a 4 day HRT (78.2%) (Table 1). The reduction of TN, NH_4, NO_3 and Ortho-P was greater than 84% at a HRT of 2 days, while removal of NO_2 (76.3%) and TP (65.9%) was poorer. The outlet concentrations of NH_4, NO_2, NO_3, and Ortho-P were all extremely low (< 0.03 mg/L) at all HRTs studied (Table 1). In particular, outlet concentrations of NO_3 (the main N species of concern) were generally less than 0.05 mg/L throughout the study period, with greater than 97% load reduction consistently being achieved. TN and TP outlet concentrations, while still quite low, were greater than 0.38 mg/L (TN) and 0.025 mg/L (TP) at all HRTs (Table 1). This suggests that the majority of N and P in the treated test solution was in the organic form (both dissolved and particulate). This is illustrated in Figures 3 and 4 which show the average monthly inlet and outlet N composition of the test solution during the study period, and Figures 5 and 6 which show the average monthly inlet and outlet P compositions. The decline in inlet TN concentration between October 1999 and February 2000 was due to an accumulation of biological activity in the Header Tanks, which resulted in conversion of N (particularly NO_3) and subsequent loss. This was rectified in January 2000. While this decline meant that inlet NO_3 concentrations fell below the target of 6 mg/L, concentrations were still within the range observed in actual nursery runoff. Of interest is the fact that, despite this fluctuation in inlet concentration, outlet TN concentrations remained stable, at less than 0.45 mg/L (Figure 3).

Table 1 Mean nutrient inlet and outlet concentrations (mg/L), and percentage load reductions for 3 different hydraulic residence times. Standard errors of the means are shown in parentheses. 5 day HRT: 13/7/99–30/12/99 ($n = 64$); 4 and 2 day HRT: 5/1/00–17/4/00 ($n = 48$)

	HRT	TN	NH_4-N	NO_2-N	NO_3-N	TP	Ortho-P
5 days	In	3.50 (0.24)	0.656 (0.08)	0.281 (0.05)	2.06 (0.25)	0.49 (0.02)	0.32 (0.01)
	Out	0.41 (0.02)	0.008 (0.002)	0.001 (5×10⁻⁴)	0.03 (0.01)	0.025 (0.002)	0.009 (9×10⁻⁴)
	% red.	86.8 (0.59)	93.8 (1.5)	94.4 (1.46)	97.7 (0.45)	94.4 (0.42)	96.7 (0.16)
4 days	In	5.70 (0.47)	0.715 (0.07)	0.249 (0.04)	3.49 (0.4)	0.55 (0.01)	0.35 (0.02)
	Out	0.38 (0.03)	0.008 (9×10⁻⁴)	0.002 (3×10⁻⁴)	0.001 (1×10⁻⁴)	0.13 (0.03)	0.007 (0.001)
	% red.	90.4 (1.0)	98.5 (0.23)	92.9 (1.6)	97.8 (0.64)	78.2 (3.1)	97.8 (0.23)
2 days	IN	5.70 (0.47)	0.715 (0.07)	0.249 (0.04)	3.49 (0.4)	0.55 (0.01)	0.35 (0.02)
	OUT	0.54 (0.03)	0.014 (0.001)	0.004 (3×10⁻⁴)	0.002 (4×10⁻⁴)	0.15 (0.02)	0.007 (9×10⁻⁴)
	% red.	84.4 (1.89)	97.3 (0.35)	76.3 (7.9)	97.4 (0.66)	65.9 (3.9)	96.6 (0.51)

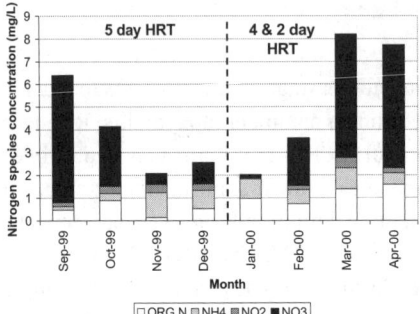

Figure 3 Mean monthly inlet TN concentration showing the proportion of each nitrogen species

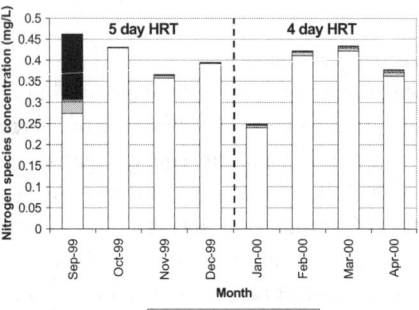

Figure 4 Mean monthly outlet TN concentrations showing the proportion of each nitrogen species

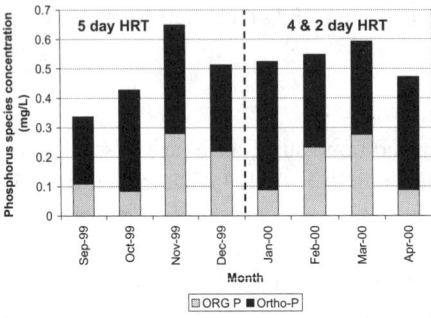

Figure 5 Mean monthly inlet TP concentrations showing the proportion of each species

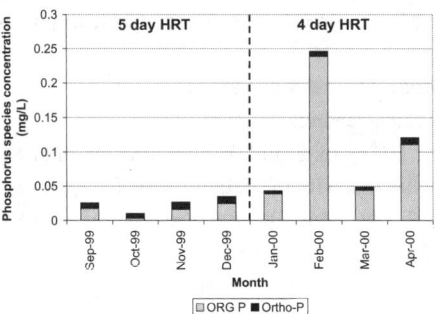

Figure 6 Mean monthly outlet TP concentrations showing the proportion of each species

Organic nutrients in the reed bed effluent would be sourced from the decay of plant root material and attached growth microorganisms, as well as phytoplankton living within the water column. This suggests that, even at very low loading rates, background levels of organic N and P will always be present in treated reed bed effluent due to internal generation within the reed bed itself. This supports the claim by Kadlec and Knight (1996) that small concentrations of organic N will always persist in the effluent from a constructed wetland, regardless of the HRT.

The data for TN and NO_3 removal at the 5 day HRT were not included in the ANOVA because the inlet TN and NO_3 concentrations for the 5 day HRT were found to be significantly different to those at the 4 and 2 day HRTs. This was due to the decline in inlet NO_3

concentrations discussed earlier (seen in Figure 3). No differences were found for Ortho-P, NO_2 or NO_3 load reduction between any of the observed HRTs ($P > 0.05$). This indicates that the HRT may be shortened to less than 2 days before any substantial decline in treatment for Ortho-P, NO_2 or NO_3 is observed. On the other hand, TN, NH_4 and TP load reduction was affected by HRT ($P < 0.05$). The reduction in TN load was greater ($P < 0.05$) for the 4 day HRT (90.4%) than the 2 day HRT (84.4%). NH_4 removal at the 5 day HRT (93.8%) was less than at the 4 day HRT (98.5%) ($P < 0.05$), while treatment at the 2 day HRT (97.3%) showed no difference to the other HRTs ($P > 0.05$). This result, with the lowest removal for NH_4 being achieved at the longest HRT (5 days), might suggest that the additional time allowed further breakdown of organic N into NH_4. If this was the case, higher outlet concentrations of NH_4 would be expected for the 5 day HRT. However, the outlet NH_4 concentrations were similar for both 5 and 4 day HRTs (0.008 mg/L for both). Moreover, because no significant difference was found for the 2 day HRT, it is more likely that factors other than HRT were important. Because the 5 day HRT treatment period occurred before the 4/2 day HRT period, a maturation effect related to reed development may have influenced the results. The reeds were only 5 months old at the commencement of the 5 day HRT period, whereas the 4/2 day HRT treatment commenced some 9 months after planting with the reeds being more fully established. Consequently, oxygenation of the root zone may have been greater by this time, facilitating more efficient nitrification of NH_4 into NO_3; an oxygen requiring process. Nitrification is often the rate-limiting process for N removal in horizontal flow reed beds due to slow oxygen diffusion rates in the predominantly anaerobic conditions (Kadlec and Knight, 1996; Davies et al., 1993; Johnston, 1993). On the other hand, the extremely low outlet NH_4 concentrations of 0.08 mg/L (Table 1) would probably persist at any HRT, and the poorer removal observed at the 5 day HRT may simply be a factor of the slightly lower inlet concentrations observed during this period.

There was a highly significant difference for TP reduction between all HRTs ($P < 0.01$). A distinct trend was observed, with mean TP reductions of 94.4%, 78.2%, and 65.9% for the 5, 4, and 2 day HRTs respectively. This declining TP removal with increasing HRT implies a strong dependency of TP treatment on HRT. Removal of Ortho-P was not found to be affected by HRT ($P > 0.05$), with around 96% removal at all HRTs studied. Therefore, the decline in TP removal at the shorter HRTs was related to increasing levels of P associated with organic substances in the reed bed effluent (Figure 6). A HRT of less than 4 days may not have been long enough for the organic P entering the reed beds to be broken down into inorganic forms and subsequently removed through plant uptake or gravel fixation. In addition to the organic P entering the reed beds within the runoff itself, substantial amounts may have been sourced from within the reed beds through the growth and decay of microorganisms and plant material. Organic P sourced in this way represents an internal cycling of P, where Ortho-P that has been previously removed through biological uptake is subsequently returned to the water as organic P (Horne and Goldman, 1994). Therefore, P removal needs to be studied over long time periods (years) in order to distinguish between short-term removal related to temporary sinks, and more permanent long-term removal.

Removal rates versus loading rates

The relationship between loading rate and removal rate (g/m²/day) for TN and TP are presented in Figures 7 and 8 respectively. The 1:1 lines represent 100% removal. Table 2 shows the corresponding loading rates of TN and TP for 4, 3, 2 and 1 day HRTs at inlet concentrations typical of nursery runoff (6.0 mg/L –TN; 0.5 mg/L –TP). Over the range of loading rates studied here (0.06 – 1.03 g/m²/day) TN exhibits a strong linear relationship, with removal rate increasing as loading rate increases ($r^2 = 0.995$). The relationship for TP

Table 2 Equivalent loading rates at different HRTs with an inlet concentration of 6.0 mg/L (TN) and 0.5 mg/L (TP)

HRT	4 days	3 days	2 days	1 day
TN	0.3 g/m²/d	0.4 g/m²/d	0.6 g/m²/d	1.2 g/m²/d
TP	0.025 g/m²/d	0.033 g/m²/d	0.05 g/m²/d	0.1 g/m²/d

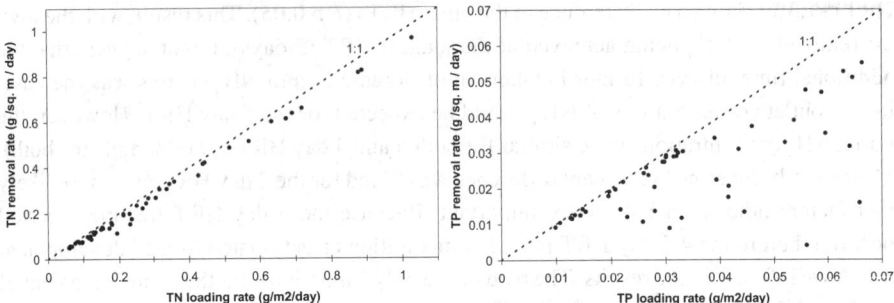

Figure 7 TN loading rate versus removal rate

Figure 8 TP loading rate versus removal rate

is weaker ($r^2 = 0.47$) and appears particularly variable at loading rates greater than 0.03 g/m²/day. TP loading rates of between 0.01 and 0.02 g/m²/day (> 4 day HRT) seem to show a fairly strong relationship to removal rate, with removal being close to 100%. This suggests that HRTs shorter than 4 days may be inefficient at removing P.

Traditionally, reed beds have been mainly employed to treat effluents richer in pollutants than in the present study (e.g. sewage, dairy effluent). Therefore, the majority of studies have investigated nutrient loading rates an order of magnitude greater than those presented here. This study provides a useful insight into nutrient removal at this lower end of the concentration range. It also highlights an important role for reed beds in polishing treated wastewaters to a high quality, particularly when that water is to be stored in open ponds for later reuse and the risk of algal blooms needs to be minimised.

System design

A preliminary design for a reed bed system to treat nursery runoff is presented based on a HRT of 2 days. For a nursery area of 1 ha (10,000 m²) on the sub-tropical northern coast of NSW (summer evaporation = 6.2 mm/day) an irrigation runoff volume of 20,000 L/day is estimated based on previous studies of irrigation efficiency (Huett, 1999). Using 10 mm gravel with a porosity of 0.4, a 2 day HRT would require a reed bed volume of 100,000 L (100 m³). Using a water depth of 0.5 m, the reed bed surface area would need to be 200 m² (approx 7 m × 30 m). Outlet concentrations of around 0.5 mg/L TN and 0.15 mg/L TP can be expected from a reed bed with a 2 day HRT, depending on a range of other factors, including the N and P concentration of the nursery runoff being dealt with.

Algal occurrence in storage dams

Although we can predict with some confidence that reed beds can achieve outlet TN and TP concentrations of below 0.5 mg/L and 0.15 mg/L respectively (HRT > 2 days), the question is whether these levels will support algal blooms in storage dams. The *Australian Water Quality Guidelines for Fresh and Marine Waters* (1992) state that algal problems in lakes and reservoirs are known to occur at or above 0.1–0.5 mg/L TN and 0.005–0.05 mg/L TP depending on a range of other factors such as light intensity and temperature. It seems that reed beds may not achieve concentrations lower than these, particularly for TP, due to internal loadings of organic nutrients. Therefore, algal problems may still occur in storage dams, although the risk is greatly diminished when compared to untreated runoff. Further studies will use algal assays to identify the range of nutrient concentrations in which algal blooms are a threat in local conditions. One possible solution may be to utilise reed beds to not only reduce the nutrient load in storage dams, but also to remove any algae that does occur by recirculating dam water through the reed bed. In this way, only freshly treated water low in algae would be used on the nursery.

Conclusion

The reed beds in this study achieved a consistently high level of nutrient removal (> 84% TN, > 65% TP) with outlet concentrations consistently below 0.54 mg/L TN and 0.15 mg/L TP. Load removal of NO_3 and Ortho-P was greater than 96% with mean outlet concentrations less than 0.03 mg/L (NO_3-N) and 0.009 mg/L (Ortho-P). The majority of TN and TP in treated runoff was organic in form. This indicates that low levels of organic N and P may always persist due to internal generation with the reed bed. Background levels identified in this study are around 0.45 mg/L TN and 0.15 mg/L TP. NO_2, NO_3 and Ortho-P removal were not affected by HRT (5,4 or 2 days) ($P > 0.05$) indicating that effective removal of these nutrients may be achieved with a HRT of less than 2 days. TN, NH_4 and TP load reductions were affected by HRT ($P < 0.05$). TN and TP removal was poorest at a HRT of 2 days, mainly due to internal loading of organic N and P. Removal of NH_4 was poorest at the longest HRT (5 days), although very low outlet concentrations were observed regardless of HRT (< 0.014 mg/L). A strong linear relationship occurred between TN loading rate and removal rate ($r^2 = 0.995$) over the range of loading rates studied (0.06–1.03 g/m²/day). This relationship was not as strong for TP ($r^2 = 0.47$) between the observed loading rates of 0.01 and 0.07 g/m²/day. It was estimated that a 1 ha nursery on the northern coast of NSW would require a reed bed surface area of 200 m² to achieve a 2 day HRT with outlet concentrations of 0.54 mg/L (TN) and 0.15 mg/L (TP). Although these levels may pose a minor algal bloom risk, this may be overcome by recirculating storage dam water through the reed bed.

References

Atkinson, I. (1998). Lies, dammed lies and statistics. *The Nursery Papers*, **1998**(016), Nursery Industry of Australia.

Australian Water Quality Guidelines for Fresh and Marine Waters (1992). Australian and New Zealand Environment and Conservation Council, Melbourne, Aust.

Bastian, R.K. and Hammer, D.A. (1993). The use of constructed wetlands for wastewater treatment and recycling. In: *Constructed Wetlands for Water Quality Improvement*, G.A. Moshiri (ed.), Lewis Publ., Florida, pp. 59–68.

Cooper, P.F. (1993). The use of reed bed systems to treat domestic sewage: the european design and operations guidelines for reed bed treatment systems. In: *Constructed Wetlands for Water Quality Improvement*, G.A. Moshiri (ed.), Lewis Publ., Florida, pp. 203–217.

Davies, T.H., Cottingham, P.D. and Hart, B.T. (1993). Application of constructed wetlands to treat wastewaters in Australia. In: *Constructed Wetlands for Water Quality Improvement*, G.A. Moshiri (ed.), Lewis Publ., Florida, pp. 577–584.

Horne, A.J. and Goldman, C.R. (1994). *Limnology*. 2nd edn, McGraw-Hill, Singapore.

Huett, D.O. (1999). *Improved irrigation and Fertiliser Strategies for Containerised Nursery Plants Through Commercial Demonstrations and Further Research*, Final Report NY 95025, Horticultural Research and Development Corporation, Sydney, Aust.

Johnston, C.A. (1993). Mechanisms of wetland-water quality interaction. In: *Constructed Wetlands for Water Quality Improvement*, G.A. Moshiri (ed.), Lewis Publ., Florida, pp. 293–299.

Juwarkar, A.S., Oke, B., Juwarkar, A. and Patnaik, S.M. (1995). Domestic wastewater treatment through constructed wetlands in India. *Wat. Sci. and Tech.*, **32**(3), 291–294.

Kadlec, R.H. and Knight, R.L. (1996). *Treatment Wetlands*. Lewis, Michigan.

Ottova, V., Balcarova, J. and Vymazal, J. (1997). Microbial characteristics of constructed wetlands. *Wat. Sci. and Tech.*, **35**(5), 117–123.

Reed, S.C., Crites, R.W. and Middlebrooks, E.J. (1995). *Natural Systems for Waste Management and Treatment*. 2nd edn, McGraw-Hill, NY.

Standard Methods for the Examination of Water and Wastewater (1995). 19th edn, American Public Health Association / Water Environment Federation, Washington DC, USA.

Vymazal, J. (1995). Constructed wetlands for wastewater treatment in the Czech Republic – state of the art. *Wat. Sci. and Tech.*, **32**(3), 357–364.

Removing filterable reactive phosphorus from highly coloured stormwater using constructed wetlands

M.A. Lund, P.S. Lavery and R.F. Froend

Centre for Ecosystem Management, Edith Cowan University, 100 Joondalup Drive, Joondalup, 6027, Western Australia.

Abstract A constructed wetland design, consisting of 16 repeating cells was proposed for Henley Brook (Perth, Western Australia) to optimise the removal of FRP from urban stormwater. Three replicate experimental ponds (15 × 5 m), were constructed to represent at a 1:1 scale a single cell from this design. Three 5 m zones of each pond were sampled: shallow (0.3 m) vegetated (*Schoenoplectus validus*) inflow and outflow zones and a deeper (1 m), V-shaped central zone. In 1998/99, inflows and outflow waters were intensively sampled and analysed for FRP and Total P. In addition, all major pools of P (plants, sediment) within the ponds, and important P removal processes (benthic flux, uptake by biofilm and *S. validus*) were quantified.

A removal efficiency of 5% (1998) and 10% (1999) was obtained for FRP. Initial uptake was mainly in plant biomass, although the sediment became an increasingly important sink. Benthic flux experiments showed that anoxia did not cause release of P from sediments, indicating that most of the P was bound as apatite rather than associated with Fe or Mn. The highly coloured waters were believed responsible for the very low biofilm biomass recorded (<1 $g.m^{-2}$). We have demonstrated that constructed wetlands can be effective for removing FRP immediately after construction, although their long-term removal capacity needs further research.

Keywords Amended sediment; biofilm; constructed wetlands; phosphorus; stormwater

Introduction

Around Australia wetlands are being constructed for stormwater pollution control, and a variety of manuals for their design have been produced (e.g. Lawrence and Breen, 1998; DLWC, 1998; WRC, 1998). These typically focus on the engineering aspects of the design rather than on optimising the biological/chemical processes that are responsible for the nutrient removal. The prevailing view appears to be that constructed wetlands are little more than water treatment plants rather than the reality that they are constructed ecosystems. Western Australia has experimented with several constructed wetlands, early designs were essentially retrofitted groundwater recharge basins and more recent examples have been purpose built. The largest and most studied of these newer examples is the Bartram Road Buffer lakes (BRB). The BRB system became operational in 1993 and consists of 5 cells (3.2 ha) designed to reduce P by 30% (Braid, 1995). It has been speculated that constructed wetlands in Perth face greater difficulties in treating stormwater than elsewhere in the world due to the high DOC and FRP levels (WRC, 1997). Perth has a Mediterranean climate and as a result stormwater and drainage flows are largely restricted to between May and November, therefore constructed wetlands (and associated plants) have to cope with prolonged dry periods.

Over 80% of Western Australia's population live on the Swan Coastal Plain (SCP) with the majority congregated in the city of Perth. The SCP consists of a series of parallel dunal systems of varying ages between the Indian Ocean to the west and the Darling Scarp to the east. As urban expansion pushes the limits of the city, many new developments are taking place on the older Bassendean sand dune system. This area was once considered only suitable for semi-rural development as it is largely situated over large groundwater mounds.

These mounds are used to supplement (up to 50%) Perth's drinking water supplies and so have been traditionally protected from intensive urban development. In the 1990s the southern mound was opened up for high density urban development and drainage schemes were put in place to reduce the flooding risk. The BRB system was constructed as part of this scheme. Bassendean sands are highly leached and so have very poor nutrient retention capacities. As a result stormwater and groundwater in these areas typically carry high concentrations (up to 80% of Total P) of filterable reactive P (FRP), nitrates/nitrites and ammonium. Initially, as the development commences, most of the baseflow within the drains is actually groundwater. Groundwater in Bassendean sands is often highly coloured by humic and fulvic acids (dissolved organic carbon (DOC)) leached from vegetation. This colour inhibits algal growth either through light limitation or more likely by forming complexes with P and trace elements (Lund and Ryder, 1998).

Braid (1995) highlighted a number of problems with constructed wetlands in Perth that reduced their performance and evaluation. A common feature was that construction often did not match design, resulting in short-circuiting. The criteria against which performance was evaluated were often poorly developed and the methods used for measuring performance of these wetlands were often inadequate.

A simple conceptual model for P removal in Australian constructed wetlands was proposed by DLWC (1998). The model (Figure 1) suggests that overall contaminant removal efficiency (percentage of incoming nutrient load retained by the wetland (CRE)) is related to three key processes, sedimentation, short-term uptake (sediments and macrophytes) and long term uptake (biofilm, filtration and litter/peat accumulation). The model does not indicate the magnitude of the processes or the timeline for each. This project aims to validate the model and quantify timelines and magnitudes where possible. This aim will be addressed through measuring the efficacy of a "state of the art" design for FRP removal in highly coloured stormwater through intensive in and outlet monitoring, and quantification of P sinks and the processes responsible for any accumulation.

Methods
Study site
A conceptual design for Henley Brook was proposed in 1997. This design represented "state of the art" when this project commenced (Jim Davies & Associates, 1997); more

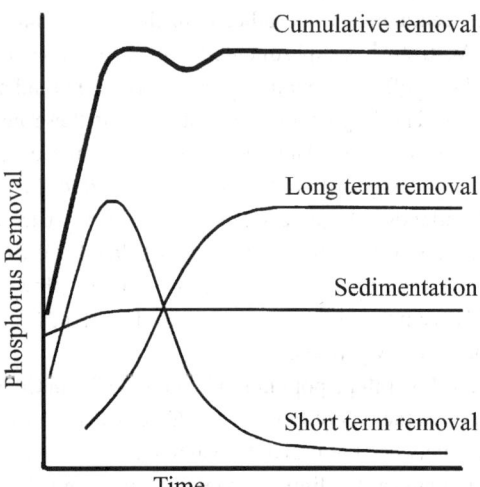

Figure 1 Conceptual model of the development of key processes responsible for P removal in constructed wetlands (adapted from DLWC, 1998)

information on the rationale for the project can be found in Bayliss (1998). The design consisted of a series of repeating cells, designed specifically to facilitate P removal (Figure 2).

This project constructed three experimental ponds (15 m long by 5 m wide) next to the Thomsons Lake Main Drain in Jandakot (Figure 3). This drain supplies the BRB system downstream of the site. Unlike the Henley Brook design (HBD) the ponds were isolated from the groundwater by a concrete shell covered with a PVC liner. The ponds were built (June 1998) to represent a single HBD cell at 1:1 scale. The liner was covered by 0.4 m of Bassendean sands (from the site), covered with a 0.1 m layer of red sand. Bauxite residues neutralised by gypsum have been tested as soil amendment to improve P retention on SCP farmlands with reasonable success (see Summers *et al.*, 1996 a and b). The residue is available in two forms, commonly called red mud (<150 µm particle size fraction) and red sand (≥150 µm particle size fraction); red mud has been used in several constructed wetlands as a sediment amendment, although in the BRB system problems were encountered with clumping and resuspension. This study chose to use red sand which has a lower P Retention Index (WRC, 1998) but is less prone to resuspension and redistribution. Each pond was divided along its length into three 5 m wide zones. The inlet and outlet zone was vegetated with the native rush *Schoenoplectus validus*, these zones slope (1:7.5; 0.2 to 0.5 m water depth) towards the central zone. A 1:5 sloped V-shape occupies the central zone (maximum water depth ~1 m). The slope was determined to minimise sediment slippage and limit plant encroachment of the open water section. The V-shape was used to increase the volume of water in the wetland to increase the hydraulic residence time (HRT). In addition, during periods of no flow (i.e. summer), the V-shape retained water to support the plants and facilitated the establishment of natural convection circulation patterns that should ensure the bottom of the V remains oxygenated. This was seen as essential to preventing a redox driven release of P following reestablishment of flow conditions. Inlet and outlet structures were designed to facilitate plug flow through the ponds. The HRT was 8 h and 24 h in 1998 and 1999 respectively. Eight hours more accurately reflects the HRT of individual cells of the HBD. Flow rates were constant throughout the experiment. When the Thomsons Lake Main Drain was flowing (June to November), a sump pump was used to lift the water into a small header tank where flow was regulated into each pond, outflow was directed back into the drain downstream of the sump pump. In addition, to extend the flow

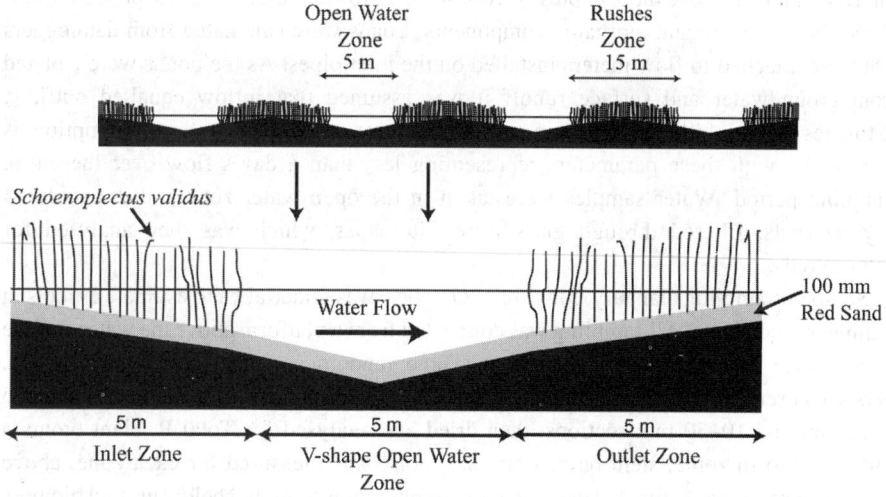

Figure 2 Illustration of the Henley Brook Design showing the repeating cell design, the lower diagram shows the section used in the experimental ponds

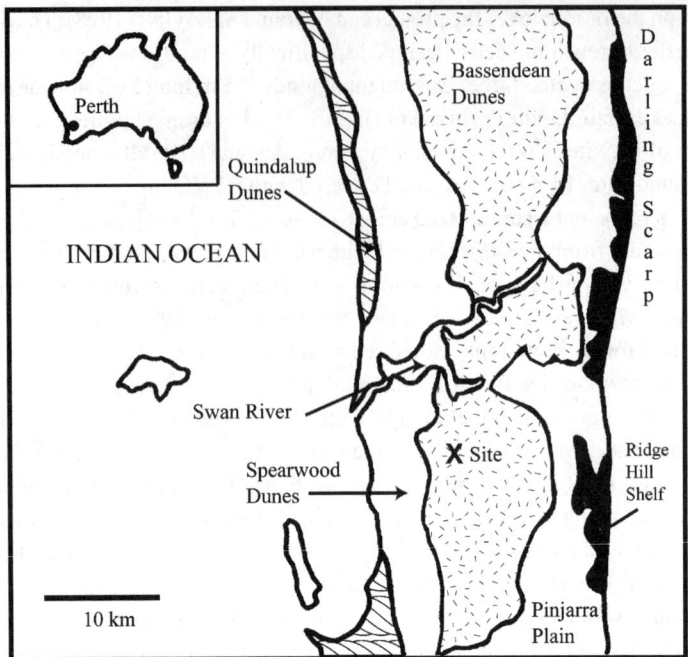

Figure 3 Location map of the study site and the dunal system of the Swan Coastal Plain. The City of Perth covers the majority of the plain shown east of the Darling Scarp

period, drain water could be supplemented with groundwater, which was added upstream of the sump to allow aeration.

Sampling methods

In 1998, the ponds were sampled between 1/10/98 and 28/11/98 and in 1999, between 20/7/99 to 23/11/99. Autosamplers (ISCO®) were used to collect samples composited at 6 hour intervals over 2 or 3 days from the inlet and each pond outlet. Water samples were frozen and later analysed for Total P and FRP. In 1998, these samples were also analysed for DOC and in 1999, the samples were also used to determine total suspended solids (TSS), both organic and inorganic components. Loads were calculated from dataloggers that were attached to flow meters installed on the inlet pipes. As the ponds were isolated from groundwater and surface runoff it was assumed that inflow equalled outflow. Estimates of rainfall, transpiration and evaporation suggest that this assumption is reasonable, with these parameters representing less than a day's flow over the entire sampling period. Water samples were taken in the open water zone and drain at 3–4 day intervals, filtered through glass fibre filterpaper, which was then analysed for chlorophyll a.

Seasonal sampling (January, May, July, October) was undertaken to estimate P pools in sediments and plants. All sampling was conducted from a platform above the water surface to minimise sediment/plant disturbance. Three random sediment cores (44 mm dia. Perspex corer) were collected from each zone in each pond. Cores were divided into 0–10 mm and 10–50 mm sections, then dried and analysed for Total P. Plant biomass (divided into rhizome, stem base, stem and roots) was measured for each zone, above ground biomass was estimated using length-weight regressions and below ground biomass was estimated by taking sediment cores. All plant material was washed, dried and analysed for Total P. Biofilm was measured using glass slides (0.0375 m^2), two slides were scraped, dried and analysed for Total P, one slide was scraped for Chlorophyll *a* determination and one slide was used to measure organic and inorganic biomass.

The benthic flux for FRP was measured in July and October using intact cores taken from 3 sites in each zone (as per Lavery *et al.*, in press). Phosphorus isotherms for the sediment were measured seasonally from 3 sites in each zone by the Government Chemistry Laboratory (Perth).

Total P, FRP and chlorophyll *a* were analysed according to *Standard Methods for the Examination of Water and Wastewater* (1998) at the Marine and Freshwater Research Analytical Laboratory (Murdoch University) or at a nutrient analysis laboratory (Edith Cowan University).

Results and discussion

The HBD had a wetland area of 5.1 ha, an average HRT of 11.9 days (minimum 4 days), an annual load of 68 kg of P, and was predicted to reduce P by 40–60% per year (Jim Davies & Associates, 1997). In the two years of operation (July 1998 to January 2000) the experimental ponds each received loads of 4.8 to 5.4 kg of P, of which 3.2 to 3.7 kg were in the form of FRP. In this study and for the HBD the ratio of FRP:TP was very high (~70%), although in the HBD, the bulk of the FRP was from groundwater inputs, which suggests that baseflow in the Thomsons Lake Drain has a large groundwater input. In the ponds, concentrations of Total P appeared substantially lower in 1998 (347 ± 57 µg l^{-1}) than 1999 (648 ± 87 µg l^{-1}), although FRP concentrations were very similar (1998, 273 ± 28 µg l^{-1}; 1999, 265 ± 20 µg l^{-1}). It was suspected that sampling technique may have been responsible for underestimating the Total P in 1998, however a comparison of different methodologies failed to indicate a problem. In 1998, the CRE for FRP ranged between 1.5–4.5%, increasing to 11.4–12.1% in 1999. These figures suggest that the ponds are performing extremely well, certainly in the predicted range for HBD when scaled to an operational size. The loading rates in the ponds were substantially higher than that predicted for HBD (HBD: 13 kg P/ha/yr; Ponds: 320–360 kg P/ha/yr). The CRE for Total P in the ponds was substantially poorer at –1.8 to 2.4% in 1998 and 4.2 to 7.5% in 1999. It should be noted that in both years the net amount of FRP removed frequently exceeded the net amount of Total P removed due to increases in particulate/organic P at the outlet. The ponds were not designed to promote sedimentation, as this is a reasonably well understood process and is typically incorporated into constructed wetlands prior to the macrophyte cells.

In 1999, only 16% of the suspended solid (SS) load into the ponds was inorganic, which reflects the sandy nature of the catchment. The ponds generated suspended solids with organic loads increasing from 5.8–5.9 kg to 7.5–9.5 kg and inorganic loads from 1.1 kg to 1.5–2.6 kg. Interestingly, the highest inorganic loads are associated with the lowest CRE for Total P. Resuspension of sediment is the most likely source of the increased inorganic SS load. We found evidence of slumping of the red sand fraction and sampling of plants and sediments also appeared to cause small amounts of resuspension.

The short HRT of the ponds would have prevented the development of detectable phytoplankton loads, and the highest chlorophyll *a* concentration recorded in the ponds was 9.9 µg l^{-1}, although the majority were <1 µg l^{-1}. It is assumed that the phytoplankton within the ponds was washed in from the drain, although there does appear on average to be a slightly higher concentration of chlorophyll leaving the ponds than entering them (Drain 0.7 ± 0.1 µg l^{-1}; Ponds 1.3 ± 0.1 µg l^{-1}), a slight increase is also seen in Phaeophytin (Drain 0.4 ± 0.1 µg l^{-1}; Ponds 1.1 ± 0.1 µg l^{-1}). A possible source for this increase may be turnover of biofilm. Biofilm was estimated to have a standing crop per pond (assuming a 485 m^2 surface area in the vegetated zone and 30 m^2 in the open water) in 1999 (Sept to Nov) of 28–216 g inorganic and 67–233 g organic components. These estimates suggest that if there was a rapid turnover of the biofilm that it could be a major source of the SS load, but not of particulate P, as the pool of P in the biofilm was estimated to be only 8.3–31 mg at any one time in each pond. Lawrence and Breen (1998) and DLWC (1998) suggest that a key component for the long term removal of P from constructed wetlands is through biofilm accumulating FRP and during turnover adding that P to the sediment. Although there is a paucity of data on biofilm biomass, results from Cronk and Mitsch (1994) suggest that biofilm biomass in the ponds is an order of magnitude below that recorded in their constructed wetlands. The inflow water was highly coloured with a mean DOC concentration of 50.8 ± 1.6 mg C l^{-1} and this is probably responsible for the low biomass recorded. The reasons are twofold with rapid attenuation of PAR, such that below 0.4 m there is effectively no light available for photosynthesis and chelation of essential elements (Lund and Ryder, 1998).

Another potential source of organic SS is from the decomposition of macrophyte tissue. The growth of *S. validus* was substantial and by January 2000 biomass had reached 1.6–5.1 kg DW m^{-2} above ground and 1.2–2.2 kg DW m^{-2} below ground depending on the area. This was aided by bird netting to prevent birds uprooting the tube stock. As the two years of the study represent the establishment phase of the stand, the stem turnover rates are therefore likely to be low. This indicates that macrophytes were unlikely to be making a substantial contribution to the organic SS. The V-slope proved very successful in preventing *S. validus* encroaching into the open water area, which would simplify maintenance.

In 1999, the amount of P retained for FRP was extremely consistent between ponds, with a similar result for Total P. In 1998, CRE was highly variable for both FRP and Total P, this may reflect the maturation of the ponds or reduced HRT in 1999. Lantzke *et al.* (1999) suggest that there is not a linear relationship between CRE and HRT, and this was also apparent

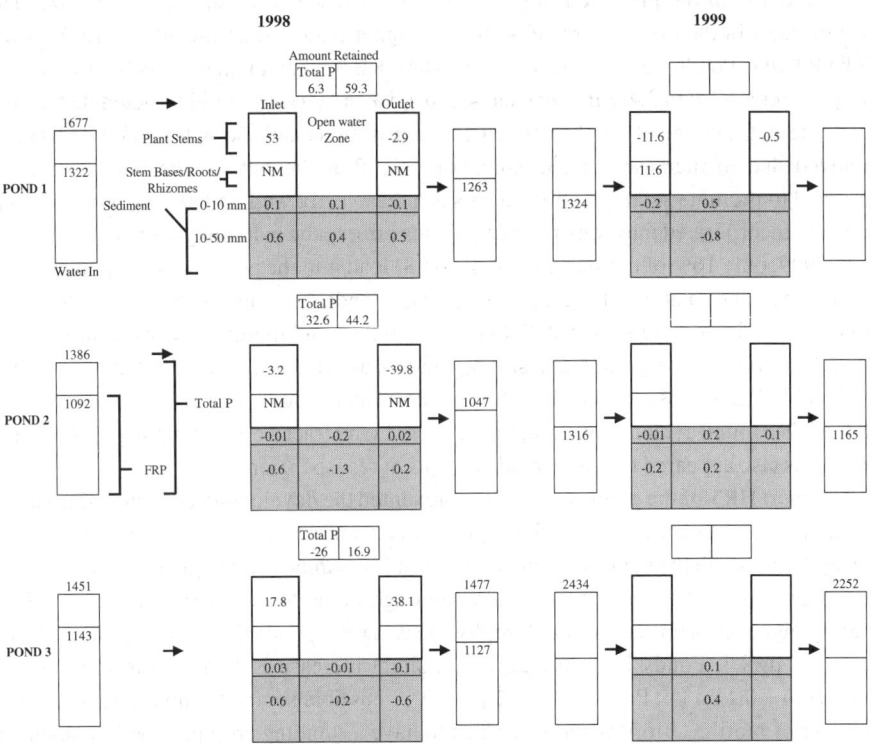

Figure 4 Pools (net change) and loadings of P (in grams) in 1998 and 1999 over the monitoring period for each experimental pond (negative values indicate a decline in the pool; NM = not measured)

here. The P isotherms indicate that by January 2000, that the sediments still had a reasonable capacity for P adsorption. The sediments accumulate little P during the monitoring period (Figure 4). One unexpected result was the apparent migration of P deep (50 mm) into the sediment; the mechanism responsible for this is currently unknown but suggests that all of the top 50 mm of sediment is available for P uptake. Benthic flux experiments suggest that P is primarily bound to Ca in the sediment, as under anaerobic conditions P was still being taken up. The remainder of the uptake can be accounted for by uptake into plant biomass. The pools of P in *S. validus* were highly variable, with a large proportion found in the above ground biomass. It is suspected that now the stand has matured, uptake by plants will become less important for the CRE than in the first two years. Lantzke *et al.* (1999) suggested that the major pathways for removal (in the order of months) were macrophyte>sediments>>biofilm, and this appears to have been the case here.

It is suggested that the short term removal curve (Figure 1) should be divided into plants and sediments, with a similar shape. It appears that at 2 years, the uptake by macrophytes may have peaked in the systems. It is unlikely that sediment uptake has yet peaked. It currently appears that high DOC concentrations will limit the development of long term removal pathways through biofilm. This suggests that the long term efficiency of constructed wetlands receiving highly coloured waters may be lower than indicated during the macrophyte establishment phase. Substantial improvements in the efficiency of the HBD appear possible by reducing resuspension of sediment and/or enhancing sedimentation processes.

Conclusions

The experimental ponds proved extremely effective at removing FRP from incoming water in their second year of operation. The sediment accounted for little of the P removal, with P incorporation into plant biomass responsible for the majority of removal. As the plant stand has now become established it is anticipated that plant uptake will become less significant in subsequent years. Biofilm development appears limited due to the highly coloured waters, this suggests that biofilm may not be able to make a substantial contribution to long term P removal. This in turn suggests that the long term removal capacity of constructed wetlands receiving coloured waters needs further investigation.

Acknowledgements

This project has relied on the valuable research assistance of Peter Bayliss, Kelli Sargent and Scott Ranford and the technical assistance of Dr Carolyn Oldham. The project was funded through an Australian Research Council Collaborative Grant with contributions from Water and Rivers Commission (WA), Water Corporation (WA) and City of Cockburn (WA).

References

Bayliss, P.R. (1999). Optimising Design Features of Water Pollution Control Ponds in Western Australia: General Applications in Pond Habitat Creation, Restoration and Protection. In: *Ponds and Pond Landscapes of Europe*, J. Boothby (ed.) Proceedings of the International Conference of the Pond Life Project, The Pond Life Project, Liverpool John Moores University, Liverpool, UK. pp. 17–30.

Braid, A.J. (1995). *Constructed Wetland Design Criteria: A study of their role in contaminant removal from urban stormwater runoff*. Honours thesis, School of Natural Sciences, Edith Cowan University, Perth, Western Australia.

Cronk, J.K. and Mitsch, W.J. (1994). Periphyton productivity on artificial and natural surfaces in constructed freshwater wetlands under different hydrologic regimes. *Aquatic Botany* **48**(3/4), 325–331.

DLWC (1998). *Constructed Wetlands Manual*. Department of Land and Water Conservation, New South Wales, Australia.

Jim Davies and Associates (1997). *Henley Brook Drive Water Pollution Control Pond Conceptual Design.* Report No. J294bi.doc, JDA Consultant Hydrologists, Subiaco, Western Australia.

Lavery, P., Oldham, C. and Ghizalberti, M. (in press). The use of Fick's 1st Law for predicting porewater nutrient fluxes under diffusive conditions. *Hydrological Processes.*

Lantzke, I.R., Mitchell, D.S., Heritage, A.D. and Sharma, K.P. (1999). A model of factors controlling orthophosphate removal in planted vertical flow wetlands. *Ecological Engineering* **12**, 93–105.

Lawrence, I. and Breen, P. (1998). *Design guidelines: Stormwater pollution control ponds and wetlands.* Cooperative Research Centre for Freshwater Ecology, Canberra, Australia.

Lund, M.A. and Ryder, D.S. (1998). Can artificially generated gilvin (g440 m^{-1}, Gelbstoff) be used as a tool for lake restoration. *Verhandlungen der Internationalen Vereinigung fur Theoretische und Angewandte Limnologie* **26**, 731–735.

Standard Methods for the Examination of Water and Wastewater (1998). 20th edn., American Public Health Association/American Water Works Association/Water Environment Federation, Washington DC, USA.

Summers, R.N., Guise, N.R., Smirk, D.D. and Summers, K.J. (1996a). Bauxite residue (red mud) improves pasture growth on sandy soils in Western Australia. *Australian Journal of Soil Research* **34**, 569–581.

Summers, R.N., Smirk, D.D. and Karafilis, D. (1996b). Phosphorus retention and leachates from sandy soil amended with bauxite residue (red mud). *Australian Journal of Soil Research* **34**, 555–567.

WRC (1997). *Evaluation of Constructed Wetlands in Perth.* Perth, Water and Rivers Commission. Water Resources Technical Series: WRT20, Water and Rivers Commission, Perth, Western Australia.

WRC (1998). *A Manual for Managing Urban Stormwater Quality in Western Australia.* Water and Rivers Commission, Perth, Western Australia.

Environmental impacts to the Everglades ecosystem: a historical perspective and restoration strategies

M.J. Chimney* and G. Goforth*

* Ecological Technologies Department and ** Environmental Engineering Section, South Florida Water Management District, 3301 Gun Club Road, West Palm Beach, FL 33406, USA

Abstract The Everglades is a vast subtropical wetland that dominates the landscape of south Florida and is widely recognized as an ecosystem of great ecological importance. As a result of anthropogenic disturbances over the past 100 years (i.e., agricultural and urban development, eutrophication resulting from stormwater runoff, changes in hydrology and invasion of exotic species), the biotic integrity of the entire Everglades is now threatened. To protect this valuable resource, the state of Florida and the Federal Government, in cooperation with other interested parties, have developed a comprehensive restoration strategy that addresses controlling excess nutrient loading and reestablishment of a more natural hydrology. These efforts include building approximately 17,000 ha of treatment wetlands, referred to as Stormwater Treatment Areas, to treat surface runoff before it is discharged into the Everglades. We briefly discuss the history of the Everglades in the context of environmental disturbance and outline the steps being taken to ensure its survival for future generations.

Keywords Constructed wetland; Everglades; phosphorus; restoration; stormwater runoff

Introduction

The Everglades is a vast freshwater wetland that dominates the landscape of south Florida. Before the 1900s, the Everglades extended unbroken from the south shore of Lake Okeechobee to Florida Bay (Figure 1) and encompassed more than 10,000 km^2 (Gunderson and Loftus, 1993; Light and Dineen, 1994). Agricultural and urban development have since reduced the present-day Everglades to only 50% of its original extent, of which approximately 3,500 km^2 is impounded within shallow, diked reservoirs known as Water Conservation Areas (WCAs) (SFWMD, 1992a). The wetland that remains (i.e., the WCAs, the Holeyland and Rotenberger Wildlife Management Areas and Everglades National Park [ENP]) still supports unique biotic communities containing many threatened or endangered plant and animal species (USCOE and SFWMD, 1996) and is widely regarded as an ecosystem of immense regional, national and international importance. Everglades National Park has been designated as an International Biosphere Reserve, a United Nations World Heritage site and a Wetland of International Importance under the 1987 Ramsar Convention, one of only three wetlands in the world to receive such recognition (Maltby and Dugan, 1994). Water Conservation Area 1 is part of the Arthur R. Marshall Loxahatchee National Wildlife Refuge (LNWR). Both ENP and LNWR are federally protected wetlands.

History of Everglades impacts

Efforts to manage surface water in south Florida began in the late-1800s. The primary goal was to drain the land and take advantage of its rich organic soils and subtropical climate for agricultural purposes (Anderson and Rosendahl, 1998; Snyder and Davidson, 1994). Today, the hydrology of the region is managed by the South Florida Water Management District (District), which operates one of the world's largest and most complex drainage systems. Much of this infrastructure was built (or upgraded) by the U.S. Army Corps of

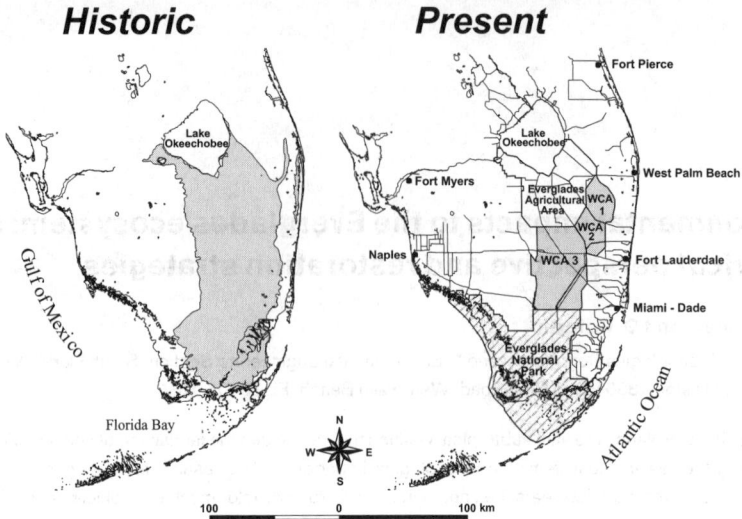

Figure 1 Comparison of areal extent of the historic Everglades with the present-day ecosystem. See text for details

Engineers (USCOE) from 1953 to 1967 as part of the Central and Southern Florida Project (C&SF Project). Management objectives for the C&SF Project have changed over time. Throughout most of its history, the project was operated primarily for regional flood protection during the wet season (May–October) and alternatively, to supply water for farm irrigation and domestic use during the dry season (November–April). Within the last 15 years, preservation and restoration of the remaining Everglades ecosystem has become a top priority for the District.

Although relatively few water quality data exist for the Everglades before 1940, the wetland is thought to have been oligotrophic throughout its history. This inference is based on: (1) rainfall and dry deposition were the main nutrient sources to the system; because nutrient concentrations in contemporaneous deposition are low (Brezonik *et al.*, 1983), historic atmospheric sources are presumed to have delivered relatively low levels of nutrients (McPherson *et al.*, 1976); (2) oligotrophic conditions still exist at interior sites in ENP and the WCAs (minimum values for total phosphorus [TP] ≤ 10 g L^{-1}) (Bechtel *et al.*, 1999); and (3) undisturbed Everglades sediments are nutrient-poor and the native vegetation has low nutrient requirements (Steward and Ornes, 1975; Swift and Nicholas, 1987); the characteristics of sediments and the vegetation community change quickly in response to nutrient enrichment and their persistence in the present-day ecosystem suggests a history of low nutrient conditions. Much of the Everglades today is considered to be P limited (Craft *et al.*, 1995).

Everglades plants and animals are adapted to the hydrologic and physico-chemical conditions (e.g., low dissolved oxygen and nutrients) that are characteristic of the region (Gunderson and Loftus, 1993). The timing, distribution, quantity and quality of water entering the Everglades are the most important factors influencing marsh ecology. Changes in water quality and other environmental disturbances were detected in ENP as early as 1938 (Beard, 1938). Operation of the C&SF Project exacerbated these problems. The improved drainage system permitted a 2,830-km^2 area immediately south of Lake Okeechobee, known as the Everglades Agricultural Area (EAA; Figure 1), to be developed for agriculture. Further degradation of water quality in the region was documented in the 1960s and 1970s (McPherson *et al.*, 1976). Most EAA runoff today flows directly into the WCAs through a system of canals (Figure 2) and carries elevated levels of nutrients and other constituents (e.g., total suspended solids and pesticides; Nearhoof, 1992). Runoff from urbanized basins also enters the Everglades at a number of locations along the eastern boundary of the system. Pollutant loads in stormwater runoff can be exceptionally high. Excessive P loading has caused eutrophication in parts of the WCAs (Belanger et al., 1989). In addition, enclosing the WCAs within levees and operating them as impoundments to meet flood control and water supply needs has altered flow paths (Figure 2) and the timing of water delivery throughout the system. This has caused excessive flooding in some areas, overdrainage of other areas and periodic reversals in the seasonal fluctuation in water depth. Eutrophication and changes in hydrology, in turn, have been linked to widespread changes to the ecology of the Everglades as evidenced by dramatic declines in the size of wading bird populations, intrusion of cattail into more than 10,000 ha of native sawgrass and slough habitat, and the widespread invasion of exotic plant species (Rader and Richardson, 1992; Davis and Ogden, 1994; Thayer et al., 2000).

Everglades restoration legislation

The environmental and scientific communities were alarmed over deteriorating conditions in the Everglades. It was clear that the impacts described above had damaged the Everglades to the extent that the biotic integrity of the remaining ecosystem was threatened. These effects were unwanted consequences of what otherwise was widely regarded as a beneficial public works project. Any actions to remedy this situation needed to substantially reduce nutrient loads in EAA and urban runoff and restore the natural hydrology of the region. The District became involved with Everglades environmental issues in the early 1970s and began evaluating the treatment efficacy of both natural and constructed wetlands in 1976 (Davis et al., 1985). By the 1980s, there was a growing consensus that treatment wetlands could effectively reduce nutrient levels in stormwater runoff and therefore might play an important role in any strategy to restore the Everglades.

Concern over the Everglades prompted the Florida Legislature to enact the Everglades Protection Act (EPA) in 1991 (§ 373.4592, F.S.). This bill was intended to help resolve long-standing litigation between the Federal government and the District and State of Florida related to environmental degradation in WCA-1 and ENP. The EPA, together with the legal agreement resulting from settlement of the Federal lawsuit in 1992, required the

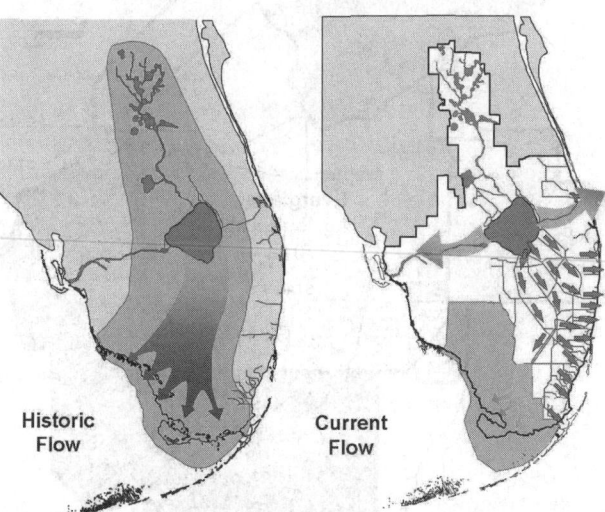

Figure 2 Comparison of major historic and current flow paths in the Everglades (USCOE and SFWMD, 1999)

District to adopt a Surface Water Improvement and Management (SWIM) Plan for the Everglades (SFWMD, 1992b) and initiate design of stormwater management systems that would bring all District facilities into compliance with water quality standards. The resulting conceptual design (Burns and McDonnell, 1992) proposed that three large treatment wetlands covering approximately 13,200 ha be built and operated to reduce P loads in runoff entering the Everglades. These treatment wetlands became known as the Stormwater Treatment Areas (STA) (Figure 3). The basis of design for the STAs was predicated on the success of other treatment wetlands in Florida and the fact that WCA-2A continued to reduce P in surface water even after decades of continuous nutrient loading (Burns and McDonnell, 1992; Kadlec and Newman, 1992; Walker, 1995). While the STAs are sometimes called "filter marshes," this is a misnomer since filtration is only a minor part of the treatment process. The STAs operate primarily by promoting biological uptake, chemical absorption/adsorption and gradual settling and accumulation of nutrients in the sediments.

The 1992 Everglades restoration plan and associated permits were challenged with several legal actions. Between December 1992 and December 1993, the District and other stakeholders engaged in mediation that produced an improved plan, which was incorporated into the 1994 Everglades Forever Act (EFA) (§ 373.4592, F.S.). The EFA was a revision of the 1991 legislation and required restoration of a significant portion of the remaining Everglades through a program of construction projects, research and regulatory controls. The goals and legislated requirements of the EFA call on many state and federal agencies to address water quality, water quantity (including hydroperiod restoration), and exotic species issues. Most of this work will be the primary responsibility of the District,

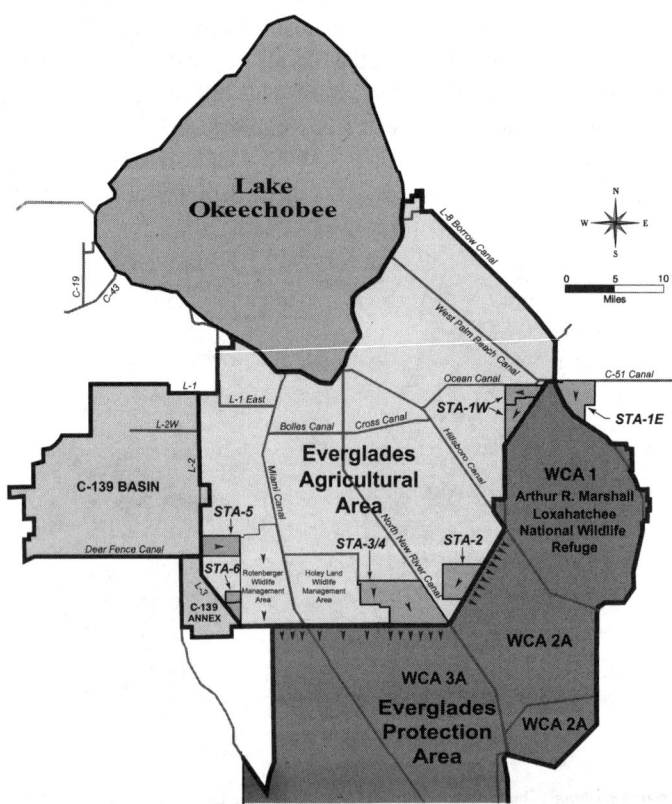

Figure 3 Location of the Stormwater Treatment Areas within the Everglades Agricultural Area

with substantial involvement by the Florida Department of Environmental Protection (FDEP) in more than half of the projects.

Everglades construction project

The EFA established the funding mechanisms and construction timetable for a more comprehensive program of six STAs that now encompass almost 17,000 ha of wetted surface area (see Figure 3). The STAs are designed to treat the annual runoff ($\sim 1.23 \times 10^9$ m^3) from seven of the 16 basins that discharge into the Everglades and regulatory releases from Lake Okeechobee. The District is building five of the STAs and associated infrastructure as part of its Everglades Construction Project (ECP), which has a capital budget of approximately $695 million. The USCOE is responsible for funding, design and construction of the remaining STA (STA-1 East). To date, over 4,720 ha of treatment wetlands have become operational (STA-1 West, STA-5 and STA-6), another 2,600 ha are under construction or in start-up (STA-2) and the remaining areas are in design (STA-1 East and STA-3/4). Even before full completion, the STAs comprise the largest system of treatment wetlands in the world.

The EFA requires the District to conduct water quality research and monitoring programs that, among other things, will seek to optimize nutrient removal (primarily P) by the STAs. In addition to mandating construction of the STAs, the EFA directs the District to (1) conduct research into improving Best Management Practices (BMPs) within the EAA to further reduce nutrient loads coming off farm fields, (2) investigate "alternative" treatment technologies that might be used in conjunction with, or in place of, the STAs to improve their performance, (3) initiate efforts to restore the region's natural hydrology and (4) deal with the invasion of exotic plant species. The water quality research and monitoring, ecosystem-wide planning and regulatory programs that make up the ECP are intended to ensure that the District has a sound foundation for science-based decision-making in its restoration efforts.

The EFA set both interim and long-term water quality goals for the Everglades and recognized that additional measures may be required to achieve compliance with long-term standards. The STAs, in conjunction with BMPs in the EAA, constitute the District's Phase I restoration efforts and are designed to produce effluent that meets an interim standard of 50 µg P L^{-1} on an average basis. The ultimate objective is to combine point-source, basin-level and regional solutions to ensure that by December 31, 2006 all runoff discharged into the Everglades has nutrient levels that do not cause an imbalance in populations of Everglades flora or fauna. A concentration TP limit, i.e., the "threshold" concentration, that achieves this goal, and the methodology to be used in determining compliance with this limit, will be set by December 2003 based on research being conducted by the District (McCormick et al., 1999, 2000) and other parties. The EFA stipulates that the TP standard will be 10 µg P L^{-1} by default if this multi-party research effort is inconclusive or consensus cannot be reached on a single TP concentration. To date, the STAs have produced effluent ranging from 18 to 25 µg P L^{-1} on a long-term basis. STA design considerations and initial treatment performance will be discussed further in companion papers presented during this conference.

Other restoration activities

Despite the massive scale and cost of Phase I restoration, the environmental benefits that will be achieved by these efforts alone cannot meet all the water quality and hydrologic needs of the Everglades. The following sections outline other activities that are critically important to the future health of the Everglades.

Advanced treatment technologies

The EFA mandated the District to investigate "alternative" P reduction technologies that could be used in combination with, or in place of, the STAs to treat stormwater runoff. All currently known drinking and wastewater treatment technologies, ranging from low-intensity management of constructed wetlands to full-scale chemical treatment plants were screened for possible use (PEER Consultants, P.C./Brown and Caldwell, 1996). Various combinations of the highest ranked technologies were evaluated based on nutrient removal performance, implementation costs and environmental criteria. These comparisons confirmed that the STAs are indeed the best interim step towards achieving long-term restoration goals. The most promising alternative technologies were identified, and any remaining performance and/or operational uncertainties were documented to guide future research. The USCOE dredge and fill permit to build the STAs required that the District investigate eight technologies in what is now referred to as the Advanced Treatment Technologies (ATT) Research Program. This program is designed to obtain critical design information about nutrient removal performance, hydrologic operating characteristics, land requirements, initial and annual costs and potential environmental impacts. Additional details on the ATT research projects will be provided in companion papers presented at this conference. The ATTs constitute the District's Phase II restoration efforts and, in combination with the STAs and BMPs, are intended to produce effluent that will achieve the P threshold concentration. There is not enough information at this time to estimate costs for full implementation of Phase II restoration.

Everglades stormwater program

The District has initiated a separate effort, the Everglades Stormwater Program (ESP) to address water quality problems in basins that discharge into the Everglades and are not covered by the ECP (Bearzotti et al., 2000). As with the ECP, the ESP has established basin-specific schedules and strategies for regulation, water quality monitoring, construction and other measures and has a target date of December 31, 2006 for full compliance with state water quality standards. Other components of the ESP include inter-governmental cooperative projects, an education campaign and development of a mechanism for reimbursement of restoration expenditures. Cost estimates to fully implement the ESP are not available at this time.

Comprehensive Everglades restoration plan

The District is a partner in a Federal initiative, known as the Comprehensive Everglades Restoration Plan (CERP), which is evaluating modifications to the infrastructure and operation of the C&SF Project to restore, to the greatest extent practicable, the natural hydrology of south Florida while continuing to fulfill flood protection, water supply and all other project objectives. A draft report detailing restoration alternatives was written by an inter-governmental agency team and submitted to Congress in July 1999 (USCOE and SFWMD, 1999). Additional work is needed to (1) determine the total water storage capacity required to achieve hydrologic restoration and (2) define requirements for temporal and spatial distribution of flows. The projected cost for implementing CERP is approximately $7.8 billion, which is an order of magnitude greater than the anticipated cost for the ECP.

Lower east coast regional water supply plan

In May 2000, the District finalized a water supply plan for Florida's lower east coast (the LEC Plan), the region that has the highest concentration of urban development in the District. The LEC Plan defined the hydrologic requirements (e.g., quantity, discharge locations, timing of delivery, etc.) necessary for sustainable future water use and defines how the District will manage the regional water supply over the next 20 years. Because the LEC Plan balances the water needs of urban, agricultural and environmental interests, it is directly linked to implementation of Everglades restoration activities.

Current status of design and implementation

As sufficient information from the various ongoing research programs and other restoration activities discussed above becomes available, basin-specific feasibility studies and conceptual designs will be started to determine the optimal combination of treatment technologies required to achieve long-term water quality goals. For planning purposes, the District is assuming an end-of-pipe discharge limit of 10 µg P L^{-1}. If the final TP discharge limit is significantly different from this value, the optimal long-term solutions may be altered, with possible adverse impacts on final costs and the time required for implementation.

Acknowledgements

The manuscript was improved based on comments from Greg Coffelt, Susan Gray, Jennifer Jorge, Jana Newman and Sharon Trost. We thank Drew Campbell, Debra Case and Cheri Craft for figure preparation.

References

Anderson, D.L and Rosendahl, P.C. (1998). Development and management of land/water resources: the Everglades, agriculture and south Florida. *J. Am. Wat. Res. Assoc.* **34**(2), 235–249.

Beard, D.B. (1938). *Wildlife Reconnaissance, Everglades National Park Project*. Report to the U.S. Department of Interior, National Park Service, Washington, D.C., USA.

Bearzotti, R., Smith, L., Marshall, A., Meiers, D. and Bell, M. (2000). Chapter 11: The Everglades Stormwater Program. In: *Everglades Consolidated Report*, South Florida Water Management District, pp. 11–1 to 11–32.

Bechtel, T., Hill, S., Iricanin, N., Jacobs, K., Mo, C., Mullen, V., Pfeuffer, R., Rudnick, D. and Van Horn, S. (1999). Chapter 4: Status of compliance with water quality criteria in the Everglades Protection Area and tributary waters. In: *Everglades Interim Report*, South Florida Water Management District, pp. 4–1 to 4–132.

Belanger, T.V., Scheidt, D.J. and Platko, J.R. (1989). Effects of nutrient enrichment on the Florida Everglades. *Lake Res. Manag.* **5**(1), 101–111.

Brezonik, P.L., Hendry, C.D., Edgerton, E.S., Schulze, R.L and Crisman, T.L. (1983). *Acidity, Nutrients, and Minerals in Atmospheric Precipitation Over Florida: Deposition Patterns, Mechanisms, and Ecological Effects*, EPA/600/3-84/004, U.S. Environmental Protection Agency, Corvallis, OR, USA.

Burns and McDonnell (1992). *Everglades Protection Project Conceptual Design – Stormwater Treatment Areas*. Report prepared for the South Florida Water Management District, West Palm Beach, FL, USA.

Craft, C.B., Vymazal, J. and Richardson, C.J. (1995). Response of Everglades plant communities to nitrogen and phosphorus additions. *Wetlands* **15**(3), 258–271.

Davis, F.E., Federico, A.C., Goldstein, A.L. and Davis, S.M. (1985). *Use of Wetlands for Water Quality Improvements*. Technical Memorandum, South Florida Water Management District, West Palm Beach, FL, USA.

Davis, S.M. and Ogden, J.C. (eds.). (1994). *Everglades – The Ecosystem and Its Restoration*. St. Lucie Press, Delray Beach, FL, USA.

Gunderson, L.H and Loftus, W.F. (1993). The Everglades. In: *Biodiversity of the Southeastern United States Lowland Terrestrial Communities*, W.H. Martin, S.G. Boyce and A.C. Echternacht (eds.), John Wiley & Sons, New York, NY, USA, pp. 199–255.

Kadlec, R.H. and Newman, S. (1992). *Phosphorus Removal in Wetland Treatment Areas – Principles and Data*, DOR 106, Report prepared for South Florida Water Management District, West Palm Beach, FL, USA.

Light, S.S. and Dineen, J.W. (1994). Water control in the Everglades: a historical perspective. In: *Everglades – The Ecosystem and Its Restoration*, S.M. Davis and J.C. Ogden (eds.), St. Lucie Press, Delray Beach, FL, USA, pp. 47–84.

Maltby, E. and Dugan, P.J. (1994). Wetland ecosystem protection, management and restoration: An international perspective. In: *Everglades – The Ecosystem and Its Restoration*, S.M. Davis and J.C. Ogden (eds.), St. Lucie Press, Delray Beach, FL, USA, pp. 29–46.

McCormick, P.V., Newman, S., Miao, S., Reddy, R., Gawlik, D., Fitz, C., Fontaine, T. and Marley, D. (1999). Chapter 3: Ecological needs of the Everglades. In: *Everglades Interim Report*, South Florida Water Management District, West Palm Beach, FL, USA, pp. 3–1 to 3–66.

McCormick, P.V., Newman, S., Payne, G., Miao, S. and Fontaine, T.D. (2000). Chapter 3: Ecological effects of phosphorus enrichment in the Everglades. In: *Everglades Consolidated Report*, South Florida Water Management District, West Palm Beach, FL, USA, pp. 3–1 to 3–72.

McPherson, B.F., Hendrix, G.Y., Klein, H. and Tysus, H.M. (1976). *The Environment of South Florida – a Summary Report*, U.S. Geological Survey Professional Paper 1011.

Nearhoof, F.L. (1992). *Nutrient-induced Impacts and Water Quality Violations in the Florida Everglades*. Water Quality Technical Series, Vol. 3, No. 24, Florida Department of Environmental Protection, Tallahassee, FL, USA.

PEER Consultants, P.C./Brown and Caldwell (1996). *Desktop Evaluation of Alternative Technologies*. C-E008-A3. Final Report prepared for the South Florida Water Management District, West Palm Beach, FL, USA.

Rader, R.B. and Richardson, C.J. (1992). The effects of nutrient enrichment on algae and macroinvertebrates in the Everglades: A review. *Wetlands* **12**(2), 121–135.

SFWMD (1992a). *Surface Water Improvement and Management Plan for the Everglades – Supporting Information Document*, South Florida Water Management District, West Palm Beach, FL, USA.

SFWMD (1992b). *Surface Water Improvement and Management Plan for the Everglades – Planning Document*, South Florida Water Management District, West Palm Beach, FL, USA.

Snyder, G.H. and Davidson, J.M. (1994). Everglades agriculture: past, present, and future. In: *Everglades – The Ecosystem and Its Restoration*, S.M. Davis and J.C. Ogden (eds.), St. Lucie Press, Delray Beach, FL, USA, pp. 85–115.

Steward, K.K. and Ornes, W.H. (1975). The autecology of sawgrass in the Florida Everglades. *Ecology* **56**, 162–171.

Swift, D.R. and Nicholas, R.B. (1987). *Periphyton and Water Quality Relationships in the Everglades Water Conservation Areas*, Technical Publication 87–2, South Florida Water Management District, West Palm Beach, FL, USA.

Thayer, D., Ferriter, A., Bodel, M., Langeland, K., Serbesoff, K., Jones, D. and Doren, B. (2000). Chapter 14: Exotic plants in the Everglades. In: *Everglades Consolidated Report*, South Florida Water Management District, West Palm Beach, FL, USA, pp. 14–1 to 14–48.

USCOE and SFWMD (1996). *Florida's Everglades Program Everglades Construction Project. Final Programmatic Environmental Impact Statement*, U.S. Army Corps of Engineers, Jacksonville District, South Atlantic Division, Jacksonville, FL and South Florida Water Management District, West Palm Beach, FL, USA.

USCOE and SFWMD (1999). *Central and Southern Florida Project Comprehensive Review Study. Final Integrated Feasibility Report and Programmatic Environmental Impact Statement*, U.S. Army Corps of Engineers, Jacksonville District, South Atlantic Division, Jacksonville, FL and South Florida Water Management District, West Palm Beach, FL, USA.

Walker, W.W. (1995). Design basis for Everglades stormwater treatment areas. *Wat. Res. Bull.* **31**(4), 671–685.

Phosphorus removal from Everglades agricultural area runoff by submerged aquatic vegetation/limerock treatment technology: an overview of research

B. Gu*, T.A. DeBusk**, F.E. Dierberg**, M.J. Chimney*, K.C. Pietro* and T. Aziz ***

* Ecological Technologies Department, South Florida Water Management District, 3301 Gun Club Road, West Palm Beach FL 33406, USA
** DB Environmental, 414 Richard Road, Suite 1, Rockledge, FL 32955, USA
*** Florida Department of Environmental Protection, Twin Towers Office Building, 2600 Blair Stone Road, Tallahassee FL 32399, USA

Abstract The 1994 Everglades Forever Act mandates the South Florida Water Management District and the Florida Department of Environmental Protection to evaluate a series of advanced treatment technologies to reduce total phosphorus (TP) in Everglades Agricultural Area runoff to a threshold target level. A submerged aquatic vegetation/limerock (SAV/LR) treatment system is one of the technologies selected for evaluation. The research program consists of two phases. Phase I examined the efficiency of SAV/LR treatment system for TP removal at the mesocosm scale. Preliminary results demonstrate that this technology is capable of reducing effluent TP to as low as 10 µg/L under constant flows. The SAV component removes the majority of the influent soluble reactive P, while the limerock component removes a portion of the particulate P. Phase II is a multi-scale project (i.e., microcosms, mesocosms, test cells and full-size wetlands). Experiments and field investigations using various environmental scenarios are designed to (1) identify key P removal processes; (2) provide management and operational criteria for basin-scale implementation; and (3) provide scientific data for a standardized comparison of performance among advanced treatment technologies.
Keywords Advanced treatment technology; Everglades; limerock filter; phosphorus; submerged aquatic vegetation; wetlands

Introduction

The Everglades is an internationally recognized ecosystem that covers approximately two million acres in South Florida and represents the largest subtropical wetland in the United States. However, the biotic integrity of the Everglades ecosystem has been endangered by the alterations of hydrological and nutrient regimes due to urban and agricultural development. Reduction of total phosphorus (TP) from the Everglades Agriculture Area (EAA) runoff is a prerequisite to restoring and protecting the remaining Everglades natural resources. The 1994 Everglades Forever Act (EFA, Section 373.4592, Florida Statutes) mandates South Florida Water Management District (District) and the Florida Department of Environmental Protection to evaluate a series of advanced treatment technologies that may be used to protect and restore the Everglades ecosystem.

One of the promising advanced treatment technologies being evaluated is a system using submerged aquatic vegetation and limerock (SAV/LR) to remove P from the water column. Removal of P is accomplished by plant uptake as well as by adsorption to (or co-precipitation with) calcium carbonate, which precipitates from the water column due to photosynthesis-related pH elevations. The downstream limerock further removes a small amount of particulate P (PP).

This research project is divided into two phases. For Phase I of this project, mesocosms operated under steady state conditions were used to evaluate P removal performance of

SAV and limerock under various hydraulic retention times (HRT), water depths, and harvesting regimes (DBEL, 1999). The Phase II SAV/LR research program addresses a number of system processes, as well as management and operational issues at various scales (Table 1). This paper reports findings mainly from the Phase I research effort.

Methods

This research project is being performed at Stormwater Treatment Area-1 West (STA-1W), a 2,610 hectare treatment wetland in western Palm Beach County, Florida (Figure 1). STA-1W currently provides P removal from nutrient-rich water of the EAA before discharge into the Loxahatchee National Wildlife Refuge. The STA-1W inflow has high nutrient concentrations characteristic of post-BMP farm runoff (~70 to 220 µg/L TP) and low nutrient concentrations at its outflow (~10 to 40 µg/L). This paper describes experiments using SAV mesocosms located at the North Supplemental Technology site (with high influent TP), as well as performance data from a large, SAV-dominated wetland cell (Cell 4 of the STA-1W).

For the HRT study, SAV mesocosms (internal dimensions: 4.66 m L × 0.79 m W × 1.0 m D) were constructed of lumber and fiberglass and then plumbed with PVC pipe. Muck was placed in the bottom of each mesocosm to a depth of 15 cm and the water depth maintained at 76 cm. Source water (i.e., Post-BMP runoff) was distributed to the mesocosms by gravity. The mesocosms were inoculated with SAV collected from STA-1W. *Ceratophyllum demersum* and *Najas guadalupensis* were the dominant species stocked. Crushed limerock with a nominal size range of 1.5 to 2.5 cm was placed in black plastic barrels (208 and 113 L). The barrels were plumbed in series after each SAV tank to serve as upflow limerock beds. STA-1W inflow waters were fed to three sets of triplicate mesocosms at different flow rates, providing HRT of 1.5, 3.5 and 7.0 days. The downstream limerock beds were sized to provide a uniform HRT of 5 hours among all treatments (DBEL, 1999). For the water depth and plant harvest studies, we deployed slightly smaller mesocosms, without a downstream limerock bed. Phosphorus removal performance was evaluated using SAV mesocosms operated at water depths of 0.4, 0.8 and 1.2 m. In a separate study, we compared P removal performance of non-harvested SAV systems with those from which macrophytes were intermittently harvested (DBEL, 1999). In all experiments, water samples were typically collected once weekly and analyzed for TP and SRP.

Table 1 An overview of research activities for the Phase II SAV/LR advanced treatment technology

Topics	Platform	Experiments
Process Issues	Microcosms, Mesocosms, ENR Test Cells and Cell 4	• Characterization of DOP and PP • Co-precipitation with Soluble Reactive P (SRP) • Effects of Calcium Concentrations, pH and Alkalinity • Sediment Characterization • Biomass Turnover and Decomposition • P Mass Balances • Diurnal P Cycle
Operation Issues	Mesocosms, ENR Cell 5	• SAV Inoculation • Lime Additions • Effects of SAV Harvesting • Effects of Dry-down/Reflooding
Management Issues	Mesocosms, ENR Test Cells and Cell 4	• Effects of Pulse-loading • Effects of Stagnation • Effects of Velocity • Effects of Substrates • Effects of Filtration Media • Hydraulic Optimization

Each sample consisted of a composite of two grab samples, collected during the week. Temperature and pH measurements were performed in the field at the time of water sample collection.

Figure 1 also shows the location of the full-scale (147 ha) SAV-dominated treatment wetland (Cell 4 of STA-1W). Since mid-1994, Cell 4 has received influent waters that were pre-treated in an emergent wetland (Cell 2) to TP concentrations of approximately 44 µg/L (Chimney et al. 2000). Since startup, the District has measured flow velocities in inflow culverts and outflow culverts at 15-minute intervals with ultrasonic velocity meters (UVMs). Calculation of daily Cell 4 hydraulic loading rates (HLRs) was based on average daily flow velocities, total culvert areas, and Cell 4 surface area. The District also sampled influent and effluent water quality in Cell 4. TP samples are collected weekly at inflow and outflow stations, with each sample representing a composite of 21 grab samples (3x/daily, 7 days/week).

Results and discussion

Effects of hydraulic retention time

This experiment was designed to examine the P removal performance by the SAV/LR treatment system under various HRT (1.5, 3.5, and 7.0 days). Results indicate that most of the P removal was provided by the "front-end" SAV unit process. Outflow TP concentrations from the 1.5, 3.5 and 7.0 day HRT SAV mesocosms averaged 53, 28 and 23 µg/L during the experimental period (Figure 2). These values were further reduced in the downstream limerock beds to 40, 19, and 15 µg/L, respectively. Influent TP values to the SAV mesocosms during this period averaged 108 µg/L. It is notable that increasing the HRT from 1.5 to 3.5 days markedly improved the P removal performance, yet doubling the HRT from 3.5 days to 7 days had little additional effect. In both the 3.5 and 7.0-day HRT mesocosms, average SRP levels were reduced to the detection limit (3 µg/L and 2 µg/L, respectively) (DBEL, 1999; 2000). Most studies of treatment wetlands have shown that long HRTs typically result in better treatment performance. These data confirm that at a constant depth, superior SAV P removal performance can be achieved at longer HRTs.

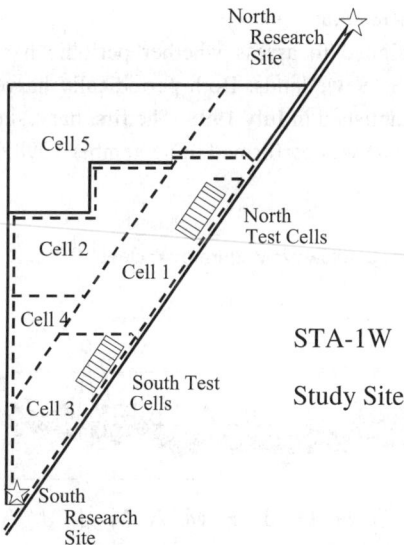

Figure 1 Site map of SAV/LR research location within STA-1W, Palm Beach County, Florida

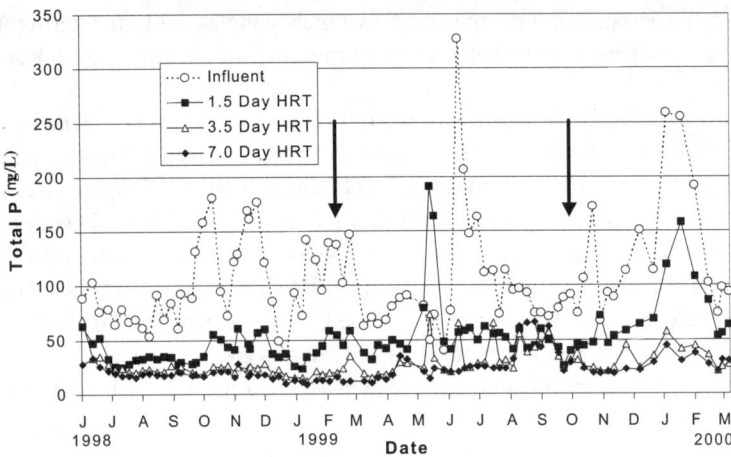

Figure 2 Mean TP concentrations in the influent and effluents of triplicate mesocosms operated at 1.5, 3.5 and 7.0 day HRT since June 1, 1998. Data from only one mesocosm of each HRT is presented for the period between February 10 and September 29, 1999 (arrows)

Effects of water depth

The SAV mesocosms operated at shallow (0.4 m), moderate (0.8 m) and deep (1.2 m) water depths received a comparable hydraulic loading rate of approximately 10 cm/day. This resulted in HRTs of 4.0, 8.1, and 12.1 days for the shallow, moderate and deep mesocosms, respectively. Results from the first year of the study indicated that despite their shorter HRT, the 0.4 m depth mesocosms performed slightly better than either the moderate or deep systems (DBEL, 1999; 2000). More recently, effluent TP concentrations have been equal among all depth treatments, averaging about 20 µg/L (Figure 3).

Because runoff from the nearby EAA is rainfall generated, the hydraulic loadings to the large STA wetlands being constructed in south Florida will vary widely on a seasonal basis. These variable hydraulic loadings will also result in varying water depths within the STAs. To date, these mesocosm findings suggest that SAV will provide effective P removal over the depth range of 0.4 to 1.2 m.

Effects of SAV harvesting on P removal

This experiment was designed to assess whether periodic harvesting can enhance P removal performance of SAV wetlands. Both periodically harvested and nonharvested SAV mesocosms were established in July 1998. The first harvest (partial removal of both SAV shoot and root biomass) was performed in September 1998. Effluent TP levels were

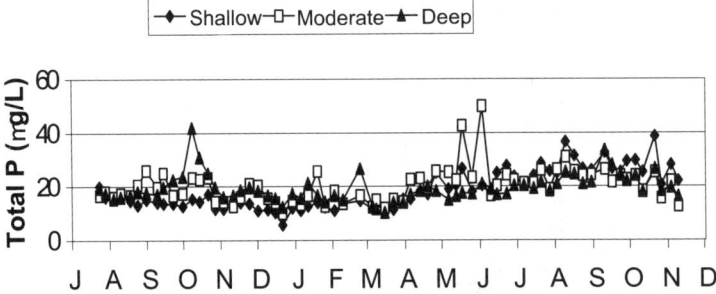

Figure 3 Mean TP concentrations in SAV mesocosms operated at shallow (0.4 m), moderate (0.8 m), and deep (1.2 m) depths from July 1998 to December 1999

considerably higher following this harvest, but then returned to pre-harvest levels after several weeks (DBEL, 1999). The second harvest (clipping and discarding the top half of the SAV) was performed on August 19, 1999. The P removal efficiency in the mesocosms, which had been considerably reduced after this "second" harvest for a 7-week "recovery" period, returned to near pre-harvest levels by the second week in October (Table 2). To date, harvesting of SAV biomass has provided no long term enhancement of P removal performance. Additionally, short term increases in outflow TP levels following harvest suggest that this management practice is not useful for SAV systems designed for low-level P removal (DBEL, 2000).

Effects of substrate types

In this study, SAV is being cultured with muck, limerock and sand substrates using STA-1W outflow (Post-STA) waters. To date, the SAV in the muck is providing slightly lower effluent TP concentrations than SAV cultured on the other substrates (Table 3). During the November to January quarter, more P was removed (relative to the influent loading) in the muck substrate mesocosms than in previous months, which may have been a direct result of increased P loadings in the source water (DBEL, 2000).

The SAV standing crop on the muck substrate appears more robust than the SAV in either the sand or limerock substrate mesocosms. The apparent superior nutritional status of the SAV in the muck systems may enhance the water column P removal either directly (through plant P uptake) or indirectly (altering water chemistry), although there is no significant difference between treatments. This study indicates that muck removal to a limerock substrate (or limerock placement over muck) does not enhance Post-STA SAV P removal (DBEL, 2000).

Sequential SAV/LR treatment system

Because of the initial superior performance of the shallow depth SAV systems, during fall 1998 we established a sequential SAV/LR treatment system (DBEL, 1999). This system consisted of the following unit processes in sequence: a SAV mesocosm operated at a 0.8m depth with a HRT of 3.6 days; a second mesocosm at a depth of 0.4 m with a HRT of 1.8 days; and a LR bed with 1 hour HRT. During the first 6 months (November 1998-April 1999) of operation, the average influent TP was reduced from 130 to 24 µg/L in the 80 cm

Table 2 Mean influent and effluent TP concentrations of triplicate harvested SAV mesocosms before harvest, and during and after a "recovery" period. Harvesting was performed on August 19, 1999

	Sampling Events	Influent TP (µg/L)	Effluent TP (µg/L)	P Removal (%)
Pre-Harvest Period (July 1–Aug 18, 1999)	7	98	29	70
Recovery Period (Aug 19–Oct 5, 1999)	7	83	84	−1
Post-Recovery Period (Oct 6–Jan 31, 2000)	13	144	37	74

Table 3 Mean and range of TP concentrations (µg/L) from July 1999 to February 2000 for the influent and effluents of mesocosms containing limerock, muck and sand substrates. Each substrate treatment consists of duplicate mesocosms operated at a 1.3 day HRT

	Influent	Limerock	Muck	Sand
mean	17	14	13	15
minimum	9	6	6	10
maximum	30	27	21	25

mesocosm, to 11 µg/L in the 40 cm mesocosm, and finally down to 9 µg/L in the limerock beds (Figure 4). However, TP concentrations at all monitoring locations of the sequential system began to increase in May of 1999, and have since been quite variable. The more erratic performance of the sequential system after the first 6 months is likely due to the senescence of *Chara*, a macroalga that dominated the SAV community early in the study. It is not clear why *Chara* became the dominant species in this experiment. *Chara* is not a major SAV in STA-1W. The results from this experiment point to the importance of SAV management in full-scale wetland systems.

Evaluation of Cell 4 performance

Since mid-1994 a stable SAV ecosystem has colonized and established in Cell 4 of STA-1W, with minimal management effort. *C. demersum* and *N. guadalupensis* are the dominant SAV species. Between mid-1994 and 1999, the average HLR in Cell 4 was 15 cm/day (standard deviation = 10 cm/day) and the mean water depth in Cell 4 was 0.64 m (standard deviation = 0.13 m).

Figure 5 depicts a time history of Cell 4 influent and effluent TP concentrations. For the first three years of operation, effluent TP levels were dynamic and exhibited seasonal

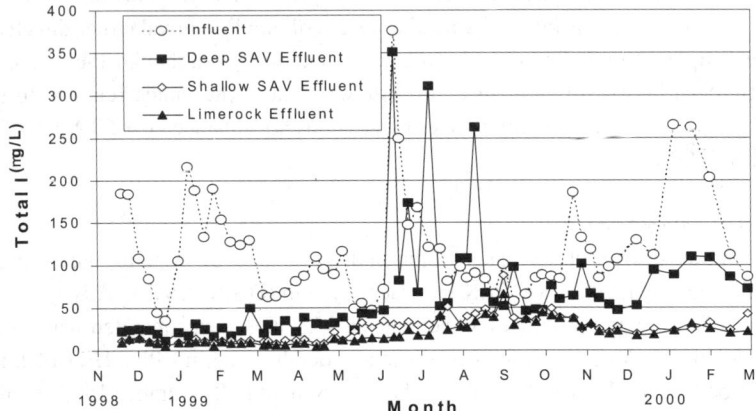

Figure 4 Weekly TP concentrations at four locations in the sequential system treatment train from November 18, 1998 to March 1, 2000

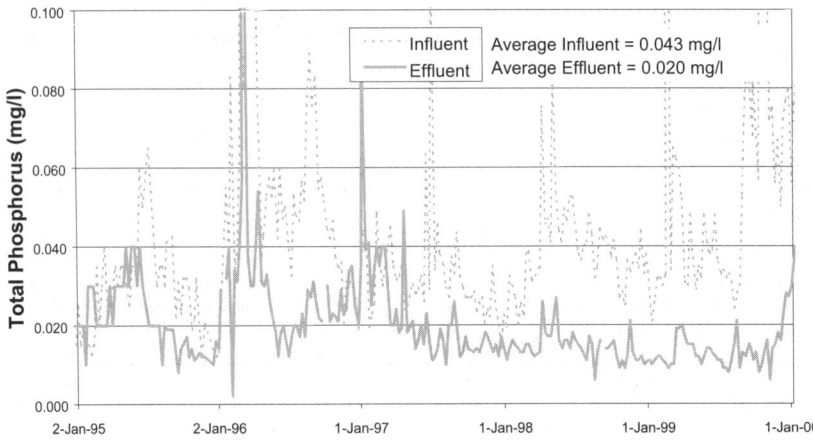

Figure 5 Cell 4 influent and effluent TP concentrations from samples collected weekly, with each sample representing a composite of 21 grab samples (3×/daily, 7 days/week)

Table 4 Summary of Cell 4 TP removal performance

Year	Hydraulic Loading Rate (cm/day)	Effluent TP Concentration (µg/L)	Mass Removal Rates (g/m²/yr)	TP Settling Rate, k (m/yr)	Removal Efficiency (%)
1995	16	21	0.7	28	26
1996	21	29	1.7	49	31
1997	13	21	0.7	26	30
1998	13	14	1.2	44	70
1999	11	13	1.9	55	61
Average	15	20	1.2	40	44

(winter) increases in concentration. However since mid-1997, effluent concentrations have been stable and have averaged 14 µg/L. During the 1999 calendar year, 25% of measured effluent concentrations in Cell 4 were less than or equal to 10 µg/L.

Table 4 summarizes Cell 4 TP removal performance on an annual basis in terms of mass removal rates, TP settling rates, and removal efficiency. During the 1998-99 calendar years when effluent concentrations averaged 14 µg/L, the HLR was slightly lower than historic averages, but mass removals, settling rates, and removal efficiencies were above average values for those years (DBEL, 2000). The average 5-year TP settling rate of 40 m/yr is approximately 3–4 times greater than exhibited by emergent macrophyte treatment wetlands within STA-1W (Chimney *et al.*, 2000) and elsewhere within Florida (Kadlec, 1994).

Future research

Several experiments and field surveys have been initiated to address key system processes, operational and management issues related to the scale-up of SAV/LR technology (Table 1). Additionally, the District has also started or plans to initiate several studies to provide supplemental information on long-term SAV performance and key system P removal processes. These studies include investigation of techniques for DOP removal, SAV decomposition, and sediment and water quality analyses from natural aquatic systems with a history of long-term SAV dominance.

Conclusions

Preliminary findings from Phase I research indicate that the SAV/LR treatment system displays high TP removal rates. Direct plant uptake and co-precipitation of SRP with calcium carbonate are likely the major mechanisms for P removal in SAV systems. In our Phase II efforts, we will attempt to define the relative importance of these P removal mechanisms. The results from short-term mesocosm experiments are largely supported by the P removal performance of a 5-year old SAV-based wetland system (STA-1W Cell 4). Additional experiments are being conducted under ambient hydraulic conditions to provide a full evaluation of this low-level P removal system, so that it can be compared to other conventional and wetland-based treatment technologies.

References

Chimney, M.J., Nungesser, M., Newman, J., Pietro, K., Germain, G., Goforth, G., Lynch, T. and Moustafa, M.Z. (2000). Stormwater Treatment Areas-Status of Research and Monitoring to Optimize Effectiveness of Nutrient Removal and Annual Report on Operational Compliance. Everglades Consolidated Report. South Florida Water Management District, West Palm Beach, FL. USA.

DBEL (1999). *A Demonstration of Submerged Aquatic Vegetation/Limerock Treatment System Technology for Removal Phosphorus from Everglades Agricultural Area Water: Final Report.* Report prepared for SFWMD, West Palm Beach, FL. USA.

DBEL (2000). *Demonstration of submerged aquatic vegetation/limerock treatment system technology for phosphorus removal from Everglades agricultural area waters: Follow-on Study. First Quarterly Report.* Report Prepared for South Florida Water Management District and Florida Department of Environmental Protection.

Kadlec, R.H. (1994). Phosphorus uptake in Florida marshes. *Water Science and Technology* **30**(8), 225–234.

Evaluation of phosphorus retention in a South Florida treatment wetland

M.K. Nungesser and M.J. Chimney

Ecological Technologies Department, South Florida Water Management District, 3301 Gun Club Road, West Palm Beach, FL 33406, USA. (E-mail: *mnunges@sfwmd.gov, mchimney@sfwmd.gov*)

Abstract The Everglades Construction Project of the South Florida Water Management District (District) will employ large constructed wetlands known as Stormwater Treatment Areas (STAs) to reduce phosphorus concentrations in runoff entering the Everglades. The District built and operated a prototype STA, the 1,545 ha Everglades Nutrient Removal Project (ENRP), to determine the efficacy of subtropical wetlands for improving regional water quality with a focus on reducing total phosphorus (TP). In five years of operation, the ENRP has consistently exceeded its performance goals of TP outflow concentrations <50 µg P/L and a 75% TP load reduction.

Since August 1994, the ENRP has retained 70.3 metric tons of TP that otherwise would have entered the Everglades. When corrected for surface area and inflow TP load, TP removal efficiency was highest in the inflow buffer cell and decreased generally in a downstream fashion through the wetland. High TP removal efficiency in treatment cell 4 was attributed to superior performance of its submerged aquatic vegetation community relative to the emergent and floating macrophyte community in the other cells. Controlled experiments in the District's STA Optimization Research Program will help clarify what effect vegetation and operational conditions may have on nutrient removal in the STAs.

Keywords Constructed wetlands; phosphorus; subtropical wetlands; water quality; water treatment

Introduction

Biotic integrity of the Everglades is endangered due to urban and agricultural development, flood-control efforts that have disrupted the region's hydrologic patterns, and influx of nutrient-rich water from a number of sources (Chimney and Goforth, 2000). For example, nutrient enrichment has altered Everglades plant and periphyton communities, which are very sensitive to phosphorus (P) availability (McCormick *et al*., 1999). The Everglades Construction Project of the South Florida Water Management District (District) includes building large treatment wetlands, referred to as Stormwater Treatment Areas (STAs), to reduce nutrients (primarily total P [TP]) in runoff entering the Everglades. The District operated a prototype STA, the Everglades Nutrient Removal Project (ENRP), from August 1994 through April 1999 (period of record) as part of a research and monitoring program designed to gain the operational data necessary to optimize nutrient removal performance in the STAs (Chimney and Goforth, in prep). Performance objectives for the ENRP were 1) to achieve a TP mass reduction of 75% in water processed through the wetland, and 2) to discharge water with an annual flow-weighted TP concentration no greater than 50 µg P L^{-1}. This paper focuses on operational performance of the ENRP from May 1995, when the wetland was fully instrumented, through April 1999 (study period), after which it was incorporated into one of the STAs, i.e., STA-1 West (Figure 1).

Study site

The ENRP (Figure 1) was a 1,546 ha constructed wetland designed to operate as a once-through treatment system with the capacity to process up to 5.3×10^6 m^3 of runoff annually (~ 4.3×10^5 ac-ft yr^{-1}). Water was pumped from the ENRP supply canal into the buffer cell and then routed to one of two parallel flow-ways separated by a longitudinal interior levee

(Figure 1). The eastern flow-way consisted of treatment cells 1 and 3; the western flow-way included treatment cells 2 and 4. Water moved from treatment cell 1 into 3 and from treatment cell 2 into 4 via gravity through culverts in separating interior levees. A perimeter canal captured groundwater seepage from the northern and western project boundaries and returned it to the inflow pump station. Additional groundwater seepage entered the ENRP along its eastern border with Water Conservation Area 1 (WCA-1). The outflow pump station removed water from the ENRP and pumped it into WCA-1. Physical characteristics of the ENRP, including size, depth, vegetation types, and hydraulic retention time, are summarized in Table 1.

The vegetation community differed markedly among the interior cells of the ENRP. Naturally occurring cattail (*Typha domingensis* and *T. latifolia*) dominated treatment cells 1, 2, and the buffer cell. The plant community in treatment cell 3 was a mixture of cattail and 164 ha planted with wetland species common to south Florida, including arrowhead (*Sagittaria latifolia* and *S. lancifolia*), spikerush (*Eleocharis interstincta*), maidencane (*Panicum hemitomon*), pickerelweed (*Pontederia cordata*), and sawgrass (*Cladium jamaicense*). Treatment cell 4 was maintained as a submerged aquatic vegetation (SAV) community dominated by coontail (*Ceratophyllum demersum*) and southern naiad (*Najas guadalupensis*). Portions of treatment cells 1, 2 and 3 also supported dense stands of SAV.

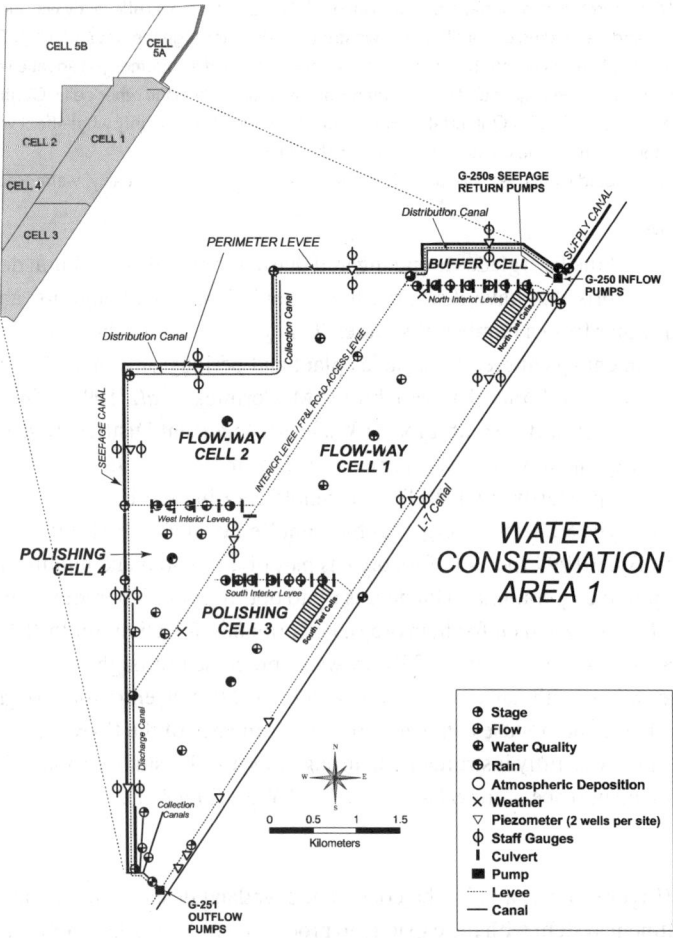

Figure 1 Sampling locations in the Everglades Nutrient Removal Project (ENRP). Insert shows location of the ENRP relative to Stormwater Treatment Area-1 West. Cell size is reported in Table 1

Table 1 Select characteristics of the Everglades Nutrient Removal Project and its component buffer and treatment cells

	Surface Area (ha)	Mean Depth[a] (m)	Range in Vegetation Coverage (%)[b]				Median HRT[a] (days)
			Cattail	Other Emergents	Floating	Open Water/SAV	
ENRP	1,545	0.56	—	—	—	—	20.2
Buffer	54	0.56	40.3–44.8	8.4–16.1	28.3–45.3	2.5–14.8	0.5
Cell 1	527	0.54	29.1–43.4	15.2–21.9	2.2–12.4	34.8–46.4	15.1
Cell 2	413	0.75	45.8–82.8	1.0–9.2	0.0–10.6	7.8–46.7	11.1
Cell 3	404	0.37	1.2–47.3	40.4–49.2	0.0–2.0	10.1–57.8	7.2
Cell 4	147	0.64	0.0–4.8	0.1–5.5	0.0–2.2	93.3–99.9	4.5

[a] Data summarized for the period May 1995 through April 1999
[b] Data summaries from overflights conducted from October 1993 through November 1998

Methods

Vegetation coverage

Aerial photographs of the ENRP were taken at regular intervals from October 1993 through November 1999. Photo interpretation, together with field ground-truthing, was used to determine the extent of four vegetation community types: cattail, other emergent vegetation, floating vegetation and open water/SAV (Table 1). Changes in the coverage of each vegetation type were monitored over time.

Water budgets

Separate water budgets were computed for the entire ENRP, the buffer cell and each treatment cell (Chimney et al., 2000). Water budgets were based on estimated daily inflow and outflow water volumes and changes in wetland storage capacity using a water mass balance equation of the general form: $I - O = \Delta S + r$, where I = inflow water volume; O = outflow water volume; ΔS = change in storage capacity; and r = remainders to the water budget. Inflow to the entire ENRP included pumped inflow, returned perimeter seepage, rainfall, and estimated groundwater seepage coming from WCA-1; while outflow consisted of pumped outflow, estimated evapotranspiration for each major vegetation type, and seepage losses. Daily change in storage for the entire ENRP was estimated as the sum of the daily storage changes for each of the five internal cells. Cell-level water budgets were based on flows into and out of each cell, evapotranspiration adjusted for the composition of the cell's vegetation community, rainfall proportional to the cell's surface area, and seepage adjusted for head differences and levee length between adjacent cells. Surface outflow from treatment cell 3 was not monitored directly as in other cells, but was estimated instead as the difference between the measured daily flow at the outflow pump station and outflow from treatment cell 4. Chimney and Moustafa (1999), Chimney et al. (2000), and Nungesser et al. (2001) describe the methodology used to calculate ENRP water budgets.

Water loading rates into the ENRP averaged 4.2 cm d^{-1} for the four-year period from May 1995 through April 1999 (Table 2). By far the highest loading rate, 91.1 cm d^{-1}, was in the buffer cell, into which all water entered before distribution into the two flow paths. The next highest loading rate, 15.6 cm d^{-1}, was into treatment cell 4. The small surface area of these cells contributed largely to the high loading rates. Average annual inflow also was highest in the buffer cell, 176.2×10^6 m^{-3} yr^{-1}.

Phosphorus budgets

Phosphorus budgets were calculated for the entire ENRP, the buffer cell and each treatment cell. Phosphorus budgets accounted for the TP mass associated with flow from the inflow,

Table 2 Hydrologic features of the Everglades Nutrient Removal Project treatment cells[a]

	Average HLR (cm d^{-1})	Average Flow (hm^3 yr^{-1})[b]	
		Inflow	Outflow
ENRP	4.2	218.3	218.0
Buffer	91.1	176.2	157.8
Cell 1	4.0	78.9	70.6
Cell 2	6.3	98.5	108.7
Cell 3	5.2	76.7	70.5
Cell 4	15.6	87.5	86.4

[a] Data summarized for the period May 1995 through April 1999
[b] 1 hm^3 = 1,000,000 m^3

outflow and seepage return pumps, rainfall, groundwater seepage and atmospheric dry deposition (Chimney et al., 2000). Total P loads were calculated by multiplying daily water volumes with corresponding TP concentrations and then summing these values on a temporal basis. Monthly flow-weighted mean TP concentrations were computed by first weighting flow-proportioned sample concentrations by their corresponding flow volume. All inflow and outflow TP concentrations are reported as flow-weighted values. The methodology used to calculate TP budgets for the ENRP is detailed in Chimney and Moustafa (1999), Chimney et al. (2000) and Nungesser et al. (2001). In this paper, P removal (i.e., load reduction) is computed as the net difference between inflow and outflow TP mass expressed on an annualized unit-area basis (R_{UL}; g m^{-2} yr^{-1}) to account for differences in cell surface area. Annualized unit-area TP load reduction is commonly reported for treatment wetlands. In addition, to account for differences in TP loading among cells, we normalized aerial TP load reduction over the entire study period (mg m^{-2}) by the TP mass (kg) that entered each cell to produce what we term "area-load" TP reduction (mg m^{-2} kg^{-1}).

Because the interior cells were loaded differently, differences in nutrient removal performance were evaluated with a one-way analysis of covariance (ANCOVA) of R_{UL} (3-month moving values) that used inflow TP load as a covariate (SAS, PROC ANCOVA; SAS Institute Inc, Cary, NC). The ANCOVA analysis was followed by a *post hoc* means separation test (Ryan-Einot-Gabriel-Welsch [REGW]). The REGW procedure groups mean values for each treatment class according to the probability that they are not statistically different from each other (Sokal and Rohlf, 1981), and it controls the maximum experiment-wise error rate (SAS, 1990).

Results

Vegetation coverage

Except for treatment cell 4, the interior cells were dominated by emergent (including cattail) and/or floating vegetation (Table 1). The northern portion of the ENRP (buffer cell and portions of treatment cells 1 and 2) was flooded in 1989 and had developed a substantial plant community before the first photographic overflight (and therefore was not seen as predominately open water in the first photographs). There were relatively large changes in vegetation coverage in some cells, such as shifts in cattail and open water/SAV coverage in treatment cells 2 and 3. Treatment cell 4 shifted from being almost entirely open water to complete coverage by a SAV community within several years after flooding.

Water budgets

Although the western flow-way accounted for only 36.2 % of the ENRP's surface area (Table 1), it received 55.9% of outflow from the buffer cell during the study period, whereas the larger eastern flow-way processed the remaining buffer cell outflow (Table 2). Monthly water depths in the ENRP ranged from 0.2 m to 0.9 m, while mean depths in each cell for the entire study period varied from 0.37 m to 0.75 m (Table 1). The western flow-way was consistently 0.1 to 0.3 m deeper than the other cells. Monthly storage volume for the entire ENRP ranged from 5.5 to 11.4 hm^3. However, the net ΔS for the study period was only a fraction of these volumes. Daily total inflows to the project varied greatly over the study period (~ 0.03 to 2.68 hm^3). Median hydraulic retention times (HRT) ranged from 0.5 days in the buffer cell to 15.1 days in treatment cell 1. Median HRTs for the eastern flow-way, western flow-way, and the entire ENRP were 22.3, 15.6 and 20.2 days, respectively (Table 1).

Phosphorus budgets

Monthly flow-weighted mean TP concentrations at the inflow pump station ranged from 57 to 201 µg P L^{-1}, while corresponding concentrations at the outflow pump station ranged from 10 to 39 µg P L^{-1}. The cumulative flow-weighted outflow TP concentration during the study period was 22 µg P L^{-1}. The ENRP removed a total of 70.3 metric tons (mt) of TP from inflow waters during the period of record and 52.4 mt (= 77.1 % load reduction) during the study period (Table 3). The interior cells within the ENRP varied in their individual TP removal performance. The greatest TP mass was removed by the buffer cell (18,408 kg), which largely represented those P fractions most easily retained, specifically soluble reactive P and P bound to particles subject to mechanical settling. The western flow-way retained almost twice the TP mass (21,393 kg) as the eastern flow-way (11,028 kg) (ratio = 1.9:1.0), even though it had a smaller surface area (Table 1) and was not loaded in the same proportion (ratio = 1.7:1.0). Total P retention rates in the buffer cell far exceeded that observed in any other cell, both on a annualized unit-area and an area-load basis (8.544 g m^{-2} yr^{-1} and 0.536 mg m^{-2} kg^{-1}, respectively). This was expected because removal efficiency is usually greatest at the headwaters of both natural and constructed wetlands and decreases with linear distance through the system (e.g., Kadlec and Newman, 1992; Kadlec, 1994). Area-load TP reduction in treatment cell 4 (0.323 mg m^{-2} kg^{-1}) was markedly higher then in the other treatment cells (0.0075 to 0.128 mg m^{-2} kg^{-1}). Throughout its operation, the ENRP consistently exceeded its TP outflow concentration (< 50 µg P/L) and load reduction (75%) performance objectives. The ANCOVA of TP retention detected significant differences among treatment cells (Table 4). The means separation test identified four statistically similar groups of cells: (1) the buffer cell, (2) the eastern flow-way, (3) the western flow-way and (4) treatment cells 1 and 2. Note that the regression model resulting from the ANCOVA accounted for a relatively large amount of variance in TP retention ($R^2 = 0.8686$).

Table 3 Summary of phosphorus budget for the Everglades Nutrient Removal Project and its component buffer and treatment cells for the period May 1995 through April 1999

	TP Inflow (kg)	TP Outflow (kg)	TP Retention		
			(%)[a]	(g m^{-2} yr^{-1})[b]	(mg m^{-2} kg^{-1})[c]
ENRP	67,972	15,557	77.1	0.848	0.050
Buffer	63,792	45,384	28.9	8.544	0.536
Cell 1	17,530	9,980	43.1	0.358	0.082
Cell 2	28,972	13,675	52.8	0.925	0.128
Cell 3	11,424	7,946	30.4	0.215	0.075
Cell 4	12,820	6,725	47.5	1.037	0.323

[a] Relative TP retention = (inflow − outflow)/inflow
[b] Annualized unit-area TP retention rate
[c] Area-load TP retention rate = unit-area TP retention for entire study period adjusted for TP inflow load

Table 4 Summary of analysis of covariance for total phosphorus retained (3-month moving unit-area values in the interior cells of the Everglades Nutrient Removal Project from May 1995 through April 1999

Source	DF	Sum of Squares	Mean Square	F value	P > F
Model	5	2,816.0	563.2	296.4	0.0001
Error	224	425.7	1.9		
Corrected Total	229	3,241.6		$R^2 = 0.8686$	
REGW Procedure: $\alpha = 0.05$, df = 224, MSE = 1.900					
Number of means		2	3	4	
Critical Range		0.6719	0.7341	0.7440	

REGW[a] Grouping	Mean	N	Cell
A	8.874	46	buffer
B	1.102	46	4
C B	0.966	46	2
C D	0.366	46	1
D	0.225	46	3

[a] Means with the same letter are not significantly different.

Discussion

The mechanisms that control short- and long-term nutrient removal in freshwater treatment wetlands are not completely understood. However, nutrient cycling by plants (both algae and macrophytes) is an important process that regulates net retention (Richardson and Craft, 1993; Vymazal, 1995; Kadlec and Knight, 1996). One of the design features of the ENRP was to establish different vegetation communities in the cells and evaluate the effect that vegetation type might have on nutrient removal. If vegetation type is an important factor regulating TP retention, one might expect to see this effect reflected in the TP retention data based on the similarity of vegetation communities, that is, similar performance in treatment cells 1, 2 and 3 and differences in treatment cell 4. Alternatively, if vegetation type is not important, one would expect cell differences associated only with a downstream decrease in removal efficiency consistent with the performance observed in other wetlands. To some degree, we found evidence in the ENRP data for both scenarios.

Inspection of the rank order of area-load TP retention rates indicated a decrease in removal efficiency that would be expected based on increasing distance from the ENRP inflow: buffer cell > treatment cells 1 and 2 > treatment cells 3 and 4. The similarity of area-load retention rates in treatment cells 1 and 2 (0.082 and 0.128 mg m^{-2} kg^{-1}, respectively) is consistent with their position at the upper portion of each flow-way and having comparable vegetation communities. However, the area-load retention rate for treatment cell 4 (0.323 mg m^{-2} kg^{-1}) was higher than expected considering that this cell is located at the bottom of the western flow-way. The fact that treatment cell 4 was dominated by a SAV community and was very effective in P retention suggests that vegetation type contributed to its greater P removal performance.

The ANCOVA of unit-area TP retention found significant differences between the eastern and western flow-ways even after considering differences in TP loading. The inverse relationship between smaller size and higher removal efficiency is at first counterintuitive, although similar findings have been reported for other treatment wetlands (e.g., Li, 2000). These results may be attributable to differences between flow-ways in HRT, operating depths, vegetation characteristic and/or the influence of seepage into the eastern flow-way from WCA-1.

Our data suggest that management strategies involving manipulating water depth, HRT, vegetation type, nutrient loading and other operational variables affect P sequestration in the STAs. Further analysis of data from the District's STA Optimization Research Program will help clarify these relationships. Controlled experiments are under way in two sets of fifteen 0.2 ha test cells in treatment cells 1 and 3 (Figure 1) to determine the roles of these variables in phosphorus retention (Chimney and Goforth, 2000; Nungesser et al., 2001).

Acknowledgements

We appreciate comments and improvements in this manuscript based on reviews by Susan Gray, Binhe Gu, Jennifer Jorge, and Jana Newman, as well as two anonymous reviewers.

References

Chimney, M.J. and Goforth, G. (2000). Environmental impacts to the Everglades ecosystem: a historical perspective and restoration strategies. Preprint of the 7th International Conference on Wetland Systems for Water Pollution Control, Institute of Food and Agricultural Science, University of Florida, Lake Buena Vista, FL, USA, pp. 159–167.

Chimney, M.J. and Goforth, G. (In preparation). History and description of the Everglades Nutrient Removal Project. To be submitted to *Ecological Engineering*.

Chimney, M.J. and Moustafa, M.Z. (1999). Chapter 6: Effectiveness and optimization of stormwater treatment areas for phosphorus removal. In: *Everglades Interim Report*, South Florida Water Management District, West Palm Beach, FL, USA, pp. 6–1 to 6–45.

Chimney, M.J., Nungesser, M., Newman, J., Pietro, K., Germain, G., Lynch, T., Goforth, G. and Moustafa, M.Z. (2000). Chapter 6: Stormwater Treatment Areas – status of research and monitoring to optimize effectiveness of nutrient removal and annual report on operational compliance. In: *Everglades Consolidated Report,* South Florida Water Management District, West Palm Beach, FL, USA, pp. 6–1 to 6–127.

Kadlec, R.H. (1994). Phosphorus uptake in Florida marshes. *Wat. Sci. Tech.* **30**(8), 225–234.

Kadlec, R.H. and Knight, R.L. (1996). *Treatment Wetlands*. Lewis Publishers, Boca Raton, FL, USA.

Kadlec, R.H. and Newman, S. (1992). *Phosphorus removal in wetland treatment areas – principles and data*, DOR 106, Report prepared for South Florida Water Management District, West Palm Beach, FL, USA.

McCormick, P.V., Newman, S., Miao, S., Reddy, R., Gawlik, D., Fitz, C., Fontaine, T. and Marley, D. (1999). Chapter 3: Ecological needs of the Everglades. In: *Everglades Interim Report*, South Florida Water Management District, West Palm Beach, FL, USA, pp. 3–1 to 3–66.

Li, X. (2000). *Purification function of wetlands: spatial modeling and pattern analysis of nutrient reduction in the Liaohe Delta*. Wageningen University, the Netherlands.

Nungesser, M.K., Newman, J.M., Combs, C., Lynch, T., Chimney, M.J. and Meeker, R. (2001). Chapter 6: Optimization research for the stormwater treatment areas. In: *2001 Everglades Consolidated Report*, South Florida Water Management District, West Palm Beach, FL, USA, pp. 6–1 to 6–44.

Richardson, C.J. and Craft, C.B. (1993). Effective phosphorus retention in wetlands: Fact or fiction. In: *Constructed Wetlands for Water Quality Improvement*, G.A. Moshiri (ed.), Lewis Publishers, Boca Raton, FL, USA, pp. 271–282.

SAS (1990). *SAS/STAT Users Guide*, Version 6, 4th Edition, Volume 2. SAS Institute, Inc. Cary, NC, USA.

Sokal, R.R. and Rohlf, F.J. (1981). *Biometry, the Principles and Practice of Statistics in Biological Research*. 2nd Edition. W. H. Freeman and Company, New York, USA.

Vymazal, J. (1995). *Algal and Element Cycling in Wetlands*. Lewis Publishers, Boca Raton, FL, USA.

The Everglades Nutrient Removal Project test cells: STA optimization – status of the research at the north site

J. Majer Newman and T. Lynch

Ecological Technologies Department, South Florida Water Management District, 3301 Gun Club Road, West Palm Beach, FL 33406, USA

Abstract The Everglades is an oligotrophic ecosystem that is being adversely impacted by hydrologic changes and nutrient-rich runoff generated from urban and agricultural sources. The Stormwater Treatment Area (STA) Optimization Research and Monitoring program is mandated by the 1994 Everglades Forever Act and will assist the South Florida Water Management District in developing operational strategies that maximize performance of emergent macrophyte STAs. The primary objective of this research is to examine how hydrologic conditions may influence STA performance. The study was conducted in 0.2 ha, shallow, fully lined test cells located within the perimeter of the Everglades Nutrient Removal Project. Experiments were designed to examine the effect of increased and decreased hydraulic loading rate (HLR) on wetland performance and to determine, if possible, the HLR at which STA treatment fails to reduce outflow total phosphorus concentration to the interim target of 50 µg-P/L. To date, two HLR experiments have been completed at the north site. Preliminary data indicated at all HLRs tested that particulate phosphorus and dissolved organic phosphorus ratios remained virtually unchanged from inflow to outflow. The dissolved organic and particulate compounds within these test cells are extremely recalcitrant, and are not easily assimilated within the system. High HLRs may not result in detention times long enough to mineralize these forms into easily assimilated inorganic compounds, resulting in mean TP concentrations greater than 50 µg-P/L.

Keywords Constructed wetland; Everglades; hydrology; mass retention; nitrogen; phosphorus

Introduction

The Everglades is an oligotrophic ecosystem that is being adversely impacted by hydrologic changes and nutrient-rich runoff generated from urban and agricultural sources (Light and Dineen, 1994; Chimney and Goforth, 2000). The Everglades Forever Act (EFA) requires the South Florida Water Management District (District) to construct a series of large treatment wetlands (ca. 17,000 ha) called Stormwater Treatment Areas (STAs) to reduce nutrient levels in runoff to an interim target of 50 µg-P/L before it reaches the Everglades Protection Area (Chimney and Moustafa, 1999). The EFA also requires the District to conduct research to optimize nutrient removal performance by the STAs. The STA Optimization research and monitoring program will assist the District in developing operation strategies that maximize performance of the STAs (Chimney *et al.*, 2000). One part of this Program involves conducting hydrologic research in the Everglades Nutrient Removal Project (ENRP), a prototype STA built and operated by the District, within which test cells were built. The test cells (Figure 1) are shallow, fully lined wetlands, about 0.2 ha in size. STA Optimization research utilizes the test cells to examine how hydrologic conditions may influence STA performance; i.e., what water management scenarios will promote maximum TP removal efficiency in these systems and conversely, under what hydrologic conditions will TP removal efficiency fail to meet mandated requirements.

Methods

The test cells are arranged into two groups of 15 cells each; one group is located at the northern end of the ENRP and the other at the southern end. Inflow concentrations at the

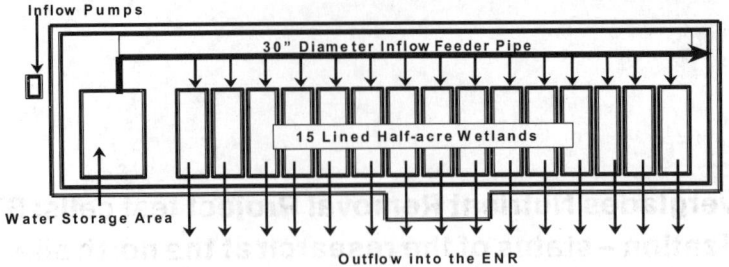

Figure 1 Design of the Everglades Nutrient Removal Project test cells

north and south sites averaged about 0.0771 µg-P/L and 0.029 µg-P/L, respectively. Inflows to each set of test cells come from within the ENRP itself.

The District has dedicated 10 test cells, six at the north and four at the south, for STA Optimization experiments. Two test cells at both the north and south sites are being used as controls, and they are operated at a mean hydraulic loading rate (HLR) of 2.7 cm/day and nominal depth of 0.6 m approximating the average design conditions for the STAs (Walker, 1991). While holding depth constant, the HLR for two north test cells and one south test cell is being incrementally decreased by 50% (thereby increasing hydraulic residence time) every 15 weeks from the initial value of 2.7 cm/day to approximately 0.3 cm/day (low HLR experiments). Concurrently, the HLR in the remaining north and south test cells is being incrementally increased by 50% (conversely, decreasing hydraulic residence time) every 15 weeks to approximately 20 cm/day (high HLR experiments) (Table 1). Following completion of the HLR experiments, all cells will be returned to a HLR of 2.7 cm/day, after which water depth will be incrementally increased step-wise in all experimental cells to a maximum of 1.2 m over three 15-week periods. At this time, only one HLR experiment has been completed at the south site, therefore, this paper will focus on the results of HLR experiments at the north site.

Weekly grab and/or composite water quality samples were collected at storage cell outlets (inflow source water) and the outflow of each STA Optimization test cell and analyzed for 30 parameters (Chimney et al., 2000). Previous testing verified that there was no change in water quality due to the delivery system (Chimney et al., 2000). Here, we will discuss the results of the phosphorus, nitrogen and carbon water quality analyses. All sample collection and analyses have been conducted in accordance with South Florida Water Management District Comprehensive Quality Assurance Plan (SFWMD, 1998), which mandates the use of either Environmental Protection Agency (USEPA, 1983) or American Public Health Association (APHA, 1989) approved analytical methods. Phosphorus, carbon, and nitrogen values less than the minimum detection limit were reported as the detection limit value and were used in calculations. Organic nitrogen, particulate and dissolved organic phosphorus were calculated using the means of measured parameters.

Table 1 Implementation dates, mean hydraulic loading rates, and nominal depths for the STA Optimization research at the ENRP Test Cells

Exp.	North test cells	South test cells	HLR (cm/d)			Depth (m)
			Low	Control	High	
1	May 19, 1999	November 2, 1999	1.27	2.7	4.92	0.6
2	September 1, 1999	February 14, 2000	0.72	2.7	10.72	0.6
3	February 14, 2000	July 3, 2000	0.27	2.7	19.04	0.6
4	July 3, 2000	October 23, 2000	–	2.7	–	1.07
5	October 23, 2000	February 5, 2001	–	2.7	–	0.1
6	February 5, 2001	May 28, 2001	–	2.7	–	low to high

Results

Concentration

Mean influent concentrations were generally higher than outflow concentrations for phosphorus, nitrogen, and carbon, except for total and dissolved organic carbon at the higher HLRs (Figure 2). Mean outflow total phosphorus (TP) concentrations were less than the TP interim target concentration of 50 µg-P/L except when with increased mean HLR of 10.72 cm/d. Mean outflow TP concentrations during the low HLR trials were significantly less than mean outflow TP values from the high HLR trials during experiments 1 (Exp1) and 2 (Exp2), and outflow TP concentrations generally increased as HLRs increased within each experimental phase (Figure 2). Mean total dissolved phosphorus (TDP) represented about 50% of the inflow TP concentrations for Exp1 and Exp2 (Figure 2). Although mean TDP concentrations were substantially reduced for all HLRs, TDP outflow concentrations continued to represent about 50% of the outflow TP concentration, and did not increase with increased HLRs. Mean inflow soluble reactive phosphorus (SRP) reduced by about 80% for all HLRs, and mean outflow concentrations were not affected by changes in the HLR (Figure 2). SRP comprised about 70% and 50% of the influent TDP (organic and inorganic)

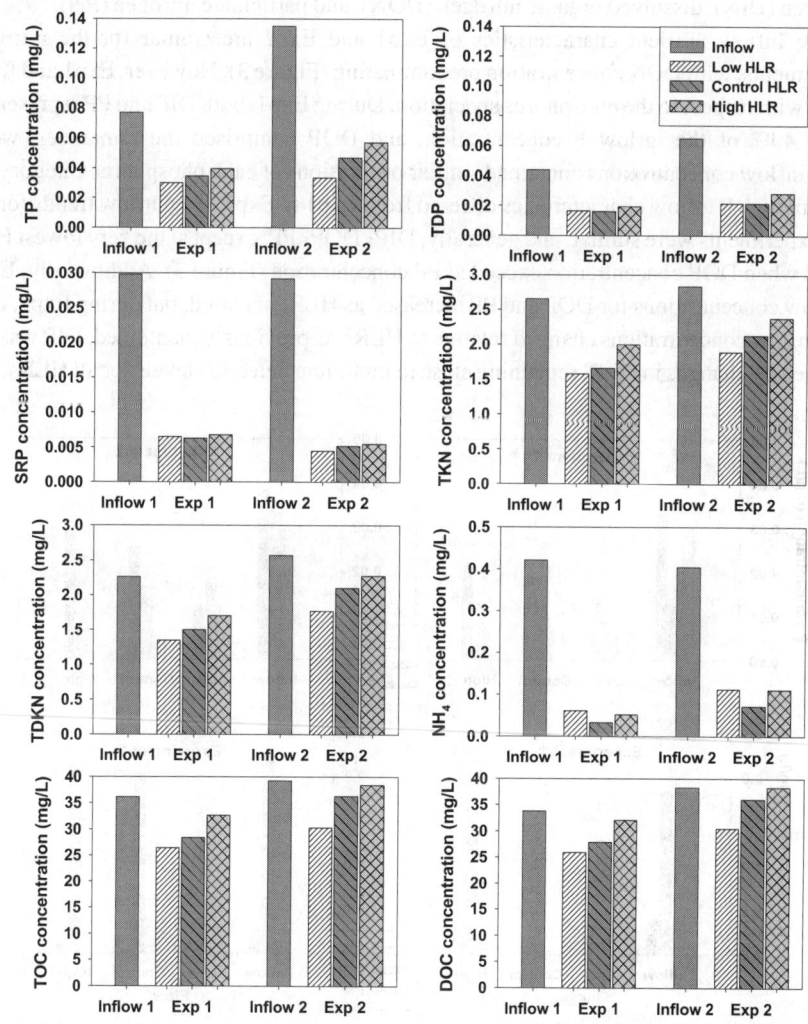

Figure 2 Mean inflow and outflow concentrations for various forms of nitrogen, phosphorus, and carbon during HLR Experiments 1 and 2 at the ENR north test cells

for Exp1 and Exp2, respectively, but only 25% of the outflow TDP concentration for all experimental trials.

Total Kjeldahl nitrogen (TKN) concentration accounted for over 99% of the mean inflow and outflow nitrogen concentration in these test cells. Mean outflow TKN concentrations were generally less than inflow concentrations and ranged from 1.577 mg/L to 2.379 mg/L with TKN outflow concentrations generally increasing as HLRs increased within each experimental phase (Figure 2). In contrast, NH_4-N concentrations were reduced by about 80% regardless of the mean HLR, and represented about 5% of the effluent nitrogen. Mean outflow total dissolved Kjeldahl nitrogen (TDKN) concentrations were less than inflow concentrations and generally increased as HLRs increased (Figure 2). Additionally, TDKN represented the majority of the inflow and outflow TKN concentrations.

While total nutrients can be categorized into four main fractions (dissolved inorganics, dissolved organics, particulate inorganics and particulate organics), however, the particulate constituents are lumped into one category. Therefore, total phosphorus is comprised of dissolved inorganic phosphorus (DIP), dissolved organic phosphorus (DOP), and particulate phosphorus (PP). Likewise, nitrogen is compartmentalized into dissolved inorganic nitrogen (DIN), dissolved organic nitrogen (DON), and particulate nitrogen (PN).

The inflow nutrient characteristics of Exp1 and Exp2 are similar for the nitrogen constituents, with DON concentration predominating (Figure 3). However, Exp1 and Exp2 differ with respect to the phosphorus speciation. During Exp1, both DIP and PP represented about 40% of the inflow P concentration, and DOP comprised the remainder, while Exp2 inflow concentrations contained similar proportions of each phosphorus category.

Although P inflow characteristics differed from Exp1 to Exp2, the outflow trends for the two experiments were similar, and generally, DIP<DOP<PP, except at the very lowest HLR tested when DOP concentration exceeded PP concentration (Figure 3). Additionally, Exp1 outflow concentrations for DOP and PP increased as HLR increased, but during Exp2, only PP outflow concentrations changed relative to HLR. As previously mentioned, DIP was not affected by changes in HLR, remaining close to minimum detection levels for all HLRs.

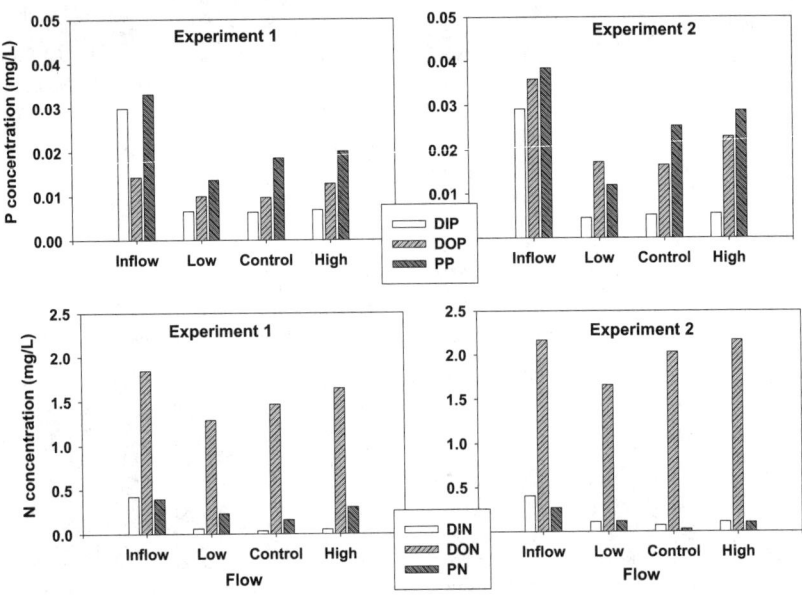

Figure 3 Mean phosphorus and nitrogen constituents for the inflow and outflow for the controls, low, and high HLR experiments 1 and 2 at the north test cells

During Exp1 and Exp2, generally DIN<PN<DON, with DON representing more than 75% of the inflow and outflow nitrogen concentration, and increased as HLR increased (Figure 3). This is in contrast to phosphorus, where PP outflow concentrations were greater then DOP, and represented about 40 and 50% of the outflow concentration, respectively. As with DIP, DIN was not affected by changes in HLR and comprised less than 5% of the outflow nitrogen concentration.

The mean inflow concentrations for total organic carbon (TOC) and dissolved organic carbon (DOC) were greater than mean outflow concentrations for low and control HLRs during Exp1 and Exp2, but percent reductions were less than or equal to 30% (Figure 2). At high HLRs, the mean outflow TOC and DOC concentrations during both Exp1 and Exp2 were about equal to inflows, with percent reductions less then 1%. DOC predominated the inflow and outflow TOC concentrations, indicating that particulate carbon was not an important constituent in the test cells.

Mass

The water year for these experiments extended from 1 May 1999 through 30 April 2000. Separate water balances were calculated for each test cell based on daily inflows, outflow and change in storage capacity using the general water mass balance equation:

$$I - O + RO - ET = \Delta S + r$$

where: I = inflow water volume to the test cell (m^3); O = outflow water volume from the test cell (m^3); RO = runoff water volume to the test cell (m^3); ET = evapotranspiration (m^3); ΔS = change in storage capacity within the test cell (m^3); and r = residuals to the water budget (m^3).

A calibrated tipper bucket was used to verify the inflows for each HLR (Chimney et al., 1999). For the north test cells, the residuals (i.e., unmeasured components and measurement error) ranged from 16.27 to –1.29% for the water year (Table 3). Surveys of various lake and wetland water balance studies showed that it is not uncommon for residuals to range between 10–20% (Winter, 1981).

Mean TP mass percent retention ranged between 30 and 52%, with retention during the low HLR experiments slightly higher than during the high HLR experiments. At the higher HLRs, the TP mass exported from the system was generally related to changes in the inflow TP concentration (ranging between 0.04 µg-P/L and 0.20 µg-P/L) peaking when influent TP concentrations spiked (Figure 4). However, at the low HLRs the TP mass exported was independent of the influent TP concentrations, indicating that the test cells at the low HLR were able to process all the TP regardless of the influent TP concentrations.

Conclusions

These preliminary data indicate that in the test cells outflow TP concentrations were reduced from a mean concentration of 77.1 µg-P/L to less than 50 µg-P/L except at mean HLR of 10.72 cm/day. Of the three main nutrient constituents, the dissolved inorganic faction was readily attenuated in these systems at all HLRs. Except at the lowest HLR,

Table 3 Water balance at ENR north test cells extending May 1, 1999 to April 30, 2000

Test Cell Number	Inflow (m³)	Outflow (m³)	Residuals (%)
TC 07N Low	8418	9130	–8.45
TC 08N Low	8119	8224	–1.29
TC 05N Control	25331	22479	11.26
TC 10N Control	26953	31230	–15.87
TC 06N High	93572	78347	16.27
TC 09N High	88890	76395	14.06

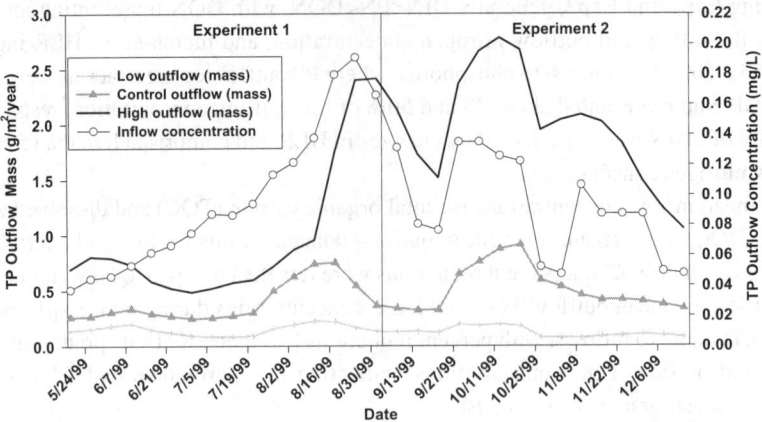

Figure 4 Mean TP inflow concentration and outflow mass during the HLR experiments at the ENRP north test cells

outflow PP concentrations exceeded DOP, but PP and DOP predominated the outflow concentrations at all HLRs. Organic nitrogen outflow concentration increased with increasing HLR, and was the predominant form of nitrogen exiting these wetland systems. DOC was the predominant form of carbon entering and exiting the system, indicating that there was no net attenuation of DOC in these systems at any HLR tested. The dissolved organic compounds within this system are extremely recalcitrant, and are not easily assimilated within the system. High HLRs may not result in detention times long enough to break down these organics into easily assimilated inorganic compounds, resulting in mean TP concentrations greater than the interim target of 50 µg-P/L. Additionally, the TP mass exported from these systems increased in response to TP inflow concentration spikes at higher TP loading levels but not at lower HLRs. This operational response was not observed for SRP. Mean outflow SRP concentration values for all experiments at all the sites were very close to the minimum detection limit, indicating that SRP is limited within these systems, and that even at low nominal residence times of about 5 days, the biota are able to take up almost all the available dissolved inorganic phosphorus.

References

American Public Health Association. (1989). *Standard Methods for the Examination of Water and Wastewater*, 17th ed. American Public Health Association, Washington, DC, USA.

Chimney, M.J. and Goforth, G. (2000). Environmental Impacts to the Everglades Ecosystem: A Historical Perspective and Restoration Strategies. In: Preprints for the 7th International Conference on Wetland Systems for Water Pollution Control, November 11–16, 2000, Lake Buena Vista, FL, USA.

Chimney, M.J. and Moustafa, M.Z. (1999). Chapter 6: *Effectiveness and Optimization of Stormwater Treatment Areas for Phosphorus Removal*. In: Everglades Interim Report, South Florida Water Management District, FL, USA, pp. 6–1 to 6–45.

Chimney, M.J., Nungesser, M., Newman, J., Pietro, K., Germain, G., Lynch, T., Goforth, G. and Moustafa, M.Z. (2000). Chapter 6: *Stormwater Treatment Areas – Status of Research and Monitoring to Optimize Effectiveness of Nutrient Removal and Annual Report on Operational Compliance*. In: Everglades Consolidated Report, South Florida Water Management District, FL, USA, pp. 6–1 to 6–127.

Light, S.S. and Dineen, J.W. (1994). *Water Control in the Everglades: a historical perspective. In: Everglades*. The ecosystem and its restoration, S.M. Davis and J.C. Ogden (eds.), St. Lucie Press, Boca Raton, FL, USA, pp. 47–84.

SFWMD (1998). *SFWMD Quality Assurance Plan*. West Palm Beach, FL, South Florida Water Management District, FL, USA.

USEPA (1983). *Methods for Chemical Analysis of Water and Wastes*. Cincinnati, OH, USA, Environmental Monitoring and Support Laboratory.

Walker, W.W. (1991). Water quality trends at inflows to Everglades National Park. *Wat. Res. Bull.* **27**(1), 59–72.

Winter, T.C. (1981). Uncertainties in estimating the water balance of lakes. *Wat. Res. Bull.* **17**(1), 82–115.

Progress in the research and demonstration of Everglades periphyton-based stormwater treatment areas

J.S. Bays*, R.L. Knight**, L. Wenkert***, R. Clarke**** and S. Gong*****

* CH2M HILL, 4350 W. Cypress St., Tampa, FL 33607-4178 U.S.A.
** Environmental Scientist, 2809 NW 161 Court, Gainesville, Florida 32609, U.S.A.
*** South Florida Water Management District, 3301 Gun Club Road, West Palm Beach, FL-33406 U.S.A.
**** CH2M HILL, 3011 SW Williston Rd., Gainesville, FL 32608 U.S.A.
***** CH2M HILL, 800 Fairway Dr. Suite 350, Deerfield Beach, FL 33441-1831 U.S.A.

Abstract The South Florida Water Management District (District) is conducting research focused on potential advanced treatment technologies to support reduction of phosphorus (P) loads in surface water entering the remaining Everglades. Periphyton-based stormwater treatment areas (PSTA) are one of the advanced treatment technologies being researched by the District. This detailed research and demonstration project is being conducted in two phases. Basic research in field-based mesocosm experiments was conducted during the first phase within the District's Everglades Nutrient Removal Project (ENR). Studies were conducted in 24 portable PSTA mesocosms and three of the south ENR test cells. Phase 1 studies addressed the effects of system substrate (shellrock, organic peat, or sand), water depth, hydraulic loading rate, vegetation presence, depth:width ratio, and inhibition of algal growth on total phosphorus removal performance of the PSTA mesocosms. A second phase of research is currently under way, during which PSTA feasibility will be evaluated further in four field-scale constructed mesocosms totaling about 2 ha, and follow up studies within the ENR test cells and portable mesocosms will be conducted to further investigate the effects of other inorganic substrates, shallow water depth, and velocity on treatment performance. Phase 1 monitoring has determined that periphyton-dominated communities can be established in constructed wetlands within 5 months. The algal component of these periphyton plant communities is characteristic of natural Everglades periphyton. High macrophyte densities resulted from use of peat soils in PSTA mesocosms, while shellrock and sand soils promoted more desirable sparse macrophyte stands. P removal rates under the conditions of this research were relatively high considering the low influent total P concentrations tested (average 23 µg/L). PSTA mesocosms on shellrock soils were able to attain long-term average outflow total P concentrations as low as 11 µg/L. The maximum one-parameter TP first-order removal rate constant (k_1) measured was 27 m/y. Minimum attainable outflow total P concentrations and mass removals appear to be the result of a balance between internal P loading from antecedent soils, uptake and burial processes in new sediments, and rainfall inputs. A different soil type (limerock) will be tested for effectiveness during Phase 2. Selected existing treatments will also be continued to look for trends over a second growing season.

Keywords Everglades; periphyton; phosphorus removal; restoration; substrate; treatment wetlands

Introduction

The South Florida Water Management District (District) is conducting research focused on determining the effectiveness and design criteria of potential advanced treatment technologies to support reduction of phosphorus (P) loads in surface waters entering the remaining Everglades (South Florida Water Management District, 2000). Particular focus is being placed on the treatment of surface waters from the Everglades Agricultural Area (EAA) as well as Lake Okeechobee water, which is diverted through the primary canal system to the Lower East Coast of Florida.

Periphyton-based stormwater treatment areas (PSTAs) are one of the advanced treatment technologies being researched by the District. The PSTA concept was proposed for phosphorus removal from EAA waters by Doren and Jones (1996) and further described and evaluated by Kadlec (1996) and Kadlec and Walker (1996). Prior to initiation of the District's PSTA project in July 1998, detailed research to evaluate treatment performance issues and the long-term viability of the PSTA approach to phosphorus reduction in Everglades Agricultural Area surface waters had not been performed.

A two-phased approach has been taken to meet the stated objectives of the PSTA concept evaluation: an Experimental Phase (Phase 1), and a Validation/Optimization Phase (Phase 2). Some tasks have been completed in Phase 1, while several tasks will continue from Phase 1 into Phase 2. The two phases, and the types of activities that are included in each, are described as follows.

- Phase 1 (Experimental Phase) included development of the work plan and experimental design, initial research in three experimental test cells (PSTA Test Cells) located at the southern end of the Everglades Nutrient Removal (ENR) project, and construction and start up/monitoring of research using 24 portable experimental mesocosms (Porta-PSTAs). The Phase 1 experimental studies have yielded critical information needed to plan for field-scale mesocosm (Field PSTAs) design and construction in Phase 2. Development of a forecast model and associated predictive tools has occurred, along with preliminary model calibration with the Phase 1 experimental data.
- Phase 2 (Validation/Optimization Phase) will include continued research in the ENR PSTA Test Cells, in the Porta PSTAs, and in the Field PSTAs currently under construction immediately west of STA 2. During Phase 2, the expanded database will be used to validate the performance forecast model. This phase of the project will be used to assess the feasibility of full-scale implementation of PSTAs as a successful advanced technology. It will focus on developing design criteria for a full-scale PSTA system. Additional research will be conducted and the forecast model will be applied to provide projections of the long-term cost of implementing PSTAs to meet ultimate phosphorus reduction goals under the Everglades Forever Act.

The PSTA Research and Demonstration Project is designed to develop defensible conclusions related to specific hypotheses that are relevant to key research questions described in the PSTA Research Plan (CH2M HILL, 1999). The research addresses whether a desired periphyton-dominated ecological community can be established in a full-scale PSTA, whether that community can be maintained over time, and whether such a constructed PSTA can consistently and predictably reduce total phosphorus to concentrations targeted to support Everglades restoration. This paper provides an interim summary of the progress made through March 2000. More detailed findings from Phase 1 are presented in CH2M HILL (2000).

Methods
Experimental mesocosms
Table 1 provides a summary of the treatments evaluated during Phase 1 of the PSTA Research and Demonstration Project. A more detailed description of the Porta PSTA (PP) and south Test Cell (STC) mesocosm scales is provided below.

South ENR Test Cells. The STC mesocosms consist of 15 rectangular, 0.2 ha cells receiving flows from a single Head Cell. Water pumped into the Head Cell from the ENR flows by gravity through a distribution manifold into each of the Test Cells. The District assigned three south ENR Test Cells to the PSTA Research Project. During final construction, substrate within these PSTA Test Cells was modified by the District by placing the following layers of substrate over the cell liner:
- **Test Cell 13** – about 80 cm of sand surcharge plus 30 cm of locally mined shellrock plus 30 cm of peat taken from a local, unflooded, and former agricultural lands area

Table 1 Summary of PSTA Phase 1 design criteria and experimental treatments

Treatment	Cells	Area (m²)	Soil Type	Water Depth (cm)	HLR (cm/d)	Depth:Width Ratio	Other Considerations
Portable PSTA (PP) Mesocosms							
PP-1	9, 11, 18	6	peat	60	6	0.6	Macrophytes
PP-2	4, 7, 8	6	shellrock	60	6	0.6	Macrophytes
PP-3	12, 14, 17	6	peat	30	6	0.3	Macrophytes
PP-4	3, 5, 10	6	shellrock	30	6	0.3	Macrophytes
PP-5	2, 13, 16	6	shellrock	60	12	0.6	Macrophytes
PP-6	1, 6, 15	6	shellrock	0–60	0–12	0–0.6	Macrophytes
PP-7	19	6	sand	30	6	0.3	Macrophytes
PP-8	20	6	sand	60	6	0.6	Macrophytes
PP-9	21	6	peat	60	6	0.6	aquashade; no macrophytes
PP-10	22	6	shellrock	60	6	0.6	aquashade; no macrophytes
PP-11	23	18	shellrock	30	6	0.1	Macrophytes
PP-12	24	18	peat	30	6	0.1	Macrophytes
South Test Cell (STC) Mesocosms							
STC-1	13	2240	peat	60	6	0.0214	Macrophytes
STC-2	8	2240	shellrock	60	6	0.0214	Macrophytes
STC-3	3	2240	shellrock	0–60	0–12	0–0.0214	Macrophytes

Note: mesocosms with variable depth and HLR values were "dry out" treatments subjected to varying seasonal conditions tracking a natural Everglades wetland hydroperiod. Depth:width ratio provides a measure for assessing the potential effect of mesocosm size on performance

- **Test Cell 8** – about 1 m of sand surcharge plus 30 cm of locally mined shellrock
- **Test Cell 3** – about 1 m of sand surcharge plus 30 cm of locally mined shellrock.

Figure 1 provides a photograph of PSTA Test Cell 8 (treatment STC-2), with shellrock substrate after nearly one year of colonization.

Porta-PSTA Mesocosms. Twenty-four Porta-PSTA mesocosm units were fabricated of fiberglass offsite and delivered to the South ENR Supplemental Technology Research Compound. Twenty-two of the fiberglass tanks are 6 m long by 1 m wide by 1 m deep. The remaining two tanks are 3 m wide to allow assessment of mesocosm configuration effects.

Figure 1 Photograph of PSTA test cell 8 (treatment STC-2) following about 12 months of colonization. This photo is looking upstream from the outfall standpipes towards the inflow at the far end of the cell. Monitoring walkways are located at 1/3 and 2/3 points along the flow path

Figure 2 is a photograph of the Porta-PSTA experimental setup showing the layout of typical 1 and 3 m wide mesocosms in relation to the constant-head tank and inlet manifolds.

Sampling methodology

Weekly water samples were taken at the mesocosm inlets and outlets. Periphyton biomass, algal species composition, biovolume, and pigment concentration samples were collected monthly from the entire water column using a 15 cm diameter coring device. Soils were sampled for P analysis monthly from the top 10 cm. Sediment samples were taken quarterly from each mesocosm for analysis of reactive and non-reactive phosphorus levels. Samples were analyzed according to *Standard Methods* (APHA, 1998).

Results

Plant community establishment

A total of 361 algal taxa were identified in the PSTA periphyton samples. Table 2 provides a summary of the dominant species (defined as those that most frequently comprised more than 10% of the cell count or biovolume totals). Filamentous green algae were observed to occupy the front end of the mesocosms in areas of measurable dissolved reactive P, while filamentous blue greens and diatoms dominated floating and benthic periphyton mats throughout the majority of the test systems.

Ash-free dry weight biomass rose to sustained levels (typically between 100 and 600 g/m^2) within 1 to 2 months from startup. Chlorophyll *a* (corrected for pheophytin), and algal biovolume continued to rise throughout the study period. Average chlorophyll *a* concentrations were between 40 and 200 mg/m^2.

Eleocharis cellulosa (spike rush) and *Utricularia* spp. (bladderwort) were purposely added to most of the PSTA mesocosms. These macrophytes are present in natural Everglades

Figure 2 Photograph of Porta-PSTA Tank 23 (Treatment PP-11) after 11 months of colonization. This 6 × 3 m tank has shellrock soils and was operated at a 30 cm water depth. Floating periphyton mats are visible among the sparse emergent macrophytes. Narrow tanks can be seen in the background as well as the constant head tank used to feed all mesocosms at this site

Table 2 Summary of dominant algal species in PSTA mesocosms during the Phase 1 research period. Principal algal groups on all substrates were blue greens and diatoms

TAXON	PHYLUM	Percent of Samples > 10% of Total Cell Counts	Percent of Samples > 10% of Total Biovolumes	Long-Term Averages Cell Counts (# cells/m²)*10⁶	Long-Term Averages Biovolumes (cm³/m²)
Mastogloia smithii	Bacillariophyceae		26.3%	861	2.994
Lyngbya limnetica	Cyanobacteria	15.9%	5.2%	65617	1.640
Oscillatoria angustissima	Cyanobacteria	14.8%		19043	0.038
Oscillatoria limnetica	Cyanobacteria	13.7%		46492	0.325
Rhopalodia gibba	Bacillariophyceae		10.4%	87	2.197
Scytonema sp?	Cyanobacteria	1.1%	7.8%	6650	9.211
Lyngbya lagerheimii	Cyanobacteria	7.4%		14333	0.086
Mastogloia smithii v.lacustris	Bacillariophyceae		7.4%	765	1.229
Cylindrospermum sp.	Cyanobacteria	5.9%	0.7%	7355	0.299
Aphanothece stagnina	Cyanobacteria	4.4%	0.7%	12988	0.312
Oedogonium punctatostriatum	Chlorophyta	0.4%	4.4%	237	1.905
Oscillatoria formosa	Cyanobacteria	3.3%	1.5%	8893	0.703
Aphanocapsa delicatissima	Cyanobacteria	4.1%		7787	0.008
Mastogloia lanceolata	Bacillariophyceae		3.7%	154	1.036
Johannesbaptista pellucida	Cyanobacteria	2.6%	0.7%	2922	0.164
Aphanothece smithii	Cyanobacteria	3.0%		7466	0.044
Amphora lineolata?	Bacillariophyceae		2.6%	111	0.606
Lyngbya sp. (small)	Cyanobacteria	2.6%		11604	0.058
Nitzschia semirobusta	Bacillariophyceae		2.6%	382	0.225
Oscillatoria princeps	Cyanobacteria		2.6%	860	5.401
Schizothrix arenaria?	Cyanobacteria	2.6%		29670	0.386

periphyton-dominated plant communities; they were introduced in the test mesocosms to provide periphyton attachment sites, and increase mat stability against wind-induced drift. *Typha latifolia* (cattail), *Hydrilla verticillata* (hydrilla), and *Chara* sp. (stonewort) invaded some of the PSTA mesocosms. Macrophyte biomass estimates document that macrophyte proliferation occurred within the peat soil mesocosms (Table 3), dominating visual plant cover estimates. Nevertheless, macrophyte cover dominance did not appear to limit the periphyton community importance as measured by chlorophyll *a* and algal biovolume (Table 3).

Table 3 Summary of average macrophyte standing crop biomass, corrected chlorophyll a, and algal cell volume in the PSTA mesocosms during the post start-up period of Phase 1 (August 1999 through end)

Treatment	Soil Type	Macrophyte Biomass (g dw/m²)	Chlorophyll a (mg/m²)	Biovolume (cm³/m²)
PP-1	Peat	75	74	13
PP-2	Shellrock	19	73	17
PP-3	Peat	215	89	15
PP-4	Shellrock	21	93	33
PP-5	Shellrock	26	107	24
PP-6	Shellrock	14	59	11
PP-7	Sand	0.0	170	37
PP-8	Sand	3.4	144	16
PP-9	Peat	0.0	113	12
PP10	Shellrock	0.0	55	8.8
PP-11	Shellrock	89	107	22
PP-12	Peat	176	42	1.9
STC-1	Peat	582	223	31
STC-2	Shellrock	61	75	12
STC-3	Shellrock	55	39	4.3

Inflow phosphorus concentrations

Inlet P concentrations were variable throughout the project period. While mean TP concentrations were similar at both sites (23 μg/L), TP concentrations were greater at the Test Cells during late summer and higher at the Porta-PSTAs during the early spring and late winter. These differences in TP were largely attributable to complex seasonal variations in the concentrations of total dissolved P (TDP) and total particulate P (TPP) in the two water supplies. On the average TDP comprised 52 to 70% of TP. Dissolved reactive P (DRP) was typically about 5 to 6 μg/L while dissolved organic P (DOP) averaged between 7 and 11 μg/L in the inflow waters.

Phosphorus removal performance

Table 4 summarizes the TP concentrations and estimated k-C^* model parameters (IWA, 2000; Kadlec and Knight, 1996) for each treatment during the post-startup period-of-record. Values for k_1 are also summarized in Table 4 and offer a normalized comparison between treatments.

The following preliminary conclusions were drawn from these Phase 1 research data.

- Estimated values for C^*, the effective background TP concentration resulting from internal and external loadings and removals ranged from 9 to 15 μg/L.
- Estimated k_{TP} values ranged from 14.5 to 89.3 m/y.
- Lowest long-term average TP outflow concentrations were about 13 μg/L and lowest monthly averages were 9 μg/L.
- TP k_1 values ranged from 0.02 to 27.0 m/y.
- There were no consistent effects of water depth (30 vs. 60 cm steady depth) on outflow TP concentration; however, higher k_1 values were recorded in mesocosms with shallower depths.
- Variable water depths resulted in reduced TP removal performance as reflected by lower k_1 values when compared to similar treatment groupings.
- Macrophyte biomass was greater in peat-based systems indicating some substrate preparation, such as limerock capping, will be necessary to prevent out-competing the periphyton.

Table 4 Summary of average performance and estimated parameters for k_1 and k-C^* models from PSTA Phase 1 mesocosm research. Data are for the period-of-record after the end of startup (STCs: July 99 – end; PPs: October 99 – end). C^* is set equal to the lowest observed monthly average. The k values are estimated by the Excel Solver routine to minimize the difference between observed and estimated monthly average outflow TP concentrations. The k_1 values are based on the period-of-record averages assuming $C^* = 0$

Treatment	TP (mg/L) In	TP (mg/L) Out	HLR (m/yr)	k_1 (m/yr)	k (m/yr)	C^* (mg/L)
PP-1	0.020	0.014	34.9	11.2	25.8	0.009
PP-2	0.020	0.013	33.4	13.9	39.1	0.010
PP-3	0.025	0.015	31.9	15.7	32.7	0.010
PP-4	0.025	0.014	32.5	18.1	33.4	0.009
PP-5	0.025	0.017	62.8	27.0	68.1	0.013
PP-6	0.026	0.015	16.5	9.1	23.4	0.010
PP-7	0.025	0.015	31.4	16.2	41.7	0.012
PP-8	0.020	0.016	33.9	7.8	89.3	0.015
PP-9	0.026	0.020	34.9	8.2	35.5	0.012
PP-10	0.026	0.015	32.4	17.7	45.4	0.012
PP-11	0.025	0.016	32.5	14.9	34.0	0.012
PP-12	0.025	0.017	32.7	13.2	36.2	0.013
STC-1	0.025	0.016	16.5	6.9	16.6	0.012
STC-2	0.024	0.013	16.9	9.6	26.3	0.011
STC-3	0.023	0.018	17.0	4.3	14.5	0.013

- Outflow TP concentrations were lower and k_1 values higher in mesocosms built with shellrock substrates than in comparable mesocosms with peat soils.
- Higher loading rates (hydraulic and TP mass) increased k_1.
- A mesocosm scale effect was observed that indicated that smaller mesocosms underestimated outflow TP values by 0 to 14% and overestimated k_1 values by 16 to 38%.
- In Aquashade control mesocosms, average outflow TP concentrations were higher but k_1 values were not consistently different from non-Aquashade treatments.

Discussion and conclusions

The periphyton communities that became established within the PSTA mesocosms attained biomass levels and replicated normal periphyton algal species assemblages typical of low-P Everglades waters (Browder et al., 1994) within the first year of operation. These plant communities display predictable community-level responses to environmental forcing functions such as sunlight and antecedent soil conditions. Based on the conditions selected for this research, these PSTA mesocosms were able to attain average TP outflow concentrations during the post-startup Phase 1 period as low as 13 µg/L. These concentrations are considerably lower than the long-term average outflow TP concentration from the ENR of 22 µg/L (Walker, 1999). Continued studies of some of these mesocosms through Phase 2 will help confirm whether the PSTA treatment performance records to date will continue or perhaps improve with a longer period of record.

Lower average TP outflow concentrations have been observed in natural periphyton-dominated communities in Water Conservation Area 2A (McCormick et al., 1996), in the southern Everglades, and in experimental mesocosms built with limerock soils (DBEL, 1999). The minimum TP values recorded during this Phase 1 research appeared related to internal P loading from antecedent soils. It is not currently known if these minimum outflow TP concentrations will continue to decline with increasing system maturity and eventual complete burial of antecedent soils. Phase 2 PSTA research includes continued testing on shellrock soils, on calcium-amended peat soils, and on limerock soils at both small and large scales to determine which of these substrates results in lower achievable TP outflow concentrations.

The k_1 values recorded in this research are comparable to or higher than values recorded for emergent macrophyte and submerged aquatic vegetation (SAV) dominated treatment wetlands in south Florida. Walker (1999) determined that the overall ENR k_1 value was about 15.5 m/y for the period from March 1995 through November 1998. The k_1 value for Cell 3 of the ENR is probably most comparable due to similar inflow water quality conditions as the PSTA research sites. This cell averaged $k_1 = 9.5$ m/y during this operational period. Cell 4 is dominated by SAV and averaged $k_1 = 17.3$ m/y during this same period. Continuing research with the PSTA mesocosms needs to be conducted to validate and refine the TP performance estimates obtained during the Phase 1 operational period.

References

APHA (1998). *Standard Methods for the Examination of Water and Wastewater.* 20th edn, American Public Health Association/American Water Works Association/Water Environment Federation, Washington DC, USA.

Browder, J.A., Gleason, P.J. and Swift, D.R. (1994). Periphyton in the Everglades: spatial variation, environmental correlates, and ecological implications. Chapter 16, pp. 379–418 in: S.M. Davis and J.C. Ogden (eds.) *Everglades. The Ecosystem and Its Restoration.* St. Lucie Press, Delray Beach, FL.

CH2M HILL (1999). Periphyton-based Stormwater Treatment Area (PSTA) Research and Development Project – PSTA Research Plan. Prepared for the South Florida Water Management District. CH2M HILL, Deerfield Beach, FL.

CH2M HILL (2000). Periphyton-based Stormwater Treatment Area (PSTA) Research and Development

Project – Phase 1 Report (February 1999–March 2000). Prepared for the South Florida Water Management District. CH2M HILL, Deerfield Beach, FL.

DBEL (1999). A Demonstration of Submerged Aquatic Vegetation/Limerock Treatment System Technology for Removing Phosphorus From Everglades Agricultural Area Waters. Report to the South Florida Water Management District and the Florida Department of Environmental Protection. DB Environmental Laboratories, Inc., Rockledge, FL.

Doren, R.F. and Jones, R.D. (1996). Conceptual Design of Periphyton-Based STAs. Memo to Col. T. Rice, COE dated January 30, 1996.

IWA (2000). *Constructed Wetlands for Pollution Control – Processes, Performance, Design, and Operation. (Scientific and Technical Report No. 8)*. IWA Specialist Group on Use of Macrophytes in Water Pollution Control. London, U.K.

Kadlec, R.H. (1996). Algal STAs for Achieving Phase II Everglades Protection. Technology Outline., Letter Report dated October 21, 1996. 9 pp.

Kadlec, R.H. and Knight, R.L. (1996). *Treatment Wetlands*. Lewis Publishers. Boca Raton, FL. 893 pp.

Kadlec, R.H. and Walker, W.W. (1996). Perspectives on the Periphyton STA Idea. Draft letter report dated December 26, 1996. 26 pp.

McCormick, P.V., Rawlik, P.S., Lurding, K., Smith, E.P. and Skylar, F.H. (1996). Periphyton-water quality relationships along a nutrient gradient in the northern Florida Everglades. *J. N. Am. Benthol. Soc.* **15**(4), 433–449.

South Florida Water Management District (2000). Everglades Consolidated Report. South Florida Water Management District. West Palm Beach, FL.

Walker, W.W. (1999). Contributions to workshop on STA/polishing cell design. Unpublished data analysis for the U.S. Department of Interior, March 18, 1999.

Performance of a recirculating wetland filter designed to remove particulate phosphorus for restoration of Lake Apopka (Florida, USA)

M.F. Coveney, E.F. Lowe, and L.E. Battoe

Department of Water Resources, St. Johns River Water Management District, P.O. Box 1429, Palatka, FL, USA

Abstract Operation of a 14-km^2 wetland filter for removal of total phosphorus (TP) from lake water is part of the restoration program for hypereutrophic Lake Apopka, Florida. This system differs from most treatment wetlands because 1) water is recirculated back to the lake, and 2) the goal is removal of particulate phosphorus (P), the dominant form of P in Lake Apopka. The operational plan for the wetland is maximization of the rate rather than the efficiency of P removal. The St. Johns River Water Management District operated a 2-km^2 pilot-scale wetland to examine the capacity of a wetland system to remove suspended solids and particulate nutrients from Lake Apopka. TP in the inflow from Lake Apopka ranged from about 0.12 to 0.23 mg l^{-1}, and hydraulic loading rate (HLR) varied from 6.5 to 42 m yr^{-1}. The performance of the pilot-scale wetland supported earlier predictions. Mass removal efficiencies for TP varied between about 30% and 67%. A first-order, area-based model indicated a rate constant for TP removal of 55 m yr^{-1}. We compared actual removal of P with model predictions and used modeled performance to examine optimal operational conditions. Correspondence between observed and modeled outflow TP was not good with constant variable values. Monte Carlo techniques used to introduce realistic stochastic variability improved the fit. The model was used to project a maximal rate of P removal of about 4 g P m^{-2} yr^{-1} at P loading 10–15 g P m^{-2} yr^{-1} (HLR 60–90 m yr^{-1}). Data from the pilot wetland indicated that actual rates of P removal may prove to be higher. Further operation of the wetland at high hydraulic and P loading rates is necessary to verify or modify the application of the model.

Keywords Florida; P; phosphorus; lake; nutrient; removal; restoration; USA; Wetland

Introduction

Wetlands are effective in removal of suspended particles and can function to remove and sequester nutrients from inflowing water. A large number of wetland treatment projects involve use of natural or constructed wetlands to remove pollutants from wastewater. Growing numbers of projects use wetlands in treatment of stormwater runoff from agricultural lands and other non-point sources (Kadlec and Knight, 1996).

Lake Apopka is a large, shallow, hypereutrophic lake where phosphorus (P) loading from agriculture has been the primary cause of eutrophication (Battoe *et al.*, 1999; Schelske *et al.*, 2000). A 14 km^2 treatment wetland will be used by the St. Johns River Water Management District (District) to accelerate the restoration of Lake Apopka (Lowe *et al.*, 1992; Battoe *et al.*, 1999). The function of the treatment wetland (Marsh Flow-Way) is the removal of algae and resuspended sediments and the associated P from lake water. This treatment wetland differs from typical systems because water is recirculated between the wetland and the lake and because the goal is removal of particulate P, the dominant form of P in Lake Apopka. The operational goal for the wetland is maximization of the areal rate of P removal (g P m^{-2} yr^{-1}). In other words, the intent is to maximize power (P removed per unit time) and capacity (permanent P storage) rather than efficiency (P removed in a single pass) (Lowe *et al.*, 1989).

The District operated a 2-km^2 pilot-phase wetland (Marsh Flow-Way Demonstration Project) in periods from 1990 until 1997 to examine the capacity of a wetland system to remove suspended sediments and particulate nutrients from Lake Apopka. A detailed discussion of performance of the pilot project during one complete 29-mo operational cycle (start-up to drawdown) was presented elsewhere (Coveney et al., in preparation). Our focus in the present work was limited to P removal during a 10-mo portion of the operational period. We compared observed removal of P with model predictions and used modeled performance to examine optimal operational conditions of the system for P removal.

Methods

Lake Apopka is a large (125 km^2), shallow (mean depth 1.6 m) lake located near Orlando in central Florida, U.S.A. The lake is hypereutrophic, primarily because of heavy P loading from floodplain farms since the 1940s. The lake water is characterized by high concentrations of chlorophyll a (0.092 mg l^{-1}), suspended solids (79 mg l^{-1}), total nitrogen (5.14 mg l^{-1}), and total P (0.204 mg l^{-1}) (Battoe et al., 1999).

The Marsh Flow-Way Demonstration Project was located on the shore of Lake Apopka on former sawgrass (*Cladium jamaicense*) marsh that was drained for farming. The demonstration project consisted of two wetland cells in series. Water flowed or was pumped from Lake Apopka into the project and was pumped back to the lake. We focused on performance data from the first wetland cell (0.73 km^2) because most P removal occurred in this area.

We report here on P removal during December 1991 through October 1992 (10.2 mo), a period when evaluation of performance was not complicated by the initial flux of P from flooded soils or by later hydraulic short-circuiting. During this period, mean flows varied from 1.3×10^4 to 8.4×10^4 m^3 d^{-1}, depth in the first cell varied from 0.48 to 0.70 m, hydraulic retention time in the first cell was between 5.6 and 27 d, and hydraulic loading rate for the first cell was between 6.5 and 42 m yr^{-1}. The plant community was dominated by *Typha latifolia*, with *Hydrocotyle ranunculoides*, *Pontederia cordata*, *Sagittaria lancifolia*, *Alternanthera philoxeroides*, *Salix caroliniana*, and *Ludwigia peruviana* as important subdominants (Stenberg et al., 1997).

Methods for data and sample collection and analysis were described elsewhere (Coveney et al., in preparation). Briefly, inlet flow was measured daily with mechanical flow meters, and outlet flow from the first cell was estimated from a water budget. Water was sampled at the inflow and outflow twice weekly for chemical and physical analyses. Nutrient species were measured using standard automated techniques (Kopp and McKee, 1983). Particulate organic P (POP) was calculated as total P (TP) minus dissolved TP, and dissolved organic P (DOP) was calculated as dissolved TP minus soluble reactive P (SRP). Interpolated daily values for concentration of constituents at the inlet and outlet were multiplied by daily flows to calculate daily fluxes. Mass fluxes were averaged over consecutive 15-d periods for reporting and analysis. In this manner 21 data points were derived for each variable for the 10-mo period considered here.

Results and discussion

Behavior of P fractions in the wetland

Compared to surface-flow wetlands used to treat wastewater, this system operated at lower inflow concentration of TP but higher HLR (Kadlec and Knight, 1996). TP in the inflow from Lake Apopka ranged from about 0.12 to 0.23 mg l^{-1}. In contrast, HLR varied from 6.5 to 42 m yr^{-1}, a factor of almost 6.5. Thus, variations in loading rate of P largely were due to changes in flow rather than changes in inflow P concentration. POP made up 86% to 98% of TP in the inflow from Lake Apopka, and both soluble inorganic and organic forms of P were present in low concentrations. POP was removed effectively in the wetland, but concentrations of SRP and DOP tended to remain unchanged or to increase from inflow to outflow. Mass removal efficiencies based on TP varied between about 30% and 67% (Figure 1).

When the wetland first was inundated, concentrations of SRP were high in the outflow because of leaching from the soils and dead vegetation. These elevated SRP values masked the simultaneous removal from the inflowing water of POP. SRP release rates declined over the first 6 months of operation and averaged 0.44 mg P m^{-2} d^{-1} during the period considered here. DOP was released in small and variable amounts in the wetland throughout the period, with a mean release rate of 0.54 mg P m^{-2} d^{-1}.

Models for removal of TP

We evaluated two models to describe the removal of TP in the wetland. Model 1 was the first-order, area-based "k-C*" model (Kadlec and Knight, 1996) applied directly to TP.

The model can be represented as

$C_o = C^* + (C_i - C^*) \exp(-kA/Q)$, with
- C_o = Flow-weighted mean concentration in the outflow, g m^{-3}
- C^* = Background concentration in the wetland, g m^{-3}
- C_i = Flow-weighted mean concentration in the inflow, g m^{-3}
- k = First-order areal rate constant for removal, m yr^{-1}
- A = Area of the wetland, m^2
- Q = Water flow, m^3 yr^{-1}.

This model describes the decrease in concentration of a constituent in the wetland from the level at the inlet towards a background level C* set by internal processes, rainfall, or other factors (Kadlec, 1999). We used the lowest outlet concentrations observed to estimate C* for TP as a curvilinear function of temperature. Estimated C* values lay between 0.06 and 0.09 mg P l^{-1} and were high compared with other systems (Kadlec, 1999). High C* for TP was expected because the wetland was constructed on P-rich organic soils and received hypereutrophic lake water.

We also evaluated a model (Model 2) which separated different fractions of TP because of their differing behavior in the wetland. Whereas POP was removed, concentrations of SRP and DOP either were unchanged or were augmented. In Model 2, we used the k-C* model to calculate POP but assumed constant release rates in the wetland for SRP and DOP (TP = POP + SRP + DOP). These model rates were taken as the mean release rates observed

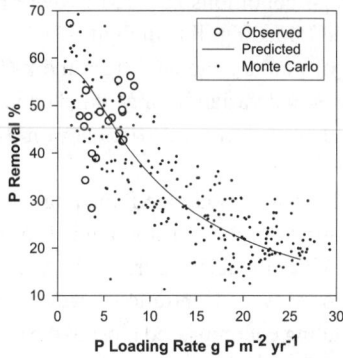

Figure 1 Phosphorus removal efficiency (%) as a function of phosphorus loading rate in the first wetland cell. Open circles are observed values. Line shows k-C* model prediction under average conditions (C_i, T, k). Solid circles show predicted values with Monte Carlo techniques used to introduce realistic stochastic variability

during the period, 0.44 mg P m^{-2} d^{-1} for SRP and 0.54 mg P m^{-2} d^{-1} for DOP. C* for POP was estimated in the same manner as for TP.

We calibrated each model to the 21 observations for outflow TP concentration. Calibration was accomplished by varying the first-order areal rate constant k to minimize the sum of squares of deviations between observed and predicted outflow TP. Despite the different mechanisms represented by the two models, the mean square deviations after calibration were close (2.71×10^{-4} for Model 1 and 4.87×10^{-4} for Model 2). Because of its slightly better fit, we selected Model 1 to characterize P removal in the wetland and to examine performance at P loading rates (i.e. HLR) higher than those achieved in the pilot project.

One limitation in using this model for prediction was that outflow P concentration was calculated as a function of HLR at constant (mean) values for inflow P concentration and temperature (i.e. C*). To consider better the stochastic nature of wetland conditions, we used a Monte Carlo technique to calculate a set of 250 outflow P concentrations at randomly-selected values for inflow P, temperature, and k. Here, inflow P concentration and temperature were distributed normally according to the actual mean and standard deviation (SD) for the data set (C_i, mean 0.174 mg P l^{-1}, SD 0.025; T, mean 22.3 C, SD 4.6), k was distributed normally with the fitted mean 55 m yr^{-1} and an assumed SD 5.5, and HLR was distributed randomly between 5 and 150 m yr^{-1}.

P removal performance of the wetland

Calibration of Model 1 resulted in a first-order areal rate constant (k) for removal of TP of 55 m yr^{-1}. This value was high in comparison with a set of 92 wetlands where k averaged 10 m yr^{-1} (SD 5) (Kadlec, 1999). However, as opposed to most systems in that data set, almost all P in Lake Apopka water was particulate, and particles were removed at a high rate in the wetland. Also, for most of the other systems k was calculated under the assumption that C* = 0, which underestimated k if background P was present; k values can be inflated during initial operation of wetlands because of rapid storage of P on soil sorption sites and in growing vegetative biomass (Kadlec and Knight, 1996). These phenomena were not major factors at the Marsh Flow-Way, where the soil was a source rather than a sink for P during initial operation, and the change in mean plant biomass was moderate (August 1991, 990 g dwt m^{-2}; March 1994, 1,500 g dwt m^{-2}; live and standing dead, above and below ground) (Stenberg et al., 1997).

The observed efficiency of TP removal varied between 30% and 67% and showed no relationship to P loading rate (Figure 1). This behavior was at odds with that predicted by the k-C* model under average conditions (C_i, T, k) where removal efficiency declined asymptotically with increased P loading (line in Figure 1). However, most of the observed values lay within the cloud of points generated from the k-C* equation by Monte Carlo modeling. We conclude that actual variation in wetland conditions may have contributed to the apparent lack of dependence of removal efficiency on P loading and to the poor fit of the k-C* model.

Most of our data were collected at moderate P loading rates (3–7 g P m^{-2} yr^{-1}). In this range, observed P concentrations in the outflow tended to increase with P loading (i.e. HLR), as predicted by the k-C* model (Figure 2A). Again, agreement was not close between observed values for outflow concentration and the model prediction, but the points derived by Monte Carlo modeling encompassed most of the observations.

The potential rate of P removal in the wetland is critical to the successful operation of this project. The purpose of the Marsh Flow-Way is removal of P stores from Lake Apopka (Battoe et al., 1999), and lake water is recycled through the project. Therefore, the operational goal is to maximize the long-term rates of P removal and storage (power and capaci-

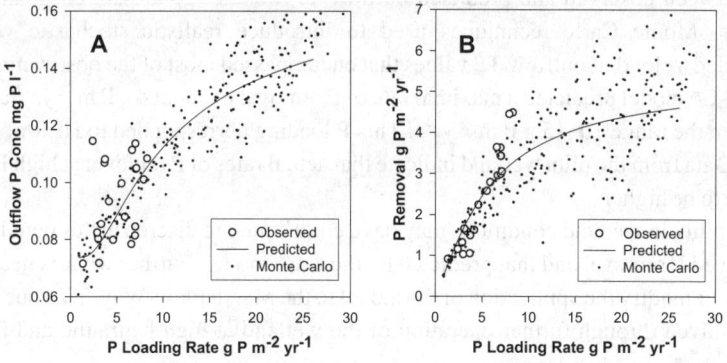

Figure 2 Outflow phosphorus concentration (A) and rate of phosphorus removal (B) as functions of phosphorus loading rate in the first wetland cell. See Figure 1 for explanation of symbols

ty) rather than to maximize the efficiency of removal. Lowe *et al.* (1989) predicted that the rate of P removal of the wetland would increase with increasing P loading rate due to the asymptotic decline in efficiency of P removal. Based on observations in the pilot project and on the k-C* model results, this prediction was substantially correct (Figure 2B).

The actual rate of P removal increased linearly with P loading (Figure 2B). The k-C* model indicated that further increases in P loading (i.e. increased HLR) would begin to yield diminishing returns in TP removal. Considering the effects of variation in wetland conditions (Monte Carlo model), P loading rates in the range 10–15 g P m^{-2} yr^{-1} should give a maximal P removal rate averaging about 4 g P m^{-2} yr^{-1} (Figure 2B). One important caveat is the fact that observed rates of P removal did not show the curvilinear behavior predicted by the model at the highest loading rates tested (around 8 g P m^{-2} yr^{-1}). Instead, removal rates continued to increase almost linearly (Figure 2B). Thus, higher P removal rates than those predicted by the k-C* model may be possible.

The practical limit to P removal likely will be set by the cost of moving increasing amounts of water, as suggested by Lowe *et al.* (1989) rather than by a loss of P removal efficiency at high HLR. The optimal range in P loading of 10–15 g P m^{-2} yr^{-1} would correspond to a range in HLR of 60–90 m yr^{-1} at the mean inflow concentration of TP (0.174 mg P l^{-1}). Kadlec (1999) concluded, based on hydraulic considerations and on P removal performance, that operation of a wetland at HLR greater than about 100 m yr^{-1} is not feasible. The HLR necessary to maximize P removal rate in the Marsh Flow-Way appears to be high but still feasible.

Based on the Monte Carlo modeling, we conclude that variation in wetland conditions may have contributed to the discrepancies noted between observed P removal and that predicted by the k-C* model. Further work is necessary to verify or modify the application of the k-C* model to the Marsh Flow-Way. This question will be resolved most convincingly through further operation of the wetland at high hydraulic and P loading rates.

Conclusions

- Compared to most surface-flow treatment wetlands, inflow to the pilot wetland was characterized by low TP, a predominance of particulate P, and high HLR.
- Particulate P was removed effectively in the wetland; mass removal efficiencies calculated for TP varied between about 30% and 67%.
- A "k-C*" model (Kadlec and Knight, 1996) could be fit to observed data for TP removal, although other models should not be excluded.
- The calibrated first order areal rate constant was 55 m yr^{-1}. High k was expected because P was largely in particulate form.

- Fit between observed and predicted outflow TP was not good with constant variable values. Monte Carlo techniques used to introduce realistic stochastic variability provided a cloud of outflow TP values that encompassed most of the observations.
- The k-C* model predicted a maximal rate of P removal of about 4 g P m^{-2} yr^{-1} at loading rates in the range 10–15 g P m^{-2} yr^{-1}. This P loading corresponded to HLR of 60–90 m yr^{-1}. Data from the pilot wetland indicate that actual rates of P removal at high HLR may prove to be higher.
- Variability in wetland conditions may have contributed to discrepancies noted between observed P removal and that predicted by the k-C* model. Further work is necessary to verify or modify the application of the model to the Marsh Flow-Way. This question will be resolved through further operation of the wetland at high hydraulic and P loading rates.

Acknowledgements

D.L. Stites and R. Conrow collaborated in all phases of the Marsh Flow-Way Project. E.J. Olson, B.R. Cooper, and W.W. Griffy conducted the field work. We acknowledge the assistance of colleagues in the Divisions of Engineering, Hydrologic Data Services, and Laboratory Services, St. Johns River Water Management District.

References

Battoe, L.E., Coveney, M.F., Lowe, E.F. and Stites, D.L. (1999). The role of phosphorus reduction and export in the restoration of Lake Apopka, Florida. In: *Phosphorus Biogeochemistry of Subtropical Ecosystems*, K.R. Reddy, G.A. O'Connor, and C.L. Schelske (eds), Lewis Publishers, Boca Raton, Florida, pp. 511–526.

Coveney, M.F., Stites, D.L., Lowe, E.F., Battoe, L.E. and Conrow, R. (In preparation). Nutrient removal from eutrophic lake water by wetland filtration.

Kadlec, R.H. (1999). The limits of phosphorus removal in wetlands. *Wetlands Ecology and Management*, **7**, 165–175.

Kadlec, R.H. and Knight, R.L. (1996). *Treatment Wetlands*. Lewis Publishers, Boca Raton.

Kopp, J.F. and McKee, G.D. (1983). Methods for Chemical Analysis of Water and Wastes, Report No. EPA-600/4-79-020, United States Environmental Protection Agency.

Lowe, E.F., Battoe, L.E., Stites, D.L. and Coveney, M.F. (1992). Particulate phosphorus removal via wetland filtration: an examination of the potential for hypertrophic lake restoration. *Environ. Manage.*, **16**(1), 67–74.

Lowe, E.F., Stites, D.L. and Battoe, L.E. (1989). Potential role of marsh creation in restoration of hypertrophic lakes. In: *Constructed Wetlands for Wastewater Treatment*, D.A. Hammer (ed.), Lewis Publishers, Chelsea, Michigan, pp. 710–717.

Schelske, C.L., Coveney, M.F., Aldridge, F.J., Kenney, W.F. and Cable, J.E. (2000). Wind or nutrients: Historic development of hypereutrophy in Lake Apopka, Florida. *Arch. Hydrobiol. Spec. Issues Advanc. Limnol.*, **55**, 543–564.

Stenberg, J.R., Clark, M.W. and Conrow, R. (1997). Development of natural and planted vegetation and wildlife use in the Lake Apopka Marsh Flow-Way Demonstration Project: 1990–1994, Special Publication SJ98-SP4, St. Johns River Water Management District, Palatka, Florida.

Nitrogen removal in a combined system: vertical vegetated bed over horizontal flow sand bed

S. Kantawanichkul*, P. Neamkam* and R.B.E. Shutes**

* Department of Environmental Engineering, Chiang Mai University, Chiang Mai 50200, Thailand
** Urban Pollution Research Centre, Middlesex University, Bounds Green Road, London, N11 2NQ, UK

Abstract Pig farm wastewater creates various problems in many areas throughout Thailand. Constructed wetland systems are an appropriate, low cost treatment option for tropical countries such as Thailand. In this study, a combined system (a vertical flow bed planted with *Cyperus flabelliformis* over a horizontal flow sand bed without plants) was used to treat settled pig farm wastewater. This system is suitable for using in farms where land is limited.
The average COD and nitrogen loading rate of the vegetated vertical flow bed were 105 g/m^2.d and 11 g/m^2.d respectively. The wastewater was fed intermittently at intervals of 4 hours with a hydraulic loading rate of 3.7 cm/d. The recirculation of the effluent increased total nitrogen (TN) removal efficiency from 71% to 85%. The chemical oxygen demand (COD) and total Kjeldahl nitrogen (TKN) removal efficiencies were 95% and 98%. Nitrification was significant in vertical flow *Cyperus* bed, and the concentration of nitrate increased by a factor of 140. The horizontal flow sand bed enhanced COD removal and nitrate reduction was 60%. Plant uptake of nitrogen was 1.1 g N/m^2.d or dry biomass production was 2.8 kg/m^2 over 100 days.
Keywords Combined system; *Cyperus fabelliformis*; nitrogen; pig farm wastewater; plant uptake; recirculation

Introduction

Livestock farming has recently undergone rapidly development in Thailand. The growth of large scale pig farms has created a need for suitable technology for waste disposal. The majority of these farms do not manage their wastes appropriately and their wastewater discharges have an adverse impact on receiving water. Many producers are considering the use of economic technologies to treat their wastes and control odor emission. Treatment technologies, including anaerobic digestion, lagoons and composting have been used for several years but the effluent quality is still unsatisfactory. Constructed wetland systems have recently been introduced to treat pig farm effluent following anaerobic digestion but due to insufficient knowledge and experience, these systems have not yet been effective. The high concentration of ammonia in the wastewater could be a major cause of their failure.

Constructed wetlands have the potential to remove nitrogen from pig farm wastewater (Szogi *et al.*,1999). According to Cooper and de Maeseneer (1996), the system with a vertical flow followed by horizontal flow bed can be smaller and can remove significant amounts of nitrate in the horizontal flow stage. A vertical flow bed showed 80–95% nitrification and a combined horizontal flow bed 75–80% denitrification rate in a study by Platzer (1996). Platzer (1999) also found that a recirculation rate up to 200% can be used without hydraulic problems for vertical flow beds. More research is needed to evaluate the suitable operating conditions for pig farm wastewater treatment.

Methods

The experiment was conducted in a tank constructed of iron plate, 1.2 m × 1.2 m × 1.2 m (W × L × H). The tank was separated into two sections by a plastic sheet : the upper section was 80 cm deep and the lower section was 30 cm deep. The upper section was filled with gravel (1–2 cm) to a depth of 10 cm, followed by 70 cm depth of course sand (1–2 mm). Four 2 cm diameter influent PVC pipes with 0.5 cm holes and separated by 5 cm were laid on the surface of the upper section. The wastewater was fed intermittently (4 hours on and 4 hours off) at the surface of the media and flowed vertically to the bottom of the upper section. A drainage pipe of the same diameter collected water from the bottom of the upper section (vertical flow) and passed to the top of the lower section which was filled with coarse sand. The water flowed horizontally and across to the bottom of the lower section (Figure 1). In the second run, 50% of the effluent was recycled to the upper vertical flow section.

The upper section was planted with *Cyperus flabelliformis*. The height of the plants was measured at weekly intervals. Nitrogen concentration in the plants were also analyzed both before and after the operation of each run.

Pig farm wastewater from the faculty of Agriculture, Chiang Mai University, was used in the experiment. It was flushed from the pig pens into the settling tank with a retention time of about 5 days. Only clear supernatant was used in the experiment.

The influent and effluent from vertical (effluent1) and horizontal flow sections (effluent2) were collected and analyzed for COD, TKN, NH_3-N, NO_x-N, SS, Alkalinity, Coliform bacteria, pH and temperature according to standard methods for the examination for water and wastewater (1995).

Results and discussion

Run 1: without recirculation of the effluent

The analysis commenced when the plants were 2 weeks old and were acclimatized to the wastewater used in the experiment. The raw wastewater was fed intermittently with a hydraulic loading rate of 3.7 cm/d. The average of COD concentration in the raw wastewater was 2,800 mg/l or 105 g COD/m^2.d. The COD in the raw wastewater varied from 1,725 to 3,210 mg/l. The experiment was conducted for 60 days from June to August 1999 and the average temperature of the raw wastewater was 25°C. COD removal efficiency reached 97% after passing through the vertical flow section and at the end of horizontal flow part, the removal efficiency further increased to 98%. The average COD in the effluent was 43 mg/l (Figure 2).

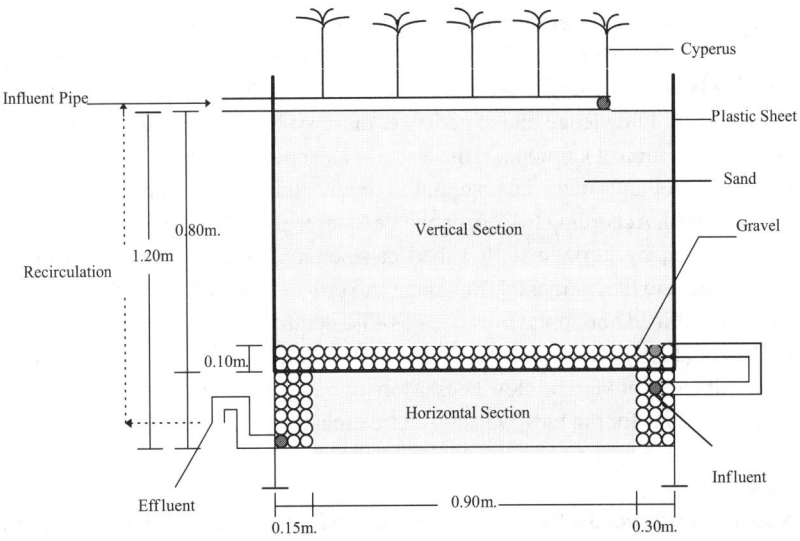

Figure 1 Experimental design

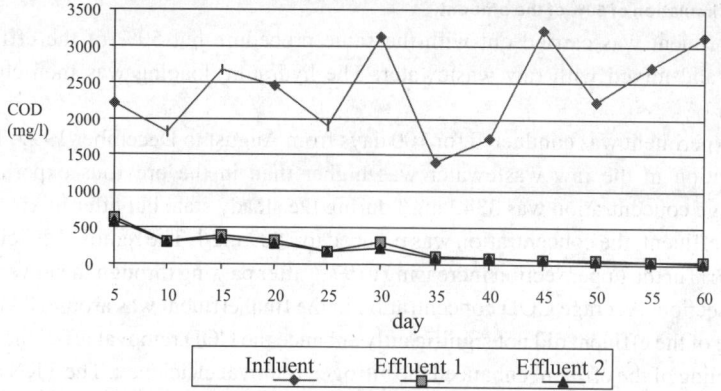

Figure 2 Total COD profile without recirculation of the effluent

Nitrification and denitrification were measured and nitrification was satisfactory. The average concentration of TKN in the raw wastewater was 240 mg/l or 9 gTKN/m^2.d.

The TKN reduction in the vertical flow section was 97% and the TKN in the effluent was reduced to 6.6 mg/l. The reduction was increased to 98% in the horizontal flow section resulting in 3.9 mg/l TKN effluent concentration. The same pattern was observed for NH_3-N with a concentration of 168 mg/l in the raw wastewater, which was reduced in the final effluent to 3.1 and 1.9 mg/l, respectively (Figure 3). There was minimal loss of ammonia due to volatilization as the pH of the wastewater in both sections was neutral. The nitrate concentration was increased significantly in the final effluent. There was only a trace amount of nitrate in the raw wastewater (0.27 mg/l) but after passing vertically through the upper section, nitrification resulted in an increase in the concentration of nitrate to 72.8 mg/l followed by a reduction to 63.2 mg/l in the horizontal flow section due to denitrification.

The reduction of alkalinity after passing through the vertical flow section (93.8%) was due to nitrification. The concentration of alkalinity increased again from 53.4 to 79.4 mg/l in the horizontal flow section due to the denitrification reaction. Denitrification was not significant though the removal efficiency in terms of total nitrogen reached 70.8%. The following experiment aimed to increase denitrification in the horizontal flow section by the recycling of 50% of the effluent to the raw wastewater.

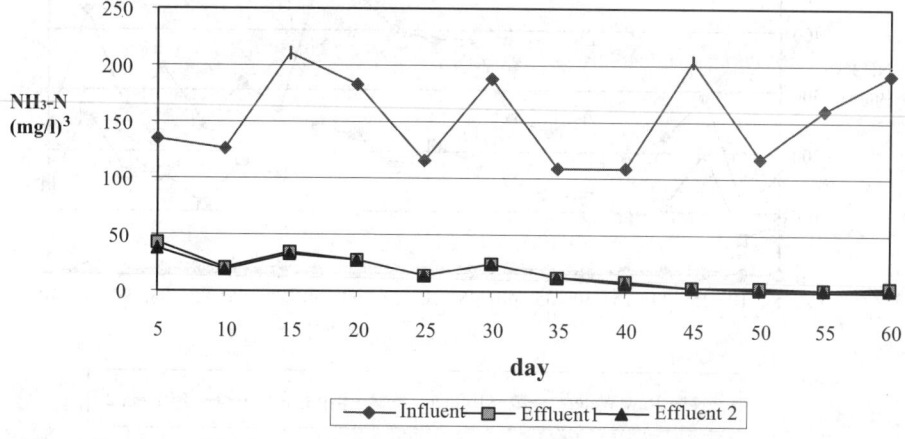

Figure 3 NH_3-N profile without recirculation of the effluent

Run 2: Recirculation of 50% of the effluent

The experiment was carried out with the same procedure but 50% of the effluent was recycled and mixed with raw wastewater. The hydraulic loading was then changed to 5.6 cm/d.

The experiment was conducted for 100 days from August to December 1999. The COD concentration in the raw wastewater was higher than in the previous experiment and the average concentration was 3342 mg/l during the steady state but after mixing with the recycled effluent, the concentration was reduced to 1809 mg/l. The removal efficiency was around 96% in the upper section increasing to 97% after passing through the lower horizontal flow section. Average COD concentration in the final effluent was around 83 mg/l. The recycling of the effluent did not significantly enhance the COD removal efficiency.

Recycling of the effluent enhanced the nitrogen removal efficiency. The TKN and NH_3-N concentrations in the raw wastewater were also higher than the previous run. The average concentration of TKN and NH_3-N were 406 and 302 mg/l, respectively. After mixing with 50% effluent, the concentrations of TKN and NH_3-N were diluted to 230 and 170 mg/l (Figure 4). Both TKN and NH_3-N were reduced by 99%. The average nitrate concentration in raw wastewater was 1.1 mg/l and after mixing with the recycled effluent, the concentration increased to 32 mg/l. The increase of nitrate to 140 mg/l in the vertical flow bed was significant (Figure 5). The reduction of TKN and NH_3-N was caused by nitrification, and the denitrification occurred in the horizontal flow section reducing nitrate to 57 mg/l or approximately 60%.

The total nitrogen removal efficiency was higher when 50% of the effluent was recycled. In the first run, without recycling, the TN removal was around 70% but in the second run the TN removal reached 85%. Laber *et al.* (1997) also found that a recirculation rate of 50–60% and intermittent feeding could increase the nitrification/denitrification and total nitrogen removal was 53% in a combined constructed wetland system.

Evaporation in a vertical flow section was observed since the experimental unit was located outdoors and exposed to sun light. The water lost was approximately 63%.

The growth and nitrogen accumulation in the plants were measured and analysed at the end of each run. The dry biomass of *Cyperus* at the end of the first and the second runs were 1.55 and 2.79 kg/m^2 (15.5 and 27.9 ton/ha) respectively and the growth rate was reduced with time. The nitrogen content was maximum in the stems then leaves and roots as shown in table 1. The total nitrogen accumulation were 0.81 g/m^2.d and 1.1 g/m^2.d at the end of

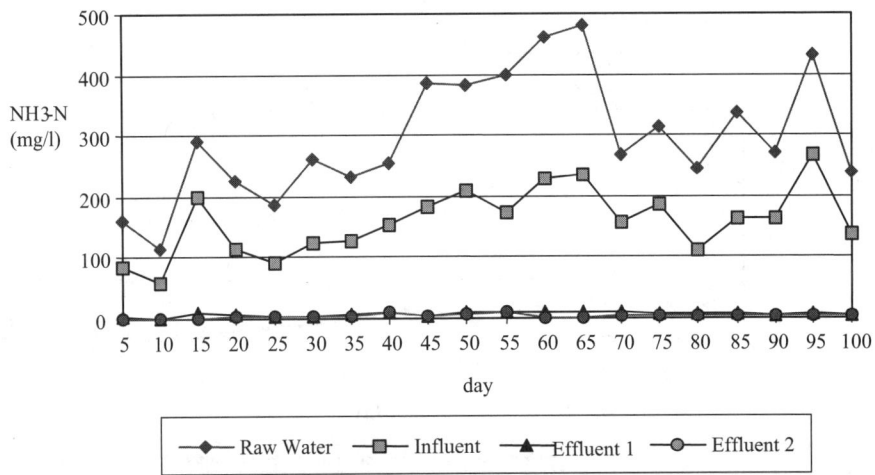

Figure 4 NH_3-N profile with 50% recirculation of the effluent

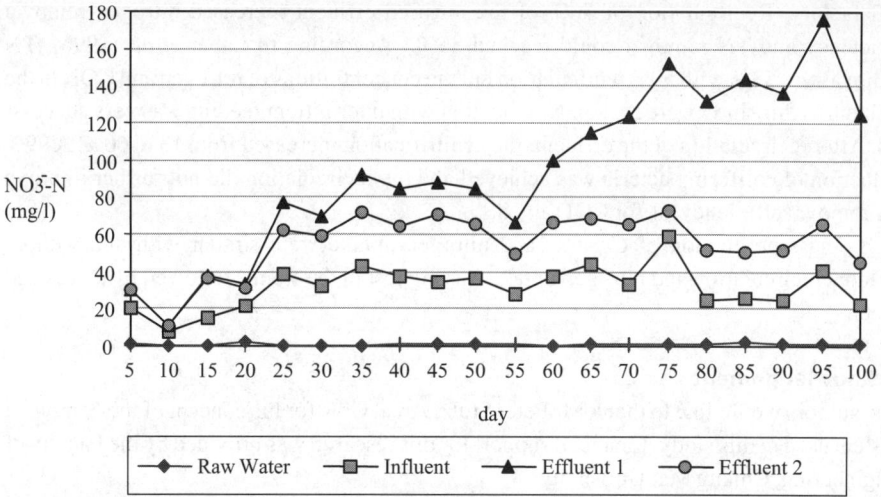

Figure 5 Nitrate profile with 50% recirculation of the effluent

each run, respectively. Nitrogen take up by plants was equivalent to 9% and 7% of nitrogen loading.

Koottatep and Polprasert (1997) reported N uptake by cattails in the tropics using domestic wastewater was 3 kg/ha.d and the increment of dry biomass was 16.1 ton/ha in 90 days in comparison to 8.13 kg/ha.d and 15.5 ton/ha in 60 days and 11.2 kg/ha.d and 27.9 ton/ha. in 100 days of *Cyperus* in this study.

The experiment is continuing in order to determine the optimum recycling ratio for nitrogen removal.

The reduction of coliform bacteria was also measured. In the raw wastewater, total coliform (TC) and fecal coliform (FC) bacteria were 25×10^{11} and 21×10^{11} MPN/100 ml respectively. In the first run, the reduction was very high and the effluent from the vertical flow bed showed TC and FC concentration of approximately 16×10^4 and 20×10^4 MPN/100 ml., the reduction increased in the horizontal flow bed and 15×10^2 TC and 12×10^2 FC MPN/100 ml were recorded in the final effluent. Recycling of 50% of the effluent showed a similar reduction in the bacteria. The TC and FC in the raw wastewater were 26×10^{11} and 13×10^{11} MPN/100 ml. and after passing through the vertical flow section, the number of TC and FC were 16×10^4 and 11×10^4 MPN/100 ml and reduced to 14×10^2 and 10×10^2 MPN/100 ml. in the final effluent, respectively. The concentration of SS in the effluent was 1,000 mg/l on average (37.5 g/m^2.d). The removal efficiency was 97–98% in both runs. Clogging was observed at the end of the second run and geotextile membrane which was used to cover the effluent pipes was then removed.

Conclusions

The use of a combined system with a vegetated vertical flow bed constructed wetland over horizontal flow sand bed is an alternative method for wastewater treatment where land area is limited. The efficiency of the system, especially for COD and nitrogen removal was

Table 1 Nitrogen accumulation in plants with recirculation of the effluent

Constituent	initial period			after 100 days			
	leaf	stem	root	leaf	stem	root	total
N content, gN/g	0.0074	0.0083	0.0037	0.0261	0.0139	0.0107	
	N plant uptake, gN/m^2.d			0.414	0.491	0.197	1.102

satisfactory. Recirculation of 50% of the nitrified effluent increased nitrogen removal efficiency and TN removal could reached 85% . According to Laber *et al.* (1996), TN removal was 53% with 50% recirculation but the concentrations of nitrogen and COD in the influent in this study were higher. Nitrification with intermittent feeding was very successful. After recirculation of the effluent, the denitrification increased from 13 to 60%. A 99% reduction of coliform bacteria was achieved and the recirculation did not further improve the removal efficiency, as for COD and SS.

Biomass production of *Cyperus* and nitrogen uptake were similar with and without effluent recirculation and nitrogen uptake was highest in the stems, followed by leaves and roots.

Acknowledgement

The authors would like to thank Mr Peter Stolz, Areal GbR for his concept of the combined system used in this study. Financial support for this research was provided by the Faculty of Engineering, Chiang Mai University.

References

Cooper, P.F. and de Maeseneer, J. (1996). Hybrid system – What is the best way to arrange the vertical and horizontal-flow stages? *IAWQ Specialist Group on Use of Mesophytes in Water Pollution Control*, No.15, December: 8–13.

Koottatep,T. and Polprasert, C.(1997). Role of plant uptake on nitrogen removal in constructed wetlands located in the tropics. *Wat.Sci.Tech.* **36**(12), 1–8.

Laber, J., Perfler, R. and Raimund, H. (1997). Two strategies for advanced nitrogen elimination in vertical flow constructed wetlands. *Wat. Sci. Tech.* **35**(5), 71–77.

Platzer, C. (1996). Enhance nitrogen elimination in subsurface flow artificial wetlands- a multistage concept. In: *Preprints Volume 1 of the 5th International Conference on Wetland Systems for pollution control,* Vienna, Sept. 15–19, 1996. I/7–1–I/7–9.

Platzer, C. (1999). Design recommendations for subsurface flow constructed wetlands for nitrification and denitrification. *Wat. Sci. Tech.* **40**(3), 257–263.

Standard Methods for the Examination of Water and Wastewater (1995). 19th edn, American Public Health Association/American Water Works Association/Water Environment Federation, Washington, DC., USA.

Szogi, A.A., Rice, J.M., Humenik, F.J., Hunt, P.G. and Stem, G. (1999). Constructed wetlands for confined swine wastewater treatment. *Proceedings of 1999 Animal Waste Management System Symposium*. North Carolina, 379–383.

Nitrogen and phosphorus budget in rewetted fens

A. Lenz* and U. Wild**

* Ingenieurbüro Lenz, Lusenstr. 6, D-94160 Ringelai, Germany. (E-mail: *LenzAnton@aol.com*)
** Vegetation Ecology, TU München, Am Hochanger 6, D-85 350 Freising, Germany.
(E-mail: *wild@weihenstephan.de*)

Abstract A former dewatered fen was flooded for a multi-purpose landuse system including cattail production, fen protection, and water purification. These research plants with an area of 6 ha consist of three constructed surface-flow wetlands. The inflowing water is polluted by non-point sources due to intensive agriculture. The focus of this paper is the estimation of the potential of rewetted fens to reduce phosphorus and nitrogen. The dominating forms of nitrogen in the inflow are organic nitrogen and nitrate. The reduction rate is higher for nitrate than for organic nitrogen, although the nitrate reductions occur only during the summer season. If no nitrate is available for denitrification, there is a release of ammonia from the peat into the water. The main form of phosphorus in the in- and outflow is ortho-phosphate. In contrast to the values of nitrate, the concentrations of phosphorus are very regular with no significant seasonal pattern.

When nitrate isn't available in the water any more, the release of phosphorus begins and the rewetted fens change from a sink for phosphorus to a source of it. Rewetted fens can be a sink for phosphorus and nitrogen with nitrate as the limiting factor. Only if denitrification can occur, can the release of ammonia and phosphorus from the peat layer be prevented.

Keywords Cattail; fen; nitrogen reduction; phosphorus reduction; rewetting

Introduction

Most fens in middle Europe were drained for agricultural use which led to a degradation of the peat soil (Göttlich *et al.*, 1993; Pfadenhauer and Klötzli, 1996). Another consequence was that these peatlands converted from a sink for nutrients and carbon to a source of it (Glenn *et al.*, 1993). A restoration of the sink function is only achievable by rewetting.

Pfadenhauer and Klötzli (1996) and Brülisauer and Klötzli (1998) described several examples of rewetting experiments in middle European wetland systems. These often small scale experiments have been installed generally for nature conservation purposes. Despite the functioning of wetlands as nutrient-traps being well known, there are only a few experiments in middle Europe concerning this feature. The aim of the project is to combine the restoration of a landscape with an economically viable production of cattail, a non-food crop. On the other hand, reconstructed wetlands can be an important factor for water purification in intensive agricultural catchments by reducing the non-point pollution. This is the focus of this paper.

The two cattail species (*Typha latifolia* and *T. angustifolia*) were selected because of their high productivity, in particular under eutrophic wetland conditions. The leaves of cattails seem especially suited to serve for an industrial raw material production. In addition cattails endure continuous flooding. As a result the cultivation areas provide extensive water retention possibilities and allow the simulation of a fen typical water regime.

Site description and experimental design

Experimental site

The experiments have been performed in the Donaumoos, Germany, approximately 80 km north of Munich in the valley of the river Danube (48°42′N, 11°11′E). It is a percolating mire and the largest peat deposit in southern Germany. The mean annual air temperature is

7.6°C and the mean annual precipitation is about 700 mm. The experimental site is deeply drained; the groundwater table can reach down to 1.5 m beneath land surface in dry summer months. The peat thickness is 0.8 to 1.1 m; the degree of decomposition in the top soil layer is H9 (terric histosol) with a pH ($CaCl_2$) of 6.6 and a C:N ratio of 13.0, the total organic N is 2.1%.

Experimental design

In the northwest of the Donaumoos the demonstration plant was established in spring, 1998. The plant consists of three constructed wetlands with a net area of 6 ha. The sites were formerly used as meadows without wetland vegetation. For construction, the areas were graded by bulldozers, so every area has a uniform level. Parts of the soil were used to construct the dams around the areas; the heights of the dams are 0.8 or 1.0 m.

The water source for flooding and permanent inflow consists of two ditches which drain catchments with, in the main parts, intensive agriculture. Ditch water is pumped by two borehole pumps through pressure lines into Area 1 and Area 2. For Area 3 the inflow occurs by the outflow of Area 2. Water was first pumped into the constructed wetlands at the end of June, 1998. The water levels in these surface flow wetlands are 0.2 m for the Areas 1 and 3

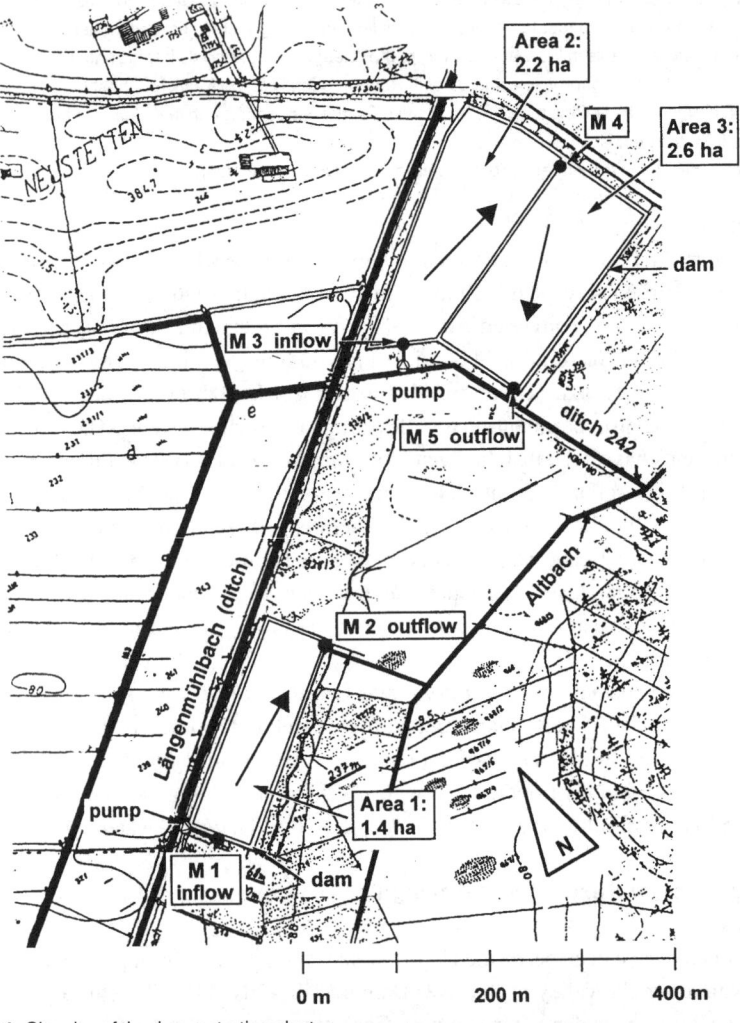

Figure 1 Site plan of the demonstration plants

and 0.4 m for Area 2. The water regulation is constructed by riser pipes at the sampling locations M2, M4 and M5. The inflows at the points M1 and M3 are built as multiple-point discharges by a pipe with orifices, at point M4 as a point discharge. The cattail stands were established in May and June, 1998, by planting.

This project is supported by Deutsche Bundesstiftung Umwelt (DBU).

Methods

Water quantity and quality were measured continuously at 5 sampling locations (M1-5, Figure 1), as was electrical conductivity (WTW conductivity meter), oxygen content (amperometric), pH, and temperature (glass combination electrode). Additional samples for water quality analysis were collected every three weeks in a three day-series. These samples were 24-hour composite samples taken from the 5 sampling locations. The analyses included settleable solids, TKN, NH_4-N, NO_3-N, NO_2-N, total-P, PO_4-P, and TOC (German Industrial Standard-methods).

Results and discussion

Water budget

Water pumping into the demonstration plants started in June, 1998. In the first year of research the average hydraulic loading rate was 30 mm/d in Area 1, 36 mm/d in Area 2, and 25 mm/d in Area 3. For the second year the load is similar. The pH-values at the five sampling locations ranged from 7.0 to 8.5, the conductivity from 600 to 1,200 µS cm^{-1}. These values are characteristic for alkaline drained fens like the Donaumoos.

The periods of water temperatures lower than 10°C ranged from November 1998, to March 1999, and from October to December 1999.

Nitrogen

On an annual and a half-annual basis, the role of the wetlands as sinks or sources for nutrients were investigated.

Organic nitrogen and nitrate were the prevailing nitrogen forms in the in- and the outflow. In the Areas 1 and 2 the reduction rates of organic nitrogen were lower than for nitrate, not only in the first year (Table 1), but also in the following half year. From July to December 1999 in Area 1 the reduction rate of nitrate-N was 65%, in Area 2 92%, and in Area 3 69%. During the same period organic nitrogen was reduced in the Areas 1 and 2 in a range of 32 and 37%. In Area 3 more organic-N left the wetlands than entered it. From July to December 1999, the tendency to release ammonia in Area 3 was increasing to a rate of 385%. Area 3 is also a source of total nitrogen in a range of 18%.

The main concentrations of organic nitrogen ranged from 2 to 10 mg/l with a peak in spring 1999. In summer the wetlands changed temporarily from a sink to a source of organ-

Table 1 Nitrogen mass load from middle of June 1998 to 30 June 1999

	Org. N (kg)	NH_4-N (kg)	NO_3-N (kg)	Total N (kg)
Area 1 (1.4 ha):				
Reduction in %	24	67	62	42
Reduction in kg/ha	121	31	234	391
Area 2 (2.2 ha):				
Reduction in %	32	74	49	41
Reduction in kg/ha	241	38	265	547
Area 3 (2.6 ha):				
Reduction in %	–3	–37	26	7
Reduction in kg/ha	–13	–4	61	45

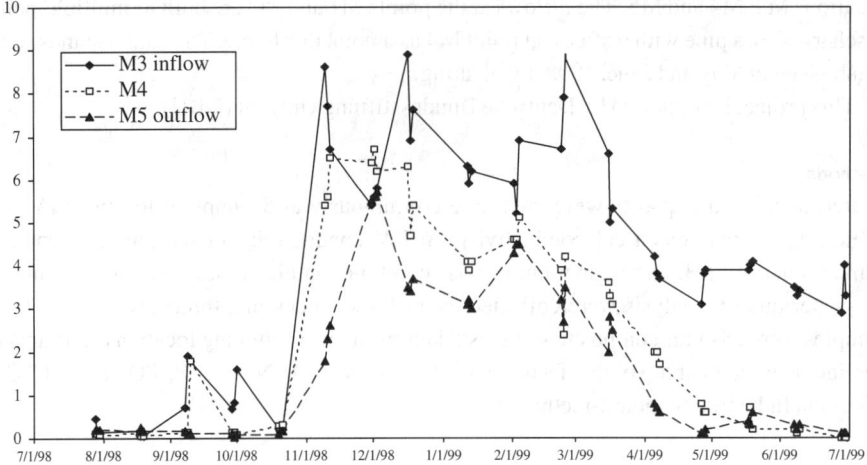

Figure 2 Nitrate nitrogen in the seasonal pattern (Areas 2 and 3)

ic nitrogen. The values had a high fluctuation without a discernable pattern, similar to results of the Des Plaines Project (Phipps and Crumpton, 1994), but at a higher level. In surface flow constructed marshes with comparable hydraulic and mass load, but fed by wastewater, the reduction of organic nitrogen ranged between 30 and 60% (Kadlec and Knight, 1996).

In the first year of research the average concentrations of ammonia ranged from 0.44 mg/l (sampling point M1, inflow) and 0.36 mg/l (M3, inflow) to 0.18 mg/l (M5, outflow), 0.16 mg/l (M2, outflow) and 0.13 mg/l (M4, in- and outflow). In contrast to the wetlands of the Areas 1 and 2, Area 3 is a source of ammonia.

The second important form of nitrogen was nitrate with a clear peak in winter for the inflowing and outflowing water. This pattern of the inflow is typical for streams polluted by non-point sources (Phipps and Crumpton, 1994). Nitrification processes in the plants were not restricted by low oxygen values. Nitrate concentrations in water, exiting the wetlands, were consistently reduced relative to inlet concentration, more effective in Area 1 and 2 than in Area 3. Dates of Area 1 and 2 are similar, so here it may be sufficient to present the results of Area 2 and 3 (Figure 2). During periods of temperatures higher than 10°C the nitrate was removed to average concentrations lower than 0.1 mg l^{-1} in the Areas 1 and 2. Hence an additional reduction in Area 3 was not possible. During winter the removal was restricted by a low denitrification caused by low temperatures and the lack of uptake by plants.

Phosphorus

In the three wetlands PO_4-P was the dominating form of phosphorus which corresponds to results of the Des Plaines Project (Hey *et al.*, 1994).

For the following research period (July to December, 99), there was no significant change of the reduction rates in the Areas 1 and 2. Area 3 became an increasing source of phosphorus with rates of 128% of the inflow for PO_4-P and 54% for total-P. Most values in the inflow of the Areas 1 and 2 range from 0.1 to 0.3 mg/l with a peak in summer.

The total-P loads of 3 g $(m^2 a)^{-1}$ in the Areas 1 and 2 and 1 g $(m^2 a)^{-1}$ in Area 3 in the first year of research were in a range where reduction in other wetlands occured (Mitsch and Gosselink, 1993). It is remarkable that, in spite of the lower load in Area 3, this wetland is a source of phosphorus.

Table 2 Phosphorus mass load from middle of June 1998 to 30 June 1999

	PO$_4$-P (kg)	Total P (kg)
Area 1:		
Reduction in %	78	74
Reduction in kg/ha	20	25
Area 2:		
Reduction in %	75	70
Reduction in kg/ha	16	22
Area 3:		
Reduction in %	7	−22
Reduction in kg/ha	0	−2

Discussion

The main reduction of total nitrogen occurs by the reduction of nitrate. This process runs very effectively when the water temperature is higher than 10°C, but during this time the concentrations in the inflow are lower than during the winter season. So this is the prevailing restriction for the effectness of these systems. Organic nitrogen is reduced in the Areas 1 and 2 at a lower level than nitrate. Sometimes the wetlands change from a sink to a source of organic nitrogen. The prolongation of the detention time, which happens in Area 3, is not suitable for an increasing of the reduction rate. This wetland is a source of organic nitrogen. Ammonia, a less important load than organic nitrogen or nitrate, is reduced in the Areas 1 and 2 at a higher rate than the other nitrogen fractions. In contrast to the Areas 1 and 2, Area 3 is a source of ammonia with a tendency to increase this effect. For total nitrogen Area 1 and 2 are sinks. Although the load of N in Area 3 is lower than in the other wetlands, the reduction effect is very low. In the second year of research it even becomes a source because of the increase of ammonia in the outflow.

For phosphorus we can see an analogous tendency as for ammonia. The Areas 1 and 2 are effective sinks, and Area 3 is a source fed with a lower load. The reduction is not very different during the year, so the main factor cannot be the uptake by plants, but must be the precipitation by calcium and magnesium (Kadlec and Knight, 1996).

The fact that Area 3 is a source of nutrients is not caused by the load, but must have another reason. Our explanation is that the reason therefore is the lower redox potential in this wetland. The combination of ammonia- and phosphorus release in wetlands is pointed out in other publications (Wetzel, 1975; Fenchel and Blackburn, 1979; Bryant and Bauer, 1987; Reddy and D'Angelo, 1994). The redox potential can be lower in Area 3 because denitrification, as a stabilisation factor, occurs on a much lower level (Gumbricht, 1993).

Conclusion

Rewetted fens, designed as constructed surface flow wetlands, can be sinks for nutrients from non-point sources, but only under some special conditions. After 1.5 years of research, we can present our results. A release of ammonia and phosphorus can be avoided by denitrification. So the ideal form of nitrogen in the inflow is nitrate. Denitrification guarantees that the redox potential in the soil cannot decrease to a level where a release of nutrient occurs. The detention time of the inflowing water should be adapted to this condition. The effectivity of the fen wetlands as a nitrogen sink depends on the nitrate reduction. Other processes are of lower importance. During periods with water temperatures higher than 10°C, no limit for denitrification can be seen for our mass load.

For phosphorus the water temperature is of lower importance. Here no limit for the retention over the whole year can be seen.

References

Brülisauer, A. and Klötzli, F. (1998). Notes on the ecological restoration of fen meadows, ombrogenous bogs and rivers: definitions, techniques, problems. *Bulletin of the Geobotanical Institute ETH*, **64**, 47–61.

Bryant, C.W. and Bauer, E.C. (1987). A simulation of benthal stabilisation. *Water Science and Technology*, **19**(12), 161–168.

Fenchel, T. and Blackburn, T.H. (1979). *Bacterial and Mineral Cycling*. Academic Press, London.

Glenn, S., Hayes, A. and Moore T.R. (1993). Carbon dioxide and methane fluxes from drained peat soils. *Global Biochem. Cycles*, **7**, 247–258.

Göttlich, K.H., Richard, K.-H., Kuntze, H., Eggelsmann, R., Günther, J., Eichelsdörfer, D. and Briemle G. (1993). Mire Utilisation. In: Heathwaite, A.Z. and Göttlich, K.H. (Eds), *Mires. Process, Exploitation and Conservation*, John Wiley & Sons, Chichester, pp. 325–415.

Gumbricht, T. (1993). Nutrient removal processes in freshwater submerged macrophyte systems. *Ecological Engineering*, **2**, 1–30.

Hey, D.L., Kenimer, A.L. and Barrett, K.R. (1994). Water quality improvement by four experimental wetlands. *Ecol. Eng.*, **3**, 381–397.

Kadlec, R.H. and Knight, R.L. (1996). *Treatment Wetlands*. Lewis Publishers, Boca Raton.

Mitsch, W.J. and Gosselink, J.G. (1993). *Wetlands*. Van Nostrand Reinhold, New York.

Pfadenhauer, J. and Klötzli, F. (1996). Restoration experiments in middle European wet terrestrial ecosystems: an overview. *Vegetatio*, **126**, 101–115.

Phipps, R.G. and Crumpton, W.G. (1994). Factors affecting nitrogen loss in experimental wetlands with different hydraulic loads. *Ecological Engineering*, 3/4, 399–408.

Reddy, K.R., D'Angelo and E.M. (1994). Soil processes regulating water quality in wetlands. In: Mitsch, W.J. (Ed.): *Global Wetland Sold World and New*. Elsevier, Amsterdam, pp. 309–324.

Wetzel, R. (1975). *Limnology*. W.B. Saunders, Philadelphia.

Nutrient removal in subsurface flow constructed wetlands for application in sensitive regions

H. Rustige* and Chr. Platzer**

* Akut Umweltschutz Ingenieurgesellschaft mbH, Sydower Feld 4, 16359 Biesenthal, Germany
** COBAS – Consultaria Brasil-Alemanha de Saneamento ltda., Rua Padre Roma 281, 88015-100 Florianópolis, Brazil

Abstract One of the most interesting sites for research on CWs in Germany has been established in Wiedersberg (Saxonia). The multi-stage concept with primary settling, vertical and horizontal flow reed bed followed by UV-disinfection and a special phosphorus filter bed, allows numerous ways of operation and investigations. Denitrification can be improved by recirculation through VF bed and sedimentation tank or by means of adding carbonaceous water from the primary stage to a second level within the VFB or directly to the following HF bed. In order to investigate the efficiency of P-elimination four kinds of natural sands containing different amounts of iron have been used. To maintain a long-term capacity for P-reduction an additional filter bed is filled with gravelly sand which had been used for the precipitation of iron from drinking water before. After saturating with P this filter medium can be exchanged easily.
A result of more than one year of operation is the high performance rate for adsorption of phosphorus by enriched iron on drinking water filter sand. At a total loading rate of 350 g P/m³ filter medium 250 g P/m³ have been adsorbed. Design considerations can not be given yet. The median denitrification rate at VFB is 1.3 g N m^{-2}d^{-1} and at HFB is 0.25 g Nm^{-2}d^{-1}. The low denitrifcation rate of HFB might be due to a very high quota of wastewater dilution by storm- and ground-water of 100 to 200 percent. The investigations on this wastewater treatment plant will be continued until June 2001 and experiments with filter columns will be added.
Keywords Iron sand; nutrient removal; phosphorus filter; recirculation; subsurface flow wetlands

Introduction

In regions with many lakes and very slow rivers there is an increasing interest in elimination of phosphorus and nitrogen from domestic wastewater. This is the case in East Germany, where there exist a lot of small villages without the possibility of discharge. For these villages constructed wetlands can be a convenient solution if they are able to meet high standards in performance.

In order to show the efficiency of constructed wetlands in practice a research program on "planted soil filter beds" (PSF) has been started by the German Federal Environmental Foundation (DBU) in 1998 and is being carried out by seven cooperating companies, universities and the German Federal Environmental Agency. Detailed information is being provided at: www.bodenfilter.de. The study carried out by the authors as part of the PSF-program focuses on the removal of phosphorus and nitrogen. Preliminary results of this investigation are presented in this paper. The data were collected on a full-scale reed bed system which was constructed in 1998.

Parallel investigations on microbiological parameters and soil clogging are being made at the same site by cooperating institutions. First results on retention of microbiological organisms are described in another paper at this conference (Hagendorf *et al.*, 2000).

Wiedersberg (Saxonia) is a small village with 120 persons situated in a formerly prohibited area near the East German border. After a short distance of about three kilometres the local river flows into a drinking water reservoir (Figure 1). Until 1998 there have been two aerobic ponds and preliminary anaerobic settling tanks for treating the wastewater of a combined sewer system. The separation of stormwater runoff and upgrading of the wastewater treatment by means of planted soil filter was the only accepted solution for maintaining discharge into this river system.

It was assumed that the use of a reed bed system would provide the best conditions for achieving a good and long-term performance in nutrient reduction and hygienic acceptable standards for such a small size system with a total of 145 pe.

Methods

System design

General considerations. The design of the wastewater treatment system follows the requirements of protecting the drinking water reservoir. This means limiting the concentration and load of oxygen consuming substances, nutrients and pathogenic microorganisms (Table 1). The design had to be adapted to the proportions of the formerly existing parts of the wastewater treatment and integrated into the free space of this site. After separation of the sewer a maximum flow of about 19 m^3/d and about 11 m^3/d at dry weather was presumed.

Because of the high efficiency of vertical flow beds for nitrification on a rather small area and the capability of denitrification at low concentrations in horizontal flow beds a combination of both was chosen. In order to minimize the area of HFB a recirculation into the primary settling tank was planned. Some other options for distributing the load during the cold winter period or for improving the C/N-ratio had to be included. As a final stage a third filter bed for phosphorus removal is installed in order to guarantee long-term efficiency. The use of physical UV-disinfection was required by the authorities of the drinking water reservoir.

Basis for design. The dimensioning of the VFB and HFB was made according to the design recommendations of Platzer (1998). For VFB this is based on oxygen transfer and oxygen demand. It leads to a maximum load of about 25 g COD $m^{-2}d^{-1}$ and can here be achieved with a specific area of 3.2 m^2/pe if a COD-reduction of about 30 percent within the primary stage is assumed. In order to reach a complete nitrification an oxygen transfer of about 90 g O_2 $pe^{-1}d^{-1}$ into the soil filter would be necessary. An oxygen yield of about 9 g O_2 $pe^{-1}d^{-1}$ can be taken into account if a denitrification rate of 30 percent in the VFB is being considered.

The total amount of oxygen demand can be covered by convection into soil through intermittent loading and diffusion by the surface of the VF bed. If a diffusion time of 15 hours a day is calculated (depending on the number and time of intermittent loading) the theoretical input is 15 g O_2 $m^{-2}d^{-1}$ or 36 g O_2 $pe^{-1}d^{-1}$ assuming three of four sections of the VFB are in use at a time. This means the remaining amount of 45 g O_2 $pe^{-1}d^{-1}$ has to be put in by convection. With an oxygen concentration of 300 mg O_2/l in the air a volume of 150 l/pe would be necessary. In this case a minimum hydraulic loading rate on the used area of 63 mm/d equivalent to a flow rate of 22 m^3/d is sufficient. Presuming a dry weather flow of

Table 1 Official permit of site

Parameter	Concentration limit [mg/l]	Maximum daily discharge [g/d]
BOD_5	20	580
COD	90	2610
TN (>12°C)	18	522
NH_4-N (>12°C)	10	290
TP	2	58

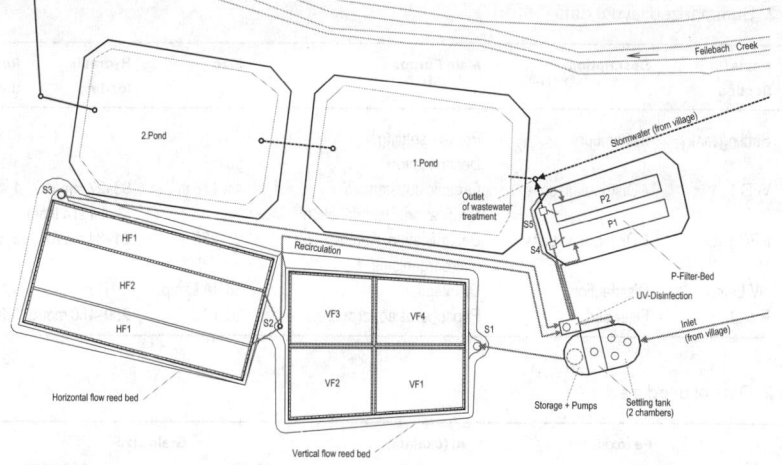

Figure 1 Diagrammatic presentation of the Wetland system at Wiedersberg/ Saxonia

11.3 m³/d a minimum recirculation rate of 94 percent of inflow would be required for optimum oxygen input.

Denitrification. The third stage HFB was expected to have a denitrification capacity of 0.65 g N m^{-2}d^{-1} if loaded with 1 g N m^{-2}d^{-1} (Platzer, 1998). The available area of 3.75 m²/pe is then able to remove 2.4 g N pe^{-1}d^{-1}. Added to the performance capacity of 3 g N pe^{-1}d^{-1} of the VFB the total elimination rate of nitrogen within the planted soil filters sums up to 5.4 g N pe^{-1}d^{-1}. This would mean an effluent concentration of 46 mg/l without recirculation. To achieve a maximum rest concentration of 18 mg TN/l or a maximum remaining load of 3.6 g N pe^{-1}d^{-1} a removal of 1 g N pe^{-1}d^{-1} by means of primary recirculation is required. This means a recirculation rate of at least 10 per cent.

Without recirculation the effluent concentration of VFB would be 70 mg TN/l considering a flow rate of 100 l pe^{-1}d^{-1} and a 30 per cent removal within the planted soil. With a recirculation rate of 50 per cent a denitrification rate of 25 per cent or a total of 2.5 g N pe^{-1}d^{-1} is expected. This leads to a maximum N-reduction of 55 per cent in the first treatment stages and to a calculated concentration of 45 mg TN/l. After the subsequent HF stage a rest concentration of 16 mg TN/l or a rest load of 1.6 g N/pe d could be achieved.

Retention of Phosphorus. For long-term removal of phosphorus by subsurface flow wetlands no well founded dimensioning proposal could be found in the literature. Nonetheless many publications show very high performance of phosphorus removal by planted or unplanted soil filters in the first years of operation. Rettinger (1992) for example cites 10 authors finding an elimination rate of 75 to 99 per cent within the soil. Börner (1992) reports a mean retention performance for a high number of HF reed beds of more than 80 per cent after some years of operation.

Almost every author dealing with P names the role of precipitation and adsorption in the soil. The role of plants is regarded as subordinate (Reddy and D'Angelo, 1996). It is generally agreed that the concentrations of iron, aluminium or calcium are the main factors of influence. Since the process of adsorption is finite, it is important to know the cumulative loading at which a required phosphorus concentration breaks through.

According to several studies at laboratory or half-technical scale the use of such filter sands is very efficient for phosphorus removal. Starck and Lennartz (1996) found a cumulative loading capacity of 233 mg P/kg filter medium until a breakthrough concentration of

Table 2 Summarized layout data

Stage	Name/Section	Description	Main Purpose	Size	Hydraulic loading	Retention time
I.	Settling tank	2 chambers	Primary settling, Denitrification	55 m³		> 1 d
II.	VFB 1..4	4 Distribution units	Organic substances, Nitrification, Denitrification	4 × 116 m²	63–97 mm/d (3 of 4 at a time)	1,4–2,2 d
III.	HFB 1..2	2 Distribution units	Denitrification, Retention of Microorganisms	2 × 270 m²	21–34 mm/d	3,3–5,4 d
IV.	UV-Lamp	Disinfection	Disinfection	36 VA Lamp	< 1,1 m_/h	> 10 sec
V.	P 1..2	Filter bed	Phosphorus adsorption	20 m³	280–460 mm/d	0,3–0,5 d

Table 3 Data of used media

Section	Fe (oxalate) (mg/kg)	Al (oxalate) (mg/kg)	Grain size d 10 (mm)	Grain size d 60 (mm)
VF 1 and HF 1	85	53	0,21	0,86
VF 2 and HF 2	731	170	0,24	0,82
VF 3	964	403	0,26	1,22
VF 4	741	307	0,41	3,34
P 1 and P 2	20600	742	0,58	1,2

1 mg P/l starting with an influent concentration of 3 mg/l. Güldner and Hegemann (1995) found a cumulative loading capacity of about 1000 mg P/kg filter medium at the same breakthrough concentration but with an influent concentration of about 4 mg/l. The efficiency of retention was about 90 per cent.

At the site of the water treatment system two basins with an area of each 40 m² were used for "P-Filtration". They were filled each with 20 m³ of filter medium at a height of 0.5 m. Based on an approximate loading capacity of 400 to 1,500 g P/m³ cited above, the possible running time for each filter bed could be 230 to 600 days if an influent concentration of 3 mg P/l and an effluent concentration of 1 mg P/l was assumed.

General layout. The main layout data of the wastewater treatment system and media used are summarized in Tables 2 and 3. Hydraulic data correspond to the presumed dry and wet weather flow of 11.3 to 18.5 m³/d.

Three gravelly sands with higher amounts of Fe(ox) and one ordinary sand were selected for comparison. As filter medium at stage V, gravelly sand from a drinking water processing plant in Leipzig was used.

Monitoring

The monitoring program has not finished yet at the time of writing and will be carried out until June 2001. Most physical and chemical parameters are monitored weekly. Samples are taken at every stage. Most parameters are measured by photometry using pre calibrated tests and standards. Only some of the TP and PO_4-P samples were below the detecting limits of this method. Parallel monitoring by the Federal Environmental Agency using standard methods showed comparable values and were used for interpretation as well. The flow rates are measured by an inductive flow counter behind HFB, by water meters and pumping before the HFB and the recirculation rate and mass loading of each section of VFB is determined by calculation from pump operation time.

Results and discussion

General operating results

Hydraulics. Ever since the new separated sewer system of the village had been in use a much higher flow rate than predicted was detected. The long-term average flow rate was 26 to 28 m^3/d. Only several times the flow rate decreased to the designed maximum of 19 m^3/d. In consequence the concentration of domestic wastewater was very low and the intended recirculation rates could not be achieved. The reason for this high amount of water was illegal connections of drainages to the wastewater sewer mainly on private grounds.

Organic substances. The average loading rate of COD on to the VF bed was 22 g m^{-2}d^{-1} ranging from 6 to 46 g m^{-2}d^{-1}. That is about 88 per cent of that predicted. This corresponds to an estimated number of only 130 pe connected to the treatment system. The measured BOD$_5$/COD ratio is 0.51 and the average ratio of DOC/ COD is 0.31. The mean performance rate of VFB is 90 per cent (range: 80 to 96) and does not vary with the COD loading rate. At HFB another 27 per cent of remaining COD is removed.

Nitrogen removal

Total nitrogen. The median removal of TN within the soil filters can be estimated to be 37 per cent at VFB (up to 68 per cent) and at HFB to be 18 per cent (7 to 43 per cent) (Figure 2). In relation to TKN the HFB eliminates about 32 (24–60) per cent. The mean loading rate of HFB is 1.4 g TN m^{-2}d^{-1}. The ratio of C/N is estimated to be 0.25 (0.1–0.6) and C/TKN ratio is 0.66 (0.2–2.5). The median loading of VFB is 3.5 (1.9– 6.6) g TN m^{-2}d^{-1}.

Due to the high hydraulic load the recirculation was only used at low rates and only for a short time so far. At other times uncontrolled recirculation happened and the flow rate was not measured. The performance of denitrification at the primary settling stage will be tested in the next period of operation.

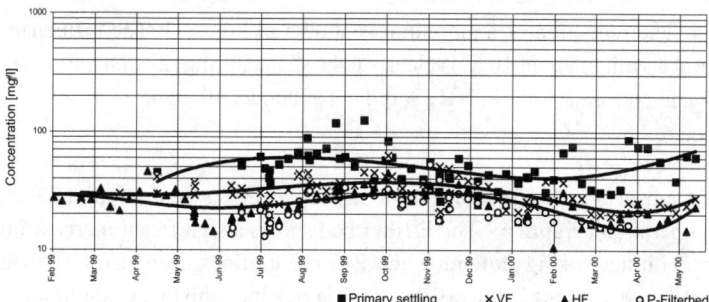

Figure 2 TN concentrations at sequencing stages

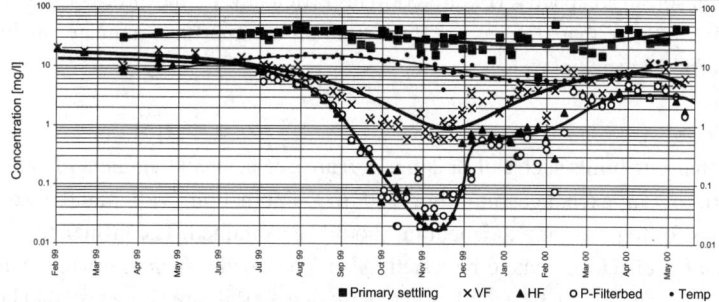

Figure 3 NH$_4$-N concentrations at sequencing stages

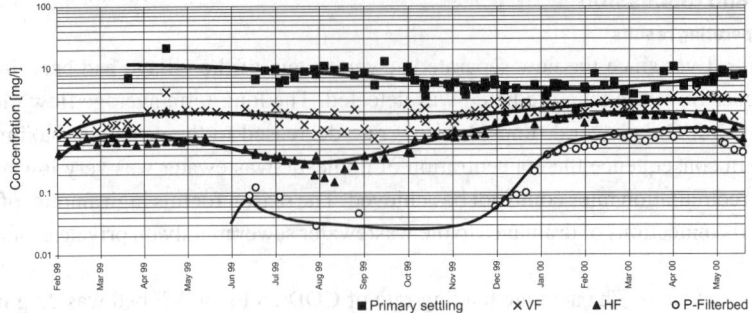

Figure 4 TP concentrations at sequencing stages

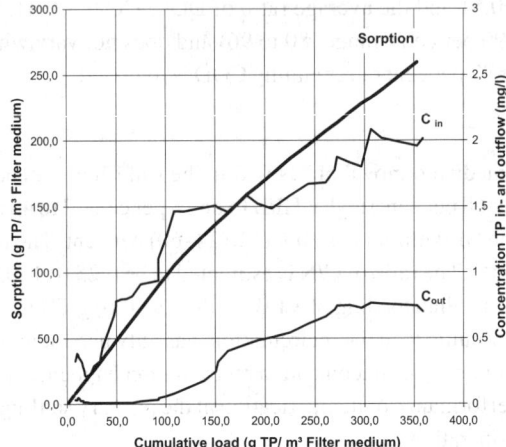

Figure 5 Cumulative loading of P-Filter medium

Nitrification. The reduction of ammonium is shown in Figure 3. The influence of water temperature at a limit of about 10°C is clearly to be seen. During the first half year of operation the effluent concentrations of NH_4-N did not get under 10 mg/l.

Phosphorus removal

All stages of subsurface flow wetland show an increasing concentration of TP after about one year of operation (Figure 4). The P-filter bed shows a significant increase after about half a year. With decreasing flow rates the P-concentrations drop again. Although there were lower P concentrations in the effluent of iron rich medium in the beginning, they now seem to converge. A final loading capacity can not be calculated yet.

The loading of the first P-filter bed has reached about 50 per cent of the minimum predicted absorbing capacity (Figure 5). The efficiency of the process has decreased already. In spite of that decreasing flow rates in summertime indicate an increasing retention efficiency again.

Conclusions

If low discharging limits such as 1 or 2 mg TP /l are given, small treatment plants could use the benefit of an exchangeable filter bed in combination with planted soil filters. Possibilities for regenerating the medium should be examined. The influence of filtration velocity on the efficiency has to be tested. Also other media such as crushed concrete or granulated iron sludge can be useful. Some of these materials are being tested in laboratory scale columns. Design recommendations can not be given yet.

The Wiedersberg wetland system shows that a high flow rate of water from combined sewer systems does not affect the performance of BOD/COD removal. For nitrogen and P-removal, retention time seems to be of more importance. The HF bed only shows a very low denitrification rate which might be due to high flow rates and channeling effects, as may be suggested by tracer response curves. The concept of recirculation from VFB to primary settling has its advantages rather with high strength domestic wastewater.

References

Börner, T. (1992). Einflußfaktoren für die Leistungsfähigkeit von Pflanzenkläranlagen. Dissertation, Schriftenreihe WAR Bd. 58, TH Darmstadt (Hrsg.). Darmstadt.

Güldner, C. and Hegemann, W. (1995). Ermittlung von Bemessungsverfahren und Erprobung einfacher Verfahren zur weitergehenden Phospatelimination für kleine, dezentrale Kläranlagen. Technische Universität Berlin, FB 06, Fachgebiet Siedlungswasserwirtschaft. Abschlußbericht (unpublished).

Hagendorf, U. and Diehl, K. and Feuerpfeil, I. and Szewzyk, R. (2000). Retention of microbiological organisms in constructed wetlands. 7th Int. Conf. On Wetland Systems for W.P.C. Lake Buena Vista, USA.

Platzer, Chr. (1999). Design recommendations for subsurface flow constructed wetlands for nitrification and denitrification. *Wat. Sci. Tech.,* **40**(3), 257–263.

Reddy, K.R. and D'Angelo, E.M. (1997). Biochemical indicators to evaluate pollutant removal efficiency in constructed wetlands. *Wat. Sci. Tech.,* **35**(5), 1–10.

Rettinger, S. (1992). *Wasser und Stoffdynamik bei der Abwasserperkolation.* Berichte aus der Wassergüte- und Abfallwirtschaft, Nr. 97. TU München.

Starck, H.-G. and Lennartz, B. (1996). Wiederverwendung ausgewechselten Wasserwerkskieses in der Abwasserbehandlung. Christian-Albrecht -Universität Kiel. Institut für Wasserwirtschaft und Landschaftsökologie. DBU-Abschlußbericht.

Distribution of ammonium-N in the water-soil interface of a surface-flow constructed wetland for swine wastewater treatment

A.A. Szögi* and P.G. Hunt**

* Washington State University, Cooperative Extension, 128 N Second Street, Yakima, WA 98901, USA
** USDA-ARS, Coastal Plains Soil, Water and Plant Research Center, 2611 W Lucas Street, Florence, SC 29501, USA

Abstract Most livestock wastewaters treated in constructed wetlands are typically rich in ammonium N. The objective of this study was to evaluate the soil-water ammonium distribution and the diffusive flux through the soil-water interface. Wetland system 1 (WS1) was planted to rush and bulrushes, and wetland system 2 (WS2) was planted to bur-reed and cattails. Nitrogen was applied at a rate of 2.5 g m^{-2} d^{-1}. Interstitial soil water was sampled at 9, 24, 50, and 70 m from the inlet. In both wetlands, we found that NH$_4^+$ diffusion gradient and N losses were highest in the wetland system with lowest water depth. From other studies, we knew that shallower depths may have promoted a more effective interfacing of nitrifying and denitrifying environments. In turn, this N reduction in the water column may be the reason for steady NH$_4^+$-N upward diffusion fluxes. The assumed mechanism for N removal has been nitrification and denitrification but ammonia volatilization could also have occurred. Although diffusion may explain a significant portion of the material transport between the soil-water interface, the large differences in concentrations between outlet and inlet need further explanation.

Keywords Ammonia; diffusion; denitrification; hogs; surface-flow wetland; wastewater

Introduction

Confined swine production generates large amounts of wastewater that are typically treated and stored in anaerobic lagoons. Lagoon effluents are rich in ammonia/ammonium (NH$_4^+$/NH$_3$) nitrogen and customarily land applied for terminal treatment. However, over application of nitrogen can occur in operations when land is limiting. An alternative to land application of liquid manure is the use of constructed wetlands. It is believed that constructed wetlands can be part of a farm-wide waste management plan that could minimize the adverse environmental impact to water resources (Cronk, 1996; Szögi *et al.*, 2000). Hunt *et al.* (1999) report consistent removals in surface-flow wetlands of at least 80% of the added N with loading rates ranging from 0.3 to 2.5 g NH$_4^+$-N m^{-2}d^{-1} (3 to 25 kg ha^{-1} d^{-1}). These authors also found that at the lower loading rates, plant and soil accumulation constituted a significant portion (~ 30%) of the total amount applied, but at the higher loading rates, microbial transformations were likely the more dominant treatment factors. Although these results were very encouraging, denitrification enzyme assays indicated that nitrate was the limiting factor. Moreover, the denitrification values were not exceptionally high, which indicated that ammonia volatilization might have been significant. The objective of this study was to evaluate the soil-water ammonium nitrogen distribution along a constructed wetland system used for swine wastewater treatment and estimate ammonium flux across the soil-water interface that may be contributing to N gaseous losses.

Materials and methods

The study site was located in Duplin Co., NC. The study was performed in two wetland systems that consisted of parallel sets of two 4-m × 33.5-m cells connected in series with a total length, from inlet to outlet, of 82 m (Hunt *et al.*, 1999). Wetland system 1 (WS1) was planted to a mixture of rush (*Juncus effusus*) and bulrushes (*Scirpus americanus, Scirpus cyperinus* and *Scirpus validus*). Wetland system 2 (WS2) was planted to a mixture of bur-reed *(Sparganium americanum)* and cattails (*Typha latifolia and T. angustifolia*). Both wetlands were used to treat the effluent from an anaerobic lagoon that stored the wastewater generated by a confined hog production unit. Wastewater was applied during 1998 at a mean loading rate of 2.5 g NH_4^+-N m^{-2} d^{-1} (2.5 kg NH_4^+-N ha^{-1} d^{-1}) to both wetland systems. The hydraulic retention time was about 12 days. The loading rate was obtained by diluting anaerobic lagoon wastewater with fresh water. Wastewater characteristics were monitored on a weekly basis during the plant growth season (33 weeks), April–October 1998 (Table 1).

Eight Plexiglas soil pore water equilibrators were used to sample ammonia-N, and nitrate-N concentrations in interstitial soil water and the overlying water column. Each equilibrator had two parallel sets of 3 mL compartments spaced at 1 cm intervals, with a total of 23 compartments per set (Simon *et al.*, 1985). Once each compartment was filled with distilled-deionized water, both sides of the equilibrator were covered with a rectangular 0.2 µM Nucleopore polycarbonate membrane and sealed with a Plexiglas cover. The equilibrators were stored in plastic containers, also filled with distilled-deionized water, and bubbled with N_2 for 24 h. After the N_2 supply was disconnected, the plastic containers were covered with a lid and sealed. Then, the equilibrators were carried in containers to the field. The equilibrators were installed in WS1 and WS2 at 9, 24, 50, and 70 m from the wastewater inlet. The equilibrators were inserted in the soil leaving two sets of 8 compartments in the water column and allowed to equilibrate in the field for 14 days (July 24 to August 8, 1998). Water level in each wetland system was maintained at least 9 cm above the soil surface. Average daily water temperature varied from 23 to 26°C. Immediately after the equilibrators were taken out of the constructed wetland, the compartments were sampled with a syringe. Samples were placed in 4 mL plastic vials, acidified (1 µL of 50% H_2SO_4) to < pH 2 and transported with ice to the laboratory. Ammonium and nitrate plus nitrate (NO_{2+3}-N) were analyzed with a Technicon Auto Analyzer II using USEPA Methods 350.1, and 363.2 (USEPA, 1983).

Four soil cores were taken along each wetland system at about 9, 24, 50, and 70 m from the wastewater inlet. The soil cores were sectioned in the field at four depths at 5 cm increments. A 3 cm layer of muck was underlain by loamy sand (86% sand, 10% silt and 4% clay) in both wetland systems. Samples were transported with ice to the laboratory and pH was measured in wet samples. Air-dried samples were digested using Kjeldahl N digestion and analyzed for total nitrogen (TKN) following the procedure described by Gallaher *et al.* (1976).

The NH_4^+-N profiles were used to calculate steady-state diffusive flux according to Fick's law (Berner, 1980):

$$J_i = -\phi D_i \theta^{-2} \, dc/dz$$

Table 1 Characteristics of the swine wastewater Duplin Co., NC (April–October 1998)

Parameter	Unit	Mean	Sample number	Standard Error
Total N	mg L^{-1}	249	33	18
Ammonium-N	mg L^{-1}	225	33	19
Nitrate-N	mg L^{-1}	3	33	1
pH		8.2	33	0.1

where J_i is the flux of the dissolved species i per unit area and time; ϕ is the porosity of the soil; D_i is the difussion coeficient of species i; θ is the tortuosity factor, and dc/dz is the concentration gradient with depth. The concentration gradient was estimated by linear regression between –4 to +4 cm depth. The porosity was assumed to be close to 1 since soil bulk density was very low (0.10 g cm^{-3}). A diffusion coefficient of 19.8×10^{-6} cm^2 s^{-1} at 25°C was used for NH$_4^+$, according to Li and Gregory (1974). Analysis of variance (proc ANOVA), regression (proc REG), means and standard errors (proc MEANS) were determined using SAS software (SAS Institute, 1988).

Results and discussion

The NH$_4^+$-N pore water profiles peaked just below the sediment-water interface (0–5 cm) and decreased with depth at all sites in both WS1 and WS2 (Figures 1 and 2). These NH$_4$-N peak levels in the 0 to 5 cm layer were likely related to the simultaneous supply of N from plant uptake, soil adsorption, microbial assimilation and mineralization of sediment organic matter, and none of these supplies would have acted as the ultimate source of N. These profiles are consistent with the distribution of total soil N with depth (Table 2).

In flooded soils, conditions exist under which both nitrification and denitrification can proceed at the same time and NH$_4^+$-N levels are greatly influenced by the presence of aerobic and anaerobic soil layers (Reddy and Patrick, 1984). Reducing soil conditions found at 20 mm depth in WS1 and WS2 were consistently below 100 mV. This indicated that nitrification was likely limited and that denitrification was predominant. This limitation was previously tested by denitrification enzyme assay in order to ascertain that nitrate was the most limiting factor for denitrification in the two-wetland systems (Hunt et al., 1999). Analysis of the soil pore water showed no traces of NO$_{2+3}$–N, indicating that anaerobic conditions and limited nitrification and denitrification were prevalent in the wetland soils. Under these conditions, solution chemistry (pH and alkalinity) and environmental conditions (temperature and wind) could promote N losses via ammonia volatilization (Vlek and Stumpe, 1978).

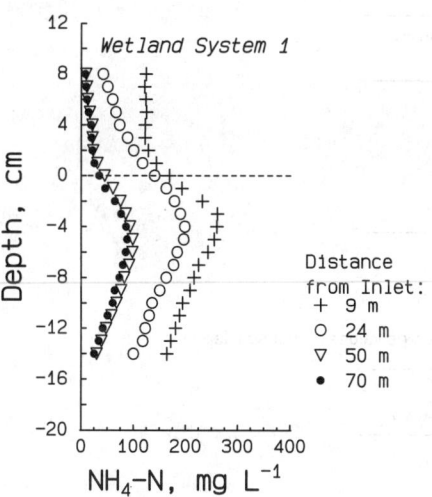

Figure 1 Mean ($n = 2$) surface water and soil pore water profiles of ammonia with depth (1-cm intervals) in Wetland System 1 (rush-bulrush plants) at four distances from the inlet

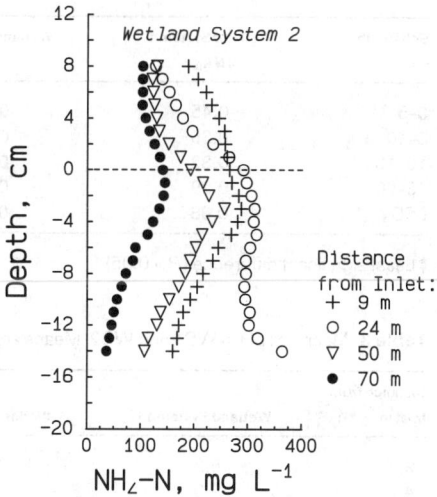

Figure 2 Mean ($n = 2$) surface water and soil pore water profiles of ammonia with depth (1-cm intervals) in Wetland System 2 (cattail-bur reed plants) at four distances from the inlet

Ammonium-N levels decreased in the interstitial soil-surface water along WS1 and WS2. Ammonium-N levels were greatest at the closest distance from the inlet (9 m) for both WS1 and WS2 (Figures 1 and 2). Concurrently, the lowest NH_4^+-N concentrations were observed at the farthest distance from the inlet (70 m) for both wetland systems. In the surface water, NH_4^+-N levels were lower in WS1 than WS2 at all sampling sites. WS1 had a 7-fold reduction in average NH_4^+-N levels (130 versus 18 mg L^{-1}) in the overlying water column between 9 and 70 m from the inlet, while WS2 had only a 2 fold reduction (237 versus 115 mg L^{-1}). The pH in surface water, from the inlet to the outlet, ranged from 7.3 to 8.2 in WS1, and from 7.4 to 7.9 units in WS2. Similar pH values were found in soils of both wetland systems (Table 3). Within these pH ranges at 25°C, the estimated amount of free NH_3 (gas) present in the wastewater is 5 to 7% of the total soluble N according to Anthonissen *et al.* (1976). Similar percentage losses of NH_3 were measured in the field by Hunt *et al.* (2000). Although a larger proportion of NH_3 could have been lost due to rising water temperature, wind speed and water turbulence, ammonia volatilization could not explain the higher N losses in WS1.

In both wetland systems, distinct gradients in the pore water NH_4^+-N profiles were found at the soil-water interface (Table 4). The diffusion flux was not calculated for WS2 sites 24 and 70 m from the inlet because NH_4^+-N was almost at equilibrium at the soil-water interface. All positive fluxes indicated that NH_4^+-N moved from the water-soil interface upward. Diffusive fluxes were higher in the WS1 than the WS2 system except at the 50 m site where the flux in WS2 was almost two orders of magnitude higher than in WS1. These NH_4^+-N flux differences between WS1 and WS2 could be explained by the fluctuation of water levels. During the 14 days of the experiment, WS2 had much higher (> 15 cm) water levels than WS1 (9 to 10 cm). This may have promoted a more effective interfacing of nitrifying and denitrifying environments in WS1. Hunt *et al.* (2000) found that denitrification potential was highest in the shallower portion of the wetland. In turn, this N loss could have increased the NH_4^+-N diffusion gradient. At the highest diffusion gradient in Table 4, NH_4^+ would diffuse at a 0.35 g m^{-2} d^{-1} rate from the water-soil interface into the surface

Table 2 Mean total soil nitrogen concentrations ($n = 4$) in Wetland Systems 1 and 2

Soil Depth cm	Wetland System 1 g N kg^{-1}	Wetland System 2
0–5	0.45	0.63
5–10	0.36	0.42
10–15	0.32	0.39
15–20	0.29	0.37
$LSD_{0.05}$*	0.08	0.12

* Least significant difference ($P > 0.05$)

Table 3 Mean soil pH in WS1 and WS2. Means are average values at four soil depths

Distance from Inlet m	Wetland System 1	Wetland System 2
9	8.2	7.7
24	7.6	7.6
50	7.3	7.4
70	7.2	7.3
$LSD_{0.05}$*	0.6	0.2

Least significant difference ($P > 0.05$)

Table 4 Ammonium flux at the soil-water interface based in concentration gradients. NS indicates gradient had non-significant regression line ($P > 0.05$)

Distance from Inlet m	Wetland System	Gradient dc/dz	R^2	Diffusive Flux mg m^{-2} d^{-1}
9	1	−20.4	0.94	+349
	2	−4.8	0.92	+83
24	1	−16.7	0.99	+287
	2	−8.9	NS	NS
50	1	−10.2	0.95	+175
	2	−16.4	0.94	+280
70	1	−9.1	0.92	+155
	2	−2.6	NS	NS

water column. This diffusion rate is about 14% of the 2.5 NH$_4^+$g m^{-2} d^{-1} application rate. Although diffusion may explain a significant portion of the material transport between the soil-water interface, it alone cannot explain the large differences in concentrations between outlet and inlet.

Conclusions

We found that NH$_4^+$ diffusion gradient and N losses were highest in the wetland system with lowest water depth. From other studies, we knew that shallower depths may have promoted a more effective interfacing of nitrifying and denitrifying environments. In turn, this N reduction in the water column may be the reason for steady NH$_4^+$ upward diffusion fluxes. The assumed mechanism for N removal has been nitrification denitrification, but ammonia volatilization could have occurred. However, solution chemistry and environmental conditions did not support the assumption that major gaseous losses occurred due to ammonia volatilization. Diffusion explained a significant portion of the material transport between the soil-water interface. However, large differences in concentrations between outlet and inlet still need to be explained by other mechanisms.

References

Anthonissen, A.C., Loher, R.C., Prakasam, T.B. and Srinath, E.G. (1976). Inhibition of nitrification by ammonia and nitrous acid. *Journal WPCF*, **48**(5), 835–852.

Berner, R.A. (1980). *Early Diagenesis: A Theoretical Approach*. Princeton Univ. Press, Princeton, NJ.

Cronk, J.K. (1996). Constructed wetlands to treat wastewater from dairy and swine operations: a review. *Agric. Ecosystems Environ.*, **58**(2/3), 97–114.

Gallaher, R.N., Weldon, C.O. and Boswell, F.C. (1976). A semi automated procedure for total nitrogen in plant and soil samples. *Soil Sci. Am. J.*, **40**(6), 887–889.

Hunt, P.G., Szögi, A.A., Humenik, F.J. and Rice, J.M. (1999). Treatment of animal wastewater in constructed wetlands. In: *Proc. of the poster presentations of the 8th International Conference on the FAO ESCORENA Network on Recycling of Agricultural, Municipal and Industrial Residues in Agriculture*. 26–29 May 1998, Rennes, France, pp. 305–313.

Hunt, P.G., Szögi, A.A., Humenik, F.J., Reddy, G.B., Poach, M.E., Sadler, E.J. and Stone, K.C. (2000). Treatment of swine wastewater in wetlands with natural and agronomic plants. In: *Proc. of the 9th International Conference on the FAO ESCORENA Network on Recycling of Agricultural, Municipal and Industrial Residues in Agriculture*. Italy (in press).

Li, Y.H. and Gregory, S. (1974). Diffusion of ions in sea water and deep-sea sediments. *Geochim. Cosmochim. Acta*, **38**(5), 703–714.

Reddy, K.R. and Patrick, W.H. (1984). Nitrogen transformations and loss in flooded soils and sediments. *CRC Crit. Rev. Env. Control*, **13**(4), 273–309.

SAS/STAT User's Guide (1988). Rel. 603. SAS Institute Inc., Cary, NC.

Simon, N.S., Kennedy, M.M. and Massoni, C.S. (1985). Evaluation and use of a diffusion-controlled sampler for determining chemical and dissolved oxygen gradients at the sediment-water interface. *Hidrobiologia*, **126**(2), 135–141.

Szögi, A.A., Hunt, P.G.and Humenik, F.J. (2000). Treatment of swine wastewater using a saturated-soil-culture soybean and flooded rice system. *Trans. ASAE*, **43**(2), 327–335.

Vlek, P.L.G. and Stumpe J.M. (1978). Effects of solution chemistry and environmental conditions on ammonia volatilization losses from aqueous systems. *Soil Sci. Soc. Am. J.* **42**(3), 416–421.

U.S. Environmental Protection Agency (1983). Methods for chemical analysis of water and wastes. EPA-600/4-79-020. Environmental Monitoring and Support Lab. Office of Research and Development, USEPA, Cincinnati, OH.

Denitrification in free water surface wetlands receiving carbon supplements

P.S. Burgoon

Water Quality Engineering Inc., 103 Palouse Street, Suite 2, Wenatchee, WA 98801, USA

Abstract Wetlands may be used as fixed film reactors for removal of NO_3-N in wastewater. Two 1 hectare free water surface wetlands were constructed for denitrification of nitrified effluent from intermittent sand filters in Connell, Washington. The wetlands were designed to remove NO_3-N from wastewater prior to land application. The design flow as 5300 m^3/d (1.4 mgd). Primary effluent from a potato processing facility was used as a carbon supplement for detnitrification. Addition of primary effluent (COD = 2800 mg/L) resulted in a COD:NO_3-N mass load ratio that ranged from 10 to 25. The total hydraulic retention time in the two wetlands varied from 1–2 days in the summer and winter. The NO_3-N load ranged from 10 to 110 kg/ha d. The NO_3-N mass removal rate ranged from 50 to 99% of the influent load. During the five months of data presented, January to May 1998, average monthly effluent NO_3-N was 1.6 mg/L; monthly average influent NO_3-N was 20.5 mg/L. An average of >85% of the NO_3-N influent load was removed. The NO_3-N removal rate coefficient (K_{20} = 358 m/yr) was higher than that measured in wetlands without carbon supplements and was independent of temperature above 12°C.

Keywords Constructed wetlands; denitrification; potato processing wastewater; nitrogen removal

Introduction

Constructed wetlands for wastewater treatment are used around the United States for treating municipal and industrial wastewater. Over the last 20–30 years there has been worldwide research, development and implementation of wetlands for all aspects of wastewater treatment (Cooper, 1999; Kadlec and Knight, 1996; Reed *et al.*, 1995).

Wetlands are naturally effective "reactors" for denitrification since they are generally anoxic, contain large amounts of surface area, and have a naturally occurring population of denitrifying bacteria. The plant denitrus and stems that accumulate in the wetlands provide high amounts of surface area for denitrification. Denitrifying cateria also reside in the wetland soil and denitrify nitrate that diffuses into the soil. In general, the critical design constraints for denitrification in wetlands are the influent mass load, carbon avaialbility, and temperature (Metcalf and Eddy, 1991).

In the absence of dissolved oxygen, the aerobic heterotrophic bacteria utilize nitrates instead of oxygen in their metabolism. The bacterial populations reduce the nitrates to nitrites and then to nitrogen gas according to the following denitrification stoichiometric equation.

$$10NO_3 + C_{10}H_{19}O_3N \rightarrow 5Nv2 + 10CO_2 + 3H_2O + NH_3 + 10OH^- \tag{1}$$

The denitrifying bacteria utilize organic compounds as electron donors in denitrification. Equation 1 shows a "model" septic tank wastewater compound as the carbon source for denitrification. Denitrifying bacteria use a variety of carbon sources. The rate of denitrification will proceed faster with the simple organic carbon compounds (Lee, 1984). Methanol is a common but expensive source of carbon. Other sources that may be used are effluent from municipal or industrial primary clarifier and acetic acid. This paper discusses the use of effluent from a primary clarifier treating potato processing wastewater.

Summary of constructed wetland/natural system design

A wastewater treatment system consisting of a series of integrated natural systems is located in the northwestern United States in the central portion of Washington State, in Connell, Washington. The integrated natural system consists of wetlands, intermittent sand filters, and a storage lagoon. The design wastewater flow is 5300 m^3/d (1.4 mgd) of primary clarifier effluent from a potato processing facility. The average influent water quality is shown in Table 1. The free water surface wetlands prior to the intermittent sand filters remove COD, TSS and some NH_4-N (Burgoon *et al.*, 1999). The intermittent sand filters nitrify the wastewater and further reduce COD and TSS. Following the sand filters are 2 free water surface wetlands designed to reduce the NO_3-N prior to storage and land application of treated wastewater. This paper will focus on the performance of the denitrifying wetlands receiving carbon supplementation via slipstream of primary effluent. The denitrifying wetlands consisted of two 1 hectare (2.4 acre) cells in series, designated DN1 and DN2 (Figure 1). The hydraulic retention time in the denitrification wetlands varied from 1 to 2 days in the summer (depth = 30 cm) and winter (depth = 45 cm).

The free water surface wetlands were constructed and planted with cattail *Typha latifolia* and bulrush *Scirpus* sp. in summer and fall 1995. Plants were acclimated to wastewater and then loading with full strength wastewater began in September of 1996. The data set presented was collected from January 25 through May 30, 1998. Samples were collected 1 to 3 times per week from the sand filter effluent, and from the effluent of each of the denitrification wetland cells. Flow meters monitored influent flow to the sand filters and the flow rate of the primary clarifier effluent used as the supplemental carbon source. Flow balances and calculation of removal rate coefficients for the wetlands do not account for evapotranspiration or precipitation.

Results and discussion
Carbon supplementation
The slipstream of primary effluent was mixed with the effluent from the intermittent sand filters prior to entering the wetlands. Monthly average concentrations of COD, TKN,

Table 1 Monthly average concentrations (mg/L) and characteristics of primary effluent

Month	T, °C	pH	COD	TKN	NH_4-N	TSS
January 98	34.5	5.5	3489	191	120	265
February	33.0	5.9	2950	169	109	Na
March	33.7	6.1	3125	191	112	Na
April	31.7	6.0	2104	104	56	Na
May 98	Na	Na	2488	152	95	Na

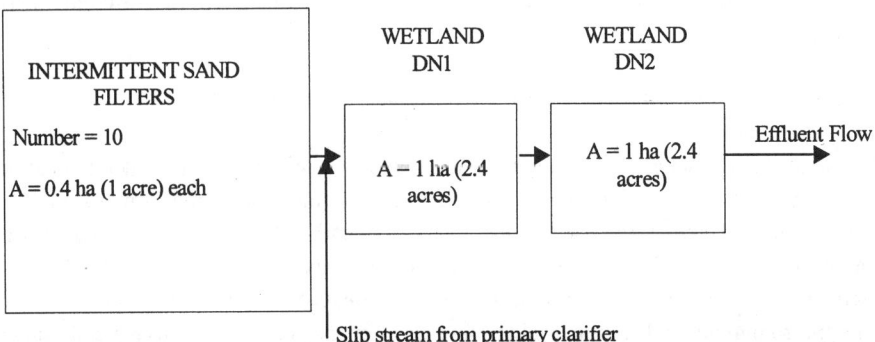

Figure 1 Basic plan view of intermittent sand filters and denitrification wetlands

NH_4-N, TSS in the slipstream are shown in Table 1. The hydraulic loads of nitrified sand filter effluent and slipstream are presented in Table 2.

The resulting COD and NO_3-N mass loads are also shown in Table 2. The COD mass load does not include COD in the nitrified effluent, since it was considered unusable by the denitrifying bacteria if it resisted degradation in the previous anaerobic and aerobic treatment processes. The COD/NO_3-N mass load ratio varied between 13 and 28. This was influenced by the influent hydraulic loads, concentration and, operator changes in the field.

Recommended COD/NO_3-N ratios for activated sludge systems may range from 2–10 depending on the carbon source (Lee, 1984; WPCF, 1983). The ratio used in the denitrifying wetlands was much higher than the ratio of 4–8 reported for subsurface flow wetlands (Gersberg et al., 1984). The COD was clearly in excess since COD mass removal was always incomplete (Table 2); also the wetland effluent COD was always greater than the sand filter effluent COD (data not shown).

Water quality

Effluent concentrations of COD, NH_4-N, NO_3-N and organic nitrogen from the wetlands are shown in Figures 2–5. The influent concentration for all constituents is flow weighed based on flow into and concentration out of the intermittent sand filters, and flow and concentration of slipstream. Addition of the primary clarifier effluent to the sand filter effluent elevated the influent concentration of all constituents except NO_3-N. The warm primary effluent increased the influent temperature in the first wetland. The temperature approaches the original temperature of the sand filter effluent when it is discharged from the second wetland (Figure 6).

The majority of the COD was removed in wetland DN 1 (Figure 2). The average mass COD removal for DN1 was 1156 kg/ha d versus 245 kg/ha d for DN2. The lower rate of removal in the second wetland may be due to a less available form of COD for the denitrifying bacteria. The effluent COD from CN2 was always greater than the COD of the sand filter effluent. This implied that the wetlands always had excess carbon for denitrification. The effluent concentration decreased as the water warmed, implying a significant effect of temperature on COD removal (Figures 2 and 6).

The majority of the NO_3-N was also removed in wetland DN 1 (Figure 3). The lower rate of removal in DN2 was due to the lower loads and may be due to a less available form of COD for the denitrifying bacteria (Figure 3 and Table 2).

The two wetlands combined, received a maximum average NO_3-N load of 58 kg/ha d in April. The loads and removal rates to DN1 were significantly higher than the total wetland area. The average mass NO_3-N removal for DN1 was 65 kg/ha d versus 18 kg/ha d for DN2.

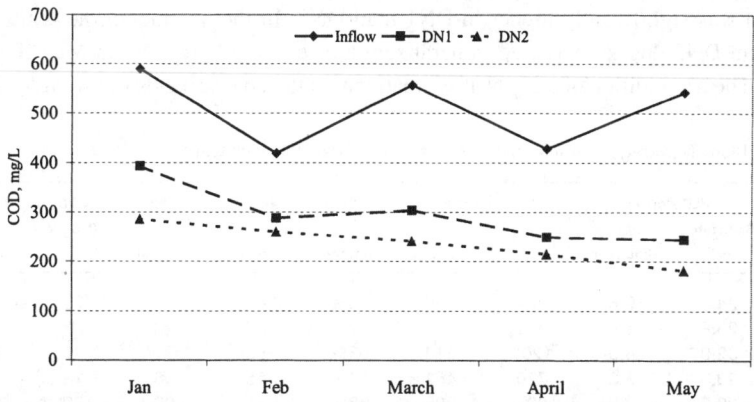

Figure 2 COD removal from denitrification wetlands

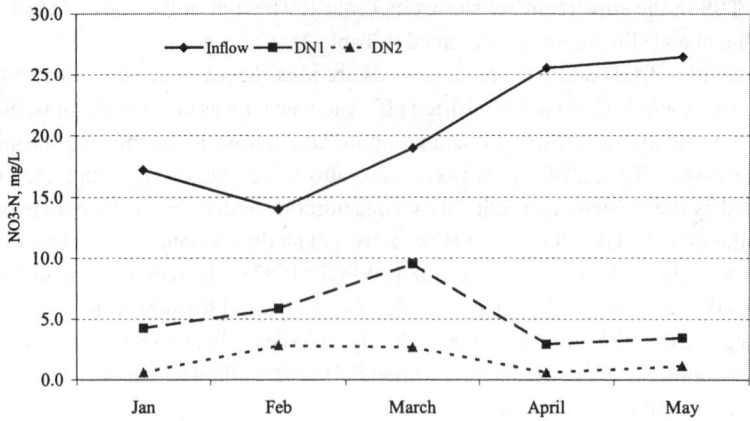

Figure 3 NO$_3$-N removal in denitrification wetlands

The weekly loads to DN1 were greater than 111 kg/ha d with a peak of 164. The peak removal rate for DN1 was 161 kg/ha d. Throughout April and May, an average of 89% of the NO$_3$-N load was removed in DN1. In comparison an average of 48% of the NO$_3$-N load to DN2 was removed during April and May.

The reduced nitrate removal in February and March, less than 61% removal (Table 2), can not be explained but may have been due to inconsistent loading of COD to the W4s. Averaging loads results in rates that appear similar but may actually have been applied inconsistently over the time period. It is not unusual to review the daily data and see very high daily loads followed by several days of no flow. Often this is beyond the control of the operator, such as when the processing plant shuts down, resulting in limited supply of carbon supplement. We assume that the performance in April and May was indicative of improved operations.

Denitrification is expected to increase pH due to alkalinity production during NO$_3$-N reduction. This was indicated by the increase in pH after passing through each denitrification wetland. The pH of the effluent from the sand filter, DN1, and DN2 averaged 6.9, 7.08, and 7.22, respectively. This pH increase may also have been due to minimeralization of the organic matter in the slipstream (Burgoon *et al.*, 1999).

Addition of the slipstream as a carbon supplement resulted in addition of organic and NH$_4$-N nitrogen. The influent concentrations are shown in Figures 4 and 5. The average increase, above the sand filter effluent, was 8 mg/L NH$_4$-N and 3 mg/L of organic nitrogen. Both forms of nitrogen were removed in the DN wetlands. During the first 4 months NH$_4$-N removal was significantly higher in DN1 than DN2. In the last month the removal was greater in DN2 due to increased mineralization of organic N in DN1 in May (Figures 4 and 5). The net addition of NH$_4$-N after treatment in the DN wetlands was 4 mg/L. The net

Table 2 Monthly average hydraulic and mass loads to the denitrification wetlands

1998	HLR, cm/d Sand filter eff.	slipstream	COD, kg/ha d load	removal	% COD removal	NO$_3$-N, load	kg/ha d removal	COD /NO$_3$-N	% NO$_3$-N removal
January	24.0	2.6	924	790	85%	41	38	23	93%
February	23.8	1.8	538	425	79%	32	20	17	61%
March	32.0	4.0	1207	813	67%	55	31	22	57%
April	23.5	3.3	720	461	64%	58	56	13	96%
May	20.5	3.5	894	785	88%	41	39	28	96%

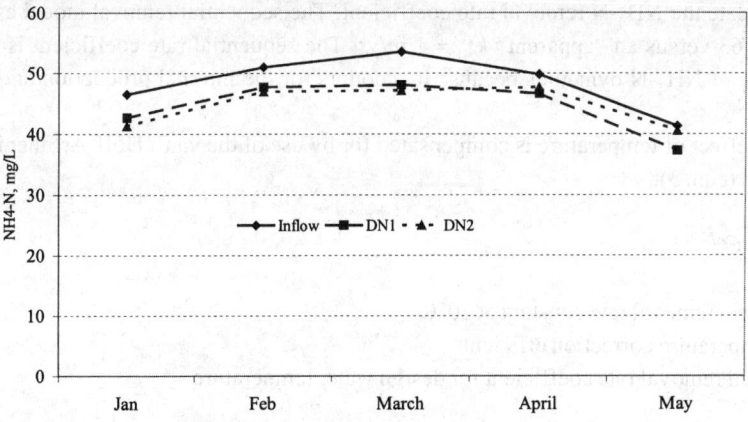

Figure 4 NH$_4$-N removal in denitrification wetlands

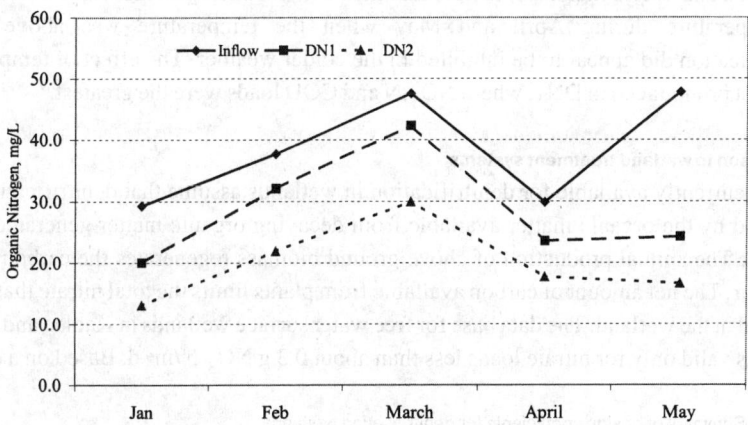

Figure 5 Removal of organic nitrogen in the denitrification wetlands

removal of NO$_3$-N averaged 22 mg/L. Therefore net N removal in the DN wetlands averaged 5.5 – NO$_3$-N removed per NH$_4$-N added in the carbon supplement.

Design models for denitrification wetlands

The first order area model, solved for effluent concentration (Kadlec and Knight, 1996) is shown below:

$$C_{out} = C^* + (C_{in} - C^*)exp(-kA/0.0365Q) \qquad (2)$$

C_{out} = expected effluent concentration
C_{in} = influent concentration
C^* = background concentration in wetland (lowest concentration achievable), 0.0 mg/L for NO$_3$-N
k = first order areal rate constant, m/yr
Q = flow rate, m^3/d

Equation 2 is used for estimation of removal rate coefficients for COD, organic nitrogen, and NO$_3$-N. Use of equation 2 for estimating removal coefficients for nitrogen species results in "apparent" removal rate coefficients. A sequential model accounting for mineralization of organic nitrogen and nitrification of NH$_4$-N (Kadlec and Knight, 1996) is used to calculate the NH_4-N removal rate coefficient. The sequential removal model estimates a $k_{20} = 63$ versus an "apparent" $k_{20} = 1$ m/yr. The sequential rate coefficient is a better estimate of NH_4-N dynamics because it accounts for the internal production and loss of NH_4-N.

The effect of temperature is compensated for by use of the van't Hoff-Arrnhenius relationship (eqn. 3).

$$k_T = k_{20} \Theta^{(T-20)} \tag{3}$$

k_{20} = areal removal rate constant at 20°C
Θ = temperature correction efficient
k_T = areal removal rate coefficient for design water temperature

Apparent removal rate coefficients for both DN1 and DN2 combined were calculated using equation 2 and 3. It is important to note that the denitrification process was not influenced by temperature during April and May when the temperature was above 12°C. Denitrification did appear to be inhibited in the colder weather. The effect of temperature was most pronounced in DN1, where NO_3-N and COD loads were the greatest.

Comparison to wetland treatment systems
Models currently available for denitrification in wetlands assume that denitrification will be fueled by the organic matter available from decaying organic matter generated in the wetland. The annual production of above ground biomass regenerates the carbon source each year. The net amount of carbon available from plants limits the total nitrate that can be removed in the wetland. The data base for free water surface wetlands in Kadlec and Knight (1996) is valid only for nitrate loads less than about 0.3 g NO_3-N/m² d. Based on a conser-

Table 3 Summary of design coefficients for denitrification wetlands

Parameter	k_{20}, m/yr	C*, m/yr	Theta
COD	150	100	1.04
Org N	159	10	1.05
NH_4-N	63	0.25	1.03
NO_3-N			
< 12°C	381	0.0	1.01
> 12°C	358	0.0	1.00

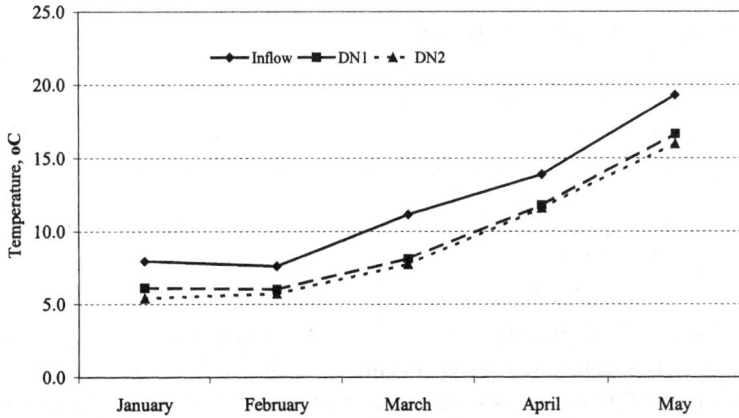

Figure 6 Temperature of water in the denitrification wetlands

vative estimate for primary productivity in a wetland (Reed et al., 1995; Wetzel, 1983) and a limiting C/N ratio of 5 required for denitrification (Ingersoll and Baker, 1998), approximately 0.3 g/m^{2}d of NO$_3$-N can be denitrified with available plant biomass.

Although there is very little information available on supplementation of wetlands with organic matter to enhance denitrification, research has shown that denitrification rates are significantly different in wetlands with and without supplemental carbon (Kadlec et al., 1997; Kadlec and Knight, 1996; Gersberg et al., 1984). Carbon may be supplemented by providing ground plant matters (Gersberg et al., 1984; Bachand, 1996), methanol or primary sewage effluent (Gersberg et al., 1984). Kadlec and Knight (1996) calculated apparent removal rate coefficients for mulched subsurface flow wetlands of 74 m/yr, and 215 m/yr for substrate flow wetlands supplemented with methanol.

Gersberg et al. (1984) achieved greater than 90% denitrification with a COD/NO$_3$-N mass loading ratio of 6.8. Lower ratios will improve net nitrogen removal and will depend on metabolic efficiency of the denitrifying bacteria using the carbon source. Other studies with industrial waste stream as carbon supplement for activated sludge have shown greater than 95% NO$_3$-N reduction with ratio less than 10 (Water Pollution Control Federation, 1983). Comparison to other systems implies that carbon supplementation to the DN wetlands was in excess.

This full-scale system has clearly shown that wetlands can be used effectively for removal of NO$_3$-N. When supplemented with exogenous carbon the removal rates are an order of magnitude higher than those seen in natural wetlands.

References

Burgoon, P.S., Kadlec, R.H., and Henderson, M. (1999). Treatment of potato processing wastewater with engineered natural systems. *Wat. Sci. Tech.* **40**(3), 211–215.

Bachand, P.A.M. (1996). Effects of managing vegetative species, hydraulic residence time, wetland age and water depth on removing nitrate from nitrified wastewater in constructed wetland marcocosms in Prado Basin, Riverside County, California. PhD Dissertation, UMI Microform. Ann Arbor, Michigan.

Cooper, P. (1999). Wetland Systems for Water Pollution Control 1998. *Wat. Sci. Technol.*, 40(3), 1–363.

Gersberg, R. M., Eklins, B. V., and Goldman, C. R. (1984). Use of artificial wetlands to remove nitrogen from wastewater. *JWPCF*, 56(2), 152–156.

Ingersoll, T.I. and Baker, L.W. (1998). Nitrate removal in wetland microcosms. *Water Research*, **32**(3), 677–684.

Kadlec, R.H., and Knight, R.L. (1996). *Treatment Wetlands.* Lewis Publishers.

Kadlec, R.H., Burgoon, P.S., and Henderson, M. (1997). Integrated natural systems for treating potato processing wastewater. *Water Sci. Technol.*, **35**(3), 262–270.

Lee, B.Y. (1984). Denitrification with Wastewater Organics. A Thesis presented to the Graduate School of the University of Florida. Master of Science.

Reed, S., Crites, R. and Middlebrooks (1995). *Natural Systems for Waste Management and Treatment.* McGraw-Hill, Inc.

Water Pollution Control Federation (1983). *Nutrient Control: Manual of Practice FD-7.* Water Pollution Control Federation, Washington, DC.

Wetzel, R.G. (1983). *Limnology.* Saunders College Publishing, Harcourt Brace Jovanovich College Publishers, New York, 767.

Removal of nutrients from combined sewer overflows and lake water in a vertical-flow constructed wetland system

L. Gervin* and H. Brix**

* Copenhagen Water, Department of Environment, Studiestræde 54, 1554 Copenhagen, Denmark
** Department of Plant Ecology, University of Aarhus, Nordlandsvej 68, 8240 Risskov, Denmark

Abstract Lake Utterslev is situated in a densely built-up area of Copenhagen, and is heavily eutrophicated from combined sewer overflows. At the same time the lake suffers from lack of water. Therefore, a 5,000 m^2 vertical flow wetland system was constructed in 1998 to reduce the phosphorus discharge from combined sewer overflows without reducing the water supply to the lake. During dry periods the constructed wetland is used to remove phosphorus from the lake water. The system is designed as a 90 m diameter circular bed with a bed depth of c. 2 m. The system is isolated from the surroundings by a polyethylene membrane. The bed medium consists of a mixture of gravel and crushed marble, which has a high binding capacity for phosphorus. The bed is located within the natural littoral zone of the lake and is planted with common reed (*Phragmites australis*). The constructed wetland is intermittently loaded with combined sewer overflow water or lake water and, after percolation through the bed medium, the water is collected in a network of drainage pipes at the bottom of the bed and pumped to the lake. The fully automated loading cycle results in alternating wet and dry periods. During the initial two years of operation, the phosphorus removal for combined sewer overflows has been consistently high (94–99% of inflow concentrations). When loaded with lake water, the phosphorus removal has been high during summer (71–97%) and lower during winter (53–75%) partly because of lower inlet concentrations. Effluent phosphorus concentrations are consistently low (0.03–0.04 mg/L). Ammonium nitrogen is nitrified in the constructed wetland, and total suspended solids and COD are generally reduced to concentrations below 5 mg/L and 25 mg/L, respectively. The study documents that a subsurface flow constructed wetland system can be designed and operated to effectively remove phosphorus and other pollutants from combined sewer overflows and eutrophicated lake water.
Keywords Combined sewer overflow; constructed wetland; nitrogen; phosphorus; stormwater; urban runoff; vertical-flow system; water treatment

Introduction

Lakes located in densely built-up areas have a high recreational value for the public. However, because of the intensive urban development and human activities in the catchment of the lakes, the water quality is often poor and reduces the recreational values. Discharge of uncontrolled urban stormwater and combined sewer overflows may be major contributors to the deterioration of the water quality. Stormwater runoff originates as runoff from parking lots, roads, roofs and other impervious surfaces, and as runoff across bare or vegetated soils. Combined sewer overflows occur during rain events when large amounts of rainwater are added to the normal domestic flows of the sewerage systems. Upgrading and extending the sewerage system in the catchment of the lakes may reduce these sources of pollution, but often this will also lead to an undesirable diversion of a major part of the freshwater source for the lakes. Another option is to clean the runoff water and the combined sewer overflows before discharge to the lake.

Lake Utterslev is an urban marsh system located in a densely built-up area about 10 km northwest of the city centre of Copenhagen. The marsh covers 91 ha and has a mean water depth of c. 1 metre. The marsh has at present no natural inflow, but water is pumped to the lakes through a channel at the west end of the marsh. In addition, the marsh receives urban runoff and combined sewer overflows. The residence time of the water in the lakes is three to thirteen months during summer. Until 1970 the lakes in Utterslev marsh received large discharges of untreated and poorly treated wastewater and, in spite of diversion and treatment of the sewage water, the lakes are heavily eutrophicated. The yearly mean phosphorus concentration of the lake water is 0.3–0.4 mg/L, and because of the large pool of nutrients accumulated in the sediments, release of phosphorus in the warm summer months periodically raises the concentrations of phosphorus in the lake water to more than 1 mg/L (Municipality of Copenhagen, 2000). Hence, there is no submerged aquatic vegetation in the lake, and water quality is generally poor because of algal blooms. The marsh is however a significant habitat for birds and other fauna, and has great recreational value for the public. Therefore, during recent decades large investments have been made in order to improve the environmental conditions in the marsh.

Constructed wetland systems have been successfully established to treat combined sewer overflows and urban stormwater (e.g. Shutes *et al.*, 1997; Green *et al.*, 1999; Scholes *et al.*, 1999). In 1998 the Municipality of Copenhagen developed and constructed a wetland system to treat the combined sewer overflow before discharge into the marsh (Seehuusen and Gervin, 1999). This was done in order to reduce the loading of phosphorus in particular to the marsh, and at the same time retain the supply of water to the marsh. Besides treating combined sewer overflows, the constructed wetland is used to clean lake water in "dry" periods when there are no sewer overflows. This paper describes the system design and operational schemes and presents results from the initial two years of operation.

Location and design of the constructed wetland

The constructed wetland is located in the natural littoral zone of one of the lakes in an existing reed swamp (Figure 1). The system is designed as a 90 m diameter circular bed partly buried in the swamp and is separated from the rest of the swamp by a polyethylene membrane (Figure 2). The membrane is placed 3 m beneath the water level of the lake on the underlying clay layer. The two metre deep bed substrate consists of a mixture of gravel and crushed marble, which has a high binding capacity for phosphorus (Brix *et al.*, 2001). The surface level of the bed medium is located one metre beneath the water level of the lake. The plant is surrounded by a dike of earth, and is planted with reeds (*Phragmites australis*). The surface area of the plant is 5,000 m^2, and the storage capacity for water in the system, including the void volume of the bed medium, is 6,500 m^3. The inlet-pipes are

Figure 1 Location of the circular constructed wetland system (CW) in Utterslev marsh

buried in a dike leading to the centre of the plant, from where the influent water is distributed over the surface of the bed (Figure 3). The system is intermittently loaded with combined sewer overflow water or lake water, and after percolation through the bed medium, the water is collected in a network of drainage pipes at the bottom of the bed and pumped to the lake.

A buried in-line retention basin built into the sewerage system is located immediately upstream of the constructed wetland system. The basin withholds the "first flush" of the combined sewer overflow, and therefore the water going into the wetland during overflow events is rather dilute. During periods with no overflow events, the system is used to clean lake water. The fully automated loading cycles result in alternating wet and dry periods.

Several built-in attributes of the system contribute to enhance the nutrient removal performance. The bed medium is relatively deep (2 m) and contains crushed marble, which has a high binding capacity for phosphorus (Brix *et al.*, 2001). The vertical-flow regimen secures a good contact between the water and bed medium, and the intermittent loading creates alternating wet and dry periods which increase oxidation of organic compounds and nitrification in the bed medium.

The routine operation of the system is automated via data acquisition and real time control. The control is based on flow transmitters measuring influent and effluent flow rates, and level transmitters placed in the retention basin and in the constructed wetland itself. Furthermore, on-line monitoring equipment is installed in the system to measure oxygen, temperature and pH in the inlet and the outlet water, in different depths of the free water on the surface of the bed during loading periods, as well as in different depths within the bed medium. Automatic samplers are installed at the inlet and outlet of the bed taking samples proportional to volume flow.

Treatment performance of the system

The initial period of operation has been a running-in period, where the focus has been to study the effects of residence time on removal performance. In the period from October 1998 to January 1999 only lake water with low contents of nutrients were treated in the system as there were no overflows from the sewer system until June 1999. In the period from October 1998 to the first of January 2000 a total of 125,000 m^3 of water were treated in the system (115,000 m^3 in 1999). The inflows consisted of c. 50 loadings of lake water (42 loadings in 1999), and 16 loadings of sewer overflows. The performance when treating lake water was tested by analysis of 24 in- and outlet samples. The performance when treating sewer overflows was tested by analysis of 12 inlet and 16 outlet samples. At two of the overflow events samples were taken of the outlet water after 3, 5 and 7 days, respectively, to

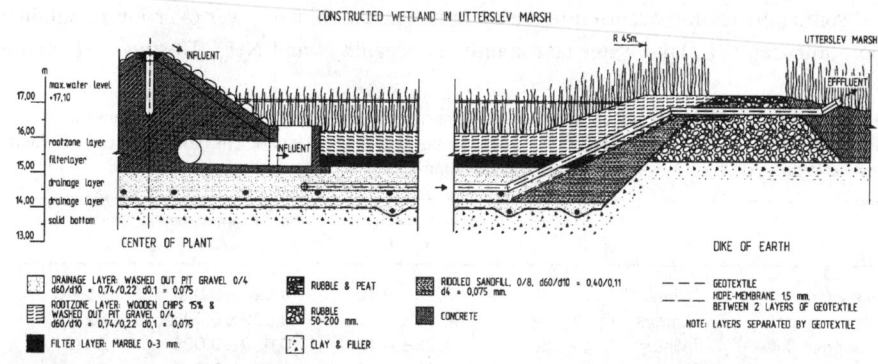

Figure 2 Cross-section through the circular constructed wetland system at Utterslev marsh

Figure 3 Photograph of the constructed wetland system at Utterslev marsh showing a boardwalk leading to the center of the circular bed

evaluate the effects of residence time. All samples were analysed using standard methods by a certified laboratory.

Phosphorus

Inlet concentrations of phosphorus in the lake water varied between 0.08 and 0.37 mg/L during winter, but were higher (0.12 to 0.78 mg/L) during summer because of release of phosphorus from the lake sediments during anoxic conditions in the lake. However, effluent concentrations from the constructed wetland system were consistently low (0.03–0.04 mg/L) resulting in high removal rates that were largely independent of inlet concentration (Table 1). Concentrations of phosphorus in the sewer overflows were higher (0.9 to 3.2 mg/L) than concentrations in lake water, but effluent concentrations were still low, and removal efficiencies greater than 97%. The retention times studied (3, 5 and 7 days) did not have any consistent effect on effluent phosphorus concentration. A total mass balance showed that 86 kg of phosphorus, corresponding to 96% of the loading, were removed by the constructed wetland system during 1999.

Nitrogen

The concentrations of nitrogen in the lake water were generally low and occurred nearly exclusively as organic nitrogen (Table 2). The total nitrogen content was reduced by c. 25% and some nitrification occurred in the system. The combined sewer overflows contained more nitrogen than lake water and mainly as organic-N and NH_4. This was effectively

Table 1 Average (±SD) inlet and outlet concentrations of total-phosphorus and percentage removal (% of inlet concentration) in the constructed wetland at Utterslev marsh when loaded with lake water or combined sewer overflows (Winter: November–May; summer: June–October)

		n	Inlet concentration (mg/L)	Outlet concentration (mg/L)	Removal (%)
Lake water	Winter	7	0.15 ± 0.10	0.048 ± 0.020	65 ± 8
	Summer	17	0.53 ± 0.18	0.032 ± 0.006	92 ± 6
Sewer overflow	Winter	2	1.27 ± 0.13	0.029 ± 0.005	98 ± 1
	Summer	10173 & 14	1.79 ± 0.83	0.041 ± 0.013	97 ± 1

Table 2 Average (±SD) inlet and outlet concentrations of total-nitrogen (Total-N), ammonium-nitrogen (NH_4-N) and nitrite + nitrate-nitrogen (NO_{2+3}-N) and percentage removal (% of inlet concentration) in the constructed wetland at Utterslev marsh when loaded with lake water or combined sewer overflows

		n	Inlet conc. (mg/L)	Outlet conc. (mg/L)	Removal (%)
Lake water	Total-N	21	1.64 ± 0.63	1.19 ± 0.55	25 ± 27
	NH_4-N	20	0.03 ± 0.05	0.03 ± 0.03	–
	NO_{2+3}-N	20	0.01 ± 0.02	0.83 ± 0.49	–
Sewer overflow	Total-N	16	7.26 ± 3.69	1.08 ± 0.65	81 ± 12
	NH_4-N	15	2.74 ± 0.79	0.02 ± 0.03	99 ± 1
	NO_{2+3}-N	15	0.58 ± 0.19	0.71 ± 0.71	–

removed in the system as a consequence of nitrification-denitrification processes (Table 2). The denitrification rate was higher when treating overflows compared to lake water, probably because of the higher nitrogen contents in the sewer overflow. However, the higher content of organic matter, and the degradability of the organic matter may also have affected the denitrification process. It was not possible to observe a significant relation between the residence time and the denitrification rate for any of the types of water. The effluent concentrations of ammonium were consistently low, and as a consequence of the nitrification process in the system some nitrate was present in the effluent.

COD and TSS

The COD (Chemical Oxygen Demand) concentrations in the lake water were very low in winter, but increased during summer as a consequence of the presence of algae (Table 3). This could also be seen in the contents of TSS (Total Suspended Solids), which varied between <5 mg/L during winter to 48 mg/L in August (yearly average 13 mg/L). The COD contents in the sewer overflows were consistently higher (up to 294 mg/L on one occasion) as were the contents of TSS (average 90 mg/L). Effluent concentrations of COD were generally low, but were higher during summer than during winter. Effluent concentrations of TSS were below the detection limit (5 mg/L) in 22 out of 24 samples.

Discussion

The present paper describes the initial experiences from the constructed wetland system. The initial year has been an experimental period to document performance and to experiment with loading practice, retention time, etc. Therefore, the constructed wetland system has not been loaded according to its expected capacity. In 1999 a total of 115,000 m³ of water was treated in the constructed wetland, of which 38,000 m³ was combined sewer overflows, and during summer (May to October) 75,000 m³ of water was treated. A total of 86 kg of phosphorus was removed by the constructed wetland system of which two thirds were removed from the combined sewer overflows.

It is anticipated that the total capacity of the constructed wetland will be 180,000 m³ per year, of which 135,000 m³ will be treated during the summer period. The summer period is

Table 3 Average (±SD) inlet and outlet concentrations of COD and percentage removal (% of inlet concentration) in the constructed wetland at Utterslev marsh when loaded with lake water or combined sewer overflows (Winter: November–May; summer: June–October)

		n	Inlet conc. (mg/L)	Outlet conc. (mg/L)	Removal (%)
Lake water	Winter	2	9 ± 3	2 ± 1	68 ± 16
	Summer	12	59 ± 8	25 ± 12	58 ± 19
Sewer overflow	Winter	2	116 ± 20	11 ± 8	90 ± 8
	Summer	9 & 13	99 ± 77	27 ± 9	63 ± 23

relatively dry and therefore only a few sewer overflows are expected to occur. Therefore, the constructed wetland will mainly be treating lake water during summer. It is estimated that 105,000 m^3 out of the expected 135,000 m^3 treated during summer will be lake water. Hence, the constructed wetland will mainly be used to treat lake water when the phosphorus levels in the lake water are high as a consequence of release of phosphorus from the lake sediments. It is anticipated that in the future a total of 130 kg of phosphorus will be removed per year from sewer overflows and lake water.

The constructed wetland system clearly reduces the discharge of phosphorus and other pollutants contained in the sewer overflows to the Utterslev marsh. Furthermore, the constructed wetland removes phosphorus from the lake water when no overflows occur. Therefore, the establishment of the wetland no doubt contributes to improve the environmental conditions of Utterslev marsh. However, additional measures must be taken to restore the former environmental quality of the marsh. The heavily polluted sediments of the lakes, which release large amounts of phosphorus every summer, are of prime concern (Municipality of Copenhagen, 2000). However, a few minor combined sewer overflows and stormwater discharges located around Utterslev marsh also contribute in sustaining the poor water quality of the lakes.

The initial experiences with the constructed wetland at Utterslev marsh document that a constructed wetland can be designed and operated to effectively remove phosphorus and other pollutants from polluted water. The lifetime of the system is not known, but estimates based on the phosphorus sorption capacity of the bed medium of the constructed wetland indicate that the phosphorus removal can be sustained for several decades.

Acknowledgements
We thank Morten Seehuusen, construction engineer of the wetland system, Bent Petersen and Jørgen Kristensen, members of the construction project group, and Gert Jørgensen and Poul Erik Jøhncke, service staff of the wetland system. Jacob Harboe is thanked for preparing Figure 2 and Dr Brian K. Sorrell for linguistic improvements of the manuscript.

References
Brix, H., Arias, C.A. and del Bubba, M. (2001). Media selection for sustainable phosphorus removal in subsurface-flow constructed wetlands. *Wat. Sci. Tech.* **44**(11–12) 46–53 (this issue).
Green, M.B., Martin, J.R. and Griffin, P. (1999). Treatment of combined sewer overflows at small wastewater treatment works by constructed reed beds. *Wat. Sci. Tech.* **40**(3), 357–364.
Municipality of Copenhagen (2000). Lake Utterslev, 1999. Aquatic Environment Nationwide Monitoring Programme, NOVA-2003 (in Danish).
Scholes, L.N.L., Shutes, R.B.E., Revitt, D.M., Purchase, D. and Forshaw, M. (1999). The removal of urban pollutants by constructed wetlands during wet weather. *Wat. Sci. Tech.* **40**(3), 333–340.
Seehuusen, M. and Gervin, L. (1999). Europas største rodzoneanlæg. Stads- og Havneingeniøren 2000(1), 4–9 (in Danish).
Shutes, R.B.E., Revitt, D.M., Mungur, A.S. and Scholes, L.N.L. (1997). The design of wetland systems for the treatment of urban run off. *Wat. Sci. Tech.* **35**(5), 19–25.

Microbial indicator removal in onsite constructed wetlands for wastewater treatment in the southeastern U.S.

E.C. Barrett*, M.D. Sobsey**, C.H. House*** and K.D. White****

* 1308 Aggie Lane, Austin, TX 78757, USA (Research performed at University of North Carolina at Chapel Hill) (Email: ebarrett@texas.net)
** University of North Carolina at Chapel Hill, Department of Environmental Science and Engineering, Chapel Hill, NC 27599-7400, USA
*** NC State, Department of Forestry, Raleigh, NC 27695-8008, USA
**** University of South Alabama, Department of Civil Engineering, Mobile, AL 36688, USA

Abstract Seven onsite constructed wetlands for wastewater treatment in the coastal plains of Alabama and North Carolina were studied from September 1997 to July 1998. Each site was examined for its ability to remove a range of fecal contamination indicators from settled wastewater. Indicator organisms include total and fecal coliforms, enterococci, *Clostridium perfringens*, and somatic and male-specific (F+) coliphages. Four identical domestic wastewater treatment sites in Alabama were evaluated. In these sites the Log_{10} geometric mean reductions ranged between 0.5 and 2.6 for total and fecal coliforms, 0.1 and 1.5 for enterococci, 1.2 to 2.7 for *C. perfringens*, −0.3 and 1.2 for somatic coliphages, and −0.2 and 2.2 for F+ coliphages. Three unique designs were examined in North Carolina. Log_{10} geometric mean reductions ranged between 0.8 and 4.2 for total and fecal coliforms, 0.3 to 2.9 for enterococci, 1.6 to 2.9 for *C. perfringens*, −0.2 and 2.8 for somatic coliphages, and −0.1 and 1.5 for F+ coliphages. Somatic and F+ coliphage detection was highly variable from month to month.
Keywords *C. perfringens*; coliforms; coliphages; constructed wetlands; enterococci; wastewater

Introduction

In the U.S. today, 75 to 90 million people are not serviced by municipal sewage treatment systems (Moeller, 1997). These individuals must use onsite wastewater treatment to control nutrient and pathogen contamination of groundwater and surface water. Usually the onsite treatment of choice is a septic tank with a drain field; however, this technology is not appropriate for certain soils and in regions with a high water table. In fact, one quarter of soils in the U.S. are not suitable for drain field use (U.S. EPA, 1980). One inexpensive and attractive alternative to septic systems is the constructed wetland. A broad range of designs for these systems have been studied for removal of indicator bacteria, especially total and fecal coliforms, which are usually reduced by at least 1 log (Kadlec and Knight, 1996). Viral indicator organisms like bacteriophages have been shown to be removed by as much as 34% to 99.9% (Gersberg *et al.*, 1989; Karpiscak *et al.*, 1996). Protozoan organisms like *Giardia lamblia* have been shown to be removed by as much as 99.9% (Quiñónez *et al.*, 1997). However, there is little research to profile the treatment of constructed wetland onsite wastewater treatment in "real-life" nonmunicipal field situations. In this study, seven constructed wetlands were studied for their ability to remove four bacterial indicator species and two viral indicator species, including total and fecal coliforms, enterococci, *C. perfringens*, and somatic and male-specific (F+) coliphages.

Total and fecal coliforms are the indicator organisms typically used by regulators to assess water quality. They have come under increasing criticism because it is widely accepted that protozoan cysts and viruses are hardier in treatment processes than these organisms. Enterococci is one alternative indicator group that has been shown to correlate with gastroenteritis in swimmers and is used by EPA as a bacterial criterion for the acceptability of swimming waters (Cabelli et al. 1983). *C. perfringens* is another bacteria that forms spores and may better model how protozoan cyst formers like *Giardia* and *Cryptosporidium* are treated. Somatic and F+ (male-specific) coliphages have been suggested as viral indicators. Somatic coliphages are numerous in wastewater. Although they vary as a group and are shaped differently than most disease-causing viruses, they are easier to detect than the F+ coliphages. The F+ coliphages have a size and shape similar to disease-causing organisms but are not found in the majority of the population's waste. They may therefore be difficult to detect in small flow systems like septic systems.

Methods

All wastewater in this study had been settled (using a septic tank) before wetland treatment.

Site designs

The four constructed wetland systems in Alabama treat domestic wastewater from individual residences, are of the subsurface horizontal flow type, and are based on design criteria published by the TVA (Steiner and Watson, 1993) with minor modifications. The systems consist of two 0.3-m-deep basins containing gravel as a substrate for emergent vegetation. The first gravel basin is lined and designed for treatment. The second gravel basin is unlined and is designed to allow for infiltration. The treatment cell area is about 33 m^2 with a design flow of 1.7 m^3/day. Hydraulic retention time (HRT) in the treatment cell is about 2.25 days and the hydraulic loading rate (HLR) is 51.5 L/m^2/day.

The NC School Site serves a population of approximately 350 people. The average measured flow is 15,140 L/day. The wastewater is pressure dosed into the system three to four times per day. The wastewater is first treated by two parallel 9.1 m × 13.7 m × 0.6 m vertical flow cells filled with a mixture of coarse sand, gravel, and crushed brick, which is planted with *Juncus effusus*. The HLR for the vertical flow cells alone is 101 L/m^2/day and the HRT is approximately 3 days. The water is collected from the bottom of the vertical flow cells and half is pumped back to the septic tank for recirculation through the vertical cells. The other half is treated by one of two 9.8 m × 13.7 m × 0.6 m horizontal flow cells with a 5 to 7-day HRT. The horizontal cells have a substrate of coarse sand and gravel and are planted with *Scirpus validus*. After passing through the horizontal flow cells, the water is disinfected by an ultraviolet chamber and released as discharge (House et al., 1997).

The NC Ranger Station Site serves a ranger station with five to six employees plus a ranger's quarters and a day care. It is a subsurface horizontal flow system with an average measured flow of 1,514 L/day. A cell with a gravel substrate treats half of the flow (757 L/day). The HLR of the gravel cell is 31.2 L/m^2/day and the HRT is 3 days. A sand-filled cell with a 3-day HRT and a HLR of 27 L/m^2/day treats the other half of the flow. The effluent from both cells is combined and treated by conventional tile field treatment.

The NC Home Site is another subsurface horizontal wetland that treats wastewater from a three-bedroom home (1,363 L/day). The wastewater is split and half is routed to a 4.6 m × 6.1 m × 0.6 m cell with a 7.4 day HRT, and the other half is routed to a cell with a 4.6 m × 6.1 m × 0.3 m cell with a 3.4 day HRT. Both cells have a gravel substrate and are planted with a combination of shrubs and emergent wetland vegetation. The effluent from the two cells is spray-irrigated in a wooded/brushy area behind the home (Spooner et al., 1998).

Microbiological techniques

Grab samples for microbiological analysis were taken every 1 to 2 months from September 1997 to July 1998. All bacteria were enumerated from water samples by membrane filtration using 47 mm Gelman GN-6 (0.45 µm) membranes. Total and fecal coliform techniques have been described previously (APHA, 1995). A modification of the membrane filtration method developed by Levin et al. (1975) was used for enterococci isolation. This modification amounted to the addition of 75 mg of indoxyl β D glucoside per 100 mL of mCp agar. C. perfringens was isolated by the technique developed by Bisson and Cabelli (1979).

Both kinds of virus were enumerated with the single-agar assay described by Grabow and Coubrough (1986). The bacterial host used for the somatic coliphage assay was E. coli CN-13. The bacterial hosts used for the male-specific assay were Salmonella typhimurium WG-49 (September to January) and E. coli F-Amp (February to June). E. coli F-Amp was selected for the remainder of the study due to less variability in host quality. In June the adsorption elution method described by Sobsey et al. (1990) was used for F+ coliphage isolation in order to analyze larger volumes. Here water was filtered through a 0.45 µm filter and eluted with 3% beef extract. The effluent was plated just as the sample water was for the single agar layer.

Results and discussion

A Wilcoxon Signed Rank Test (a nonparametric, paired test) was used to determine the significance of Log_{10} geometric mean reductions of bacterial indicators with the following results: (1) 73% (35 out of 48) were significant at p ≤ 0.05, (2) 14% were marginally significant with a p between 0.06 and 0.1, and (3) 12% were not significant (p > 0.1). Most of the systems/cells profiled provide significant reductions of bacterial fecal contamination indicators.

Overall the NC School Site provides the best bacterial treatment (Figure 1). This may be due to the redirecting of some of the water through the vertical cell twice, the combining of two types of flow (vertical and horizontal), and the comparatively long retention time (5–7 days). Mann-Whitney tests showed that reductions of total and fecal coliforms were significantly greater when horizontal cells were added to the vertical cells, while the addition did not influence the treatment of the other indicators. The improved treatment provided by the horizontal cells may be due to the increased HRT from 3 days in the vertical cells to 6 days total in the combined system.

The AL sites were not sampled as often as the NC sites. This reduces the statistical power of the data and may explain in part why there are more marginally significant and nonsignificant reductions at the AL sites (Figure 2). It is assumed that wastewater is treated to some extent in the cells designed for infiltration; however, this treatment was not measured in this study.

The somatic coliphage detection at the AL sites was highly variable. At AL Site 1, phage was detected in October, March, and July with Log_{10} reductions of 0.51, 0.13, and 0.11 respectively. At AL Site 2, somatic coliphage was detected in February, March, May, and July with Log_{10} reductions of ≥2.8, ≥1.3, ≥ 0.3, and ≥0.3 respectively. No coliphage was detected in AL Site 3 in the influent or the effluent. At AL Site 4, there were more coliphage in the treated water leaving the wetland (effluent) than in the wastewater entering the wetland (influent) in September and October. The increase was 1 Log_{10} and 0.5 Log_{10} respectively. F+ coliphage detection in the AL sites was also highly variable (Table 1). At AL Site 1 in February, reduction was 0.75 Log_{10}. AL Site 2 results were more reliable with 5 months of data and average reduction of ≥2.2 Log_{10}. Wetlands 3 and 4 also demonstrated high variability (Table 1) with negative average reductions of –0.2 and –1.1 Log_{10} (Figure 2). These negative reductions indicate that coliphage may be accumulating in the wetland substrate, only to be flushed out when the water chemistry changes.

Somatic coliphage detection in the NC sites was more reliable than in the Alabama sites (Figures 1–3). The F+ coliphage detection, however, was highly variable (Table 1). F+

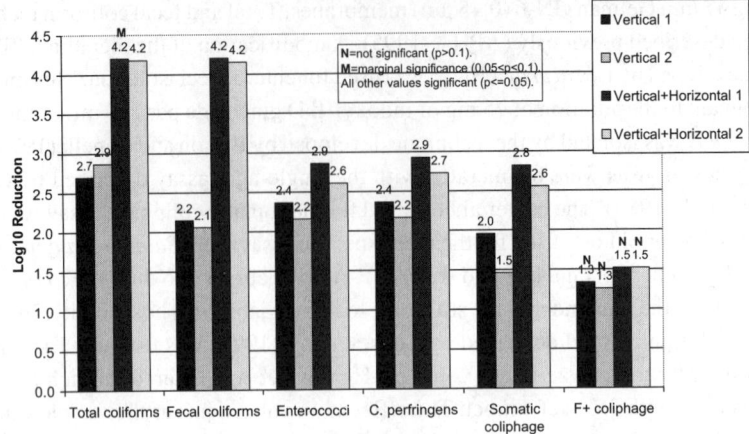

Figure 1 Log$_{10}$ Geometric Mean Indicator Reductions in the NC School Site (Vertical/Horizontal Flow Combination Design)

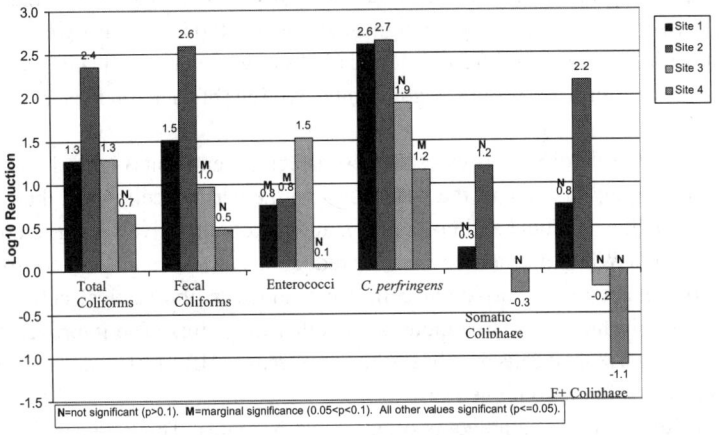

Figure 2 Log$_{10}$ Geometric Mean Indicator Reductions in the Alabama Sites (Horizontal Subsurface Flow Design)

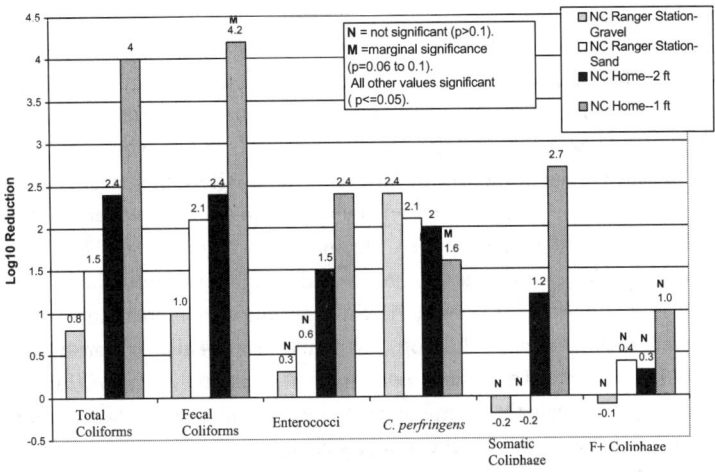

Figure 3 Log$_{10}$ Geometric Mean Bacterial Reductions of the NC Ranger Station and Home Sites (Horizontal Subsurface Flow Designs)

Table 1 Log 10 Concentrations of F+ Coliphage per 100 ml

Site	Sept	Oct	Nov	Jan	Feb	Mar	Apr	May	June	July
NC School-Influent	ND	ND	>=3.00	ND	ND	0.29	ND	NS	0.78	NS
NC School-Vertical 1	ND	ND	<=0.22	ND	ND	0.85	ND	NS	<=-1.00	NS
NC School-Vertical 2	ND	ND	1.30	ND	ND	0.00	ND	NS	<=-1.00	NS
NC School-Horizontal 1	ND	ND	<=0.22	ND	ND	0.30	ND	NS	<=-1.00	NS
NC School-Horizontal 2	ND	ND	<=0.22	ND	ND	0.30	ND	NS	<=-1.00	NS
NC Ranger Station-Influent	ND	ND	>=3.00	<=0.39	<=0.10	ND	<=0.00	NS	1.07	ND
NC Ranger Station-Gravel	ND	ND	2.64	0.40	1.66	ND	0.30	NS	-0.18	ND
NC Ranger Station-Sand	ND	ND	1.12	1.26	<=0.00	ND	0.30	NS	<=-0.18	ND
NC Home-Influent	NS	ND	ND	ND	0.00	<=-0.01	<=0.00	NS	NS	3.82
NC Home-2 ft	NS	ND	ND	ND	<=0.1	<=0.00	1.54	NS	NS	0.90
NC Home-1 ft	NS	ND	ND	ND	<=0.1	0.00	0.30	NS	NS	-0.70
AL 1-Influent	ND	ND	NS	ND	1.65	ND	NS	ND	NS	ND
AL 1-Effluent	ND	ND	NS	ND	0.90	ND	NS	ND	NS	ND
AL 2-Influent	3.10	ND	NS	ND	2.63	2.35	NS	2.90	NS	3.36
AL 2-Effluent	<=1.15	ND	NS	ND	1.54	0.48	NS	<=0.00	NS	0.00
AL 3-Influent	ND	ND	NS	ND	0.60	0.00	NS	ND	NS	ND
AL 3-Effluent	ND	ND	NS	ND	1.04	0.00	NS	ND	NS	ND
AL 4-Influent	ND	ND	NS	ND	0.60	ND	NS	ND	NS	ND
AL 4-Effluent	ND	ND	NS	ND	1.68	ND	NS	ND	NS	ND

ND = No Data; NS = No Sample Taken

values from the NC School Site indicate a Log_{10} reduction of ≥2.8 and ≥1.7 during November for vertical flow cells 1 and 2, respectively.

Both horizontal flow cell values indicate a Log_{10} reduction of ≥2.8. March results suggest a pulsing effect with greater numbers of plaque-forming units (pfus) in the wetland effluent than the wastewater influent. June values indicate ≥1.8 Log_{10} reductions for each vertical and horizontal cell.

At the NC Ranger Station Site, November showed F+ coliphage Log_{10} reductions of ≥0.4 for the Gravel cell and ≥1.9 for the Sand cell. In June, Log_{10} reductions were 1.3 for the Gravel cell and ≥1.3 for the Sand cell. January, February, and April all exhibit the above mentioned pulsing (Table 1). The NC Home Site also exhibited pulsing in April, with more pfus in the effluent than the influent.

The high variability in coliphage detection is assumed to be a combination of infrequent sampling (sampled every 1–2 months) and low-volume sampling (except for June the sample volume was usually 100 ml). Weekly or biweekly sampling should be conducted at on onsite constructed wetlands so "pulsing" from storm events can be better predicted and documented.

Conclusions

Onsite constructed wetlands are capable of significantly reducing total and fecal coliforms, enterococci, *C. perfringens*, and somatic and F+ coliphages in septic wastewater effluent. The addition of the horizontal cells in series with the vertical cells significantly improved the reductions of total and fecal coliforms compared to the vertical cells alone. Negative reduction values for somatic and F+ coliphages indicate that wetlands may be accumulating and retaining viruses, and later releasing them into the wastewater. Coliphages are likely desorbing from wetland cell substrate due to a change in the chemistry or hydraulics of the system. More field studies on constructed wetlands for residential wastewater treatment are needed to determine design parameters that can improve microbial indicator and pathogen treatment.

References

Bisson, J.W. and Cabelli, V.J. (1979). Membrane filter method for *Clostridium perfringens*. *Appl. Environ. Microbiol.*, **37**(1), 55–66.

Cabelli, V.J., Dufour, A.P. and McCabe, L.J. (1983). A marine recreational water quality criterion consistent with indicator concepts and risk analysis. *J. Water Poll. Con. Fed.* **55**, 1306–1314.

Gersberg, R.M., Gearheart, R.A. and Ives, M. (1989). Pathogen removal in constructed wetlands. In: *Constructed Wetlands for Wastewater Treatment*, D.A. Hammer (ed.), Lewis Publishers, Chelsea, Michigan, pp. 431–444.

Grabow, W.O.K. and Coubrough, P. (1986). Practical direct plaque assay for coliphage in 100-ml samples of drinking water. *Appl. Environ. Microbiol.* **52**, 430–433.

House, C.H., Broome, S.W., Frederick, D.J. and Rubin, A.R. (1997). Vertical flow-horizontal flow constructed wetlands combined treatment system design and performance. In: *Proceedings of the 70th Annual Water Environment Federation Conference and Exposition*, October 21, 1997, Chicago, Illinois.

Kadlec, R.H. and Knight, R.L. (1996). *Treatment Wetlands*. Lewis Publishers, Boca Raton.

Karpiscak, M.M., Gerba, C.P., Watt, P.M., Foster, K.E. and Falabi, J.A. (1996). Multi-species plant systems for wastewater quality improvements and habitat enhancement. *Wat. Sci. Tech.* **33**(10–11), 231–236.

Levin, M.A., Fischer, J.R. and Cabelli, V.J. (1975). Membrane filter technique for enumeration of enterococci in marine waters. *Applied Microbiology* **30**, 66–71.

Moeller, D.W. (1997). *Environmental Health*. Harvard University Press, Cambridge, Massachusetts.

Quiñónez, M.J., Karpiscak, M. and Gerba, C.P. (1997). Constructed wetlands for wastewater quality improvement. In: *Proceedings of the 97th General Meeting of the American Society for Microbiology*, May 4–8, 1997, Miami Beach, Florida.

Sobsey, M.D., Schwab, K.J. and Handzel, T.R. (1990). A simple membrane filter method to concentrate and enumerate male-specific RNA coliphages. *J. Am. Water Works Assoc.* **82**, 52–59.

Spooner, J., House, C.H., Hoover, M.T., Rubin, A.R., Silverthorne, R., Steinbeck, S.J., Harris, V., Uebler, R.L. and Martin, B. (1998). Long term performance evaluation of innovative and alternative onsite wastewater systems for four soil resource groups in Craven County, North Carolina. In: *Proceedings of the 8th National Symposium on Individual and Small Community Sewage Systems*, March 8–10, 1998, Orlando, Florida.

Standard Methods for the Examination of Water and Wastewater (1995). 19th edn, American Public Health Association/American Water Works Association/Water Environment Federation, Washington, DC, USA.

Steiner, G.R. and Watson, J.T. (1993). *General Design, Construction, and Operation Guidelines: Constructed Wetlands Wastewater Treatment Systems for Small Users including Individual Residences*, 2nd Edition, TVA/WM-93/10.

U.S. Environmental Protection Agency (1980). *Design Manual: Onsite Wastewater Treatment and Disposal Systems*. EPA 625/1-80-012.

Removal of bacterial indicators and pathogens from dairy wastewater by a multi-component treatment system

M.M. Karpiscak*, L.R. Sanchez**, R.J. Freitas*** and C.P. Gerba**

* Office of Arid Lands Studies, University of Arizona, 1995 East 6th St., Tucson Arizona 85719 USA. (E-mail: karpisca@ag.arizona.edu)
** Department of Soil, Water and Environmental Science, University of Arizona, Shantz Bldg. 429, Tucson Arizona 85721 USA. (E-mail: sanchez–none; gerba@ag.arizona.edu)
*** Department of Agricultural Biosystems Engineering, University of Arizona, Shantz Bldg. 403, Tucson Arizona 85721 USA. (E-mail: bobf@ag.arizona.edu)

Abstract Microbial removal by a multi-component treatment system for dairy and municipal wastewater is being studied in Arizona, USA. The system consists of paired solids separators, anaerobic lagoons, aerobic ponds and constructed wetlands cells. The organisms under study include: total coliform, fecal coliform, enterovirus, *Listeria monocytogenes*, *Clostridium perfringens*, coliphage, *Giardia lamblia* and *Cryptosporidium parvum*. Organism removal rates from dairy wastewater varied from 13.2 per cent for fecal coliform to 94.9 per cent for coliphage. It appears that the much higher turbidity of the dairy wastewater, nearly 1,300 NTU, decreased the treatment systems' ability to remove some microbial indicators and pathogens. Information from this study can be used to determine the adequacy of multi-component treatment systems for the control of wastewater-borne pathogens, both in municipal treatment systems as well as in confined animal feeding operations (CAFO). This information also can assist municipalities and the CAFO industry in the implementation of rational and efficient treatment strategies for appropriate reuse of wastewaters.
Keywords Constructed wetlands; dairy wastewater; indicator bacteria; pathogens

Introduction

Discharges from dairies are of increasing concern because the discharges carry waste materials, potentially affecting public health and causing other environmental problems (Hollon *et al.*, 1982). The wastewater originates from washing of equipment, leaks and spillage, processing losses, deliberate waste of low value products, sanitizers, lubricants, chemicals from boilers, and cow washing (Bull *et al.*, 1981; Viraraghavan and Kikkeri, 1990). Manure and dirt tracked into the milk room or washed from worker's boots and utensils are added to the wastewater (Graves, 1987). Dairy wastewater typically is more concentrated than domestic sewage because of the lactose, fats and proteins originating from the milk carbohydrates (Fang,1990; Perle *et al.*, 1995).

Manure from domestic animals is produced nationally at a rate of approximately 1.8×10^{12} kg/yr (Ross *et al.*, 1978). The number of liquid manure handling systems being used continues to increase, especially on dairy and swine operations, because of the convenience, the availability of equipment to handle large volumes of material and the ability to apply the manure immediately in the fields. Liquid manures traditionally are spread on top of the soil or cover crop as a method of fertilization (Ross *et al.*, 1978). The strength and volume of dairy wastewater can vary from one facility to another because of its intermittent production (Rico Gutierrez *et al.*, 1991). The concentration of biochemical oxygen demand in dairy wastewater can vary from 40–48,000 mg/L (mean strength value of 2,300 mg/L) (Rico Gutiérrez *et al.*, 1991).

Background

Dairy wastewater contains indicator bacteria and may contain pathogenic microorganisms such as *Escherichia coli*, *Salmonella* spp., *Listeria monocytogenes*, *Clostridium perfringens* (Cullor, 1995), as well as parasites (i.e., *Giardia* and *Cryptosporidium*) and viruses (Gersberg *et al.*, 1989). Some of these pathogens (i.e., bacteria) are part of the normal intestinal microflora of man and other warm-blooded animals. Other pathogens (i.e., viruses and parasites) are of enteric origin (Bitton, 1994).

Escherichia coli is the predominant fecal coliform. Fecal samples from 6,894 calves from 1,068 herds in 28 states were examined for the presence of *E. coli* O157:H7. The verotoxigenic strain was found in 25 calves sampled (representing 19 herds) (Hancock *et al.*, 1993). A similar study on Washington State dairy herds showed 0.28 per cent of the 3,750 fecal samples or 8.3 per cent of the 60 herds examined contained *E. coli* O157:H7 (Hancock *et al.*, 1993).

Salmonellae are widely distributed in the environment and are potentially pathogenic. *S. dublin* is an invasive pathogen, causing septicemia with high morbidity and mortality in calves. *Salmonella* also can cause bacteremia in bovine neonates if they are infected with the serotypes *S. dublin* or *S. typhimurium*. Animals that recover from *Salmonella* infections become carriers and shed the organism in their feces for weeks after recovery (Smith *et al.*, 1989).

Listeria can live in a variety of hosts, including domestic and wild animals (Gray and Killinger, 1966; Cranfield *et al.*, 1985; Skovgaard, 1987; Inoue *et al.*, 1991). Listeriosis may cause abortion, stillbirth or neonatal sepsis in humans and mammals. It is likely to contaminate dairy wastewater and the environment through feces, urine, and bodily secretions such as conjunctival, oral, nasal and uterine fluids of carriers (Cullor, 1995). The organism has been found in the milk from asymptomatic cows and goats (Loken *et al.*, 1982).

Enterococci are a subgroup of the fecal streptococcus group. Studies involving marine and fresh water beaches indicated that swimming-associated gastroenteritis was related to the quality of the water and that enterococci was the most efficient bacterial indicator (Cabelli, 1983; Dufour, 1984). *Clostridium* is the most widely occurring pathogenic bacterium (Smith and Williams, 1984) and is found within human and animal intestinal tracts.

Enteric viruses can be responsible for diseases such as a skin rash throuh gastroenteritis and paralysis. Enteric viruses occur in small numbers in the environment, which makes them very difficult to culture (Bitton, 1994). Bacteriophages are more easily detected because they occur in higher numbers (Bitton, 1980). Phages infecting *E. coli* are called coliphages, which can be used as indicators for assessing the removal efficiency of water and wastewater treatment plants (Bitton, 1987) and the efficacy of water treatment processes (Payment, 1991).

Giardia and *Cryptosporidium* are the parasites of primary concern in drinking water and wastewater (AWWA, 1988). *Cryptosporidium* can be transmitted to humans through contact with the feces of cows, rodents, dogs and cats and may be responsible for up to 30 per cent of the diarrheal illness in undeveloped countries (Lippy and Waltrip, 1984). *C. parvum* is the major species responsible for infections in humans and animals (Rose *et al.*, 1985; Current, 1987; Rose, 1990). The duration of symptoms and the outcome depends on the immunological status of the patient (Bitton, 1994). The diarrhea generally lasts 1 to 10 days and may eventually cause death in immunosuppressed patients.

Methods

A dairy wastewater treatment system was constructed at a dairy in Glendale, Arizona. This facility requires approximately 250 m³/day of water for cow and equipment washing as well as facility cleanup. Details of the system's design and operation are discussed in

Karpiscak *et al.* (1999). A schematic diagram of the overall system is presented in Figure 1. Water samples were taken from the collection sump, solids separators, anaerobic lagoons, aerobic ponds and wetlands (8 surface flow wetland cells). The sampling locations are indicated in Figure 1 by uppercase letters; for example, water exiting the wetland cells [E].

The performance of the overall treatment system was evaluated by measuring the physical, chemical, and biological parameters in grab samples taken from the selected locations. Chemical and physical parameters included pH, biochemical oxygen demand, nitrogen species, electrical conductivity, and turbidity. Microbes monitored included coliforms (total and fecal), *Cryptosporidium*, *C. perfringens*, *Salmonella*, coliphages, enterococci and *L. monocytogenes*.

Water samples were collected in 1 L plastic bottles from October 1996 through March 1999. After collection, the samples were packed in ice for the trip to the laboratory. Samples for indicator bacteria (coliform), coliphages, *Salmonella*, enterococci, *C. perfringens*, *L. monocytogenes*, and turbidity were processed within 6 h of collection and parasite samples (*Giardia* and *Cryptosporidium*) were concentrated within 48 h. Data on biochemical oxygen demand, electrical conductivity, total suspended solids and nitrogen compounds were presented in Karpiscak *et al.* (1999). This paper will discuss only the microbial parameters and turbidity.

Total coliform was identified using mEndo agar and fecal coliform was detected using mFC agar using the spread plate or membrane filtration methods according to *Standard Methods* (APHA, 1995). Coliphages were detected using the double layer agar method technique described by Adams (1959). *Escherichia coli*, strain ATCC 15597 (American Type Culture Collection), was used as the host bacterium for the analysis.

Salmonella spp. were isolated by first placing the sample in dulcitol selenite enrichment broth and incubating at 37°C for 24 h. This was followed by spread plating on Hektoen enteric agar and incubation at 37°C for 24 h as outlined by APHA (1995). Suspected *Salmonella* colonies were stabbed into Lysine Iron Agar (LIA) and incubated at 37°C for 24 h. Black colonies resulting from the LIA agar were further isolated after being streaked onto TSA and confirmed with an API 20E biochemical test strip specific for *Salmonella*.

The enterococci group was determined using the multiple-tube technique. Azide dextrose broth was used for isolation and Bile-Esculin Agar (BEA) for specific identification

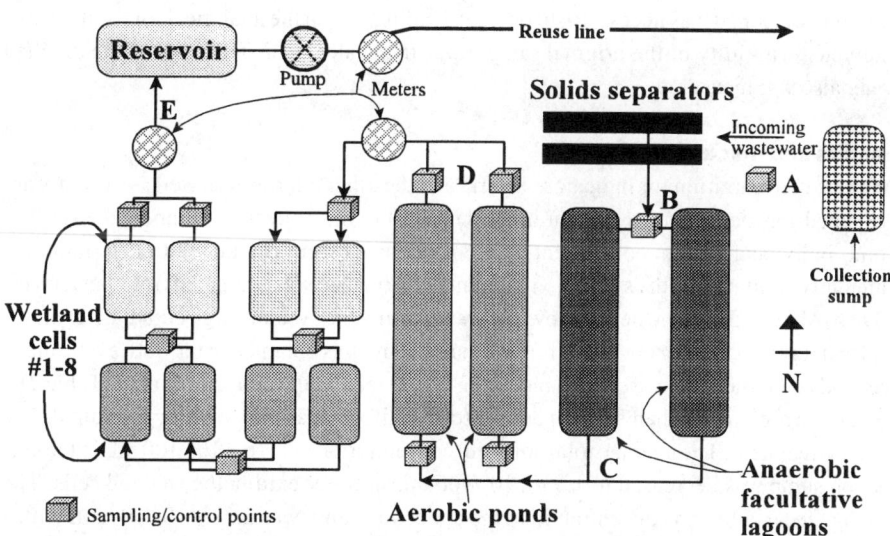

Figure 1 Dairy wastewater treatment reuse elements and sampling points. (adapted from Karpiscak *et al.*, 1999)

of enterococci. To estimate the enterococci density from the number of tubes, the combination of positives tubes was computed for most probable number (MPN) using Table 9221.IV as outlined in *Standard Methods* (APHA, 1995).

Clostridium perfringens was identified by using the method described by Armon and Payment (1988). The sample was heat shocked to kill vegetative bacteria. The sample was then membrane-filtered using a basal medium called m-CP. The plates were inverted and incubated in an anaerobic jar at 45°C for 18 h. Yellow colonies were counted and exposed to ammonium hydroxide vapors. Colonies that turned red and dark pink were scored as acid phosphatase (positive) colonies of *C. perfringens*.

A modified procedure from Lovett (1988) and Fraser and Sperber (1988) was used to isolate and identify *L. monocytogenes*. The method consisted of using two warm enrichment media based on inhibition of indigenous microflora through the addition of inhibitory agents. The first step consisted of placing samples into *Listeria* Enrichment Broth (LEB)-FDA (primary enrichment) and incubating at 30°C for 24–48 h. After incubation, turbid samples were transferred from the primary enrichment broth to Fraser broth and incubated at 30°C for 24–48 h. Turbidity obtained in this medium (black-dark brown) was streaked onto *Listeria* Plating Medium (LPN). Plates were incubated at 35°C for 48 h. Presumptive blue-grayish colonies of *L. monocytogenes* were streaked onto Tryptic Soy Agar (TSA) and incubated at 35 ± 0.5°C for 24 h, after which a catalase test was performed. The catalase positive colonies were picked by a loop and stabbed into motility media. The motility tubes were incubated at room temperature for 48 h. A red-like uniform brush appearance confirmed the motility of *L. monocytogenes*. In addition, an API 20S biochemical test strip was performed on the colonies that were motile. *Listeria* densities were calculated in terms of MPN using Table 9221.IV from Standard Methods (APHA, 1995).

The analysis of *Giardia* and *Cryptosporidium* included three major steps: (1) parasite concentration into a pellet, (2) pellet flotation to clarify the samples, and (3) antibody staining for detection of the parasite using a microscope with fluorescent light. After staining, the filters were examined microscopically. Cysts and oocysts observed were identified according to specified criteria such as immunofluorescence, size, shape, and internal morphological characteristics described in Standard Methods (APHA, 1995).

Extremely high turbidities were found in the dairy wastewater effluent. Therefore, before analysis could take place, each sample was diluted with distilled water. Successive dilutions were made, as necessary, to obtain a reading within the analytical limits of the turbidimeter. Turbidity of the original sample was then calculated. Turbidity of the distilled water also was measured.

Results and discussion

Average concentration for indicator bacteria and the other microbes as well as turbidity are presented for the treatment system in Table 1. Using the collection sump as the starting point, individual system component per cent removals were calculated to determine the cumulative removal by the system. Treatment performance was evaluated from January 13, 1998 to March 25, 1999 when wastewater was entering the system. High cumulative microbial and turbidity per cent removal for most organisms was obtained in the outflow from the wetland cells when compared with incoming wastewater in the collection sump (Table 2).

A decrease greater than 99 per cent for total coliform was obtained for the cumulative system treatment. The average total coliform concentration of 8.16×10^8/100 mL in the collection sump was decreased to 1.05×10^6/100 mL in water exiting the wetland cells. The average fecal coliform concentration in the collection sump was 3.31×10^8/100 mL with a cumulative reduction of 99.96 per cent to 1.38×10^5/100 mL in water exiting the wetland cells.

Table 1 Average concentration of indicator bacteria and pathogens, and turbidity in a multi-component dairy wastewater treatment facility

Parameter (bacteria concentration in 100 mL)	Collection sump	Solids separators	Anaerobic lagoons	Aerobic ponds	Wetland cells
Total coliform	8.16×10^8	6.26×10^8	1.02×10^7	8.36×10^5	1.05×10^6
Fecal coliform	3.31×10^8	2.76×10^8	1.01×10^6	1.59×10^5	1.38×10^5
Coliphage	9.44×10^5	7.35×10^5	2.99×10^4	1.67×10^4	8.51×10^2
Enterococci	2.50×10^7	1.94×10^7	3.97×10^5	6.94×10^4	1.82×10^4
Listeria monocytogenes	1.47×10^7	8.96×10^6	1.74×10^5	2.93×10^4	2.00×10^4
Clostridium perfringens*	5.16×10^6	6.97×10^6	3.03×10^6	1.63×10^4	1.31×10^6
Cryptosporidium	1.79×10^3	2.64×10^3	<1	<1	<1
Turbidity (NTU)	3,891	4,447	1,644	1,277	1,295

*Clostridium perfringens assays discontinued after July 14, 1998 because no meaningful differences were observed

Table 2 Reduction in indicator bacteria, pathogens and turbidity in a multi-component dairy wastewater treatment facility.

Parameter	Solids separators (per cent)	Anaerobic lagoons (per cent)	Aerobic ponds (per cent)	Wetland cells (per cent)	Cumulative treatment system removal (per cent)
Total coliform	23.3	98.4	91.8	+20.4	99.87
Fecal coliform	17	99.6	84.3	13.2	99.96
Coliphage	22.4	95.9	44.1	94.9	99.9
Enterococci	22	97.95	82.5	73.8	99.9
Listeria monocytogenes	39.1	98.1	83.2	31.7	99.86
Clostridium perfringens	+35.6*	52.7	50.6	19.6	74.6
Cryptosporidium	+47.5	99.99	n/a	n/a	>99.99
Turbidity (NTU)	+14.3	63	22	+1	67

*'↑' indicates increase in parameter concentration

A high coliphage removal of 99.9 per cent was observed for the system that reduced the average concentration in the collection sump from 9.44×10^5/100 mL to 8.51×10^2/100 mL in the wetland cells. An enterococci removal rate of more than 99.9 per cent was observed with the average concentration of 2.50×10^7/100 mL in the collection sump reduced to 1.82×10^4/100 mL at the wetland cells. An overall system removal of 99.86 per cent of *L. monocytogenes* was observed. The collection sump average concentration for *C. perfringens* was 5.16×10^6/100 mL that was reduced only 74.6 per cent across the system. Since no meaningful differences in *C. perfringens* were observed between the sampling locations, analysis for this parameter was stopped in July 1998.

Although there was a large increase in *Cryptosporidium* between the system's inflow point and the exit from the solids separators, the cumulative system reduction was greater than 99.99 per cent, with virtually all *Cryptosporidium* removed by the end of the anaerobic lagoons. Turbidity was reduced by only 67 per cent between the inflow and outflow from the overall system, with the largest removal occurring after transit through the anaerobic lagoons (Table 2).

The highest total and fecal coliform average concentrations were found in the collection sump. This is to be expected since the sump is the collection point for all water used in the sanitation of cows and stormwater runoff. Moreover, large amounts of manure were added to the wastewater stream by the flush system from corrals during grooming. The large cumulative system reductions for total and fecal coliform observed may be attributed primarily to removal in the anaerobic lagoons as a result of cell die-off and settling of the materials, degradation of organic matter, competition of microorganisms for limiting nutrients or trace elements, membrane desiccation, and membrane enzyme degradation due to UV-light. In addition, filtration, adsorption, aggregate formation, predators, bacteriophages and toxins emitted by other microorganisms may have had bactericidal effects. The long detention time of 60 days in the anaerobic lagoons also suggests that settling may have been a major contributor to the removal of some microorganisms.

The greatest per cent removal of coliphages was observed in the anaerobic lagoons (95.9 per cent) and wetland cells (94.9 per cent) indicating that the longer residence time within the anaerobic lagoons and UV light within the wetland cells as well as microbial activity most likely had a positive effect on the removal of coliphage. Another factor that may have contributed to the large removal observed in the anaerobic lagoons could have been the adsorption of coliphages onto large settable solids.

Although the aerobic ponds did not yield large reductions for most of the parameters studied, reductions were observed for total coliform (91.8 per cent), fecal coliform (84.3 per cent), coliphage (44.1 per cent), *L. monocytogenes* (83.2 per cent) and enterococci (82.5 per cent) in comparison to the outflow from the anaerobic lagoons. The *C. perfringens* concentration, however, did not change significantly in the aerobic ponds. Turbidity also was not reduced significantly by the aerobic ponds.

The wetland cells achieved small reductions for most of the studied organisms with the exception of total coliform. This system component was most effective in the removal of coliphage and enterococci (94.9 and 73.8 per cent, respectively), but neither of these organisms was reduced significantly in the short detention time of four days. Overall removals of coliphage greater than 99 per cent and nearly 95 per cent observed in the outflow from the wetland cells were very significant when compared with coliphage reductions observed by Thurston *et al.* (1996) of about 35 per cent in the effluent from a duckweed pond and 94 per cent from a multispecies wetland in Arizona.

The information obtained from this study can be used to determine the adequacy of current wastewater treatment for the control of wastewater-borne pathogens, especially from confined animal feeding operations (CAFO). The information also can assist the CAFO industry in the implementation of rational and efficient treatment strategies for reuse of wastewater. While the anaerobic lagoons appear to be effectively reducing or removing the microbial species of interest, further improvements in the aerobic ponds and wetland components need to be considered in order to improve overall wetland system treatment capability.

Acknowledgements

Major funding for construction of the lower subsystem was obtained from the Phoenix Active Management Area of the Arizona Department of Water Resources. The NRCS provided engineering design assistance and partial funding through long-term cost-sharing agreements with the dairy for construction of the upper subsystem. The Arizona Department of Environmental Quality provided funding from the 319 (h) pass through funds from the U.S. Environmental Protection Agency to monitor the components of the upper subsystem. The Bureau of Reclamation, U.S. Department of the Interior, supplied in-kind assistance by drilling the on-site monitoring well. The Pima County Wastewater Management Department's support of the CERF facility provided a logistic base from which to conduct some of the research activities. Major contributions of time, equipment and personnel were also provided by the dairy management.

References

Adams, M.H. (1959). *Bacteriophages*. Interscience Publisher, Inc., New York.

American Public Health Association. American Water Works Association, and Water Environment Federation (1995). *Standard Methods for the Examination of Water and Wastewater*. 19th Edition. American Public Health Association. Washington DC.

American Water Works Association (1988). *Cryptosporidium* round table. (Nov 18, 1987). Baltimore, Maryland. *Journal of the American Water Works Association*, **80**, 14–27.

Armon, R. and Payment, P. (1988). A modified m-CP medium for enumeration of *Clostridium perfringens* from water samples. *Canadian Journal of Microbiology*, **34**, 78–79.

Bitton, G. (1980). *Introduction to Environmental Virology*. Wiley, New York.

Bitton, G. (1987). Fate of bacteriophages in water and wastewater treatment plants. In: *Phage Ecology*, S.M. Goyal, C.P. Gerba and G. Bitton (eds.), Wiley Interscience, New York, pp. 181–195.

Bitton, G. (1994). *Wastewater Microbiology*. John Wiley & Sons Inc., New York.

Bull, M., Sterrit, R. and Lester, A. (1981). Some methods available for treatment of wastewater in the dairy industry. *Journal of Chemical Technology and Biotechnology*, **31**, 579–583.

Cabelli, V.J. (1983). Health Effects Criteria for Marine Water. EPA-600/1-80-031. U. S. Environmental Protection Agency, Cincinnati, Ohio.

Cranfield, M., Eckhaus, M.A., Valentin, B.A. and Strandbe, J.D. (1985). Listeriosis in Angolian Giraffes. *Journal American Veterinary Medical Association*, **187**, 1238–1240.

Cullor, J.S. (1995). Common pathogens that cause foodborne disease: Can they be controlled on the dairy? *Veterinary Medicine*, **90**, 185–86, 188, 190, 192–194.

Current, W.L. (1987). *Cryptosporidium*: Its biology and potential for environmental transmission. *Critical Reviews in Environmental Control*, **17**, 21–51.

Dufour, A.P. (1984). Health Effects Criteria for Fresh Recreational Waters. EPA-600/1-84-004. U. S. Environmental Protection Agency, Cincinnati, Ohio.

Fang, H.H.P. (1990). Aerobic treatment of dairy wastewater. *Biotechnology Techniques*, **4**, 1–4.

Fraser, J.A. and Sperber, W.H. (1988). Rapid detection of *Listeria* spp. in food and environmental samples by esculin hydrolysis. *Journal of Food Protection*, **51**, 762–765.

Gersberg, R.M., Gearheart, R.A. and Ives, M. (1989). Pathogen removal in constructed wetlands. In: *Constructed Wetlands for Wastewater Treatment. Municipal Industrial and Agricultural*. D.A. Hammer. (ed.), Lewis Publishers, Chelsea, Michigan, pp. 431–445.

Graves, R.E. (1987). *Animal Manure-Milking Center*. Agricultural Engineering Department, Penn State University, College Station, Pennsylvania.

Gray, M.L. and Killinger, A.H. (1966). *Listeria monocytogenes* and Listeric infections. *Bacteriological Review*, **30**, 309–382.

Hancock, D.D., Besser, T.E., Kinsel, M.L., Tarr, P.I., Rice, D.H. and Paros, M.G. (1993). The prevalence of *Escherichia coli* O157:H7 in dairy and beef cattle in Washington State. *Epidemiology and Infection*, **113**, 199–207.

Hollon, B.F., Owen, J.R. and Sewell, J.I. (1982). Water quality in a stream receiving dairy feedlot effluent. *Journal of Environmental Quality*, **11**, 5–9.

Inoue, S., Iida, T., Tanikawa, T., Maruyama, T. and Morita, C. (1991). Isolation of *Listeria monocytogenes* from roof rats (*Rattus rattus*) in buildings in Tokyo. *Journal of Veterinary Medicine Science*, **53**, 521–522.

Karpiscak, M.M., Freitas, R.J., Gerba, C.P., Sanchez, L.R. and Shamir, E. (1999). Management of dairy waste in the Sonoran desert using constructed wetland technology. *Water Science and Technology*, **40**(3), 57–65.

Lippy, E.C. and Waltrip, S.C. (1984). Waterborne disease outbreaks: 1946–1980. A thirty-five year perspective. *Journal of the American Water Works Association*, **76**, 2–60.

Loken, T., Aspoy, E. and Gronstol, H. (1982). *Listeria monocytogenes* excretion and humoral immunity in goats in a herd with outbreaks of listeriosis and in a healthy herd. *Acta Veterinaria Scandinavia*, **23**, 392–399.

Lovett, J. (1988). Isolation and enumeration of *Listeria monocytogenes*. *Food Technology*, **42**, 172–175.

Payment, P. (1991). Fate of human enteric viruses, coliphages, and *Clostridium perfringens* during drinking-water treatment. *Canadian Journal of Microbiology*, **37**, 154–157.

Perle, M., Kimchie, S. and Shelef, G. (1995). Some biochemical aspects of the anaerobic degradation of dairy wastewater. *Water Research*, **29**, 1549–1554.

Rico Gutierrez, J.L., Garcia Encina, P.A. and Fdz-Polanco, F. (1989). Anaerobic treatment of cheese-production wastewater using a UASB reactor. *Bioresource Technology*, **37**, 271–276.

Rose, J.B. (1990). Occurrence and control of *Cryptosporidium* in drinking water. In: *Drinking Water Microbiology*. G.A. McFeters. (ed.), Springer-Verlag, New York, pp. 294–321.

Rose, J.B., Musial, C.E., Arrowood, M.J., Sterling, C.R. and Gerba, C.P. (1985). Development of a method for the detection of *Cryptosporidium* in drinking water. Water Technology Conference. American Water Works Association, Houston, Texas, pp. 8–11.

Ross, I.J., Sizemore, S., Bowden, J.P. and Haan, C.T. (1978). Effects of soil injection of liquid dairy manure on the quality of surface runoff. National Technical Information Service, Technical Report 113, Springfield, Virginia.

Skovgaard, N. (1987). *Listeria*: Major sources and routes of human infection, environment and plant. In: Listeriosis. Joint WHO/ROI Consultation on Prevention and Control, Berlin (West). Compiled by A. Schonberg. *Veterinary Medicine Hefte*, **5**, 86–87.

Smith, B.P., Oliver, D.G., Singh, P., Dilling, G., Marvin, P.A., Ram, B.P., Jang, L.S., Sharkow, N., Osborn, J.S. and Jackett, K. (1989). Detection of Salmonella-dublin mammary-gland infection in carrier cows, using an enzyme-linked immunosorbent-assay for antibody in milk or serum. *American Journal of Veterinary Research*, **50**, 1352–1360.

Smith, L. and Williams, B.L. (1984). *The Pathogenic Anaerobic Bacteria*, 3rd ed, C.C. Thomas, Springfield, Illinois.

Thurston, J.A., Falabi, J.A., Gerba, C.P., Foster, K.E. and Karpiscak, M.M. (1996). Fate of indicator microorganisms, *Giardia* and *Cryptosporidium* in two constructed wetlands. Fifth International Conference on Wetland Systems for Water Pollution Control. Vienna, Austria. Poster 30–1.

Viraraghavan, T. and Kikkeri, S.R. (1990). Anaerobic filter treatment of dairy wastewater at low temperatures. Proceedings of the 44th Purdue Industrial Waste Conference. Purdue University, West Lafayette, Indiana, May 9–11, 1989, pp. 199–208.

Protozoan predation as a mechanism for the removal of *cryptosporidium* oocysts from wastewaters in constructed wetlands

R. Stott*, E. May**, E. Matsushita*** and A. Warren****

* Department Civil Engineering, University of Portsmouth, Portsmouth, PO1 3QL, UK
** School of Biological Sciences, University of Portsmouth, PO1 2DY, UK
*** Biologie Cellulaire et Physiologie, Université Pierre et Marie Curie, Paris VI, France
**** Dept Zoology, Natural History Museum, London, SW7 5BD, UK

Abstract The removal of the protozoan parasite, *Cryptosporidium parvum*, from wastewaters is becoming of increasing importance in the UK, especially since contamination of raw waters by sewage effluents has been implicated in major waterborne outbreaks of cryptosporidiosis in recent years. Compared to conventional wastewater-treatment processes, constructed wetlands have demonstrated favourable removal rates for *Cryptosporidium* oocysts. The removal mechanisms, however, remain unknown. Predation by free-living ciliated protozoa, which are commonly found in constructed wetlands, was investigated as a possible mechanism for oocyst removal. In laboratory feeding experiments, ciliates (*Euplotes patella*, *Stylonychia mytilus*, *Paramecium caudatum* and an unidentified wetland ciliate species), were exposed to doses ranging from 10 to 10^6 oocysts/ml for between 5 and 60 minutes. Ciliate predatory activities were assessed by enumerating fluorescently labelled ingested oocysts using epifluorescence microscopy. Oocysts were found to be ingested by all species investigated. *Paramecium* demonstrated the highest mean ingestion rates (up to 170 oocysts/hr) followed by *Stylonychia* (up to 60 oocysts/hour). *Euplotes* and the wetland ciliate had lower mean grazing rates (4 and 10 oocysts/hr respectively). These results indicate that protozoan predation may be an important factor in the removal of *Cryptosporidium* oocysts from wastewaters in constructed wetlands.

Keywords Ciliates; constructed wetlands; *Cryptosporidium*; oocyst removal; predation; protozoa

Introduction

Cryptosporidium parvum is an enteric parasitic protozoan that causes severe diarrhoeal illness which may be fatal in infants, immunocompromised and immunosuppressed people (O'Donoghue, 1995). It is a highly transmissible disease with the ingestion of 30 oocysts sufficient to establish an infection (Dupont *et al.*, 1995). The principal route of transmission is waterborne through faecally contaminated drinking and recreational waters (Smith and Rose, 1998; Kramer *et al.*, 1998)). Discharge of wastewaters into receiving waters and runoff from agricultural wastes applied to land have both been implicated as the major sources of oocyst contamination in raw waters. However, there is evidence that recent major waterborne outbreaks of cryptosporidiosis in the UK were specifically associated with wastewater-derived oocysts (Patel *et al.*, 1998). The removal of *C. parvum* oocysts from wastewaters is therefore becoming of increasing importance.

Cryptosporidium oocysts may be plentiful in sewage. Madore *et al.* (1987) reported 850–5280 oocysts per litre of raw sewage, but in a sewage receiving waste from a slaughterhouse, almost 14,000 oocysts per litre were recorded. Conventional wastewater treatment systems are usually not effective in the removal of *Cryptosporidium* and oocysts are frequently reported in effluents. In the UK, 26% of sewage effluent samples were positive for oocysts (up to 60 per litre) (Bukhari *et al.*, 1997); other studies have reported concentrations of oocysts ranging from 0.02 to ~4000 oocysts per litre in treated wastewaters (Smith and Rose, 1998). Most oocyst removal is reported to occur during secondary aerobic treatment rather than primary sedimentation and conventional tertiary treatment but rates of removal may be negligible (Robertson et al., 2000). Compared to conventional wastewater treatment systems, constructed wetlands have demonstrated favourable removal rates for protozoan (oo)cysts (Stott et al., 1997; Oswald et al., 2000). However, whilst the removal of oocysts during wastewater treatment in constructed wetlands can minimise the release of oocysts into the aquatic environment, the removal processes remain unknown.

Predation by free-living ciliated protozoa is an important mechanism for the removal of bacteria from wastewaters in biological wastewater-treatment processes including constructed wetlands (Decamp and Warren, 1998). However, although it is known that predatory protozoa can ingest bacterial pathogens (Curds, 1992) and viruses (Gonzalez and Suttle, 1993) there is a lack of information about protozoan predation on parasite pathogens. *Cryptosporidium* oocysts are known to be ingested by a variety of aquatic invertebrates including rotifers (Fayer et al., 2000) and unidentified zooplankton (Medema et al., 1997). There are no previous reports that *Cryptosporidium* oocysts are ingested by free-living protozoa.

The aim of this study was therefore to investigate predation by free-living ciliated protozoa as a potential mechanism for the removal of *Cryptosporidium* oocysts from wastewaters treated in constructed wetlands.

Methods
Free-living ciliated protozoa
Four species of filter-feeding ciliated protozoa commonly found in constructed wetlands used for wastewater treatment were selected for the current investigation: *Stylonychia mytilus, Euplotes patella, Paramecium caudatum* and an unidentified ciliate species isolated from a wetland (probably a *Paramecium* sp.). *Stylonychia mytilus, E. patella* and *P. caudatum* were obtained as monocultures from Sciento (Manchester, UK). Specimens were measured with an ocular micrometer and mean sizes were as follows; *S. mytilus* (149.6 µm × 62.9 µm), *E. patella* (114.6 µm x 74.7 µm) and *P. caudatum* (166.8 µm × 64.8 µm). The wetland ciliate (163.9 µm × 56 µm) was isolated from a horizontal subsurface flow Gravel Bed Hydroponic (GBH) constructed wetland planted with *Phragmites australis* and treating primary settled domestic wastewater (bed length 56 m, width 30 cm, depth 15 cm; gradient 1:40; flow rate 2 L/min). Ciliate specimens were collected from the wetland at 10 m from the inlet from the rhizospheric zone and surrounding bulk water.

Cultures of *E. patella* and *P. caudatum* were maintained at room temperature in Petri dishes containing sterile mineral water ("Volvic") and a boiled rice grain. *Stylonychia mytilus* was maintained on a mix of dried and live algae (13 µm × 10 µm) and sterile Volvic water. The species isolated from the wetland was kept in water from the GBH system diluted with Volvic water.

Cryptosporidium oocysts
Oocysts of *Cryptosporidium parvum* were obtained from Moredun Research Institute (Scotland). Oocysts were labelled with fluorescein isothiocyanate (FITC) conjugated to IgM monoclonal antibodies (A400; Waterborne Inc, USA) and detected and enumerated using epifluorescence microscopy (Nikon E800 Eclipse equipped with FITC filter). Prior to feeding, the oocyst suspension was microcentrifuged at 7,000 g for 5 minutes and resuspended in sterile Volvic water. Suspensions of FITC-labelled oocysts were enumerated using a Neubauer haemocytometer and prepared to final concentrations of 10^4 and 10^6 oocysts/ml. Control slides were prepared to check fluorescence of oocyst suspensions and to confirm oocyst exposure doses in the predation trials.

Predation trials

Ciliated protozoa were exposed to high (10^6 oocysts/ml) and low numbers (*ca.* 10) of FITC-MAb stained *Cryptosporidium* oocysts for varying time intervals between 5 and 60 minutes.

High dose exposure. A 3 µl aliquot was taken from the 10^6/ml FITC-oocyst suspension and placed in a 96-well plate to which was added 3 µl of sterile Volvic water. For each ciliate, three cells were added per well using a micropipette and the protozoa incubated with the oocyst suspensions for 5, 10, 15, 20, 25, 30, and 60 minutes. After incubation, individuals were removed from the oocyst suspension and transferred to a microscope slide. A coverslip mounted on Vaseline was used to immobilise the live protozoa by applying gentle downward pressure. Slides were examined by epifluorescence microscopy at 460–500 nm in order to observe ingested oocysts and thus determine ciliate grazing rates.

Low dose exposure. For low dose exposure of around 10 oocysts, a 1–3 µl aliquot of a 10^4 ocyst/ml suspension was placed on a microscope slide and the total number of oocysts counted under epifluorescence using a ×20 objective. An individual ciliate specimen of *S. mytilus* or *P. caudatum* was then added to the oocyst suspension on the slide and a coverslip supported on Vaseline was placed over the protozoa. After 10, 15, 30 or 60 minutes, numbers of oocysts internalised within the protozoa were determined as before.

Results and discussion

Rates of ingestion

High dose exposure. The predation rates of the four ciliates at high exposures (10^6/ml) are shown in Figure 1. The highest mean rate of oocyst ingestion was observed in the largest ciliate, *P. caudatum* (2.8 oocysts/cell/minute) followed by *S. mytilus* (1.0 oocysts/cell/min) and the wetland ciliate (0.2 oocysts/cell/min). *Euplotes patella* had the slowest rate of ingestion of 0.07 oocysts/cell/minute. A significant difference in the median rates of ingestion was observed between ciliate species (Kruskal Wallis, $H = 134.30$, DF = 3, $P = <0.001$).

The numbers of oocysts internalised by individual protozoan cells also varied significantly between ciliate taxa (Kruskal Wallis, $H = 133.38$, DF = 3, $P = <0.001$) and ranged on average from 1 to >30 oocysts/cell (Figure 2). The mean number of ingested oocysts was significantly correlated to protozoan mean size (Pearson correlation coefficient $r = 0.976$, $p = 0.024$). The frequency distribution of ingested oocysts within ciliates is shown in Figure 3.

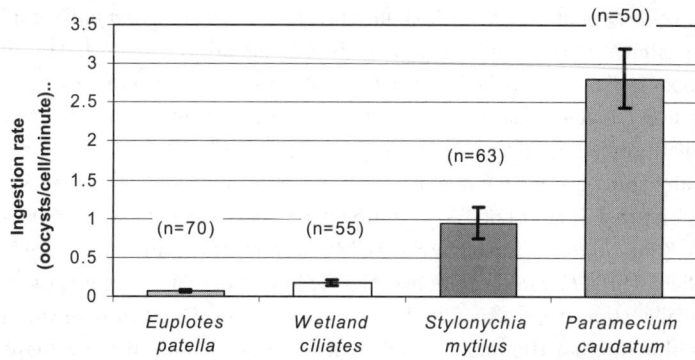

Figure 1 Rates of oocyst ingestion by ciliates exposed to high concentrations of *Cryptosporidium* oocysts (10^6 oocysts/ml) (Mean and SEM shown)

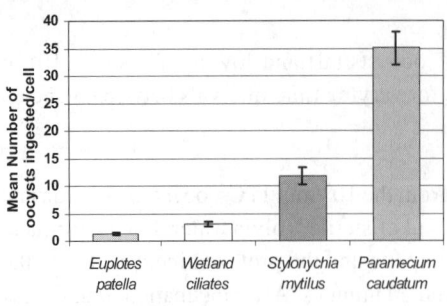
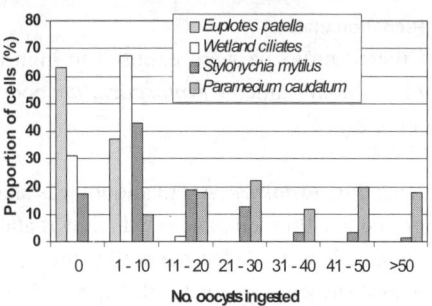

Figure 2 Numbers of oocysts ingested by ciliated protozoa (mean and SEM shown)

Figure 3 Distribution of ingested oocysts within ciliates (relative frequency shown)

Although most individual ciliates had ingested oocysts, the proportion of cells containing oocysts was found to be highly associated with ciliate species (Chi sq. $\chi^2 = 61.85$, DF = 3, $P = <0.001$). The majority of *E. patella* cells (63%) did not contain oocysts. However, 37% of *E. patella* cells had ingested between 1 and 10 oocysts and on average, *E. patella* contained 1 oocyst/cell during the study. The wetland ciliate demonstrated slightly greater predatory activity with the majority of cells (67%) ingesting between 1 and 10 oocysts and a few (2%) containing between 11 and 20 oocysts; on average, the wetland ciliate ingested 3 oocysts/cell. Higher numbers of oocysts were ingested by *S. mytilus*; the majority of cells (43%) ingested between 1 and 10 oocyst, 32% of cells contained between 11 and 30 oocysts and 8% of cells had ingested more than 30 oocysts. However, some individuals had accumulated greater numbers; on several occasions high numbers of oocysts, estimated at between 20–50 but too numerous to count, were observed in individual specimens. Overall, the mean oocyst ingestion for *S. mytilus* was 11 oocysts/cell. Greatest predatory activity was found in *P. caudatum;* all cells contained oocysts but 50% of cells had ingested between 1 and 30 oocysts and the remaining 50% of cells had ingested more than 30 oocysts. Average predation by *Paramecium* was 35 oocysts/cell.

Differences in ciliate ingestion rates and the proportion of actively ingesting individuals may be due to a variety of prey and predator factors. Filter feeding in ciliates is largely determined by prey particle size (Fenchel, 1980). *Stylonychia mytilus* and *P. caudatum* are perhaps more likely to ingest oocysts whose size of 4–6 µm falls within their prey particle range of 0.2 to 20 µm (Posch and Arndt, 1996; Pfister and Arndt, 1998) than *E. patella* which usually feeds on small bacteria.

Filter feeding ciliated protozoa in this study demonstrated a higher predacious activity on oocysts than that reported for the other filter feeders (such as rotifers) commonly found in wastewater-treatment systems. Maximum levels of oocysts within individual cells varied between ciliates and ranged from around 10 oocysts/cell (*E. patella* and the wetland ciliate) to 50 oocysts/cell (*S. mytilus*) and up to 90 oocysts/cell (*P. caudatum*). In comparison, lower ingestion capacities have been reported in rotifers ranging from 5–15 oocysts per individual with some ingesting up to 25 oocysts (Fayer *et al.*, 2000).

The relationship between the rate of oocyst ingestion and exposure time for each ciliate is shown in Figure 4. Rates of oocyst ingestion varied between taxa, with predation activity greatest for *P. caudatum* followed by *S. mytilus*. Lower rates of ingestion were found in the wetland ciliate whilst *E. patella* demonstrated the lowest predation activity over time.

Rates of ingestion were also found to vary with time. Greatest ingestion of oocysts occurred within the first 10 minutes but thereafter ingestion rates then decreased. Highest rates of ingestion were found after 5 minutes incubation for all ciliate taxa with mean rates of ingestion varying from 5.1 oocysts/cell/min (*P. caudatum*), 2.9 oocysts/cell/min

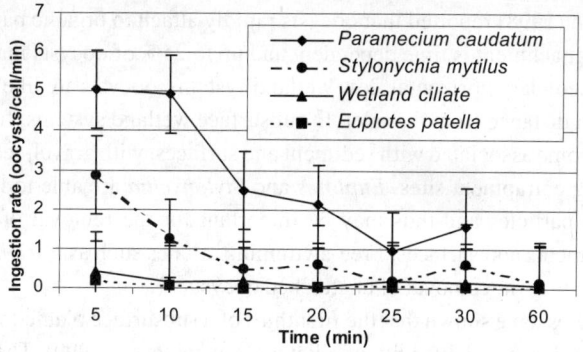

Figure 4 Relationship between ingestion rate (oocysts/cell/min) and time for ciliated protozoa. (Mean and SEM shown, $n = 8-10$)

(*S. mytilus*), 0.4 oocysts/cell/min (wetland ciliate) to 0.2 oocysts/cell/min (*E. patella*). After the initial intense feeding activity, ingestion rates in *S. mytilus* and *P. caudatum* decreased and appeared to settle at much lower levels. After 30 minutes, mean rates of oocyst ingestion were 1.6 oocysts/cell/min (*P. caudatum*) while after 60 minutes, rates of ingestion were reduced to 0.2 oocysts/cell/min (*S. mytilus*) and less than 0.1 oocysts/cell/min for *E. patella* and the wetland ciliate.

Low dose exposure. Rates of ingestion at low dose exposures were similar for *P. caudatum* and *S. mytilus*. Mean rates of ingestion varied between 0 and 0.03 oocysts/cell/minute with maximum individual ingestion rates of 0.2 oocysts/cell/minute observed for both *P. caudatum* and *S. mytilus*. Grazing activities were not observed for *P. caudatum* cells until after 30 minutes incubation (0.01 oocysts/cell/min; 0.33 oocysts/cell). Ingestion rates then increased to 0.02 oocysts/cell/minute at 60 minutes (1.22 oocysts/cell). In contrast, grazing activity was seen in *S. mytilus* after 10 minutes (0.02 oocysts/cell/min); rates of ingestion then increased to 0.03 oocysts/cell/min after 20 minutes. Similarly numbers of oocysts ingested by *S. mytilus* increased from 0.2 oocysts/cell to 0.6 oocysts/cell after 20 minutes when greatest numbers of oocyst were ingested. However, rates of ingestion and numbers of oocysts ingested then decreased thereafter to 0.01 oocysts/cell/min and 0.2 oocysts/cell respectively. All *S. mytilus* cells were empty after 60 minutes.

Interestingly, comparable rates of ingestion were observed in *S. mytilus* at low oocyst concentration and after short term feeding (30–60 min) at high oocyst concentrations of 10^6/ml with a rate of ingestion of 0.2 oocysts/cell/min. Rates of ingestion by protozoa have been found to increase with prey concentration (Choi, 1994). Although the fate of ingested oocysts within protozoa was not determined, it is possible that oocysts retained within ciliates during the short term (1 hr) feeding trials, may have been digested and/or excreted. In some *P. caudatum* cells, diffused fluorescence was seen in food vacuoles due possibly to dissociated fluorescein labelling or fragments of oocyst cell wall, and a few individual *S. mytilus* cells were seen to egest food particle debris that included several oocysts. The effect of ingestion on oocyst viability is the subject of ongoing investigation.

Protozoan removal of oocysts in constructed wetlands

Stylonychia, *Euplotes* and *Paramecium* species have been found routinely in constructed wetlands (Decamp, 1996). In these studies, *P. caudatum* demonstrated the greatest predation on oocysts in suspension. However, *in situ*, the impact of ciliate predation on removal of oocysts from wastewaters may depend on ciliate feeding habit, population diversity and density, contact time and distribution of oocysts.

Medema *et al.* (1998) reported that oocysts rapidly attach to organic particles in wastewater although attachment is time dependent and up to 25% of oocysts may remain freely suspended. In secondary horizontal flow wetland systems, oocysts are likely to be deposited within a short distance from the inlet. In subsurface wetland systems, free and attached oocysts may become associated with sediment and surfaces, with gravel, reed roots and rhizomes providing entrapment sites. *Euplotes* and *Stylonychia* are able to feed on surface-associated food particles and thus may be important for the removal of oocysts lightly attached to sediments and surfaces. Free-swimming ciliates such as *Paramecium* are more likely to contribute to the removal of unattached oocysts.

Previous studies have shown that the first third of a subsurface planted wetland (length) has a higher abundance and diversity of ciliates (Decamp *et al.*, 1999). The wetland ciliate used in the predation trials was isolated from this section at 10 m along a 56 m reed bed; the abundance of this species was approximately 450 ciliates/ml. Tracer dye studies indicate that the retention time of wastewater within the first 10 m of the wetland used in this study is around 30 minutes (Hughes *et al.*, 1995). The ingestion rate for the wetland ciliate after 30 minutes contact time was observed to be 4.09 oocysts/cell and thus population grazing activities could lead to ingestion and removal of 1,840 oocysts/ml. However, only 80% of the wetland ciliate cells in the trial were actively ingesting. If feeding cells only are considered, rates of ingestion increase to 5.6 oocysts/cell and a total of 2,531 oocysts/ml could be removed by ciliate predatory activities. Furthermore, rates of ingestion may vary with time. Average oocysts predation by the wetland ciliate over one hour was 10 oocysts/cell/hour (including non-feeders) at high oocyst concentration indicating that wetland ciliates may be capable of removing up to 4,670 oocysts/ml/hour. Predation activity observed at the low oocyst concentration is more likely representative of that *in situ* when treatment systems usually receive a constant flow of wastewater containing relatively low numbers of oocysts, although oocyst concentrations in wastewaters can range considerably. In general, raw wastewater usually contains <200 oocysts/L (Bukhari *et al.*, 1997). However, up to around 5,300 oocysts/L and 14,000 oocysts/L have been reported in sewage and slaughterhouse wastewaters respectively (Madore *et al.*, 1987). High numbers of oocysts have also been reported in backwash waters (Oswald *et al.*, 2000). Ciliates are thus theoretically capable of complete oocyst removal from wastewaters treated in constructed wetlands although it is likely other processes are also involved.

Conclusions

Oocysts of *Cryptosporidium parvum* were effectively ingested by free-living, filter-feeding ciliated protozoa commonly found in constructed wetlands. High rates of ingestion were observed with *Stylonychia mytilus* and *Paramecium caudatum* that preferentially feed on surface-associated and suspended prey respectively. *Euplotes patella* and an unidentified ciliate isolated from a Gravel Bed Hydroponic constructed wetland also demonstrated oocyst grazing potential. Rates of oocyst ingestion varied widely between ciliate species and time. Although oocyst predation by protozoa *in situ* is unknown, and other mechanisms may also play a role in removal, these results indicate that protozoan predation may be an important factor in the removal of oocysts from wastewaters in constructed wetlands.

Acknowledgements

The authors would like to thank Mrs S. Hope for technical assistance during the preliminary studies, Mr D. Maund, Dr. J. Chernin and Dr. M. Guille (School of Biological Sciences) for provision of microscope facilities and the Department of Civil Engineering for funding support.

References

Bukhari, Z., Smith, H.V., Sykes, N., Humphreys, S.W., Paton, C.A., Girdwood, R.W.A. and Fricker, C.R. (1997). Occurrence of *Cryptosporidium* sp. oocysts and *Giardia* sp. cysts in sewage influents and sewage effluents from sewage treatment plants in England. *Water Science and Technology*, **35**(11–12), 385–390.

Choi, J.W. (1994). The dynamic nature of protistan ingestion response to prey abundance. *Journal of Eukaryotic Microbiology*, **41**, 137–146.

Curds, C.R. (1992). *Protozoa in the Water Industry*. Cambridge University Press, UK.

Decamp, O. (1996). *The microbial ecology of the rootzone method of wastewater treatment*. PhD Thesis, University of Leicester.

Decamp, O. and Warren, A. (1998). Bacterivory in ciliates isolated from constructed wetlands (reedbeds) used for wastewater treatment. *Water Research*, **32**(7), 1989–1996.

Decamp, O., Warren, A. and Sanchez, R. (1999). The role of ciliated protozoa in subsurface flow wetlands and their potential as bioindicators. *Water Science and Technology*, **40**(3), 91–98

Dupont, H.L., Chappell, C.L., Sterling, C.R., Okhuysen, P.C., Rose, J.B. and Jakubowski, W. (1995). The infectivity of *Cryptosporidium parvum* in healthy volunteers. *New England Journal of Medicine*, **332**, 855–859.

Fayer, R., Trout, J.M., Walsh, E. and Cole, R. (2000). Rotifers ingest oocysts of *Cryptosporidium parvum*. *Journal of Eukaryotic Microbiology*, **47**(2), 161–163.

Fenchel, T. (1980). Suspension feeding in ciliated protozoa: structure and function of feeding organelles. *Archive fur Protistenkunde*, **123**, 239–260.

Gonzalez, J.M. and Suttle, C.A. (1993). Grazing by marine nanoflagellates on viruses and virus-sized particles – ingestion and digestion. *Marine Ecology – Progress Series*, **94**(1), 1–10.

Hughes, C.M., Loveridge, R.F., Ford, M.G. and Butler, J.E. (1995). Recovery of dosed copper from planted and unplanted gravel sewage treatment beds. In: Proceedings of a workshop on *Natural and Constructed Wetlands for Wastewater Treatment and Reuse – Experiences, Goals and Limits*, Perugia, Italy, 26–28th October 1995.

Kramer, M.H., Sorhage, F.E., Goldstein, S.T., Dalley, E., Wahlquist, S.P. and Herwaldt, B.L. (1998). First reported outbreak in the United States of cryptosporidiosis associated with a recreational lake. *Clinical Infectious Diseases*, **26**(1), 27–33.

Madore, M.S., Rose, J.B., Gerba, C.P., Arrowood, M.J. and Sterling, C.R. (1987). Occurrence of *Cryptosporidium* oocysts in sewage effluents and selected surface waters. *Journal of Parasitology*, **73**, 702–705

Medema, G.J., Bahar, M. and Schets, F.M. (1997). Survival of *Cryptosporidium parvum*, *Escherichia coli*, faecal enterococci and *Clostridium perfringens* in river water: influence of temperature and autochthonous microorganisms. *Water Science & Technology* **35**(11–12), 249–252.

Medema, G.J., Schets, F.M., Teunis, P.F.M. and Havelaar, A.H. (1998). Sedimentation of free and attached *Cryptosporidium* oocysts and *Giardia* cysts in water. *Applied and Environmental Microbiology*, **64**(1), 4460–4466.

O'Donoghue, P.J. (1995). *Cryptosporidium* and cryptosporidiosis in man and animals. *International Journal of Parasitology*, **25**, 139–195.

Oswald, A.M., Gerba, C.P. and Karpiscak, M.M. (2000). Removal of enteric microorganisms from secondary effluent and backwash filter water by artificial wetlands. Proceedings of the IWA 3rd International Symposium on *Wastewater, Reclamation, Recycling and Reuse*, 3–7th July 2000, Paris, France.

Patel, S., Pedraza-Diaz, S., McLaughlin, J. and Casemore, D.P. (1998). Molecular characterisation of *Cryptosporidium parvum* from two large suspected waterborne outbreaks. Outbreak Control Team South and West Devon, 1995 Incident Management Team and Further Epidemiological Studies Subgroup North Thames 1997, *Communicable Disease and Public Health*, **1**(4), 231–233.

Pfister, G. and Arndt, H. (1998). Food selectivity and feeding behaviour in omnivorous filter-feeding ciliates; a case study for *Stylonychia*. *European Journal of Protistology*, **34**(4), 446–457.

Posch, T and Arndt, H. (1996). Uptake of sub-micrometre and micrometre-sized detrital particles by bacterivorous and omnivorous ciliates. *Aquatic Microbial Ecology*, **10**, 45–53.

Robertson, L.J., Paton, C.A., Campbell, A.T., Smith, P.G., Jackson, M.H., Gilmour, R.A., Black, S.E., Stevenson, D.A. and Smith, H.V. (2000). *Giardia* cysts and *Cryptosporidum* oocysts at sewage treatment works in Scotland, UK. *Water Research*, **34**(8), 2310–2322.

Smith, H.V. and Rose, J.B. (1998). Waterborne cryptosporidiosis. *Parasitology Today*, **14**(1), 14–22.

Stott, R., Jenkins, T., Shabana, M. and May, E. (1997). A survey of the microbial quality of wastewaters in Ismailia, Egypt and the implications for wastewater reuse. *Water Science and Technology*, **35**(11–12), 211–217.

Distribution and retention of faecal coliforms in the Nakivubo wetland in Kampala, Uganda

F. Kansiime* and J.J.A. van Bruggen*

* Makerere University Institute of Environment & Natural Resources, P.O. Box 7062, Kampala, Uganda
** IHE Delft, P.O. Box 3015, 2601 DA Delft, The Netherlands

Abstract Nakivubo wetland, which has been receiving wastewater from the capital of Uganda for more than 40 years is a tropical wetland dominated by *Cyperus papyrus* and *Miscanthidium violaceum*. Field, pilot and laboratory studies were carried out to assess the distribution of faecal coliforms and factors responsible for their retention in different compartments of the two macrophytes in the wetland. There were higher coliform numbers in the free water column below the mat of zones dominated by *Miscanthidium* ($1.1 \pm 0.6 \times 10^5$ MPN/100 ml) compared to those dominated by papyrus ($8.9 \pm 3.1 \times 10^4$ MPN /100 ml). The thick (1.3 m) and compact mat of *Miscanthidium* restricts vertical transport of wastewater into the mat, resulting in flow-through of wastewater under the mat. The papyrus mat is loose, open and thin (0.5 m) and allows easy vertical penetration of wastewater into the mat. The unrestricted interaction between the wastewater in the water column and that in the mat of papyrus in addition to plant debris and detritus continuously sedimenting out of the mat to the wetland bottom are responsible for the retention of coliforms in the papyrus dominated zones. Attachment, sedimentation and natural die-off were found to be important mechanisms responsible for the retention of coliforms in the Nakivubo wetland.

Keywords *Cyperus papyrus*; faecal coliforms; *Miscanthidium violaceum*; Nakivubo wetland; Uganda

Introduction

Wetlands offer an attractive low cost wastewater treatment especially for medium and small sized communities. However, the processes and factors responsible for the retention of pathogens are not well documented, especially for natural floating tropical wetlands. Factors that are considered to affect the survival of pathogens in wetlands include sedimentation, aggregation, inactivation by UV light, exposure to biocides excreted by plants, adsorption to organic matter, grazing by protozoa and attack by lytic bacteria and viruses (Lijklema *et al*., 1987; Scheuerman *et al*., 1989; Brettar and Höfle, 1992). Besides the modest number of studies that have examined the reduction of indicator bacteria in wetlands, there is little information on the long-term survival of these organisms in wetlands.

Nakivubo wetland is a natural wetland located at the northern shores of Lake Victoria and has been receiving secondary effluent from the Bugolobi sewage works for more than 40 years, as well as storm water run-off from the Nakivubo channel and raw sewage from the Luzira prisons. The wastewater finally enters Lake Victoria at the Inner Murchison Bay, where the water supply for Kampala City is abstracted at Gaba water works, just 4 km South East. Some previous studies concentrated on the wetland inflow and on the Inner Murchison Bay (Kizito, 1986). Kansiime *et al*., 1994 reported preferential water flows in the Nakivubo wetland. No study has been carried out to assess the fate of faecal coliforms as the water flows through the wetland into the bay and for possible causes of the preferential flows. Since pathogens could eventually be transported to Gaba water works, the pathogen and / or disinfecting aspect of pollution is very important. As a result this study was carried out to assess: the longitudinal distribution of faecal coliforms in the wetland; the vertical distribution of faecal coliforms in different compartments of the dominant macrophytes; and the major mechanisms responsible for the removal of faecal coliforms in the wetland by conducting bioassays under controlled conditions in the laboratory.

Materials and methods
Study area
This study was carried out in the Nakivubo wetland in 1996. This tropical wetland is located on the northern shores of Lake Victoria at an altitude of 1,135 m. The wetland is co-dominated by papyrus (*Cyperus papyrus*) and *Miscanthidium violaceum*. Papyrus covers a large expanse of the wetland from the landside edge up to the shores of the Inner Murchison Bay. At the wetland edge, papyrus is rooted in the sediment, whereas in the middle parts and towards the lake, the plants are rooted in a floating mat. *Miscanthidium* is restricted to the centre of the wetland and is held in a thick mat floating over a column of 1.5–2 m of water. An overview of the vegetation in the wetland and root mat structure is reported in Kansiime and Nalubega 1999 and Azza *et al.*, 2000). To access the wetland four transects (1 to 4) were made by cutting tracks across the wetland. Sampling sites are designated by transect number and distance along the transect as measured from the western edge (Bukasa side) of the wetland. For instance, T2-100 refers to 100 m along transect 2 (Figure 1).

Methods
The Multiple Tube Fermentation Technique was used to detect faecal (thermotolerant) coliforms as described in Standard Methods (APHA, 1992). Lauryl sulphate (lactose) broth was used as a test media for the presumptive phase and EC media for a confirmatory test. Samples were collected from the mat, free water column below the mat and from the top of the peat sediment. Sampling was done by gently driving a pole (outside diameter 15 cm) into the floating root mat to the desired depth. By withdrawing the pole an opening was created, and a plastic tube connected to a battery operated electric pump was inserted and the sample was pumped into 120 ml sterile glass bottles. After collection, samples were kept in a cold icebox before being transported to the laboratory. Samples were analysed

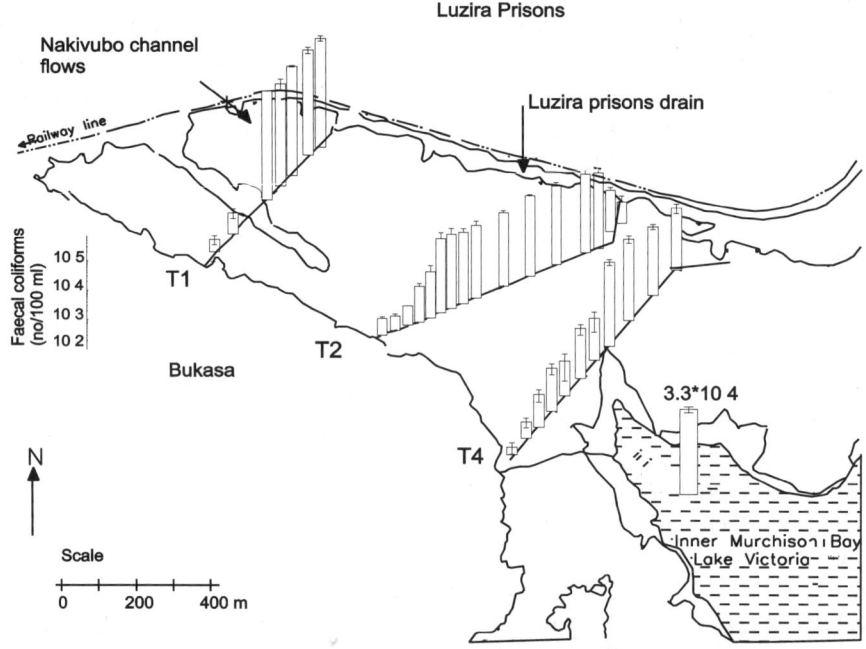

Figure 1 Distribution of faecal coliforms (log units/100 ml) in the Nakivubo wetland. (10 5 in the figure means 10^5, etc)

within 4 to 6 h of sampling. Electrical conductivity, pH and dissolved oxygen were measured in situ. Statistical analysis was performed with the package MINITAB Release 10 for Windows and included analysis of variance (ANOVA), Barlett's and Levene's test for homogeneity of variance and Tukey's multiple comparisons for differences between means. Linear regression was used to estimate the decay rates.

To assess the longitudinal distribution and retention of faecal coliforms in the wetland, samples were taken from the major inflow into the wetland, the Nakivubo channel, and from the main flow path through *Miscanthidium* (Inlet, T2-400 and T4-400) and through papyrus (Inlet, T2-200 and T4-200). Sampling was done between 9.00 and 12.00 am for 3 days during a dry season. For each sampling day, samples were taken from the water column and simultaneously. To get an insight into the possible impact of discharged effluent on the Inner Murchison Bay with the intake for the water supply of Kampala, samples were taken in the bay at approximately 500 m, 1,500 m and 2,500 m perpendicular to the wetland lake interface at T4-400.

To determine the vertical distribution of faecal coliforms and to assess the influence of dominant macrophytes on their distribution, samples were taken from different compartments (mat, water column and peat-sediment) of the papyrus and *Miscanthidium* areas. Sampling locations were T2-100, T4-395 (papyrus), T2-400 and T4-410 (*Miscanthidium*). The sites at T4 were close to each other (15 m apart) at the wetland lake interface. These two sites were selected to make a comparison between papyrus and *Miscanthidium* under more or less similar conditions. Location T2-100 is a typical papyrus site far from *Miscanthidium* and with little influence from the lake. For each sampling date, samples from the locations on a given transect were taken on the same day in the dry season (for six weeks in June–July 1996), with an hour difference between the two locations.

Laboratory simulation experiments

Attachment to plant roots. To determine the number of faecal coliforms that are attached to plant surfaces in contact with wastewater, root samples of papyrus and *Miscanthidium* were collected from the wetland at T4-395 and T4-410, respectively. In the field, whole plant units were cut out of the mat. Loose mat peat was removed by gentle shaking, and the rhizome and its roots were put in a polythene bag. In the laboratory, roots were separated and 10 g (wet-weight) of root material was transferred to 95 ml of a saline diluent made up of 8.5 g NaCl/l and neutralised bacteriological peptone (1 g/l) and this resulted in a 10^{-1} dilution. Subsequent dilutions were prepared from this dilution. Sonication was applied to detach faecal coliforms from plant roots and the peat-sediment layer as described in Kansiime and Nalubega (1999).

Sedimentation. To assess the role of suspended material (fresh peat from the mat in the Nakivubo wetland) on settling out of coliforms from the water column, two batch reactors of 1 litre were compared. One was filled with plain wetland water (R1) and in the second one wetland water was mixed with 250 ml of fresh peat (R2). They were left to settle undisturbed for 17 days. Electrical conductivity, pH, temperature and dissolved oxygen were also monitored. The number of coliforms was determined daily. The decay rate of faecal coliforms was determined according to Chick's Law (Chick, 1908), from the plot of the logarithm of the fractional decrease in coliform concentration over time, versus time. The slope of the line (fitted by linear regression) is equal to the decay constant.

Results

Longitudinal distribution of coliforms in the Nakivubo wetland and Murchison Bay

An overview of the levels of faecal coliforms at different points of the transects is given in

Figure 1. The levels in the Murchison Bay decreased rapidly and at 4,000 m from transect 4 the numbers were only 16/100 ml. The ratio of faecal coliforms and the conductivity of the same sample decreased in the direction of the flow, indicating that faecal coliform levels were decreasing independently of dilution effects (conductivity was assumed to be conservative; Kansiime and Nalubega, 1999). Considering the ratio of conductivity to faecal coliforms, 2 log units of coliforms were removed from the water flowing through the papyrus zone (Inlet, T2-200 and T4-200), compared to 1 log unit retained in the *Miscanthidium* zone (Inlet, T2-400 and T4-400). The removal of coliforms was higher (92%) between T2 and T4 for the papyrus zone compared to (78%) between T1 and T2. For the wastewater flowing through the *Miscanthidium* zone, the highest removal of coliforms (69%) took place between T1 and T2, whereas it was lower between T2 and T4 (54%). The zones before T2 are dominated by papyrus.

Vertical profiles of faecal coliforms

The vertical distribution of faecal coliforms and conductivity in wetland compartments along transect 2 is depicted in Figure 2. The faecal coliforms numbers in the papyrus mat at T2-100 m (papyrus site), were not significantly different from those in the free water column below the mat, but were significantly lower than those recorded in the peat-sediment top layer (Figure 2A). Conductivity did not differ significantly in the compartments. The mat was thin (0.4 m), loose and soggy. The water in the water column had a lot of detritus, which was sedimenting out of the mat. At location T2-400 *(Miscanthidium* site), faecal coliform numbers were significantly lower in the *Miscanthidium* mat compared to the water column and peat-sediment layer (Figure 2B). The number of coliforms in the water column was lower than in the peat sediment top layer though the difference was not significant. Electrical conductivity had the same pattern as coliforms, having the lowest value in the mat. The *Miscanthidium* mat was compact and thick (1.2 m) with no visible signs of water on the surface, though the mat surface was soggy. The water below the *Miscanthidium* mat was always clear and very little detritus was observed in the water column. The water in the free water column was greyish-black in colour with a foul smell of wastewater, similar to that at the wetland inlet.

The vertical distribution of faecal coliforms and conductivity along transect 4 is depicted in Figure 3. Faecal coliform numbers did not differ significantly between papyrus compartments at T4-395 (p = 0.315). Electrical conductivity had the same distribution as coliforms and was also not significantly different among all the compartments (p = 0.40). The mat was also thin (0.5 m) and its bulk density was 14.7 ± 4.7 kg/m^3.

Similar to the observations made along transect 2, faecal coliform numbers were lowest in the mat of *Miscanthidium* at T4-410 m (Figure 3B). The coliform numbers in the water

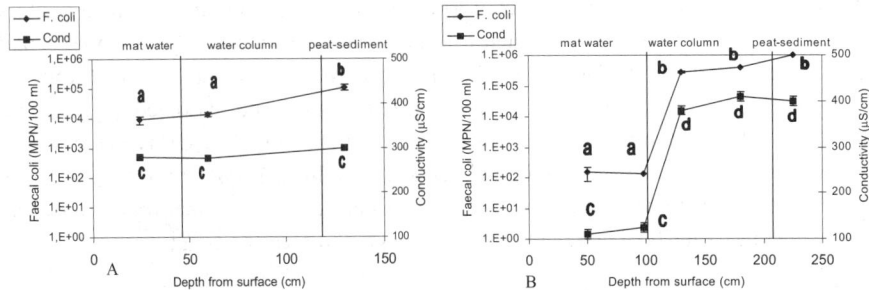

Figure 2 Vertical profile of faecal coliforms and conductivity. Bars indicate standard error of mean and n = 6. A = papyrus zone at 100 m at transect 2. B = *Miscanthidium* zone at 400 m at transect 2. Values in the graph with the same letters are not significantly different

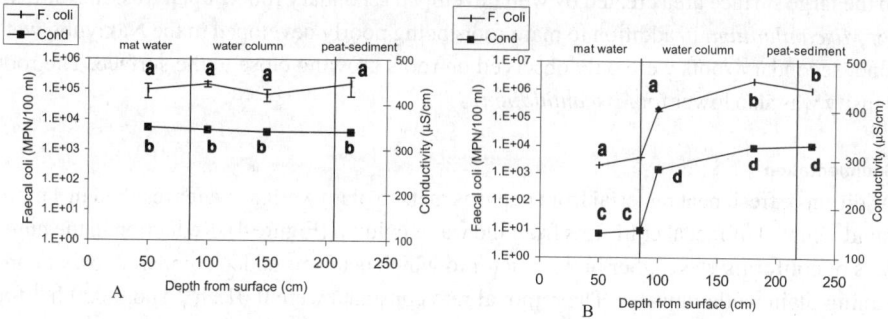

Figure 3 Vertical profile of faecal coliforms and conductivity. Bars indicate standard error of mean and n = 8. A = papyrus zone at 395 m at transect 4. B = *Miscanthidium* zone at 410 m at transect 4. Values in the graph with the same letters are not significantly different

column and peat sediment top layer were not significantly different from each other. Electrical conductivity depicted the same distribution as coliforms, and was again significantly lower in the mat. At this site the mat of *Miscanthidium* was 0.8 m thick and the bulk density was 69.7 ± 5.5 kg/m^3.

Comparison of papyrus and Miscanthidium zones

For the measurements made at the two sites at T4, faecal coliforms were significantly lower in the *Miscanthidium* mat liquid at T4-410 than other compartments. The number of coliforms in the water column and the peat-sediment layer of papyrus (T4-395) were significantly lower than those recorded in the corresponding *Miscanthidium* compartments. Since the sites at T4-395 and T4-410 are close to each other (15 m apart), low numbers of coliforms recorded at the former site may be attributed to the higher removal in the papyrus zone. In Table 1 information is given about the pH at the various sites. The *Miscanthidium* mat liquid had the lowest pH. The pH values for other compartments were all in the same range. The mat compartments were hypoxic whereas the water column and peat sediment layer were anoxic. In general the gradients (differences) of the measured variable were higher in *Miscanthidium* than in papyrus as a consequence of differences in bulk density of the mat.

Faecal coliform attachment to root surfaces

High numbers of faecal coliforms were detached from the root surfaces of papyrus ($18,000 \pm 4,000$ MPN/g DW of root, DW = dry weight) compared to *Miscanthidium* roots (220 ± 95 MPN/g DW). The high density of secondary roots observed on papyrus roots could have provided a high surface area, explaining the high numbers of attached coliforms per unit mass. The roots of *Miscanthidium* were generally less developed and secondary roots were only observed close to the surface, whereas inside the mat, root development was poor without noticeable secondary roots. Furthermore, the surface of *Miscanthidium* roots was smooth compared to those of papyrus. In addition to easy penetration of water into the mat, the high numbers of coliforms detached from papyrus roots may be attributed

Table 1 Variation of pH at the various sites

Site	Species	pH		
		mat	Water column	peat-sediment
T2-100	Papyrus	6.5	6.4	6.3
T2-400	*Miscanthidium*	5.0	6.3	6.4
T4-395	Papyrus	6.9	6.8	6.7
T4-410	*Miscanthidium*	5.7	6.5	6.6

to the large surface area created by well developed secondary roots of papyrus. In contrast, for *Miscanthidium*, in addition to main roots being poorly developed in the Nakivubo wetland, secondary roots were only observed on roots growing close to the surface. The root density was also lower for *Miscanthidium*.

Sedimentation

Addition of fresh peat material from papyrus mats to plain wetland water resulted in a more rapid removal of faecal coliforms from the water column (Figure 4). Reduction in the numbers of coliforms was higher in the reactor to which peat was added compared to that containing plain wetland water. The removal rate constants were 0.018 h^{-1} and 0.029 h^{-1} for the respective reactors containing plain wetland water and wetland water with peat.

Discussion

Guidance for this discussion is the conceptual model given in Figure 5. The longitudinal distribution of coliforms in Nakivubo wetland indicates that more coliforms were removed from the water in the papyrus dominated vegetation compared to that of *Miscanthidium* (Figure 1). This may be due to the thick and compact mat of *Miscanthidium* with little detritus falling out of it resulting in a clear water column beneath. Under these conditions, faecal coliforms remain in suspension rather than being attached to organic detritus and subsequent sedimentation to the wetland bottom. The combination of less resistance to hydraulic flow and little detrital material in the water column results in shorter retention and less die-off and settling of coliforms in this portion of the wetland. This results in a fast flow below the mat and hence a short hydraulic retention time. The coliforms would eventually be transported to the Inner Murchison Bay through this portion of the wetland. This was confirmed by the presence of high numbers of coliforms as far as 1500 m offshore in the Bay. This explains why the faecal coliform numbers in the region dominated by *Miscanthidium* along T2 to T4 were more or less the same.

The flow-through was also confirmed in storm events, during which storm water (brown in colour) was seen dispersing into the Inner Murchison Bay from this part of the wetland. The morphological difference between papyrus and *Miscanthidium* was further manifested in differences in the vertical distribution of faecal coliforms in different wetland compartments dominated by these aquatic macrophytes. The similarity in the number of coliforms in the mat and free water column below, in the papyrus sites, is attributed to easy vertical transport of wastewater into the mat. The high number of coliforms in the peat-sediment layers would be due to the high flux of suspended solids falling out of the mat, to which coliforms have been attached. The detritus and peat and attached coliforms eventually settle from the

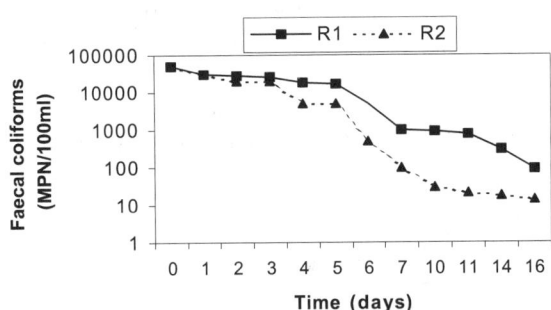

Figure 4 Reduction of faecal coliforms by die-off and settling under laboratory conditions. Reactor 1 (R1) had only swamp water, whereas reactor 2 (R2) contained peat and swamp water. (Temp. = 23°C, dissolved oxygen = 2.5 ± 0.8 mg/l)

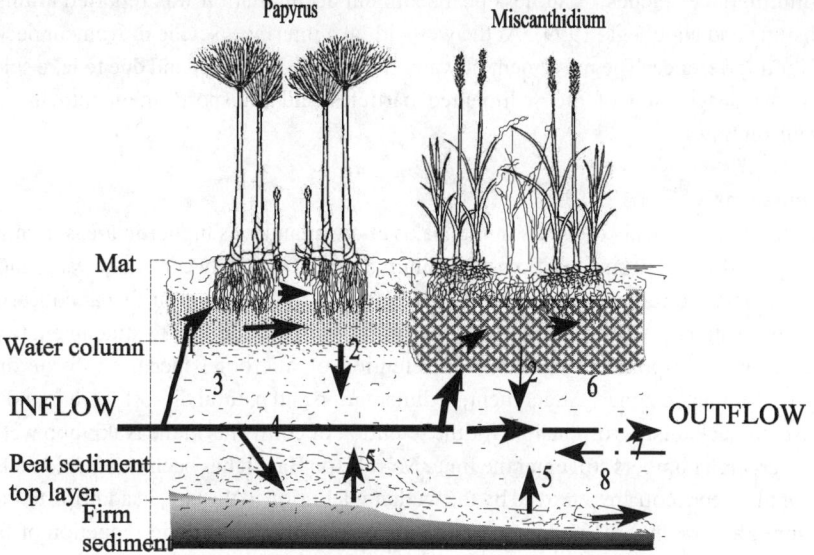

Figure 5 Conceptual model of faecal coliform transport in the Nakivubo wetland. Key: 1 = water penetration into the mat; 2 = detritus falling out of the mat with attached coliforms; 3 = flow through the water column; 4 = attachment to suspended particles in the water column; 5 = resuspension of sedimented particles; 6 = outflow of the wetland; 7 = back-flow from the lake and 8 = export of peat/sedimented matter

mat through the water column to the wetland bottom. Faecal coliforms, as reported for other organisms in the microscopic range, can adhere to particles dispersed in the water and wastewater (Brettar and Höfle, 1992). The detritus and peat deposited from the mat to the wetland bottom have been reported to be a sink of nutrients in floating papyrus wetlands (Howard-Williams, 1985) and this is likely to be true also for faecal coliforms in the Nakivubo wetland and especially in the floating zones of papyrus as observed at T2-100.

The low pH in the *Miscanthidium* mat could also result in a faster die-off of faecal coliforms in this compartment. Faecal coliforms have been reported to survive better at pH values between 6–7 with a rapid decline above and below this range (Solic and Kortulovic, 1994). Since some of these factors act synergetically, it was not possible in this study, to decidedly identify which of the investigated factors was responsible for the low numbers of coliforms in the *Miscanthidium* mat. The high correlation between faecal coliforms and conductivity along the vertical profiles in both papyrus and *Miscanthidium* indicates that coliforms and ions contributing to the conductivity are transported together in the Nakivubo wetland. Therefore, conductivity and coliforms are complementary indicators of water flow in the wetland.

The higher removal rate of coliforms (0.029 h^{-1}) in wetland water to which peat was added compared to plain wetland water (0.018 h^{-1}), demonstrates that the presence of settling material acted as seed material for sedimentation. This was confirmed by the presence of more coliforms in the settled material than in the water column in the reactor to which peat was added. This may explain why high numbers of coliforms were recorded in the peat sediment top layer compared to the water column of papyrus dominated areas especially at T2-100. Faecal coliforms have been reported to accumulate in the sediment of various aquatic systems (Lijklema *et al.*, 1987; Scheuerman *et al.*, 1989). According to Brettar and Höfle (1992), survival may also be due to reduced grazing pressure in the sediment and even regrowth. Finally, settled particles to which coliforms are attached may prolong the survival of coliforms. The similarity in numbers of coliforms in the peat-sediment and other compartments in the papyrus site at T4-395, is probably due to the influence of the diluting lake seiches. Also, less peat/sediment accumulation was reported at this site (Kansiime and Nalubega, 1999). At the wetland/lake interface, seiche movement becomes a major interference. The movement of water in and out of the wetland due to lake seiches, leads to resuspension of the sedimented particles and transports them into the Inner Murchison Bay.

Conclusions

The retention of faecal coliforms in the Nakivubo wetland was higher in areas dominated by papyrus. Less resistance to flow and minimal interaction between wastewater and mat (where plants are rooted) may be responsible for the low removal of faecal coliforms in areas dominated by *Miscanthidium*. Peat and plant debris in addition to attachment to plant surfaces, were responsible for the high retention of microbial faecal coliforms in the papyrus dominated zones. Attachment, sedimentation and natural die-off were found to be important mechanisms responsible for the retention of coliforms in the Nakivubo wetland. However, high numbers still enter the Inner Murchison Bay. It has been calculated that only 91% of the faecal coli are removed by the wetland. Dilution in the bay, in addition to factors like sunlight, grazing, pH and temperature are responsible for a rapid reduction of faecal coliforms in the Bay with distance. Very low numbers of faecal coliforms were recorded at a site close to Gaba Water Works.

Acknowledgements

We extend our sincere thanks to both the International Institute for Infrastructure, Hydraulics and Environment, and the Agricultural University of Wageningen, The Netherlands, who sponsored this study through the Ecotechnology Project. Further thanks go to Danielly Orozco, who participated in part of the fieldwork. Our thanks also go to Professor Patrick Denny who critically read this manuscript and made constructive suggestions.

References

APHA (1992). *Standard Methods for Examination of Water and Wastewater*. American Public Health Association, Washington. D C.

Azza, N.G.T., Kansiime, F., Nalubega, M. and Denny, P. (2000). Differential permeability of papyrus and *Miscanthidium* root mats in Nakivubo swamp, Uganda. *Aquat. Bot.* **67**, 169–178.

Brettar, I. and Höfle, M. (1992). Influence of ecosystematic factors on survival of *Escherichia coli* after large-scale release into lake water mesocosms. *Appl. Environ. Microbiol.* **58**(7), 2201–2210.

Chick, H. (1908). Investigations of the law of disinfection. *J. Hyg.*, **8**, 92.

Howard-Williams, C. (1985). Cycling and retention of nitrogen and phosphorus in wetlands: A theoretical and applied perspective. *Freshwater Biology* **15**, 391–431.

Kansiime, F and Nalubega, M. (1999). *Wastewater treatment by a natural wetland: the Nakivubo swamp, Uganda. Processes and implications*. PhD thesis, A.A. Balkema Publishers, Rotterdam, The Netherlands.

Kansiime, F., Nalubega, M., Bugenyi, F.W.B. and Tukahirwa, E.M. (1994). The role of Nakivubo swamp in maintaining the water quality of Inner Murchison Bay, Lake Victoria. *J. Trop. Hydrobiol. Fish.* **5**, 79–87.

Kizito, S. (1986). *The evaluation of pollution levels in Nakivubo Channel, Kampala-Uganda*. M.Sc. thesis, Makerere University, Uganda.

Lijklema, L., Habekotte, B., Hooijmans. C.M., Aalderink, R.H. and Havelaar, A.H. (1987). Survival of indicator organisms in a detention pond receiving combined sewer overflow. *Wat. Sci. Tech.* **19**(3/4), 547–555.

Scheuerman, P., Bitton, G. and Farrah, S. (1989). Fate of microbial indicators and viruses in a forested wetland. *In*: Hammer, A.D. (ed.) *Constructed Wetlands for Wastewater Treatment*. Lewis Publishers Chelsea, Michigan. pp 657–663.

Solic, M. and Kortulovic, N. (1992). Separate and combined effects of solar radiation, temperature, salinity and pH on the survival of faecal coliforms in sea water. *Marine Pollution Bulletin* **24**, 411–416.

Bacterial dynamics in the sub-surface constructed wetland

J. Vymazal*, J. Balcarová** and H. Doušová**

* Ecology and Use Wetlands, Říčanova 40, 169 00 Praha 6, Czech Republic.
(E-mail: vymazal@yahoo.com)
** Institute of Chemical Technology, Department of Water and Environmental Technology, Technická 3, 160 00 Praha 6, Czech Republic. (E-mail: jarmila.balcarova@vscht.cz)

Abstract Constructed wetlands have been shown to be capable of removing a wide variety of contaminants, including bacterial pollution. However, only limited information exists on the distribution of bacteria on roots of macrophytes growing in constructed wetlands. Constructed wetland with sub-surface horizontal flow at Nučice near Prague, Czech Republic, was put in operation in 1996. The system treats municipal sewage from 650 PE and the total area of the beds, planted with *Phalaris arundinacea* and *Phragmites australis* in alternate stripes perpendicular to the flow direction, is 3,224 m² (2 beds 62 × 26 m each). Pea gravel (8/16 mm) was used as a filtration material. During the period 1998–1999, distribution of total aerobic and anaerobic bacteria, coliform bacteria and fecal streptococci was monitored in wastewater as well as on roots of both macrophyte species. Counts of bacteria on root surface in the system at Nučice indicate that there is a steep decrease in bacterial numbers within the first few metres of the bed and that there is significantly more bacteria on roots of *Phragmites* as compared to *Phalaris*. There was no statistically significant influence of the season on the bacterial counts on roots of macrophytes.
Keywords Aerobic bacteria; anaerobic bacteria; fecal stroptococci; *Phalaris arundinacea*; *Phragmites australis*; roots; subsurface flow

Introduction

Constructed wetland with sub-surface horizontal flow at Nučice near Prague, Czech Republic, was put in operation in 1996. The system treats municipal sewage from 650 PE and the total area of the beds, planted with *Phalaris arundinacea* and *Phragmites australis* in alternate stripes perpendicular to the flow direction, is 3,224 m² (2 beds 62 × 26 m each). Pea gravel (fraction 8/16 mm) was used as a filtration material. Treatment effect of the constructed wetland at Nučice is presented in Table 1.

Methods

Removal of bacteria in the constructed wetland at Nučice was determined during the period of 1998–1999 by sampling inflowing and outflowing water on a monthly basis. Psychrophilic (PB), mesophilic (MB), anaerobic (AB) and total coliform (TC) bacteria

Table 1 Removal of organics and suspended solids in constructed wetland at Nučice. Average values for the period 1998–1999

Parameter	Concentration (mg L⁻¹)			Loading (kg ha⁻¹ d⁻¹)	
	Raw sewage	Inflow*	Outflow*	Inflow*	Outflow*
BOD_5	153	98	15.6	25.1	4.0
COD	328	204	44	52.1	11.4
SS	132	90	18.7	22.9	5.0

* Inflow = after pretreatment, i.e., inflow to the vegetated beds, outflow = outflow from the vegetated beds, i.e., final effluent

together with fecal streptococci (FS) were determined. The samples were serially diluted and plated onto pepton-agar (PB, MB), Endo-agar (TC), VL-agar (AB) and Slanetz-Bartley agar (FS). Plates were incubated at 20°C for 72 h (PB), 37°C for 24 h (TC) and 48 h (MB, FS, AB) (Doušová, 1999).

In the system at Nučice total coliform bacteria and fecal streptococci were also determined on roots of *Phragmites australis* (Common reed) and *Phalaris arundinacea* (Reed canarygrass). In this system, stripes of Reed canarygrass and Common reed are planted alternately perpendicular to wastewater flow. There are 7 stripes (4x *Phalaris* and 3x *Phragmites*) about 8 metres wide. Roots were sampled at three locations in the bed: 1) in the beginning of the first strip at metre 1 of the bed (*Phalaris*), 2) at the border between the first (*Phalaris*) and second strip (*Phragmites*) at metre 8 of the bed, and 3) at the last border between *Phragmites* and *Phalaris* at metre 52 of the bed. For each sample, roots (rhizomes were not included) from three plants were mixed. The size of the composite sample was about 30 g fresh mass. In the laboratory, the below-ground plant material was washed for 30 minutes in a shaker in sterile 0.85% physiological solution on NaCl and then the solution was filtered. After the root biomass was washed it was oven-dried at 70°C to a constant weight. The bacterial counts were expressed per unit of dry mass (Doušová, 1999).

Results and discussion

The removal of studied groups of bacteria is presented in Table 2. The results indicate high rate of removal, especially for fecal streptococci. Soto *et al.* (1998) reported removal of FS in experimental systems up to 99.9% but the inflow counts were 7–8 order higher than in our survey. Stott *et al.* (1996) reported the FS removal in the 200 m experimental bed to be 88.9, 99.2, 99.8 and 99.9% after 50, 100, 150 and 200 m passage with the final effluent count of 23 CFU ml^{-1}.

The removal of coliform bacteria amounted to 99.3%.. The high removal efficiency is in good agreement with literature data where efficiency of coliform removal is usually >97% (Gersberg *et al.*, 1987, 1989a, b; Cooper *et al.*, 1996; Stott *et al.*, 1996; Soto *et al.*, 1998) and many times exceeds 99.9% (e.g., Christian, 1990; Soto *et al.*, 1998). Outflow counts of total coliforms (80–740 CFU ml^{-1}) were lower than those reported for secondary treatment in the literature. Gersberg *et al.* (1989b) found an average of 5770 CFU ml^{-1} in Santee, California, Stott *et al.* (1996) reported 3400 and 3900 CFU ml^{-1} for secondary treatment systems in Egypt and England, respectively.

The counts of bacteria on roots of macrophytes are listed in Table 3. Presented results indicate that there is significantly more ($p < 0.05$) bacteria on roots of *Phragmites* as compared to *Phalaris* per unit of dry matter at the same distance from the inlet. It is surprising to some extent as *Phragmites* is considered to be a producer of root excrements which can effectively kill fecal indicators (Vincent *et al.*, 1994). Also, the decrease of bacteria numbers on *Phalaris* roots between metres 8 and 52 is much greater for mesophilic, psychrophilic and anaerobic bacteria as compared to numbers on *Phragmites* roots.

Table 2 Removal of bacteria from wastewater in constructed wetland at Nučice. Mean numbers in CFU ml^{-1} (min. and max. values in parentheses), removal in %

	Inflow	Outflow	Removal
Mezophilic bacteria	30,620 (1,400–104,500)	424 (100–1,900)	98.6
Psychrophilic bacteria	87,380 (1,900–350,000)	2,220 (570–4,000)	97.5
Total coliform bacteria	48,420 (1,000–137,500)	323 (80–740)	99.3
Total anaerobic bacteria	33,000 (500–85,000)	545 (220–1,300)	98.3
Fecal streptococci	9,280 (100–22,500)	20 (1–110)	99.8

Table 3 Mean numbers (CFU g^{-1}) of bacteria growing on roots of *Phalaris arundinacea* and *Phragmites australis* at different distances from the inflow

	Phalaris			Phragmites	
	1 m	8 m	52 m	8 m	52 m
Mesophilic bacteria	1,093	721	141	1,619	1,204
Psychrophilic bacteria	11,168	7,261	3,914	17,455	10,691
Total coliform bacteria	701	76	29	3,470	653
Total anaerobic bacteria	2,530	582	142	2,089	2,994
Fecal streptococci*	8,900	35	10	420	120

* Numbers in CFU 100 ml^{-1}

References

Christian, J.N.W. (1990). Reed bed treatment systems: experimental gravel beds in Gravesend – the Southern Water experience. In: *Constructed Wetlands in Water Pollution Control*, P.F. Cooper and B.C. Findlater (eds.), Pergamon Press, Oxford, pp.309–319.

Cooper, P.F., Job, G.D., Green, M.B. and Shutes, R.B.E. (1996). *Reed Beds and Constructed Wetlands for Wastewater Treatment*, WRc Swindon, U.K.

Doušová, H. (1999). *Dynamika některých skupin mikroorganismů ve vegetační čistírně odpadních vod* (Dynamics of selected groups of microorganisms in constructed wetland for wastewater treatment). M.Sc. thesis, Department of Water and Environmental Technology, Institute of Chemical Technology, Prague, Czech Republic (in Czech).

Gersberg, R.M., Brenner, R., Lyons, S.R. and Elkins, B.V. (1987). Survival bacteria and viruses in municipal wastewater applied to artificial wetlands. In: *Aquatic Plants for Water Treatment and Resource Recovery*, K.R. Reddy and W.H. Smith (eds.), Magnolia Publishing, Orlando, FL. Pp. 237–245.

Gersberg, R.M., Lyon, S.R., Brenner, R. and Elkins, B.V. (1989a). Integrated wastewater treatment using artificial wetlands: a gravel marsh case study. In: *Constructed Wetlands for Wastewater Treatment*, D.A. Hammer (ed.), Lewis Publishers, Chelsea, MI, pp. 145–152.

Gersberg, R.M., Gearheart, R.A. and Ives, M. (1989b). Pathogen removal in constructed wetlands. In: *Constructed Wetlands for Wastewater Treatment*, D.A. Hammer (ed.), Lewis Publishers, Chelsea, MI, pp. 431–445.

Soto, F., García, M., de Luís, E. and Bécares, E. (1998). Role of *Scirpus lacustris* in bacterial and nutrient removal from wastewaters. *Wat. Sci. Tech.*, **40**(3), 241–247.

Stott, R., Jenkins, T., Williams, J. Bahgat, M. May, E. Ford, M. and Butler, J. (1996). *Pathogen Removal and Microbial Ecology in Gravel Bed Hydroponic (GBH) Treatment of Wastewater*. Research Monographs in Wastewater Treatment and Reuse in Developing Countries No. 4, University of Portsmouth, U.K.

Vincent, G., Dallaire, S. and Lauzer, D. (1994). Antimicrobial properties of roots exudate of three macrophytes: *Mentha aquatica* L., *Phragmites australis* (Cav.) Trin. and *Scirpus lacustris* L. In: *Proc. 4th Internat. Conf. on Wetland Systems for Water Pollution Control*, ICWS Secretariat, Guangzhou, P.R.China, pp. 290–296.

Biota participating in wastewater treatment in a horizontal flow constructed wetland

J. Vymazal*, V. Sládeček** and J. Stach**

* Ecology and Use Wetlands, Říčanova 40, 169 00 Praha 6, Czech Republic. (E-mail: *vymazal@yahoo.com*)
** Havlovického 3, 140 00 Praha 6, Czech Republic
*** Hydroeko, Nová hospoda 135, 261 01 Příbram, Czech Republic

Abstract During the period 1996–1997, three constructed wetlands with sub-surface horizontal flow were investigated. All systems are designed to treat municipal sewage from small villages (150, 200 and 300 PE). The survey included microscopical identification of organisms in both wastewater and filtration substrate. The organisms were used as an indication of oxygen conditions (aerobic, anoxic and anaerobic) in the particular microenvironment. Saprobiological terms characterizing different levels of saprobity were employed to characterize inflowing wastewater, filtration bed and outflowing water. The occurrence of organisms was correlated with BOD_5 values in particular profiles. It has been found that the biocenosis in the inflowing wastewater differs from those found in the filtration bed and water outflowing from the vegetated beds. The organisms were grouped into those living under anaerobic and anoxic conditions and those living under aerobic conditions. More than 70 species of bacteria, amoebae, ciliates, rotifers, colorless flagellates, cyanobacteria and algae were found and the most important 45 species were figured in a plate together with saprobiological information for each species. Biota of the inflowing water is usually restricted to bacteria, ciliata and colorless flagellata while the organisms found in outflowing water as well as in periphyton growing on outflow structures indicate 2–3 levels better quality.
Keywords Bacteria; BOD_5; ciliates; saprobic index; subsurface flow

Introduction

There is a substantial lack of information on the biota found in wastewater and vegetated beds of constructed wetlands and, therefore, biota participating in the treatment process. The individual organisms may also indicate the aeration conditions and could be related to organic pollution (BOD_5). This relationship may be described by means of saprobic index (S_i) (Sládeček and Tuček, 1975; Sládeček, 1985). It has been shown that the BOD_5 value of 50 mg L^{-1} is an important boundary for biochemical processes because it divides aerobic and microaerobic conditions from anaerobic ones (Sládeček, 1985). The species have different indication value (I_s): the best indicators are found only at one saprobic level (indication value 10) and the worst indicators are found at four or more levels. In Table 1, the complete scale of saprobity is presented together with saprobity index values, examples of biotops, typical organisms and approximate BOD_5 values for each saprobic level.

Methods

During 1996–1997, three constructed wetlands with subsurface horizontal flow were investigated. All three systems are designed to treat municipal sewage from small villages. Major design parameters of surveyed constructed wetlands are given in Table 2. The survey included microscopical identification of organisms in both wastewater and filtration substrates. Samples were taken into small vials and analysed within the same or the next day. It was necessary to analyse the samples quickly because changes in the sample have been observed even within several hours after sampling. This was caused mostly by depletion

Table 1 Scale of saprobity with saprobity index and approximate BOD_5 values

Saprobity Index S_i	Saprobic level		Examples of biotops	Typical organisms	Approximate BOD_5 (mg/L)
−1.5			distilled water	stygobionts	
−1.0	k	katarobity	ground water		
−0.5			drinking water	stygophyls	0.0
−0.5			springs	stygoxens	0.0
0.0	x	xenosaprobity			
0.5			brooks	Rhodophyta	1.0
1.0	o	oligosaprobity	small streams	Xanthophyta	
1.5			lakes	Chrysophyta, desmids	2.5
2.0	b	β-mesosaprobity	rivers	Chlorococcales	
2.5			ponds		5.0
3.0	a	α-mesosaprobity	pollution	*Leptomitus*	
3.5					
4.0	p	polysaprobity	pollution	*Sphaerotilus natans*	10.0
4.5					50.0
Anaerobic conditions					
4.5					50.0
5.0	I	isosaprobity	sewage	ciliata	
5.5					400
6.0	m	metasaprobity	H_2S zone	colorless flagellata	
6.5					700
7.0	h	hypersaprobity	CH_4 zone	bacteria or fungi	
7.5					1,200
8.0	u	ultrasaprobity	abiotic zone	without life	
8.5			(towards solid phase)	(spores, cysts)	120,000

Table 2 Major design parameters of monitored constructed wetlands

	Kotenčice	Svaty	Jan Zbenice
Starting date	10/94	10/95	6/96
Type of wastewater*	S+R	S	S+R
Population equivalent	326	147	200
Bed area (m²)	1,800	770	1,000
Substrate	rock	rock	pea gravel
Fraction (mm)	4–8	8–16	4–8
Bed depth (m)	0.6–0.8	0.6–0.8	0.6–0.8
Vegetation	*Phragmites*	*Phragmites*	*Phragmites Phalaris*

* S = sewage, R = stormwater runoff

of dissolved oxygen in the sample. The samples were analysed live in a Sedgwick-Rafter counting cell.

Results

In Table 3 and Figure 1, major organisms found in wastewater, vegetated beds and outflowing water are presented. There is a clear difference between sampling locations. The saprobic index (S_i) of major organisms found in raw wastewater, in the middle of the vegetated bed and in outflowing water varied between 4.0–5.0, 1.9–4.2 and 0.8–3.2, respectively. The presence of a particular species reflects the BOD_5 concentrations at sampling points. For example, the average BOD_5 of raw sewage at Zbenice was 101 mg L^{-1}, the average BOD_5 in the outflowing water was 6.9 mg L^{-1}.

Biota of the inflowing water is usually restricted to bacteria, ciliata and colorless flagellata. In particular, colorless flagellata (Table 3, Nos. 14–18), indicators of anaerobic conditions, are very abundant. Biota found in vegetated beds is more variable and reflects the aerobic-microaerobic-anoxic mosaic of the filtration bed. The organisms found in outflow-

Table 3 Major organisms found in wastewater and vegetated beds of monitored constructed wetlands

Taxon	x	o	b	a	p	i	m	I_s	S_i
Organisms found in raw wastewater									
1. Zoogloea ramigera	–	–	–	+	4	5	1	2	4.7
2. Leucothrix mucor	–	–	–	2	6	2	–	3	4.0
3. Sphaerotilus natans	–	–	–	+	7	3	–	4	4.3
4. Beggiatoa alba	–	–	–	1	5	4	–	2	4.3
5. Thiothrix nivea	–	–	–	+	4	6	–	3	4.6
6. Mycophyta g. sp.	–	–	–	+	+	+	–	–	4.0
7. Paramecium putrinum	–	–	–	+	6	4	–	3	4.4
8. Tetrahymena pyriformis	–	–	–	+	5	5	–	3	4.5
9. Dexiostoma campylum	–	–	–	+	6	4	–	3	4.4
10. Colpidium colpoda	–	–	–	+	5	5	–	3	4.5
11. Glaucoma scintillans	–	–	–	+	7	3	–	4	4.3
12. Acineria incurvata	–	–	–	–	5	5	–	3	4.5
13. Vorticella microstoma	–	–	–	–	5	5	–	3	4.5
14. Bodo caudatus	–	–	–	+	3	7	–	4	4.7
15. Polytoma uvella	–	–	–	–	+	10	–	5	5.0
16. Cercobodo longicauda	–	–	–	+	3	7	–	4	4.7
17. Hexamitus inflatus	–	–	–	+	1	9	–	5	4.9
18. Trepomonas rotans	–	–	–	–	+	10	–	5	5.0
Organisms found in the substrate of the bed									
19. Paramecium caudatum	–	–	–	2	7	1	–	3	3.9
20. Trithigmostoma cucullulus	–	–	2	5	3	–	–	2	3.1
21. Uroleptus piscis	–	–	3	6	1	–	–	3	2.8
22. Stentor roeseli	–	1	4	5	–	–	–	2	2.4
23. Stentor polymorphus	–	1	5	4	–	–	–	2	2.4
24. Chilodonella uncinata	–	–	2	6	2	–	–	3	3.0
25. Leptothrix ochracea	–	3	5	2	–	–	–	2	1.9
26. Euglena viridis	–	–	–	2	5	3	–	2	4.1
27. Vahlkampfia limax	–	–	–	+	8	2	–	4	4.2
28. Arcella vulgaris	–	1	8	1	–	–	–	4	2.0
29. Cinetochilum margaritaceum	–	1	3	3	3	–	–	1	2.8
30. Carchesium polypinum	–	–	2	7	1	–	–	3	2.9
31. Nematoda g. sp.	–	–	+	+	+	–	–	–	3.0
32. Chironomus cf. thummi	–	–	–	1	9	–	–	5	3.9
33. Limnodrilus hoffmeisteri	–	–	–	2	8	–	–	4	3.8
Organisms found in water outflowing from vegetated beds									
34. Lecane clara	–	4	4	2	–	–	–	2	1.8
35. Colurella adriatica	1	3	4	2	–	–	–	1	1.8
36. Dicranophorus grandis	–	6	3	1	–	–	–	3	1.4
37. Lepadella patella	+	6	4	–	–	–	–	3	1.4
38. Rotaria rotatoria	–	+	1	6	3	–	–	3	3.2
39. Chlamydomonas sp.	–	–	+	10	+	–	–	5	3.0
40. Navicula sp.	–	+	5	5	–	–	–	3	2.5
41. Ulothrix zonata	–	–	1	9	–	–	–	5	2.9
42. Klebsormidium flaccidum	1	3	4	2	–	–	–	1	1.7
43. Phormidium spp.	–	+	+	10	–	–	–	5	3.0
44. Meridion circulare	3	6	1	–	–	–	–	3	0.8
45. nauplius Cyclops	–	+	8	2	–	–	–	4	2.2

1–5: bacteria, 7–13: Ciliata, 14–18: colorless flagellata, 19–24: Ciliata, 25: bacterium, 26: alga, 27–28: Rhizopoda, 29–30: Ciliata, 32: Chironomidae, 33: Oligochaeta, 34–38: Rotatoria, 39–44: algae including blue-green algae (cyanobacteria), 45: Copepoda

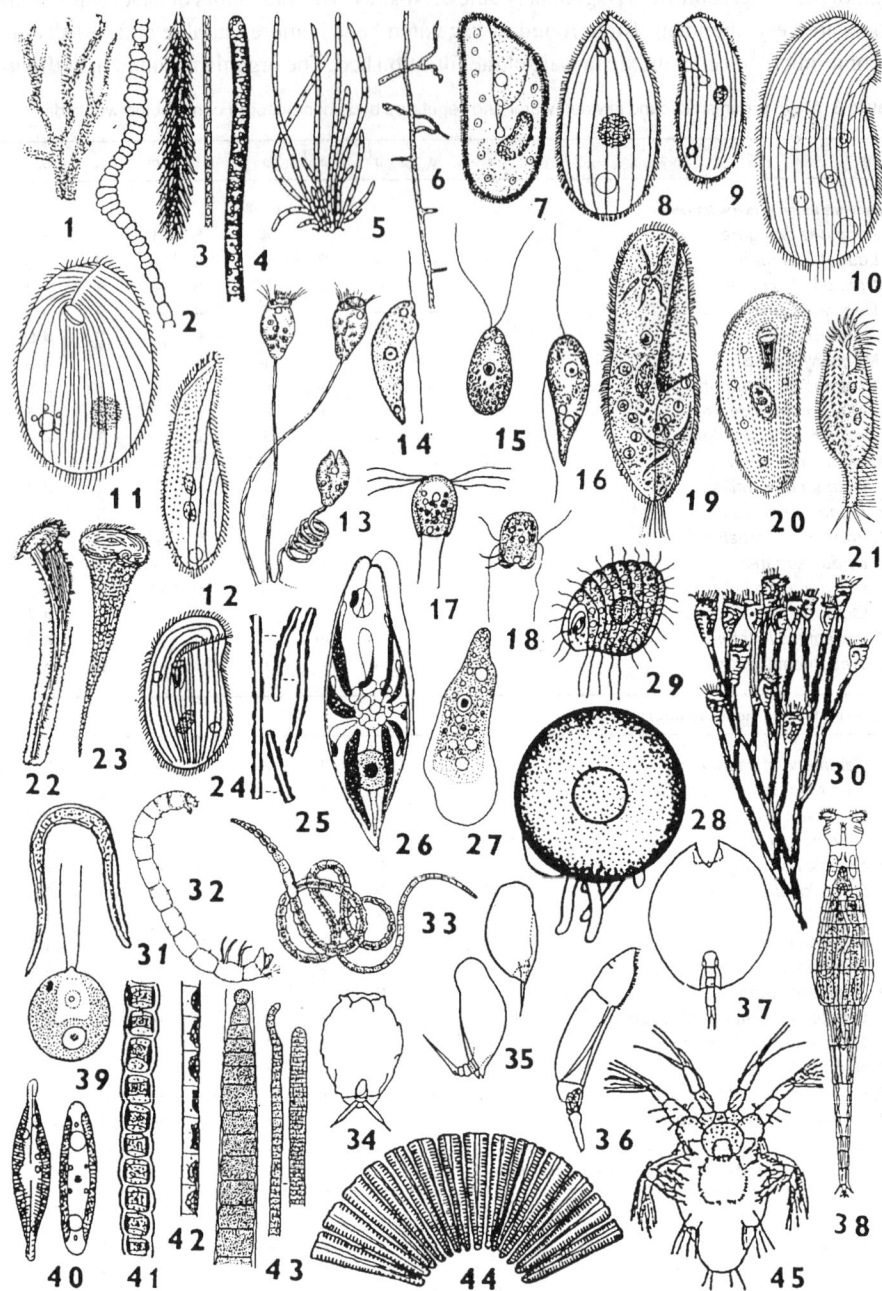

Figure 1 Major organisms found in wastewater and vegetated beds of monitored constructed wetlands

ing water as well as in periphyton growing on outflow structures indicate a 2–3 levels better quality of water from a saprobiological point of view compared to inflow.

References

Sládeček, V. (1985). Scale of saprobity. *Verh. Internat. Verein. Limnol.*, **22**, 2337–2341.
Sládeček, V. and Tuček, F. (1975). Relation of the saprobic index to BOD_5. *Water Research*, **7**, 791–794.

Removal of *Salmonella* and microbial indicators in constructed wetlands treating swine wastewater

V.R. Hill and M.D. Sobsey

Department of Environmental Sciences and Engineering, School of Public Health, University of North Carolina at Chapel Hill, CB#7400, Chapel Hill, North Carolina 27599, USA

Abstract Reductions of *Salmonella* bacteria and enteric microbial indicator organisms were measured in swine wastewater treated by a field-scale surface flow (SF) constructed wetland at a commercial hog nursery in North Carolina and in laboratory-scale SF and subsurface flow (SSF) constructed wetland reactors. Overall reductions of *Salmonella*, fecal coliforms and *E. coli* were 96, 98 and 99%, respectively, in the two-cell field-scale wetland. Somatic and F-specific coliphage viral indicators were reduced by 99 and 98%, respectively. Reductions of *Salmonella*, fecal coliforms and *E. coli* were similar in the first cell of the field system and in the laboratory-scale SF wetland operated at a TKN loading of 25 kg ha^{-1} d^{-1} and 30°C (approximately 70, 90 and 90%, respectively). In the SSF wetland reactor, *Salmonella* and fecal coliform reductions were 80 and 98%, respectively, at a 40 kg TKN ha^{-1} d^{-1} loading and 99.8 and 99.99%, respectively, at a 10 kg TKN ha^{-1} d^{-1} loading. These results show that SF constructed wetlands can be effective for reducing enteric pathogens in swine wastewater and that greater removals can be achieved using SSF designs and lower TKN loading rates.

Keywords Constructed wetlands; pathogens; salmonella; swine waste; wastewater treatment

Introduction

Manure and wastewater from animal feeding operations (AFOs) are potential sources of a wide range of pollutants, including pathogens, that can be transported to environmental resources near farms. Removal or inactivation of enteric pathogens in swine wastewater is important because potentially infectious human pathogens (e.g., *Salmonella, Yersinia, Cryptosporidium parvum*, emerging viruses like swine hepatitis E virus) may be present and transported to nearby water resources by surface water runoff and groundwater infiltration (Cole *et al.*, 1999). Fecal microbes have been found at high concentrations in flushed swine waste (Hill and Sobsey, 1998). In the US, waste generated on swine farms (as well as dairy farms) is often stored in lagoons from which the liquid is periodically applied to land application fields (i.e., sprayfields). Public pressure has been increasing to develop alternative treatment systems to lagoons due to concerns regarding potential public health and environmental risks of lagoon-sprayfield waste management systems. Constructed wetlands represent a promising alternative or additional treatment system for wastewaters generated by animal feeding operations (AFOs). Over 100 constructed wetland systems are treating livestock wastewater in the US (CH2M Hill and Payne Engineering, 1997). Most of these are surface flow (SF) systems, but some use subsurface flow (SSF) designs.

Previous research indicates that enteric microbe removal efficiency in constructed wetlands can be affected by changes in hydraulic loading rate (HLR) and resultant hydraulic residence time (HRT) (Tanner *et al.*, 1995), the presence of vegetation (Soto *et al.*, 1999), and whether the systems are SF or SSF (Kadlec and Knight, 1996). Enteric bacteria have been reported to be removed by 90-99.9% (Tanner *et al.*, 1995; Ottová *et al.*, 1997; Gerba *et al.*, 1999) and viruses by 90–99% (Gersberg *et al.*, 1989; Gerba *et al.*, 1999) in SF and SSF constructed wetlands. Although the potential for substantial reductions of enteric microbes in constructed wetlands is suggested by previous research, much of this research has been limited to the analysis of fecal coliforms and other bacterial indicators which may not be indicative of the removal of other microbes, such as viruses, protozoan parasites or helminths. Additionally, little research has focused on the removal of frank pathogens, such as *Salmonella*, in constructed wetlands. Available data indicate that protozoan pathogens such as *Cryptosporidium parvum* and *Giardia lamblia* may be less effectively removed than enteric bacteria and viruses in wetland systems, with reported removal efficiencies of less than 90% (Gerba *et al*., 1999). Research on the removal of free-living amoebae indicates that these microorganisms can be removed by 75-95% in vegetated SSF wetlands (Rivera *et al*., 1995). Other research indicates that helminth ova such as *Ascaris lumbricoides* can be removed by 80-90% in SSF wetlands (Stott *et al*., 1997). The spore-forming bacterium, *Clostridium perfringens*, was investigated in this study as an indicator for the removal of environmentally stable protozoan and helminth parasites like *C. parvum* and *Ascaris* spp., respectively.

The objectives of this study were to (1) quantify reductions of the pathogen, *Salmonella*, and enteric bacterial and viral indicator microbes in a two-cell SF wetland operating at a commercial swine farm and in laboratory-scale SF and SSF wetland reactors, (2) investigate the effect of loading rate on enteric microbe reductions in the laboratory-scale wetland reactors, (3) evaluate the effect of vegetation on enteric microbe removals by comparing removals in vegetated and non-vegetated SSF reactors, and (4) investigate correlations between microbial indicator and *Salmonella* reductions in constructed wetlands.

Methods and materials
Operation and sampling of constructed wetland systems
The field-scale constructed wetland system was installed at a 2,600-head swine nursery in North Carolina, USA in 1992 (Hunt *et al*., 1998). The SF system contained two cells (each 3.6 m × 33.5 m) in series, planted with bur-reed (*Sparganium americanuum*) and cattails (*Typha angustifolia* and *Typha latifolia*). Wastewater from the anaerobic lagoon at the farm was diluted 1:1 with water and pumped through the system at a total nitrogen loading rate of 25 kg ha^{-1} d^{-1}. The hydraulic loading rate (HLR) of lagoon liquid to the field system averaged 2.0 cm d^{-1}.

Three laboratory-scale reactors (76 cm × 30 cm × 61 cm polyethylene tanks) were installed in a walk-in incubator in October 1998: a SF reactor with soft-stem bulrush (*Scoenoplectus validus*) planted in sandy loam (30 cm deep); a SSF reactor with soft-stem bulrush planted in 30-cm-deep expanded-slate gravel (9.5 mm average diameter) (Carolina Stalite Co.; Salisbury, North Carolina); and an non-vegetated 30-cm deep SSF expanded-slate gravel reactor. Full spectrum "Sunshine" plant grow-lights (General Electric) were suspended above each reactor. Beginning in February 1999, lagoon liquid diluted 1:1 with tap water was pumped from a refrigerated central distribution tank into each reactor using peristaltic pumps. The incubator temperature was set at 30°C to model Summer temperature conditions. Between September 1999 and March 2000, the reactors were studied at an initial total Kjeldahl nitrogen (TKN) loading rate of 40 kg ha^{-1} d^{-1} (3.8 cm d^{-1} HLR), an intermediate TKN loading rate of 25 kg ha^{-1} d^{-1} (2.3 cm d^{-1} HLR), and a final TKN loading rate of 10 kg ha^{-1} d^{-1} (1.1 cm d^{-1} HLR). Tracer tests were conducted at each loading rate using sodium fluoride to measure the hydraulic residence time (HRT) in each reactor. *Salmonella typhimurium*, isolated from a commercial hog farm lagoon, was spiked into the influent tank to maintain an approximate influent concentration of 1000 000 MPN per 100 mL.

Sample collection and analysis
Grab samples were collected from the field system influent and effluent, as well as between

the two cells. Between March 1997 and May 2000, 18 sets of samples were collected from the field-scale system and analyzed for fecal coliforms, *Eschericia coli*, enterococci, *Clostridium perfringens* spores, somatic coliphages and male-specific (F-specific) coliphages. Samples were also analyzed for *Salmonella* spp., chemical oxygen demand (COD), pH and total suspended solids (TSS) during 9 rounds of sampling conducted between December 1998 and May 2000. For the laboratory study, samples were collected from the influent tank and the effluent from each of the three reactors. At least four rounds of sampling were conducted at each TKN loading rate. Samples were analyzed for fecal coliforms, *E. coli*, enterococci, *C. perfringens* spores, somatic coliphages, F-specific coliphages, *Salmonella* spp., pH, COD, and TSS.

Fecal coliforms and *E. coli* were enumerated by membrane filtration as described in *Standard Methods for the Examination of Water and Wastewater* (1999). Enterococci were enumerated by incubating membrane filters on modified mE agar (Difco) for 48 h at 41°C (Levin *et al.*, 1975). *C. perfringens* spores were analyzed by filtering heat-treated (60–70°C for 20 minutes) samples and incubating membranes on mCP agar (Acumedia®) in an anaerobic jar for 18–24 h at 41°C (Bisson and Cabelli, 1979). Viral indicators were enumerated using the single-agar layer, pour plate plaque technique (Grabow and Coubrough, 1986), with somatic and F-specific coliphages detected using host bacteria *E. coli* CN-13 and *E. coli* Famp, respectively. *Salmonella* were enumerated using the most probable number (MPN) technique as follows: pre-enrichment for 20–24 h at 37°C in buffered peptone water (Difco) (Edel and Kampelmacher, 1973); enrichment for 24 h at 43°C in Rappaport-Vassiliadis R10 broth (Difco) (Vassiliadis, 1983); parallel isolation on Salmonella-Shigella agar (Difco) and Rambach® agar (CHROMagar Microbiology); and biochemical testing of a subset of presumptive positives using BBL® Enterotube™ II media (Becton Dickinson). COD was measured using the Hach COD System and a Spectronic 1201 spectrophotometer (Milton Roy) set at $\lambda = 620$ nm. TSS was measured using Standard Method 2540 D (1999). For tracer tests, fluoride ion concentration was measured using an Accumet™ ion-selective electrode. First-order volumetric and areal rate constants (k) were calculated using standard exponential decay equations (Kadlec and Knight, 1996).

The nonparametric Wilcoxon Matched Pairs test was used to (1) compare influent and effluent data for individual microbes to determine whether the treatment systems achieved significant reductions of these microbes; (2) evaluate whether differences in measured treatment effectiveness between the various microbes were significant; and (3) investigate differences in treatment effectiveness for the laboratory-scale wetland reactors. Correlation analysis of the $\log_{10}$ reductions measured for the different microbes was performed using the Spearman Rank Order nonparametric method. Linear regression analysis was used to determine if loading rate was significantly associated with enteric microbe removals. For all statistical analyses, significance is considered to be a p value ≤ 0.05. All statistical analysis was performed using *Statistica* software (StatSoft, Inc.).

Results and discussion

Diluted lagoon liquid pumped through the SF constructed wetland at the hog nursery had an average COD of 620 mg L^{-1}, TSS of 190 mg L^{-1} and pH of 7.9 (Table 1). COD and TSS were reduced by 71 and 92%, respectively, in effluent from the system (Cell 2 effluent).

The influent to the wetland system had geometric mean concentrations of fecal coliforms and *E. coli* of 240 000 and 180 000 colony forming units (CFU) per 100 mL, respectively (Figure 1). These bacterial indicators were significantly reduced in each cell of the wetland system: 1.0 $\log_{10}$ (91%) and 1.1 $\log_{10}$ (92%), respectively, in Cell 1 and 0.7 $\log_{10}$ (80%) and 0.8 $\log_{10}$ (84%), respectively, in Cell 2. Overall reductions for fecal coliforms and *E. coli* were 1.7 $\log_{10}$ (98%) and 1.9 $\log_{10}$ (99%), respectively.

Table 1 Average wastewater characteristics in surface flow constructed wetland at commercial swine nursery, North Carolina, USA

Sample Location	COD (mg L^{-1})	TSS (mg L^{-1})	pH
System Influent	620	190	7.9
Cell 1 Effluent	250	33	7.8
Cell 2 Effluent	180	15	7.8

Enterococci were less effectively reduced in the wetland system than were fecal coliforms and *E. coli*, with measured reductions of 0.7 $\log_{10}$ (80%) in Cell 1 and 0.9 $\log_{10}$ (87%) in effluent from the system. Enterococci concentrations in wetland system effluent were significantly lower than in system influent, but the concentrations in Cell 2 effluent were not significantly different than in Cell 1 effluent. Overall, enterococci reductions in the constructed wetland were significantly lower than for all the other enteric microbes studied. The reductions of the other enteric microbes were not significantly different from each other in the wetland system. Enterococci (as well as other fecal streptococci) are generally thought to be more resistant to environmental degradation than fecal coliforms, including *E. coli*. These data suggest that enterococci may be a good indicator for more environmentally stable bacterial pathogens. It is also possible, however, that the low reductions of enterococci during some sampling rounds may reflect the reproduction of these organisms in the wetland system. On three occasions, enterococci concentrations in effluent from the wetland system were higher than in system influent, while fecal coliform and *E. coli* concentrations were reduced by 1.5 to 2 $\log_{10}$. Research has shown that enterococci can exist naturally and reproduce on some plant species (Clausen *et al.*, 1977; Anderson *et al.*, 1997).

Salmonella were measured at far lower concentrations than the indicator bacteria in influent to the wetland system (Figure 1), although they were readily detectable in 100 mL volumes. *Salmonella* were reduced from an influent geometric mean of 350 MPN/100 mL to a mean of 130 MPN/100 mL in Cell 1 (a 0.4 $\log_{10}$, or 63% reduction) and a mean of 12 MPN/100 mL in effluent from the system (a 1.5 $\log_{10}$, or 96% reduction). *Salmonella*

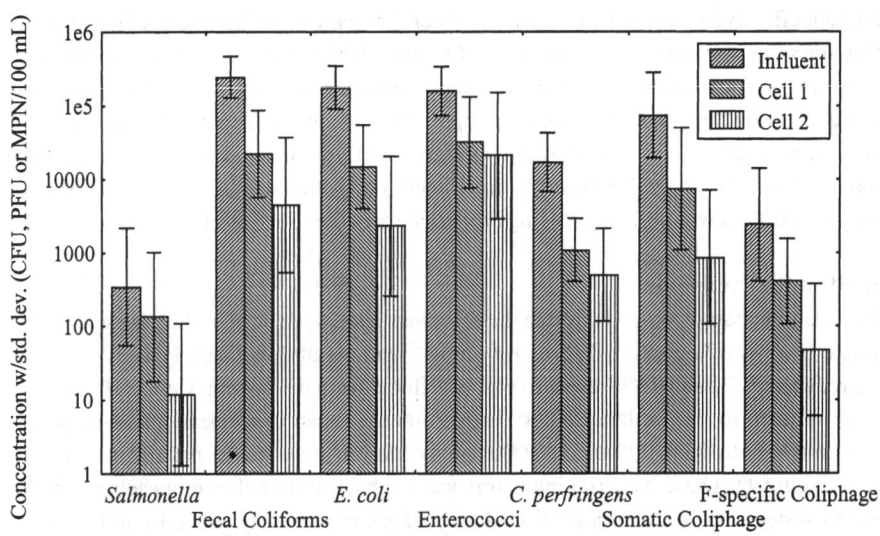

Figure 1 Geometric mean concentrations of Salmonella and microbial indicators in surface flow constructed wetland treating swine lagoon liquid

reductions in Cell 1 were not significant ($p = 0.08$), but were significant in system effluent. *Salmonella* concentrations varied greatly in the wetland system, which may have been due to seasonal and climatic conditions as well as differences in the prevalence of infection and fecal excretion levels in the numerous groups of animals that moved through the nursery facility during the study. Variations in constructed wetland treatment performance for *Salmonella* reduction were significantly correlated with reductions of fecal coliforms ($R = 0.85$), *E. coli* ($R = 0.82$) and *C. perfringens* spores ($R = 0.71$), but were not significantly correlated with reductions in enterococci or the coliphages. *Salmonella* reductions in the SF wetland were not significantly different than fecal coliform or *E. coli* reductions, but were significantly higher than enterococci reductions.

C. perfringens spore concentrations were reduced by 1.2 $\log_{10}$ (93%) in Cell 1 effluent and by 1.5 $\log_{10}$ (97%) in system effluent. The 95% confidence limits for *C. perfringens* spore reductions overlap slightly between Cell 1 and Cell 2 effluent. This was likely due to highly variable seasonal and climatic conditions that affected both the constructed wetlands performance as well as the treatment performance of the lagoon that was used as the source of wastewater for the study. When the data were analyzed using paired nonparametric statistics it was determined that the effluent concentrations of *C. perfringens* spores from Cell 2 were significantly lower than corresponding concentrations in influent to the cell. Reductions of *C. perfringens* spores were significantly correlated with reductions of *Salmonella* ($R = 0.71$), fecal coliforms ($R = 0.64$), *E. coli* ($R = 0.68$) and somatic coliphages ($R = 0.59$). These results suggest that more environmentally-stable enteric microbes (e.g., *Cryptosporidium parvum* oocysts, *Giardia lamblia* cysts and helminth ova) may also be significantly removed from wastewater by similarly designed and operated SF constructed wetland systems. Because bacterial spores and parasites are relatively stable in the environment, release of these microbes from the wetland treatment system is a possibility, especially during system perturbations (e.g., precipitation events).

Somatic and F-specific coliphages were reduced to a similar, and significant, extent in each cell of the SF constructed wetlands system: 1.0 $\log_{10}$ (90%) and 0.8 $\log_{10}$ (83%), respectively, in Cell 1 and 0.9 $\log_{10}$ (87%) and 1.0 $\log_{10}$ (90%), respectively, in Cell 2. Overall reductions of these coliphages in the wetland system were 1.9 $\log_{10}$ (99%) and 1.8 $\log_{10}$ (98%), respectively. $\log_{10}$ reductions of these two viral indicator microbes were more strongly correlated with each other ($R = 0.83$) than with the other microbes studied. Somatic coliphage reductions were also significantly correlated with fecal coliforms ($R = 0.54$), *E. coli* ($R = 0.51$), enterococci ($R = 0.50$) and *C. perfringens* spores ($R = 0.59$). Other than with somatic coliphages, F-specific coliphage reductions were only significantly correlated with enterococci ($R = 0.63$).

The first-order areal rate constants (k_a) for inactivation/removal of enteric microbes in the two-cell SF constructed wetland (as cm/d) were calculated to be: 6.7 for *Salmonella*, 8.0 for fecal coliforms, 8.7 for *E. coli*, 4.0 for enterococci, 8.9 for somatic coliphages, 7.8 for F-specific coliphages, and 7.0 for *C. perfringens* spores. The calculated rate constant for fecal coliforms is within the range of reported first order areal decay rates for municipal SF constructed wetland systems (Kadlec and Knight, 1996).

In the laboratory-scale SF constructed wetland reactor, *S. typhimurium* was reduced by 0.5 $\log_{10}$ (71%) at a TKN loading of 25 kg ha^{-1} d^{-1} (Figure 2). This result is similar to the 0.4 $\log_{10}$ reduction measured in Cell 1 of the field-scale SF constructed wetland operated at a total nitrogen loading of 25 kg ha^{-1} d^{-1} (50 kg ha^{-1} d^{-1} if considering the loading to Cell 1 only). Fecal coliform and *E. coli* reductions in the laboratory-scale SF reactor at a TKN loading of 25 kg ha^{-1} d^{-1} (0.9 and 1.0 $\log_{10}$, respectively) were also similar to corresponding reductions in the first cell of the field system (1.0 and 1.1 $\log_{10}$, respectively). In the laboratory-scale SF reactor, treatment effectiveness for the enteric bacteria was lowest at

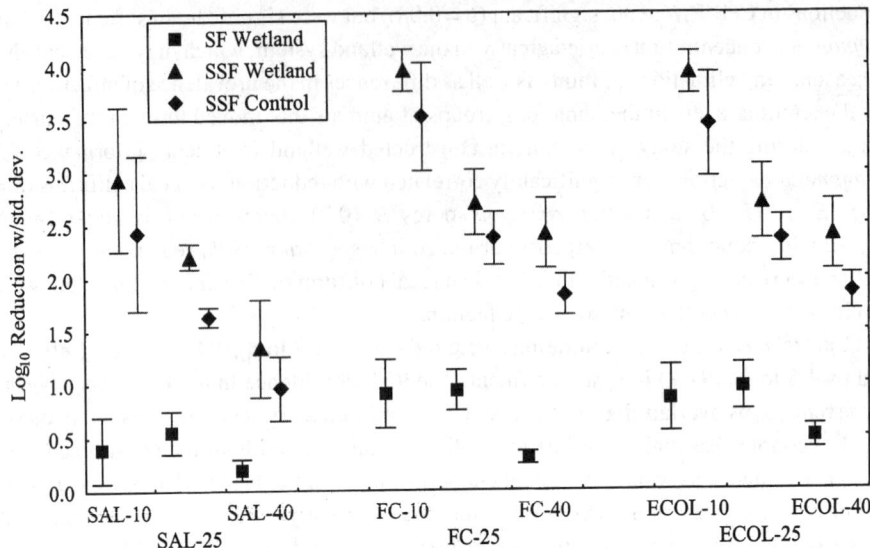

Figure 2. Reductions of *Salmonella typhimurium*, fecal coliforms, and *E. coli* in surface flow wetland, subsurface flow wetland, and subsurface flow control reactors at TKN loading rates of 10, 25 and 40 kg ha^{-1} d^{-1}

the highest loading rates, but the association between loading rate and microbe removal was significant only for fecal coliforms. In the SSF reactors, loading rate was significantly associated with $\log_{10}$ reductions of *S. typhimurium*, fecal coliforms and *E. coli*.

Salmonella, fecal coliform and *E. coli* reductions were significantly greater in the SSF reactors than in the SF reactor at the same TKN loading rates (Figure 2). Reductions of these microbes decreased in the vegetated SSF reactor as the TKN loading rate was increased: 2.9 $\log_{10}$ to 0.7 $\log_{10}$ (99.8 to 80%) for *Salmonella*, 4.0 $\log_{10}$ to 1.7 $\log_{10}$ (99.99 to 98%) for fecal coliforms, and 3.9 $\log_{10}$ to 1.6 $\log_{10}$ (99.98 to 97%) for *E. coli*. Geometric mean reductions of these microbes in the non-vegetated SSF control reactor were significantly lower than in the vegetated SSF reactor, indicating that the presence of vegetation in the SSF wetland had a positive effect on bacterial reductions. These data support the conclusions of previous research on the effects of wetlands vegetation on enteric microbe reductions in SSF treatment wetlands (Gersberg *et al.*, 1989; Soto *et al.*, 1999; Warren *et al.*, 2000).

HRT in the laboratory-scale reactors varied from 2 to 4 days in the SF reactor, 4 to 15 days in the SSF wetland reactor, and 4 to 17 days in the SSF control reactor. Corresponding volumetric decay rates ($k_{v,30}$) for fecal coliforms varied from 0.4 to 0.9 d^{-1} in the SF wetland, 0.6 to 1.4 d^{-1} in the SSF wetland and 0.5 to 1.1 d^{-1} in the SSF control reactor. Areal decay rates ($k_{a,30}$) for fecal coliforms were 2.9 to 6.3 cm d^{-1} in the SF wetland, 9.1 to 20 cm d^{-1} in the SSF wetland and 8.1 to 15 cm d^{-1} in the SSF control. Reduction rates for *E. coli* and fecal coliforms were similar. *S. typhimurium* volumetric decay rates were lower than those for *E. coli* and fecal coliforms, varying from 0.2 to 0.5 d^{-1} in the SF wetland, 0.5 to 0.8 d^{-1} in the SSF wetland, and 0.3 to 0.6 d^{-1} in the SSF control. Areal decay rates for *S. typhimurium* were 1.4 to 3.6 cm d^{-1} in the SF wetland, 6.8 to 11 cm d^{-1} in the SSF wetland and 5.6 to 8.0 cm d^{-1} in the SSF control. The areal decay rates calculated for *Salmonella*, fecal coliforms and *E. coli* were higher for the field SF wetland than in the laboratory-scale SF wetland reactor, possibly due to the effects of environmental conditions not present during the laboratory study (e.g., sunlight irradiance) or differences in wetland design (e.g., two-cell vs. single-cell, multiple vegetation species vs. monoculture).

Conclusions

Significant reductions of enteric microbes can be achieved in SF treatment wetlands. The field-scale and laboratory-scale surface flow systems of the present study were operated as secondary treatment systems (i.e., receiving wastewater from primary treatment anaerobic lagoons). As such, the results of this study show that a secondary treatment system using surface flow constructed wetlands can be effective for achieving significantly greater pathogen reductions in swine wastewater than would be achieved using a single-stage lagoon system. Of the microbial indicators studied, fecal coliforms and *E. coli* appeared to be the best indicators for removal of *Salmonella* in the SF wetland system. The viral indicators and *C. perfringens* spores were removed to a similar extent as the enteric bacteria studied, indicating that SF constructed wetlands may be as effective for reducing concentrations of viral and parasitic pathogens as for bacterial pathogens.

The results from the laboratory study show that loading rate is an important variable to consider when designing surface or subsurface flow wetland treatment systems to remove pathogens from wastewater. The data also indicated that SSF systems, whether vegetated or not, can achieve greater pathogen reductions than similarly sized SF wetlands operated at the same loading rates. The presence of vegetation in the SSF wetland significantly improved the removal of *Salmonella*, fecal coliforms and *E. coli* compared to the non-vegetated SSF control reactor, thus supporting previous research reporting positive effects of vegetation on enteric microbe removal in treatment wetland systems.

Acknowledgments

We thank the North Carolina Water Resource Research Institute (Project #70173), National Pork Producers Council (Project #99-112) and United States Department of Agriculture (Project #99-35102-8178) for providing funding to support this research. We also thank Frank Humenik and Mark Rice, North Carolina State University, Biological and Agricultural Engineering Department, for their assistance in studies of the wetland field system.

References

Anderson S.A., Turner S.J. and Lewis G.D. (1997). Enterococci in the New Zealand environment: Implications for water quality monitoring. *Wat. Sci. Tech.*, **35**(11-12), 325–331.

Bisson J.W. and Cabelli V.J. (1979). Membrane filter enumeration method for *Clostridium perfringens*. *Appl. Environ. Microbiol.*, **37**(1), 55–66.

CH2M Hill and Payne Engineering (1997). *Constructed Wetlands for Livestock Wastewater Management*.

Clausen E.M., Green B.L. and Litsky W. (1977). Fecal streptococci: Indicators of pollution. In: *Bacterial Indicators/Health Hazards Associated with Water*, A.W. Hoadley and B.J. Dutka (eds.), American Society of Testing and Materials, Philadelphia, pp. 247–264.

Cole D.J., Hill V.R., Humenik F.J. and Sobsey M.D. (1999). Health, safety, and environmental concerns of farm animal waste. *Occupational Medicine*, **14**(2), 423–448.

Gerba C.P., Thurston J.A., Falabi P.M., Watt P.M. and Kapiscak M.M. (1999). Optimization of artificial wetland design for removal of indicator microorganisms and pathogenic bacteria. *Wat. Sci. Tech.*, **40**(4–5), 363–368.

Gersberg R.M., Gearhart R.A. and Ives M. (1989). Pathogen removal in constructed wetlands. In: *Constructed Wetlands for Wastewater Treatment*, D.A. Hammer (ed.), Lewis Publishers, Chelsea, MI, pp. 431–445.

Grabow W.O.K. and Coubrough P. (1986). Practical direct plaque assay for coliphages in 100-mL samples of drinking water. *Appl. Environ. Microbiol.*, **52**(3), 430–433.

Hill V.R. and Sobsey M.D. (1998). Microbial indicator reductions in alternative treatment systems for swine wastewater. *Wat. Sci. Tech.*, **38**(12), 119–122.

Hunt P.G., Szögi A.A., Humenik F.J. and Rice J.M. (1998). Treatment of animal wastewater in constructed wetlands. In: *Proc. Eighth International Conf. of the FAO European Research Network on Animal Waste Management*, Rennes, France, 26–28 May.

Kadlec R.H. and Knight R.L. (1996). *Treatment Wetlands*. New York: CRC Press, Inc.

Levin M.A., Fischer J.R. and Cabelli V.J. (1975). Membrane filter technique for enumeration of enterococci in marine waters. *Appl. Environ. Microbiol.*, **30**(1), 66–71.

Ottová V., Balcarová J. and Vymazal J. (1997). Microbial characteristics of constructed wetlands. *Wat. Sci. Tech.*, **35**(5), 117–123.

Rivera F.A., Warren A., Ramirez E., Decamp O., Bonilla P., Gallegos E., Calderon A. and Sánchez J.T. (1995). Removal of pathogens from wastewaters by the root zone method (RZM). *Wat. Sci. Tech.*, **32**(3), 211–218.

Soto F., Garcia M., de Luis E. and Bécares E. (1999). Role of *Scirpus lacustris* in bacterial and nutrient removal from wastewater. *Wat. Sci. Tech.*, **40**(3), 241–247.

Standard Methods for the Examination of Water and Wastewater (1999). 20th edn, American Public Health Association/American Water Works Association/Water Environment Federation, Washington, DC, USA.

Stott R., Jenkins T., Shabana M. and May E. (1997). A survey of the microbial quality of wastewaters in Ismailia, Egypt and the implications for wastewater reuse. *Wat. Sci. Tech.*, **35**(11–12), 211–217.

Tanner C.C., Clayton J.S. and Upsdell M.P. (1995). Effect of loading rate and planting on treatment of dairy farm wastewaters in constructed wetlands— I. Removal of oxygen demand, suspended solids and faecal coliforms. *Wat. Res.*, **29**(1), 17–26.

Warren A., Decamp O. and Ramirez E. (2000). Removal kinetics and viability of bacteria in horizontal subsurface flow constructed wetlands. In: *Preprint. Seventh International Conf. Wetland Systems for Water Pollution Control*, International Water Association, Lake Buena Vista, Florida, 11-16 November.

Occurrence and die-off of indicator organisms in the sediment in two constructed wetlands

T.A. Stenström and A. Carlander

Department of Water and Environmental Microbiology, Swedish Institute for Infectious Disease Control, S-171 82 Solna, Sweden. (E-mail: *thor-axel.stenstrom@smi.ki.se, anneli.carlander@smi.ki.se*)

Abstract The interest in constructed wetlands for municipal wastewater and stormwater treatment has recently increased but data for the reduction efficiency of indicator organisms are often restricted to the water phase. In a full-scale wastewater wetland in Sweden fecal coliforms and enterococci were reduced by 97–99.9% and coliphages by approximately 70%. The factors affecting the reduction are however less well understood.

In two full-scale wetlands, for stormwater and wastewater treatment, an assessment has been done of the particle associated fraction of indicator organisms. No significant differences in the particle-associated numbers were seen between the inlet and the outlet of the wetlands, but the amounts of sedimenting particles varied between the two sites. In the stormwater wetland the amount of sedimenting particles at the outlet was 3% of the amount at the inlet, while the wastewater wetland had much lower particle removal efficiency. The reduction of suspended particles seems to be the main factor for bacterial elimination from the water phase, governed by vegetation and design. In the sediment, survival of presumptive *E.coli*, fecal enterococci, *Clostridium* and coliphages were long with T_{90}-values of 27, 27, 252 and 370 days, respectively. The organisms can however be reintroduced by resuspension. Viruses in the water phase may be of main concern for a risk assessment of receiving waters.

Keywords Coliphages; constructed wetlands; die-off; indicator organisms; sediment; stormwater; wastewater

Introduction

Surface-flow constructed wetlands have received an increasing interest as an alternative or supplementary treatment step for municipal wastewater or stormwater during the last decades, mainly due to their potential reduction of nitrogen, other chemicals and suspended matter. The main interest has been directed towards the water phase and the effects of hydrology and vegetation regarding the nutrients and particle removal efficiency while less is understood about the reduction mechanisms of fecal pathogens and indicator organisms introduced with the influent. This has been of less concern in several countries due to the common practice of using chlorination or other means of disinfection of the effluents before entering the wetlands but is not practiced in Sweden partly due to the negative side effects on the aquatic organisms.

Several combining factors necessitate further evaluations of the fate of fecal organisms in surface-flow wetlands in the temperate region, the actual reduction mechanisms occurring and the potential secondary transmission of pathogens. This includes the lack of disinfections, the common practice of promoting constructed wetlands as official recreational areas, the increasing awareness of the role of animals and birds as carriers, the prolonged survival times of fecal organisms in water and sediments in temperate regions and the common design criteria of much shorter retention periods then is even practiced in wastewater stabilization ponds in the tropical region.

Inflow and outflow densities and the treatment reduction efficiencies of fecal indicator bacteria in the water phase from pilot and full-scale wetland systems have been summarized by Kadlec and Knight (1996). In general, the reduction values are in the range of

85–99.9% and depicted as a first order function against detention time. Variability may be due to the impact of animals and birds as well as land run-off (Girts and Knight, 1989). Information about the reduction efficiency of viruses varies; some authors report a slower reduction (Scheuerman *et al.*, 1989) while others report a similar or more rapid reduction correlated with suspended solid removal (Gersberg *et al.*, 1989). A similar conclusion was reached by Gerba *et al.* (1999) for the reduction of *Giardia* and *Cryptosporidium* in a duckweed pond and for viruses and indicator bacteria in a subsurface flow wetland. One main factor responsible for the reduction of organisms from the water-phase and the sedimentation of particles and associated organisms which merits further attention is the optimization of wetland treatment. Furthermore the die-off of pathogens and fecal indicators in the associated sediment phase under different environmental situations may provide further insight into the risk of secondary transmissions due to resuspension and to birds.

Thus, the aim of this study was to investigate the occurrence and survival of fecal indicator microorganisms in the sediment of two constructed surface-flow wetlands in Sweden. Oxelösund, a wetland constructed for treatment of municipal wastewater and Flemingsberg, a wetland constructed for treatment of stormwater were sampled. Both wetland areas are also used as recreational areas for the public and the water is not disinfected.

Material and methods
Description of field sites
The 22 ha wastewater wetland receives its load from a 15,200 p.e municipality. The inlet water is mechanically and chemically pretreated. The wetland consists of two parallel pond systems (Figure 1) with a daily load of 5,000–6,000 m^3 and with a retention time of 7 days (Wittgren *et al.*, 1996). One of the parallel systems was sampled, where each of the ponds hold 20–25,000 m^3 and are intermittently filled and emptied. The vegetation consists of *Carex sp, Phragmites australis* and *Elodea canadensis*. Large areas of the wetland are open ponds with little vegetation.

The 18 ha stormwater wetland (Figure 2) receives run-off water from housing and industrial areas. The total drainage area is 7 km^2 and the lake after the wetland receives approx. 1.8×10^6 m^3 stormwater each year. The wetland area consists of a sedimentation pond, a surface overflow area and a denitrification pond and has a total retention time of 3–5 days. The dominating vegetation in the wetland consists of *Lemna sp, Potamogeton sp*, *Alisma plantago-aquatica, Typha sp* and *Myrophyllium sp*.

From each wetland, water sedimenting particulates were captured in sediment traps placed in transects across the wetlands, (Figure 1 and 2). Each trap was 70 mm in ∅ and had a height of 295 mm, giving a height/diameter ratio ≥3 (Blomqvist and Håkansson, 1979). The traps were emptied 3 times; each representing 1–2 months, during the Swedish spring to autumn period. The amounts of material were measured and the fecal organisms quantified. At the two first sampling transects (dotted lines, Figure 1 and 2) at each wetland the sediment traps were analyzed separately, while at the preceding transects the samples where pooled. For each water sample, three bottles of 500 ml were taken and pooled. The

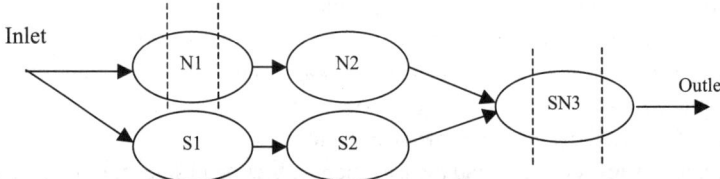

Figure 1 Constructed municipal wastewater wetland in Oxelösund, Sweden. Dotted lines indicate transects with sediment traps

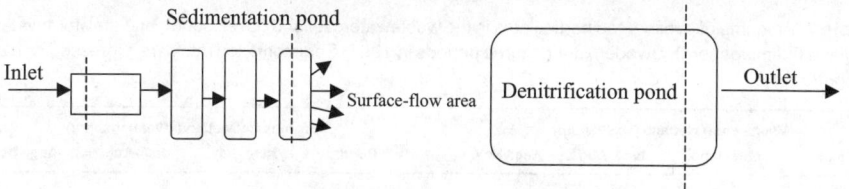

Figure 2 Constructed stormwater wetland in Flemingsberg, Sweden. Dotted lines indicate transects with sediment traps

samples were stored in +4°C and analyzed within 24 hours. All samples (water, bottom sediment and sedimentary material) were analyzed in triplicate for total coliforms, presumptive *E.coli*, fecal enterococci, sulfite-reducing anaerobic sporeformers and coliphages

A sediment core from each wetland was placed in glass bottles. The survival studies were performed at room temperature, 20–25°C, with sub-samples taken at day 0, 1, 4, 7, 11, 19, 25, 34, 50 for organism quantification.

The concentrations of indicator organisms in the sediment were quantified from 10 g weighed sub-samples. These were mixed with 80 ml of phosphate buffer and 10 ml 1% Tween 80-solution, shaken for 2 min on a Whirl-mixer and refrigerated for 10–12 minutes to let the material sediment. Thereafter 40 ml of the liquid was taken off, diluted in ten-fold steps and analyzed. The dry substances were measured and the concentrations of organisms were calculated as cfu or pfu per gram dry substance.

For the bacteria the Swedish Standard methods were used. For total coliforms (mEndoagar-LES, 44h, 35°C), presumptive *E.coli* (MFC agar, 24h, 44°C with further verification) and the fecal enterococci (mEnterococcus agar, 44h, 35°C) the plate-spread method was used. For sulfite-reducing anaerobes the samples were heated (75°C, 15 min) and thereafter analyzed (Perfringens agar base, 44 h, 37°C) with the pour plate method. Coliphages were analyzed according to the method described by Adams (1959) with the host *E.coli* C ATCC 13706.

Results

Survival studies

The die-off of total coliforms, *E.coli*, enterococci, *Clostridium* and coliphages were followed during 50 days in a laboratory study of the sediment samples and the results given as the logarithmic reduction (T_{90}-values) in Table 1.

The bacterial indicators were reduced much more rapidly than the coliphages and the anaerobic sporeformers. The variation in die-off of coliphages between the sediment from the two sites was due to a much lower quantity of indigenous phages from the stormwater site. Thus, this value may be too low.

Sedimentation

In Oxelösund the amounts of deposited particulates were similar over the wetland, except during the first period with 27% in the outlet region as compared to the inlet (Table 2). The same comparison at the two other sampling periods gave values of 80–83%.

Table 1 T_{90}-values in days for fecal indicator organisms in the sediment in laboratory survival study

T_{90} values	Total coliforms	Presumptive *E.coli*	Fecal enterococci	*Clostridium perfringens*	Coliphages
Oxelösund	16	27	27	252	370
Flemingsberg	17	24	53	396	51

Table 2 Amounts of sedimented particulates in the wastewater wetland (Oxelösund), and the stormwater wetland (Flemingsberg), Sweden, during three periods in 1997. Sedimentation rates are expressed as g dry weight/day m^2

Pond	Wastewater wetland (Oxelösund)			Pond	Stormwater wetland (Flemingsberg)		
	May–June	June–Aug.	Aug.–Nov.		June–July	July–Aug.	Aug.–Oct.
N1 in	74.9 ± 39.9	24.2 ± 5.1	20.4	1 in	103.9 ± 18.1	54.3 ± 12.1	74.6 ± 20.1
N1 out	49.3	17.4 ± 6.9		3 out	41.6 ± 36	12.4 ± 4.6	7.6 ± 2.9
SN3 in	17.3	19.6	8.5	5 out	2.9	1.7	3.8
SN3 out	20.0	20.2	16.4				

In Flemingsberg the reduction of particulate matter over the wetland was much higher at all three periods (Table 2), with just 3–5% deposition as compared to the inlet.

Occurrence of indicator organisms

Water. A separate initial assessment of the reduction of indicator organisms was made in the wastewater wetland system with reduction values given in Table 3 (Wittgren *et al.*, 1996). The corresponding values for control measurements are given for this investigation (Table 3). A slightly lower reduction, although not significant, was obtained during the autumn sampling except for the coliphages. In the study performed by Wittgren *et al.* (1996), the reduction capacity of the coliphages were lower over the wetland system compared with the rest of the analyzed indicator organism.

The stormwater wetland had much lower quantities of fecal enterococci and coliphages in the inlet and these parameters were reduced to below the methodological detection limit applied at the outlet. The coliforms and *E.coli* were reduced to a similar extent, 88 and 99% respectively, while an 83% reduction was obtained for the sulfite reducing anaerobic sporeformers.

Wastewater wetland. The variation in the quantities of indicator organisms in the sediment samples representing individual traps within each transect was high on all occasions. At the inlet (N1 in), values (Figure 3) are an average of the individual analyzed traps. The coliforms showed a high reduction in June, 99.5%, while in August and November the concentrations were constant or increased over the wetland. E.coli showed a similar pattern but with lower concentrations. *Clostridium* had during all three samplings a low reduction over the system, with a maximum reduction in June, 88%. Coliphages were the organisms with the highest reduction. In June and August the coliphages were below the methodological detection limit after N1.

Table 3 Concentrations and reduction of indicator organisms in the water phase in the wastewater wetland during spring and autumn. Concentrations are given as cfu or pfu per 100 ml. N.D = Not Detected. 1) Reduction values from Wittgren *et al.* (1996), for March and November

	Colif.	E. coli	Fecal enter.	Clost.	Coliphages
May					
Inlet	1.1×10^4	1.9×10^3	4.8×10^2	1.6×10^3	8.3×10^4
Reduction	87.9%	99.8%	99.98%	98.7%	N.D
Reduction[1]	99.99%	99.5%	99.9%	50%	
November					
Inlet	1×10^4	4×10^2	2.4×10^3	2.3×10^3	6.7×10^3
Reduction	92%	>97.5%	99.9%	95.9%	98.1%
Reduction[1]		98.6%	97.5%	+55%	95.4%

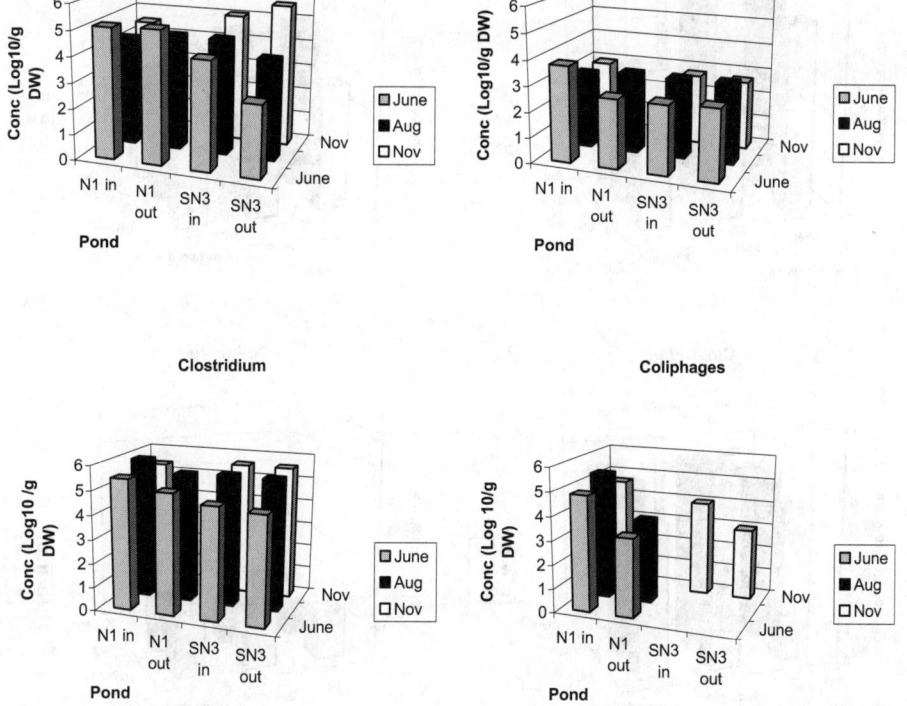

Figure 3 Concentrations of coliforms, *E.coli*, *Clostridium* and coliphages in the deposited material in the wastewater wetland. The values are given as $\log_{10}$/ g DW

The quantities of fecal enterococci were generally low in the inlet part and often below the methodological detection limit applied, in the preceding parts of the wetland sediments. These values have therefore been omitted.

Stormwater. At the two first sampling transects the traps were analyzed separately with large variations between the different traps. At the final transects, the traps were pooled and analyzed as one sample. The reduction of organisms in the sediment over the wetland was similar between the three sampling occasions (Figure 4) with a low reduction from the inlet to the outlet. At the first sampling, the only notable reduction over the wetland was among the total coliforms with a reduction of approximately 1 $\log_{10}$. On all three sampling occasions *Clostridium* had a very low reduction varying between 18–44%. The coliphage concentrations were very low in the stormwater wetland, less than 100 pfu/g dry weight on all occasions.

Discussion

Survival studies and the effect of sedimentation

The sediment environment is favourable for most of the organisms, with prolonged persistence for several of them. For presumptive *E. coli* the time for 90% reduction was 24–27 days and for fecal enterococci somewhat longer as judged from the die-off experiments in the sediment. Compared with the time for total reduction in freshwater or sewage of between < 30 to < 60 days used by Feachem *et al.* (1983) and <50 days at 20–30°C, the reduction is much lower in these sediment environments. However, the low quantities of

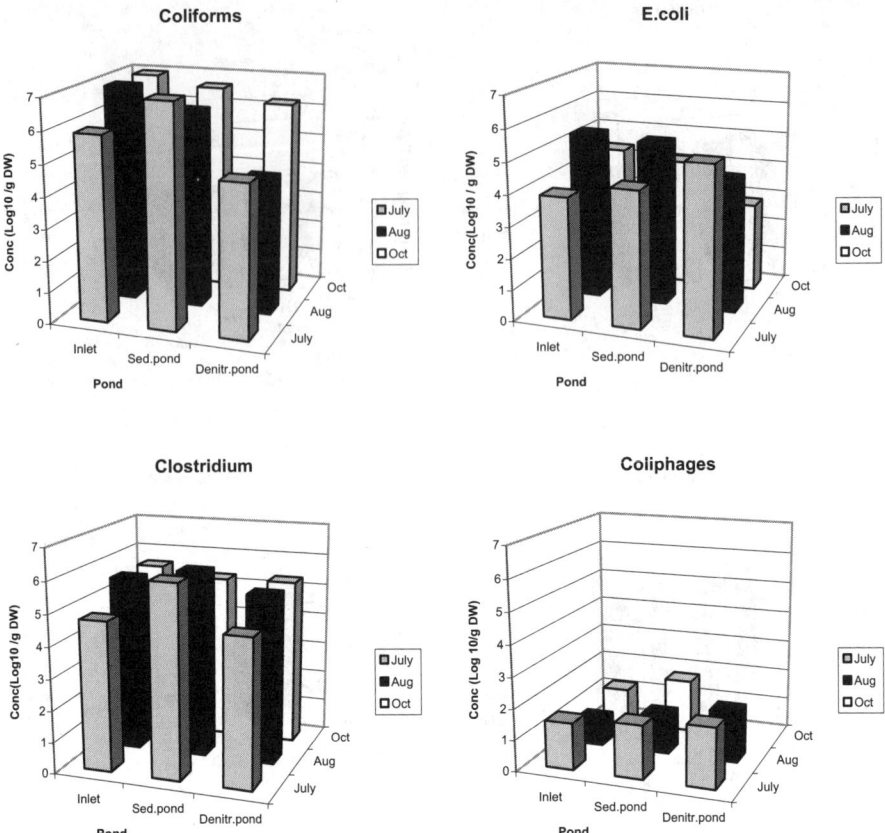

Figure 4 Concentrations of coliforms, *E.coli*, *Clostridium* and coliphages in the sedimentary material in the stormwater wetland. The values are given as $\log_{10}$/ g DW

enterococci in the deposited particulates in the sediment traps are in contradiction with the die-off expected reduction in the sediment and can presently not be explained. The die-off experiments were performed at room temperature, 20–25°C. This also corresponds with the temperature in the water phase and in the top sediment layer during the summer period, while for the rest of the year the temperature will be much lower, 5–10°C or below, which will further prolong the survival time for most organisms. For the sulfite reducing anaerobic sporeformers the reduction was as expected very slow in both wetlands, with T_{90}-values of 252 (wastewater wetland) and 396 (stormwater wetland) days respectively. A slow reduction was also seen for the coliphages in the wastewater wetland with a T_{90}-value at 370 days. The accumulation of organisms and especially viruses can give high concentrations in the sediment and if re-suspended in the water high concentrations can be seen. Such an accumulation was not evident from the sediment trap samples.

Sedimentation of particles and associated organisms seems to be a main important factor for the reduction of microorganisms from the water phase in the wetland. Thus, an optimization of the reduction of particulate matter and a minimization of the resuspension is governing factors also to reduce the potential health impact. In both wetlands most deposition occurred at the inlet and then decreased further down-flow. In the stormwater wetland and independent of the amount of inflowing particulates, the sedimenting quantities were reduced by 95–97%. For the wastewater wetland, the reduction was much more variable and less efficient over the wetland even though it consisted of several parts in sequence. In this study we have mainly focused on the relative amounts of deposits at different distances from the inlet. This can not directly be compared with the traditional reduction of turbidity normally reported. Linker (1989) reported a 77% reduction and Daukas *et al.* (1989) between 80–95% turbidity reduction. However, great variability may occur due to local conditions. The reduction of particulate matter over the stormwater system was high. One important factor for high sedimentation in the stormwater wetland may be the deeper open water areas (the sedimentation pond, Figure 2) as discussed by Fennessy *et al.* (1994) as a more conductive area for sediment accumulation than shallower open water areas. Other factors are areas of overland flow, the actual design of the system and the effects of vegetation. All these factors are of importance in treatment units for stormwater, where large variations in amounts of inflowing particulates could be expected due to differences in rainfall and seasonality. A good design and construction of the wetland with established high retention of particulate matter will give both a high reduction of organisms in the water phase and lessen the seasonal variability.

The coliphages, here used as indicators for enteric viruses reduction, need to be particle associated to sediment. Karpiscak *et al.* (1996) showed that the degree of removal of indicator organisms from the water in a duckweed pond was related to their size, with the lowest grade of removal for the coliphages. In a study reported by Gerba *et al.* (1999) three different wetlands were compared, where the reduction of coliphages varied between 40% in the aquatic pond and 95% in a sub-surface wetland.

In this study the concentrations of coliphages differed between the two wetlands, both in the water and the sediment phase, with 100–1000 higher concentrations in the wastewater wetland. The explanation for this may be due to differences in excreted amounts of somatic coliphages between humans and animals/birds, with higher concentrations excreted in the human feces compared with the animal. However, the amounts of fecal sterols (data not shown) indicated a significant fecal impact also in the stormwater system. For the other indicator organisms the concentrations were at the same levels or even higher in the stormwater wetland. In the sedimentation pond in the stormwater wetland, several of the organisms increased which may be a result of the bird-life in this part of the wetland.

Conclusions

Prolonged survival times in the sediment were found for the fecal microorganisms analyzed. Coliphages and spore forming bacteria showed a much longer survival time than the vegetative bacteria. The survival times in the sediment were similar or in a similar range in the two wetlands.

The reduction efficiency of indicator organisms in the water phase is related to the reduction of particular matter over the wetland system. The reduction efficiency of particulate matter is mainly due to construction of the wetland and not to retention time.

The quantities of coliforms were higher, clostridia in the same range and coliphages lower in the stormwater wetlands as compared to the wastewater wetland.

Seasonal variations did occur but were in the same range.

References

Adams, M.H. (1959). *Bacteriophages*. Interscience Publishers, Inc, New York.
Blomqvist, S. and Håkansson, L. (1979). *Sedimentfällor i akvatisk miljö*. Statens Naturvårdsverk, rapport 1229. Solna. (In Swedish).
Daukas, P., Lowry, D. and Walker, W. Jr. (1989). Design of wet detention basins and constructed wetlands for treatment of stormwater runoff from a regional shopping mall in Massachusetts. In: *Constructed Wetlands for Wastewater Treatment: Municipal, Industrial and Agricultural*. D.A Hammer (ed.), Lewis Publishers, Chelsea, MI, pp 686–694.
Feachem, R.G., Bradley, D.J., Garelick, H. and Mara, D.D. (1983). *Sanitation and disease. Health Aspects and Wastewater Management*. John Wiley & Sons, Chichester.

Fennessy, M.S., Brueske, C.C. and Mitsch, W.J. (1994). Sediment deposition patterns in restored freshwater wetlands using sediment traps. *Ecological Engingeering* **3**, 409–428.

Gerba, C.P., Thurston, J.A., Falabi, J.A., Watt, P.M. and Karpiscak, M.M. (1999). Optimization of artificial wetland design for removal of indicator microorganisms and pathogenic protozoa. *Wat. Sci. Tech.*, **40**(4–5), 363–368.

Gersberg, R.M., Gearheart, R.A. and Ives, M. (1989). Pathogen removal in constructed wetlands. In: *Constructed Wetlands for Wastewater Treatment: Municipal, Industrial and Agricultural*, D.A Hammer (ed.), Lewis Publisher, Chelsea, MI, pp 431–445.

Girts, M.A. and Knight, R.L (1989). Operation optimization. In: *Constructed Wetlands for Wastewater Treatment: Municipal, Industrial and Agricultural*, D.A Hammer (ed.) Lewis Publisher, Chelsea, MI, pp 417–429.

Kadlec, R.H and Knight, R.L (1996). *Treatment wetlands*. CRC Press. Boca Raton, FL.

Karpiscak, M.M., Gerba, C.P., Watt, P.A., Foster, K.E. and Falabi, J.A. (1996). Multi-species plant systems for wastewater quality improvements and habitat enhancement. *Wat. Sci. Tech.*, **33**(10–11), 231–236.

Linker, L.C. (1989). Creation of wetlands for the improvement of water quality: a proposal for the joint use of highway right-of way. In: *Constructed Wetlands for Wastewater Treatment: Municipal, Industrial and Agricultural*. D.A Hammer (ed.),. Lewis Publishers, Chelsea, MI, pp 695–701.

Scheuerman, P.R., Bitton, G. and Farrah, S.R. (1989). Fate of microbial indicators and viruses in a forested wetland..In*: Constructed Wetlands for Wastewater Treatment: Municipal, Industrial and Agricultural*. Hammer, D.A (ed.), Chelsea, MI: Lewis Publisher, pp 657–663

Wittgren, H.B., Stenström, T.A. and Sundblad, K. (1996). Removal of indicator microorganisms in surface-flow treatment wetlands. In: Preprints from the 5th International Conference on Wetland Systems for Water Pollution Control, September 15–19, 1996, Vienna, Austria, pp.I/19 1–8.

Performance modeling of subsurface-flow constructed wetlands systems

M.F. Dahab,* R.Y. Surampalli,** and W. Liu*

*Department of Civil Engineering, University of Nebraska-Lincoln, Lincoln NE 68588-0531, USA
**US Environmental Protection Agency, Region VII, Kansas City, KS 66101, USA

Abstract A subsurface flow constructed wetlands (CW) system, located at a neighborhood consisting of a small housing development and golf courses outside of Lincoln, NE, was studied for its effectiveness as a small community wastewater system. Extensive monitoring was conducted biweekly between June 1996 and December 2000. Prediction models for soluble $CBOD_5$ NH_3-N, and TP removal in CW were employed for comparison with the field data. It was found that the disappearance of BOD_5 and NH_3-N could be approximated using first-order kinetics, but the kinetics of TP removal were unclear. The reduction rate constants regressed from the field data were found to be lower than literature reported values.
Keywords Constructed wetlands; kinetics; modeling; subsurface-flow; wastewater treatment

Introduction

Natural wetlands have been used for wastewater treatment for a long time. However, uncontrolled discharge of wastewater often led to irreversible degradation in natural wetlands. The first experiment aimed at the possibility of wastewater treatment by wetland plants was undertaken in Germany in 1952 at the Max Planck Institute in Ploen (Seidel 1955). It took more than 20 years of research before the first operational full-scale constructed wetland for municipal sewage was built in Germany (Kickuth 1977). The use of CWs for domestic wastewater treatment has been increasing in the United States since the first intentionally engineered CW treatment pilot systems constructed in 1973 at the Brookhaven National Laboratory near Brookhaven, NY. In the Midwestern United States, numerous facilities have been reported for domestic wastewater treatment, including South Dakota, North Dakota, Kentucky, Iowa and Missouri (Vanier, 1997; Vanier and Dahab, 1997). The constructed wetland treatment system investigated in this study, located in a neighborhood of a small housing development and a golf course outside of Lincoln, Nebraska, is the first constructed wetland in the state.

Constructed wetland treatment systems are engineered systems that have been designed and constructed to utilize the natural processes involving wetland vegetation, soils, and their associated microbial assemblages to assist in treating wastewater (Vymazal *et al.* 1998). The pollutants in such systems are removed through a combination of physical, chemical, and biological processes including sedimentation, precipitation, adsorption to soil particles, assimilation by the plant tissue, and microbial transformations (Brix 1993).

Organic compounds are removed through settling, entrapment of particulate matter in the void spaces, and microbial growth on the media surface and plant roots and rhizomes. Nitrogen is removed by nitrification/denitrification, volatilization, adsorption and plant uptake. But, the major removal in constructed wetlands is nitrification/denitrification (Vymazal *et al.* 1998). Phosphorus removal in wetland treatment systems occurs through adsorption, plant absorption, complexation, and precipitation (Watson *et al.* 1989). However, the pathways by which available P is utilized are not fully understood.

Several prediction models have been established to simulate the removal mechanisms of carbonaceous, nitrogenous and phosphorous pollutants in subsurface-flow constructed wetlands. The rates at which constituents move through wetlands have been reported from experimental or operational experiences. The purpose of this paper is to compare existing prediction models with our operational experience with emphasis on soluble $CBOD_5$, NH_3 and TP.

Materials and methods
Site and sampling
The CW treatment system investigated during this study serves a residential community of about 110 households and a golf club east of Lincoln, NE. Construction of this wetland was completed and macrophyte sprigs were planted in November of 1995 and operation began in December 1995. The treatment system consists of four parallel gravel-based cells, each 50m × 25m, with horizontal subsurface flow through a depth of 0.6m. Each CW cell was planted with three equal vegetative zones of common cattails (*Typha latifolia*) and woody cattails (*Typha domingensis*), alkali bulrush (*Scripus acutus*), and common reeds (*Phragmites communis*). Influent into the wetland cells is the effluent from a series of pre-treatment tanks acting as combined primary settling basins and sludge digestion units. The design flow rate of the treatment system was calculated at 225 m³/d with a current typical flow rate in the range of 115 to 125 m³/d.

Sampling and analysis were carried out on a biweekly basis from June 1996 to present. The locations of the sampling sites were selected at five points along the length of one cell at fractional distances of 0 (inlet), 0.15, 0.40, 0.67, and 1.00 (outlet). Each set of samples were analyzed for temperature, pH, DO, TSS and VSS, total and soluble $CBOD_5$, total and soluble COD, nitrogen (TKN, NH_3-N, and NO_3-N), total phosphorus (TP), fecal coliform, and Salmonella. All samples were collected as grab samples and analyzed according to Standard Methods (APHA, 1995).

Performance models
Several design models describing the removals of soluble $CBOD_5$, NH_3-N, and total P in CW wetlands can be found in the wetlands literature. One of the widely used models is an area-based, first-order equation from Kadlec and Knight (1996). The area-based, first-order model, if considering irreducible background concentration, can be written as:

$$\ln\left[\frac{Ce - C^*}{Ci - C^*}\right] = -\frac{k}{q}y \quad (1)$$

where k is the area-based, first-order rate constant (m/yr), q is the hydraulic loading rate, m/yr, y is the fractional distance through the wetland, Ce is the effluent concentration (mg/L), Ci is the influent concentration (mg/L), and C^* is the irreducible background wetland concentration (mg/L).

The area-based, first-order rate constant, k, and the irreducible background wetland concentration, C^*, will vary depending on different water quality parameters. In describing $CBOD_5$ removal, according to Kadlec and Knight (1996), several studies on horizontal subsurface flow wetlands produced data from which k_{BOD} and C^*_{BOD} values may be regressed. The average rate constant, k_{BOD}, for SSF systems was approximately 180 m/yr at 20°C and the irreducible background wetland BOD_5 concentration, C^*_{BOD5}, was approximately 9.8 mg/L (Kadlec and Knight, 1996). Temperature adjustment was found to be negligible and thus was ignored by Kadlec and Knight (1996). In the prediction of ammonia removal, the suggested area-based, first order NH_3-N rate constant, k_{NH3-N}, of 34 m/yr was used with a temperature correction factor of 1.04, according to Kadlec and Knight (1996). There appears to be a very small ammonia nitrogen concentration in background for subsurface flow wetlands: in the range of 0.05 to 0.10 mg/L. For this reason, C^*_{NH3-N} was assumed to equal zero (Kadlec and Knight, 1996). The rate constants for phosphorus removal, reported by Kadlec and Knight (1996), for SSF systems had k_P values of approximately 11.7 ± 4.2 m/yr. Background phosphorus levels in the model were considered to be insignificant. Temperature effect was neglected in TP removal performance in the wetland (Kadlec and Knight, 1996).

Besides the area-based, first-order model from Kadlec and Knight (1996), analysis of subsurface flow CW system data verifies that soluble $CBOD_5$ removal is reasonably approximated by a first-order, irreversible, plug flow relationships (Reed and Brown, 1995; USEPA, 1993; Cooper and Finlater, 1990). The commonly used first-order plug flow model is represented as follows:

$$\frac{Ce}{Co} = e^{(-k_T t)} \qquad (2)$$

where k_T is the temperature dependent rate constant (d^{-1}), $k_T = k_{20}(1.06)^{(T-20°)}$, ($k_{20}$ is the rate constant at $20°C(d^{-1})$, T is the temperature of liquid in the system (°C)), t is the hydraulic residence time (d), C_e is the effluent BOD_5 (mg/L), and C_o is the influent BOD_5 (mg/L).

In utilizing Equation (2), a conservative rate constant k_{20} of 0.86 d^{-1} was used in the model for gravel sand media. According to the USEPA (1988), different k_{20} values were associated with different characteristics of media. For gravel sand, coarse sand or medium sand, k_{20} values were reported as 0.86 d^{-1}, 1.35 d^{-1} or 1.84 d^{-1}, respectively. The rate was modified with a temperature adjustment of 1.06 for both summer and winter conditions. The "t" or hydraulic residence time (HRT) in equation (2) can be defined as:

$$t = \frac{LWnd}{Q} \qquad (3)$$

where L is the length of the system, parallel to flow path (m), W is the width of system (m), n is the porosity of the system, d is the average depth of the system (m), and Q is the average flow rate through the system (m^3/d).

In the equations above, where appropriate, a porosity of 0.4 and a treatment depth of 0.45m were used. The surface area of wetland cell was 1250 m^2 (50m × 25m). The measured average summer and winter CW wastewater temperature of 19.5 °C and 3.9 °C were used in the analysis. The flow rate was 26.4 m^3/d under summer and 31.4 m^3/d under winter average conditions.

Results and discussion

As indicated earlier, two prediction models were used to estimate the $CBOD_5$ removal in the CW wetlands systems. The comparisons of the field data with the predictions using the area-based, first-order model (Kadlec and Knight, 1996) and the first-order, irreversible, plug flow model (Reed and Brown, 1995; USEPA 1993; Cooper and Finlater, 1990) are presented in Table 1. According to the results of the area-based, first-order model, Equation (1), the predicted summer and winter $CBOD_5$ effluents values are identical at 9.80 mg/L, which actually equaled the assumed irreducible background wetland $CBOD_5$ concentration in the model. Meanwhile, Equation (2) predicted final effluent concentrations of 6.40 mg/L in the summer and 7.52 mg/L in the winter, respectively. Comparing to the measured results, which were 29.1 mg/L under summer condition and 19.0 mg/L under winter condition, it can be seen that both prediction models overestimated the soluble $CBOD_5$ removal capacity in the wetland. The field data resulted in low removal rates compared to data from the literature. Calculated from the measured data, the area-based, first-order rate constant was 38 m/yr (temperature independent) as opposed to 180 m/yr in Equation (1). Additionally, the observed plug flow rate constants were 0.31 d^{-1} as opposed to 0.84 d^{-1} (Equation (2)) under summer condition and 0.17 d^{-1} as opposed to 0.34 d^{-1} under winter condition. As can be seen from Table 1, the exponential regression of the CBOD$_5$ disappearance with the distance using Equations 1 or 2, provided a rather high (all measured values were >0.6) coefficients of determination.

The area-based, first-order model (Kadlec and Knight, 1996) also was used in predicting ammonia reductions along the fractional length of the CW system in summer and winter (Table 1). The measured performance data indicated that the exponential decay of the ammonia concentration along the length of the wetland cell was such that it supports the first-order model assumption. The exponential regression of the ammonia disappearance with the distance from the discharge provided relatively high coefficients of determination of 0.84 in the summer and 0.74 in the winter, respectively. However, from Table 1, it can be seen that the Kadlec and Knight (1996) model overestimated the removal efficiency of ammonia nitrogen; 14.3 mg/L (measured) versus 0.29 mg/L (predicted) in the summer and 16.4 mg/L (measured) versus 2.94 mg/L (predicted) in the winter. This is probably because the measured area-based, first-order ammonia reduction rate constants of 11.0 m/yr in the summer and 7.91 m/yr in the winter, calculated from Equation (1), were much smaller than the reduction rate constants of 33.3 m/yr in the summer and 18.1 m/yr in the winter used by Kadlec and Knight (1996).

Measured and predicted TP concentration variations with the fractional distance are presented in Table 1. The area-based, first-order model (from Kadlec and Knight (1996)) overestimated the removal efficiency in the realistic CW treatment. The predicted TP effluent concentration was 2.03 mg/L compared to 3.98 m/L from the operational field site. The coefficients of determination of 0.37 from the exponential regression of the performance would not support the assumption that TP removal in CW followed an area-based, first-order reduction model. If the phosphorus removal mechanisms could be understood clearly, prediction of the removal efficiency in the CW wetland would probably be vastly improved.

In general, the comparison of the monitoring results in the CW wetland with prediction models commonly used in the wetlands literature revealed that the general form of the area-based, first-order model (Kadlec and Knight 1996) and the first-order, irreversible, plug

Table 1 Comparisons of Model Prediction Results with Field Data

Water	Quality	Model 1[a]				Model 2[b]			
	Parameters	Calculated Value		Measured Value		Calculated Value		Measured Value	
		Summer	Winter	Summer	Winter	Summer	Winter	Summer	Winter
	Effluent[c]	9.80	9.80	29.1	19.0	6.40	7.52	29.1	19.0
CBOD$_5$	Rate Cons.[d]	180	180	38	38	0.84	0.34	0.31	0.17
	R^{2}[e]	1	1	0.64	0.85	1	1	0.61	0.74
	Effluent	0.29	2.94	14.3	16.4				
NH$_3$-N	Rate Cons.	33.3	18.1	11.0	7.91	—	—	—	—
	R^2	1	1	0.84	0.74				
	Effluent	2.03		3.98					
TP	Rate Cons.	7.5		—		—	—	—	—
	R^2	1		0.37					

a. Model 1 is the area-based, first-order model (Kadlec and Knight 1996).
b. Model 2 is the first-order, plug flow model (Reed and Brown, 1995; USEPA 1993; Cooper and Finlater, 1990).
c. Effluent concentration unit is mg/L.
d. Rate constant unit for model 1 is m/yr. and for model 2 is d^{-1}.
e. R^2 is coefficient of determination.

flow model (Reed and Brown, 1995; USEPA 1993; Cooper and Finlater, 1990) can be used to evaluate the performance of the wetland (not including phosphorus removal). However, the literature-reported rate constants for BOD and ammonia nitrogen appeared to be quite large when applied to our wetland system. As a consequence, these models tended to overestimate the capability of sub CW treatment. It should be noted that phosphorus removal mechanisms in subsurface-flow CW are not truly understood and thus, both the performance and the predictions of total phosphorus are not satisfactory.

References

American Public Health Association (APHA). 1995. *Standard Methods for Examination of Water and Wastewater.* 19th Edition, Washington D.C.

Brix, H. 1993. Wastewater treatment in constructed wetlands: system design, removal processes and treatment performance. In: Moshiri, G.A. (ed.), *Constructed Wetlands for Water Quality Improvement*, pp. 9–22. CRC Press, Boca Raton, Florida.

Cooper, P.F., and Finlater, B.C., Eds. 1990. *Constructed Wetlands in Water Pollution Control.(Adv Wat. Poll.Control)* Oxford, U.K.: Pergamon Press/IAWPRC.

Dahab, M.F., and Surampalli, R.Y, 1999, Predicting subsurface-flow constructed wetlands performance a comparison of common design models, Proceedings, Annual Conference of the Water Environment Federation, New Orleans, LA, Oct 9–13.

Dahab, M.F. and S.M. Vanier, 1998, "Temperature effects on subsurface flow constructed wetlands performance in the Midwest," Proceedings, the ASCE National Conference on Wetlands and River Systems, Denver, CO, March 23–25.

Kadlec, R.H. and Knight R.L. 1996. *Treatment Wetlands.* Boca Raton, FL: Lewis Publishers.

Kickuth, R. 1977. Degradation and incorporation of nutrients from rural wastewaters by plant hydrosphere under limnic conditions. In: *Utilization of Manure by Land Spreading*, pp. 335–343. Comm. Europ. Commun., EUR 5672e, London.

Liu, W., M.F. Dahab and R.Y. Surampalli, 2000, Subsurface-flow constructed wetlands performance evaluation using area-based first-order kinetics, Proceedings, Annual Conference of the Water Environment Federation, Anaheim, CA, Oct. 15–18.

Reed, S.C., and Brown, D.S. 1995. Subsurface flow wetlands – a performance evaluation. *Water Environ. Res.* **67**, 244–248.

Seidel, K. 1955. Die Flechtbinse *Scirpus lacustris.* In: *Okologie, Morphologie und Entwicklung, ihre Stellung bei den Volkern und ihre wirtschaftliche Bedeutung*, pp. 37–52. Sweizerbart'sche Verlgsbuchhandlung, Stuttgart.

U.S. EPA. 1993. *Subsurface flow constructed wetlands and aquatic systems for municipal wastewater treatment.* EPA/832/R-96/008. Washington, D.C.

U.S.EPA. 1988. *Design Manual: constructed Wetlands and Aquatic Systems for Municipal Wastewater Treatment.* EPA/625/1-88/022. Cincinnati, Ohio.

Vanier, S.M. 1997. Evaluation of a subsurface flow constructed wetland system for small-community wastewater treatment in southeastern Nebraska. Master of Science Thesis, University of Nebraska Libraries, Lincoln, NE.

Vanier, S.M. and M.F. Dahab. 1997. Evaluation of a subsurface flow constructed wetlands for small community wastewater treatment in the plains. In Water Environment Federation 70th Annual Conference and Exposition. Vol. 6. Chicago, Illinois.

Vymazal J., H. Brix, P.F. Cooper, M.B. Green and R. Haberl. 1998. *Constructed Wetlands for Wastewater Treatment in Europe*, pp. 1–15. Backhuys Publishers, Leiden, The Netherlands.

Watson, J.T., Reed, S.C., Kadlec, R.H., Knight, R.L. and Whitehouse, A.E. 1989. Performance expectations and loading rates for constructed wetlands. In: Hammer, D.A. (ed.), *Constructed Wetlands for Wastewater Treatment. Municipal, Industrial and Agricultural.* pp. 319-358. Lewis Publishers, Chelsea, Michigan.

Modelling nitrogen transformations in surface flow wastewater treatment wetlands in Sweden

S. Kallner* and H.B. Wittgren**

* Department of Biology, Linköping University, SE-581 83 Linköping, Sweden
** Department of Water and Environmental Studies, Linköping University, SE-581 83 Linköping, Sweden

Abstract The purpose of this study was to describe and compare the fate of nitrogen (N) in two Swedish wastewater treatment wetlands in the cities of Oxelösund and Hässleholm. Specifically, we wanted to see if a fairly simple model, developed with regard to common data availability, could satisfactorily describe the concentration dynamics at the outlet from the wetlands. A first-order area-based model, with two alternative expressions for temperature dependence, was set up to describe three major processes: ammonification, nitrification and denitrification. The N concentration dynamics at the outlet of the Oxelösund wetland was not satisfactorily described, $R^2(NH_4^+-N) = 0.33$ and $R^2(NO_3^--N) = 0.10$, while the modelled concentrations corresponded quite well with measured concentrations in the Hässleholm wetland, $R^2(NH_4^+-N) = 0.83$ and $R^2(NO_3^--N) = 0.58$. The NO_3^--N concentrations, in both wetlands, could be slightly better described when introducing a temperature coefficient as an additional free parameter. The explained variances reported above were achieved when the model was calibrated individually for the two wetlands, when the resulting (optimised) reaction rate coefficients for each of the three processes were quite different between the two wetlands. To improve model performance, the rate equations may have to be changed to include factors in addition to concentration and temperature, such as dissolved oxygen and hydraulic efficiency. It may also be important to include other processes, such as plant uptake/decay and ammonia volatilisation.
Keywords Cold climate; constructed wetland; modelling; nitrogen transformation; wastewater

Introduction

The biogeochemical cycle of nitrogen (N) in wetlands is complex and comprises many processes. In addition, several factors, e.g., temperature, nutrient concentrations and oxygen status, affect these processes (Kadlec and Knight, 1996). More or less complex models have been used to describe N transformations in wetlands (Martin and Reddy, 1997; Knight et al., 2000). The transformation rates are usually described by first-order concentration dependence, which is related to either wetland area or to wetland volume (Kadlec, 2000).

The purpose of this study was to describe and compare the fate of N in two Swedish wastewater treatment wetlands in the cities of Oxelösund and Hässleholm. Specifically, we wanted to see if a fairly simple model, developed with regard to common data availability, could satisfactorily describe the concentration dynamics at the outlet from the wetlands. A first-order area-based model was set up to describe three major processes: ammonification, nitrification and denitrification. The simulations of N transformations in the two wetlands, along with measurements of N content in the sediment of the Oxelösund wetland, were used to discuss the fate of N in the two different wetlands.

Wetland descriptions

The wetland of Oxelösund consists of five basins with connecting canals and overland flow areas (Figure 1). It is divided into two parallel systems, the north and the south, and a final common basin with an outlet to the Baltic Sea. This design allows intermittent filling and emptying of the two first basins in each system while the last basin is constantly filled

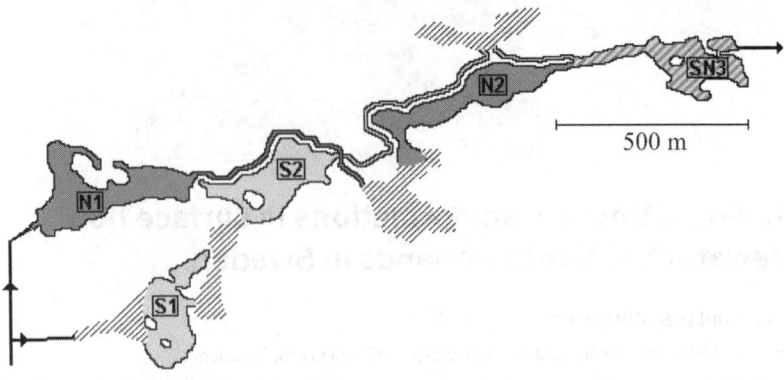

Figure 1 Map of the wetland system in Oxelösund

(Andersson et al., 2001). The total wetland area is about 23 ha. The catchment area has been estimated to be 105 ha including the wetland area and the residence time of the wetland is about eight days. During 1994–98, the average wastewater flow to the wetland was 4,900 m³ d⁻¹, the average total N concentration in incoming water was 23 mg l⁻¹, with NH_4^+-N (17 mg l⁻¹) as the major N compound, and annual total N retention varied between 35 and 48% (Andersson et al., 2001).

The wetland of Hässleholm consists of five basins, one initial, followed by four parallel basins leading to one small final common basin (Figure 2). The wetland area is 20 ha, surrounded by a small catchment area of 10 ha, and the residence time of the wetland is about six days. During 1996–99, the average wastewater flow to the wetland was 9,600 m³ d⁻¹, the average total N concentration in incoming water was 21 mg l⁻¹, with NO_3^--N and NH_4^+-N (14 and 7 mg l⁻¹, respectively) as the dominant N compound, and annual total N retention was approximately 30% (Hässleholms kommun, 1999).

In conclusion, a major difference between the two wetlands is that Oxelösund received wastewater with almost no NO_3^--N, while this was the major compound in Hässleholm.

Methods

Nitrogen transformation model

The transformations of organic N, NH_4^+-N and NO_3^--N were assumed to be due to the processes ammonification (organic N transformation to NH_4^+-N), nitrification (NH_4^+-N transformation to NO_3^--N) and denitrification (NO_3^--N transformation to N_2), respectively. These processes were further assumed to take place at the water-sediment

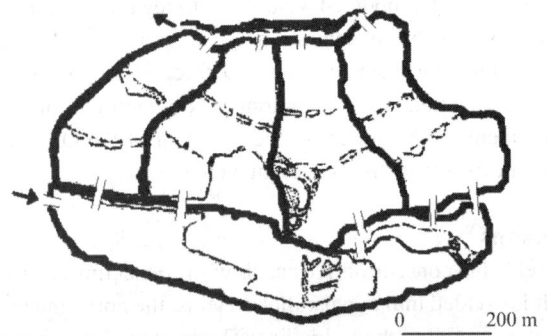

Figure 2 Map of the wetland system in Hässleholm

interface (area-based), and be affected by the concentrations (first-order) of the different N compounds, and by temperature. In Oxelösund, each of the wetland basins was modelled as a completely mixed batch reactor, each of which was connected as in Figure 1. The wetland of Hässleholm was modelled as one completely mixed batch reactor. The following area-based first-order equation was applied with a one-day time step for each of the three N transformation processes:

$$C_{end} = C_{start} \cdot \exp\left(\frac{-k_{aT} \cdot A \cdot \Delta t}{V}\right) \qquad (1)$$

where C_{end} was the daily end and outflow nitrogen concentration (mg l^{-1}), C_{start} was the daily initial nitrogen concentration after discharge, filling and mixing (mg l^{-1}), k_{aT} was the area-based reaction rate coefficient at temperature T (m d^{-1}), A was the wetland (basin) area (m^2), V was the wetland (basin) water volume (m^3) and Δt was the time step (d). The reaction rate coefficients were assumed to be temperature dependent according to:

$$k_{aT} = k_a \cdot T \qquad (2)$$

where k_a was the area-based reaction rate coefficient (m d^{-1} °C^{-1}) and T was the water temperature (°C). To determine if the transformations could be better described by including one more free parameter, the modified Arrhenius temperature dependence was used as an alternative to Eq. 2:

$$k_{aT} = k_{a20} \cdot \theta^{(T-20)} \qquad (3)$$

where k_{a20} was the area-based reaction rate coefficient at 20°C (m d^{-1}) and θ was the temperature coefficient (dimensionless).

Hydrological data and modelling

In Oxelösund, the flow of treated wastewater piped to the wetland was measured continuously with an inductive flow meter. The outflow from the wetland was calculated from measurements with an ultrasonic level recorder (1994–1997) and a pressure head level recorder (1998) (Andersson et al., 2001). The flow recorder at the outlet did not produce reliable data during some periods and a water balance model, HBV (Bergström, 1992), was used in addition to measurements to estimate the flow at the outlet from the wetland. The HBV-model was also used to calculate water balances for each of the wetland basins. Air temperature, precipitation and potential evapotranspiration were measured at the climate station in Oxelösund (Swedish Meteorological and Hydrological Institute, Norrköping).

In Hässleholm, water flow was measured with an electromagnetic flowmeter at the inlet of the wetland. At the outlet, water flow was calculated from measurements with a pressure head level recorder. Precipitation was measured at the wastewater treatment plant close to the wetland.

Water chemistry and temperature data

The wetland of Oxelösund has been monitored during 1994–98 (Andersson et al., 2001). During this period, inlet and outlet weekly samples (1994–95), or samples taken every second week (1996–98), were analysed for total N, NH_4^+-N, NO_3^--N, total P, dissolved oxygen (less frequently) and pH. Concentrations of organic N were calculated as the difference between total N and NH_4^+-N + NO_3^--N. Inlet measurements of NO_3^--N were abandoned when it was established that concentrations were constantly below 0.1 mg l^{-1}. Additional monitoring within each basin was performed during short periods of 1994 and 1995 (Wittgren and Tobiason, 1995; Andersson *et al.*, 2001). Water temperature was measured during 1994–97. Water temperature in 1998 was estimated from regression of measured water temperature and average air temperature during the last ten days (Water temp. = 1.8 + 0.89 × Air temp.; R^2=0.87).

The wetland of Hässleholm has been monitored during 1995–99. (Hässleholms kommun, 1999). Samples were analysed for total N, NH_4^+-N, NO_3^--N, total P and pH. Water temperature was measured, but during some periods (Jan 1996–Aug 1996, Nov 1996–Feb 1997) values were estimated as in Oxelösund (Water temp. = 2.7 + 0.96 × Air temp.; R^2=0.90).

Calibration, validation and sensitivity analysis of the nitrogen transformation model

In Oxelösund, the model was calibrated manually against observed concentrations at the outlet of the wetland. The reaction rate coefficients, k_a (Eq. 2), of each the three processes were calibrated in the interval 0.0001 to 0.9. The coefficients for ammonification and nitrification were calibrated simultaneously against concentrations of NH_4^+-N at the outlet of the wetland, while the coefficient for denitrification was calibrated against the NO_3^--N concentrations after the calibration of the two other coefficients. The ammonification coefficient was not calibrated against organic N concentrations, as they were low and their dynamics not possible to model with the chosen approach. When calibrating the model with temperature coefficients, θ (Eq. 3), as additional free parameters, the reaction rate coefficients, k_{a20}, of the three processes were changed in the interval 0.001 to 0.9 m d^{-1} while θ was set to the values 1.0, 1.072, 1.116, 1.175 or 1.259, corresponding to Q_{10}-values of 1, 2, 3, 5 and 10, respectively. The model parameters were optimised by maximising the explained variance according to Nash and Sutcliffe (1970).

In the wetland of Oxelösund, the model was calibrated against measured outlet concentrations from 1994–95 and validated against data from 1996–98. The concentrations within basins, obtained from the calibrated model, were also compared with the corresponding measured concentrations from the additional monitoring done in 1994–95. The calibrated model was applied in Hässleholm, but the model was also calibrated against the whole Hässleholm data set (1996–1999).

A sensitivity analysis was performed on the calibrated parameters by changing one parameter at a time. The reaction rate coefficients (k_a) were changed in the interval 50% to 150% of its calibrated value (100%). The sensitivity of the ammonification coefficient was analysed against the NH_4^+-N concentrations. Furthermore, a correlation analysis was performed to investigate if the differences between observed and simulated concentrations (the residuals) of NH_4^+-N and NO_3^--N were correlated.

Sediment chemistry

In Oxelösund, sediment samples were collected in 1994 (April–May) and 1998 (September–October) to estimate N accumulation in or release from the sediment. The basins S1, S2, N1 and N2 were divided into 3, 3, 5 and 4 sub-areas, respectively. Sediment cores were sampled with a steel tube (2.2 cm inner diameter). In each sub-area, 20 sediment cores were taken randomly at 0–10 cm and 10–20 cm depth. These were gathered randomly to four composite samples (of five cores) from each depth. The samples were analysed for total N and total C content at the Swedish University of Agricultural Sciences by a high-temperature combustion method. An analysis of variance was performed on the data, with "year of sampling" and "wetland basin" as fixed variables.

Results and discussion
Oxelösund

Modelled NH_4^+-N and NO_3^--N concentrations corresponded only weakly with measured concentrations, and for organic N there was no correspondence (Table 1 and Figure 3). The explained variance (R^2) varied, however, between years: from –0.02 to 0.61 for NH_4^+-N and from –0.16 to 0.27 for NO_3^--N. The best fit for both N compounds occurred in 1996. The NO_3^--N fit was slightly improved when using two free parameters (Eq. 3), as compared to the model with only one free parameter (Eq. 2). The k_{a20}-value for denitrification of NO_3^--N was high compared to values reported by Kadlec and Knight (1996) and Knight *et al.* (2000). This shows either a truly high denitrification rate in the Oxelösund wetland, or an overproduction of NO_3^--N in the model. Overproduction of nitrate by nitrification could occur in the model if plant uptake and ammonia volatilisation are significant alternative pathways for NH_4^+-N transformation in the Oxelösund wetland, since these processes were not included in the model.

When modelled NO_3^--N concentrations were compared to measured concentrations in the different basins, the north system showed good correspondence to the model: $R^2 = 0.80$ in N1, $R^2 = 0.86$ in N2 and $R^2 = 0.51$ in SN3 (Figure 4). The modelled concentrations in the south system were lower than the measured and thus showed a poor fit to the model. Higher measured concentrations in the south than in the north system might be explained by the fact that oxygen concentrations were also higher in the south system (ANOVA, $p < 0.05$, Figure 4), favouring nitrification and disfavouring denitrification.

The model was sensitive to changes of the reaction rate coefficients for nitrification and denitrification. Decreasing k_a for nitrification to 50% of the calibrated value (Figure 3) resulted in a decrease of R^2 from 0.33 to –0.05. With an increase to 150% of the calibrated value, R^2 decreased to 0.18. Correspondingly for denitrification, R^2 decreased from 0.10 to –0.89 and –0.01 when k_a was set to 50 and 150% of the calibrated value, respectively. Changing the reaction rate coefficient for ammonification had less impact on model performance.

The differences between modelled and measured concentrations (the residuals) for NH_4^+-N and NO_3^--N were correlated (corr. coeff. = –0.42). A negative correlation between

Table 1 Reaction rate coefficients, k_a (Eq. 2) and k_{a20} (Eq. 3), and temperature coefficients, θ (Eq. 3), for the processes ammonification, nitrification and denitrification in the wetland of Oxelösund, and associated explained variances (R^2) concerning concentrations of organic N, NH_4^+-N and NO_3^--N during calibration (1994–1995) and validation (1996–1998)

	k_a (m d⁻¹ °C⁻¹)	R^2 (calib.)	R^2 (valid.)	k_{a20} (m d⁻¹)	θ	R^2 (calib.)
Ammonification/Organic N	0.0011	–0.95	–0.94	0.014	1	–0.58
Nitrification/NH_4^+-N	0.0014	0.33	0.34	0.023	1.072	0.30
Denitrification/NO_3^--N	0.023	0.10	0.10	0.65	1.116	0.16

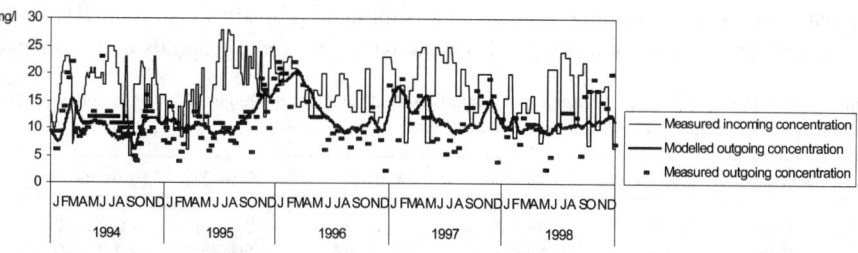

Figure 3 Modelled and measured NH_4^+-N concentrations in the wetland of Oxelösund during the calibration (1994–1995) and validation (1996–1998) periods

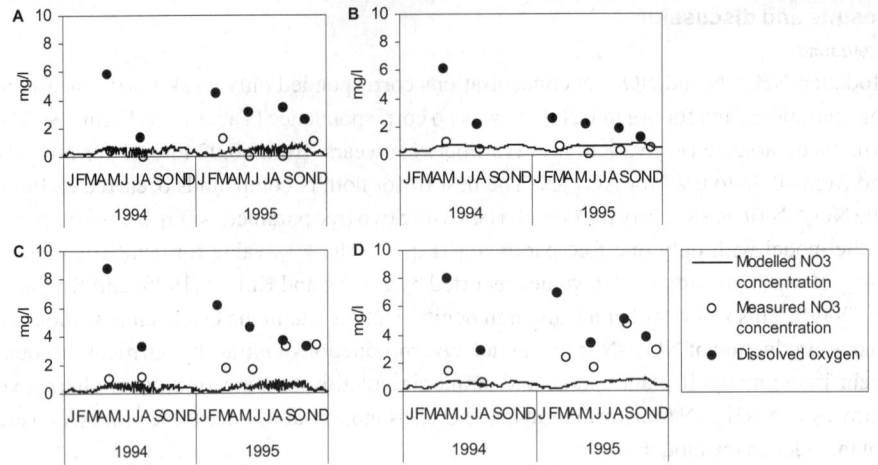

Figure 4 Modelled and measured NO_3^--N concentrations and measured dissolved oxygen concentrations in the north (A = N1, B = N2) and the south system (C = S1, D = S2) of the wetland of Oxelösund in 1994–1995

the residuals was expected since the NO_3^--N concentration at the inlet was less than 0.1 mg l^{-1}, and thus all NO_3^--N was generated from transformation of NH_4^+-N within the wetland. No seasonal pattern was observed.

Sediment chemistry. The N content of the sediments in Oxelösund decreased significantly from 1994 to 1998 (Table 2). No significant difference in carbon content was shown over time and the C/N-ratio increased in all basins from 1994 to 1998 ($p < 0.01$, interaction >0.05). The decrease in N content from 1994 to 1998 might be explained by uptake in vegetation, which increased dramatically over the years (Andersson *et al.*, 2001).

Hässleholm

The model, with reaction rate coefficients calibrated in the wetland of Oxelösund, was applied in the wetland of Hässleholm. The modelled NH_4^+-N concentrations showed a relatively good correspondence with the measured values (Table 3), while organic N and NO_3^--N showed no correspondence. After calibration against concentrations at the outlet of the wetland of Hässleholm (Table 3, Figure 5), both NH_4^+-N and NO_3^--N concentrations showed a good fit for all years: $R^2 = 0.78–0.85$ for NH_4^+-N and $R^2 = 0.41–0.66$ for NO_3^--N. There was only a small effect on explained variance when using two free parameters (Eq. 3) instead of one (Eq. 2) (Table 3). As in Oxelösund, organic N showed no correspondence.

In contrast to Oxelösund, there was no strong correlation between the residuals of NH_4^+-N and NO_3^--N concentrations (corr. coeff. = –0.08) which might be explained by the fact that NO_3^--N concentrations were much less influenced by nitrification of NH_4^+-N in Hässleholm. The analysis of residuals also showed that the model generally overestimated

Table 2 The accumulation (+) or release (–) of N and C (kg ha^{-1} yr^{-1}) from 1994 to 1998 in the sediment of the basins (N1, N2, S1 and S2) in the wetland of Oxelösund

	Depth (cm)	N1 (s.d.)	N2 (s.d.)	S1 (s.d.)	S2 (s.d.)	p (interaction)
Total N	0–10	–73 (75)	+98 (70)	–55 (75)	–28 (68)	<0.05 (>0.1)
	10–20	–154 (68)	–10 (58)	–181 (120)	+30 (85)	<0.01 (>0.1)
Total C	0–10	–553 (1,045)	+1,399 (670)	–261 (1,022)	–234 (800)	>0.1 (>0.01)
	10–20	–1,337 (517)	+143 (1,013)	–1675 (1,346)	+339 (1,029)	<0.05 (>0.1)

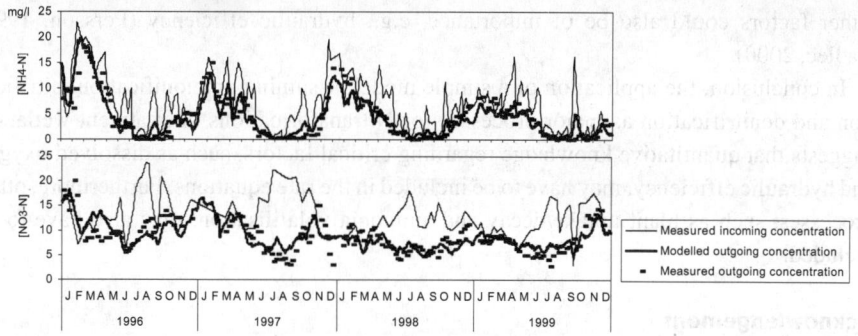

Figure 5 Modelled and measured NH_4^+-N and NO_3^--N concentrations in the wetland of Hässleholm during the calibration period (1996–99)

Table 3 Reaction rate coefficients, k_a (Eq. 2) and k_{a20} (Eq. 3), and temperature coefficients, θ (Eq. 3), for the processes ammonification, nitrification and denitrification in the wetland of Hässleholm, and associated explained variances (R^2) concerning concentrations of organic N, NH_4^+-N and NO_3^--N during calibration (1996–1999) and when using k_a as obtained from calibration in Oxelösund (Table 1)

	k_a (m d^{-1} °C^{-1})	R^2 (calib.)	R^2 (coeff. of Oxelösund)	k_{a20} (m d^{-1})	θ	R^2 (calib.)
Ammonification/Organic N	0.0001	–0.96	–0.98	0.0001	1	–0.91
Nitrification/NH_4^+-N	0.0044	0.83	0.67	0.10	1.116	0.83
Denitrification/NO_3^--N	0.0033	0.58	–3.05	0.08	1.116	0.60

both the NH_4^+-N and NO_3^--N concentrations during summer, and underestimated these concentrations during winter. Overestimation during summer conditions might indicate plant uptake or ammonia volatilisation, while underestimation during winter conditions might indicate leakage from plants or sediment during decomposition.

Concluding discussion

Calibration in the wetland of Oxelösund resulted in a high reaction rate coefficient for denitrification compared to the value estimated for Hässleholm and literature values (Kadlec and Knight, 1996; Knight et al., 2000). If the Hässleholm denitrification rate coefficient was applied in Oxelösund, the nitrification rate coefficient would have to be very low (k_{a20} = 0.002), and 50–450 kg N ha^{-1} yr^{-1} would have to be removed by processes not included in the model. Plant uptake could be of some importance. The vegetation is not harvested in Oxelösund, but the standing biomass has increased dramatically over the years. Concerning below ground biomass, the sediment samples did not show any indication of increased N content between 1994 and 1998. However, it is possible that roots and rhizomes were not included in a fully representative way in sediment sampling, and that storage in below ground biomass could be significant for the N budget. Ammonia volatilisation is not considered quantitatively important below pH 8 (Reddy and Patrick, 1984), while the highest measured pH was 7.9 and normal values were around 7.0 in the wetland of Oxelösund. Due to daily variations in pH, however, ammonia volatilisation could be of importance. NO_3^--N concentrations in the wetland of Oxelösund were low compared to Hässleholm. If NO_3^--N transformations at low and/or high concentrations were limited by other factors, in addition to concentration and temperature, then the reaction rate coefficients are lumped parameters that can not easily be compared between wetlands. The results in Figure 4 indicated that dissolved oxygen concentrations could be one important factor, which was not included in the model due to shortage of data.

Other factors could also be of importance, e.g., hydraulic efficiency (Persson, 1999; Kadlec, 2000).

In conclusion, the application of a simple model, assuming ammonification, nitrification and denitrification as major processes for N transformations in treatment wetlands, suggests that quantitative knowledge regarding critical factors, such as dissolved oxygen and hydraulic efficiency, may have to be included in the rate equations. Furthermore, other processes, such as plant uptake/decay and ammonia volatilisation, may also have to be included.

Acknowledgement

The authors wish to acknowledge Ingrid Hägermark and Hans Wallin at WRS AB and Per-Åke Nilsson at the Hässleholm Municipality for providing data. We also thank Katarina Losjö at the Swedish Meteorological and Hydrological Institute for hydrological modelling and David Bastviken for valuable comments. The study was part of the Swedish Water Management Research Program (VASTRA), which is financed by the Swedish Foundation for Strategic Environmental Research (MISTRA).

References

Andersson, J.L., Wittgren, H.B., Kallner, S., Ridderstolpe, P. and Hägermark, I. (2001). Wetland Oxelösund, Sweden – the first five years in operation. In: Mander Ü. and Jenssen P. D. (eds). *Natural Wetlands for Wastewater Treatment in Cold Climate Areas*. WIT Press, Southampton, UK (in press).

Bergström, S. (1992). *The HBV model – its structure and applications*. SMHI Report Hydrology (RH) No 4. Swedish Meteorological and Hydrological Institute, Norrköping.

Hässleholms kommun. (1999). Miljörapport 1999, Hässleholms avloppsreningsverk. Tekniska kontoret, Hässleholm, Sweden (in Swedish).

Kadlec, R.H. (2000). The inadequacy of first-order treatment wetland models. *Ecol. Eng.*, **15**, 105–119.

Kadlec, R.H. and Knight, R.L. (1996). *Treatment Wetlands*. CRC Press/Lewis Publishers, Boca Raton, Florida.

Knight, R.L., Payne, Jr. V.W.E., Borer, R.E., Clarke, Jr. R.A. and Pries, J.H. (2000). Constructed wetlands for livestock wastewater management. *Ecol. Eng.*, **15**, 41–55.

Martin, J.F. and Reddy, K.R. (1997). Interaction and spatial distribution of wetland nitrogen processes. *Ecol. Model.*, **105**, 1–21.

Nash, J.E. and Sutcliffe, J.V. (1970). River flow forecasting through conceptual models. Part I. A discussion of principles. *J. Hydrol.*, **10**, 282.

Persson, J. (1999). Hydraulic Efficiency in Pond Design. Ph. D. thesis, Dept. of Hydraulics, Chalmers University of Technology, Göteborg, Sweden.

Reddy, K.R. and Patrick, W.H. (1984). Nitrogen transformations and loss in flooded soils and sediments. *CRC Crit. Rev. Environ. Control*, **13**, 273–309.

Wittgren, H.B. and Tobiason, S. (1995). Nitrogen removal from pretreated wastewater in surface flow wetlands. *Wat. Sci. Tech.*, **32**(3), 69–78.

The influence of water table fluctuations on nutrient dynamics in the rhizosphere of common reed (*Phragmites australis*)

O. Urbanc-Berčič and A. Gaberščik

National Institute of Biology, Večna pot 111, 1101 Ljubljana, Slovenia

Abstract Lake Cerkniško jezero is an intermittent lake and thus it is an ecosystem with permanent water level fluctuations. The lake area is covered mainly by wetland vegetation, with a common reed (*Phragmites australis*) as the prevailing species. The present research evaluated the influence of water table fluctuations on nutrient dynamics in the root zone of the reed. The content of nutrients (nitrate, nitrite, ammonium, soluble phosphorus) was monitored in the pore water, using ground samplers. The vitality of roots was determined measuring terminal electron transport (ETS) activity of root tissue on the vertical profile of the root system. The content of organic matter ranged from 33% up to 48% on a soil depth profile of 1 m. During the year nutrients were present on the whole profile. Water table fluctuations influenced the form, concentration and availability of nutrients in pore water and the vitality of roots. The measurements of potential respiration of root tissue (ETS activity) revealed the highest vitality in the upper layer of the rhizosphere.

Keywords Nutrients; *Phragmites australis*; reed; terminal electron transport system (ETS) activity; water table fluctuations

Introduction

Lakes with extensive wetlands are sites where increasing human-induced eutrophication is buffered by the self-purification processes in densely vegetated littoral. Reed stands play an important role in nutrient dynamics (Wetzel, 1990; Lakatos *et al.*, 1998). During the last two decades authors discussed the syndrome of reed stand decline in Europe. They pointed out eutrophication as its main cause (Urbanc-Berčič and Gaberščik 1995; Gaberščik and Urbanc-Berčič 1996; Čičkova *et al.*, 1996; Lakatos *et al.*, 1998; Gaberščik *et al.*, 2000). On the other hand reed is the preferable species in constructed wetlands where high nutrient loadings show hardly any harmful effect on reed stand condition. Therefore additional causes are expected to be involved in the process of reed decline.

The site of our research – the intermittant Lake Cerkniško jezero is a unique karstic phenomenon. Dry and wet periods are exchanging during the year. In wet periods 80 billion m^3 of water could accumulate in the area (Habič, 1976). Water from the lake sinks underground through numerous gorges and gullets, which are connected, to a large underground cave system. In dry periods, usually in the summer, traditional agricultural activities take place at the edge of the area, while the water gathers in the main stream Stržen, meandering through the area. The majority of the area is covered by reed stands. The more or less regular water level fluctuations and exchanging of dry and wet periods are an outstanding feature of the ecosystem, the benefit of which is accelerated recycling of nutrients. Water fluctuations act also as a physical factor transporting and disintegrating particulate organic matter, which is annually accumulated in the area. In dry periods intensive aerobic mineralization occurs. Lowering of the water table enables transfer of nutrients in the deeper layers. The timing and the frequency of these processes influence annual primary production on the lake. Any disturbance, mechanical or physiological, detected in extensive reed stands would be a serious threat to this unique ecosystem, where reed vegetated areas present a main trap for loadings before the water sinks underground.

The present investigation was set up in the reed stand at Dujice which is the most vulnerable due to the lack of water in dry periods. The aim was to follow the dynamics of nutrients, namely nitrate, nitrite, ammonium and soluble phosphorus in pore water on the depth profile of soil in relation to water table fluctuations and to measure the vitality of roots of reed.

Material and methods
The Lake Cerkniško jezero is about 5 km wide and 10 km long. It is located in the transition area between Dinaric and Alpine regions (45°45'N, 14°20'E). The interference of the influences of Mediterranean and continental climate results in a specific precipitation regime (in average 1,700 mm/year). At highest water level, about 26 km^2 of Cerkniško Polje area is flooded (Habič, 1976). Young pleistocenic and holocenic deposits (clay, sand, dolomite and limestone gravel) cover the bottom of the lake. From the ecological point of view the Lake Cerkniško jezero is characterised as a water fluctuating ecosystem. During the rainy period, a large amount of water springs from numerous caves on the edge of Cerkniško Polje area and from the gorges. Waters gather in a wide riverbed and it takes only few days to flood the whole area.

Reed plants, which dominate the area, are found at locations with a different hydrological regime, therefore terrestrial stands on the water/land interface and littoral stands could be distinguished. The present investigation was set up in the terrestrial stand at Dujice which could be the most affected due to the lack of water in dry periods.

Chemical analyses
Pore water for nutrient analyses was sampled with the ground samplers (Soil Moisture Equipment, USA). They were placed on three sites within the reed stand at depths of 25, 50 and 90 cm. The concentrations of nitrate, nitrite, ammonium, and soluable phosphorus were detected *in situ* with Merck kits using spectrophotometer Spectroquant NOVA 60. Samples were collected from May 1999 to June 2000.

Terminal electron transport system (ETS) activity of roots
The terminal ETS activity was determined on the last third of the lateral roots in the depth profile of the vertical rhizome. Plant material was washed thoroughly and weighed, homogenised in an ice-cold homogenisation buffer in a potter (3–4 minutes; 500 rpm) and sonicated (20–30 seconds; 40W). After this treatment homogenate was centrifuged (4 minutes; 10,000 rpm; 0°C). Within 10 minutes triplicates of homogenate were incubated for 40 minutes in substrate solution (NADP, NADPH, Triton-x-100) with iodonitrotetrazolium chloride (INT). After stopping the reaction formazan production was determined spectrophotometrically and the ETS activity was calculated (Christensen and Packard, 1979) and expressed per fresh weight of sample (FW) as mg O_2/g FW/ h.

Results
The fluctuations of water table during the period from May 1999 to June 2000 were representative changing from saturated soil to dry soil, when the water table dropped under 1 m depth (Figure 1). In 1999 the drying began at the end of May while in 2000 it occurred one month earlier. Changing of the water regime was reflected in the amount and the form of nutrients. The highest concentrations of 2.1 mg/l ortho-phosphate were found in June 1999 in the deepest layer (Figure 2). The annual dynamics of soluble phosphorus is highly correlated with the amount of ammonium and nitrite, the correlation being 0.86 and 0.81, respectively. The results indicated that conditions for mineralization are favourable in the entire depth profile. Maximum values were determined in June 1999.

The amount of inorganic nitrogen varied annually and by depth (Figure 3). Nitrogen in

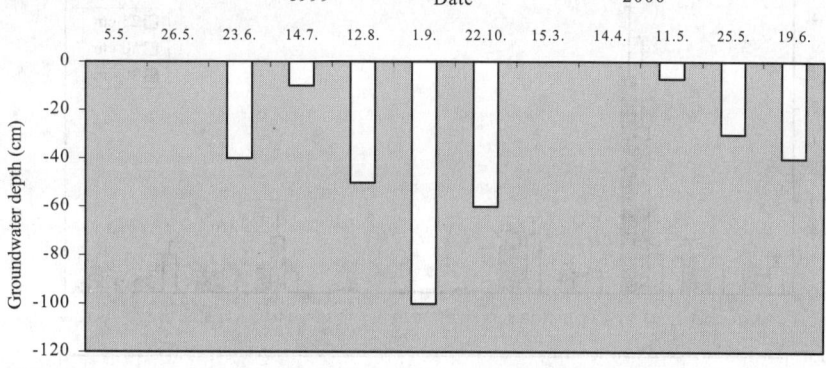

Figure 1 Water table fluctuations in the reed stand at Dujice (May 1999–June 2000)

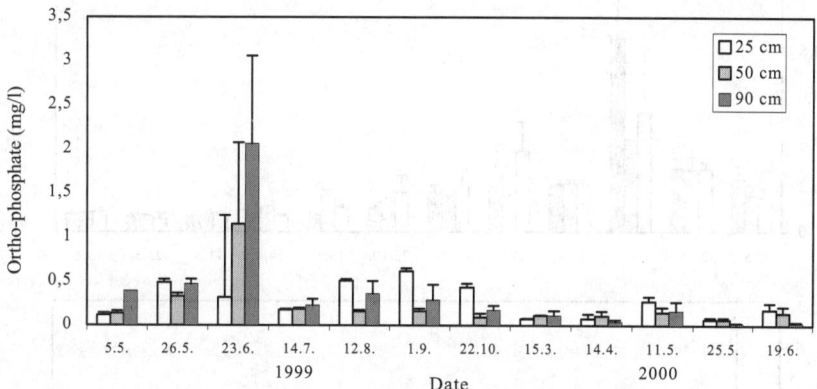

Figure 2 Pore water concentrations of ortho-phosphate at the depth profile in samples in the reed stand at Dujice (May 1999–June 2000, vertical bars represent standard error (SE), n = 3–4)

the form of ammonium is changing with no respect to ground water level. Maximal concentrations of ammonium and nitrite were detected in June 1999, being 1.45 mg/l and 0.86 mg/l, respectively. Average values of ammonium were between 0.2 mg/l to 0.4 mg/l except in the dry and cold periods, when the microbial activity is supressed either by oxygen or by lower temperatures. The contents of nitrate in the pore water ranged from 0.3 mg/l to 4 mg/l. Nitrogen in the form of nitrite was present in all samples. Forms of inorganic N and their depth distribution showed oscillations, which are highest in the lower layers.

Uniform conditions for mineralization on the soil depth profile are revealed in Figure 4 which presents the content of organic versus inorganic matter. Obtained ratios are similar on the entire profile.

Water table fluctuations induced the depth penetration of rhizomes. Vertical rhizomes spread down to 1 m, while a few horizontal rhizomes developed. In autumn 1999 ETS activity was measured at different layers to get an insight into the vitality of reed roots. Values varied from 0.18 mg O_2/g FW h at 0–20 cm to 0.04 mg O_2/g FW h at 80–100 cm (Figure 5). The vitality decreased gradually with the depth.

Discussion

Wetlands are efficient in removing and transforming nutrients even under high nutrient loading. This ability is successfully applied in constructed wetlands for wastewater treatment where hydraulic loading and permeability of media play an important role. In the Lake Cerkniško jezero nutrients leaching from the litter and flushing from the watershed

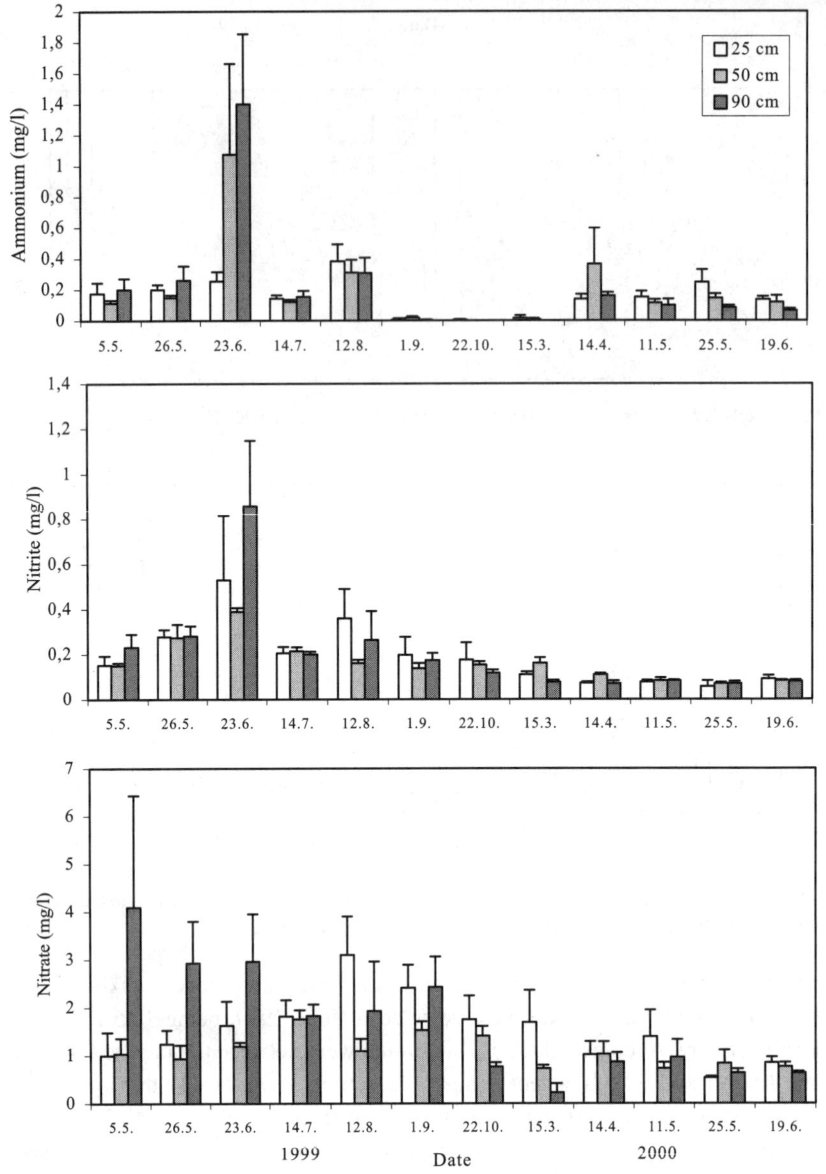

Figure 3 Pore water concentrations of ammonium, nitrite, and nitrate at the depth profile in samples in the reed stand at Dujice, (May 1999–June 2000, vertical bars represent SE, n = 3–4)

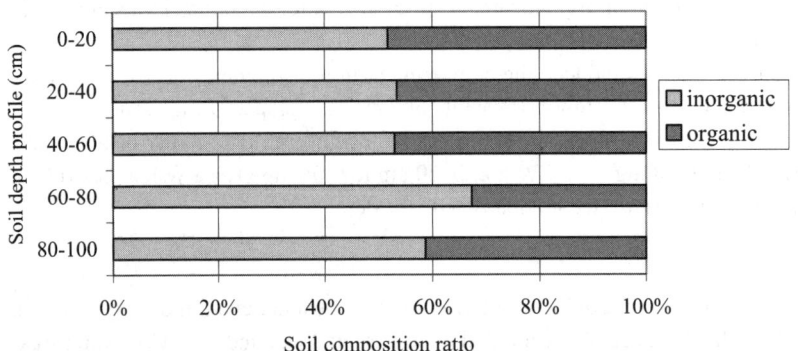

Figure 4 The ratio between organic and inorganic soil components at the depth profile at Dujice (sampled in autumn 1999)

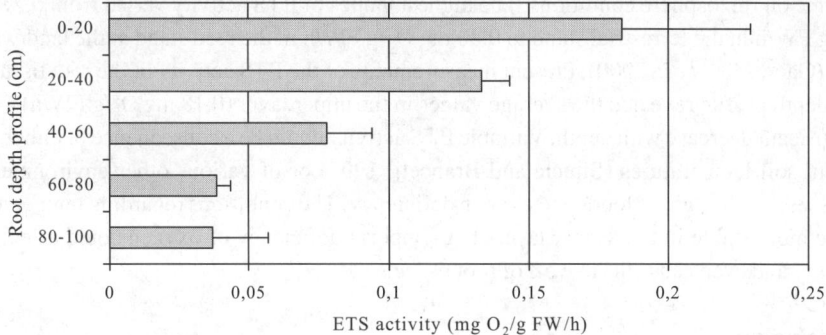

Figure 5 Terminal electron transport (ETS) activity on the depth profile of the root system in *Phragmites australis* in the location Dujice in autumn 1999. Columns present the means of three samples and SE

accumulate only briefly in reed stands. The reason for that is intensive nutrient cycling driven by short-term water level fluctuations and exchanging of dry and wet periods. These conditions enable the aeration processes which took place occasionally in the entire rhizosphere. The exchanging of dry and wet periods provide different environmental conditions in various reed stands on the lake and consequently affects their annual primary production (Gaberščik et al., 2000).

Monitoring of ground water table in the stand during two subsequent vegetative periods showed that in dry periods the water table could drop under a depth of 1 metre. These conditions accelerated mineralization of organic matter in the entire rhizosphere, which could assure a constant supply of nutrients for plant uptake as was reported by other authors (Wetzel, 1990). The intensive mineralization was evident from the organic matter content in the soil profile, which ranged from 32 to 47% (Figure 4). A well-developed rhizosphere of reed, spreading to a depth of 1 metre, was an additional indication of favourable conditions at the location. This aspect is also important during floods, when roots could provide a significant amount of oxygen in the rhizosphere (Armstrong et al., 1990; Brix, 1993).

Present research revealed that the amount and forms of nutrients in pore water at a selected site varied comparing dry and wet periods. Lowering of the water table resulted in the release of nutrients and their translocation to the deeper layers. Nitrite, which was present in low concentrations even in dry periods, reflected the exchanges of aerobic and anaerobic conditions. The concentrations of nitrate were relatively high throughout the entire growth period, which indicated permanent releasing of nitrogen and the presence of oxygen. This was not the case in the research of Starink et al. (1999) who detected no nitrate in the pore water during a high vegetative period. Our findings showed that in a reed stand, nitrification and denitrification processes were permanently going on. Concentrations of ortho-phosphate showed high correlation with nitrite and ammonium. In reed stands with water table fluctuation the availability of nutrients is assured by efficient cycling from litter to soil and back to reed plants. Nutrient dynamics in such systems is much more intensive in comparison to open water stands (Vaithiyannathan and Richardson, 1998), where anaerobic forms of nitrogen prevail and constant water saturated conditions retard mineralization of organic matter.

To get an insight into the vitality of root systems and their supportive role in processes of mineralization we estimated the respiratory potential by measuring terminal electron transport system (ETS) activity. The ETS activity measured on roots of the vertical rhizomes of reeds grown in a constructed wetland ranged from 0.06 to 0.80 mg O_2/g FW/ h (Urbanc-Berčič and Gaberščik, 1997). Our preliminary results obtained on three different reed stands in the Lake Cerkniško jezero in 1997 pointed out the influence of a different water regime on rhizosphere conditions. The highest values of ETS activity varied from 0.27 mg O_2/g FW/h in the terrestrial stand to 0.40 mg O_2/g FW/h in the reed stand at the land/water site (Gaberščik et al., 2000). Present measurements of the ETS activity of the root tissue in the depth profile revealed the average values in the upper layer (0.18 mg O_2/g FW/h) and a significant decrease with depth. Variable ETS activity could be a consequence of either different soil temperatures (Simčič and Brancelj, 1997) or of various other environmental stresses, i.e. droughts, floods and oxygen deficiency. The conditions regarding temperature were more stable in the deeper layers but temporal deficiency of oxygen could affect the vitality and even cause the dye off of root tissue.

Conclusion

The reed stands on the Lake Cerkniško jezero thrive under outstanding conditions which are driven by water level fluctuations and exchanging of dry and wet periods. Results of our study revealed that the amount of nutrients on the depth profile in rhizosphere depends on short-term water level fluctuations. Decreasing of the water table in the dry periods supports aerobic mineralization and growth of vertical rhizomes. The extended root system supplies the rhizosphere with oxygen and accelerates mineralization in the whole depth profile. Outcomes of this research provide a good basis for designing and operating constructed wetland systems, especially with subsurface flow (SSF) where short cuts and clogging present serious disturbance, lowering the efficiency of the systems.

Acknowledgement

This research was part of project No. J1-8738-0105-99 of the Ministry of Science and Technology of the Republic of Slovenia.

References

Armstrong, W., Armstrong, J. and Beckett, P.M. (1990). Measurement and Modelling of oxygen release from roots of *Phragmites australis*. In: *Constructed Wetlands in Water Pollution Control*, P.F. Cooper and B.C. Findlater (eds.), Pergamon Press, pp. 41–52.
Brix, H. (1993). Macrophyte-Mediated Oxygen Transfer in Wetlands: Transport Mechanisms and Rates. In: *Constructed Wetlands for Water Quality Improvement*, G.A. Moshiri (ed.), pp. 391–398.
Christensen, J.P. and Packard, T.T. (1979). Respiratory electron transport activities in phytoplankton and bacteria: comparison of methods. *Limnol. Oceanogr.*, **24**(3), 576–583.
Čičkova, H., Strand, J.A. and Lukavska, J. (1996). Factors associated with reed decline in an eutrophic fishpond, Rožmberk (South Bohemia, Czech Republic). *Folia Geobotanica Phytotax.*, **31**, 73–84.
Gaberščik, A. and Urbanc-Berčič, O. (1996). Monitoring approach to evaluate water quality of the intermittent Lake Cerkniško jezero. *Wat. Sci. Tech.*, **33**(4–5), 357–362.
Gaberščik, A., Urbanc-Berčič, O. and Martinčič, A. (2000). The influence of water level fluctuation on the production of reed stands (*Phragmites australis*) on intermittent lake Cerkniško jezero. Proceedings of the International Workshop and *10th Macrophyte group meeting IAD/SIL*, 24.–28. Aug. 1998, Danube Delta, Romania; pp. 29–32.
Habič, P. (1976). Investigations in Ljubljanica river basin. Geomorphologic and hydrologic characteristics. In: *Underground water tracing*, R. Gospodarič and P. Habič (eds.). Third International Symposium of Underground Water Tracing, Ljubljana – Bled, pp. 12–27.
Lakatos, G., Grigorszky, I. and Biro, P. (1998). Reed-periphyton complex in the littoral of shallow lakes. *Verh. Internat. Verein. Limnol.*, **26**, 1852–1856.
Simčič, T. and Brancelj, A. (1997). Electron transport system (ETS) activity and respiration rate in five Daphnia species at different temperatures. *Hydrobiologia*, **360**, 117–125.
Starink, M. van Rijswijk, J., Middelburg, J. and Cappenberg, T. (1999). Nitrification – denitrification in the rhizosphere of emergent macrophytes; fiction or facts. *Inter. Conf. on Phragmites-dominated wetlands, their functions and sustainable use*. Trebon, Czech Republic.
Urbanc-Berčič, O. and Gaberščik A. (1995). Potential of the littoral area in Lake Bled for reed stand extension. In: Ramadori, R., Cingolani, L. and Cameroni, L. *Natural and Constructed Wetlands for Wastewater treatments and reuse* (eds.). Centro Studi Provincia di Perrugia, Italy, pp. 95–100.
Urbanc-Berčič, O. and Gaberščik, A. (1997). Reed stands in constructed wetlands: "Edge effect" and photochemical efficiency of PS II in common reed. *Wat. Sci. Tech.*, **35**(5), 143–147.
Vaithiyannathan, P. and Richardson, C.J. (1998). Biogeochemical characteristics of the Everglades Sloughs. *Journal of Environmental Quality*, **27**(6), 1439–1450.
Wetzel, G.R. (1990). Land-water interfaces: Metabolic and limnological regulators. – *Verh. Internat. Verein. Limnol.*, **24**, 6–24.

Thermal environments of subsurface treatment wetlands

R.H. Kadlec

Wetland Management Services, 6995 Westbourne, Chelsea, MI, USA

Abstract Treatment wetlands are solar powered ecosystems, resulting in annually cyclic temperatures. This paper reports data and models for temperatures and energy flows for subsurface flow wetlands. The water temperature seasonal cycle follows the air temperature during unfrozen conditions, with small hysteresis. Winter under-ice water temperatures are approximately 2°C. The energy balance is dominated by radiation to and from the wetland, and evaporative losses. Sensible heat flows, conduction and convection are of smaller magnitude. Lateral energy losses were measured to be small. Vertical gains and losses were also small, but of importance in winter conditions. A simple model for ice formation shows that ice formation may be held to an acceptable minimum by addition of mulch or by early snow accumulation.

Keywords Conduction; constructed wetlands; convection; cycles; energy; soil; subsurface wetlands; temperature

Introduction

Treatment wetlands are used in a wide variety of climatic conditions, ranging from tropical to sub-arctic. Some, but not all, of the water quality improvement processes are sensitive to water and soil temperatures. In cold climates, there is a need to know the extent and duration of ice formation. This paper reports data on temperatures and energy balances for subsurface flow wetlands. Emphasis is on measurements and models for three wetlands located near Duluth, Minnesota, USA, one at Grand Lake (GL) and two at the Northeastern Region Correctional Center (NERCC). Details of these wetlands have been set forth in several papers (e.g., Axler et al., 1999). The Grand Lake treatment wetland consists of a two cell, subsurface flow system. Dimensions of lined cell 1 are 10.3 m × 17.7 m (L × W)(182 m^2), and for unlined cell 2 are 14.6 m × 20.1 m (L × W)(293 m^2). The design nominal residence time of cell 1 was 10 days at a hydraulic loading rate of 2.19 cm/day. Temperatures were measured at inlet and outlet of cell 1, and at each of two depths at 18 interior locations, 9 in each cell. Soil temperatures were measured at six additional locations, inside and outside the wetlands, at four depths: 0, 30, 60 and 90 cm.

The two replicate constructed wetlands at NERCC are lined two-cell, subsurface flow systems. The dimensions of each wetland cell are 7.0 m × 5.3 m (L × W), with a bed depth of 46 cm pea rock (1.0–1.2 cm) was used in cell 1 and limestone (crushed and screened to 1.2–1.9 cm) was used in cell 2 of each wetland train. The wetlands were lined with 1.0 mm LDPE. Design hydraulic residence time was 13 days with a hydraulic loading rate of 1.27 cm/d (0.95m^3/day). Temperatures were measured at the inlet and outlet of cells 1 and 2 in each train, and at each of two depths at 8 interior locations, 2 in each cell. Soil temperatures were measured at four locations, one inside each wetland cell, at four depths: 0, 30, 90 and 150 cm; and additionally at each train inlet, at 8, 15, 22 and 30 cm, and at two locations adjacent to the wetlands. Snow depths and snow temperatures were periodically monitored during the winter at both GL and NERCC.

Movement of energy to and from adjacent soil
Horizontal energy movement
The possibility of lateral heat leakage was investigated at the Minnesota sites. There was

spatial uniformity of water temperatures on any given date in three directions for GL and NERCC wetlands, for at least the last 75% of the flow path, to less than one degree standard deviation (GL: 66 dates over four years at 18 locations, sd = 0.9°C; NERCC: 51 dates over four years at 10 locations in each of two wetlands, sd = 0.8°C). The vertical uniformity of wetland water temperature indicates no stratification occurs over the very small vertical dimension, which is normally in the range of 0.3–0.9 metres.

Lateral heat losses and gains occur when the wetland water temperature differs from that of the horizontally adjacent soils. This energy flow was assessed at the GL site, by acquiring data for soil temperatures at several depths for two locations each inside and outside the wetland, separated by about nine metres. The horizontal temperature gradients were in the range $-0.3 < \partial T/\partial y < 0.3$°C/m. The thermal conductivity of the soil was estimated as 0.048 MJ/m d °C. The perimeter area of the wetland, extending to 1.82m deep, was 102m². Thus, using Fourier's law of heat conduction, there was a maximum heat loss of 1.5 MJ/d, prorated over 182 m² of wetland area, for an areal loss of 0.008 MJ/m² d. This is much less than the vertical energy flows, and may be considered negligible.

Vertical energy movement

The temperature profiles $T(z,t)$ in the (unfrozen) soils below a wetland are governed by the unsteady state heat conduction equation together with the boundary condition:

$$\frac{\partial^2 T}{\partial z^2} = \frac{1}{\alpha}\frac{\partial T}{\partial t} \qquad T(\infty,t) = T_s \tag{1,2}$$

For a sinusoidal surface temperature, the solution to this periodic, dynamic heat balance is:

$$T(z,t) = T_s + A\exp\left(-\frac{z}{H}\right)\cos\left[\omega(t - t_{max}) - \frac{z}{H}\right] \tag{3}$$

where $H = \sqrt{\frac{2\alpha}{\omega}}$ and $\alpha = \frac{k}{\rho_s c_s}$ (4,5)

and where A = amplitude of surface temperature cycle, °C; c_s = soil heat capacity, MJ/kg °C; k = soil thermal conductivity, MJ/m d °C; t = time, Julian day; t_{max} = time of maximum surface temperature, Julian day; T = temperature, °C; T_s = mean annual temperature of the soil surface, °C; z = vertical depth, m; α = thermal diffusivity of soil, m²/d; ρ_s = soil density, kg/m³; ω = annual cycle frequency = $2\pi/365$ = 0.0172 d^{-1}. The penetration depth (H) is the depth at which the mean annual temperature swing is 63.2% of that at the soil surface (A). The heat flux into the water from the soil is then:

$$G = \left[\frac{kA}{H}\right]\cdot\{\cos[\omega(t - t_{max})] - \sin[\omega(t - t_{max})]\} \tag{6}$$

It may be shown that the heat flux (G) achieves a maximum 46 days (one-eighth of an annual cycle) before the day of minimum water temperature, which is 136 days after the day of maximum water temperature. It may also be shown that the total heat gain from the soil over the 182-day heating half cycle (G_{half}) is:

$$G_{half} = (2\sqrt{2})\frac{kA}{\omega H} \tag{7}$$

The maximum daily heat gain may be shown to be a factor $\pi/2 = 1.57$ times greater than the average rate over the heating half of the year. This model provides an accurate description of the temperature gradients below the Grand Lake and NERCC treatment wetlands (Figure 1, Table 1), and of the heat fluxes to and from soils to the wetland water. The amounts of

Table 1 Regression parameters for the soil temperature heat conduction model

	NERCC 1	NERCC 2	Grand Lake
Surface Temperature Amplitude, °C	8.23	8.23	8.02
Surface Temperature Maximum, Julian Day	213	213	217
Penetration Depth, m	2.05	2.24	2.17
Thermal Diffusivity, m^2/d	0.0361	0.0432	0.0407
Correlation Coefficient, R^2 (4 depths, 4 years)	0.87	0.89	0.88
Upward Heat Flux Maximum, Julian Day	350	349	353
Maximum Heat Flux, $MJ/m^2 \cdot d$	0.274	0.250	0.250
Half-year Heat Gain, MJ/m^2	31.8	29.1	28.9

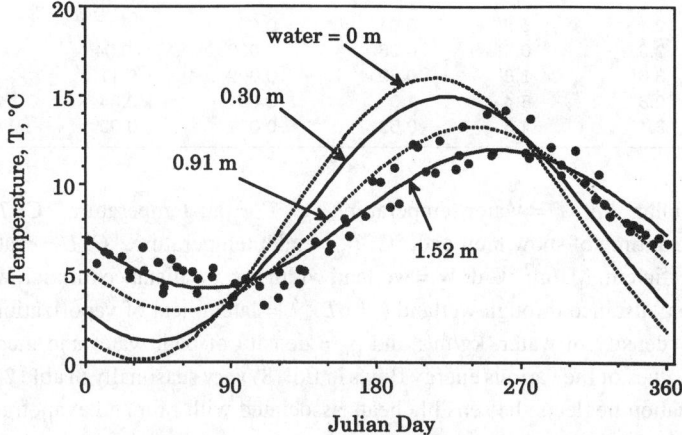

Figure 1 Soil temperatures below the Grand Lake wetland. Lines are the model fit. Data are shown only for the 1.52 m depth for clarity; scatter was comparable at all depths

these fluxes are small relative to other energy flows in the system. As a consequence of this information, the wetland water, media and plants may be considered as a one-dimensional flow-through system, with weak interactions with the energy storage in soils below, and strong interactions with the atmosphere above.

Unfrozen energy balance and temperatures
Energy balance

The temperature profiles $T(x,t)$ in the water passing through the wetland are governed by the unsteady state heat conduction equation. For a rectangular wetland, this energy balance may be expressed as a one dimensional, transient continuum model:

$$-\rho c \frac{\partial(qT)}{\partial y} + R + G - E + U(T_a - T) = \rho_\beta c_\beta h \frac{\partial T}{\partial t} \qquad (8)$$

together with the boundary and initial (early spring) conditions:

$$T(0,t) = T_{wi} \qquad T(y,t_m) = T_m(y) \qquad (9,10)$$

where c = water heat capacity, MJ/kg °C; c_b = bed, water and media, heat capacity, MJ/kg °C; E = energy loss due to ET (= λET), MJ/m^2 d; ET = evapotranspiration rate, m/d; G = energy gain from soil (= $C_V - C_L$), MJ/m^2 d; h = bed depth, m; H_a = energy gain from air (= $U(T_a - T)$, MJ/m^2 d; L = wetland length, m; q = hydraulic loading (= Q/WL), m/d; R = energy gain from net radiation (= $(1-\alpha)R_S - R_b$), MJ/m^2 d; t_m = time of snow

Table 2 Approximate components of the wetland energy budget for the Grand Lake treatment wetland (units: MJ/m² d). The amount of absorbed radiation is computed by difference

Energy Inputs	Undepleted Radiation	Absorbed Radiation	Convective (into water)	Sensible (with inflow)	Conduction (vertical from soil)	Total Input
Fall	20.0	7.2	−0.04	0.57	0.10	7.9
Winter	15.0	4.1	−0.30	0.10	0.20	4.1
Spring	30.0	6.5	0.84	0.41	−0.10	7.7
Summer	40.0	8.8	2.81	0.95	−0.20	12.3
Annual	26.3	7.3	0.29	0.51	0.00	8.1

Energy Outputs	Back (Heat) Radiation	ET Loss	Sensible (with outflow)	Heat Loss (lateral to soil)	Storage (in wetland)	Total Output
Fall	2.1	5.2	0.61	0.001	−0.11	7.9
Winter	3.6	0.5	0.08	0.003	−0.04	4.1
Spring	3.6	3.6	0.39	0.002	0.11	7.7
Summer	2.8	8.4	1.01	0.001	0.04	12.3
Annual	3.1	4.5	0.53	0.002	0.00	8.1

melt end, Julian day; T = water temperature, °C; T_a = air temperature, °C; T_m = water temperature at time of snow melt end, °C; T_{wi} = inlet temperature, °C; U = water-air heat transfer coefficient, MJ/m² °C d; W = wetland width, m; x = distance through wetland, m; y = fractional distance through wetland (= x/L); λ = latent heat of vaporization of water, MJ/m³; ρ = density of water, kg/m³; and ρ_b = density of bed, water and media, kg/m³. Estimated values of the various energy flows in Eq. (8) vary seasonally (Table 2).

This equation neglects the sensible heats associated with rain and evapotranspiration, which are small in comparison to other terms. In many situations, it may be shown that the heating or cooling of the bed is a slow process, and consequently the energy storage term may be omitted (this is the quasi-steady state approximation). Therefore, (8) is rewritten:

$$\rho c q \frac{\partial T}{\partial y} = R + G - E + U(T_a - T) \tag{11}$$

In a steady flow situation, the temperature of the water at large wetland lengths approaches a limiting value (Kadlec and Knight, 1996) and $\delta T/\delta y = 0$:

$$0 = R + G - E_b + U(T_a - T_b) \tag{12}$$

where T_b = water balance point temperature, °C; and E_b = evaporative energy loss at balance condition, MJ/m² d. The balance point temperature cannot be predicted from (12) with any reasonable accuracy, because of its sensitivity to a number of meteorological variables. However, the balance point temperature is easily determined from experiment, as the asymptotic water temperature reached along the gradient from wetland inlet to outlet (Figure 2). The approach to the balance point is rapid, often resulting in most of the wetland length being at that temperature (Figure 2). Data on water temperatures, together with the physical properties of water, allow calculation of the heat transfer coefficient (U). Subtraction of (12) from (11), together with the assumption that $E = E_b$ gives:

$$\rho c q \frac{\partial T}{\partial y} + U(T - T_b) = 0 \tag{13}$$

Eq. (13) predicts an exponential change in water temperature, approaching the balance temperature as the water progresses through the wetland. The overall heat transfer coefficient is calculated from the change in water temperature over the inlet section of the

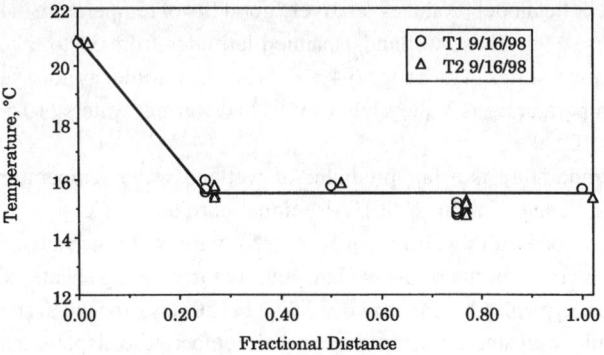

Figure 2 Water temperatures along the flow direction at the two NERCC wetlands T1 and T2

wetland. For the NERCC wetlands, there was a mean annual change over the first half of the first cell of 8.6°C, (range 3.0–14.0°C), resulting in a lower bound estimate $U = 0.31 \pm 0.03$ MJ/m^2 d °C. This value of U is consistent with the widely accepted value of the heat transfer coefficient in stagnant air above evaporating vegetated surfaces, which is $U = 0.37$ MJ/m^2 d °C (ASCE, 1990).

Wetland water temperatures

The water temperatures vary seasonally, as do air temperatures (Figure 3). Data from several SSF wetlands were regressed to a truncated, sinusoidal time series model of the form:

For the unfrozen season $(t_1 < t < t_2)$: $T = T_{avg}(1 + A \cos[\omega(t - t_{max})])$ (14)
For the frozen season $(t_2 < t < t_1)$: $T = T_o$

where A = fractional amplitude of the annual temperature, °C; t = time, Julian day; t_1 = ice-out time, Julian day; t_2 = freeze up time, Julian day; t_{max} = time of annual maximum temperature, Julian day; T = water temperature, °C; T_{avg} = annual average water temperature, °C; and T_o = under-ice water temperature, °C. In cold climates, the winter water temperatures under ice are just above freezing. Regression of 17 years data from four wetlands produced $1.92 < T_o < 2.22$°C (Table 3). Accordingly, a water temperature of 2°C appears appropriate for under-ice, or near freezing, conditions. Nearly as good a data fit may be obtained by ignoring a frozen period, and constraining the water temperature to be at or above zero

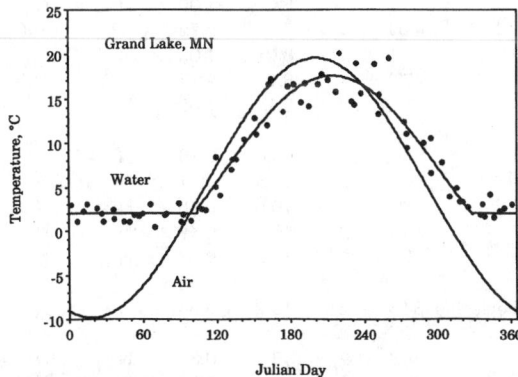

Figure 3 Balance point water temperatures in the Grand Lake wetland

(Table 3). The cyclic model produces relatively good fits of temperature in many wetlands, with $0.83 < R^2 < 0.96$. These wetlands spanned latitudes from 34 to 60°, and had mean annual water temperatures in the range $6.4 < T < 18.5°C$. Latitude is not a sufficient predictor of water temperatures, as Valleyfield (56°N) had warmer waters (10.5°C) than Grand Lake (47°N, 8.0°C).

Mean air temperature is a fair predictor of wetland water temperature for unfrozen conditions (Kadlec and Knight, 1996). Air temperature also undergoes a seasonal cycle, and may be described quite well by equation (15), with R^2 in the range of 0.80 to 0.95. Because of the thermal inertia of the wetland and its surroundings, water temperatures lag air temperatures, typically by about two weeks (11–20 days for the systems in Table 1). Therefore, the plot of water temperature versus air temperature displays a loop, with water temperatures exceeding air temperatures in fall, and the reverse in spring (Figure 4).

Frozen energy balance and ice thickness

Ice-forming conditions cause changes to the energy balance for the wetland. The presence of a layer of ice prevents evapotranspiration and precipitation is collected on top of the wetland as snow. The flow rate is therefore spatially uniform in the direction of flow. Freezing introduces a latent heat term into Eq. (11). The overall heat transfer coefficient (U) includes additional components due to the ice layer and the snow layer, which greatly decrease the heat loss. Eq. (11) is modified accordingly:

$$\rho c q \frac{\partial T}{\partial y} = R + G - E + U(T_a - T) + \lambda_f \frac{dh_i}{dt} \qquad (15)$$

where h_i = ice thickness, m; and λ_f = latent heat of fusion of water, MJ/m^3. Radiation and evaporation (sublimation) continue to dominate the wetland energy budget, but typically result in a snow surface temperature that is only slightly cooler than the mean air temperature. An energy balance at the snow surface may be used to solve for the snow surface temperature, but the supporting meteorological data are not easily available. As an approx-

Table 3 Annual SSF wetland water temperature sinusoidal cycle parameters for systems in several geographic regions. Systems with freezing conditions all regressed to winter water T = 2.0°C. Julian days at south latitudes are advanced to correspond to northern latitudes

Site	Lat. Deg.	Years	T mean °C	Amplitude	Freeze-up Julian Day	Thaw Julian Day	t_{max} Julian Day	R^2
Haugstein, Norway Maehlum, 1999	60N	5	6.4	3.07	320	100	209	0.94
			6.9	1.00	N	N	212	0.85
Grand Lake, Minnesota	47N	4	8.0	2.73	330	100	215	0.94
			8.0	1.00	N	N	217	0.87
NERCC 2, Minnesota	47N	4	7.9	2.72	330	100	215	0.96
			8.2	1.00	N	N	213	0.85
NERCC 1, Minnesota	47N	4	8.0	2.77	325	100	214	0.95
			8.2	1.00	N	N	213	0.85
Valleyfield 2, Scotland Cooper et al., 1996	56N	2	10.0	0.49	N	N	208	0.85
Valleyfield 3, Scotland	56N	2	10.5	0.47	N	N	211	0.85
Valleyfield 4, Scotland	56N	2	10.5	0.45	N	N	211	0.84
Valleyfield 1, Scotland	56N	2	10.6	0.47	N	N	205	0.83
Benton, Kentucky TVA, 1990	37N	1	13.9	0.68	N	N	195	0.88
Richmond, NSW Schoenoplectus Bavor et al., 1988	34S	2	18.2	0.34	N	N	214	0.86
Richmond, NSW Typha	34S	2	18.3	0.32	N	N	208	0.88
Richmond, NSW Gravel Only	34S	2	18.5	0.38	N	N	212	0.86

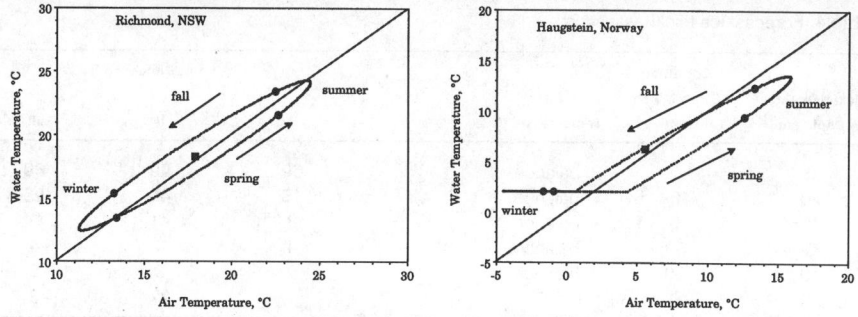

Figure 4 Annual course of air and water temperatures at two sites

imation, it may be assumed that snow surface temperature equals air temperature, and an empirical air-side heat transfer defined. For the balance condition:

$$\lambda_f \frac{dh_i}{dt} = G - U(T_a - T_b) \tag{16}$$

The overall heat transfer coefficient is comprised of several layer components:

$$\frac{1}{U} = \frac{1}{U_{water}} + \frac{h_i}{k_i} + \frac{h_s}{k_s} + \frac{h_m}{k_m} + \frac{1}{U_{air}} \tag{17}$$

where h_m = mulch (straw) thickness, h_s = snow thickness, m; k_i = thermal conductivity of ice, MJ/m d °C; k_m = thermal conductivity of mulch; k_s = thermal conductivity of snow, MJ/m d °C; U_{water} = water to ice heat transfer coefficient, MJ/m² d °C; U_{air} = snow to air heat transfer coefficient, MJ/m² d °C. Eqs (16) and (17) were solved for the temperatures and soil return heat flux measured in the GL and NERCC studies. The period of freezing temperatures extended from November 10 through March 31. The combined air and water heat transfer coefficient ($1/U_{water} + 1/U_{air}$) was assumed to be that calculated for summer conditions, 0.31 MJ/m² d °C. Three different presumed patterns of snow accumulation were considered:

I. Snow building up over the first 20 days, persisting until March, then melting over 20 days.
II. Snow building up over the first 60 days, followed by the remainder of the pattern in I.
III. No snow for 60 days, followed by the pattern in I.

Maximum snow depths of 0, 20 and 40 cm were considered, as well as the effect of 30 cm of straw added to the wetland surface. Thermal conductivities were: k_s = 0.015, k_i = 0.19 and k_{straw} = 0.009 MJ/m d °C. Both balance point and warm inlet temperatures were considered. The maximum ice thicknesses calculated ranged downward from the 69 cm predicted for zero snow or mulch (Table 4). Ice formation is greatly hindered by early snow, but the late snow of sequence II causes early freezing from which there is no recovery. A modest amount of heat in the incoming water (10°C) greatly assists in ice prevention, but cannot compensate for late snow accumulation. Mulch is effective under all three snow scenarios. The Grand Lake and NERCC sites displayed no freezing when straw-mulched, or when an early snow blanket developed, despite extreme cold. Average daily temperatures ranged down to –34°C. However, lack of early snow caused ice formation at both sites. Hydraulic problems were not encountered under flow conditions, but loss of flow caused problematic ice accumulations at NERCC.

Summary and conclusions

Water temperatures in subsurface flow treatment wetlands follow annual cycles that lag the

Table 4 Forecast ice thicknesses, cm

Maximum Snow Depth cm	Condition Straw Mulch cm	Inlet Temperature °C	Snow Sequence		
			I	II	III
0	0	balance	69	69	69
20	0	balance	18	46	25
40	0	balance	8	41	14
0	30	balance	4	4	4
10	30	balance	2	3	2
20	0	10	1	27	3

solar cycle by about four weeks. During warm seasons, water temperatures are approximately equal to air temperatures, with a slight hysteresis of cooler water temperatures in spring and warmer temperatures in fall. Under-ice water temperatures are nominally 2°C. Negligible lateral heat losses or gains were measured, and only small vertical losses or gains from deeper soils beneath the wetland. Heat associated with the incoming wastewater is dissipated in the entrance region of the wetland, with the exit regions achieving a balance temperature resulting from a combination of radiation and evapotranspiration, which are dominant; and convection and conduction, which are of lesser importance. All processes may be modeled accurately via spatially distributed energy balances. Ice formation is probable in extreme climates such as Duluth, Minnesota, although lessened by the presence of snow or mulch. The extreme case is early cold accompanied by lack of early snow.

Acknowledgements
R. Axler, B. McCarthy and J. Henneck of the Natural Resources Research Institute at the University of Minnesota-Duluth carried out the experiments. St. Louis County, the State of Minnesota and the Electric Power Research Institute provided funding.

References
ASCE (1990). *Evapotranspiration and Irrigation Water Requirements*, ASCE Manuals and Reports on Engineering Practice No. 70, American Society of Civil Engineers, 345 East 47th Street, New York, NY.
Axler, R., Henneck, J., Nordman, D., McCarthy, B. and Monson Geerts, S. (1999). Operation and maintenance experiences with constructed wetlands in Minnesota. *Proceedings for the National OnSite Wastewater Recycling Association (NOWRA) 8th Annual Conference*, Jekyll Island, Georgia, pp. 219–223.
Bavor, H.J., Roser, D.J., McKersie, S.A. and Breen, P. (1988). *Treatment of Secondary Effluent*, Report to Sydney Water Board, Sydney, NSW, Australia.
Cooper, P.F., Job, J.D., Green, M.B. and Shutes, R.B.E. (1996). *Reed Beds & Constructed Wetlands for Wastewater Treatment*, Book, Bibliography and UK Database Diskette.
Kadlec, R.H. and Knight, R.L. (1996). *Treatment Wetlands*, CRC Lewis, Boca Raton, FL, 893 pp.
Maehlum, T. (1998). Doctor Scientarum Thesis 1998:9, Agricultural University of Norway, Ås.
TVA (Tennessee Valley Authority) (1990). Unpublished data from Benton, KY and Pembroke, KY.

Cold climate wetlands: design and performance

S. Wallace*, G. Parkin** and C. Cross*

* North American Wetland Engineering, P.A. 20920 Keewahtin Ave, Forest Lake MN 55025, USA
** Department of Civil & Environmental Engineering, University of Iowa, 4016 SC, Iowa City, IA 52242-1527, USA

Abstract Constructed wetlands are gaining widespread use as a simple, low cost means of wastewater treatment. Introduction of constructed wetlands technology into the northern United States has been limited by the ability of conventional wetland systems to operate without freezing during the winter. A design approach using subsurface-flow constructed wetlands covered with an insulating mulch layer has been demonstrated to prevent freezing. However, introduction of a mulch layer will affect oxygen transfer rates, pollutant removal performance, and plant establishment. These factors must be addressed for successful application of constructed wetlands technology in cold climates.
Keywords $CBOD_5$; cold climate; constructed wetlands; insulation; mulch, nitrogen; oxygen transfer

Introduction

Constructed wetlands have many unique benefits as a wastewater treatment process, including the ability to operate on ambient solar energy, self-organize and increase treatment capacity over time, create wildlife habitat, produce oxygen and consume carbon dioxide, and achieve high levels of treatment with minimal maintenance (Wallace, 1998). The authors' primary interest has been in developing simple constructed wetland treatment systems as an alternative to more complex mechanical systems with an emphasis on household and small community wastewater treatment. This is driven in part by the demonstrated need for better wastewater alternatives, even in the United States. On-site septic systems serve approximately 25% of the US population (USEPA, 1997), and in 1995 alone, over 2.5 million septic systems malfunctioned (NODP). Contrary to the belief that regional wastewater facilities are solving the nation's problems, more Americans are using septic systems now than in 1990 (NODP).

Initial work on subsurface flow wetlands was developed in Germany (Seidel, 1973). Subsurface flow wetlands have the primary benefit that water is not exposed during the treatment process, minimizing energy losses through evaporation and convection. This makes horizontal subsurface flow (and vertical flow) wetlands more suitable for winter applications.

Adaptation of constructed wetlands technology to sub-freezing environments requires some type of insulation strategy. Leaf litter is often suggested as one source of insulation; however leaf litter is often spotty in distribution, which allows heat to escape. To be effective, insulation must be uniform in coverage, which requires that it be designed as an integral part of the wetland system. Initial use of mulch as a cover in subsurface flow constructed wetlands was suggested by the Tennessee Valley Authority (Steiner and Watson, 1993) as a means to prevent odors and sunscald in warm climates. Mulch was used as an insulation medium on the constructed wetland built at the Indian Creek Nature Center in Cedar Rapids, Iowa in 1994 and found to be highly effective in preventing the system from freezing (Wallace and Patterson, 1996). Computer modeling of insulated wetland systems has suggested that adequate insulation would be effective in preventing systems from freezing at temperatures as low as –20°C (Jenssen *et al.*, 1996).

Insulation design

Designing an effective insulation layer for the constructed wetland requires a knowledge of the basic elements of heat transfer, how the wetland will respond under cold conditions, the effect the mulch material will have on wetland performance, and what plant species are compatible with the mulch layer.

Heat transfer

Factors affecting heat transfer in aquatic and plant systems are discussed at length in several sources (Kadlec and Knight, 1996; ASCE, 1990). In considering the winter energy balance condition, the situation can be simplified to the following:

$$E_{loss} = G + (U_i - U_o)$$

where:
E_{loss} = energy lost to the atmosphere, MJ/m²/d
G = conductive transfer from ground, MJ/m2/d
U_i = energy entering with water, MJ/m²/d
U_o = energy leaving with water, MJ/m²/d

Successful design of cold climate wetlands requires that E_{loss} be "throttled down" so that the energy inputs, $G + (U_i - U_o)$ can replace the energy lost. The basic design strategy is to minimize E_{loss}, through the following methods.

- Avoid open water. This minimizes heat loss through evapotranspiration and convection.
- Do not depend entirely on surface ice. Contrary to popular belief, ice is a very poor insulator, and has a thermal conductivity (0.19 MJ/m/d/°C) almost four times greater than liquid water (0.05 MJ/m/d/°C).
- Use subsurface flow and vertical flow wetland systems. These systems have a smaller area footprint per unit flow (concentrating the incoming heat, U_i), can be substantially disposed of in the earth (maximizing G), and can be designed to avoid open water.
- Insulate the system. Placing layers with greater thermal resistance on top of the wetland reduces E_{loss}.

Performance history

One of the authors (Wallace) has designed a number of constructed wetlands since 1997 with mulch insulation. Based on data collected during quarterly (or monthly) sampling events and interviews with owners on 28 systems located in Minnesota, none of these systems froze. However, several wetland systems monitored by the University of Minnesota froze (causing hydraulic failure) during the winters of 1998/1999 and 1999/2000 (McCarthy, 2000 personal communication). These wetland systems had performed well during previous winters (McCarthy *et al.*, 1997). The primary difference was that there was ample snow cover during the previous winters but snow cover was lacking during severe cold temperatures in the 1998/1999 and 1999/2000 winters. Clearly, one of the primary benefits of the mulch insulation approach is that it provides a "safety factor" for those winters when it is severely cold without adequate snow cover.

Lutsen Sea Villas Case Study. The benefits of a mulch insulation approach can be clearly illustrated using a real-life example. A subsurface flow constructed wetland was built in 1997 to treat domestic wastewater from 27 town home units located on the north shore of Lake Superior in Lutsen, Minnesota. The fall and early winter were mild, with no snow cover on the ground. On December 19, 1998, temperatures began to drop rapidly, reaching −28°C by December 21, 1998 (Figure 1).

The system was insulated with 15 cm of mulch. A 5 cm air gap was present under the

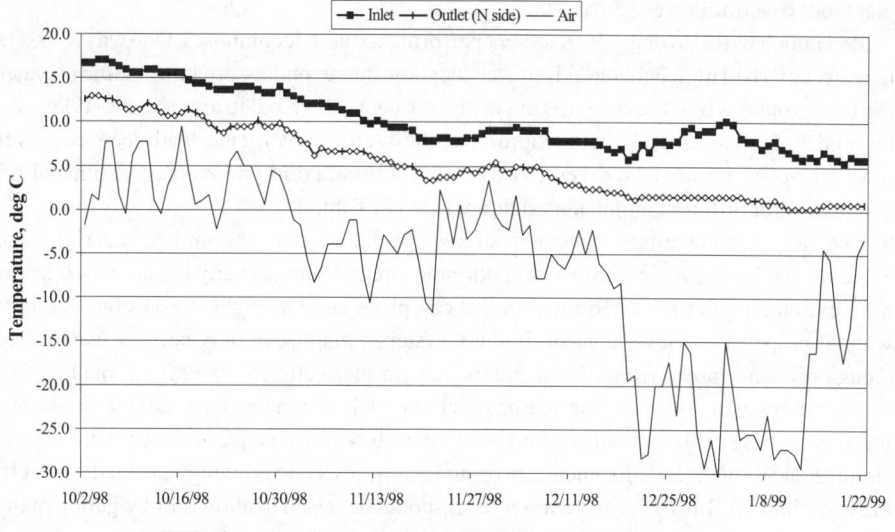

Figure 1 Temperature response of mulch-insulated subsurface flow wetland at Lutsen Sea Villas during extreme freezing event in December 1998 and January 1999

mulch insulation. Based on elapsed time meter readings on the pumps, the system was operating at 0.88 cm/day. Based on the inlet and outlet water temperatures, the heat input associated with the water $(U_i - U_o)$ is calculated as 0.24 MJ/m²/d. Based on a standard value (Kadlec and Knight, 1996) for the thermal conductivity of the mulch (0.0052 MJ/m/d/°C), the air gap (0.0021 MJ/m/d/°C), and the average thermal gradient (29.5°C), the heat lost to the atmosphere, (E_{loss}) is estimated at 0.56 MJ/m²/d.

Subtracting $(U_i - U_o)$ from E_{loss} results in an estimate of G at 0.32 MJ/m²/d (Wallace, 2001), which is very close to the transfer value of 0.31 MJ/m²/d determined for the Houghton Lake wetland system (Kadlec and Knight, 1996). If a 5 cm ice cap and 5 cm air gap had been used instead of the mulch insulation, the resulting heat loss would have been 1.22 MJ/m²/d (almost 4 times greater); an unsustainable heat loss that would result in freezing.

Mulch effects

Early references to potential mulch use in constructed wetlands suggested that a wide variety of materials such as bark, pine straw, wood chips, etc. would be suitable (Steiner and Watson, 1993). After trying a variety of mulch types, preferred materials used by the authors include reed-sedge peat (ASTM, 1969) and high quality yard waste compost. Good mulch material must meet the following characteristics.

- Be substantially decomposed, and not exert a secondary organic loading on the system.
- Have a balanced nutrient composition and a circumneutral pH.
- Have a fluffy structure with high fiber content to provide good thermal insulation and not wash down into the gravel bed.
- Be fine enough so that there is good contact between the seed coat and the mulch for germination (if seeding is used as part of plant establishment).
- Have good moisture holding capacity so that seedlings are not subjected to drought stress.

Bad mulch will adversely affect plant establishment. Mulch materials such as wood chips that have a high carbon: nitrogen ratio will cause nitrogen deficiency problems during plant establishment. Material that is chipped (rather than ground) packs tightly, making plant root penetration very difficult.

Bad mulch will also degrade treatment performance as it decomposes. One way to assess the effect of a bad mulch material is to consider how the secondary organic loading elevates the background $CBOD_5$ concentration of the effluent, C^* (Kadlec and Knight, 1996). As the mulch decomposes, C^* will improve as the secondary organic load decreases over time. However, it may take several years to see substantial improvements. Estimated C^* parameters for different mulch materials are listed in Table 1:

The rate of plant establishment is strongly influenced by the mulch material used. Systems that have a mulch layer with poor moisture holding capacity (or no mulch layer) have extremely poor seed germination and can place large drought stresses on seedlings without proper water level control. In these systems, plants can only become established though rhizome spread from mature plants. In a northern climate like Minnesota this will take a minimum of three growing seasons. Better mulch design results in surface conditions much more hospitable to plant seedlings and also allows for seed germination. Under these circumstances, plant establishment can occur in as little as one growing season, despite climatic limitations. In all cases, the water level should be raised to allow sub-irrigation of the mulch layer for the first growing season.

Treatment performance

A subset of six Minnesota constructed wetlands built in 1997 and 1998 were selected for data comparison purposes. Initial design was based on generic wetland parameters (Kadlec and Knight, 1996), which were modified based on prior design experience and temperature-corrected by one of the authors (Wallace). These wetlands all treat domestic wastewater at full design load, and were designed using the same hydraulic and organic loading rates.

Carbonaceous biochemical oxygen demand

All six systems were monitored under County or State operating permits. Monitoring requirements were generally quarterly or monthly, depending on the operating permit. Data collection generally consisted of inlet and outlet concentrations. Although the six systems reviewed had largely similar loading parameters, actual performance varied widely, as shown in Figure 2:

In general, systems with good mulch, featuring substantially decomposed material with a low C^*, achieved a $CBOD_5$ reduction of 75 per cent in the first year, with treatment performance generally exceeding 90% in the second year. The four good mulch systems shown in Figure 3 (SA Wyoming, Fields of St. Croix, Cloverdale, and Happiness Resort) had a high proportion of reed-sedge peat in the mulch layer, although two of the systems (Fields of St. Croix and Cloverdale) initially had wood chips that were replaced in late summer 1998 with reed-sedge peat.

Table 1 Estimated $CBOD_5$ C^* parameters for various as-constructed mulch materials

Material	Year 1	Year 2
Wood Chips	40 mg/L	20 mg/L
Poplar Bark ("hog fuel")	60 mg/L	20 mg/L
Wood Chips buried under Sand	120 mg/L	80 mg/L
Reed-Sedge Peat	5 mg/L	3 mg/L
High Quality Yard Waste Compost	5 mg/L	5 mg/L

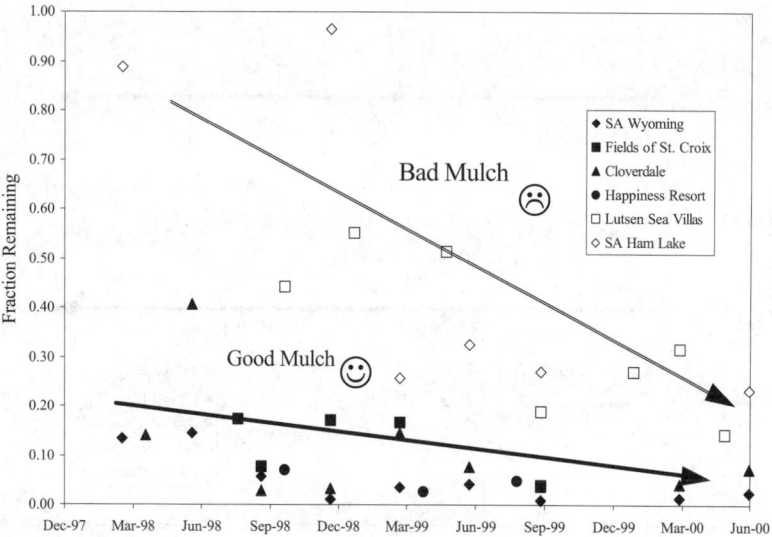

Figure 2 Summary of long-term trends in $CBOD_5$ removal for six mulch-insulated subsurface flow wetland systems located in Minnesota. Systems represented by black data points were constructed with good mulch. Systems represented by white data points were constructed with bad mulch

Improvements in $CBOD_5$ removal over time is attributed to development of a mature, stable microbial population and plant establishment, with the most dramatic improvements occurring after the conclusion of the first growing season. Systems subjected to a late fall "cold start" at full design load will generally perform poorly over the first winter, and may not improve until the end of the first growing season, almost a full year after initial start up.

Total nitrogen
Five of the six systems were also monitored for nitrogen. Reductions in total nitrogen varied widely, with no discernable improvement over time, as shown in Figure 3:

In general, good mulch systems performed better than bad mulch systems. In all of the five systems, nitrogen removal was limited due to a failure to convert ammonia to nitrate. Limited nitrogen removal in other Minnesota subsurface flow wetlands (that do not have mulch insulation) has also been observed (Kadlec *et al.*, 2000).

Methods to improve treatment performance
While effective in CBOD reduction, cold-climate subsurface flow wetlands have demonstrated limited Total Nitrogen removal. This could be attributed to either cold temperatures (that inhibit bacterial action) or limited oxygen transfer (not enough oxygen is available for ammonia oxidation).

A pilot-scale wetland (97 square metres) was installed at the Jones County, Iowa landfill to demonstrate the use of constructed wetlands as a low-cost treatment alternative for leachate generated at small rural landfills. The basic reactor is a horizontal subsurface flow wetland insulated for cold-climate operation, with an aeration system (Wallace, 2001) to eliminate any oxygen transfer limitations. The system was placed into operation in August 1999. Due to the waste strength (influent CBOD of approximately 200 mg/L, influent ammonia of approximately 500 mg/L), the initial loading was set at 4 mm/day. Monitoring

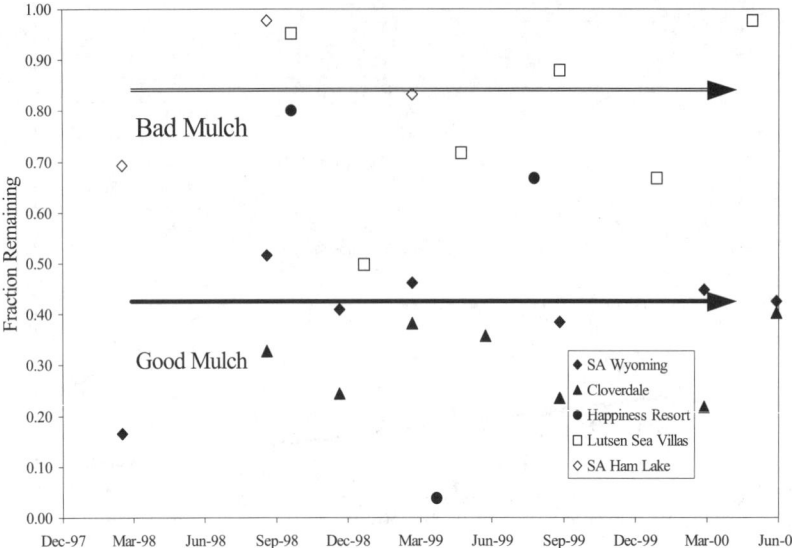

Figure 3 Summary of long-term trends in Total Nitrogen removal for five mulch-insulated subsurface flow wetland systems located in Minnesota. Systems represented by black data points were constructed with good mulch. Systems represented by white data points were constructed with bad mulch. None of the systems were aerated

is ongoing and data interpretation is still preliminary at this time. However, the system has consistently achieved very high ammonia removal rates (approximately 90%), despite the very high ammonia influent concentration (500 mg/L), with concurrent production of nitrate (approximately 100 mg/L), under winter operating temperatures very close to 0°C (Parkin and Cross, 2000).

This low-temperature nitrification is consistent with other wetland reactor designs that provide a high degree of oxygen transfer (Lemon et al., 1996).

Conclusions

Use of constructed wetlands in sub-freezing winter environments imposes a number of unique design requirements. Lessons learned from early wetland designs could be applied to other cold-climate wetlands in the future. Based on the performance history of constructed wetlands in Minnesota and Iowa, the following conclusions can be drawn.

- Properly designed insulation of the wetland bed is effective in preventing freezing and resulting hydraulic failure. Relying on snow and ice cover does not provide reliable insulation during cold periods with limited snow pack.
- The type of mulch insulation used can strongly affect the performance of the system. Only well decomposed organic materials can be used without degrading treatment efficiency.
- Properly designed cold climate insulated wetlands can achieve high levels of CBOD removal. Treatment performance will improve after the first growing season.
- In order to achieve high levels of nitrogen removal, adequate oxygen must be available. Standard horizontal subsurface flow wetlands do not transfer enough oxygen to satisfy both the carbonaceous and nitrogenous oxygen demands. Alternative reactor configurations that have higher levels of oxygen transfer are necessary for nitrogen removal. For

cold-climate wetlands, temperatures below 4°C are not a barrier to nitrification, provided the wetland is designed to prevent freezing.

References

ASCE (American Society of Civil Engineers) (1990). *Evapotranspiration and Irrigation Water Requirements*. Jensen, M., Burman, R. and Allen, R. (Eds). ASCE Manuals and Reports on Engineering Practice No. 70. ASCE, New York.

ASTM (American Society for Testing and Materials) (1969). Standard Classification of Peats, Mosses, Humus, and Related Products D2607-69. ASTM, Philadelphia, Pennsylvania.

Jenssen, P., Maehlum, T. and Zhu, T (1996). Design and Performance of Subsurface Flow Constructed Wetlands in Norway. Presented at Constructed Wetlands in Cold Climates: Design, Operation, Performance Symposium, The Friends of St. George, Niagara-on-the-Lake, Ontario.

Kadlec, R. and Knight, R. (1996). *Treatment Wetlands*. CRC Press, Boca Raton, Florida.

Kadlec, R., Axler, R., McCarthy, B. and Henneck, J. (2000). (submitted). Subsurface Treatment Wetlands in the Cold Climate of Minnesota. Constructed Wetlands for Wastewater Treatment in Cold Climates. *Advances in Ecological Sciences* (in press).

Lemon, E., Bis, G., Rozema, L. and Smith, I. (1996). SWAMP Pilot Scale Wetlands – Design and Performance, Niagara-On-The-Lake, Ontario. Presented at Constructed Wetlands in Cold Climates: Design, Operation, Performance Symposium, The Friends of St. George, Niagara-on-the-Lake, Ontario.

McCarthy, B., Axler, R., Monson-Geerts, S., Henneck, J., Crosby, J., Nordman, D., Weidman, P. and Hagen, T. (1997). Development of Alternative On-site Treatment Systems for Wastewater Treatment: A Demonstration Project for Northern Minnesota: Final Report Submitted to Minnesota Technology Inc., Legislative Commission for Minnesota Resources, Electric Research Power Institute. Natural Resources Research Institute, University of Minnesota, Duluth, Minnesota.

NODP (National Onsite Demonstration Project) (No Date). SepticStats™ An Overview. Environmental Services & Training Division, West Virginia University, Morgantown West Virginia.

Parkin, G. and Cross, C. (2000). Jones County Iowa constructed wetland operating data (unpublished data). Department of Civil and Environmental Engineering, The University of Iowa, Iowa City Iowa.

Seidel, K. (1973). United States Patent 3,770,623. United States Patent Office, Washington DC.

Steiner, G. and Watson, J. (1993). General Design, Construction and Operation Guidelines; Constructed Wetlands Wastewater Treatment Systems for Small Users Including Individual Residences, Second Edition. Tennessee Valley Authority Resource Group Water Management TVA/WM - 93/10.

USEPA (1997). Response to Congress on Use of Decentralized Wastewater Treatment Systems. United States Environmental Protection Agency Office of Water & Office of Wastewater Management, Washington DC.

Wallace, S. (1998). Putting Wetlands to Work. Civil Engineering, 98-007-0057. American Society of Civil Engineers, New York.

Wallace, S. and Patterson, R. (1996). Indian Creek Nature Center Constructed Wetlands for Wastewater Treatment. Presented at the Constructed Wetlands in Cold Climates: Design, Operation, Performance Symposium, The Friends of St. George, Niagara-on-the-Lake, Ontario.

Wallace, S. (2001). United States Patent 6,200,469. United States Patent Office, Washington DC.

Behavior of organic carbon during subsurface wetland treatment in the Sonoran Desert

D.M. Quanrud*, M.M. Karpiscak**, K.E. Lansey* and R.G. Arnold***

* Department of Civil Engineering and Engineering Mechanics, The University of Arizona, Tucson, Arizona 85721, USA
** Office of Arid Lands Studies, The University of Arizona, Tucson, Arizona 85721, USA
*** Department of Chemical and Environmental Engineering, The University of Arizona, Tucson, Arizona 85721, USA

Abstract We examined the fate of organics during wetland treatment of secondary effluent and groundwater (control) flows in parallel, research-scale, subsurface-flow (SSF) wetland raceways at the Constructed Ecosystem Research Facility (CERF) located in Tucson, Arizona. The CERF facility enabled us to distinguish experimentally among effects on effluent quality due to season-dependent processes of evapotranspiration (ET) and wetlands-derived production of organics. Organics of wastewater and wetlands origin were compared in terms of their contributions to dissolved organic carbon (DOC) in wetland effluent. Elevated temperatures and associated biochemical activities increased DOC levels in wetland effluents during summer. In other words, DOC removal efficiency was negatively correlated to temperature. The contributions of ET and wetland-derived organics to elevation of DOC in wetland effluents during summer were roughly comparable. The elevation of organic carbon concentration during wetland polishing of wastewater effluent will lead to higher levels of disinfection by-products when treated waters are chlorinated prior to reuse. Results of this work are relevant to water managers in arid regions, which may incorporate wetlands into sequential wastewater treatments leading to potable reuse of reclaimed water.
Keywords Constructed wetlands; dissolved organic carbon; evapotranspiration; secondary effluent

Introduction

The number of constructed wetlands used for treatment of municipal wastewater is increasing, numbering over 600 in North America (Cole, 1998). Treatment wetlands provide an attractive method for polishing municipal wastewater due to their ancillary benefits of public use and wildlife habitat (Knight, 1997). In the Sonoran desert region of Arizona, several constructed wetland projects have been developed over the past few years and several more are in the planning stages. Active programs of research are being conducted at the Tres Rios Demonstration Wetlands (Phoenix), Sweetwater Wetlands (Tucson), and the Constructed Ecosystems Research Facility (Tucson) to determine the efficacy of wetlands treatment in a hot, arid desert environment.

Water quality improvements that occur during passage through treatment wetlands are the result of complex and poorly understood interactions between living communities of microorganisms and plants. Variations in microbial activities can occur in response to seasonal temperature fluctuations. Thus, it is reasonable to expect that attenuation of biochemical oxygen demand (BOD_5) would follow an Arrhenius-type relationship, with higher removals occurring at higher temperatures. Several studies have shown that this is not the case: BOD_5 removal during wetland treatment is not season dependent (Bahlo and Wach, 1990; George *et al.*, 1994; Manios *et al.*, 2000). Other studies have found that there are seasonal variations in attenuation of BOD_5 during wetland treatment, with higher removals occurring in warmer summer months. For example, Vanier and Dahab (1997) found that wastewater temperature was directly correlated to removal rates of BOD_5 and chemical oxygen demand in a SSF constructed wetland. At a free-water-surface (FWS) tertiary treatment wetland, Stober *et al.* (1997) found that BOD_5 removals were greater in spring than in winter. Contrary to these studies, Karpiscak *et al.* (1996) found that BOD_5 removal was greater in winter than in summer at the CERF wetland raceways in Tucson, Arizona. Results from that study suggested the correlation between temperature and BOD_5 removal during SSF wetland treatment in a hot, arid environment may be opposite to that predicted using an Arrhenius-type relationship.

While BOD_5 measurements are useful, they do not provide insight to the behavior of organics that are resistant to biochemical oxidation. In one of the few studies to examine directly the behavior of DOC during wetland treatment, Pinney *et al.* (2000) found that DOC character and removal rates were season-dependent at a FWS wetland in Kingman, Arizona. Minimum DOC removal rates occurred during summer and were attributed to elevated contributions of plant-derived DOC. In that study, it was not possible to isolate organic carbon attenuation and production mechanisms. The objectives of work reported here were to evaluate the relative importance of season-dependent mechanisms affecting DOC (namely evapotranspiration and production of wetland-derived DOC) during wetland treatment of municipal wastewater in a hot, arid desert environment and to examine the relationship between temperature and DOC attenuation.

Methods

Research was conducted at the CERF wetlands, operated by the University of Arizona's Office of Arid Lands Studies for the Pima County Department of Wastewater Management. This pilot-scale research facility began operation in 1989. CERF consists of five raceways (61 m long by 8.3 m wide by 1.4 m deep) and one additional, larger raceway (64.6 m by 11 m by 2.6 m). All raceways are operated in parallel and are underlain by 30-mil 3-ply Hyperlon (heavy plastic sheeting) to prevent seepage losses. Raceways #2-5 are SSF wetland cells containing 0.9–1.4 m of gravel and various herbaceous shrub and tree species. The average water detention time in these raceways was 5–6 days (determined by tracer tests using bromide). Raceway #6 is an aquatic pond maintained at a depth of 0.9 m and covered with duckweed (*Lemna* spp.). Raceways received either unchlorinated secondary effluent from the Roger Road Wastewater Treatment Plant (#3 and #5) or potable water as influent (#2 and #4). The (groundwater) source water to raceways #2 and #4 contained DOC levels that were <1.0 mg/L. Raceway #6 provided pretreatment (suspended solids removal) for effluent that was sourcewater for raceways #3 and #5. Other research conducted at CERF and additional site description are available in Karpiscak *et al.* (1996) and Karpiscak *et al.* (2000).

Samples were collected at the inlets and outlets of selected CERF wetland raceways during the time period January 1997 through October 1998. For the majority of that time, samples were collected once every two weeks; during the summer of 1998, sampling frequency was increased to weekly. Additional details of sampling methodology are described in Quanrud (2000). Samples were analyzed for DOC, ultraviolet light absorbance at 254 nm (UV-254), pH, and specific electrical conductance (EC). Methods for these analyses are described in Quanrud *et al.* (1998). Specific absorbance (SUVA) was calculated using the UV-254 value normalized to the DOC concentration to provide a relative measure of aromatic carbon content. Air and water temperatures were recorded daily at the CERF facility.

Results and discussion

Seasonal trends in organic behavior

Results given here represent average values calculated using data from the two raceways receiving secondary effluent (#3 and #5) and the two raceways receiving tap water (#2 and

#4). Water quality data were similar from replicate raceways receiving the same water type. For example, per cent reductions of DOC and UV-254 from raceways #3 and #5 were highly correlated, with coefficients of determination (r^2) of 0.81 and 0.82, respectively (data not shown). This level of correlation supported lumping and averaging data from replicate raceways in order to simplify data presentation.

Trends in monthly-averaged inlet and outlet DOC concentrations and UV-254 values for raceways #3/#5 over the 22-month period of study are shown in Figure 1. Also shown in the figure are wetland-derived DOC and UV-254 measured at the outlet of the corresponding "control" raceways (#2/#4). Substantial variations (increases) in outlet DOC occurred in the raceways receiving wastewater during the summer months of 1997 and 1998. Similar behavior occurred for UV-254 (data not shown).

From the ratios of outlet-to-inlet values (C/C_o) of DOC, UV-254 and SUVA in raceways #3/#5 (Figure 2) it is apparent that there was greater attenuation of DOC than UV-254 during wetland treatment. SUVA always increased across the wetlands, indicating that residual organics were more aromatic in character as a consequence of wetland treatment.

Evaluation of mechanisms responsible for season-dependent organic behavior

Wetland outlet organic data were normalized to account for ET water losses in order to separate ET from other mechanisms (primarily the production of wetland-derived organics) that were potentially responsible for seasonal increases in outlet concentrations.

The fractional increase in EC between wetland inlet and outlet was used as an indicator

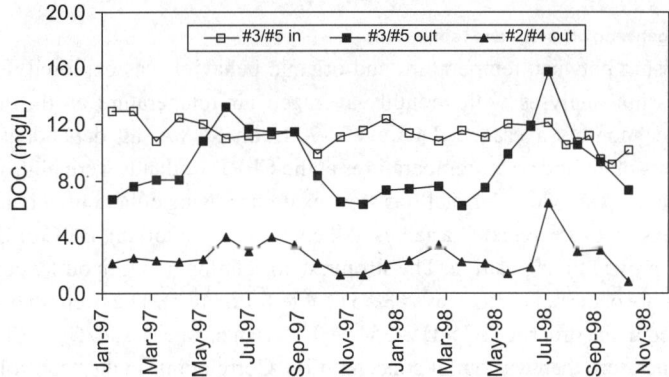

Figure 1 Trends in DOC data for raceways #3/#5 and for wetland-derived organics from corresponding control raceways #2/#4

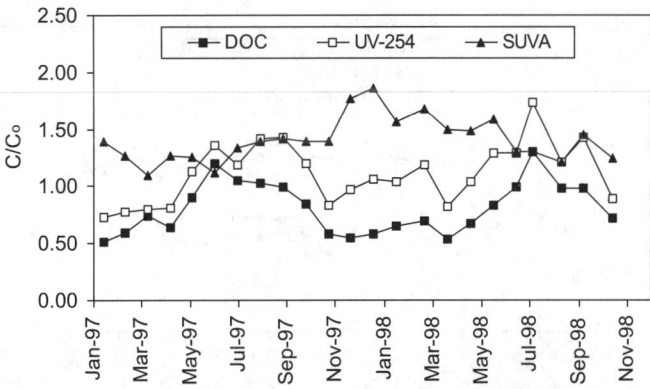

Figure 2 Behavior of organics (expressed as C/C_o) in raceways #3/#5 for the time period January 1997 to October 1998

for ET loss during wetland passage. It was hypothesized that the greater ET occurring during summer months is at least partially responsible for the elevated outlet organic concentrations during that period. To test this hypothesis, the outlet-to-inlet EC ratio was used to calculate "corrected" DOC concentrations for wetland effluents:

$$DOC_{EC\text{-}corrected\ outlet} DOC_{outlet} \times EC_{inlet}/EC_{outlet} \tag{1}$$

Thus, $DOC_{EC\text{-}corrected\ outlet}$ is the outlet DOC that would have been measured in the absence of ET. Using equation 1, corrected outlet DOC values were calculated for raceways #3/#5 (Figure 3) and raceways #2/#4 (not shown). After correcting for ET, it is apparent that outlet DOC concentrations in raceways #3/#5 were elevated during summer months by wetland-related activities. Furthermore, those increases were larger than wetland-derived organics measured in the control raceways during the same period (data not shown). It is hypothesized that the greater seasonally elevated outlet DOC in raceways receiving reclaimed water results from wastewater-related nutrient addition, which stimulates microbial activity, vegetative growth and organic decomposition.

The relative importance of ET in elevation of summertime outlet DOC concentrations in raceways #3/#5 was evaluated by integrating and comparing the areas under the curves corresponding to periods of elevated concentrations (approximately March to October) for outlet DOC (Figure 1) and ET-corrected outlet DOC (Figure 3). ET was responsible for 46% of the elevated summertime outlet DOC in 1997; in 1998, the percentage was 56%. The remainder of elevated outlet DOC was presumably due to production of wetland-derived DOC.

Relationship between organics and temperature

The relationship between temperature and organic behavior was explored using simple linear regression analyses with monthly-averaged air temperature as the independent variable and monthly-averaged DOC, UV-254, or SUVA as dependent variables. Measurements of air and water temperatures at the CERF wetlands were highly correlated ($r^2 = 0.97$; data not shown). Correlations were examined using either outlet concentrations or C/C_o values as the dependent variables. All correlations involving DOC or UV-254 and temperature proved to be positive. The strongest correlation between outlet concentration and temperature occurred in raceways #3/#5 (Table 1, Figure 4). The strength of correlation (r^2) decreased in the order: DOC > UV-254 > SUVA. In most cases, C/C_o was better correlated to temperature than was outlet concentration. Correlations in the "control" raceways (#2/#4) were much lower (Table 1), indicating a much less significant relationship between air temperature and production of wetland-derived organics.

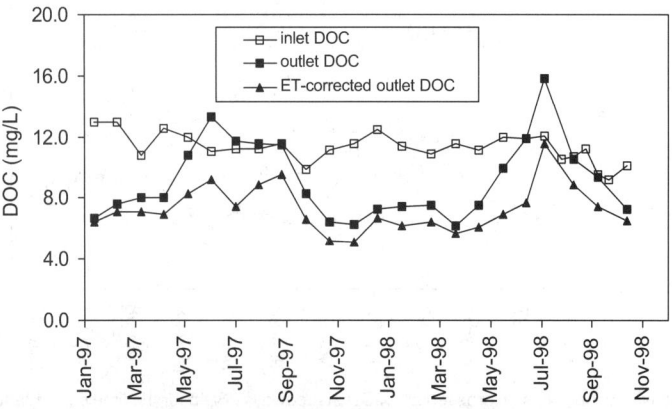

Figure 3 DOC data from raceways #3/#5 after applying corrections for ET

Table 1 Coefficients of determination (r^2) for simple linear regressions of air temperature versus organic parameters (DOC, UV-254, or SUVA) from individual CERF wetland raceways

Basis of Comparison	#2	#3	#4	#5
Outlet DOC	0.10	0.63	0.30	0.71
C/Co DOC	0.02	0.71	0.19	0.84
Outlet UV-254	0.05	0.41	0.26	0.56
C/Co UV-254	0.20	0.37	0.15	0.59
Outlet SUVA	0.01	0.01	0.02	0.13
C/Co SUVA	0.07	0.21	0.00	0.37

There was a strong positive correlation between fractional DOC residuals (C/C_o) in the raceways receiving wastewater (#3/#5) and air temperature (Figure 4). In other words, as temperature increased, the magnitude of DOC attenuation across the wetland raceway decreased. Similar behavior occurred using water temperature as the independent variable. C/C_o values in excess of 1.0 reflect a net production of DOC across the raceway. Using the regression equations obtained, it was possible to calculate the air and water temperature at which no net removal of DOC would occur ($C/C_o = 1.0$). The critical monthly-averaged air and water temperatures were 29.7 and 25.6°C, respectively.

Figure 5 shows the relationship between air temperature and outlet concentration of DOC for CERF raceways. Outlet DOC was directly related to temperature for all raceways. It is apparent that outlet concentrations of DOC in raceways receiving wastewater were more sensitive to air temperature (and hence season) than were corresponding outlet concentrations of wetland-derived DOC in raceways #2/#4. There was little significance to

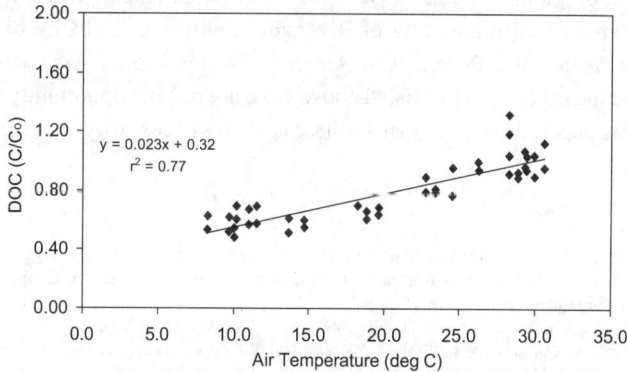

Figure 4 Relationship between air temperature and change in DOC (C/C_o) for lumped raceway #3 and #5 data

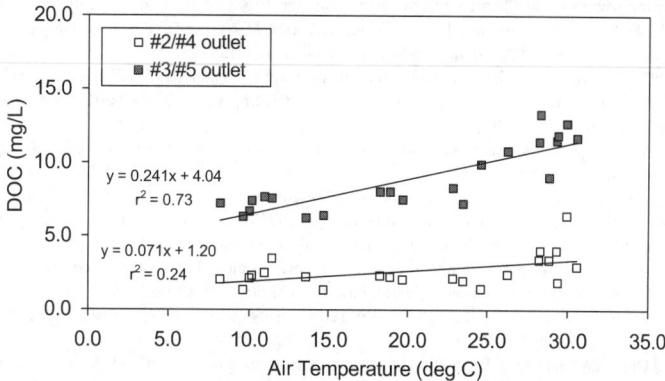

Figure 5 Relationship between air temperature and outlet DOC concentration for lumped CERF raceway data

the relationship between air temperature and outlet SUVA (Table 1). These data suggest that SUVA values were not sensitive to temperature (and hence season). That is, the proportionality of DOC and UV-absorbing materials in wetland effluent did not change as a consequence of wetland production of DOC.

Conclusions

The following conclusions regarding using subsurface-flow constructed wetlands in a hot, arid desert environment are supported by experimental results.
1. Wetland effluent DOC concentrations are strongly season dependent.
2. Mechanisms responsible for season-dependent variability in outlet DOC concentration include evapotranspiration and production of wetland-derived DOC.
3. The contributions of evapotranspiration and wetland-derived production of DOC to elevated wetland outlet DOC concentrations during summer are roughly comparable.
4. Net DOC removal efficiency was negatively correlated to temperature; greater net DOC attenuation occurred during winter.

Acknowledgements

The authors wish to acknowledge technical support from Sue Hopf and Glenn France from The University of Arizona's Office of Arid Land Studies. Financial support was provided by Tucson Water, City of Tucson, Arizona; Sanitation Districts of Los Angeles County; United States Environmental Protection Agency; American Water Works Association Research Foundation; USDA Water Conservation Laboratory; Pima County Department of Wastewater Management; Water Environment Research Foundation; City of Phoenix, Arizona; The Subregional Operators Group (Phoenix-area cities); Water Replenishment District of Southern California; City of Riverside, California; and City of Los Angeles Department of Water and Power. The American Water Works Association Research Foundation and the other agencies listed above have not had the opportunity to review and comment on this paper, therefore, none of these agencies necessarily endorses the findings presented here.

References

Bahlo, K.E. and Wach, F.G. (1990). Purification of domestic sewage with and without faeces by vertical intermittent filtration in reed and rush beds. In: *Constructed Wetlands in Water Pollution Control*, P.F. Cooper and B.C. Findlater (eds.), Pergamon Press, Oxford, UK, pp., 215–221.

Cole, S. (1998). The emergence of treatment wetlands. *Env. Sci. Tech.*, **32**(9), 218A–223A.

George, D.B., Kemp, A.S., Caldwell, A.S., Winfree, S.K. and Tsai, P.J. (1994). Design considerations for subsurface flow wetlands treating municipal wastewater. In: *Proceedings of the 67th Annual Conference of the Water Environment Federation*, vol 8, Water Environment Federation, Alexandria, Virginia, pp. 13–24.

Karpiscak, M.M., Gerba, C.P., Watt, P.M., Foster, K.E. and Falabi, J.A. (1996). Multi-species plant systems for wastewater quality improvements and habitat enhancement. *Wat. Sci. Tech.*, **33**(10–11), 231–236.

Karpiscak, M.M., Whiteaker, L.R., Artiola, J.F. and Foster, K.E. (2001). Nutrient and heavy metal uptake and storage in constructed wetland systems in Arizona. *Wat. Sci. Tech.*, **44**(11–12), 455–462.

Knight, R.L. (1997). Wildlife habitat and public use benefits of treatment wetlands. *Wat. Sci. Tech.*, **35**(5), 35–43.

Manios, T., Millner, P. and Stentiford, E.I. (2000). Effect of rain and temperature on the performance of constructed reed beds. *Wat. Env. Res.*, **72**(3), 305–312.

Pinney, M.L., Westerhoff, P.K. and Baker, L. (2000). Transformations in dissolved organic carbon through constructed wetlands. *Wat. Res.*, **34**(6), 1897–1911.

Quanrud, D.M. (2000). *Constructed wetlands and soil-aquifer treatment systems: effects on the character of effluent organic matter.* PhD Dissertation, Department of Hydrology and Water Resources, The University of Arizona.

Quanrud, D.M., Arnold, R.G., Clark, A., Massaro, M., Karpiscak, M.M. and Wilson, L.G. (1998). Efficiency and sustainability of soil-aquifer treatment leading to wastewater reclamation and reuse. In: *Proceedings, Water Reuse 98*. The American Water Works Association. Orlando, Florida, February 1–4, 1998, pp. 579–591.

Stober, J.T., O'Connor, J.T. and Brazos, B.J. (1997). Winter and spring evaluations of a wetland for tertiary wastewater treatment. *Wat. Env. Res.*, **69**(5), 961–968.

Vanier, S.M. and Dahab, M.F. (1997). Evaluation of subsurface flow constructed wetlands for small community wastewater treatment in the plains. In: *Proceedings 70th Annual Water Environment Federation Conference Exposition.* Chicago, Illinois, **6**, 75.

Development of a conceptual model for vertical flow wetland metabolism

E. Giraldo and E. Zárate

CIIA, Centro de Investigaciones en Ingenieía Civil y Ambiental. Universidad de los Andes, A.A. 4976. Santafé de Bogotá, Colombia

Abstract Four parallel vertical constructed wetlands, with a total area of 556 m², are used to treat domestic wastewater, coming from a community of 550 inhabitants. The system includes pre-treatment with an anaerobic filter and post-treatment with chlorine, before discharging the effluent to the ocean. Four native species of macrophytes were planted: *Paspalum penisetum, Typha sp, Conocarpres erectus* and *Scirpus lacustris. In situ* measurements of gas content were performed for each bed during an operation cycle. After a feeding discharge, an unaltered sample of sand from each bed was taken, and a respirometric test was implemented to measure the metabolic activity in terms of oxygen consumption kinetics, CO_2 production and organic matter degradation. The results were used to develop a conceptual model of the microbiologic metabolism for the process of organic matter removal from wastewater. Sorption in the bed is the main mechanism for organic matter removal from the wastewater, with subsequent biological oxidation during the resting period. The degradation rate for dissolved organic matter is found to be dependent on its concentration and on oxygen content in the gaseous phase. During the days of major activity, the oxygen content was not fully recovered when a new discharge occurred, finding anaerobic activity within the bed.
Keywords In situ gas measurement; metabolism; modeling; oxygen consumption kinetics; respiration rates; respirometric test; sulphide production; vertical flow wetlands.

Introduction

In 1998, four vertical flow wetlands were constructed in San Andrés, a Colombian archipelago of small islands in the Caribbean Sea, which lies about 770-km Northwest of the Colombian mainland. These wetlands were constructed as part of the domestic wastewater treatment system of a 550-inhabitant community from the island. The whole system includes the following equipment: grease traps, pre-treatment with an anaerobic filter, four parallel reed beds and post-treatment with chlorine, before discharging the effluent to the ocean. Each bed was planted with a different native plant: *Paspalum penisetum* (known as king grass), *Typha* sp., *Conocarpres erectus* (known as Mangrove: "Mangle Botón") and *Scirpus lacustris*. The effluent from the anaerobic filter exhibits high concentrations of hydrogen sulphide due to the sulphate present in the water used for flushing the toilets. Table 1 shows the parameters employed to design the wetland system. Another contribution to these proceedings presents detailed information about the performance of the system (Giraldo and Zárate, 2001).

The goals of this research were:
- to understand the mechanism of organic matter removal and mineralisation within the beds
- to characterise the kinetics of organic matter degradation and consumption
- to develop a conceptual model for the whole process. Bayor *et al.* (1995) suggest that constructed wetlands research should be focused to develop conceptual and mathematical approaches, in order to fully understand the process.

Methods

The beds were constructed 60 cm deep, being the top 20 cm coarse sand, and the bottom 40 cm gravel (between 2.4 and 4.8 mm); following the recommendations of Cooper *et al.* (1996). The hydraulic retention time was calculated as 7 minutes, by measuring the time the treated water took to leave the bed. Throughout the operation, the phenomena proved to be of critical importance: organic matter adsorption, internal oxygen transfer within the biofilm and kinetics of organic matter degradation. Several authors (Hiley, 1995; Von Felde & Kunst, 1997; Platzer, 1999) have also found that keeping a good oxygenation process within the vertical flow wetland, improves the system performance. In order to characterise each phenomenon, several measurements were performed, which are described below.

In situ measurements of gas

The gas content within the bed supplies important information about the behavior of the whole system, in terms of the oxygen required and the production of gases such as carbon dioxide (CO_2), hydrogen sulphide (H_2S) and methane (CH_4). The first measurement was taken prior to any wastewater discharge for each wetland, after a 3-day resting period. Taking into account that the beds are 60 cm deep, several measurements at different depths (20 cm, 40 cm and 60 cm) were taken along the day. Figure 1 shows a diagram of the wetland and the gas analyser apparatus.

The hydrogen sulphide measurements were taken employing the H_2S pot of the infrared gas analyser, for values up to 50 ppm. Measurements above this reading were performed employing a colorimeter device, conformed by packed glass tubes, able to determine up to 800 ppm.

Table 1 Operating parameters for the wetland system

Parameter	Units	Value	Recommended values
Wastewater average BOD,->5	mg/l O_2	290*	
Overall BOD_5 removal efficiency	%	91[(1)]	
Wastewater average COD	mg/l O_2	502*	
Wastewater average TDS	mg/l	9894*	
Wastewater average sulphide content	mg/l	93.1*	
Wastewater average salinity	‰	5.7	
Total area	m^2	556	
Length:width ratio	m/m	1.72	
Design BOD_5 loading	g $BOD_5/m^2/d$	21.4	
Operating BOD_5 loading	g $BOD_5/m^2/d$	42.3	
Design COD loading	g $COD/m^2/d$	37.0	25.0[(2)]
Operating COD loading	g $COD/m^2/d$	73.1	
H_2 loading	g $H_2S/m^2/d$	13.6	
Initial G&O loading	g $G&O/m^2/d$	1.31	
Final G&O loading	g $G&O/m^2/d$	14.6	
Hydraulic load	l/m^2/h	6.1	
Design flow	l/s	0.5	
Operating flow	l/s	0.63–0.94	
Specific area	m^2/person	1	1–5**
Resting period	Days	3	
Hydraulic retention time	Minutes	7	
Mean water temperature	°C	30	
Mean air temperature	°C	30	

*Average BOD_5, COD and TDS out of the anaerobic filter, entering the beds. **Cooper and Green (1995), Cooper (1999).
Platzer and Mauch (1997). (1) During the first year of operation. (2) For subtropical climates. G&O: Grease and oils

Respirometric test

Aiming to get information about the kinetics of organic matter degradation and oxygen consumption, a test that intended to measure the respiration rate of the microorganisms attached to the sand, was performed. As soon as the wastewater discharge occurred, and the wastewater drained off from the surface of the wetland, an unaltered soil sample was taken and put into an Erlenmeyer. It was completely sealed and connected to the infrared gas analyser, which is programmed to take samples from the headspace at selected times. The gas content is measured, and the sample is returned to the Erlenmeyer headspace. After several hours, the Erlenmeyer is opened to replace the headspace for fresh air, and it is sealed to start the process again. Figure 2 shows a diagram of the equipment employed.

The infrared gas analyser determines the percentage in volume of oxygen, carbon dioxide and methane of the gas suctioned.

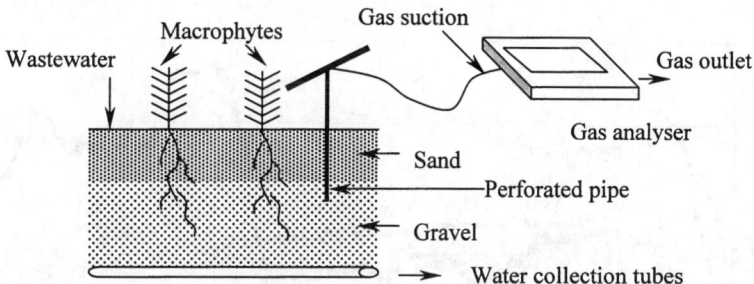

Figure 1 Apparatus of in situ gas measurement

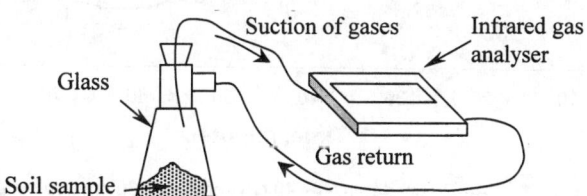

Figure 2 Respirometric test equipment

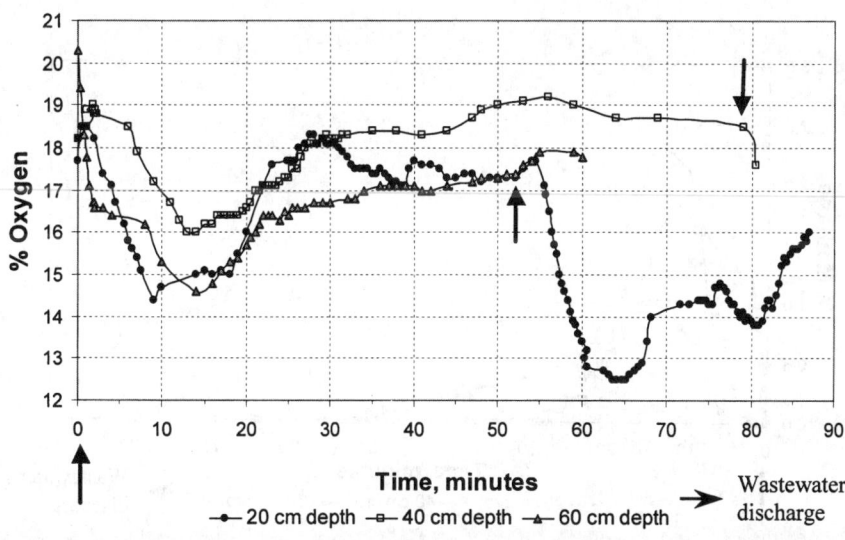

Figure 3 In situ measurement results – change in oxygen content for wetland #3

Results and discussion

In situ measurement results

Figures 3 to 6 show in situ gas content results within the soil column, for one of the wetlands (bed #3). The arrows indicate the time when a new wastewater discharge occurred. The measurements for the different depths were not taken at the same time, but all of them were started when the oxygen content was 21%, that is, after sufficient time elapsed between discharges.

Following a wastewater discharge, the oxygen concentration decreased during the first 20 minutes, slowly increasing afterwards. After 90 minutes of the discharge, the oxygen concentration was not fully recovered. During the operation, it was seen that new discharges occurred when the oxygen had not recuperated to original levels (21%), especially those days of major water consumption like laundry days. At the same time, the concentration of CO_2 increased, indicating oxidation of carbonaceous organic matter. Every time there was a new wastewater discharge, CO_2 increased.

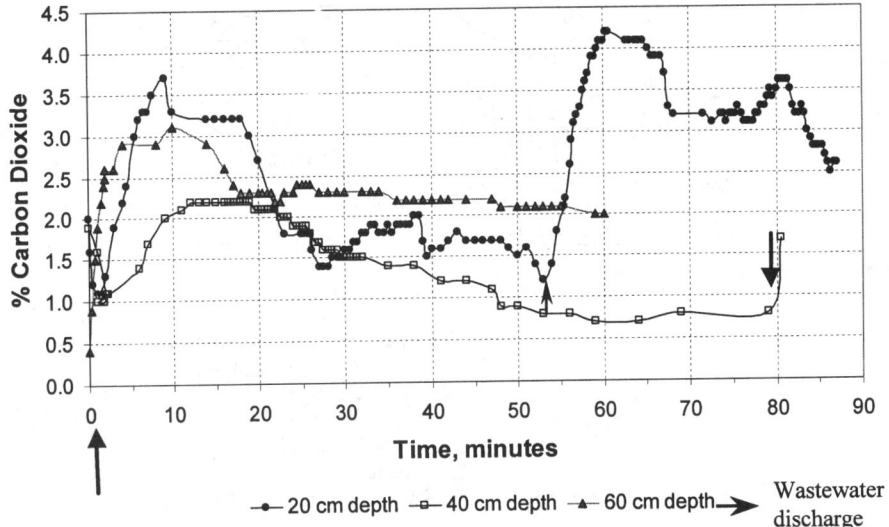

Figure 4 In situ measurement results – change in carbon dioxide content for wetland #3

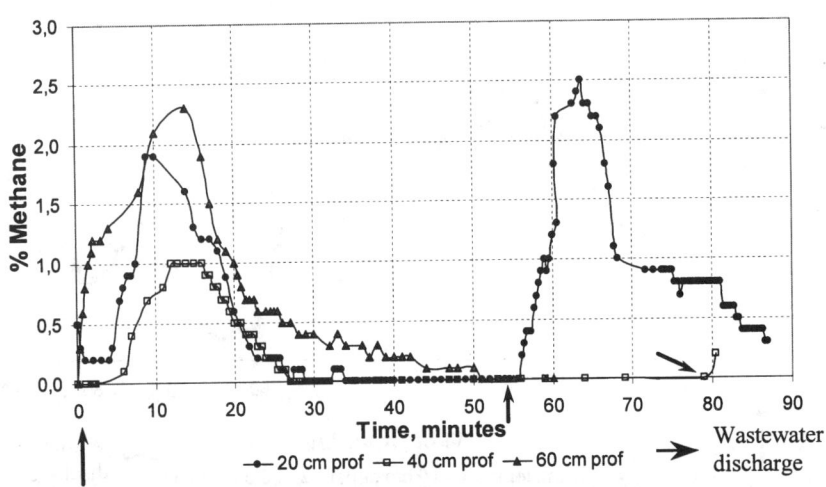

Figure 5 In situ measurement results – change in methane content for wetland #3

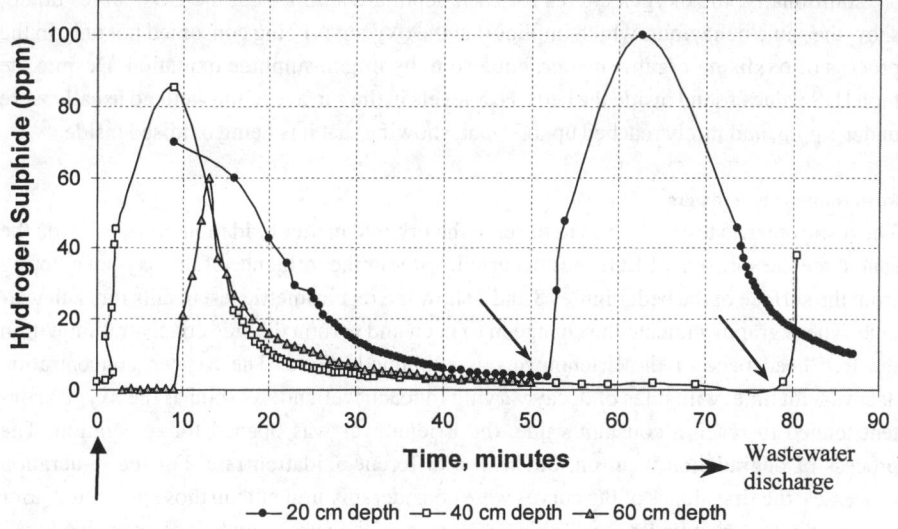

Figure 6 In situ measurement results – change in hydrogen sulphide for wetland #3

These observations indicate that there are two reactions with different kinetic rates: A fast one at the start of the discharge, and a slower one that continues for several hours after the discharge. Considering that the hydraulic retention inside the wetland bed is just of a few minutes (7 minutes were calculated for this system), and even with this short contact, high BOD_5 removal is experienced (as shown in Table 1), then it is proposed that the main mechanism for BOD removal from the water is physical. This implies that the BOD is physically transferred from the wastewater to the wetland bed and retained by sorption. Once it is sorbed inside the bed, biological oxidation takes place. This oxidation occurs in two steps: the first one is hydrolysis, in which particulate sorbed organic matter is solubilised, and the second one consists in the actual oxidation of the dissolved organic matter. The hydrolysis step is slow, while the oxidation step is fast. Figure 7 illustrates the proposed mechanism.

Readings of methane and hydrogen sulphide were also taken. As soon as the discharge occurred, there were sudden increases in the content of these two compounds, to start decreasing after some time. Part of the H_2S might come from the sulphide present in the wastewater, which volatilise during the discharge, but also the presence of methane indicates that anaerobic action is taking place inside the bed, in spite of the high oxygen concentrations in the bed (> 12%).

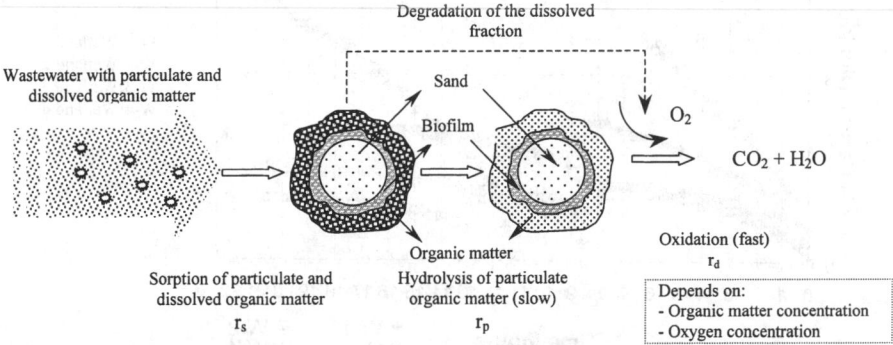

Figure 7 Mechanism for organic matter removal in the wastewater and subsequent oxidation

Additionally, the oxygen curves for each depth are similar, but the CO_2 curves do not show the same behaviour, which implies that the oxygen is being consumed not only in the process of oxidising organic matter, but also in hydrogen sulphide oxidation. Despite the high H_2S values found inside the beds, H_2S levels in the surface of the wetland usually were under 1 ppm, and rarely reached up to 7 ppm, showing that it is being oxidised inside.

Respirometric test results

The respirometric tests allow us to observe the organic matter oxidation process inside the sand once the sorption of BOD has occurred, without the influence of the oxygen entering from the surface of the bed. Figures 8 and 9 show the respirometric test results for each wetland. These graphs indicate the change in oxygen and carbon dioxide concentration within the free headspace of the Erlenmeyer, as explained above. The oxygen concentration decays with time, with rates of decay varying for each wetland. As soon as the oxygen content tended to reach a constant value, the Erlenmeyer was opened for re-aeration. The process of degradation went on, but with a different oxidation rate. For the re-aeration processes, the first slopes of the curves were considerably higher than those presented prior to opening the Erlenmeyer (see Table 2). However, the subsequent slopes remained low. This indicated that there is a clear effect in the rte of degradation, due to the oxygen concentration present. High oxygen concentrations stimulate the degradation of organic matter, as shown in Figure 8, as well as in Table 2.

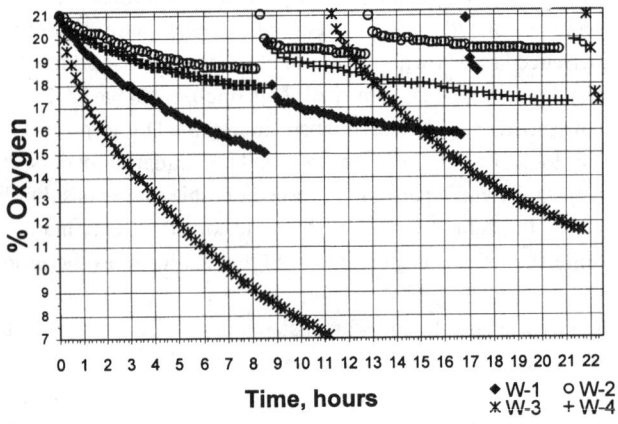

Figure 8 Respirometric test results – variation of headspace oxygen content

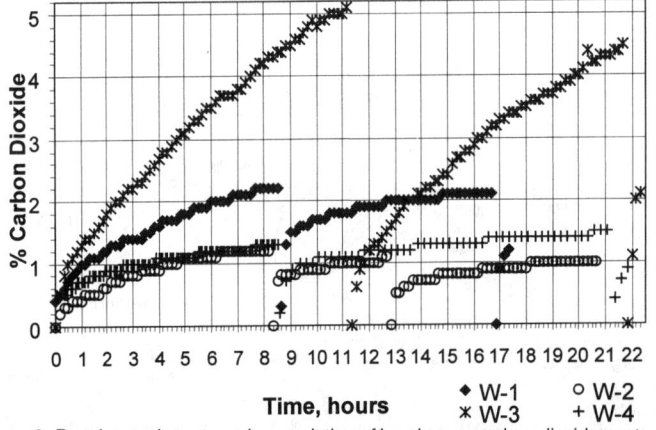

Figure 9 Respirometric test results – variation of headspace carbon dioxide content

Table 2 Slopes of the respirometric curves, from Figure 8

Reactor	Time	Slope mmol/(l*min)	O$_2$ content, %	Increase in the slope, %	Comments
Wetland 1	0 h	0.016	21.1		
	8 h, 30'	0.0041	15.1	1802	
	8 h, 40'	0.078	19.9		First re-aeration
	16 h, 40'	0.0041	15.8	1695	
	16 h, 50'	0.0736	20.9		Second re-aeration
Wetland 2	0 h	0.004	21.1		
	8 h, 10'	0.00082	18.7	5388	
	8 h, 20'	0.045	21.1		First re-aeration
	12 h, 40'	0.002	19.3	1400	
	12 h, 50'	0.03	21		Second re-aeration
Wetland 3	0 h	0.04	20.9		
	11 h, 10'	0.004	7.1	650	
	11 h, 20'	0.03	21.1		First re-aeration
	21 h 40'	0.004	11.6	1400	
	21 h 50'	0.06	21.0		Second re-aeration
Wetland 4	0 h	0.016	21		
	8 h, 30'	0.002	17.9	300	
	8 h, 35'	0.008	19.9		First re-aeration
	21 h 5'	0.0005	17.2	500	
	21 h 10'	0.003	21.0		Second re-aeration

These observations imply that once the sorption process took place, the biological mineralisation process is limited by oxygen.

Table 2 presents the slopes found at the beginning and at the end of the oxidation processes. It is seen how the slopes increase drastically when oxygen reaches 21%.

Conclusions

From the results presented above, the following conceptual model is proposed.
- Taking into account that the hydraulic retention time of the wetland system is 7 minutes approximately, and comparing it with the high removal obtained, the organic matter from the wastewater is removed by sorption to the wetland bed. Once the organic matter is sorbed inside the bed, biological mineralisation takes place.
- The biological oxidation occurs in two steps: a slow hydrolysis step, and a fast oxidation of organic matter. A mathematical model should consider two BOD fractions, a particulate and a dissolved one (see figure 7).
- The slopes increase drastically when the Erlenmeyer is re-aerated, which indicates that the organic matter degradation kinetics depend both on the amount of organic matter adsorbed and the oxygen concentration in the atmosphere. An inhibition phenomenon due to the decrease in the oxygen concentration was observed. The kinetic model for organic matter degradation should include an inhibition form depending on oxygen concentration.

The implications of the proposed mechanisms are multiple. First, the BOD removal from the wastewater is mainly determined by a physical sorption mechanism, then, the performance of the wetland system depends on the maintenance and improvement of sorption. The regeneration of the sorption capacity seems to depend on the oxidation of the sorbed organic matter, which is in turn affected by oxygen concentration. It is important to determine the effect of the resting period on the regeneration of the sorption capacity. This would help to more clearly define the optimal frequencies of wastewater discharges to the system. It is important to determine the actual significance of the oxygen limitation on the overall observed stabilisation time for the sorbed organic matter. This will help also to define the need of improved aeration in the bed or improvement in the sorption capacity, in order to increase the performance of the wetland systems. A mathematical model that considers this process would help to answer some of these questions.

Acknowledgements

Special thanks to the *Armada Nacional de Colombia* for the financial and operational support given to this project, to CARALINA, the regional regulator authority and to the personnel of the environmental laboratory at la Universidad de los Andes for the technical support.

References

Bavor, H.J., Roser, D.J., Adcock, P.W. (1995). Challenges for the development of advanced constructed wetlands technology. *Water Science and Technology*, **32**(3), 13–20.

Cooper, P. (1999). A review of the design and performance of vertical-flow and hybrid reed bed treatment systems. *Water Science and Technology*, **40**(3), 1–9.

Cooper, P. and Green, B. (1995). Reed bed treatment systems for sewage treatment in the United Kingdom – the first ten years of experience. *Water Science and Technology*, **32**(3), 317–327.

Cooper, P.F., Job, G.D., Green, M.B., Shutes, R.B. (1996). *Reed beds and constructed wetlands for wastewater treatment*. WRc Swindon, Severn Trent Water. Swindon, Wiltshire, UK.

Giraldo, E., Zárate, E. (2001). Removal of hydrogen sulphide BOD from brackish water using vertical flow wetlands in a coastal Caribbean environment. *Wat. Sci. Tech.*, **44**(11–12), 361–367 (this issue).

Hiley, P. (1995). The reality of sewage treatment using wetlands. *Water Science & Technology*, **32**(3), 329–338.

Platzer, C. (1999). Design recommendations for subsurface flow constructed wetlands for nitrification and denitrification. *Water Science and Technology*, **40**(3), 257–263.

Platzer, C., Maunch, K. (1997). Soil clogging in vertical-flow reed beds-mechanisms, parameters, consequences and solutions? *Water Science and Technology*, **35**(5), 175–181.

Von Felde, K., Kunst, S. (9197). N- and COD-removal in vertical flow systems. *Water Science and Technology,* **35**(5), 79–85.

Accumulation of organic matter fractions in a gravel-bed constructed wetland

L. Nguyen

National Institute of Water and Atmospheric Research Ltd. (NIWA), P. O. Box 11-115, Hamilton, New Zealand

Abstract The function of a gravel-bed wetland in treating wastewaters is dependent on the turn-over rate of organic matter (OM) fractions in accumulated solids. Organic deposits from a gravel-bed planted (*Schoenoplectus tabernaemontani*) wetland, which had experienced pore clogging after 5 years of receiving farm dairy wastewater were therefore collected and determined for labile (water-soluble) and stable (humic acid, fulvic acid and humin) OM fractions, total carbon (C), microbial biomass and microbial respiration rate. Over 90% of the accumulated organic solids was present as stable fractions, with humic compounds at least 2-fold higher in surface deposits and the top 100mm of the gravel bed than the lower gravel substratum. Clogging of the gravel pore spaces over a 5-year wetland operation was probably due to the accumulation of refractory (stable) organic solids, particularly in the top 100 mm of the gravel bed. Microbial respiration rate and microbial biomass were significantly correlated with stable OM fractions, suggesting that these microbial parameters may be used to predict the nature of accumulated OM fractions. Further research is required to evaluate the use of these parameters as indicators of labile and stable fractions in wetlands with a range of OM loadings and accumulation.

Keywords Constructed wetlands; dairy wastewaters; organic matter composition; organic solids; pore clogging

Introduction

Accumulation of organic solids is a typical feature in gravel–bed constructed wetlands treating organic wastes (e.g, domestic sewage and farm dairy wastes) (Burgoon *et al.*, 1995; Tanner *et al.*, 1998a). The accumulated organic solids provide a long-term storage of carbon (C) and nutrients (e.g., nitrogen and phosphorus) and a sustainable supply of C for microbial denitrification (Reddy and D'Angelo, 1996). However, they may clog wetland pore spaces, thus leading to a decline in wastewater retention time and the efficiency of nutrient removal from wastewaters (Burgoon *et al.*, 1995; Tanner *et al.*, 1998b). The breakdown (turnover rate) of organic solids and hence the overall functioning of wetlands in treating wastewaters is probably dependent on the size of microbial biomass, microbial activity and the proportion of refractory fractions in plant litter and applied organic wastewater (Reddy and D'Angelo, 1996).

The main objective of this study was to characterize organic matter (OM) fractions in a gravel-bed planted (*Schoenoplectus tabernaemontani*) wetland, which had experienced pore clogging and wastewater short-circuiting (Tanner *et al.*, 1998a) after 5 years of treating farm dairy wastewater. Biochemical indicators of accumulated OM fractions in organic solids collected from above and within gravel substratum were also investigated by relating the proportions of labile and refractory OM fractions to easily-measured soil microbial parameters (e.g., microbial biomass and activity).

Materials and methods

Deposits (sludge-like) were collected in April 1995 from a planted pilot scale subsurface flow gravel-bed constructed wetland (9.5 m long and 2m wide), which had been in operation for 5 years treating farm dairy wastewaters (pre-treated in a two-stage anaerobic and facultative waste stabilisation pond). The wetland channel was lined with butyl rubber, filled to a depth of 0.4 m with alluvial rhyolitic gravel (10–30 mm diameter and 35% initial pore space). The wetland was planted with soft-stem bulrush (*Schoenoplectus tabernaemontani* (C.C. Gmelin) Palla) and received wastewater hydraulic loading of 71.7 mm day^{-1} (retention time of approximately 2 days) and volatile suspended solids loading rate of 5.8 g m^{-2} day^{-1}. Over a 5-year wetland operation, the wetland received cumulative OM loadings from farm dairy wastewater (8.2 kg OM m^{-2}) and *in-situ* plant residues (8.4 kg OM m^{-2}) and experienced a significant pore clogging (50% reduction in pore spaces; Tanner *et al.*, 1998b).

Cores (210 mm diameter) of surface deposits (commonly 50 mm thick) and gravel substrata (0–100 and 100–400 mm depths) were collected systematically to a 450–510 mm depth (depending on the thickness of surface deposits plus 400 mm gravel substratum) at six sites along the wetland channel. These sites were located at 1, 2, 3, 5, 7 and 9 m from the inlet end of the wetland (hereafter referred to as D1, D2, D3, D4, D5, and D6, respectively). At each site, surface deposits and gravel substrata were collected from the centre of the wetland and approximately 0.75 m from either side of the centre. Surface deposits and deposits coated around the gravel (subsurface deposits) were determined for total carbon (C), microbial respiration rate (Anderson, 1982), and labile (microbial biomass and water-soluble carbon) and stable organic fractions (humic acid, fulvic acid and humin). Total C (TC) in deposits (< 150 µm) was determined using a Perkin-Elmer 2400 CHN Elemental Analyzer. Organic matter (OM) fractions were determined using the sequential extraction scheme of Murata *et al.* (1995), while microbial biomass C (BC) and biomass N (BN) were determined using the methods of Vance *et al.* (1987) and Brookes *et al.* (1985), respectively. Overall the OM fractions sequentially extracted by the Murata *et al.* scheme (1995) included:

(i) Cold water extractable C (CC).
(ii) Hot water extractable C (HC).
(ii) HCl/HF-extractable C (HCl/HF-C).
(iv) $Na_4P_2O_7$-extractable C ($Na_4P_2O_7$-C), comprising both fulvic and humic acid C fractions.
(v) NaOH-extractable C (NaOH-C), comprising both fulvic and humic acid C fractions.
(vi) Humin C.

Labile OM fractions are represented by non-humic substances, comprising BC, BN, CC and HC fractions, while stable OM fractions consist of organic C associated with soil aluminium (Al) and iron (Fe) compounds (HCl/HF-C), fulvic acids, humic acids, and humin (Murata *et al.*1995).

Results and discussion

Total C, OM fractions (except for humin C), microbial respiration rate (soil respired CO_2; mg C kg^{-1}) and microbial biomass (BC and BN) in accumulated solids significantly (at least at $P \leq 0.05$) declined with depth and with distance along the wetland channel (only BC and HCl/HF-C data shown; Figures 1–2). This decline is attributed to a higher flux of OM from plant residues and influent dairy wastewater in the upstream region of the wetland and the return of plant residues on the wetland surface (Nguyen, 2000).

Less than 10% of the TC in the accumulated solids (surface and subsurface deposits) was present as water-soluble C fractions (Table 1), while most (>90%) of the total C was composed of stable (recalcitrant) OM fractions (Table 2). Humic acid, fulvic acid and humin were the predominant stable OM fractions, accounting for 63–96% of the OM in surface deposits and the gravel substratum (Table 2).

The predominance of humic fractions (humic acid, fulvic acid and humin C) throughout

Table 1 Concentrations (%) of total carbon (TC) and the proportions of TC as microbial biomass carbon (BC), cold water soluble carbon (CC), and hot water soluble carbon (HC) in surface deposits and the gravel substratum (0–100 and 100–400 mm depths) at six distances along the wetland. Standard deviation of means ($n = 3$) and probability (p) values of F-ratios ($n = 54$) are presented in parentheses

Distance	Sampling depth	TC (%)	Proportion (%) of sediment TC as:		
			BC	CC	HC
D1	Surface deposit	16.8 (1.96)	1.62 (± 0.156)	1.53 (± 0.502)	3.15 (± 0.172)
	0–100 mm	15.9 (2.96)	1.12 (± 0.133)	1.52 (± 0.430)	2.80 (± 0.310)
	100–400 mm	3.4 (1.11)	0.74 (± 0.076)	2.45 (± 0.288)	4.11 (± 0.469)
D2	Surface deposit	15.2 (1.01)	1.41 (± 0.308)	1.40 (± 0.133)	3.56 (± 0.301)
	0–100 mm	13.5 (0.61)	0.91 (± 0.050)	1.37 (± 0.055)	2.81 (± 0.078)
	100–400 mm	3.7 (0.85)	0.68 (± 0.260)	2.14 (± 0.249)	3.59 (± 0.204)
D3	Surface deposit	15.4 (1.37)	0.78 (± 0.019)	1.13 (± 0.173)	3.05 (± 0.473)
	0–100 mm	13.4 (1.61)	0.52 (± 0.102)	1.01 (± 0.038)	2.33 (± 0.249)
	100–400 mm	2.5 (0.95)	0.74 (± 0.232)	2.68 (± 0.695)	4.28 (± 0.875)
D4	Surface deposit	14.8 (0.92)	0.68 (± 0.024)	1.15 (± 0.083)	3.12 (± 0.191)
	0–100 mm	12.7 (2.86)	0.50 (± 0.120)	1.02 (± 0.191)	2.31 (± 0.267)
	100–400 mm	4.0 (2.27)	1.62 (± 0.133)	2.18 (± 0.790)	3.02 (± 0.682)
D5	Surface deposit	13.0 (1.66)	0.92 (± 0.499)	1.09 (± 0.042)	2.68 (± 0.220)
	0–100 mm	12.5 (2.04)	0.40 (± 0.095)	0.83 (± 0.177)	1.97 (± 0.294)
	100–400 mm	4.0 (2.27)	0.83 (± 0.222)	3.02 (± 0.611)	4.39 (± 0.726)
D6	Surface deposit	13.9 (1.80)	0.67 (± 0.010)	1.06 (± 0.132)	2.91 (± 0.071)
	0–100 mm	9.2 (2.94)	0.64 (± 0.409)	1.14 (± 0.584)	2.92 (± 1.592)
	100–400 mm	1.7 (0.16)	0.60 (± 0.045)	3.28 (± 0.244)	5.61 (± 1.166)

the wetland sediment was attributed to the refractory nature of lignocellulose and their associated humic compounds originating from wetland plant litter and the influent dairy wastewater (Nguyen, 2000). Humic compounds are highly colloidal and amorphous materials with high water holding and physical binding properties (Christensen and Characklis, 1990). Thus the pore clogging of the gravel matrix and surface ponding of wastewater (extending to 2 m from the inlet) as observed by Tanner *et al.* (1998a, b), are attributed to the predominance of the refractory humic compounds in accumulated solids. Using bromide as a hydraulic tracer, Tanner *et al.* (1998a) reported a reduction of 50% in wastewater

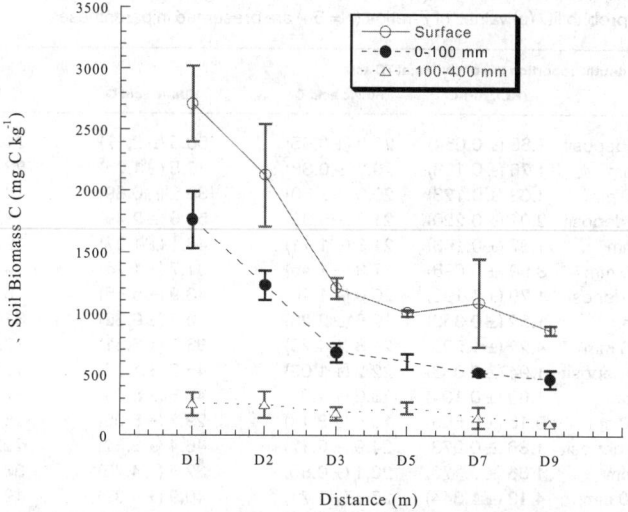

Figure 1 Amounts of microbial biomass carbon (BC) in surface deposits and gravel substratum (0–100 and 100–400 mm depths) at six sites along the wetland channel. Bars indicate standard deviation of means ($n = 3$) for comparing BC between sampling depths at each distance along the wetland

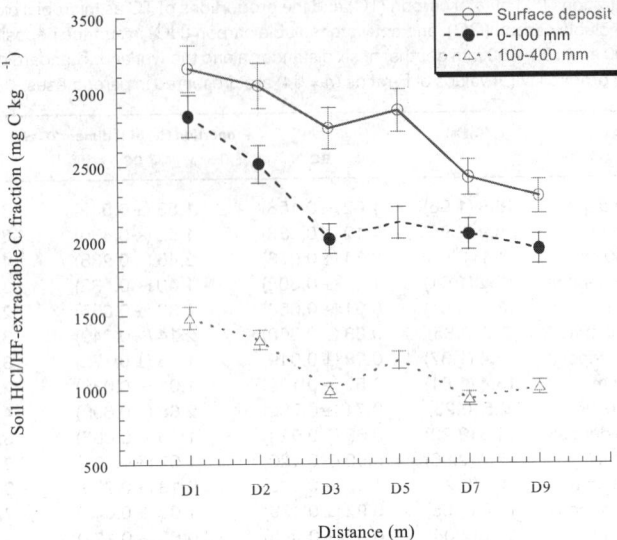

Figure 2 Amounts of carbon associated with mineral aluminium and iron compounds (extracted by hydrochloric/hydrofluoric; HCl/HF-C) in surface deposits and gravel substratum (0–100 and 100–400 mm depths) at six sites along the wetland channel. Bars indicate standard deviation of means ($n = 3$) for comparing HCl/HF-C between sampling depths at each distance along the wetland

retention time from 54.5 to 27.3 hours over a 5-year wetland operation, attributed to the clogging of gravel pore spaces over this period. This reduction in pore spaces and wastewater retention time has been implicated in the declined performance of the studied wetland to remove suspended solids and phosphorus from wastewaters after 5 years of wetland operation (Tanner *et al.*, 1998b).

Organic C associated with mineral Al and Fe compounds (HCl/HF-C) accounted for <2% of the total C in surface and subsurface deposits (Table 2). Thus mineral Al and Fe

Table 2 Proportions of total carbon in accumulated organic solids as carbon associated with mineral aluminium and iron compounds (HCl/HF-C), fulvic acid, humic acid and humin in surface deposits and the gravel substratum (0–100 and 100–400 mm depths) at six distances along the wetland. Standard deviations of means ($n = 3$) and probability (p) values of F-ratios ($n = 54$) are presented in parentheses

Distance	Sampling depth	Proportion (%) of sediment TC as:			
		(HCl/HF)-C	Fulvic acid C	Humic acid C	Humin C
D1	Surface deposit	1.88 (± 0.054)	20.3 (± 0.95)	50.5 (± 2.17)	21.0 (± 0.64)
	0–100 mm	1.76 (± 0.168)	20.7 (± 0.38)	49.5 (± 1.75)	23.6 (± 1.71)
	100–400 mm	4.33 (± 0.123)	20.0 (± 3.69)	38.2 (± 6.89)	14.9 (± 1.13)
D2	Surface deposit	2.01 (± 0.220)	21.8 (± 0.80)	51.0 (± 2.46)	19.8 (± 1.73)
	0–100 mm	1.87 (± 0.163)	21.3 (± 1.11)	48.1 (± 1.07)	19.4 (± 0.57)
	100–400 mm	3.69 (± 0.698)	17.3 (± 1.45)	31.7 (± 1.75)	14.0 (± 0.92)
D3	Surface deposit	1.79 (± 0.192)	20.0 (± 1.70)	43.9 (± 5.65)	15.4 (± 0.64)
	0–100 mm	1.52 (± 0.350)	19.6 (± 2.23)	40.2 (± 9.32)	31.4 (± 1.35)
	100–400 mm	4.22 (± 1.206)	24.8 (± 4.76)	36.7 (± 6.51)	17.5 (± 1.49)
D4	Surface deposit	1.94 (± 0.038)	22.2 (± 1.09)	46.7 (± 1.68)	19.0 (± 1.40)
	0–100 mm	1.69 (± 0.194)	23.6 (± 2.91)	45.8 (± 8.64)	37.3 (± 5.63)
	100–400 mm	3.15 (± 0.656)	18.5 (± 8.14)	29.7 (± 8.02)	16.0 (± 2.36)
D5	Surface deposit	1.86 (± 0.073)	24.9 (± 0.17)	46.4 (± 2.32)	22.2 (± 1.84)
	0–100 mm	1.66 (± 0.170)	20.1 (± 0.85)	37.5 (± 4.75)	33.7 (± 3.91)
	100–400 mm	4.13 (± 1.344)	25.1 (± 1.71)	40.9 (± 4.09)	19.0 (± 4.78)
D6	Surface deposit	1.66 (± 0.106)	23.8 (± 2.34)	46.6 (± 1.14)	20.1 (± 2.13)
	0–100 mm	2.32 (± 0.984)	23.1 (± 5.33)	39.5 (± 3.95)	33.2 (± 7.15)
	100–400 mm	6.08 (± 0.281)	26.8 (± 1.40)	41.2 (± 2.15)	22.5 (± 1.35)

Table 3 Amounts of fulvic and humic acid carbon fractions in surface deposits and the gravel substratum (0–100 and 100–400 mm depths) at six distances along the wetland. Standard deviations of means ($n = 3$) and probability (p) values of F-ratios ($n = 54$) are presented in parentheses

Distance	Sampling depth	Amount (mg kg^{-1}) Fulvic acid C	Humic acid C
D1	Surface deposit	34,127 (± 2361)	84,727 (± 7623)
	0–100 mm	32,900 (± 5608)	78,671 (± 12788)
	100–400 mm	6,614 (± 857)	12,626 (± 1578)
D2	Surface deposit	33,089 (± 1493)	77,337 (± 5073)
	0–100 mm	28,737 (± 1273)	64,927 (± 1860)
	100–400 mm	6,339 (± 1228)	11,653 (± 2129)
D3	Surface deposit	30,631 (± 1964)	67,634 (± 11130)
	0–100 mm	26,055 (± 1152)	53,154 (± 8168)
	100–400 mm	5,971 (± 947)	18,987 (± 2241)
D4	Surface deposit	32,736 (± 1335)	69,200 (± 5345)
	0–100 mm	29,645 (± 5035)	57,524 (± 11948)
	100–400 mm	6,326 (± 625)	10,871 (± 3306)
D5	Surface deposit	32,372 (± 3897)	60,101 (± 4560)
	0–100 mm	25,147 (± 4858)	46,071 (± 1883)
	100–400 mm	5,943 (± 1259)	9,685 (± 2149)
D6	Surface deposit	32,664 (± 1081)	64,502 (± 6698)
	0–100 mm	20,191 (± 2825)	37,013 (± 14220)
	100–400 mm	4,478 (± 582)	6,860 (± 487)

compounds were unlikely to play a major role in the accumulation of organic solids in the wetland. This is not surprising since gravel used as a wetland substratum contained negligible amounts of Al and Fe mineral compounds (Mann and Bavor, 1993). Furthermore, most of the mud and associated soil clay particles entering a two-stage waste stabilisation pond (WSP) from a milking parlour were probably retained within the WSP and hence unlikely to play a major role in OM stabilisation as sediment HCl/HF-C fraction.

The presence of a range of OM fractions in the accumulated organic solids (Tables 1 and 2) suggests that not only quantitative assessment of sediment OM content but also qualitative determination of sediment OM fractions is important for the understanding of OM accumulation-decomposition and pore clogging in gravel-bed wetlands. Fulvic acids are known to have a much lower molecular weight and less structural stability than humic acids (Goh and Reid, 1975). Thus sediments with similar OM content but different proportions of labile and stable OM fractions (and hence a range of molecular weights and packing density) may experience different levels of pore clogging. This could explain why differences in wastewater retention times between wetlands with a range of wastewater loading rates cannot be adequately accounted for by the amounts of OM accumulation in the wetland, as reported by Tanner et al. (1998a).

The higher fulvic and humic acid C fractions in surface deposits and the top 100 mm of the gravel bed than the lower gravel substratum (Table 3) suggests that pore clogging by these fractions was more prominent in the top layer of the gravel substratum. Since lignocellulose compounds are not readily decomposed under anaerobic conditions, pore clogging in the gravel substratum by the accumulation of these recalcitrant OM fractions and their associated humic compounds is likely to limit the operating life of gravel-bed wetlands treating dairy wastewaters. Future research is therefore required to assess the effect of periodic draining on the decomposition of lignocellulose in the gravel substratum of gravel-bed wetlands and its subsequent influence on substratum porosity.

Microbial respiration rate and microbial biomass were significantly correlated with sediment OM fractions (Table 4), suggesting that these microbial parameters may be used to predict changes in the labile and stable fractions of OM accumulation in the wetland.

Table 4 Parameters of the linear regression equations ($y = a + bx$) showing the relationships between microbial indicators and organic matter (OM) fractions in samples collected from surface deposits and gravel substratum along the gravel-bed wetland ($n = 54$, df = 52)

y variables: (Microbial indicators [a])	x variables: (OM fractions [b])	Intercept (a)	Slope (b)	r^2
	Cold water soluble carbon (CC)			
BC	CC	−306.5	0.8620	0.712****
BN	CC	−43.7	0.1178	0.769****
Respiration	CC	−30.2	0.0679	0.765****
	Hot water soluble carbon (HC)			
BC	HC	−316.0	0.3947	0.744****
BN	HC	−42.7	0.0532	0.780****
Respiration	HC	−28.6	0.0303	0.758****
	Carbon associated with mineral Al and Fe compounds (HCl/HF-C)			
BC	HCl/HF-C	−774.8	0.7952	0.704****
BN	HCl/HF-C	−107.0	0.1083	0.755****
Respiration	HCl/HF-C	−63.0	0.0606	0.708****
	Sodium pyrophosphate-extractable ($Na_4P_2O_7$) fulvic and humic C fractions			
BC	Fulvic C	−342.3	0.8307	0.681****
BN	Fulvic C	−49.2	0.1139	0.741****
Respiration	Fulvic C	−31.5	0.0644	0.707****
BC	Humic C	−580.2	0.3333	0.732****
BN	Humic C	−83.9	0.0462	0.813****
Respiration	Humic C	−49.2	0.0257	0.750****
	Sodium hydroxide (NaOH)-extractable fulvic and humic C fractions			
BC	Fulvic C	−123.7	0.0478	0.546****
BN	Fulvic C	−18.4	0.0065	0.587****
Respiration	Fulvic C	−14.3	0.0037	0.564****
BC	Humic C	−106.7	0.0233	0.697****
BN	Humic C	−15.5	0.0032	0.742****
Respiration	Humic C	−12.9	0.0018	0.717****
	Humin C fraction			
BC	Humin C	268.9	0.0247	0.261****
BN	Humin C	30.7	0.0036	0.313****
Respiration	Humin C	14.9	0.0019	0.283****

[a] Unit: mg kg^{-1} except for respiration rate in mg kg^{-1} d^{-1}; [b] Unit: mg kg^{-1}; **** = (0.0001

Conclusions

Up to 90% of the organic solids accumulated in a gravel-bed constructed wetland receiving farm dairy wastewater over a 5-year period was composed of recalcitrant organic fractions, probably originating from the lignocellulose fractions of influent wastewater and plant litter. Pore clogging as reported in previous studies in the studied wetland was therefore attributed to the accumulation of these refractory organic solids, particularly in the top 100 mm of the gravel bed. Since pore clogging limits the operating life of gravel-bed wetlands treating dairy wastewaters, future research is required to assess management factors that may enhance the decomposition of refractory organic solids in the gravel substratum and its subsequent influence on substratum porosity.

Microbial biomass and microbial respiration rates may be used as indicators of changes in stable organic fractions in the accumulated organic solids in the studied wetland. Further research is required to evaluate the use of these parameters as indicators of pore clogging status in wetlands with a range of organic loadings and accumulation. Both quantitative assessment of sediment OM content and qualitative determination of sediment OM composition are important in the understanding of OM accumulation-decomposition and pore clogging in gravel-bed wetlands.

Acknowledgements

Funding from the New Zealand Foundation for Research, Science and Technology, Dr Chris Tanner for kindly providing his study site for this study and the wetland substratum samples used, and Stu Pickmere, James Sukias and Kerry Costley for excellent technical assistance.

References

Anderson, J.P.E. (1982). Soil respiration. In: *Methods of Soil Analysis, Part 2. Chemical and Microbiological Properties*. A.G. Page, R.H. Miller and D.R. Keeney (eds.), Agronomy Monograph 9 (2nd ed.). American Society of Agronomy, Madison, WI, pp. 831–871.

Brookes, P.C., Landman, A., Pruden, G. and Jenkinson, D.S. (1985). Chloroform fumigation and the release of soil nitrogen: a rapid direct extraction method to measure microbial biomass nitrogen in soil. *Soil Biol. Biochem.*, **17**, 837–842.

Burgoon, P.S., Reddy, K.R. and DeBusk, T.A. (1995). Performance of subsurface flow wetlands with batch-load and continuous-flow conditions. *Water Environ. Res.*, **67**, 855–862.

Christensen, B.E. and Characklis, W.G. (1990). Physical and chemical properties of biofilms. In: *Biofilms*. W.G. Characklis and K.C. Marshall (eds.), Wiley, NY, pp. 523–584.

Goh, K.M. and Reid, M.R. (1975). Molecular weight distribution of soil organic matter as affected by acid pre-treatment and fractionation into humic and fulvic acids. *J. Soil Sci.*, **26**, 207–222.

Mann, R.A. and Bavor, H.J. (1993). Phosphorus removal in constructed wetlands using gravel and industrial waste substrata. *Wat. Sci. Tech.*, **27**(1), 107–113.

Murata, T., Nguyen, M.L. and Goh, K.M. (1995). The effects of long-term superphosphate application on soil organic matter content and composition from an intensively-managed New Zealand pasture. *Eur. J. Soil Sci.*, **46**, 257–264.

Nguyen, L. (2000). Organic matter composition, microbial biomass and microbial activity in gravel-bed constructed wetlands treating farm dairy wastewaters. *Ecol. Eng.*, **16**, 199–221.

Reddy, K.R. and D'Angelo, E.M. (1997). Biogeochemical indicators to evaluate pollutant removal efficiency in constructed wetlands. *Wat. Sci. Tech.*, **35**(5), 1–10.

Tanner, C.C., Sukias, J.P.S. and Upsdell, M.P. (1998a). Organic matter accumulation during maturation of gravel-bed constructed wetlands treating farm dairy wastewaters. *Wat. Res.*, **32**(10), 3046–3054.

Tanner, C.C., Sukias, J.P.S. and Upsdell, M.P. (1998b). Relationships between loading rates and pollutant removal during maturation of gravel-bed constructed wetlands. *J. Environ. Qual.*, **27**, 448–458.

Vance, E.D., Brookes, P.C. and Jenkinson, D.S. (1987). An extraction method for measuring soil microbial biomass C. *Soil Biol. Biochem.*, **19**, 703–707.

Determining ecologically acceptable nutrient loads to natural wetlands for water quality improvement

L.W. Keenan and E.F. Lowe

St. Johns River Water Management District, P.O. Box 1429, Palatka, Florida 32178, USA

Abstract Natural wetlands often function as nutrient sinks, reducing nutrient inputs into lakes and streams. P loading from anthropogenic sources has significantly affected many natural wetlands. This paper describes a method to determine an acceptable P load to natural wetlands based on ecological principles. This approach can be used to determine how much P can be assimilated without diminishing species diversity and, thereby, sets a limit for cultural eutrophication of natural wetlands. The basis for determining an acceptable load is management of risk to species diversity by determination of the maximum area of a wetland that can be put at risk while preserving biodiversity of the overall wetland system. Two cases are distinguished: 1) simple-stress, where growth of the affected area immediately increases risks for species loss, and 2) subsidy-stress, where growth of the affected area first benefits then diminishes net species diversity.

Keywords Assimilation, biodiversity, ecology, modelling, natural wetlands, nutrient balance

Introduction

Both constructed and natural wetlands can be used for improving water quality, however, typically the management goals are vastly different. The goal for a constructed wetland goal is to reduce pollutants to a defined outfall concentration, whereas for natural wetlands the primary goal is to limit pollutants to a level that prevents ecological harm. Thus constructed wetlands are engineered and sized to meet their water quality goals but with natural wetlands the pollutant load must be "sized" to meet the ecological goal. A major question facing water resource managers today is determining what portion of a natural wetland's assimilatory capacity for P can be used for treating anthropogenic nutrient load increases without unacceptable trophic state impacts.

The concept of trophic state was first applied to wetlands by Weber in 1907 (Hutchinson, 1969). However, while in lakes our understanding of eutrophication has matured greatly, for wetlands it is still in its infancy. The primary focus of studies on wetlands has been their ability to reduce nutrient concentrations instead of the nutrient effect on the wetlands (e.g. Olson, 1992; Moshiri, 1993; Kadlec and Knight, 1996). The degradation of wetlands through cultural eutrophication, however, has received considerable attention recently. Especially noteworthy in this regard are the efforts in the oligiotrophic Florida Everglades where areas of P enrichment have allowed cattail (*Typha domingensis*) to displace sawgrass (*Cladium mariscus*) as the dominant species (Davis, 1994).

Lake eutrophication management relies heavily on simple empirical models relating trophic state to phosphorus loading (OECD, 1982; Reckhow and Chapra, 1983; Henderson-Sellers and Markland, 1987; Salas and Martino, 1991). These models predict effects that are patternless and generalized throughout the lake. An acceptable degree of impact in lakes may be set as the maximum P concentration that is not expected to substantially affect the essential ecological character or function of the lake.

A comparable paradigm and methodology is needed for determining appropriate limits to loading for natural wetlands but, because wetland eutrophication is fundamentally dif-

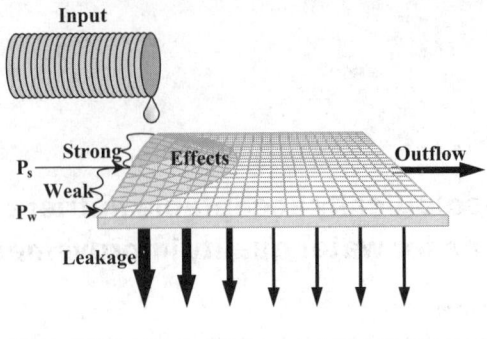

Thin Leaky Sponge Model

Figure 1 Graphical representation of conceptual model

ferent from lakes, the models developed for lakes cannot be adapted to wetlands. Lowe and Keenan (1997) presented a simple conceptual approach for eutrophication in wetlands that used the analogy of a thin leaky sponge (TLS) to represent wetlands (Figure 1).

In the TLS model, the sponge represents the water column, biomass, and soil pools of P that are actively cycling in the wetland – the biogeochemically active P (BAP). The thickness of the sponge at any point represents the total mass of BAP cycling within these short-term pools. The leakiness of the sponge represents the loss of BAP to the long-term sink of soil accretion. Both the thickness of the sponge and its leakiness are greatest proximal to the P source and decrease with distance. This conceptual model fits observations of wetland eutrophication which is localized and patterned.

Simple empirical relationships based on wetland assimilatory capacity to calculate the area of effects have been developed from data on treatment wetlands and the Everglades (Kadlec and Knight, 1996). One example of a simple relationship that could be applied was examined in our earlier paper (Lowe and Keenan, 1997). The simple equation

$$A = -\left(Q/\lambda\right)\ln\left(\frac{P_t - P_b}{P_i - P_b}\right)$$

where Q = inflow rate ($L^3 T^{-1}$)
λ = net P leakage or loss coefficient (LT^{-1})
P_i = inflow or initial P concentration at $t = 0$ (ML^{-3})
P_t = the P concentration at t or threshold conc. (ML^{-3})
P_b = the current background concentration of P (ML^{-3})
A = area with concentration above P_t (L^2).

can be used to calculate the area of effects for different threshold phosphorus concentrations using only a few variables.

Two specific areas of effects are used in this approach. The first, the strong effects area, is defined as having a concentration of P where extreme effects such as changes in species dominance or the conversion of communities are expected. The threshold concentration for these large effects is termed P_s (Figure 1). The value of P_s is difficult to accurately determine for a variety of reasons, foremost of which is that the response of the system may be extremely slow both to reach a concentration gradient equilibrium and to reach a community equilibrium even after the concentration gradient equilibrates because community changes may be slow or only triggered by infrequent, extreme events.

The second area is the weak effects zone. It is that area of the wetland between P_s and the second threshold concentration, P_w, that marks the downstream boundary of the weak effects zone (Figure 1). P_w is that concentration at which the influence of the P input approaches the natural variation of the background concentration of the wetland, i.e. where it would be difficult to definitively assert with few samples that the concentration was above the background concentration of the wetland and there are no practicably measurable effects on the community. The community in the weak effects zone is considered to be at some level of risk due to extreme events and our own lack of knowledge of long term effects. The sum of these two areas is called the total effects zone.

One crucial step needed to apply the TLS model that was not in the original paper is an ecologically based method for determining an acceptable level of impact (i.e. acceptable areas for the strong and weak effects zones), specifically, an impact that does not significantly or substantially affect the essential character or function of the wetland. Here, we present an extension of the TLS model that can be used to set limits for the zones of effects and, thereby, for P loading.

Methods

The approach to determining an acceptable area of effects builds on two strong ecological tenets. First, the continued loss of biodiversity is a major concern of conservationists and ecologists worldwide (Wilson, 1992). The need to preserve biodiversity is among the most widely accepted goals of natural resource managers (Harris, 1984; Grumbine, 1994; Noss and Cooperrider, 1994). The second tenet is that one of the most robust generalizations in ecology is the species-area relationship (Connor and McCoy, 1979; Holt *et al.*, 1999; Noss and Cooperrider, 1994). At its simplest, the species-area relationship states that the larger the area, the more species that exist there. It is generally represented by the equation

$S = CA^z$ where: S = the cumulative number of species present
C = an empirical constant
A = the cumulative area
Z = unitless exponent related to the species richness.

This simple equation has surprising good fit to data for many taxonomic levels or guilds in a variety of different communities. In wetlands, we favor using macrophyte communities because they are easy to census, have low temporal variability, are not mobile, and form the basis of much of the structure and primary production for many wetlands.

The data are plotted as the cumulative number of new species observed against the cumulative area sampled (Figure 2).

The exponent is proportional to the species richness with higher exponents reflecting the greater diversity encountered per unit area. The exponent, z, is unaffected by the units of measure of area. The exponent typically lies between 0.20–0.35, with a general tendency towards 0.25 (Shafer, 1990; Wilson, 1997).

There are two approaches that can be used to apply the species area concept to answer our central question of how to determine an acceptable impact size; 1) simple stress, where

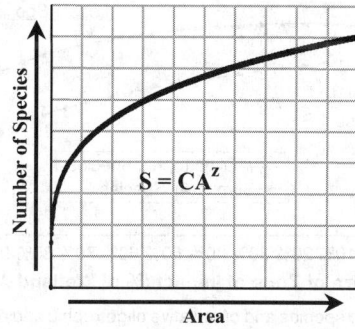

Figure 2 Example species-area curve

even a small zone of effects entails risks to overall species diversity and 2) subsidy-stress, where overall species diversity initially increases then falls with growth of the zone of effects.

In the first approach, for simple stress, the focus is only on the potential loss of species in the natural community. The variable subscript "o" represents the values for the oligotrophic community and "e" represents those for the eutrophic community.

To graphically show how this may work we have chosen a simple example where the exponent z equals 0.25 for each community and the constant C_o for the oligotrophic community equals 10 and C_e equals 2.5 for the eutrophic community. The lower line is number of species for the oligotrophic community and upper line is the sum of species for both communities. Both are plotted along an x-axis of size of the effects in per cent wetland area.

In this first case, as the area of impact (A_e), increases, the cumulative number of species decreases according to the following formula:

$$S = C_o(A_o)^{z_o}.$$

For the constants chosen in the example, we can easily see that 5 per cent of the total area must be converted before the loss of a whole integer species would be expected (Figure 3). Therefore any conversion of a smaller area would be expected to reduce the total number of species in the natural community by no more than one species. The overall essential character and function of the wetland could not be expected to be significantly or substantially compromised by the potential loss of a single rare species. Note also that this is not an insignificant area of change before an integer species reduction. This is because the species area curve tends to flatten out as area increases. As the wetland gets smaller, the acceptable area of conversion reduces at a faster rate because the curve steepens in the smaller areas. Likewise the larger the curve's exponent, z, and therefore the greater the species richness of the community, the steeper the curve and the smaller the acceptable area of impact before the loss of an integer species (see Table 1).

In contrast, the second approach also enumerates the addition of new species found in the eutrophic impact community and not previously existing in the natural wetland. This approach embodies the subsidy-stress gradient concept of Odum et al. (1979) where if the input involves usable resources there is a subsidy before stress levels are reached, in this case measured by species diversity. This is the top line in Figure 3 using the equation:

$$S = C_o A_o^{z_o} + C_e A_e^{z_e}.$$

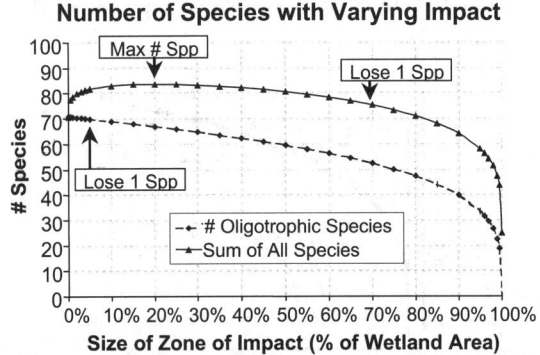

Figure 3 Cumulative oligotrophic species and cumulative oligotrophic and eutrophic species versus size of impact.

Table 1 Effects of different size wetlands and exponent values on one integer species loss area and per cent of wetland

Area Impacted to lose one spp (per cent of total area)

Total Area	z = 0.15	z = 0.25.	z = 0.35
100	19 (19%)	9 (9%)	5 (5%)
1,000	142 (14%)	52 (5%)	22 (2%)
10,000	102 (10%)	296 (3%)	100 (1%)

In this case, the number of species initially increases, the subsidy, but as the zone of effects increases beyond 20%, species number begins to decline, the stress. At about 65%, a net loss of species begins and a net loss of an interger species occurs before 70%.

Of great interest, is the large range of impacted area in the middle of the graph where the total number of species is relatively insensitive to the proportion of each community. If the management goal was to maximize species diversity, a fairly large error in the areas of the communities in this range would have only a small impact on the total biodiversity.

However, high nutrient wetlands tend to be quite common, in previously eutrophified wetlands, roadside ditches, retention ponds, and more and more, in constructed wetlands. So on a broader scale, the addition of species by the eutrophic community may not be regionally significant or beneficial, whereas the loss of species from the oligotrophic wetland may have few if any other refugia outside of the area being managed. Therefore, we suggest using the simple stress model to determine an acceptable effects zone.

Using the simple stress or only native community species approach still requires prediction of the strong effects threshold phosphorus concentration, that will cause a shift to the eutrophic community. This is the strong effects concentration (P_s) in our TLS concept. Since the value of P_s is difficult to determine, the second threshold phosphorus concentration, P_w, used in the TLS model can be of more immediate use. The area with P concentration between the input and P_w defines the total area of effects and encompasses all of the area of the wetland at any level of risk by the nutrient addition. In fact, for some portion of the fringe of the weak effects zone, there is likely to be a productivity subsidy of the native community due to increased nutrients that would provide ecological benefits with little risk of impact. This fringe therefore is a different aspect of Odum's subsidy-stress gradient concept where small increases in nutrients subsidize the production of a community but increasing levels moves towards stress by providing increasing competitive advantage to the more eutrophic species (Odum *et al.*, 1979).

We favor the simple stress model using the total zone of effects as the basis for predicting risks to overall species diversity. We believe that this is conservative and suggest that future efforts should concentrate on defining that concentration, greater than P_w, where the increased risk of conversion begins to exceed the benefits of the nutrient subsidy. This would provide a more appropriate way to calculate the area at appreciable risk. This approach may not be valid where the background concentration of the wetland has already been significantly increased from the historic background and past the subsidy level, such as through increased atmospheric loading because the entire wetland may already be at risk of eutrophication.

Conclusions

The approach outlined in this paper: defines an ecologically based method for limiting the area of impact; an easily determined threshold concentration to define where risk of impact begins; and a simple formula that would allow calculation of the phosphorus load that would correspond to the acceptable impact area at risk. Together, this provides a conservative, scientifically based method to determine the increase in nutrient load that can be safely assimilated or treated by a natural wetland with no substantial ecologically impact to the overall structure, function, or biodiversity of the natural wetland.

References

Biffi, F. (1963). Determining the time factor as a characteristic trait in the self-purifying power of Lago d'Orta in relation to a continual pollution. *1st. Ven Sci. Lettl. Arti.* **121**, 131–136.

Davis, S. M. (1994). Phosphorus inputs and vegetation sensitivity in the Everglade. In: *Everglades: The Ecosystem and its Restoration*. Davis, S.M. and Ogden, J.C. (eds.). St. Lucie Press, St. Lucie, FL, pp. 357–378.

Grumbine, R.E. (ed.) (1994). *Environmental Policy and Biodiversity*. Island Press, Washington, 415 pp.

Henderson-Sellers, B. and Markland, H.R. (1987). *Decaying Lakes: The Origins and Control of Cultural Eutrophication*. John Wiley and Sons, New York, NY, 254 pp.

Holt, R., Lawton, J., Polis, G. and Martinez, N. (1999). Trophic rank and the species-area relationship. *Ecology* **80**, 1495–1504.

Hutchinson, G.E. (1969). Eutrophication, past and present. In: *Eutrophication: Causes, Consequences, and Corrections*. Nat. Acad. Sci., Washington, D.C., pp. 17–26.

Kadlec, R.H. (1994). Phosphorus uptake in Florida marshes. *Water Sci. Tech.*, **30**(8), 225–234.

Kadlec, R.H. and Knight, R.L. (1996). *Treatment Wetlands*. CRC Press, Lewis Publishers, New York, 893 pp.

Lowe, E.F. and Keenan, L.W. (1997). Managing phosphorus-based, cultural eutrophicaton in wetlands: a conceptual approach. *Ecol. Eng.* **9**, 109–118.

Moshiri, G.A. (1993). *Constructed Wetlands for Water Quality Improvement*. Lewis Publishers, Ann Arbor, MI, 632 pp.

Noss, R. and Cooperrider, A. (1994). *Saving nature's legacy: Protecting and Restoring Biodiversity*. Island Press, 417 pp.

OECD (1982). *Eutrophication of Waters: Monitoring, Assessment, and Control*. Organisation for Economic Co-operation and Development (OECD), Paris, France, 154 pp.

Odum, E.P., Finn, J.T. and Franz, E.H. (1979). Perturbation theory and the subsidy-stress gradient. *BioScience* **29**(6), 349–352.

Olson, R.K. (ed.), (1992). The Role of Created and Natural Wetlands in Controlling Nonpoint Source Pollution. Proceedings of a U.S. EPA workshop, Arlington, Va., USA, June 10–11, 1991. *Ecol. Eng.* **1**, 1–172.

Reckhow, K. and Chapra, S. (1983). *Engineering Approaches for Lake Management. Volume 1: Data Analysis and Empirical Modeling*. Butterworth Publishers, Boston, MA, 340 pp.

Salas, H. and Martino, P. (1991). A Simplified Phosphorus Trophic State Model for Warm-water Tropical Lakes. *Water Res.* **25**(3), 344–350.

Shafer, C.L. (1990). *Nature Reserves: Island theory and Conservation Practice*. Smithsonian Institution Press, Washington, 189 pp.

Urban, N.H., Davis, S.M. and Aumen, N.G. (1993). Fluctuations in sawgrass and cattail densities in Everglades Water Conservation Area 2A under varying nutrient, hydrologic and fire regimes. *Aquatic Botany* **46**, 203–223.

Vollenweider, R.A. (1969). Possibilities and limits of elementary models concerning the budget of substances in lakes. *Arch. Hydrobiol.* **66**(1), 1–36.

Wilson, E.O. (ed.), (1988). *Biodiversity*. National Academy Press, Washington, D.C., 521 pp.

Wilson, E.O. (1988). The current state of biological diversity. In: *Biodiversity*. Wilson, E.O. (ed.). National Academy Press, Washington, D.C., pp. 3–18.

Wilson, E.O. (1992). *The Diversity of Life*. Belknap Press, Cambridge, 424 pp.

Surmounting the engineering challenges of Everglades restoration

G.F. Goforth

South Florida Water Management District, PO Box 24680, West Palm Beach, Florida, 33416, USA

Abstract The South Florida Water Management District, in partnership with other agencies and stakeholders, is undertaking one of the world's largest ecosystem restoration programs. The foundation of the nutrient control program for the Everglades is a set of six large constructed wetlands, referred to as Stormwater Treatment Areas (STAs). The initial treatment goal is to reduce phosphorus entering the Everglades to 50 parts per billion. The STAs comprise almost 17,000 hectares, with a capital cost of approximately $700 million. Approximately 4,720 hectares are currently operational, another 2,600 hectares are in the start-up phase, and construction is just getting under way on the remaining areas.

Throughout the design process, engineers and scientists collaborated to capture the best available information on wetland treatment systems, and to develop the most appropriate design criteria. Some of the more challenging issues included characterizing stormwater inflows and phosphorus loads, determining appropriate nutrient removal performance characteristics, and estimating hydraulic design parameters relating to densely vegetated systems.

The design process combined in-house staff with engineering consultants, construction contractors, external review groups and independent peer-review. This paper summarizes major design aspects and key assumptions, and sets the stage for addressing future challenges associated with achieving long-term water quality goals of Everglades restoration.

Keywords Constructed wetlands; environmental engineering; Everglades; phosphorus removal; restoration; stormwater treatment areas

Introduction

Background

The South Florida Water Management District (District), in partnership with other state and federal agencies and stakeholders, is undertaking one of the largest ecosystem restoration programs in the world. Florida's 1994 Everglades Forever Act (Act) set into action a plan for restoring a significant portion of the remaining 618,000-ha Everglades ecosystem through a program of construction, research, and regulation activities. The Act addressed water quality, water quantity (including hydroperiod), and the invasion of exotic plant species in the Everglades ecosystem. The Act also establishes both interim and long-term water quality goals to ultimately achieve restoration and preservation of the Everglades. The interim goal of the restoration program is to reduce phosphorus (P) concentrations entering the Everglades to 50 parts per billion (ppb). The foundation of the interim phosphorus control program is the Everglades Construction Project (ECP) which encompasses six strategically located constructed wetlands, referred to as Stormwater Treatment Areas, or STAs (see Figure 1). In addition to the STAs, significant phosphorus load reductions have been achieved through best management practices (BMPs) within the adjacent Everglades Agricultural Area (EAA). The long-term goal is to combine point-source, basin-level and regional solutions in a system-wide approach to ensure that all waters discharged to the Everglades Protection Area achieve final water quality goals by December 31, 2006. With respect to nutrients, the long-term goal is to reduce nutrient discharges to levels that do not cause an imbalance in natural populations of aquatic flora or fauna, however, the numerical interpretation of this narrative standard has not yet

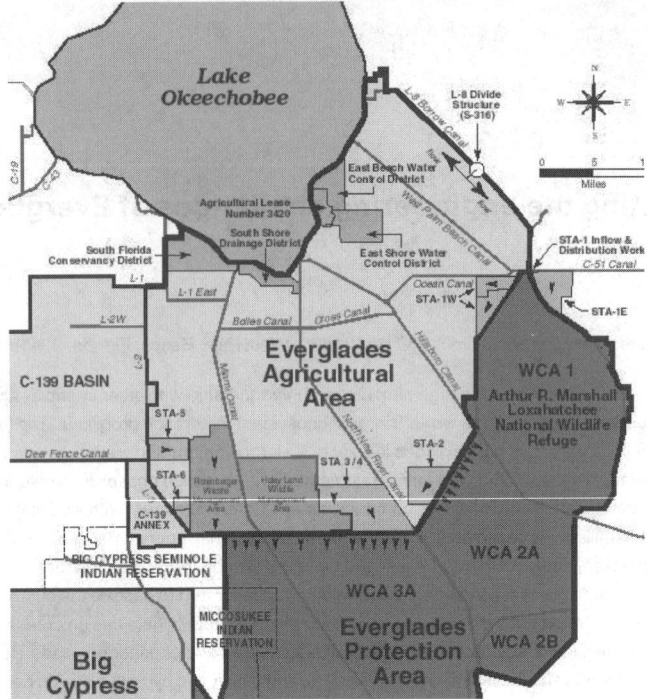

Figure 1 Overview of the Everglades Construction Project

been determined. Additional background information can be found in Chimney and Goforth (2000).

Design objectives

Through a process of scientific research and evaluation, litigation, mediation, legislation, and consensus building, the design objectives of the ECP have evolved to include the following:

1. to reduce the phosphorus concentration entering the Everglades to an interim target of 50 ppb (measured as total phosphorus); this objective will be achieved in conjunction with the BMPs of upstream landowners;
2. to increase the supply of water to the Everglades;
3. to improve the spatial distribution and timing of inflows to the Everglades;
4. to improve the flood control of an adjacent urban watershed while maintaining flood protection in the other basins;
5. to reduce harmful discharges of freshwater to coastal estuaries; and
6. to the extent possible, reduce phosphorus loading to Lake Okeechobee from local drainage districts along the southern and eastern shores.

Location

The STAs are located along the northern boundary of the Everglades Water Conservation Areas (WCAs) (see Figure 1), and in most cases are adjacent to existing major flood control pump stations that have been in operation since the late 1950s. In addition to taking advantage of the existing pump stations, these locations allow for improvement in the spatial distribution of water entering the Everglades, changing the hydraulic regime from discrete point sources to a linear distribution that approaches sheetflow. Almost 19,300 ha of agricultural lands (primarily sugar cane, citrus, sod, vegetables and pasture) and remnant wetlands were acquired for the project.

Project costs

An additional design consideration was available finances. The estimated capital cost for the ECP is approximately $700 million, with approximately $183 million for land acquisition, $498 million for design and construction, and $16.3 million for program management and related costs. Individual STA design and construction costs varied considerably, as each of the STAs has a unique configuration of existing and new components (levees, pump stations, etc.), with variations in size, prior land use and topography. Overall, the design and construction costs averaged approximately $23,800/ha, with a range of $9,600/ha to $58,000/ha. Excluding the 3,400 ha of state-owned land incorporated into the STAs, the acquisition cost ranged from $6,900/ha to $26,150/ha and averaged $14,000/ha. Revenue sources include annual agricultural privilege taxes (approximately $61/ha for the EAA and $10/ha for the C-139 Basin), *ad valorem* property taxes, state land acquisition funds, mitigation funds from a regional electric utility, and toll revenues from a highway crossing the southern Everglades. In addition, the U.S. Army Corps of Engineers (ACOE) is cost-sharing the design and construction of STA-1 East. Complete expense and revenue information is available on the ECP website: http://www.sfwmd.gov/org/erd/ecp/3_ecp.html.

Major design aspects

The design process

Numerous alternative treatment systems were evaluated during the 1980s and early 1990s leading to the decision to utilize constructed wetlands for phosphorus reduction. By 1994, alternatives investigated included source control (e.g., BMPs), algal raceways, chemical treatment, and aquifer storage and retrieval. Once the decision was made to use constructed wetlands, the design of the STAs proceeded in three phases.

1. A **Conceptual Design** for the entire ECP was completed by the consulting firm of Burns and McDonnell in March 1992 (Burns and McDonnell, 1992). This conceptual design was later revised in February 1994, as a result of mediation among the District, state and federal agencies and other stakeholders (Burns and McDonnell, 1994).
2. **General Design Memoranda** for individual ECP components were prepared by Burns and McDonnell between 1994 and 1996.
3. **Detailed Design Reports** and associated plans and construction contract specifications were prepared by multiple consulting firms, including Hutcheon Engineers (STA-1 West), Stanley Engineers (STA-1 Inflow and Distribution Works), Brown and Caldwell (STA-2), Prescott-Follett (outflow pump stations for STA-1 West and STA-2), and Burns and McDonnell (STA-5 and STA-6). In addition to these larger works, multi-discipline engineering services were provided by several consultants, including Metcalf and Eddy, Peer Consultants, Nodarse, Sverdrup Civil, Milian Swain, Weidener Surveying, and Muniz/Hazen and Sawyer. Also, District staff engineers completed the design for several of the smaller facilities. Detail design is currently under way for STA-3/4 by Burns and McDonnell and STA-1 East by the ACOE.

Throughout the design process, the District encouraged review by stakeholders and technical experts, independent peer-review, and construction contractors' value engineering as part of a formal partnering process. In addition to improving the designs, these activities led to considerable cost savings, including the use of refurbished 10-cylinder diesel engines in place of new engines for two 970-cfs pumps (cost saving: approximately $1.5 million), and resizing the outflow pump station for STA-2 (cost savings: approximately $3 million).

Inflow characteristics

The inflows to the STAs will include runoff from seven hydrologic basins that previously discharged untreated water into the northern Everglades. The following design decisions and assumptions were made.

1. The 10 yr period from October 1978 to September 1988 was selected as the basis of design for average and peak stormwater runoff and discharges from Lake Okeechobee.
2. The STAs were sized to capture and treat all basin flows that occurred during the 1978–1988 period of record without hydraulic bypass.
3. The 10 yr period of record flows was modified to reflect implementation of BMPs within the EAA, with an assumed 25% reduction in phosphorus load. With the recognition that increased on-farm retention of stormwater would be a major BMP, it was assumed that the annual volume of runoff would be reduced by 20%.
4. An amount of water equivalent to the reduction due to BMPs would need to be delivered to the Everglades during the dry season, when it was assumed that sufficient capacity would be available in the STAs.
5. It was assumed that an additional 29,100 ha-m of water would be released from Lake Okeechobee as a result of implementation of a new lake regulation schedule.
6. To optimize the size of the STAs, the transfer of water between adjacent basins would be encouraged.

The STA inflows are primarily stormwater, hence there is considerable variability on a day-to-day basis in both flow volumes and phosphorus concentrations. Examples of this variability are presented in the Figures 2 and 3 representing calendar year 1979 for STA-2 (Walker, 1999). A summary of the projected STA inflows is presented in Table 1.

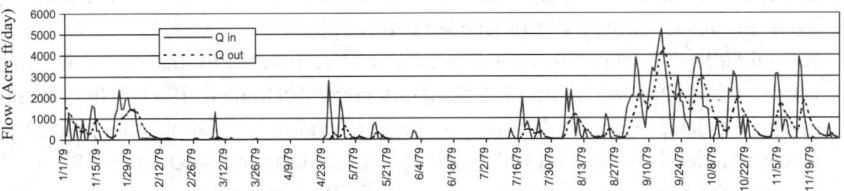

Figure 2 Simulated 1979 flow for STA-2 (1 acre ft/day = 0.1233 ha-m/day)

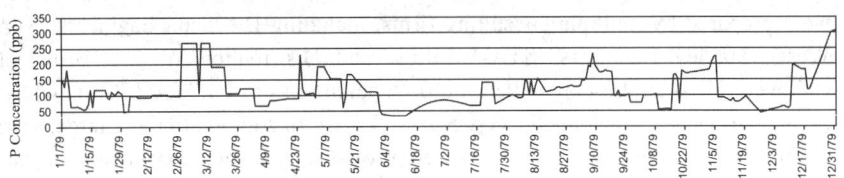

Figure 3 Simulated 1979 phosphorus concentrations for STA-2

Table 1 Summary of STA hydraulic and phosphorus inflows

STA	Average Annual flow (acre-feet)	Peak daily flow (cfs)	Average daily flow (cfs)	Ratio of peak : average flow	Average P load (metric tons/yr)
STA-1 East	124,876	4,050	185	21.8	29.4
STA-1 West	142,860	3,250	200	16.3	37.7
STA-2	174,641	3,370	244	13.8	33.8
STA-3/4	604,655	5,840	841	6.9	87.3
STA-5	78,340	2,510	110	22.9	25.3
STA-6	53,877	2,090	74	28.1	13.2
Total	1,179,240	21,110	1,629	13.0	227

Notes: 1 acre-feet = 0.1233 ha-m; 1 cfs = 0.0283 cubic metre per second

Nutrient removal performance and sizing of the STAs

The long-term phosphorus removal mechanism within the STAs is the creation of plant biomass and subsequent accretion of this organic material onto the sediment. The initial estimates of the effective treatment area required for each of the STAs were based on the work of Walker (1995) and Kadlec and Knight (1996). Phosphorus removal within the STA was assumed to be represented by a first-order equation

$$R = K A C$$

where R = removal rate, g/yr
  K = effective settling rate, m/yr
  A = effective treatment area, m^2
  C = water column concentration of phosphorus, g/m^3

Integration of the differential equations describing the water and phosphorus mass balances, with the following assumptions:
1. the flow in the STA can be represented as plug flow;
2. the STA will remain wet all year long;
3. there is negligible interaction between the STA and groundwater;
4. the apparent background phosphorus concentration within the STA is equal to zero;
5. the effective settling rate is constant and independent of hydraulic and nutrient loading rates;

and solving for area, yields the following equation for determining the effective treatment area required for each STA (Walker, 1995):

$$A = \frac{Q \left\{ \frac{(NC_i + KC_i - PC_p)}{(NC_o + KC_o - PC_p)} \right\}^{[1/(1+K/N)]} - Q}{N}$$

where C_o = target long-term average annual outflow phosphorus concentration, mg/l
  C_i = long-term average annual inflow phosphorus concentration, mg/l
  Q = long-term average annual inflow, m^3/yr
  P = long-term average annual rainfall, m^3
  N = long-term average annual difference between rainfall and evapotranspiration, m^3/yr
  C_p = long-term average annual phosphorus concentration of atmospheric deposition, mg/l
  A = area required to achieve the target outflow phosphorus concentration, m^2.

Using soil and water column phosphorus data from WCA-2A, a value of 8 m/yr was initially estimated for the effective settling rate (Burns and McDonnell, 1992). Later analysis excluded droughts from the periods in which phosphorus removal is assumed to occur, and the effective settling rate increased to 10.2 m/yr (Walker, 1995). This increase in the effective settling rate yielded smaller estimates of required treatment area, reflecting the observation that keeping the soil wet will increase phosphorus removal performance. The effective treatment areas resulting from this equation and associated loading rates are summarized in Table 2. The areas in Table 2 are effective treatment areas; an additional 2,620 ha was required for levee footprints and other ancillary components.

Table 2 Summary of STA treatment areas, phosphorus loading and hydraulic characteristics

STA	Area required to achieve 50 ppb (ha)	Average nutrient loading rate (g/m²/yr)	Estimated P removal (metric tons/yr)	Average hydraulic loading rate (cm/d)	Average hydraulic residence time (days)
STA-1 East	2,170	1.36	23.0	1.95	31
STA-1 West	2,700	1.40	30.5	1.79	34
STA-2	2,600	1.30	24.5	2.27	27
STA-3/4	6,670	1.29	53.9	3.03	20
STA-5	1,670	1.52	21.4	1.59	38
STA-6	960	1.37	10.4	1.90	32
Total	16,770	1.35	164	2.37	26

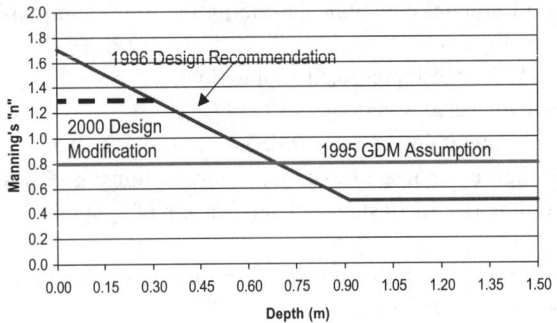

Figure 4 Influence of depth on Manning's "*n*" for a cattail marsh

STA hydraulics

Flow resistance due to vegetation. For design purposes, the hydraulic capacity of the STAs was estimated using various hydraulic routing models, all of which simulated the influence of vegetation on conveyance. The vegetation within the STAs plays a primary role in the efficacy of the treatment areas. The type and density of the vegetation within the treatment area influences the movement of water through the projects by imparting a resistance to flow, as commonly represented by Manning's coefficient of roughness, or Manning's "*n*". The value of "*n*" varies with depth and surface conditions, including soil type and vegetation. Values for Manning's "*n*" vary from 0.023 for well maintained canals to 0.045 for densely vegetated canals. During the general design of the STAs, a constant value for Manning's "*n*" of 0.8 was used. Subsequent field tests yielded values ranging from 0.2 to above 1.0 in Cell 4 of STA-1 West, suggesting the 0.8 value was a conservative design assumption for the higher flows and higher depths associated with extreme storm events (see Figure 4) (Brown and Caldwell, 1996; Kadlec, 1999). For the detail design of STA-2, Brown and Caldwell recommended the use of a depth-variant "*n*", with a value of 0.5 for depths greater than 1 metre, and a linear relationship below 1 m. As part of the STA-3/4 detail design, Burns and McDonnell modified the relationship to a constant "*n*" of 1.3 at or below 30 cm depth (Burns and McDonnell, 2000). Additional analyses are under way to differentiate the influence on "*n*" between cattail-dominated systems and areas with submersed aquatic vegetation.

Operating guidelines. In addition to normal flow conditions, the design of the STAs evaluated extreme storm events and dry periods requiring supplemental water to avoid soil dryout. As part of the levee and structure design, storms up to the Standard Project Flood, estimated as 1.25 times the 100-year flood discharge, were analyzed. In addition, to protect against the release of phosphorus from the organic sediment following exposure to the air, the STAs were designed to avoid dryout. The target minimum average depth was established at 15 cm. All but one of the STAs will have the capability to introduce supplemental water to maintain this minimum depth. Based on a review of stage exceedence curves for natural wetland systems, and the best professional judgement of biologists, a maximum operating water depth of 137 cm, for no more than 10 days, was established for the STAs.

Early results
Over 4,720 ha of treatment areas (STA-1 West, STA-5 and STA-6) are operational, while STA-2 (2,600 ha) is in the start-up phase. Construction is under way on STA-1 East, and the initial construction contract for STA-3/4 should be awarded in early 2001. The initial phosphorus removal performance of the STAs has been better than the design criterion of 50 ppb. The initial 1,515-ha treatment area in STA-1 West has been fully operational since August 1994, and has consistently produced annual phosphorus concentrations less than 25 ppb (see Figure 5). Similarly, average discharge concentrations from the 352-ha STA-6 have remained below 25 ppb, despite receiving considerably greater inflow than estimated during design.

Future challenges
Optimizing the performance of the STAs
While the performance of the STAs has exceeded design phosphorus removal expectations, much remains to be learned about how to optimize their long-term performance. One aspect receiving attention is the initial start-up period, which has varied from a couple of weeks to over a year. Investigations are under way to determine the most efficient and effective way to bring these large biological systems on-line. Factors such as prior land use, soil preparation, and initial inundation depth and duration are influential in determining the time for these systems to achieve a net improvement in phosphorus concentrations, and subsequently to achieve the design performance goal. The use of submersed aquatic vegetation (as opposed to cattails and other emergent plants) and the incorporation of periphyton-based treatment cells appears to holds promise for optimizing STA performance. In fact, the vegetation management strategy for the 930-ha Cell 5b was modified from emergent cattail vegetation to submerged vegetation (the area was inoculated with material harvested from Cells 3 and 4, and grow-in rates are being monitored). The District and other stakeholders are actively investigating these and other performance factors. In addition, a dye tracer study was conducted to identify hydraulic short circuiting in one treatment cell of STA-1 West, and corrective measures included placing earthen plugs in perimeter canals and gaps in berms adjacent to distribution canals.

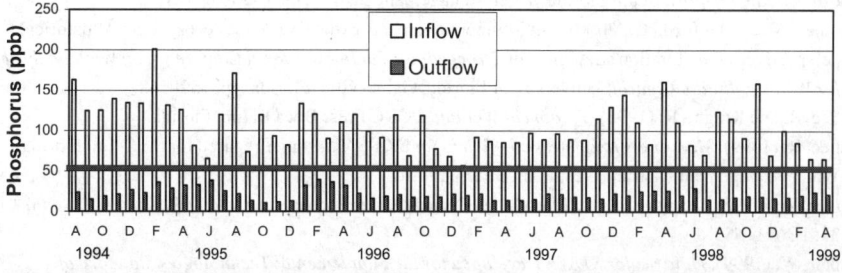

Figure 5 Initial phosphorus removal performance for STA-1 West Cells 1–4 (from Chimney et al., 2000)

Achieving long-term water quality goals

Since their inception, the STAs have been considered an interim step in achieving the long-term water quality goals of Everglades restoration. While phosphorus concentrations of 25–35 ppb may be sustainable in the STAs, discharge levels as low as 10–20 ppb may be required to halt the degradation of the Everglades ecosystem. The District and other stakeholders are evaluating the technical, environmental and economic feasibility of other technologies for achieving these long-term water quality goals. Some of the more promising systems include submersed aquatic vegetation and periphyton-based STAs, as well as chemical treatment and hybrid systems of chemical and biological systems.

Conclusions

An effective measure of the success of an engineering design is the degree to which the finished product achieves the project objectives. For the Everglades STAs, the early phosphorus removal performance has exceeded design expectations. While it is still too soon to know if the long-term performance of the STAs will replicate this early success, the lessons learned during the collaborative and open design process, in conjunction with timely integration of relevant research results, provide a solid foundation for resolving future restoration challenges, in the Everglades and elsewhere.

Acknowledgements

One objective of this paper was to adequately describe the importance of the collective contributions of numerous engineering firms and individuals to the design of the Everglades STAs. Special recognition is given to Bill Walker, Bob Kadlec and Bob Knight for their invaluable contributions to the science and engineering of both the Everglades Construction Project and the body of knowledge of treatment wetlands. Appreciation for effective peer review of this manuscript is extended to Joe Schweigart, Tracey Piccone, Laura Reilly, Jim Kunard, Randy Bushey, Jennifer Jorge and Michael Chimney of the Everglades Construction Project.

References

Brown and Caldwell (1996). *Detailed Design for STA-2 and Water Conservation Area 2A Hydroperiod Restoration*. Prepared for the South Florida Water Management District, West Palm Beach, Florida.

Burns and McDonnell (1992). *Everglades Protection Project Conceptual Design – Stormwater Treatment Areas*. Prepared for the South Florida Water Management District, West Palm Beach, Florida.

Burns and McDonnell (1994). *Everglades Protection Project Conceptual Design*. Prepared for the South Florida Water Management District, West Palm Beach, Florida.

Burns and McDonnell (2000). *Everglades Construction Project Stormwater Treatment Area-3/4 and East WCA-3A Hydropattern Restoration Task Report 5.3 2D Hydrodynamic Modeling*. Prepared for the South Florida Water Management District, West Palm Beach, Florida.

Chimney, M.J., Nungesser, M., Newman, J., Pietro, K., Germain, G., Lynch, T., Goforth, G. and Moustafa, M.Z. (2000). Chapter 6: Stormwater treatment areas – status of research and monitoring to optimize effectiveness of nutrient removal and annual report on operational compliance. In: *Everglades Consolidated Report*, South Florida Water Management District, pp. 6–1 to 6–127.

Chimney, M. and Goforth, G. (2000). Environmental impacts to the Everglades ecosystem: A historical perspective and restoration strategies. In: *Proceedings 7th International Conference on Wetland Systems for Water Pollution Control,* University of Florida, Gainesville, Florida, pp. 159–168.

Kadlec, R. and Knight, R. (1996). *Treatment Wetlands*. CRC Press Boca Raton, Florida.

Kadlec, R. (1999). *Manning's roughness coefficient for STA-3/4 design*. Prepared for the South Florida Water Management District, West Palm Beach, Florida.

Walker, W. (1995). Design basis for Everglades stormwater treatment areas. *Water Resources Bulletin* **31**(4), 671–685.

Walker, W. (1999). *Data Set for STA-2, Developed for the Supplemental Technologies Standard of Comparison*. Prepared for the South Florida Water Management District, West Palm Beach, Florida.

Changes in plant biomass and nutrient removal over 3 years in a constructed wetland in Cairns, Australia

M. Greenway* and A. Woolley**

*School of Environmental Engineering and Co-operative Research Centre for Catchment Hydrology, Griffith University, Brisbane Queensland 4111 Australia

**Department of Natural Resources, GPO Box 2454, Brisbane Qld 4000 Australia

Abstract The surface flow wetland in Cairns, Australia consists of 3 linear channels each 65 m long. Channels 1 and 2 are 5 m wide and Channel 3 is 15 m wide. The wetland was constructed in 1994 and band planted with emergent macrophyte species and alternating open water sections. The wetland was monitored for plant growth and nutrient removal until 1997. During that period HRT was 16 days in Channel 1 and 10 days in Channels 2 and 3; mass loading rates were 2.4 kg Total N and 2.0 kg Total P ha^{-1} d^{-1} in Channel 1 and 3.7 kg TN and 3.3 kg TP ha^{-1} d^{-1} in Channels 2 and 3. The aim of this work was to determine the proportion of nutrient removal that could be attributed to direct uptake by macrophytes and incorporated into plant biomass. Over the 3 year monitoring period reduction in total mass of nutrients was: Channel 1: 26% P, 85% N; Channel 2: 28% P, 87% N; Channel 3: 21% P, 81% N. Percentage reduction of FRP (Filterable Reactive Phosphorus) was similar to TP; NO$_x$ removal was 97–98%. Mass removal rates for TN and TP were higher in Channels 2 and 3 despite greater nutrient loading rates and shorter detention times. Total FRP removal was 23 kg P in Channel 1, 33 kg P in Channel 2 and 70 kg P in Channel 3 of which plant biomass accounted for 65%, 44% and 47% respectively. Total nitrogen removal was 92 kg in Channel 1, 154 kg in Channel 2 and 386 kg in Channel 3 of which plant biomass accounted for 47%, 27% and 27% respectively. Thus, in this tropical surface flow wetland supporting a mixture of emergent macrophytes and floating duckweed, vegetation is an important mechanism for direct nutrient removal.

Keywords Australia, constructed wetland; macrophytes; nutrient removal; plant biomass; surface flow wetland

Introduction

Vegetation is the dominant feature of constructed wetlands and free water surface flow systems may support a variety of macrophyte types and species including emergent macrophytes (reeds/sedges and rushes), floating leaved attached macrophytes (water lilies), free floating macrophytes (duckweed) and submerged macrophytes ("pond weeds") (Greenway and Woolley, 1997). Brix (1997) categorised the most important functions of macrophytes in the treatment of wastewater as physical and metabolic. Physical effects include filtration of particles, reduction in turbulence, stabilisation of sediments, providing an increased surface area for biofilm growth on stems, leaves, roots, rhizomes. Metabolic effects include nutrient uptake and oxygen release from the roots. Although nutrients are essential for plant growth Brix (1997) has indicated that the amount of nutrients removed and stored as plant biomass is only quantitatively important in surface flow wetland systems receiving low nutrient loadings.

As part of a pilot study on the performance efficiency of constructed surface flow wetland systems in Queensland, Australia (Greenway and Woolley, 1999), a wetland was commissioned at Cairns to polish secondary treated municipal effluent. The aim of the research presented in this paper was to monitor nutrient removal at 2 different loading rates over a 3 year period and to determine the proportion of nutrient removal that could be attributed to direct uptake by macrophytes and incorporated into plant biomass.

Materials and methods

The constructed pilot wetland at Edmonton Sewage Treatment Plant, 20 km south of Cairns, north eastern Australia, is a free water surface treatment wetland. It was constructed in April 1994 and consists of 3 linear channels: Channels 1 and 2 are 65 m long and 5 m wide (325 m^2), Channel 3 is 65 m long and 15 m wide (975 m^2), the channels are designed for a maximum water depth of 0.5 m.

The initial plant species selected were the emergent macrophytes, *Typha domingensis, Schoenoplectus validus, Eleocharis equisetina, Eleocharis sphacelata,* and the floating leaved attached macrophytes *Marsilea* (a fern) and the water lilies *Nymphaea gigantea* and *Nymphoides indica*. All species are native to Australia and were collected from local stock growing in natural wetlands and/or waterways and transplanted in the constructed wetland. The plants were band planted in sections separated by open water sections. The planting densities were: *Typha* single shoot and rhizome 0.5/m^2, *Schoenoplectus* clumps 0.5/m^2, *Eleocharis* clumps 1/m^2, *Marsilea* clumps 1/m^2 along edge, *Nymphoides* and *Nymphea* single plants 0.5/m^2. The free floating macrophytes duckweed (*Spirodela* spp, W*olffia* spp) immediately colonised the open water sections.

The constructed wetland was fully commissioned and secondary effluent introduced by the end of August. The upstream treatment process at Edmonton used an oxidation ditch to aerate the secondary effluent stream enough to oxidise most of the ammonium to nitrite/nitrate. Influent water quality was NH_4-N (0.5 mg/L), NO_x-N (5.3 mgN/L), TN (6.1 mg/L), Filterable Reactive Phosphorus (7.5 mg/L), TP (7.8 mg/L). Flow rate was 0.12 L/s, 0.22 L/s and 0.56 L/s in channels 1, 2 and 3 respectively. As part of the experimental design of this pilot wetland, Channels 2 and 3 had equivalent loading rates (median HLR 500 m^3/ha/d; NH_4–N 0.1 kg/ha/d, NO_x-N 2.7 kg/ha/d, TN 3.7 kg/ha/d, FRP 3.3 kg/ha/d, TP 3.4 kg/ha/d) and 10 days retention time. Channel 1 had a median HLR 300 m^3/ha/d (NH_4–N 0.1 kg/ha/d, NO_x-N 1.6 kg/ha/d, TN 2.4 kg/ha/d, FRP and TP 2 kg/ha/d) and 16 days retention.

Water quality monitoring

Influent and effluent monitoring was conducted by the staff at the Treatment Plant. Between November 1994 and June 1997 samples were collected on a weekly basis and analysed by Cairns City Council for TN, NO_x, NH_4, TP, FRP. Flow was monitored by daily readings of V-Notch wiers at the inlet and outlet of each channel.

Vegetation mapping

Vegetation mapping and photographic recording were conducted monthly between Aug 1994 and Sept 1996, then at 6 month intervals in order to monitor the successful growth and spread of transplanted species and the invasion and colonisation of new species. Vegetation maps were scanned and the area of each plant species determined using a GIS computer package.

Biomass

In February 1997 plant biomass was estimated for each plant species growing in different sections of each channel. 1 m^2 and 0.25 m^2 quadrats were used as the sampling units and for each plant species there were 5 replicates. All plant components (shoots, roots, rhizomes) were removed from the sampling unit. A net was used to harvest floating macrophytes (duckweed and *Azolla*) from sections of open water and amongst emergent macrophytes. Plant components were oven dried at 75°C for 72 hours and weighed. Plant nutrient concentrations were analysed for each component (Greenway and Woolley, 1997).

Harvesting

Following sampling in February 1997, harvesting of vegetation occurred in channels 1 and 2. The emergent leaves/shoots of all plants at or above the water level were harvested. Using shears the shoots of emergent macrophytes were cut at 5–10 cm above the water level. All leaves of floating leaved macrophytes were harvested. Floating rafts of *Paspalum* were totally removed. Floating duckweed and *Azolla* were removed using a net, leaving between 2–5% of the original biomass. The submerged macrophyte *Ceratophyllum* was manually removed leaving about 10% of original biomass. Channel 3 was not harvested. In June 1997 after 17 weeks regrowth, the vegetation was remapped and plant biomass determined.

Results and discussion

Monthly observations of the constructed wetland from the initial planting showed that dense plant cover was achieved within 5 to 6 months of transplanting individual plant shoots or clumps. Growth rates from the biomass harvesting study (Table 4) showed that shoot regrowth could attain preharvest biomass values within 5 to 6 months. Thus, in order to determine the proportion of nutrient removal that could be attributed to plant uptake and incorporation into plant biomass, nutrient mass balance and plant biomass was calculated over 6 discrete (5 or 6 month) periods: September 1994–January 1995 (Feb 95); February 1995–June 1995 (July 95); July 1995–January 1996 (Feb 96); February 1996–July 1996 (Aug 96); August 1996–January 1997 (Feb 97) and February 1997–June 1997 (July 97). Plant biomass was calculated based on the vegetation maps produced at the end of each period i.e. Feb 95, July 95, Feb 96, Aug 96, Feb 97 and July 97.

Water quality

Nitrogen and phosphorus loadings in the 3 channels over six consecutive periods are presented in Tables 1 and 2. Mass removal efficiency for TN was highest during the first 5 months, however NO_x was lowest. This was due to some very high NO_x concentrations (8–10mgL^{-1}) in the outlet effluent (Nov–Dec1994) possibly from analytical error. Consequently calculated loadings of NO_x are higher than TN. From January 1995 through to June 1997 the mass removal efficiency of NO_x was consistently almost 100%. Removal efficiencies for TN declined over time. Mass removal efficiency of NH_4-N in the wetland was highly variable due to low background concentrations (median NH_4-N 0.5 mg/L) and an extremely low loading rate of 0.1 kg/ha/day. Wetland nitrogen processes including ammonification would have contributed to occasional export.

Over the 3 year monitoring period mass removal efficiencies for TN were 85%, 87% and 81% in Channels 1, 2 and 3 respectively. NO_x removal (excluding the first period) was 97%, 98% and 97% respectively. By contrast NH_4 removal was lower 61%, 74% and 21% respectively. Daily mass removal rates for TN were 2.95 kg ha^{-1} d^{-1} in Channel 1, 4.93 kg ha^{-1} d^{-1} in Channel 2 and 4.12 kg ha^{-1} d^{-1} in Channel 3.

The mass removal efficiency of TP and FRP follow similar trends within each channel with highest removal occurring during the first 10 months in Channel 1 (49% Sept 94–Jan 95; 41% Feb 95–June 95) which had a P loading rate of approximately 60% of that of Channels 2 and 3. The lowest removal efficiencies (< 10%) were between Aug 96–Jan 97 in channels 1 and 3, but during this period there was no flow between mid August–end October and these wetland channels dried out. The drying out process may have facilitated the remobilisation of inorganic phosphate. Channel 2 retained more water throughout that period and maintained a 20% removal efficiency. Removal efficiency increased considerably post harvest in channel 1, but to a lesser extent in channel 2 (channel 3 was not harvested).

Over the 3 years monitoring mass removal efficiencies were 26%, 28% and 21% for TP, and 27%, 28% and 20% for FRP respectively in channels 1, 2 and 3. Daily mass removal rates for TP were 0.84 kg, 1.43 kg and 0.98 kg ha^{-1} d^{-1} in channels 1, 2 and 3.

Plant biomass standing stock

Table 3 provides information on whole plant biomass (shoots, rhizomes/roots) from the February 1997 study. Nutrient storage was determined from mean P and N content of the different plant components. *Typha* single shoot biomass was determined from the average of young, mature and old shoots, including below ground root and rhizome material. The average density of shoots was used to calculate biomass per m^2. In February 1995 shoot density averaged 6/m^2 from an initial planting density of 0.5/m^2 in August 1994. By July 1995 average shoot density was around 14/m^2 in the original planted sections. In February 1997 average shoot density was still 14/m^2 indicating a sustainable standing stock.

For the other emergent species shoot density could not be used to estimate biomass due to the clumping nature of the plants and the high variability of number of stems (leaves) arising from a single "shoot". For *Eleocharis* spp. a value of cover i.e. dense, mid dense or sparse, together with whole quadrat harvesting was used to determine biomass. For

Table 1 Nitrogen loading (kgN), efficiency of mass removal (%) and removal rate (kg ha^{-1} d^{-1}) in each wetland channel at Cairns over 3 years

Period	Days Flow	Channel	Total N In (kg)	Total N Out (kg)	Removal %	Removal kg ha^{-1} d^{-1}	NO$_x$-N Out (kg)	NO$_x$-N In (kg)	% Removal	NH$_4$-N In (kg)	NH$_4$-N Out (kg)
Feb 95*	153	1	16.65	0.61	96	3.22	5.77	2.62	55	0.23	0.08
(Sep 94–Jan 95)		2	36.54	3.60	90	6.64	12.68	9.39	26	0.51	0.30
		3	98.48	10.00	90	5.93	33.21	10.82	67	1.34	4.36
export											
July 95	150	1	16.76	1.32	92	3.16	10.71	0.42	96	1.58	0.25
(Feb 95–June 95)		2	29.97	3.40	89	5.46	18.12	1.19	93	2.80	0.46
		3	77.86	11.64	85	4.53	47.19	5.79	88	7.28	3.16
Feb 96	215	1	25.91	3.10	88	3.26	13.31	0.11	99	2.91	0.47
(July 95–Jan 96)		2	46.85	6.56	86	5.77	23.67	0.04	100	5.14	1.10
		3	107.29	22.47	79	4.07	56.10	0.36	99	12.60	8.02
Aug 96	182	1	22.84	3.90	83	3.20	10.11	0.06	99	0.80	0.44
(Feb 96–July 96)		2	30.80	4.22	86	4.50	13.42	0.03	100	0.90	0.28
		3	92.84	15.93	83	4.33	40.43	0.34	99	2.78	2.11
Feb 97** export	113	1	13.53	4.02	70	2.59	7.74	0.39	95	0.66	1.03
(Aug 96–Jan 97) 0.34		*** 2	19.34	2.39	88	4.63	11.76	0.06	100		0.95
		3	56.27	16.16	71	3.64	32.36	0.11	100	2.57	1.16
July 97	150	1	13.33	3.86	71	1.95	7.70	0.38	95	0.37	0.27
(Feb 97–June 97)		2	14.14	3.21	77	2.23	7.99	0.08	99	0.38	0.31
		3	41.79	12.15	71	2.02	23.75	0.18	99	1.19	1.41
export											
Total	963	1	109.02	16.81	85	2.95	55.34	4.02	93	6.55	2.54
Sep 94–June 97							+(49.57)	+(1.40)		+(97)	
		2	177.64	23.38	87	4.93	87.64	10.79	88	10.68	2.79
							+(74.96)	+(1.40)		+(98)	
		3	475.03	88.35	81	4.12	275.61	17.60	94	27.76	20.22
							+(242.40)	+(6.78)		+(97)	

* For the first period (Sep 94–Feb 95): Higher calculated loadings of NO$_x$ out than TN out reflects the analysis methods: TN, NO$_x$ and NH$_4$ were measured independently, and the medians calculated for each parameter independently. There were a few high NO$_x$ conc measured in the effluents in Nov and Dec.
[It is more common to measure TKN, NH$_4$ and to calculate TN from TKN and NO$_x$.]
** No flow 17 Aug–26 Oct (commissioning new STP) +Values in brackets are exclusive of the first period (Sep 94–Jan 95)

Table 2 Phosphorus loading (kgP), efficiency of mass removal (%) and removal rate (kg ha^{-1} d^{-1}) in each wetland channel at Cairns over 3 years

Period	Days Flow	Channel	Tot P in (kg)	Tot P out (kg)	%	Removal kg ha^{-1} d^{-1}	FRP In (kg)	FRP Out (kg)	%	Removal kg ha^{-1} d^{-1}
Feb 95 (Sep 94–Jan 95)	153	1	11.96	6.06	49	1.19	7.56	3.91	48	0.73
		2	26.65	20.39	23	1.27	16.82	12.93	23	0.78
		3	78.65	59.44	24	1.29	48.55	34.43	29	0.95
July 95 (Feb 95–June 95)	150	1	12.71	7.54	41	1.06	11.70	6.93	41	0.98
		2	22.46	14.46	36	1.64	20.85	13.44	23	1.52
		3	58.79	44.33	26	0.99	54.61	41.19	25	0.92
Feb 96 (July 95–Jan 96)	215	1	20.44	14.17	31	0.90	20.20	14.02	16	0.87
		2	36.88	26.62	28	1.47	36.37	26.35	35	1.43
		3	85.75	71.20	17	0.69	84.08	69.79	25	0.68
Aug 96 (Feb 96–July 96)	182	1	25.68	20.79	19	0.83	17.32	12.68	31	0.78
		2	33.90	23.01	32	1.73	22.37	15.78	28	1.11
		3	102.77	67.25	35	2.00	68.47	51.66	25	0.95
Feb 97** (Aug 96–Jan 97)	113 **	1	12.67	12.06	5	0.17	11.29	10.95	3	0.09
		2	19.74	15.54	21	1.14	17.73	13.78	22	1.08
		3	53.18	51.18	4	0.18	47.36	42.63	10	0.43
July 97 (Feb 97–June 97)	150	1	19.49	16.06	18	0.70	17.32	14.18	18	0.64
		2	20.62	15.57	25	1.04	18.30	13.58	26	0.97
		3	64.05	57.28	11	0.46	56.82	49.98	12	0.47
Total (Sep 94–June 97)	963	1	102.95	76.68	26	0.84	85.39	62.66	27	0.73
		2	160.25	115.56	28	1.43	132.97	95.86	28	1.19
		3	443.19	350.68	21	0.98	359.89	289.68	20	0.75

** No flow 17 Aug–26 Oct (commissioning new STP)

Table 3 Plant biomass (g dry wt m^2), nutrient content and storage in selected macrophytes at Cairns wetland. Mean biomass values used to calculate P and N storage

Species	Biomass (g dw/m^2)	P Content (mg P/g dw)	P Storage (g/m^2)	N Content (mg N/g dw)	N Storage (g/m^2)
Typha orientalis 1 shoot	125 ± 75		0.48/shoot		1.69/shoot
6/m^2 (Feb 95)	750	3.85	2.89	13.5	10.12
14/m^2 (Feb 97)	1,750		6.74		23.63
Eleocharis dense	1,000 ± 250	4.2	4.20	15	15.00
mid dense	500 ± 300	4.2	2.10	15	7.50
sparse	300 ± 140	4.2	1.26	15	4.50
Schoenoplectus validus	800 ± 500	3.5	2.80	14.5	11.60
S. validus (among *Typha*)	360 ± 390	3.5	1.01	14.5	5.22
Marsilea spp	270 young	9.5	2.57	27	7.29
	370 mature	9.5	3.52	27	9.99
	470 old	7.27	3.42	19.23	9.04
Nymphoides indica	83 ± 20	8.2	0.68	22	1.83
Paspalum distichum	860 ± 110	2.78	2.39	15.46	13.30
Alternathera philoxeroides	780 ± 170	3.2	2.50	16	12.48
Duckweed (open water)	40 ± 10	14.43	0.58	42.88	1.72
Azolla/Duckweed (open water)	33 ± 7	8	0.26	41.18	1.36
Ceratophyllum demersum	90 ± 30	18.95	1.71	30.47	2.74

Marsilea spp. biomass was determined for young, mature and old (senescing) stands. Although biomass was higher for old stands the tissue nutrient content was lower. Biomass for floating macrophytes – duckweed (*Spirodela* and *Wolffia*) and *Azolla* was 100% higher in sections of open water compared to where the plants grew among emergents.

Plant growth and turnover rates

By comparing plant biomass before harvesting (Feb 1997) and after 17 weeks of regrowth

(June 1997) it is possible to obtain estimates for plant growth rates (and nutrient removal) and turnover time. Table 4 provides this data for *Typha, Schoenoplectus* and *Eleocharis*. The results indicate that initial total biomass standing stock would be achieved in 5 months for *Typha*, 10 months for *Schoenoplectus* and 2 months for *Eleocharis*. Cropping appears to invigorate growth in *Eleocharis* but retard growth in *Schoenoplectus*. Harvestable shoot yield would be achieved in 6 months for *Typha* and *Schoenoplectus* and 3 months for *Eleocharis*. Stand stock and growth rates for *Typha* are within the range of values reported by International Water Association (2000).

Estimation of wetland plant biomass and nutrient removal

The total area of each plant species in each channel at the 6 periods and the biomass values in Table 3 were used to calculate total nutrient storage in plant biomass:

$$\text{Nutrient storage (kg)} = \frac{\text{Area}(m^2) \times \text{Nutrient Content (mg P or N)}}{1.0 \times 10^6 \, mg/kg}$$

These estimates assume a 5–6 month turnover of plant biomass for emergent macrophytes. Given the rate of spread of *Typha*, *Eleocharis* and *Marselia* this is probably the case, but *Schoenoplectus* did not display such rigorous growth.

Turnover rates for floating macrophytes however can be in the order of days or weeks. Under optimal conditions duckweed biomass can double in 24 hours (Landolt, 1996), if this ocurred at Cairns duckweed would produce 40 g m^{-2} d dry weight which is twice the theoretical maximum predicted by Landolt. Rejmankova *et al.* (1990) predicted optimum growth of 5.9 g m^{-2} d by removing 25% of duckweed cover every 4 days and 2.1 g m^{-2} by removing 75% of cover. Doubling times of 7–12 days have been reported in temperate (10–15°C) wastewater ponds (Ozimek, 1996). Assuming a 7 day turnover at Cairns, daily growth would be 5.71 g m^{-2} d, the optimum predicted by the Rejmankova *et al.* model at 20°C. Without any harvesting however optimal growth is unlikely, thus nutrient storage in duckweed for each period was calculated on a 14 day turnover basis.

The total amount of nutrients storage in each wetland channel was determined by adding all values for the different plant species. Table 5 shows the amount of N and P in plant biomass production for each channel over the 6 periods, and the total for 34 months. Biomass production varied between channels and periods due to the different composition of plant species and the extent of duckweed. Over the 3 year period emergent macrophytes notably *Typha* spread to colonise open water sections and out compete other species such as *Schoenoplectus* and *Marsilea*. Duckweed however always occurred amongst the *Typha*. Biomass production rates were higher in the February to July periods. Due to the wetland channels partially drying out in September and October 1996 – rates of growth could not accurately be calculated over this period. The particularly high growth rates following harvesting of channels 1 and 2 in February 1997 indicate the vigor of the regrowth.

Table 4 Comparison of initial standing stock i.e. total biomass and harvestable shoot biomass (Feb 1997) and after 17 weeks regrowth (June 1997) and growth rates, in *Typha, Schoenoplectus* and *Eleocharis*

Species	Feb 1997 biomass gm^{-2}	June 1997 biomass gm^{-2}	% regrowth	growth rate kg m^{-2}yr^{-1}	growth rate g P or g N m^{-2} yr^{-1}
Typha total biomass	1750 ± 500	1300 ± 700	74%	3.977 kg	15.3 g P; 53.7 g N
Typha shoots biomass	1120 ± 320	740 ± 125	67%	2.264 kg	8.72 g P; 30.56 g N
Schoenoplectus total	800 ± 525	330 ± 200	41%	1.009 kg	3.53 g P; 14.63 g N
Schoenoplectus shoot	300 ± 170	190 ± 136	63%	0.581 kg	2.03 g P; 8.42 g N
Eleocharis total	500 ± 140	1020 ± 230	200%	3.120 kg	13.1 g P; 46.8 g N
Eleocharis shoot	230 ± 50	300 ± 70	130%	0.918 kg	3.86 g P; 13.77 g N

Table 5 Mass balance of nutrient removal (TN and FRP) and nitrogen and phosphorus bioaccumulation in plant biomass production in each wetland channel at Cairns over 3 years

Period		Removal TN kg	Biomass Production kg N		% N in plant biomass	Removal FRP kg	Biomass Production kg P		% P in plant biomass
			Total	kg ha⁻¹ d⁻¹			Total	kg ha⁻¹ d⁻¹	
Feb 95	1	16.0	4.70	0.94	29	3.65	1.50	0.30	41.1
5 months	2	33.0	4.87	0.98	14.5	3.89	1.51	0.30	38.8
Sep94/Jan 95	3	88.5	17.01	1.14	19.2	14.12	6.61	0.44	46.8
July 95	1	15.4	9.43	1.93	61	4.77	3.83	0.68	70
5 months	2	26.6	5.10	1.05	19	7.41	1.76	0.36	24
Feb/Jun 95	3	66.2	17.98	1.23	27	13.42	6.28	0.43	47
Feb 96	1	22.8	7.82	1.12	34	6.10	2.75	0.39	45
7 months	2	40.3	6.77	0.97	17	10.02	2.42	0.35	24
Jul 95/Jan 96	3	85.3	15.17	0.72	18	14.29	4.61	0.22	32
Aug 96	1	18.9	8.39	1.42	44	4.64	2.67	0.45	58
6 months	2	26.6	8.38	1.42	32	6.59	2.98	0.50	45
Feb/July 96	3	76.9	19.48	1.10	25	16.81	6.82	0.38	39
Feb 97*	1	9.5*	6.16		64	0.34*	2.14		629
6 months	2	17.0*	7.85	NC	46	3.95*	2.93	NC	74
Aug 96/Jan 97	3	40.1*	16.21		40	4.73*	5.81		123
July 97	1**	9.5	7.14	1.46	75	3.14	2.53	0.52	80
5 months	2**	10.9	8.76	1.80	80	4.72	3.15	0.65	67
Feb/June97	3	29.6	19.22	1.31	65	6.84	6.98	0.48	102
Sept 94	1	92.21	43.64	1.39	47	22.73	14.92	0.48	65.6
July 97	2	154.26	41.73	1.33	27	33.11	14.75	0.47	44.5
34 months	3	386.68	105.07	1.12	27	70.21	37.11	0.40	47.2

* No flow 17 Aug–26 Oct 1996 (mass balance for 4.5 months only); ** post harvest biomass; NC – not calculated due to wetland partially drying out

These biomass production values can be used as an indicator to estimate the nutrient uptake capacity of the plants. Over the 3 years uptake capacity was 162 kg P, 474 kg N ha⁻¹ year⁻¹ in Channel 1, 160 kg P, 453 kg N ha⁻¹ year⁻² in Channel 2, 134 kg P, 380 kg N ha⁻¹ year⁻² in Channel 3. The uptake capacity of emergent macrophytes is in the range of 30–150 kg P ha⁻¹ y⁻¹ and 200–2500 kg N ha¹ y⁻¹ (Brix, 1997). In the current study the presence of duckweed in open water sections and among emergents would account for the higher phosphorus values, and the P:N ratios of almost 1:3.

By comparing the mass removal of N and P from the effluent with the N and P content in plant biomass it is possible to get some indication of how much N and P may have been removed directly by the plants themselves. During the first two years the percentage nitrogen removal as plant biomass was consistently higher in channel 1 which had the lowest loading rate. Between Feb–July 97 post harvest regrowth in the two harvested channels accounted for 75% (channel 1) and 80% (channel 2) of nitrogen removal, and 65% in channel 3. These high values suggest nitrogen removal is optimal in the Cairns wetland. Previous studies on nitrogen removal by emergent macrophytes suggest that biomass production rarely exceeds 10% of total removed nitrogen (IWA, 2000). Phosphorus removal by plant tissue was also higher in channel 1 and over the first 2 years ranged from 41–70% of FRP lost from the wetland. Post harvest regrowth accounted for 80% of FRP removal in channel 1 and 67% in channel 2, however during the same period plant production in channel 3 accounted for all removal of FRP. Again these results indicate the high nutrient removal capacity of wetland plants in the Cairns surface flow wetland.

Conclusions

The constructed surface flow wetland in Cairns has demonstrated that a mixture of emergent macrophytes and duckweed can contribute significantly to the removal of nitrogen and phosphorus from secondary treated effluent with concentrations of NO_x 5.3 mg N L^{-1} and loadings of 1.6–2.7 kg N ha^{-1} d^{-1} and FRP 7.5 mg P L^{-1} and loadings of 2.0–3.4 kg P ha^{-1} d^{-1}. As plant biomass increased over the 3 years, production rate and nutrient uptake capacity also increased. The gradual displacement of open water section of highly productive duckweed with larger biomass emergent macrophytes did not affect the nutrient uptake capacity of the wetland system. This study however has demonstrated that nutrient removal by wetland plants is maximised by maintaining a variety of macrophyte types and species.

Acknowledgements

Funding for this project was provided through a joint Griffith University and Queensland Dept of Natural Resources collaborative grant. The constant support of Darryl Ahlers, operator at the Edmonton Sewage Treatment Plant provided us with the water quality data. Through the diligence of Rud Westaway, Qld Dept Natural Resources, we have a comprehensive series of vegetation maps and photographic records over the 3 years.

References

Brix, H. (1997). Do macrophytes play a role in constructed wetlands? *Wat. Sci.Tech.* **35**(5), 11–17.

Greenway, M. and Woolley, A. (1997). Constructed wetlands in Queensland: Performance efficiency and nutrient bioaccumulation. *Ecological Engineering.* **12**, 39–55.

Greenway, M. and Woolley, A. (1999). Nutrient content of wetland plants in constructed wetlands receiving municipal effluent in tropical Australia. *Water. Sci. Tech.* **35**, 135–142.

International Water Association (2000). *Constructed Wetlands for Pollution Control: processes, performance, design and operation.* IWA Specialist Group on use of Macrophytes in Water Pollution Control. IWA Publishing.

Landolt, E. (1996). Duckweeds (Lemnaceae): Morphological and ecological characteristics and their potential for recycling of nutrients. Environmental Research Forum, 5–6, 289–296.

Ozimek (1996). Usefulness of *Lemna minor* in wastewater treatment in temperate climates – Myth or fact? Environmental Research Forum, 5–6, 297–302.

Rejmankova, E., Kvet, J. and Rejmanek, M. (1990). Maximizing duckweed (Lemnaceae) production by suitable harvest strategy. In Whigham, D.F., Good, R.E. and Kvet, J. (eds), *Wetland Ecology and Management: Case Studies.* pp. 39–43. Kluwer Academic Publishers, Den Haag, The Netherlands.

Performance of two macrophyte species in experimental wetlands receiving variable loads of anaerobically treated municipal wastewater

D.M.L. da Motta Marques*, G.R. Leite and S.G.T. Giovannini

* Instituto de Pesquisas Hidráulicas, Universidade Federal do Rio Grande do Sul, Avenida Bento Gonçalves, 9500, Caixa Postal 15029, CEP: 91501-970, Porto Alegre, RS, Brazil. (*E-mail: dmm@iph.ufrgs.br*)

Abstract Two highly productive emergent macrophytes, *Zizaniopsis bonariensis* and *Typha subulata* were established in experimental subsurface flow, sand-based wetlands receiving anaerobically treated municipal wastewater. The hydraulic loading rate was tested in two levels, sequentially, 6.8 cm.d^{-1} and 13.6 cm. d^{-1}, for 70 days each. In the 13.6 cm. d^{-1}-loading treatment, among all monitored variables only COD, PO_4-Total-P, and Turbidity were removed more efficiently by the planted beds in comparison to unplanted sand beds (P<0.001). When the 6.7 cm. d^{-1}-loading rate was applied no significant improvement in removal was found comparing macrophyte beds to unplanted sand beds, except for PO_4-P. *T. subulata* beds were significantly more efficient than *Z.bonariensis* beds for most of the variables. The highest significant differences (P< 0.001) were related to the main effect of the factor *hydraulic loading rate*, with decreased removal for increased load when considering the variables Total Coliforms (99.4% to 87.7%), Fecal Coliforms (100% to 89.7%), NH_3-N (95.8% to 55.2%), NO_3^--N (–54% to –396%), Total-N (90.4% to 59.6%), and TSS (86.1% to 46.1%). The performance similarity of planted and unplanted wetland beds in the lower loading condition, except for PO_4-P, indicates that plants may not be needed under low loading. However aquatic macrophytes improved wetland efficiency under high loading.

Keywords Constructed; loading; performance; *Typha subulata*; wetland; *Zizaniopsis bonariensis*

Introduction

Different aquatic macrophytes species, genera or even functional types may be considered as a fundamental factor in treatment wetlands. However, the detection of significant differences comparing macrophyte performance may be a matter of controlling and optimizing a number of other factors before. The response variation among aquatic macrophytes is related to their variable response and needs in relation to non-biotic factors – such as substrate, water, air, chemical content and physical properties, specially water vertical and horizontal movement amplitude and rate. Biotic factors in treatment wetlands may rise from the action of macrophytes as a consequence of having varying intraspecific or interspecific competition.

The influence of all these non-controlled or semi-controlled biotic and non-biotic factors makes the biological productivity highly variable among wetlands, even when they have the same aquatic macrophytes species. Different macrophytes performance can only be compared if all other factors are controlled and/or constant. Significant differences among macrophytes and/or macrophyte and unplanted beds removal efficiency in controlled experiments have been found (Burgoon *et al.*, 1989; Tanner, 1996; Soto *et al.*, 1999). Generally the significance of the results is not evaluated and an adequate data treatment is rarely used, resulting that most of the conclusions about performance come from inventory/monitoring studies (Watson *et al.*, 1989, Kadlec, 1996), with the affecting factors being presumed but not controlled.

The species used in this experiment, *Typha subulata* and *Zizaniopsis bonariensis*, can establish and survive over a wide variety of conditions although their optimum habitats are not known yet. Z. bonariensis has an ecological importance as it dominates thousands of hectares in Taim wetland; a 33,000 ha ecological preserve in Southern Brazil, that is habitat for native cayman, capybara, nutria and several native or migratory waterfowl species. Its auto-ecology and response to several factors have been studied (Motta Marques et al., 1997; Motta Marques and Giovannini, 1997, 1998; Giovannini and Motta Marques, 1998).

Comparative studies of Z. bonariensis and T. subulata have been done in relation to their responses to water level (Giovannini and Motta Marques, 1999), and to petrochemical effluent (Campagna, 1998). Individual species studies have been done to evaluate their responses to substrate and water nutrient (Giovannini and Motta Marques, 1998).

The specific objectives of this study were: (i) compare the performance of *Typha subulata* and *Zizaniopsis bonariensis* in sand substrate wetlands; (ii) Compare planted and unplanted beds; (iii) compare the performance of these species and the sand filter under different hydraulic loading rates.

Methods

The experiment was carried out in the Southern Hemisphere at the *Estação de Tratamento de Esgoto* (sewage treatment plant) *do Parque da Matriz,* Porto Alegre-RS, Brazil, in spring and summer. The local climate is Koeppen's fundamental Cfa 1. The annual mean temperatures and precipitation vary from 16.5 C to 17.4 C and from 1186 to 1364 mm, respectively (Brasil, 1973).

The experiment was carried out in twelve experimental units (1.90 m × 1.28 m each), filled with a 25 cm – sand layer (40% of initial porosity) where the treatments were allocated. The first tested factor was (A) *Species* with four qualitative levels: (a_1) unplanted, (a_2) *Zizaniopsis bonariensis* (16 cuts.m^{-2}), (a_3) *Typha subulata* (16 cuts.m^{-2}), and (a_4) mixed *Z.bonariensis* with *T.subulata* (16 cuts.m^{-2}) (Giovannini and Motta Marques, 1999). The second factor (B) was *hydraulic loading rate* at two levels: 6.8 cm.d^{-1} (b_1), and 13.6 cm.d^{-1} (b_2) tested sequentially, for 70 days each. In this experiment the factor *hydraulic loading rate* was associated with the factor *hydraulic retention time*, meaning that doubling the loading resulted in reducing to half the *hydraulic retention time*.

The experimental units were batch loaded, with anaerobically treated municipal wastewater. The average Total-N influent concentration was 66.9 mg.L^{-1}, supplying 4482.5 mg.m^{-2}.d^{-1} in the 6.8 cm.d^{-1}-hydraulic loading rate, and 8965 mg.m^{-2}.d^{-1} in the 13.6 cm.d^{-1}. The average Total-P influent concentration was 15 mg/L, supplying 1005.2 mg.m^{-2}.d^{-1} in the 6.8 cm.d^{-1}, and 2010.4 mg.m^{-2}.d^{-1} in the 13.6 cm.d^{-1}. Irrigation was done four times a day with one fourth of the total daily load applied each irrigation time. The influent was applied superficially in an inlet ditch at one side (1.28 cm wide). An outlet drainage pipe, in the bottom of the bed at the inflow opposite side, collected the outflow.

Monitoring was done at 15 days intervals, and sampling was done at the time to have 1.5 day of theoretical retention time associated to the 6.8 cm.d^{-1} hydraulic loading rate and 0.75 day to the 13.6 cm.d^{-1}. The monitored variables were Alkalinity, Cl, Fecal Coliforms, Total Coliforms, Conductivity, Color, COD, NH_3-N, NO_3-N, Total-N, Total-P, Ortho-P, Turbidity, and TSS (APHA, 1995).

The analyzed variables were removal efficiencies given by (influent concentration-effluent concentration/influent concentration)×100 of the monitored variables. The experiment had a factorial design in a split-plot treatment allocation with three whole-plot replicates. The repeated measures condition was eliminated by averaging measures collected along time instead of considering time as a factor. The variables were analyzed by ANOVA and means were compared by Least Significant Difference (LSD) procedure (Cochran and Cox, 1964). Since the experiment was run outside, and climate conditions were not controlled, the rain events were measured and used as cofactor in the analysis.

Results and discussion

The results of comparisons of treatment means for removal efficiency and significance of main effects, and interactions of factor *species* and factor *hydraulic loading rate*, are in Table 1 and Table 2, respectively. Very high significant differences (P < 0.001) were related to the main effect of the factor *hydraulic loading rate*, with decreased removal for increased load when considering the variables Total Coliforms (99.4% to 87.7%), Fecal Coliforms (100% to 89.7%), NH_3-N (95.8% to 55.2%), NO_3^--N (–54% to –396%), Total-N (90.4% to 59.6%), and TSS (86.1% to 46.1%). Fecal and Total Coliforms did not depend on factor *species* but only on *hydraulic loading*, since all macrophyte and unplanted sand had the same trend of decreasing removal when load was increased.

Nitrate (NO_3-N), and Total-N were independently affected by the factors *species* and *loading*. For these variables the increase of loading affected every macrophyte and sand wetland bed in the same way, that is, decreasing the performance. The independent effect of factor *species* was significant to Total-N, where the macrophyte beds were more efficient than unplanted sand, and to NO_3-N where *Z. bonariensis* beds removal were lower than any other treatment. NH_3-N removal was the same in any macrophytes or unplanted sand.

Species vs. *Loading* interaction occurred for Alkalinity, Cl, Conductivity, Color, COD, PO_4-P, Total-P, and Turbidity, meaning the increased loads affected differently the different macrophytes and/or sand bed. In this sense, when the 6.7cm.d^{-1}-loading rate was applied, macrophyte beds in general were not significantly more efficient for removal than unplanted sand, except for PO_4-P. On the other hand the unplanted sand bed was significantly more efficient to remove Cl, and Color than *T. subulata*, and *Z. bonariensis* wetlands. Also, it was more efficient to remove Alkalinity and Conductivity than *Z. bonariensis*.

Under the 13.6 cm.d^{-1}-loading treatment, planted beds removed COD, PO_4-P Total-P and Turbidity more efficiently than the control unplanted sand beds did. Actually, Total-P and PO_4-P removals under unplanted sand treatments were lower than in any other treatment. It shows macrophytes had a better performance than the sand filter under higher load. Tanner *et al.* (1995) have also found better performance of planted wetlands in higher loads, in their case specifically to $CBOD_5$ removal using *Schoenoplectus validus*.

Table 1 Means of removal efficiency (%) and significance of the main effect of the factors *species* and *hydraulic loading rate* over the mean differences

Variables	F Prob. Covar.	Factor *Hydraulic Loading rate*							Factor *Species*		
		6.8 cm.d^{-1}	13.6 cm.d^{-1}	F Prob.	LSD	a_1	a_2	a_3	a_1+a_2	F prob.	LSD
Alkalinity	<.001	25.8	31.9	0.081	6.87	34.9 bc	13.2 a	40.6 c	26.7 b	<.001	9.63
Cl	0.367	49.5 b	19 a	<.001	8.47	41.1 b	30.8 ab	38.7 b	26.3 a	0.054	11.87
Fecal colif.	<.001	100.2 b	89.69 a	<.001	3.57	97.76	91.93	94.34	95.38	0.146	5.01
Total colif.	0.004	99.4 b	87.7 a	<.001	5.46	97.1	88.5	94.4	94.2	0.167	7.66
Conduct.	<.001	37.6 b	29.1 a	0.001	5.19	38.9 bc	16.9 a	45.7 c	31.9 b	<.001	7.28
Color	0.288	–248 a	–77 b	<.001	96	–35 c	–276 a	–123 bc	–216 ab	0.003	134.6
COD	<.001	71.4 b	37.5 a	<.001	13.82	32.9 a	54.2 b	70.2 b	60.5 b	0.002	19.39
NH_3-N	<.001	95.8 b	55.2 a	<.001	4.41	74	75.7	76.3	76	0.873	6.18
NO_3^--N	0.772	–54 b	–396 a	0.002	213.8	–128 b	–499 a	–94 b	–179 b	0.033	299.8
Total-N	<.001	90.4 b	59.6 a	<.001	5.60	69.4	76.9	79.6	74.2	0.070	7.85
Ortho-P	0.130	95.3 b	88.9 a	0.012	4.97	71.8 a	99.3 b	97.8 b	99.6 b	<.001	6.97
Total-P	0.002	93.3 b	20.9 a	<.001	30.80	26.5	59	78.8	64	0.111	61.16
Turbidity	<.001	86.1 b	46.1 a	<.001	16.82	42.2 a	64.6 ab	81.9 b	75.7 b	0.007	23.59
SS	<.001	10 b	–89 a	<.001	56	–22	–97	–15	–24	0.137	78.5

F. prob.: probability of F from ANOVA; LSD: Least Significant difference; a_1: unplanted; a_2: *Zizaniopsis bonariensis*; a_3: *Typha subulata*; a_4: *Z. bonariensis+T.subulata*; different letters after numbers shows significant differences (p = 0.05), being a<b<c

Table 2 Means and significance of interactions among factor *species* and factor *hydraulic loading rate*

Variables	F Prob. Cov.	Unplanted Sand		Z. bonariensis		T. subulata		Z. bonariensis + T. subulata		F prob. Interac.	LSD
		L0	L1	L0	L1	L0	L1	L0	L1		
Alkalinity	<.001	39.8 cd	30 bc	5.2 a	21.3 b	33.2 bc	48 d	25 b	28.4 bc	0.031	13.63
Cl	0.367	67.3 d	19.9 a	44.5 c	17.1 ab	45.4 c	31.9 bc	40.6 c	12 a	0.015	16.80
Fecal Coliforms	<.001	100	95.12	100	83.47	100	88.52	99.11	91.65	0.109	7.09
Total Coliforms	0.004	99.7	94.4	100	77	99.6	89.1	98.1	90.4	0.103	10.84
Conductivity	<.001	49.1 e	28.7 bc	19.4 ab	14.4 a	46.2 e	45.2 de	35.8 cd	28 bc	0.055	10.31
Color	0.288	27 d	−96 bcd	−372 a	−179 bc	−258 ab	12 d	−389 a	−44 cd	0.005	190.6
COD	<.001	76.3 c	−10.4 a	67.8 bc	43.7 b	75.5 c	64.8 bc	69.1 bc	51.8 bc	<.001	27.45
NH_3-N	<.001	94.9	53.1	97.3	54.1	93.1	59.6	98	54.1	0.316	8.76
NO_3-N	0.772	−79	−177	−130	−867	7	−195	−13	−345	0.169	424.5
Total-N	<.001	89.1	49.7	91	62.8	91.7	67.6	90	58.4	0.269	11.11
PO_4-P	0.130	84.7 b	58.8 a	99.2 c	99.5 c	98.1 c	97.4 c	92.2 bc	99.9 c	<.001	9.87
Total-P	0.002	95.6 c	−42.7 a	95.6 c	22.4 b	95.4 c	62.3 bc	86.5 c	41.4 bc	0.078	61.16
Turbidity	<.001	97.1 c	−12.7 a	76.5 bc	52.7 b	84.4 c	79.3 bc	86.4 c	65 c	<.001	33.41
SS	<.001	37	−81	−7	−187	11	−42	−2	−47	0.291	111.1

F Prob. Cov: Probability of F from ANOVA related to the significance of covariable influence. L0: 6.8 cm.d^{-1}; L1: 13.6 cm.d^{-1}; letters after means represent significant differences given by LSD (Least Significant Differences), "a" goes at the side of the lowest means, being a<b<c<d... F Prob. Interac.: Probability of F from ANOVA related to effect of interaction on means treatment

Phosphorus (PO_4-P) removal did not depend on the specific kind of macrophyte or on the applied loads, considering all the macrophyte treatments worked similarly and significantly better than the sand filter, in the two loads. On the other hand, Total-P removals were very high in the lower load and decreased by increasing load. Much more decrease occurred in unplanted sand bed in comparison to *T. subulata* beds, explaining the significant *species* vs. *loading* interaction.

The same reasoning is true for Total-N removals that were very high to be mostly due to macrophytes uptake. Tanner (1996) using lower loads, then the present ones, of Total-N and Total-P (3,347, and 496 mg.m^{-2}.d^{-1}, respectively) applied to different emergent macrophyte species, found mean removals of 79–93% of Total-P and 65–92% of Total-N. These figures are similar to those we found, with no significant species differences. Also they found maximum plant uptake around 30%. The possibility of root-zone oxygen release stimulation to nitrification with further denitrification in anaerobic sites (Reddy *et al.*, 1989) may explain most of Total-N removal in planted wetlands supplied under higher loads.

It is clear (and statistically significant) that planted wetland beds were more efficient than unplanted sand beds to remove PO_4-P, under the two loads tested, and in removing Total-P in the higher load (13.6 cm.d^{-1}). The removal mechanisms, however, were not clear. The observed removals were favored in some way by the aquatic macrophytes. This difference can be partially explained by the normal prompt uptake of PO_4-P by plants. However, the Total-P loads applied were much higher than any macrophyte (Reddy *et al.*, 1987) could uptake. Orthophosphate removal in planted wetlands occurs through three parallel paths, with reaction rates as sorption to substratum being higher than biofilm assimilation, which is much higher than macrophytes uptake. On the other hand the quantity of P removed by the three paths over months is mainly due to macrophytes (70%), for a range of 120 to 24,000 mg.m^{-2}.d^{-1} of Total-P removal (Lantzke *et al.*, 1999). Aquatic macrophytes P uptake as high as 16,800 mg/m^2/d looks unreliable. The performance of planted wetland beds will depend on load, but the overall performance seems to depend also on some aquatic macrophytes work, directly or indirectly, in the systems.

Conclusions

The removal similarity, except for PO_4-P, of planted and unplanted wetland beds under lower loading condition (6.8 cm.d^{-1}) indicates that plants may not be necessary to treat lower loads as found in this experiment. Paradoxically they are expected to be more efficient under these conditions. On the other hand, the better performance of planted wetland beds, in relation to COD, PO_4-P, Total-P, and Turbidity, under higher load (13.6cm.d^{-1}), demonstrated that plants are capable of improving significantly the wetland performance. The species *Zizaniopsis bonariensis* and *Typha subulata*, like other *Typha* spp., showed a high potential to be used in constructed wetlands.

Acknowledgements

To CNPq, CAPES, FINEP, and IPH-UFRGS for their financial support to authors research activities. This work is part of the programme PROSAB II.

References

APHA (1995). *Standard Methods for the Examination of Water and Wastewater*. 19th edn. American Public Health Association, Washington, DC, USA.

Brasil. Ministério da Agricultura. Departamento de Pesquisa Agropecuária. Divisão de Pesquisa Pedológica. (1973) *Levantamento de reconhecimento de solos do Rio Grande do Sul*. Recife. 431p.

Burgoon, P.S, Reddy, K.R. and DeBusk, T. (1989). Domestic wastewater treatment using emergent plants cultured in gravel and plastic substrates. In: *Constructed Wetlands for Wastewater Treatment*, D.A. Hammer (ed.). Lewis Publishers, Chelsea, Michigan, pp. 536–541.

Campagna, A.R. (1998). Efeito de Efluente de Refinaria de Petróleo sobre Macrófitas Aquáticas em Ecossistemas Construídos – Banhados. MSc Dissertation, Curso de Pós-graduação em Ecologia do Instituto de Biociências, Universidade do Rio Grande do Sul.

Cochran, W.G. and Cox, G.M.(1964). *Experimental Designs*. 2 ed. John Wiley, New York, 614 p.

Giovannini, S.G.T. and Motta Marques, D.M.L. da. (1998). O uso de subsolo como susbtrato para construção de banhados com macrófitas aquáticas emergentes sob diferentes regimes hídricos. *Acta Limnologica Brasiliensia*, 10(2), 71–81.

Giovannini, S.G.T. and Motta Marques, D.M.L. da. (1999). Establishment of Three emergent macrophytes under different water regimes. *Water Science and Technology*, **40**(3), 233–240.

Kadlec, R.H. (1996). *Treatment Wetlands*. CRC Press, Boca Raton, FL, 893 pp.

Lantzke, I.R., Mitchell, D.S., Heritage, A.D. and Sharma, K.P. (1999). A model of factors controlling orthophosphate removal in planted vertical flow wetlands. *Ecological Engineering*, **12**, 93–105.

Motta Marques, D.M.L. da. and Giovannini, S.G.T. (1997). Hydroperiod influence in the establishment of emergent macrophytes *Zizaniopsis bonariensis* and *Scirpus californicus* in experimental constructed wetlands. In: SIL Congress, Ireland, Dublin, 1997, p.106. in press Verh. Int. Ver. Limnol.

Motta Marques, D.M.L. da, Irgang, B. and Giovannini, S.G.T. (1997). A importância do hidroperíodo no gerenciamento de água em terras úmidas (wetlands) com uso múltiplo – o caso da estação ecológica do Taim. In: Simpósio Brasileiro De Recursos Hídricos, 7, Vitória, Anais 3. São Paulo, ABRH, 1–8.

Motta Marques, D.M.L. da and. Giovannini, S.G.T. (1998). Conditioning factors for establishment of *Zizaniopsis bonariensis* in constructed wetlands. In: *Proceedings 6th International Conference on Wetland Systems for Water Pollution Control*. São Paulo, 559–563.

Reddy, K.R. and De Busk, W.F. (1987). Nutrient storage capabilities of aquatic and wetland plants. In: *Aquatic Plants for Water Treatment and Resource Recovery*, K.R. Reddy and W.H. Smith (eds.), Magnolia Publishers, Orlando, Florida, p.337–357.

Reddy, K.R., Patrick, W.H. and Lindau, C.W. (1989). Nitrification-denitrification at the plant root-sediment interface in wetlands. *Limnology Oceanography*, **34**, 261–267.

Soto, F., Garcia, M., De Luís, E. and Bécares, E. (1999). Role of *Scirpus lacustris* in bacterial and nutrient removal from wastewater. *Water Science and Technology*, **40**(3), 41–247.

Tanner, C.C. (1996). Plants for constructed wetland treatment systems – A comparison of the growth and nutrient uptake of eight emergent species. *Ecological Engineering*, **7**, 59–83.

Tanner, C.C., Clayton, J.S. and Upsdell, M.P. (1995). Effect of loading rate and planting on treatment of dairy farm wastewaters in constructed wetlands – I. Removal of oxygen demand, suspended solids and faecal coliforms. *Water Research*, **29**(1), 17–26.

Watson, J.T., Sherwood, C.R., Kadlec, R.H. and Knight, R.L. (1989). Performance expectations and loading rates for constructed wetlands. In: *Constructed Wetlands for Wastewater Treatment*, D.A. Hammer (ed.). Lewis Publishers, Chelsea, Michigan, pp. 319–351.

Ecological characteristics of a natural wetland receiving secondary effluent

J.R. Martin*, R.A. Clarke, Jr.* and R.L. Knight**

* CH2M HILL, 3011 SW Williston Road, Gainesville, Florida 32614, U.S.A.
** Environmental Scientist, 2809 NW 161 Court, Gainesville, Florida 32609, U.S.A.

Abstract The Boot wetland treatment system is a 115-acre, hydrologically altered cypress-gum wetland in Polk County, Florida. The Poinciana Wastewater Treatment Plant No. 3 has discharged secondary effluent to the bermed Boot wetland since August 1984. Before that time this natural wetland had been affected adversely by forestry, drainage, and surrounding development which contributed to dying trees and a groundcover of invasive upland plants. In accordance with the Florida Department of Environmental Protection's Wetlands Application Rule (Chapter 62–611, F.A.C.), a routine biological and water quality monitoring program has been in effect since October 1990. Components of the biological monitoring program include surveys of canopy and subcanopy, herbaceous and shrub groundcover species, benthic macroinvertebrates, fish, and nuisance mosquitoes. Effluent addition to the Boot wetland has resulted in continuous wetland inundation with a typical water depth of 2.5 to 3.0 feet for the past 15 years. Dominance and density of trees has steadily increased, upland invader species were eliminated, and stable plant, fish, and invertebrate communities were established. The long term biological data from this treatment wetland is compared to data from other natural treatment wetlands and a control wetland.
Keywords Fish; macroinvertebrates; natural wetlands; treatment wetland; vegetation; wastewater effluent

Introduction

The Boot wetland is a 115-acre natural cypress dome that has been receiving secondarily treated municipal wastewater since 1984. Because the wetland had been completely bermed to support the retention and treatment of wastewater, the entire wetland has been continuously inundated with water depths typically ranging from 2.5 to 3 feet. Before 1985, forestry, drainage canals, and surrounding land development had adversely affected this wetland. In the early 1980s, peat was rapidly oxidizing, cypress trees were leaning and dying, and groundcover vegetation had changed to invasive upland plants (Breedlove, 1986). The Boot wetland now supports a healthy and growing stand of trees, diverse wetland groundcover species, and a stable population of fish and macroinvertebrates.

The Boot wetland is located in Poinciana, Polk County, Florida (Figure 1). It lies within the Marion Creek/London Creek watershed, which is a subbasin of the Reedy Creek and Kissimmee River basin. Severn Trent-Avatar Utility Services, LLC., operates the Poinciana Wastewater Treatment Plant No. 3 (WWTP No. 3) which discharges secondarily treated municipal wastewater via a submerged discharge pipe into the north end of the Boot wetland (Station 1). The treated wastewater discharge to the Boot wetland has been nearly continuous since August 1984. The wetland seasonally overflows from its south end to the M-7 drainage canal (Station 3).

As part of the permit to operate WWTP No.3, required monitoring of the treatment wetland was conducted in accordance with Florida Administrative Code Chapter 62–611 (Wetland Application), Table 1. The permit has allowed for the continuous discharge of up to 0.35 million gallons per day (mgd) from WWTP No. 3 to the Boot wetland. CH2M HILL has generally collected monthly water quality and quarterly biological data from the Boot

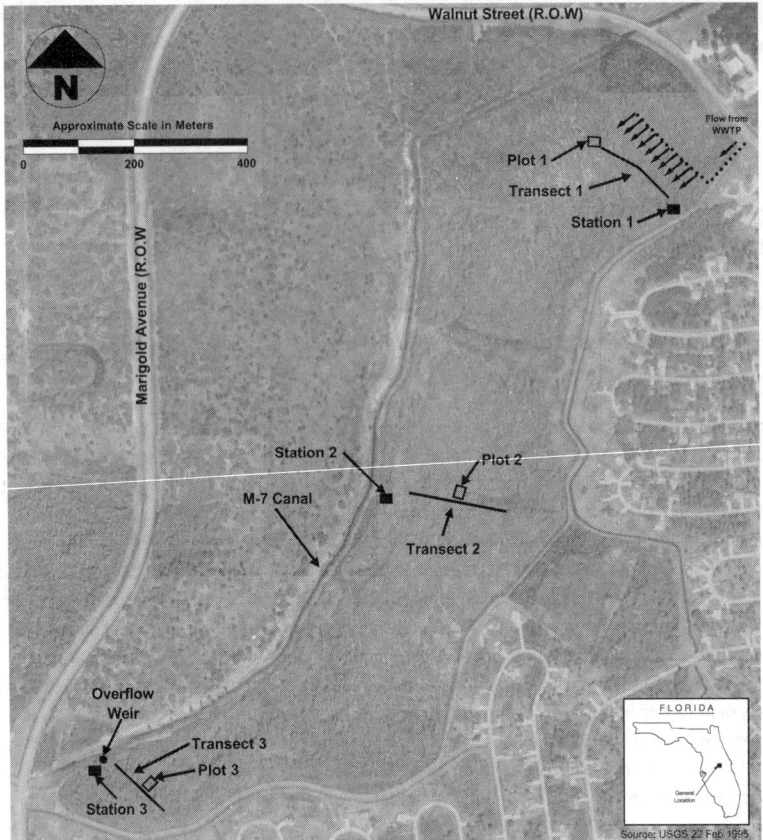

Figure 1 Boot wetland location map

wetland since October 1990. This report presents the findings of the biological monitoring in this unique wetland.

Methods

Vegetative communities

Three permanent vegetation sampling plots in the Boot wetland treatment system were established in 1990 (Figure 1). Plot 1 is located near the submerged influent distribution pipe at the northern end of the Boot wetland. Plot 2 is located in the approximate center of the wetland, and Plot 3 is located near the overflow weir at the southwest corner of the wetland.

Canopy and subcanopy strata are monitored within Plots 1, 2, and 3. Each plot measures 10 m by 30 m and is permanently marked with plastic stakes. All canopy trees (trees greater than 4 inches (10.16 cm) diameter at breast height – (dbh)) and subcanopy trees (trees between 1 and 4 inches (2.54 and 10.16 cm) dbh) within each plot are identified to species and permanently marked with numbered aluminium tags. Tree diameter measurements are conducted annually and measured at the bottom edge of the hanging tag to maintain consistency between measurements.

Canopy and subcanopy tree diameter data are used to calculate dominance and density as follows:

Dominance = Total basal area of Species A (centimetres)

Density = $\dfrac{\text{Number of individuals of Species A}}{\text{Total area sampled (hectares)}}$

Shrub and herbaceous strata are monitored quarterly within each vegetation plot. The line-intercept method is used in which a tape measure is stretched through the middle of the plot (30 metres), and the linear coverage of each plant species lying vertically over, under, or touching the tape is recorded. The shrub/herbaceous stratum is defined as all vascular plant species exclusive of the canopy and subcanopy. The line-intercept data are used to generate estimates of percent cover as follows:

Linear Cover Distance for Species A = Sum of all line-intercept distances for Species A (cm)

Per cent Cover = $\dfrac{\text{Linear cover distance of Species A} \times 100}{\text{Total transect distance}}$

Mosquitoes

The density of mosquito larvae and pupae is monitored at Stations 1, 2, and 3 (Figure 1) on a monthly basis from April to November each year. Twenty surface dips with a 0.45 litre dipper are made at each station. The total number of mosquito larvae and pupae captured in these dips is recorded.

Benthic macroinvertebrates

Benthic macroinvertebrates are sampled at Stations 1, 2, and 3 (Figure 1) in the Boot wetland using Hester-Dendy artificial substrate samplers. Five samplers with a combined surface area of 0.75 m^2 are placed on a quarterly basis at each station. Hester-Dendy samplers are suspended from a floating frame that holds the samplers within two inches of the water surface, regardless of water level fluctuations, for a period of 28 days of incubation. Taxonomic identification to the lowest practical level is completed by a subcontracted taxonomic laboratory.

Fish

Fish populations are sampled along Transects 1, 2, and 3 (Figure 1) in the Boot wetland on a quarterly basis. A Wegener ring with a surface area of 0.94 m^2 is used to quantitatively sample fish in the wetland. Five ring tosses are made at approximately 30 m intervals along each transect. After each toss, the bottom of the ring is pressed into sediment and the water column inside the ring is inventoried with at least 10 sweeps of a dip net. All collected fish are identified to species, counted, and preserved in alcohol for laboratory determination of wet weight for biomass calculations.

Control wetland

A control wetland located approximately one mile south of the Boot wetland was monitored for a period of one year, from October 1990 to September 1991. This wetland represented a typical unimpacted cypress dome in the Poinciana development. No drainage activities have taken place in this wetland, however shallow ditches are located adjacent to the wetland on its north and southwest sides. This wetland did experience seasonal drawdowns during the monitoring program. Biological monitoring in the control wetland was identical to that conducted at the Boot wetland, but at a single station. Therefore, there was a single vegetation plot, mosquito and macroinvertebrate station, and fish monitoring transect.

Results and discussion

Biological monitoring data are reviewed to provide a quantitative comparison of biological communities within the wetland and in some cases comparisons with other regional treatment wetlands. Three types of biological data are presented: vegetation, fish, and invertebrates.

Vegetation

Canopy and Subcanopy. A total of nine canopy and subcanopy species have been identified in the Boot wetland monitoring plots since 1990. They include pond cypress (*Taxodium ascendens*), blackgum (*Nyssa sylvatica* var. *biflora*), wax myrtle (*Myrica cerifera*), dahoon (*Ilex cassine*), Virginia-willow (*Itea virginica*), sweet bay (*Magnolia virginiana*), fetterbush (*Lyonia lucida*), swamp bay (*Persea palustris*), and red maple (*Acer rubrum*).

The dominance of canopy and subcanopy species has steadily increased in the wetland during the past nine years of continuous inundation. Figure 2 shows that the average canopy basal area for the entire wetland increased from 51.8 to 70.9 m^2/ha, while the subcanopy increased from 1.3 to 2.7 m^2/ha. Pond cypress and blackgum have consistently dominated the canopy during this period, with respective basal area increases of 45.2 to 61.0 and 6.5 to 9.8 m^2/ha. In the subcanopy there was a general shift in dominant species from pond cypress to wax myrtle, beginning about 1995. Dominance of subcanopy pond cypress decreased slightly from 0.7 to 0.5 m^2/ha from 1990 to 1999, while wax myrtle dominance increased from 0.09 to 1.7 m^2/ha. The cause was a gradual reassignment of five pond cypress into the canopy size class, lack of new recruitment of young pond cypress, and the new recruitment of 70 wax myrtles in recent years.

Dominance was also compared between tree plots along the surface water flow path in the Boot wetland. Plot 1 is located in close proximity to the effluent distribution pipe, Plot 2 is in the center of the wetland, and Plot 3 is the most distant from the outfall and near the wetland overflow. Canopy dominance increased steadily at all three plots and indicated no adverse trend associated with proximity to the outfall (Figure 2). Subcanopy dominance also showed an increase over time at all three plots and no apparent relationship with proximity to the outfall. The increased wax myrtle recruitment into the subcanopy affected all plots similarly with a greater rate of basal area increase beginning in 1996.

The mean annual growth rate of trees within this continuously inundated wetland is illustrated in Figure 3. Pond cypress was identified as a representative species due to its dominance throughout the wetland, and only those trees (canopy and subcanopy) which had been measured since 1990 were included. A total of 102 pond cypress were evaluated, and on average the mean annual diameter growth for all trees was 0.44 cm/year. This growth rate is higher than that measured by Straub (1984) where bald cypress (*Taxodium distichum*) in two separate wastewater receiving wetlands had annual growth rates ranging from approximately 0.20 to 0.40 cm/yr. The control wetland in this same study had a slower mean annual growth rate that ranged from approximately 0.01 to 0.20 cm/yr. In an unpublished study conducted by CH2M HILL, tree growth in an unimpacted natural wetland (Green Swamp Wildlife Management Area, Florida) was measured. Growth rates of cypress trees ranged from 0.29 to 0.67 cm/year, with an average of 0.54 cm/yr, and is comparable to the growth rate of Boot wetland cypress. Amongst the three Boot wetland plots, Plot 3 (wetland overflow area) had an annual average growth rate that was typically higher than the other two plots, and for the period of record averaged 0.63 cm per year, compared with 0.42 cm per year at Plot 1 (near effluent discharge) and 0.38 at Plot 2 (middle wetland)

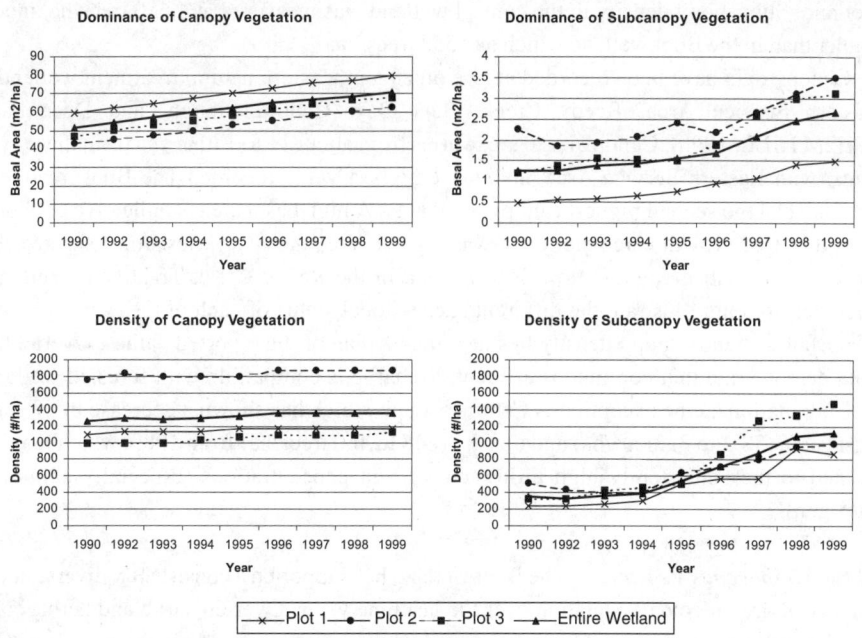

Figure 2 Summary of dominance and density of canopy and subcanopy in the Boot wetland

(Figure 3). It is apparent from these comparisons that pond cypress growth rate in the Boot wetland has not been adversely affected by the addition of treated effluent or the continuous inundation, and is similar and sometimes better than other treatment or natural wetlands.

There was no significant tree mortality recorded during the nine years of monitoring. The first vegetation monitoring event in 1990, five years after initiation of effluent discharge, identified the following standing dead trees in all plots: 11 pond cypress, 4 wax myrtle, and 4 red maple. Since 1990 only 8 subcanopy size class trees out of 222 monitored canopy and subcanopy trees died, including 1 magnolia, 3 pond cypress, 3 wax myrtle, and 1 swamp bay. Thus, there has been no significant tree mortality in the Boot wetland during the last nine years.

The nearby, unimpacted control wetland was dominated by pond cypress. In 1990, total canopy dominance in the control wetland was 19.4 m²/ha, compared to the higher 51.8 m²/ha in the Boot wetland. This corresponded with canopy density measurements of 400 trees/ha in the control wetland compared to the much higher 1,259 trees/ha for the Boot wetland. The control wetland however is a subcanopy dominated system. Subcanopy dominance was 10.3 m²/ha in the control wetland compared to 1.3 m²/ha at the Boot

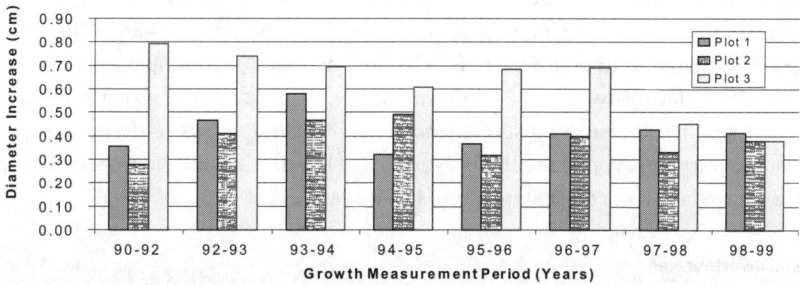

Figure 3 Mean annual diameter growth of *Taxodium ascendens* in the Boot wetland

wetland. Subcanopy density in the control wetland was measured at 4,650 trees/ha, much higher than in the Boot wetland which had 353 trees/ha.

Canopy data have been recorded at five other southeastern natural treatment wetlands: Eastern Service Area, Reedy Creek, Bear Bay, Central Slough, and Deer Park (CH2M HILL, 1998). Canopy basal area varies from about 11 to 140 m^2/ha from these systems, with Eastern Service Area having the highest value recorded. The Boot treatment wetland had the second highest canopy basal area. A high basal area is indicative of a long period of tree growth since the forest was logged. Tree density varies from 400 trees/ha in the Boot control wetland, to 3,785 trees/ha in the Reedy Creek Treatment Wetland. Tree density data illustrate the different successional status of each of these ecosystems. The Boot wetland canopy density lies near the median of the reported values. Overall the data demonstrate that continuous effluent discharge is compatible with forested wetland tree populations. One exception is Central Slough which has shown a decrease in average basal area (58.2 to 32.6 m^2/ha) and density (848 to 460 trees/ha) from 1989, which may be related to the consistently high inflow constituent concentrations (especially ammonia nitrogen).

Shrub/Herbaceous Vegetation. The Boot wetland has supported a consistently diverse community of groundcover vegetation over the last nine years. Over 60 shrub and herbaceous species have been identified during this period. There was no exposed ground surface within any of the plots due to the continuously high water level, and the resulting surface water typically had 100 per cent cover of floating aquatic plants. As a result, the dominant groundcover species occurring in the Boot wetland were duckweeds (*Lemna* spp.), frog's bit (*Limnobium spongia*), and water fern (*Salvinia rotundifolia*). Mosquito fern (*Azolla caroliniana*), water-meals (*Wolffia* spp.), and bog mats (*Wolffiella* spp.) were commonly associated with these floating species. Of the plant species attached to the bases of trees, cypress knees, and fallen logs, the most common were swamp fern (*Blechnum serrulatum*), cinnamon fern (*Osmunda cinnamomea*), chain fern (*Woodwardia virginica*), water primrose (*Ludwigia peruviana*), water hoarhound (*Lycopus rubellus*), Virginia-willow (*Itea virginica*), saw-grass (*Cladium jamaicense*), and poison ivy (*Toxicodendron radicans*).

The diversity of groundcover species within each plot and the wetland as a whole was measured quarterly since 1990. Species diversity was high at all transects, and remained relatively stable over the nine year monitoring period with seasonal fluctuations. The average number of groundcover species for Plots 1, 2, and 3 were 20, 11, and 13, respectively. For the Boot wetland as a whole, the number of groundcover species per monitoring event ranged from 21 to 36, with an average of 29. The data do not indicate any adverse effects of effluent on the diversity and density of groundcover species during the last nine years of wetland treatment.

The control wetland groundcover species community was different from the Boot wetland primarily due to the low frequency of inundation. A single vegetation plot in this wetland was monitored four times in 1990–91. Twenty-nine groundcover species were identified in the control wetland compared to over 60 in the Boot wetland. Thirteen plant species were common to both wetlands. The dominant herbaceous species in the control wetland was water hyssop (*Bacopa carolinianum*). Other common groundcover species included wax myrtle seedlings, cinnamon fern, rush (*Juncus polycephalus*) water hoarhound, and pond cypress seedlings. None of the floating aquatic plant species observed in the Boot wetland were present in the control wetland.

Fish and invertebrates

Mosquitoes. Mosquito larvae and pupae monitoring was conducted monthly during the summer months. Samples were collected in areas that typically contained about 70 per cent floating or rooted wetland plants, and about 30 per cent open water. The number of immature mosquitoes collected at each station was typically low. For the period of record (67 events), the average number of mosquitos at Stations 1, 2, and 3 was 143, 506, and 527 larvae/m^3, respectively. This is comparable to the nearby control wetland, where although only 8 sampling events were conducted during one summer, the average number of mosquitoes was 296 larvae/m^3. The average immature mosquito density for all Boot wetland stations combined was 392 larvae/m^3. These larvae densities are less than those reported at two natural treatment wetlands (791 and 4830 larvae/m^3) at Reedy Creek, Florida (CH2M HILL, 1998). As described below, the abundance of small fish species appears to provide a significant control on mosquito populations in the Boot wetland.

Benthic macroinvertebrates. Macroinvertebrate density, as measured using Hester-Dendy samplers, has been consistently low throughout the wetland during the period of record. A total of 34 sampling events have been conducted. Overall, the average density of macroinvertebrates in the wetland was measured at 39.6 organisms/m^2, and the average number of taxa was 11.1. The most common group of organisms found in the samples were of the family Chironomidae, and to a lesser extent Naididae.

Across the three wetland monitoring stations, average density was low and showed a slightly increasing trend from Station 1 near the effluent discharge to Station 3 near the wetland overflow. For the period of record, average macroinvertebrate densities for Stations 1, 2, and 3 were 15.5, 40.2, and 63.1 organisms/m^2, respectively. The corresponding average number of taxa for Stations 1, 2, and 3 were 3.0, 5.8, and 5.6. The maximum number of taxa recorded at a single station was 19 at Station 3, and the maximum density recorded was 347 organisms/m^2, also at Station 3. At Stations 1 and 3 there were many sampling events where no organisms were identified on the five Hester-Dendy samplers. The most significant factor affecting macroinvertebrate populations is the low dissolved oxygen concentration in the wetland water column. Typically dissolved oxygen within one to two inches of the water surface is between 1 to 2 mg/L, and then drops to less than 0.5 mg/L, and often less than 0.1 mg/L, at mid depth.

The nearby control wetland was monitored with Hester-Dendy samplers only once in January 1991 due to the lack of or very shallow water conditions over its annual monitoring period. The results of this single event identified a much greater diversity and density of macroinvertebrates compared to the Boot wetland, with a total of 28 taxa and 467 organisms/m^2. It is likely that the higher dissolved oxygen (3.0 mg/L) in this control wetland was a significant factor in the greater relative diversity and abundance of macroinvertebrates.

The use of Hester-Dendy artificial substrate samplers in the Boot wetland (required by permit conditions), has not provided a representative picture of the macroinvertebrate community. Qualitative observations of macroinvertebrates occupying the root structures of floating macrophytes indicate a greater density and diversity than was measured with the suspended Hester-Dendy samplers. Other sampling methods (e.g. dip net sampling) would likely measure a more productive macroinvertebrate community than was indicated by the submerged artificial substrate samplers.

Fish. The Boot wetland supports a diverse and healthy fish population. A total of six fish species were collected which included the least killifish (*Heterandria formosa*), mosquitofish (*Gambusia affinis*), sailfin molly (*Poecilia latipinna*), everglades pygmy sunfish (*Elassoma evergladei*), golden topminnow (*Fundulus chrysotus*), and pirate perch (*Aphredoderus sayanus*).

A total of 34 fish monitoring events were conducted at each of the three wetland transects during the period of record. Total fish biomass in the wetland averaged 1.2 grams/m^2 for the sampling period. This is primarily due to the small size of the species that occur in this type of natural wetland environment. No game fish species were captured during the study, although a yellow bullhead (*Ameirurus natalis*) was observed once in the wetland. Density of fish in the Boot wetland ranged from 2.8 to 20.7 fish/m^2, and averaged 8.3 fish/m^2.

The mosquitofish and least killifish were the most abundant species, ranging in average density from 3.5 to 3.8 fish/m^2, respectively. The remaining species, including the everglades pygmy sunfish, sailfin molly, golden topminnow, and pirate perch had average densities of 0.6, 0.4, 0.2, and 0.1 fish /m^2, respectively.

Fish were sampled in the control wetland twice in 1991 (two other events during the single monitoring year were dry). A total of six species were collected, including the everglades pygmy sunfish, mosquitofish, golden topminnow, flagfish (*Jordanella floridae*), a juvenile sunfish species (*Lepomis* sp.), and a shiner species (*Notropis* sp.). Average fish biomass in the control wetland was 0.3 grams/m^2, lower than in the Boot wetland (1.2 grams/m^2). The average fish density of 2.4 fish/m^2 was also less than the Boot wetland (8.3 fish/m^2). The most abundant species was the mosquitofish, averaging 1.8 fish/m^2 (also less than in the Boot wetland) while all other species averaged 0.2 fish/m^2.

Conclusions

The Boot natural wetland treatment system has received treated municipal wastewater for over 15 years. A significant biological database has been collected for the Boot wetland during the past 10 years. These data demonstrate that long-term discharge of treated municipal effluent into a continuously inundated natural cypress-dominated wetland supports healthy and diverse populations of trees, vegetative groundcover, fish, and macroinvertebrates.

The only detrimental responses identified in the wetland were the lack of new recruitment by the dominant tress (pond cypress and black gum) and low ambient dissolved oxygen conditions. No adaptation to the lack of tree recruitment is necessary in the Boot wetland due to the very long lifespan of the dominant tree species (typically many hundreds of years in the absence of logging or drainage). The Boot wetland has adapted to low dissolved oxygen conditions by compressing diverse and large populations of aquatic fauna at or near the water surface.

Compared to the impacted conditions of the Boot wetland before the addition of treated wastewater and compared to a control wetland not receiving treated effluent, the structure and function of the Boot wetland have been significantly improved.

References

CH2M HILL (1998). Treatment Wetland Habitat and Wildlife Use Assessment Project. Report prepared for the U.S. Environmental Protection Agency, U.S. Bureau of Reclamation, and City of Phoenix with funding from the Environmental Technology Initiative Program. Available on CD-rom from Ron Clarke at CH2M HILL (352)-335-7991 (USA).

Breedlove, Dennis and Associates Inc. (1986). Effects of Wastewater Discharge on the "Boot" Wetland, Poinciana Utilities, Inc., Polk County, Florida. Report prepared for Poinciana Utilities.

Straub, P.A. (1984). Effects of Wastewater and Inorganic Fertilizer on Growth Rates and Nutrient Concentrations in Dominant Tree Species in Cypress Domes. In: *Cypress Swamps*, K.C. Ewel and H.T. Odum (ed.), University of Florida Press, Gainesville, Florida, pp. 127–140.

Protection of surface water against contamination by wetland systems in Poland

H. Obarska-Pempkowiak*, T.Ozimek** and W.Chmiel***

* Faculty Hydro and Envrionmental Engineering, Technical University of Gdańsk, ul. Narutowicza 11/12, 80-952 Gdańsk, Poland. (E-mail: *hoba@pg.gda.pl*)
** Department of Hydrobiology, University of Warsaw, ul. Banacha 2, 02-097 Warszawa, Poland
*** Rural Development Foundation, ul. Obozowa 20, 01-161 Warszawa, Poland

Abstract Facilities constructed in order to protect streamS against storm water in the Gdańsk region are described. The first of them is located on the Rynarzewski Stream (water flow 25 l/s). The stream is the main tributary of the Jelitkowski Stream which in turn drains to the Baltic Sea in the area of popular beaches and hotels. Results of analyses indicate the improvement of water quality in the stream and along beaches in this region. Another facility is situated on the Swelina Stream (water flow 30 l/s). The stream is fed with storm water originating from residential districts. In order to improve water quality a pond was constructed supported by a subsurface flow filter (HF-CW type). After implementation of the system substantial improvement of water quality occurred. In order to protect drinking water intake for the city of Gdańsk against surface and point sources of contaminants a hydrophite treatment system was constructed in Bielkowo. The system consists of two subunits: wet unit (pond), filled with water all the time and dry unit (extention of the pond), designed for storm water. In the wet unit dams constructed of medium size sand are placed. The system, especially the dams, is inhibited with reed. The drainage systems collect water percolating through the dams, and directs it downstream. The system was constructed in 1997. Since then it has proven a substantial improvement of water quality discharged of inflowing loads, on average.
Keywords Constructed wetland; contamination; hydrophites; surface waters

Introduction

Deterioration of the environment in Poland particularly in the 1970s, was a result of an incorrect investment policy and indifferent attitude to the principles of the preservation of nature. Water resources, as part of the environment were affected accordingly. Shortage of water resources is caused by deterioration activity of industry, progressing urban development, over-consumption caused by low prices of water and low status of the legislation connected with the natural environment.

Up till now the law has been very liberal with respect to the quantity and the quality of wastes discharged to the natural environment. The European integration processes make it necessary to adapt the Polish standards and technological solutions in the sphere of the environmental protection to the standards required by the European Union. The destroyed natural environment cannot secure a sufficiently high standard of living. Hence the concept of ecological safety based on the conviction that "there is no high quality of life without a high quality of the surrounding environment" has more followers.

Poland's potential impact on the Baltic Sea is reflected in the following facts: about 99.7% of Poland's area belongs to the Baltic Sea drainage basin, over a half of the entire basin population live in Poland and approximately 40% of the entire basin's farmland is situated in Poland.

The rural areas in Poland, populated with 14.6 million people (38% of population) are exposed to the inflow of contaminants as sewage produced in the rural areas in Poland is drained to ground and surface water.

Up till now wetland systems were applied to removal of contaminants from point sources. They are commonly used in the rural areas, where the costs of constructing sewer systems are high. They are usually built as one stage filters with subsurface horizontal flow of water, known as Horizontal Flow Constructed Wetlands (HW-CWs). A new application of constructed wetlands is for the removal of contaminants from surface sources (Mander *et al.*, 1991). Recently three wetland systems were constructed in the Gdańsk Region in order to protect streams against contamination. Two of them contributed to improvement of marine coastal waters quality in Sopot and Oliwa. The third one was constructed in order to protect the drinking water intake for the city of Gdańsk. These facilities proved to be very efficient in removal of nutrients and, to some extent, of microbiological contaminants (*E. coli* bacteria).

Methods

In order to evaluate results of the undertaken actions, samples of water were collected and selected properties of water were measured. Composite samples were collected once a month in the period of one to three years.

Measurements of water quality included BOD_5, COD_{Mn}, total nitrogen, total phosphorus and coli Index. The analyses were carried out in accordance to Polish Standards. Description of methods of analyses was given by Banach *et al.* (1995), Obarska-Pempkowiak (1994, 1996, 2000).

Constructed wetland system along the streams

One of these systems was constructed on the Swelina Stream, which inflows to the Gulf of Gdańsk close to the beaches and bathing places in Sopot. The stream is fed with storm water from the surrounding districts. In order to improve water quality a pond was constructed (total volume 500 m^3) supported with a subsurface flow filter (HF-CW type) of the total volume of 870 m^3 (Figure 1). The overflow installation secures a stabilised flow from the retention reservoir to the filter. The filter was built as a reed bed filled with gravel of a hydraulic conductivity about 5×10^{-5} m/s. At the beginning of inlet block made of crushed stone, the 72 metres long perforated distribution pipeline was installed. At the end of the outlet block, made of the same material, a perforated drainage pipeline was installed. The drainage pipe collects filtered water and carries it away through the control well to the Swelina Stream. The mean annual flow of the stream varies in the range from 12 to 33 l/s during the dry period. During rainfall, after retaining of the first freshet, the water overflows to the Gulf of Gdańsk (Banach *et al.*, 1995).

After implementation of the system on the Swelina Stream substantial improvement of water quality occurred. During the three last years of operation, the frequency of samples included as first class has increased to 75% according to both chemical and bacteriological criteria . For this reason the bathing place and beach at the outflow of the Swelina Stream has been used during the whole summer season in the last three years.

One of the most important streams outflowing to the Gulf of Gdańsk is the Jelitkowski Stream (average flow 158 l/s). Due to its microbiological pollution and high loadings of nutrients, the stream created a severe threat to the ecological balance of the coastal region of the Gulf ofGdańsk.

Between 1991 and 1998 the Faculty of Environmental Protection of the City Council in Gdańsk undertook several actions in order to reduce the loads of nutrients and microbiological pollutants carried by the waters of the Jelitkowski Stream. In the catchment area of the Stream illegal sewage inflows were cut off and settling tanks were installed in order to treat storm waters. Furthermore 9 water reservoirs were built on the Stream to restore its ability of self-purification. On the Rynarzewski Stream, which inflows to the Jelitkowski Stream

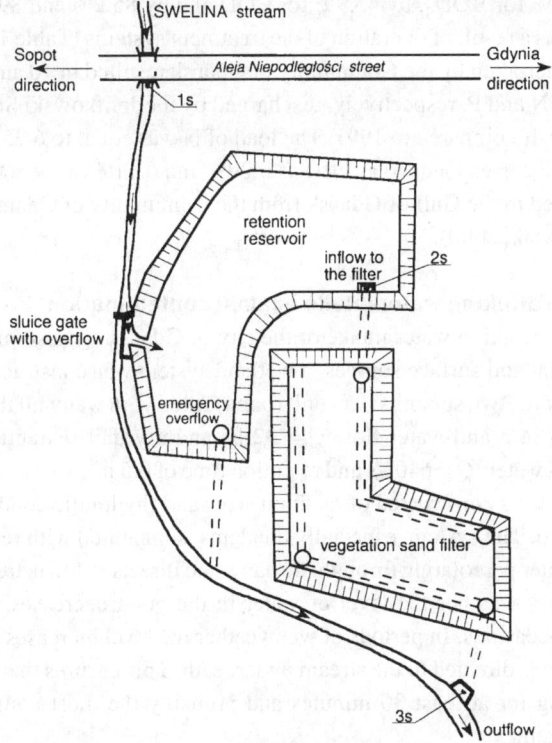

Figure 1 The schematic of the constructed wetland system situated along the Swelina Stream

after flowing across the area of the zoo in Oliwa (average flow 70 l/s), 4 sand filters with horizontal flow, of a total area of 1,650 m^2, and a natural wetland system with surface flow of water, of an area of 2,000 m^2, were built, in order to remove contaminants from point sources. To reduce the loads originating from surface sources, five buffer zones inhabited with willow were constructed, of a total area of 6,650 m^2 (Obarska-Pempkowiak, 1994). Buffer zones are strips of land parallel to the stream, planted with willow (*Salix sp.*). Their task is to separate surface sources of contaminants from the stream. The strips are cut with furrows and antislopes. Both furrows and antislopes were designed to increase the retention volume and retention time of the contaminated storm water run-off (Obarska-Pempkowiak, 1994). A schematic illustration of a buffer zone is shown in Figure 2.

The analyses of the quality of the waters of Rynarzewski Stream conducted in the years 1991–1999 showed that the average load of contaminants carried with the stream waters

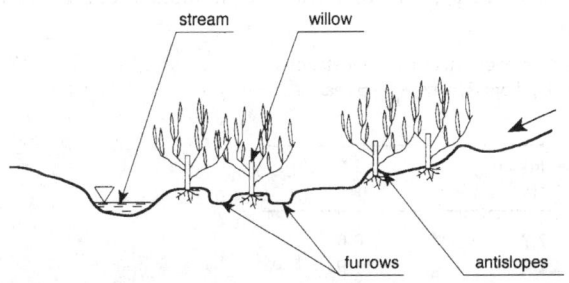

Figure 2 A buffer zone

decreased by 14.3% for BOD_5, by 45.9% for COD and by 85.2% and 89.2% for N_{tot} and P_{tot}, respectively, as a result of operation of the treatment systems (Table 1).

The actions undertaken by the Community of Gdańsk resulted in 10 and 8 fold decreasing of the loads of N and P, respectively, discharged by the Jelitkowski Stream to the Gulf of Gdańsk in 1999, if compared to 1991. The load of N was equal to 6.25 t and the load of P – to 0.23 t, which corresponds to as little as 0.29% and 0.11% of the total loads of these elements discharged to the Gulf of Gdańsk from the Community of Gdańsk (Kopeć, 1994; Obarska-Pempkowiak, 2000).

Protection of the drinking water intake against contamination

In order to protect drinking water intake of the city of Gdańsk against inflow of contaminants from the point and surface sources, a wetland system was constructed in Bielkowo. The system consists of two subunits: wet unit (pond) filled with water all the time, designed for retention time 24 h and water flow Q = 32 l/s and dry unit (extention of the pond), designed for storm water: Q = 640 l/s and retention time of 0.5 h.

In the wet unit dams constructed of medium size sand (hydraulic conductivity k = 40 – 86.4 m/d are placed. The system, especially the dams is inhabited with reed. The drainage system collects water percolating through the dams and directs it downstream (Figure 3).

In periods of dry weather the level of water in the pond decreases. The dry section emerges on such occasions. In periods of wet weather the level increases until water overflows the dams and is directed to the stream underneath. This ensures that the first surge of stormwater, lasting for at least 30 minutes and probably the most contaminated one is retained in the system.

In the design documentation of the system, it was assumed that during 8–9 months each year the whole flow of the stream (average flow Q = 32 l/s) will run through the "wet" part of the pond, with the retention time equal to 24 h. However, during the visits to the facility and collecting samples of water in 1999, it was found that the system is all the time working with the water dammed up above the overfall crest. This means that the "wet" part was covered with water all the time and there were no periods when the "dry" part of the pond emerged. In such a situation there is no retention volume for the storm water run-off in the system.

In 1997–98 due to atmospheric conditions, the hydrophyte plants (mostly common reed) were not able to take their roots on the dykes, and the facility worked as a system of filters. The system was inhabited with reed in spring, 1999. During the first two years of operation, due to mass algae blooming and lack of roots cultivating the ground, the surface of the dykes was covered with a thick mat of organisms. This was the direct reason for the changing of the permeability of the dams and flooding the elevated part with water.

The removal of contaminants in the hydrophyte system in Bielkowo occurs in the process of sedimentation, slow filtration through the dams and bioaccumulation of nitrogen and phosphorus compounds in the hydrophytes and phytoplankton. The efficiency of

Table 1 Average yearly concentrations of organic substances and nutrients in water outflowing from Rynarzewski Stream in the years 1991–1999

Parameter	Load [t/year]	
	1991	1999
BOD_5	7.7	6.6
COD_{Mn}	22	11.9
N_{tot}	22.3	3.3
P_{tot}	1.3	0.14

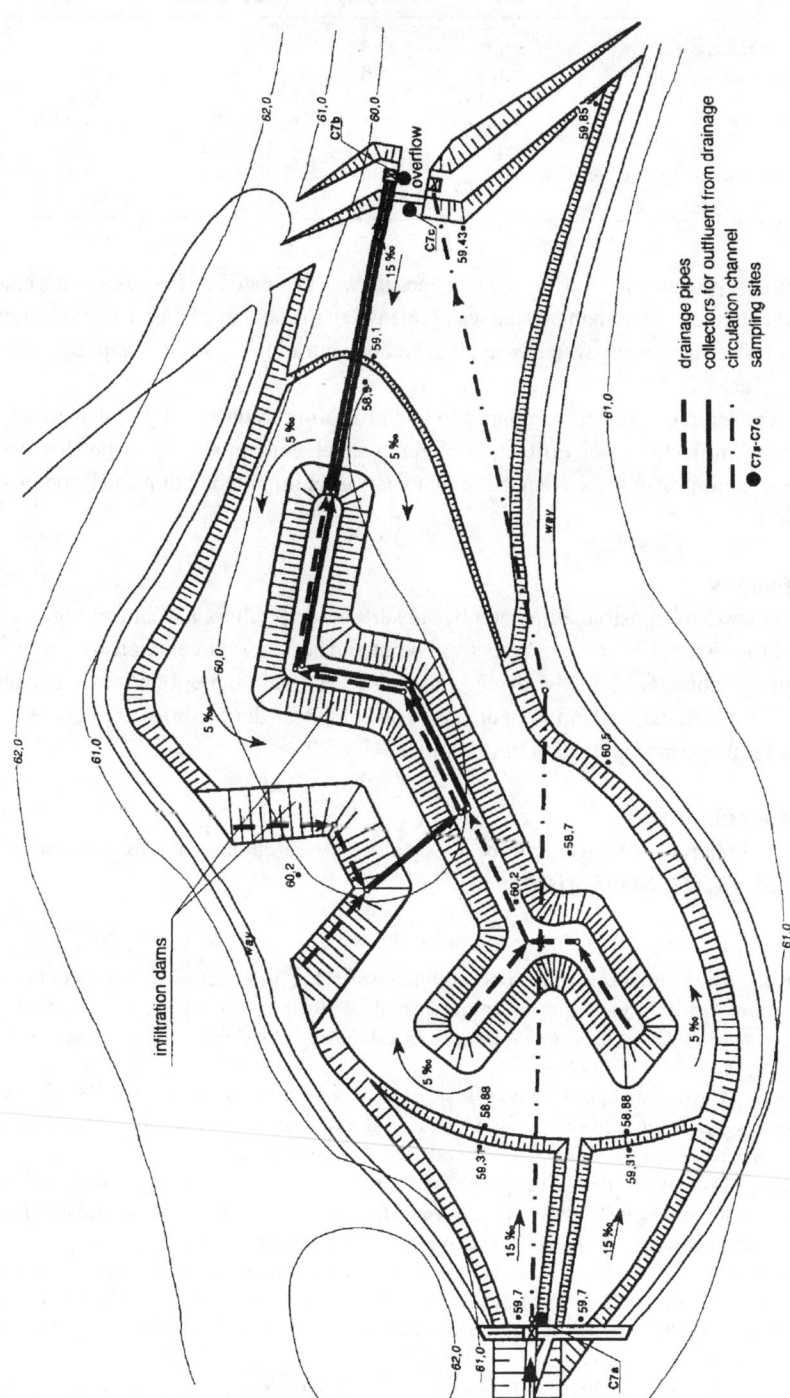

Figure 3 The schematic of the constructed wetland system in Bielkowo

Table 2 Average concentrations of characteristic contaminants in the waters inflowing and infiltrating to and outflowing from the hydrophite system in Bielkowo

Parameter	Unit	Inflowing water	Infiltrating water	Overflowing water
Suspended solids	mg/l	23.8	11.1	18.6
BOD_5	mg O_2/l	6.6	6.0	6.4
COD_{Mn}	mg O_2/l	11.7	11.6	11.0
N_{tot}	mg/l	2.1	1.7	1.8
P_{tot}	mg/l	0.11	0.08	0.08
Substances extracted with ether	mg/l	15.0	6.4	7.1
MPN *E. coli*		9,600	4,800	9,000

biochemical processes taking place in the facility was evaluated on the basis of the changes of characteristic contaminants in the water inflowing to and outflowing from the system, as well as in the infiltrating water. The average concentrations in these sampling points are given in Table 2.

The concentrations of contaminants in the infiltrating water are higher than assumed in the design project. Lower efficiency of removal of contaminants can be the result of improper operation of the system in the first years after implementation and flooding of the system.

Conclusions

Wetland systems constructed as ponds, subsurface flow filters and buffer zones can be applied for removal of contaminants from point and surface sources. Constructed wetland systems are inhabited mainly with reed and willow. High efficiency of removal of contaminants in constructed wetland systems takes place only if the facilities are designed, constructed and operated according to strict rules.

Acknowledgements

Funding support from the Committee of Scientific Research in Poland for constructed wetland study (3TO 9C 040 17) is acknowledged.

References

Banach, K., Obarska-Pempkowiak, H. and Olańczuk-Neyman, K. (1995). Chemical and bacteriological assesment of efficiency of operation constructed wetland system on Swelina Stream in Sopot. In: *Proceedings of I International Conference*, Technical University of Gdańsk, Gdańsk, Sepember 1995, 15–22.

Mander, U., Matt, O. and Nugin, U. (1991). Perspectives of vegetated shoals, ponds and ditches as extensive outdoor systems of wastewater treatment in Estonia. In: C. Etnier and B. Gutterstam (Eds.): *Ecological Engineering for Wastewater Treatment*, Bokskogen, Sweden, Gothenburg, 271–282.

Obarska-Pempkowiak, H. (1994). The application of willow and reed vegetation filters to the protection of stream passing a zoo. In: P. Arronson and K. Perttu (ed.) *Willow vegetation filters for municipal wastewaters and sludges*. Swedish University of Agricultural Sc. Uppsala, Rapport 50, 59–67.

Obarska-Pempkowiak, H. (1996). Efficiency of removal of contaminants from point souras and surface on example Rynarzewski Sream In: M. Sozański (ed.) Municipal and rural water supply and water quality. In: *Proceedings 7 of International Conference*, PZITS and Technical University of Poznań, Poznan, June 1996, 29–39.

Obarska-Pempkowiak, H. (2000). Load of contaminants discharged from Municipality of Gdańsk to Gulf of Gdańsk in 1999 year. Gdańsk, manuscript pp.15.

Natural wastewater treatment in Hungary

A. Szabó*, A. Osztoics** and F. Szilágyi***

* Department of Sanitary and Environmental Engineering, Budapest University of Technology and Economics, H-1111 Budapest, Müegyetem rkp. 3., Hungary. (E-mail: anita@vcst.bme.hu; andras@vcst.bme.hu; szilagyi@vcst.bme.hu)

Abstract Over the last few decades more and more natural wastewater treatment systems have been built in Hungary. The present study is the first step in creating a broad database on the water quality parameters and on the pollutant removal efficiency of these systems. The investigation included 78 plants out of which we analysed 16 systems in detail. Four types of natural methods are evaluated: wetlands, ponds, bio-mechanical combined oxidation (BMKO) systems, and poplar plantations. Pond systems are efficient in ammonium-nitrogen (NH_4-N) removal, reducing it with 83% (41–88%). Their chemical oxygen demand (COD_{Cr}) removal capacity is only 55% (37–81%). The only BMKO system that could be evaluated performs high COD_{Cr} (77%) and total suspended solid (TSS) (89%) removal. Removal of NH_4-N and total nitrogen (TN) declines during the years of operation giving an average value of 39% and 49%, respectively. The system is not efficient in phosphorus removal (13%). In wetlands the 71% COD_{Cr} (53–96%), and 57% TSS (33–91%) removal provides satisfactory effluent quality most of the time. Wetlands performed low nutrient removal, i.e., 17% (–21–46%) for TN and 26% (–20–92%) for phosphorus. Poplar plantations are very effective in pollutant removal. Even the average removal of each nutrient type is above 75%. Several problems have occurred in the operation of natural treatment systems. However, if carefully planned and constructed, and the required maintenance work is done properly, they can be possible alternatives for wastewater treatment.

Keywords BMKO; database; natural wastewater treatment; pond; poplar plantation; wetland

Introduction

Water supply and sewerage

In Central-Eastern European countries the drinking water supply is usually far more developed than sewerage. In Hungary more than 90% of the population are connected to the drinking water system. At the same time only 48% of the population are connected to the sewerage system (Somlyódy, 2000). This difference is due to two main reasons (Bácskai, 1996).

- In most of the villages drinking water was supplied traditionally from shallow wells. The infiltration of municipal waste (especially excreta) and the overuse of fertilizers resulted in the nitrate and pathogenic pollution of the groundwater. Further consumption from the polluted wells caused health problems. Under these circumstances it seemed imperative to ensure the supply of safe, healthy drinking water.
- Drinking water supply brings significant improvement in level of comfort while it does not make such a difference for people if sewage is disposed of in a septic tank (for storage or local infiltration) or led to the sewage system. Thus drinking water was considered as the essential need, while developing the sewage net was of secondary importance. It was also a kind of political target to please and satisfy people in order to avoid friction. In this way drinking water supply was developed more intensively than the sewage network and wastewater treatment.

The high percentage of homes using drinking water without sewers constitutes a high danger to groundwater quality, since the water consumption and consequently wastewater production has increased without solving its treatment properly. To make the situation even worse, less than half of the collected sewage is treated at all (1.1 million m³/day out of 2.4 million m³/day) (Bácskai, 1996).

Natural systems in Hungary

Natural methods in wastewater purification have been used in Hungary since the late1960s (Kárpáti et al., 1968). After several trials and research, four main types of natural wastewater treatment methods are in use today. These types are wetlands, pond systems, biomechanical combined oxidation (BMKO) systems and sewage disposal and treatment on a poplar plantation. The study is the first step in creating a broad database on water quality parameters and on pollutant removal efficiency of the systems. The investigation included 78 plants treating communal, industrial, or mixed wastewater.

The analysed systems treat settled wastewater with the exception of poplar plantations, which receive biologically treated wastewater.

Facultative ponds are among the first natural treatment systems in Hungary. Earlier, already existing lakes and ponds have been used for disposing of wastewater. Later, ponds were designed and constructed directly for this purpose. They were constructed usually with two or three basins. We can also find pond systems consisting of 5–7 basins.

BMKO systems are a mixture of ponds and wetlands. The screened or settled sewage enters the first pond, where – if necessary – wastewater is aerated mechanically. Subsequently the sewage is treated in a series of 5 or 9 facultative ponds. The last two basins have dense submerged and emergent vegetation as a polishing step.

The first experiments with sewage disposal and treatment on poplar plantations started more than 40 years ago. In this treatment method the wastewater – after a primary pretreatment and short term temporary storage – is disposed of all year round on a timber (*Populus euramericana* and *Populus robustia*) plantation. The sewage is distributed to ditches located between the rows of trees. In order to prevent groundwater pollution the infiltrated wastewater is collected by a drainage system (Vermes, 1985; Vermes, 1996).

Wetlands are the most recent type of natural wastewater treatment systems used in Hungary. After the success of the experiments in the 1980s, some small reed beds (serving a few hundred people) were put into operation in the beginning of 1990s. Most of the presently operating wetlands are subsurface vertical flow wetlands, there has been built only one with horizontal flow. The only free surface flow wetland is used as a polishing step.

Methods

Data used in the evaluation were provided by the operators of the treatment plants or by the relevant laboratories where the chemical analysis had been carried out. Usually the registration of the plants was found to be very poor. Few measurements were taken; they are not even well ordered or consistent. The objective of sampling was usually only to check whether the effluent parameters met the standard (see Table 1) or not. Removal efficiencies were seldom examined, usually only the effluent quality was measured, as this is the only standard being enforced. Because of these difficulties only 14 treatment plants were evaluated in detail.

The evaluation is based on the following aspects:
- quality of the influent and effluent (range and average);
- removal efficiency (range and average);
- changes in effluent quality and removal efficiency in terms of temperature;
- changes in effluent quality and removal efficiency in time.

Originally the emphasis in the evaluation was on the removal efficiency of the systems. As there were only a few cases where both influent and effluent were measured, the

Table 1 Emission standards for the most important water quality components in EU and in Hungary

	EU	Hungary[i]					
		I.	II.	III.	IV.	V.	VI.
chemical oxygen demand (COD_{Cr})	125	50	75	100	100	150	75
biochemical oxygen demand (BOD_5)	25	–	–	–	–	–	–
total suspended solids (TSS)	35	100	100	200	200	500	200
total nitrogen (TN)	10[ii]	2[iii]	5[iii]	30[iii]	10[iii]	30[iii]	10[iii]
total phosphorus (TP)	1.0[ii]	1.8[iv]	2.0[iv]	2.0[iv]	2.0[iv]	–	2.0[iv]

i: Area categories:
    I   – Prominent water quality area
    II   – Sources of drinking water supply and recreation area
    III – Industrial area
    IV – Sources of irrigation water
    V   – Non-priority regions along the Danube and the River Tisza
    VI – Non-priority area
ii: In case of sensitive areas, personal equivalent greater than 100 000
iii: NH_4^+ ion values
iv: If the receiving water body is standing water

evaluation of the effluent water quality completes the assessment. The data gained in this way is still insufficient for studying the plants one-by-one. Thus systems of the same kind are evaluated together, giving a picture about the operation of different system types in Hungary.

For different systems the number of sampling occasions and thus the reliability of the available data differs a lot. In some of the systems over 30 measurements were taken, while in others less than five. The intervals given in the evaluation at a certain system type (pond, BMKO, wetland or poplar plantation) show the overall minimum and maximum of the measured values. The average value describing a certain system type is calculated from the average values of each plant.

Results and discussion
Pond systems

In the case of the ponds, data was found only about chemical oxygen demand (COD_{Cr}) and ammonium-nitrogen (NH_4-N) (Table 2–3). The six ponds had an average influent concentration of 454 mg/l (75–1,850 mg/l) COD_{Cr}. Very high influent concentrations could be due to the following reasons. First, water consumption dropped in recent years as the water fee increased. Thus the same amount of pollutants are diluted with less water giving higher concentrations. Second, the number of connections to the sewerage is far below the expected number, resulting in long residence time of the raw sewage in the sewerage system. Effluent data was gained only from three systems (Balmazújváros, Biharkeresztes, Hortobágy). In the case of the other three ponds – due to leakage through the bottom of the lake – no water appeared as effluent. The average effluent value was 111 mg/l (43–280 mg/l), which would meet the Hungarian standards only in one of the categories (Table 1). The average removal rate of the three systems was 55% (25–88%), which was lower than reported in the literature (Hettiaratchi and Smith, 1989; Zhao and Wang, 1996). Where there was no effluent water (Hajdúsámson, Nyírábrány, Nyíracsád), the evaluation was based on the concentrations measured in the last pond containing water. COD_{Cr} concentration is 290 mg/l (18–889 mg/l). From the few data available it was not possible to determine whether the 889 mg/l was extreme or normal. If this high value was not considered, the average would be only 190 mg/l (18–334 mg/l).

The average NH_4-N concentration of the influent was 49 mg/l (20–121 mg/l). These high figures can be the consequence of the long residence time in the sewer net (mentioned above), where the sewage becomes anaerobic and thus the main nitrogen form is ammonium. The three ponds having effluent measured had an average effluent concentration of 13 mg/l (0–60 mg/l), which would exceed the limits in most of the categories (Table 1). Where it was possible to calculate removal efficiency, it was 71% (5–100%). If we neglect the extremely low 5%, then the average would be 83% (46–100%), which is a higher efficiency than found in the literature (Zhao and Wang, 1996). The ponds with no effluent had 19 mg/l (0.6–34 mg/l) NH_4-N concentration in their last basin. Temperature had a significant effect on the operation of the pond systems. As the temperature rose from 0°C to above 20°C the removal efficiency of NH_4-N improved with 20–50%.

BMKO systems

We found only one BMKO system that could be evaluated (Püspökladány). This treatment plant consists of a settling tank and the BMKO system. Data was available on COD_{Cr}, total suspended solid (TSS), NH_4-N, total nitrogen (TN), ortho-phosphate phosphorus (PO_4-P), and total phosphorus (TP) (Table 3).

The BMKO system performed high removal efficiency both in COD_{Cr} and in TSS. COD_{Cr} concentration was reduced from 468 mg/l (175–1,399 mg/l) to 91 mg/l (38–274 mg/l) meeting the expected 100 mg/l limit (Table 1). In many cases, wastewater with higher concentrations than the recommended 300–600 mg/l entered the system (OVH, 1984). These high values could occur because of the same reasons described for the pond systems. Effluent concentrations exceeded the expected 80–150 mg/l (OVH, 1984) in only a few cases. Average removal was 77% (38–93%). This is better than found for facultative ponds in Hungary and is comparable to other pond systems (Hettiaratchi and Smith, 1989; Zhao and Wang, 1996).

The average influent concentration of TSS was 271 mg/l (42–809 mg/l), which was reduced to 23 mg/l (0.4–109 mg/l). The effluent concentration meets the Hungarian standards and the even more strict EU standards (Table 1). The removal of TSS had a high average of 89% (63–99%).

Nitrogen forms were removed from the wastewater less efficiently. The influent 35 mg/l (0.2–56 mg/l) NH_4-N concentration was decreased to 24 mg/l (0–66 mg/l), which exceeds the standards. The average removal was 39% (24–94%), significantly lower than found in the case of pond systems. TN concentration was reduced from 63 mg/l (41–88 mg/l) to 31 mg/l (3.5–60 mg/l). The removal efficiency was 49% (16–83%).

Table 2 Performance of pond systems

Pond systems	COD_{Cr} (mg/l)			TSS (mg/l)			NH_4-N (mg/l)		
	inflow	outflow	Removal (%)	inflow	outflow	removal (%)	inflow	Outflow	removal (%)
Balmazújváros 355 (14)	68 (25)	81 (9)	–	–	–	46,9 (14)	4,8 (25)	88 (9)	
224–508	43–101	69–88	–	–	–	33–70	0–18	66–100
Biharkeresztes 228 (13)	125 (30)	37 (2)	1301 (3)	–	–	27,83 (13)	14,40 (30)	41,4 (2)
75–397	50–198	25–49	87–174	–	–	12,5–41,6	0,5–41,0	5–78
Hajdúsámson 642 (2)	334 (1) *	–	–	–	–	68,5 (2)	17,0 (1) *	–
450–833	–	–	–	–	–	60–77	–	–
Hortobágy 223 (8)	140 (50)	47 (7)	–	122 (8)	–	30,8 (8)	19,3 (50)	81 (7)
125–360	44–280	31–57	–	17–525	–	19,69–52	0,29–60	46–99
Nyírábrány 694 (4)	302 (5) *	–	–	–	–	45,8 (4)	10,0 (5) *	–
198–1850	18–889	–	–	–	–	20–75	0,6–33,5	–
Nyíracsád 579 (3)	188 (1) *	–	–	–	–	75,0 (3)	29,0 (1) *	–
369–944	–	–	–	–	–	48–121	–	–

* No effluent occurs, numbers indicate the measured values in the last pond containing water
Numbers in brackets indicate the number of measurements.
Italic numbers show the minimum and maximum measured values.

Phosphorus forms are the most problematic pollutants for this system. The average 6 mg/l (4.2–8.3 mg/l) concentration of PO_4-P in the influent increased to 7.2 mg/l (2.5–12.6 mg/l) in the effluent. The average value of PO_4-P increase was 24% (from 137% increase to 56% removal). The efficiency of total phosphorus removal was somewhat better. The 10.3 mg/l (6.2–15.8 mg/l) influent concentration was decreased to 8.4 mg/l (2.8–13.6 mg/l). The average 13% (–37–70%) removal was far below the initial 56% removal (Dévai and Woynarovich, 1981). Increase of PO_4-P and TP concentration occurred only in the past few years. This can be the consequence of the low oxygen level measured in the water resulting in the release of the formerly adsorbed phosphorus and thus causing internal load. The increase of PO_4-P can also be due to that part of TP that is transformed to PO_4-P.

Temperature had a significant effect on the system performance. NH_4-N removal was around 10% in cold weather but it improved to 80% in warm weather. Phosphorus removal also increased slightly when temperature rose.

Wetlands

The evaluation of wetlands was based on six plants (Table 4). One of them contains two parallel units, with two separate effluents (Tóalmás). For about three plants (Boldog, Kacorlak, Salgótarján) we have found data only on the raw sewage (before settling) and in all the other cases on the settled wastewater (Tóalmás, Szépalmapuszta, Szügy). In most cases we have only one to four data, due to the wetlands being rather new systems in Hungary on full scale, and pollutant concentrations being measured seldom. For this reason long-term variation and temperature dependence were impossible to study. Instead, seasonal variation was assessed in the case of two systems (Boldog and Szügy).

The examined wetlands provided satisfactory treatment of COD_{Cr} and TSS. COD_{Cr} influent concentration averaged at 363 mg/l (20–1350 mg/l). This was decreased to 66 mg/l (9–742 mg/l), which means that the effluent concentration met the standards in all except the most vulnerable areas. However, on some occasions the effluent concentration was far above the limits in every standard category. The removal efficiency was 71% (3.6–99%).

Influent TSS concentration averaged at 128 mg/l (9–766 mg/l) and was reduced to 42 mg/l (9–456 mg/l). This satisfies the Hungarian standards in all the categories and it is

Table 3 Performance of BMKO systems

BMKO plants	COD_{Cr} (mg/l)			TSS (mg/l)			NH_4-N (mg/l)		
	inflow	outflow	removal (%)	inflow	outflow	removal (%)	inflow	outflow	removal (%)
Hajdúnánás	313 (1)	96 (1)	69 (1)	–	–	–	55 (1)	44 (1)	20 (1)
	–	–	–	–	–	–	–	–	–
Püspökladány raw wastewater	648 (34) *175–1399*	91 (87) *38–274*	77 (32) *38–93*	271 (11) *42–809*	23 (21) *0.42–109*	89 (11) *62.63–99*	35.1 (34) *0.2–56.2*	24.0 (87) *0–66*	39 (32) *–24–94*
Püspökladány settled wastewater	268 (19) *133–488*	91 (87) *38–274*	54 (15) *5–92*	88.06 (4) *60–130*	23 (21) *0.4–109*	83 (3) *76–90*	31.4 (19) *0.1–47.0*	24.0 (87) *0–66*	29 (15) *–45–100*

BMKO plants	TN (mg/l)			PO_4-P (mg/l)			TP (mg/l)		
	inflow	outflow	removal (%)	inflow	outflow	removal (%)	inflow	outflow	removal (%)
Hajdúnánás	–	–	–	–	–	–	–	–	–
	–	–	–	–	–	–	–	–	–
Püspökladány raw wastewater	62.7 (13) *41.3–88.0*	30.6 (24) *3.5–60.2*	49 (12) *16–83*	5.99 (15) *4.20–8.26*	7.23 (25) *2.49–12.6*	–24 (13) *–138–566*	10.26 (17) *20–15.8*	8.35 (27) *22.81–13.6*	13 (15) *–37–70*
Püspökladány settled wastewater	56.3 (8) *41.7–72.1*	30.6 (24) *3.49–60.23*	44 (5) *–8–95*	6.27 (9) *4.7–7.5*	7.23 (25) *2.49–12.6*	–11 (5) *–68–67*	8.93 (10) *5.5–11.6*	8.35 (27) *62.81–13.6*	–0.3 (6) *–39–76*

Numbers in brackets indicate the number of measurements
Italic numbers show the minimum and maximum measured values

not far from EU standards (Table 1). However, there were very high effluent concentrations in some cases. Removal efficiency of TSS was 57% on average (–221–98%).

Influent NH_4-N concentration was 61 mg/l (1.2–204 mg/l). The wastewater had high ammonium concentrations already at the point of discharge into the sewerage. The effluent contained still 23 mg/l (0.01–162 mg/l). High effluent concentrations can be the consequence of that the systems were designed for lower influent concentrations (Szilágyi, 1996). In the literature similar high values were reported for effluent concentration, whereas the influent concentrations were lower (Watson et al., 1989; Cooper and Hobson, 1988; Conley et al., 1991). Removal efficiency of ammonium-nitrogen was 57% (–50–99.95%), which is rather good when comparing to the international experience (Watson et al., 1988; Cooper and Hobson, 1988; Conley et al., 1991; Reed et al., 1995). The 75 mg/l (13.6–182 mg/l) influent concentration of TN was reduced to 52 mg/l (16.5–159 mg/l), which is high even if we consider that the influent contains high amounts of nitrogen. The removal efficiency was 17% (–21–74%). This is much below the expected 30–40% (Brix, 1994). Nitrogen removal reflects the start-up period of the systems, which they were in at the time of assessment. The low removal efficiency of the systems shows that the vegetation and the attached microorganisms were not yet developed to the desired level.

Phosphorus was removed poorly from the wastewater in these systems. PO_4-P influent concentration of 9.3 mg/l (0–20 mg/l) was decreased to 7.7 mg/l (0–21 mg/l). It falls into

Table 4 Performance of wetlands

Wetlands	COD_{Cr} (mg/l)			TSS (mg/l)			NH_4-N (mg/l)		
	inflow	outflow	removal (%)	inflow	outflow	removal (%)	inflow	outflow	removal (%)
Boldog (raw)	170 (15)	64 (15)	64 (15)	97 (15)	47.6 (15)	47 (15)	45.9 (15)	33.0 (15)	30 (15)
	110–275	30–89	47–81	50–173	24–65	10–83	45–67.1	10.4–66.3	1–58
Kacorlak (raw)	1061(2)	37 (2)	96 (2)	310 (2)	28,5 (2)	91 (1)	165.3 (2)	13.6 (2)	92 (2)
	772–1350	9–64	92–99,3	291–328	9–48	85–97	133.6–197	7.4–19.8	90–94
Salgótarján (raw)	110 (1)	38 (1)	65 (1)	60 (1)	40 (1)	33 (1)	14.6 (2)	3.7 (2)	72 (2)
	–	–	–	–	–	–	13.2–15.9	0.1–7.3	45–99
Szépalmapuszta	593 (1)	102 (1)	83 (1)	154 (1)	48 (1)	69 (1)	69.4 (1)	31.3 (2)	47 (1)
	–	–	–	–	–	–	–	26–36.5	–
Szügy	503 (32)	170 (31)	67 (31)	221 (32)	93 (31)	50 (31)	105 (32)	78.2 (31)	23 (31)
	39–885	29–742	4–91	54–766	26–456	–221–89	2–204	2–162	–16–71
Tóalmás I.	51 (4)	27 (4)	53 (4)	28 (6)	11 (6)	54 (6)	13.7 (3)	0.7 (3)	50 (3)
	20–89	10–69	22–71	13–47	1–21	9–98	1.2–20	0.01–1.8	–50–99,9
Tóalmás II.	51 (4)	23 (4)	68 (3)	28 (6)	23 (5)	56 (4)	13.7 (3)	3,5 (2)	87.2 (1)
	20–89	5–60	50–83	13–47	6–72	50–69	1.2–20	2.56–4.4	–

Wetlands	TN (mg/l)			PO_4-P (mg/l)			TP (mg/l)		
	inflow	outflow	removal (%)	inflow	outflow	removal (%)	inflow	outflow	removal (%)
Boldog (raw)	–	–	–	–	–	–	7.2 (2)	6.75 (2)	17 (2)
	–	–	–	–	–	–	4.2–10.2	2.4–11.1	–9–43
Kacorlak (raw)	–	–	–	–	–	80 (2)	24.1 (1)	2 (1)	92 (1)
	–	–	–	–	–	69–92	–	–	–
Salgótarján (raw)	13.6 (1)	16.5 (1)	–21 (1)	7.7 (1)	11 (1)	–43 (1)	3 (1)	3.6 (1)	–20 (1)
	–	–	–	–	–	–	–	–	–
Szépalmapuszta	84.3 (1)	45.7 (1)	46 (1)	14.52 (1)	6.18 (1)	58 (1)	17.2 (1)	8.3 (1)	51 (1)
	–	–	–	–	–	–	–	–	–
Szügy	127.8 (32)	94.3 (31)	27 (31)	8.6 (32)	1.7 (31)	74 (31)	25 (32)	6 (31)	71 (31)
	45–182	29–159	–9–74	0–20	0–9.1	–50–99	7.3–92	1.2–17.8	3–92
Tóalmás I.	–	–	–	7.78 (4)	8.05 (4)	–4 (4)	3.13 (3)	3.34 (3)	–7 (3)
	–	–	–	4.6–11.8	4.8–11.6	–16–2	2.39–4.5	2.2–4.8	–26–12
Tóalmás II.	–	–	–	7.78 (4)	11.42 (4)	–54 (4)	3.13 (3)	2.33 (2)	–20 (1)
	–	–	–	4.6–11.8	4.8–21.46	–211–7	2.39–4.5	1.8–2.87	–

Numbers in brackets indicate the number of measurements
Italic numbers show the minimum and maximum measured values

the range reported in the literature (Brix, 1987; Watson et al., 1989; Conley et al., 1991). The average removal efficiency was 18.4% (–211–99%). This is lower than the reported efficiencies (Brix, 1987; Watson et al., 1988; Conley et al., 1991). TP concentration was reduced from 12 mg/l (2.4–92 mg/l) to 4.6 mg/l (1.2–18 mg/l). The removal efficiency was 26% (26–92%). Seasonal variation can be seen in wetland systems for nutrient removal. Between summer and winter performance even 40% difference occured in some cases.

Poplar plantations

The evaluation of poplar plantations is based on two systems in respect of COD_{Cr}, NH_4-N, TN, PO_4-P, and TP. Data was available mainly on removal efficiency. Influent and effluent values were given only for COD_{Cr}, TN, and TP. The results presented here were calculated from annual average values. Thus the only aspect of evaluation is the long-term efficiency.

The assessed poplar plantations were very effective in pollutant removal (Table 5). Removal of COD_{Cr} averaged at 87% (44–99.3%). TN and NH_4-N were decreased with 79% (27–97%) and 95% (66–99.9%), respectively. Phosphorus was also reduced significantly; the removal efficiency was 77% (52–94%) for TP and 93% (56–99.9%) for PO_4-P. COD_{Cr} concentration was reduced from 249 mg/l (163–376 mg/l) to 48 mg/l (39–86 mg/l), providing acceptable quality for all standard categories (Table 1). Average influent TN concentration was 59 mg/l (46–79 mg/l). It was reduced to 19 mg/l (7.9–49 mg/l). TP influent concentration was 19 mg/l (8.6–48 mg/l). This was decreased to 5 mg/l (2.5–12 mg/l), which means it is always above the standard limit (2 mg/l).

In the long-term the removal efficiency of the assessed systems usually did not change during the years of operation. In some cases the efficiency dropped at the beginning of the operation and improved somewhat later on. Only phosphate removal decreased a little with time.

Conclusion

Our results show that in spite of the many problems existing in wastewater management, the natural treatment methods can provide a possible alternative for wastewater treatment.

Pond systems are efficient in ammonium-nitrogen removal, reducing it by 83% (41–88%). Their COD_{Cr} removal capacity is only 55% (37–81%). The evaluated BMKO system performed high COD_{Cr} (77%) and TSS (89%) removal. Removal of NH_4-N and TN declines during the years of operation giving an average value of 39% and 49%, respectively. The system is not efficient in phosphorus removal (13%), often it does not remove phosphorus, but even adds to it. In wetlands the 71% COD_{Cr} (53–96%), and 57% TSS

Table 5 Performance of poplar plantations

Poplar plantation	COD_{Cr} (mg/l)			TSS (mg/l)			NH_4-N (mg/l)		
	inflow	Outflow	Removal (%)	inflow	outflow	removal (%)	inflow	outflow	removal (%)
Gyula	249 (15)	48 (15)	83 (21)	–	–	–	–	–	95 (6)
	163–376	*39–86*	*71–93*	–	–	–	–	–	*91–99*
Zalakaros	–	–	91 (62)	–	–	–	–	–	94 (62)
	–	–	*42–99*	–	–	–	–	–	*66–99.9*

Poplar plantation	TN (mg/l)			PO_4-P (mg/l)			TP (mg/l)		
	inflow	Outflow	removal (%)	inflow	outflow	removal (%)	inflow	outflow	removal (%)
Gyula	58.5 (15)	18,9 (15)	74 (20)	–	–	96 (3)	18.58 (15)	5.03 (15)	77 (21)
	46–79	*7.9–48.5*	*27–94*	–	–	*95–97*	*8.6–48.0*	*2.5–11.8*	*52–94*
Zalakaros	–	–	83 (23)	–	–	89 (51)	–	–	–
	–	–	*50–97*	–	–	*56–99.9*	–	–	–

Numbers in brackets indicate the number of measurements
Italic numbers show the minimum and maximum measured values

(33–91%) removal is not as high as expected, but provides satisfactory effluent quality most of the time. Wetlands performed low nutrient removal, giving 17% (–21–46%) for TN and 26% (–20–92%) for phosphorus. Poplar plantations are very effective in pollutant removal. Even the average removal of each nutrient type is above 75%. Still, the effluent data, where it was available, shows that effluent quality may not always meet discharge standards. As natural systems have lower costs than conventional treatment and their performance provides water that is good enough to protect recipients from the effects of raw wastewater, it might be a good idea to revise the standards and allow less stringent standards for these systems.

The assessment could have been more precise, if the data, on which the evaluation was based on, was available in greater amount. Still, the results give an overall picture of the operation of the natural treatment systems applied in Hungary. As a future task, it would be very useful to carry out a widespread measurement programme for a more detailed evaluation of the natural wastewater treatment systems.

Acknowledgement

We are very grateful for the support of the Ministry of Education, "Pro Renovanda Culture Hungarie" Foundation, The Hungarian Professional Association of Water and Sewage Companies, Budapest Sewage Works and Ökotech Ltd. We wish to thank Hajdu-Bihar County Waterworks, Public Service Company of Gyula, Hatvan Waterworks, Aqua-Mélyépterv International Ltd., Vituki Consult for providing data for the assessment.

References

Brix, H. (1987). Treatment of wastewater in the rhizosphere of wetland plants – The root zone method, *Wat. Sci. Tech.*, **19**(1/2), 107–118.

Brix, H. (1994). Use of constructed wetlands in water pollution control: Historical development, present status, and future perspectives, *Wat. Sci. Tech.* **30**(3), 325–333.

Cooper, P.F. and Hobson, J.A. (1988). Sewage treatment by reed bed systems: The present situation in the United Kingdom, In: Donald A. Hammer (ed.), *Constructed Wetlands for Wastewater Treatment: Municipal, Industrial and Agricultural*, Lewis Publisher, USA., pp. 153–171.

Conley, L.M., Dick, R.I. and Lion, L.W. (1991). An assessment of the root zone method of wastewater treatment, *JWPCF*, **63**, 239–347.

Dévai, I. and Woynarovich, E. (1981). Eutrophication and oligotrophication processes occurring in a BMKO sewage treatment plant, *Some Approaches to Saprobiological problems*, 37–47.

Hettiaratchi, J.P.A. and Smith, D.W. (1989). Lagoons and ponds, *JWPCF*, **61**(6), 810–814.

Kárpáti, I., Kárpáti, V., Szabó, I., Szeglet, P. and Tóth, I. (1968). *The use of macrophytes in municipal wastewater treatment – Lake Balaton, Final Report*, Keszthely, Manuscript.

OVH (1984). Alkalmazási engedély a BMKO biológiai szennyvíztisztítási megoldásnak (Licence of BMKO biological wastewater treatment system), *OVH 110.*, Budapest. (In Hungarian).

Reed, S.C., Crites, R.W. and Middlebrooks, E.J. (1995). *Natural Systems for Waste Management and Treatment*, McGraw-Hill, Inc., New York, pp. 423.

Somlyódy, L. (2000). A magyar vízgazdálkodás (Water management in Hungary), *Magyar Tudomány*, XLV. kötet, 6. szám, 657–672. (In Hungarian).

Szilágyi, F. (1996). *A szügyi gyökérzónás szennyvíztisztító próbaüzemének értékelése* (Evaluation of the experimental operation of the root zone wastewater treatment plant at Szügy), Final report, VITUKI, Budapest, pp. 54. (In Hungarian).

Vermes, L. (1985). Results of Poplar Plantation for Wastewater Utilization in Hungary.. In: *Agricultural Waste Utilization and Management, Proceeding of the Fifth International Symposium on Agricultural Wastes*, American Society of Agricultural Engineers, Chicago. pp. 399–403.

Vermes, L. (1996). Special Poplar Plantation for Water Pollution Control in Agricultural Areas, *Hrvaske Vode*, **4**(15), 143–147.

Watson, J.T., Reed, S.C., Kadlec, R.H., Knight, R.l. and Whitehouse, A.E. (1989). In: Donald A. Hammer (ed.), *Constructed Wetlands for Wastewater Treatment: Municipal, Industrial and Agricultural*, Lewis Publisher, USA, pp. 319–351.

Zhao, Q. and Wang, B. (1996). Evaluation on a pilot-scale attached-growth pond system treating domestic wastewater, *Wat. Res.*, **30**(1), 242–245.

Constructed wetlands for wastewater treatment in central Italy

G. Conte*, N. Martinuzzi*, L. Giovannelli**, B. Pucci* and F. Masi*

* IRIDRA S.r.l. -Via Lorenzo il Magnifico 70, 50129, Florence, Italy
** A.R.P.A.T. (Regional Environmental Protection Agency), Dept. of Prato, Via Vittorio Veneto 9, Prato, Italy

Abstract The performance of 4 constructed wetlands designed by IRIDRA Srl and operating in Tuscany (central Italy) has been monitored during the last three years. The 4 treatment systems have different sizes and characteristics: one single stage secondary treatment (150 p.e.); two secondary treatment plants with effluent reuse: one small (60 p.e) and the other big (350 p.e.); a tertiary treatment of effluents from an activated sludge plant with high hydraulic load fluctuation (5–500 p.e.). Due to geographical and economic constraints the four systems have high hydraulic and organic loading rates, nevertheless the systems show very good removal performance of COD (62–95%), especially the ones with higher inflow COD concentrations (87–95%). Interesting results concerning also removal percentage of MBAS (42–88%) and ammonium (42–85%) were obtained, even though NH_4^+ concentration in the outflows of some of the plants, doesn't always comply with Italian quality standards.
Keywords Constructed wetlands; horizontal subsurface flow; macrophytes; reed beds; secondary treatment; wastewater reuse

Introduction

The Tuscany region, is one of the first areas in Italy where constructed wetlands have been used for wastewater treatment. In fact, human settlements in Tuscany still preserve a diffuse pattern originated several centuries ago. Thus a significant part of the population lives in many small isolated areas (ranging from below 100 to one or two thousand p.e.). Some of the original rural settlements have been converted in tourist facilities, hosting up to a few hundred p.e., especially in the spring and summer period. In this situation constructed wetlands appear to be the best technology for wastewater treatment.

The company IRIDRA S.r.l., based in Florence, has designed most of the constructed wetlands for wastewater treatment operating in Tuscany. The performance of four different constructed wetlands designed by IRIDRA have been monitored during the last three years by IRIDRA and ARPAT (Agenzia Regionale per la Protezione dell'Ambiente, the regional public authority responsible for environmental monitoring).

The four treatment facilities are:
1. Moscheta (Firenze): a small village of summer residents with a big restaurant working all year round using a SFS-h (Subsurface Flow System – horizontal) reed bed;
2. Gorgona Island (Livorno): penitentiary wastewater treatment and reuse using SFS-h reed bed series followed by a wet meadow;
3. Spannocchia (Siena): lodging site with wastewater treatment and reuse using a SFS-h reed bed followed by a collecting pond;
4. Pentolina (Siena): tourist village with a high hydraulic loading fluctuation, tertiary treatment of effluents from a total oxidation activated sludge plant by a SFS-h reed bed.

Methods and materials
The dimensioning tools utilised for the design of the four systems was based on published

criteria from the following publications: USEPA (1993), Cooper (1990, 1993), Reed et al. (1995), Kadlec and Knight (1996). The superficial area and basin dimensions were based on desired effluent characteristics, the daily average hydraulic loading, organic load, fill depth and gravel size and the basin slope. Location of the plants (lack of flat areas) and/or economic costraints required the reduction of the plants area to the minimum possible size compatible with desired effluent characteristics. The largest plant in terms of surface is the Gorgona Island plant, but the tertiary system of Pentolina treats a larger population.

At the time the systems were designed, the outlet standards were established in the Italian National Law no 319/76, but, with the enforcement of a new law (D.Lgs. n.152 of 1999), which implements EC Directive 91/271 concerning municipal wastewater treatment, more stringent standards have been adopted (Table 2).

The treatment facilities have been monitored four times a year, in different seasons, since their construction. Standard analysis methods IRSA/CNR were used for both chemical and microbiological measurements in all cases (IRSA/CNR – Water Research Institute of National Research Center – is the Governmental Research Institute that provides the analytical standards for Italy: standards methods are almost the same provided by APHA 1992). Unfortunately microbiological data was not available for Gorgona, therefore only the remaining three systems could be compared in terms of microbiological performance.

Results and discussion

The average concentrations of parameters monitored at the input and at the output of the reed bed (SFS-h) of the four treatments facilities are summarised in Table 4. The performances, in percentage of pollutant removal, are indicated in the following picture.

Table 1 Main features of treatment facilities and number of monitoring samples

	Moscheta	Gorgona	Spannocchia	Pentolina
Load (p.e.)	150	350	60	500
Area SFS-h (m^2)	375	700	160	550
Area SFS-h/p.e. (m^2)	2.5	2	2,6	1.1
Area third stage (m^2)	–	450 Wet meadow	20 Pond	–
Bed depth (m)	0.5	0.6	0.5	0.6
Gravel size (mm)	10–15	5–20	5–10	5–10
HLR (cm d^{-1})	4.8	8	5.6	22.7
Organic loading rate (Kg COD ha^{-1} d^{-1})	220	200	340	450
Water use (l/p/day)	120	125	150	250
Flow (m^3/d)	18	56	9	125
Primary treatment	Imhoff + Degreaser	Grid + Imhoff	Imhoff + Degreaser	Act. Sludge Plant
Operating since	1999	1997	1997	1998
N° of samples	4	11	8	8

Table 2 Outlet criteria

Parameters	Max effluent acceptable concentration	
	Law 319 (1976)	Law 152 (1999)
B.O.D.$_5$	40 mg/l O$_2$	25 mg/l O$_2$
C.O.D.	160 mg/l O$_2$	125 mg/l O$_2$
SS	80 mg/l	35 mg/l
N-NH$_4^+$	15 mg/l	15 mg/l
P-total	10 mg/l P	10 mg/l P
N-total	35 mg/l N	35 mg/l N

Table 3 Mean concentration of inputs and outputs of the reed bed component of the four treatment facilities

		Moscheta		Gorgona		Pentolina		Spannocchia	
		in	Out	in	out	in	out	In	Out
TSS	mg/L	308.0	5.0	138.8	72.7	31.5	18.2	114.2	14.5
COD	mg/L O_2	543.0	29.7	251.5	94.8	175.3	45.9	599.9	78.3
Ammonium	mg/L NH_4^+	52.0	7.6	33.8	19.7	29.3	11.6	74.6	40.8
Nitrates	mg/L $N-NO_3^-$	0.84	0.59	4.0	0.8	16.1	9.3	1.5	3.7
Phosphates	mg/L $P-PO_4^-$	10.97	6.1	11.7	5.6	2.44	1.97	3.4	2.2
MBAS	mg/L	15.8	2.0	3.2	1.2	0.14	0.08	15.4	2.5
Total Coli	CFU/100 ml	165,150,000	40,983	–	–	28,340	1,400	2,088,000	340,100
Faec. Coli	CFU/100 ml	154,048,333	33,919	–	–	2,250	700	126,000	73,433
Faec. Strept.	CFU/100 ml	38,483	907	–	–	2,250	137	261,600	57,513
Esch. Coli	CFU/100 ml	58,335,667	20,302	–	–	100	10	73,333	3,083

Table 3 shows that, considering average results, the effluents of the two reed beds that discharge directly into the environment (Moscheta and Pentolina) are always well within the Italian quality standards regarding discharge of wastewater in the environment, even considering the more stringent standards set by Law 152/99 that were yet to be enforced when the facilities were designed. Considering single sampling, concentrations at the output of Moscheta plant appear to be always far below legal limits for all parameters except for NH_4^+, that once, in September 2000, reached 15.1 mg/l. Also in the Pentolina plant, pollutant concentrations in the output are generally below the standard, except for a sample collected in August 2000, when NH_4^+ concentration reached 50 mg/l: the bad performance was due to a malfunctioning in the activated sludge plant upstream of the reed bed (the input COD and NH_4^+ concentrations were, respectively, 520 and 128 mg/l, nevertheless COD concentration in the output remained at 85 mg/l, far below the legal standards).

The outflows of the other two reed beds (Gorgona and Spannocchia, which have a third stage before discharging into the environment), show higher concentrations: average results exceed legal limits; Gorgona for SS and NH_4^+ and Spannocchia for NH_4^+ only. The third stage treatment occurring in the wet meadow of the Gorgona facility allows a substantial reduction of SS and NH_4^+ in its final effluent, while the small third stage pond of the Spannocchia system is not always able to guarantee good water quality, especially when weather conditions prevent good water oxygenation in the pond.

In Figures 1 and 2 are provided the seasonal variation of COD and NH_4^+ concentrations (input, output of the SFS-h system, output of the collecting pond) in the Spannocchia facility. The figures show that COD at the output of the collecting pond remains always below the legal limits.

The SFS-h system shows a good removal performance: in the single case in which the output from the SFS-h system was over the standard, the pond was able to "buffer" the

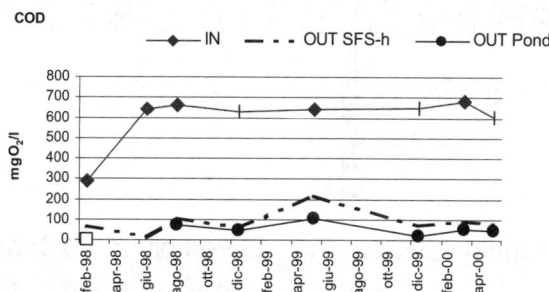

Figure 1 COD concentration values in the inlet and oulet of the Spannocchia CW plant

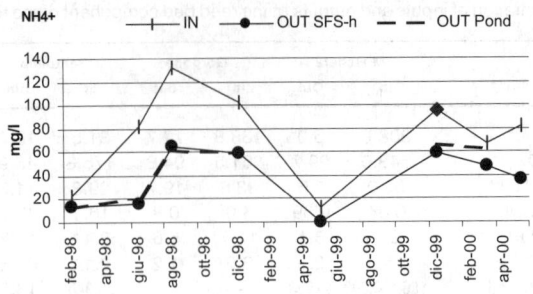

Figure 2 Ammonium concentration values in the inlet and oulet of the Spannocchia CW plant

effect. Concerning NH_4^+, removal perfomance is far lower, even though it is in good agreement with other literature results (Vymazal et al., 1998; Vymazal, 1999), while the buffer capacity of the pond is nearly negligible.

Average removal percentage of the four systems for different pollutants are summarized in Figure 3. In terms of removal efficiency the performance of the Moscheta system appears to be the best of the four, with reference to SS, COD, NH_4^+, MBAS, but the Spannocchia system also shows very interesting results for COD and MBAS. The high COD removal rate appears to be due to the high input concentration, which, in presence of good oxygenation, allows better conditions for bacterial activity. The low TSS reduction of the Pentolina plant is due to the low concentration of the inflow (which is already treated in the activated sludge plant).

Due to the good climate conditions (hot and dry for most of the year) and oxygenation, despite the small area per p.e. of all the four plants (ranging between 1.1 and 2.6 m²/p.e.), COD removal rate appears to be comparable to similar plants with larger area/p.e. ratio operating in other European countries (Cooper et al., 1996; Vymazal et al., 1998). Moscheta and Spannocchia facilities, with an HLR of 4.8 and 5.6 cm/d, show a COD removal rate that ranges around 90%: such results are achieved by plants of the same size operating in colder climates, only with smaller HLR (less than 1.5 cm/d) (Axler et al., 2001).

Two of the plants (Moscheta and Pentolina) show a high removal percentage of ammonium, compared to the literature data (Vymazal et al., 1998; Kemp and George, 1997).

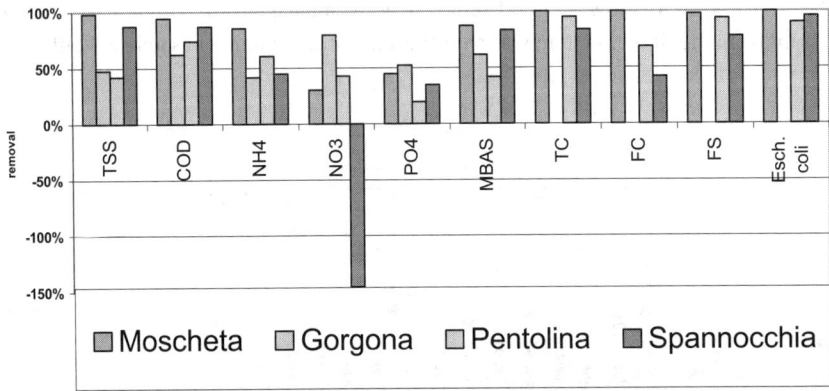

Figure 3 Mean chemical and hygienic parameters removal percentages in four SFS-h systems

Regarding of the microbiological parameters, the three plants where data are available (except Gorgona) show very high performances, often over the 99.9% of removal. These figures are comparable with values obtained in plants with lower HLR in Europe (Ottovà et al., 1996) and with results of experiments using similar HLR but performed in warmer and more constant climates (Khatiwada and Polprasert, 1999; Masi et al., 1999).

References

APHA (1992). *Standard methods for the examination of water and wastewater analysis*. 19th ed. American Public Health Association. AWWA and WPCF, Washington DC, USA.

Axler, R., Henneck, J. and McCarthy, B. (2001). Residential subsurface flow treatment wetlands in northern Minnesota. *Wat. Sci. Tech.*, **44**(11–12), 345–352 (this issue).

Cooper, P.F., Ed. (1990). *European design and operations guidelines for reed bed treatment systems*. WRc Report UI 17, presented to the *Conf. Constructed Wetlands in Water Pollution Control*. Cambridge, U.K., September 1990.

Cooper, P.F. (1993). The use of reed bed systems to treat domestic sewage: the European design and operations guidelines for reed bed treatment systems. In: *Constructed Wetlands for Water Quality Improvement*. G.A. Moshiri Ed., Lewis, Boca Raton, 203–217.

Cooper, P.F., Job, G.D., Green, M.B. and Shutes, R.B.E. (1996). *Reed beds and constructed wetlands for wastewater treatment*. WRc Seven Trent Water.

Kadlec, R.H. and Knight R.L. (1996). *Treatment Wetlands*. CRC Press/Lewis Publishers, Boca Raton, Florida.

Kemp, M.C. and George, D.B. (1997). Subsurface flow constructed wetlands treating municipal wastewater for nitrogen transformation and removal. *Wat. Environ. Res.*, **69**(7), 1254–1262.

Khatiwada, N.R. and Polprasert, C. (1999). Kinetics of fecal coliforms removal in constructed wetlands. *Water Science and Technology*, **40**(3), 109–116.

Masi, F., Martinuzzi, N., Loiselle, S., Peruzzi, P. and Bacci, M. (1999). The tertiary treatment pilot plant of PubliSer Spa (Florence, Tuscany): a multistage experience. *Water Science and Technology*, **40**(3), 195–202.

Ottovà, V., Balcarovà, J. and Vymazal, J. (1996). Microbiological Characteristics of constructed wetlands. *Proceedings of the 5th International Conference on Wetland Systems for Water Pollution Control*, I/13, IAWQ, Vienna, Austria.

Reed, S.C., Crites, R.W. and Middlebrooks, E.J. (1995). *Natural Systems for Waste Management and Treatment* – 2nd ed. McGraw Hill, New York, 173–284.

USEPA (1993). *Subsurface flow constructed wetlands for wastewater treatment*. EPA 832-R-93-001, USEPA Office of Water (WH547).

Vymazal, J., Brix, H., Cooper, P.F., Green, M.B. and Haberl, R. (1998). *Constructed wetlands for wastewater treatment in Europe*. Backhuys publ. Leiden. The Netherlands.

Vymazal, J. (1999). Removal of BOD_5 in constructed wetlands with horizontal sub-surface flow: Czech experience. *Water Science and Technology*, **40**(3), 133–138.

Residential subsurface flow treatment wetlands in northern Minnesota

R. Axler*, J. Henneck* and B. McCarthy*

* Natural Resources Research Institute, University of Minnesota, 5013 Miller Trunk Highway, Duluth, MN 55811, USA

Abstract Approximately 30% of Minnesotans use on-site systems (~500,000 residences) and >50% are failing or non-compliant with regulations due to restrictive soils and site conditions. Many sites occur near lakes and streams creating health hazards and deteriorating water quality. SSF CWs have been evaluated year-round at two northern sites since 1995. The NERCC CWs simulate single homes and the Grand Lake demonstration CW treats STE from a cluster of 9 lakeshore homes. Systems were generally able to achieve design criteria of 25 mgTSS/L and 30 mgBOD5/L and the NERCC CWs required only 0.3m of unsaturated soil to achieve consistent disinfection to <200 fecals/100 mL year round. Seeding experiments with *Salmonella* indicated removal efficiencies of 99.8% in summer and 95% in winter. High strength (~300 mgBOD/L, 95 mgTN/L) influent at NERCC probably limited system performance, particularly N-removal (mass) which was ~42% in summer and 20% in winter. The data indicate CW's are a viable, year-round treatment option for homeowners in terms of performance, ease of operation, and cost but require additional maintenance related to inconsistent vegetation growth, winter insulation, and meeting concentration-based regulatory standards since they were seasonally and annually variable due to rain events, partial freezing, spring snowmelt, and summer evapotranspiration.

Keywords Alternative technologies; cold-climate; constructed wetlands; pathogens; wastewater

Introduction

Historically, constructed wetlands (CWs) have been used world-wide at numerous locations for over 30 years, in warm and cold climates. In Minnesota, CWs have only recently begun to be used at several locations to treat wastewater from both residential and commercial establishments. In 1995, research sites were set up in northern Minnesota near Duluth at the Northeast Regional Correction Center (NERCC) and at Grand Lake (McCarthy *et al.*, 1997) and in southern Minnesota at Lake Washington, near Mankato (Anderson, 1998; Henneck *et al.*, 2001). The research is in its fifth year of testing alternative on-site treatment technologies.

About 30% of Minnesotans rely upon on-site systems for wastewater treatment (~500,000 residences). Unfortunately, >50% are estimated to be out of compliance with state standards or hydraulically failing and effective treatment options are needed for the thousands of locations with restrictive soil and site conditions. In particular, many sites occur near lakes and streams creating a health hazard and deteriorating water quality. Constructed wetlands are one option currently being evaluated, as well as sand, peat and textile filters, aerobic treatment units, and drip irrigation (McCarthy *et al.*, 1997, 1998, 1999). The use of alternative on-site technologies for wastewater treatment in Minnesota and other Great Lakes states will be limited until their seasonal performance is proven acceptable. Accurate assessment of the potential risks of these technologies requires quantification of solids, organic matter, nutrients and pathogen removal efficiencies as well as their operation and maintenance requirements and costs during the entire year.

This paper addresses subsurface flow (SSF) CWs as a viable wastewater treatment option in northern Minnesota based on our experiences from 1995–2000 and presents an overview of existing data. These are small flow subsurface flow gravel beds located at the NERCC (Northeast Regional Correction Center) research facility and at the Grand Lake, MN demonstration project. Both sites experience severe winters with extended periods of air temperatures <–20°C and occasionally to <–40°C.

Methods and system designs

The northern Minnesota research site at NERCC, near Duluth, was designed to allow side-by-side comparisons of the performance of both alternative and standard onsite systems using the same wastewater (septic tank effluent [STE]; see McCarthy *et al.*, 1997 for details) at daily flows approximating those used by single family homes in this region ~ 0.95 m^3/day (250 gal/d). One of the two replicated constructed wetlands (CW2) discharges to a standard drainfield trench monitored at three depths by pan lysimeters filled with fine silica beads. All systems were designed to achieve a *secondary* level of treatment of 25 mg TSS/L, 30 mg BOD$_5$/L, and disinfection to a recreational bathing standard of 200 fecal coliform bacteria per 100 mL. The CWs are two-cell (upper = *Typha*, lower = *Scirpus*), lined, subsurface flow systems. Additional treatment goals for the wetlands were to perform advanced wastewater treatment for nitrogen (TN <10 mg/L) during the growing season (May–Oct) and to improve phosphorus removal by using the best P-adsorbing, locally available substrates. Cell dimensions are 7.0 m L × 5.3 m W × 0.45 m D. Design hydraulic residence time is 13 days with a hydraulic loading rate of 1.3 cm/d (see McCarthy *et al.*, 1997, 1998 for details). For the period through mid-2000, the wastewater strength was higher than anticipated with typical values of BOD$_5$ >300 mg/L and NH$_4$-N ~100 mgN/L).

The Grand Lake cluster system CW was designed to correct the problems of 10 single family homes along a lakeshore just north of Duluth. The CW receives STE via a small diameter pipe to two cells in series designed for a flow of ~4m^3/day with dimensions of 10 m L × 18 m W × 0.6 m D for cell-1 (*Typha*) and 15 m L × 20 m W × 0.60 m D for the unlined, dispersal cell-2 (details in McCarthy *et al.*, 1997; Crosby *et al.*, 1998; and Axler *et al.*, 1999). Design HRT for the measured flows would be ~15 days (at an HLR = 1.3 cm/d), however, a summer bromide tracer study suggested the actual retention time was ~23 days (Kadlec *et al.*, 2001).

Nutrient analysis methods are described in detail in McCarthy *et al.* (1997, 1998) and follow standard methods (APHA, 1995; Ameel *et al.*, 1998). NERCC removal efficiencies are based on mass-removal (i.e. flow weighted) whereas Grand Lake efficiencies are concentration-based (no outflow monitoring). Salmonella seeding experiments are described in Pundsack *et al.* (2001); somatic coliphages were determined by the double agar method as per APHA (1995) and IOS (1993).

Pooled data from Years 2–4 for each CW were used to calculate areal removal rate constants assuming the first order plug flow model as derived by Brix (1998) and Cooper and Green (1998):

$$C_{out} = C_{in}\, exp\,[-k/q] \text{ and } k = k_{20}\, q^{(T-20)}$$

where C_{out} is the effluent concentration (mg/L), C_{in} is the influent concentration (mg/L), k is the first order areal rate constant (m/yr), q is the hydraulic loading (m/yr), k_{20} is the rate constant at 20°C, T (°C) is the temperature, and q is the modified Arrhenius temperature factor. A concurrent nonlinear regression was used to optimize the parameters k_{20} and q (R. Kadlec pers. comm.) for each of the water quality constituents.

Results and discussion
NERCC-TSS/BOD5/nutrients

Overall, during the growing season (May/June through September) the NERCC wetlands (2 cell-train) achieved their primary objective of treating septic tank effluent to 2° treatment standards (25 mg TSS/L, 30 mg BOD_5/L), despite higher strength wastewater than anticipated. Annual summer effluent values averaged (mean of annual means) 8 ± 2 mg TSS/L (85% removal) and 23 ± 10 mg BOD_5/L (92% removal) for 1996–1999 (Figure 1). TSS remained low in winter (mean TSS <9 mg/L), effluent BOD values rose for the same 4 years to 51 ± 17 mg BOD_5/L although the mean %-removal remained a relatively high 79%. Although initially designed to reduce growing season effluent-TN to the drinking water standard of 10 mgN/L (assuming all N converts to nitrate-N), the systems have clearly been unable to convert sufficient ammonium to nitrate in order to denitrify a major fraction of the N-load, presumably due to oxygen limitation. Alkalinities have ranged from 300–500 $mgCaCO_3$/L indicating that nitrification was not limited by available inorganic carbon. Influent TN was typically 80–100 mgN/L, >90% as NH_4-N; BOD_5 ranged from ~250 to >300 mg/L, and oxygen was rarely detected in the wetland effluent (although redox values rose steadily along the length of the systems (Axler, unpubl.). The pH has been circumneutral and unchanged across the length of the wetlands, suggesting that ammonia volatilization was a small component of the actual TN loss. The wetlands removed over 50 mgN/L during the summer of their first full year of operation but %-removal has declined annually – from 68% in 1996 to 19% in 1999 (mean of the annual means = 42% for all four years).

Phosphorus removal averaged 51% in summer and dropped to 20% in winter for the four years, again declining throughout the period of record for each season, suggesting saturation (Figure 1). The limestone substrate in the lower cell of each wetland did not enhance phosphate removal relative to the gravel used in the upper cell, as indicated by the gradual decline down the length of the cells (*cf.* Kadlec *et al.*, 2001), presumably due to pH remaining near 7.

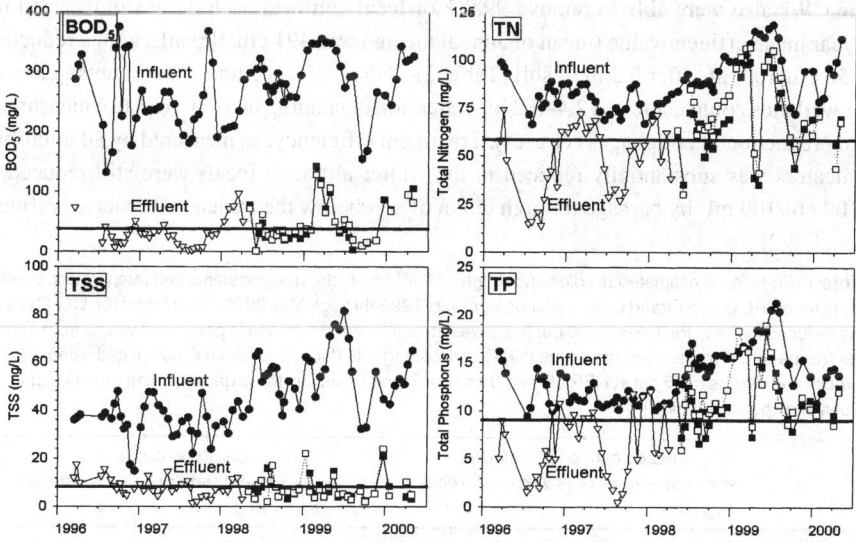

Figure 1 NERCC wetland comon influent (●), and effluent. Effluent values (▽) for 1996–1998 are the means of CW1 and CW2 when inflows were nearly idential. CW2 effluent values for 1998–2000 denoted by open boxes (□) for similar flows; CW1 effluent values for 1998–2000 denoted by closed boxes (■) when flows were reduced by 35%. The solid lines are the mean effluent values (mg/L) for 1996–2000; $BOD_5 = 39$, TN = 58, TSS = 8 and TP = 9

Vegetative growth and nutrient content (N and P) was monitored from fall 1996 (Year 1) to the present to estimate the plant nutrient uptake during the growing season. Growth was luxuriant from the initial planting in spring 1996 (unlike at Grand Lake, see below) and we estimate that the average N-uptake was ~172 mgN/m^2/day over an average 116 day growing season (shoot emergence to first hard frost). We estimate that plant uptake into above-ground biomass represented ~11–23% of the N-removal during this period and 5–8% of the annual influent-N for years 2 to 4. Although these values are low, they are not unusual in the literature (Kadlec and Knight, 1996; Reed *et al.*, 1995), and in 1999, when summer N-removal was only 19%, plant uptake could potentially have accounted for more than the measured N-removal. Note that these are likely to be conservative estimates because we could not measure root growth. Estimated vegetative P-removal was also low, 11–19% during the growing season and only 4–6% of the annual removal. Note that considerable recycling of these nutrients also must take place since plants were not harvested.

Table 1 summarizes first order plug flow reactor rate constants and temperature coefficients calculated for Years 2–4 for comparison to other published studies since this model is widely used by wetland designers. The limitations of using the first order plug flow model are recognized, in particular the assumption of constant flow. There are strong seasonal and shorter-term variations in hydraulics due to evapotranspiration, snow melt runoff, rainstorms, and freezing effects. The k's also vary with hydraulic loading, mass loading and system age (Kadlec, 2000a; Brix, 1998). Despite these shortcomings, the reaction rate constants provide a basis of comparison to previously published results. The results show that the design values based on the North American Wetland DataBase (USEPA, 1994) greatly overestimated rate constants for all parameters (1–2 orders of magnitude) and also underestimated their temperature dependence-particularly BOD$_5$. A detailed analysis of seasonal effects for both NERCC and Grand Lake may be found in Kadlec *et al.* (2001). Grand Lake k_{20} and q values were generally similar to those estimated for the NERCC wetlands.

NERCC-Pathogens

The CWs also were able to remove 99.9% of fecal coliform bacteria in summer and the 4-year mean effluent value (mean of annual means) was 491 cfu/100 mL, with a reduction to 58 cfu/100 mL after 0.3 m of soil (Table 2). *Salmonella* seeding experiments indicated the wetlands could achieve a 2.4–5.2 log reduction depending on flow. We also measured a 2 log reduction in coliphages (viruses). Treatment efficiency, as measured by all microbial indicators was substantially reduced in the winter although fecals were still reduced to <100 cfu/100 mL by passage through 0.3 m of soil below the trench that receives effluent from CW2.

Table 1 Empirically estimated (Kadlec and Knight, 1996) first order rate constants and temperature coefficients for NERCC and Grand Lake wetlands from July 1996 through May 2000. Flow from NERCC-CW1 was reduced by 35% for 1998–2000 but there was no significant difference in parameter values from CW2 over this period. Hydraulic retention time (HRT) and loading rate (HLR) are nominal, calculated using a 40% porosity and depths of 45 cm at NERCC and 60 cm at Grand Lake. L is the organic loading rate. Design values shown parenthetically [ ]

	NERCC-CW1, CW2 (2 cell-trains)		Grand Lake Cell-1	
	HLR = 0.86 m^3/d, HRT = 16–26 d, L = 2.8 g BOD/m^2		HLR = 2.31 m^3/d, HRT = 24 d, L = 2.3 g BOD/m^2	
	K_{20} (m/yr)	θ	K_{20} (m/yr)	θ
TSS	5.0, 8.1 [3000]	0.983, 1.001 [1.000]	13.2 [3000]	1.037 [1.000]
BOD	19.4, 15.0 [180]	1.071, 1.043 [1.000]	19.5 [180]	1.133 [1.000]
Fecals	27, 28 [95]	1.053, 1.049 [1.000]	45 [95]	1.028 [1.000]
TN	1.3, 1.5 [27]	1.019, 1.021 [1.050]	3.0 [27]	1.037 [1.050]
TP	1.7, 1.9 [12]	1.049, 1.035 [1.000]	3.6 [12]	1.064 [1.000]

NERCC-ice and vegetation

Substantial performance declines for all parameters in the last two winters are presumed to be a result of partial freezing in 1998/99 (reducing HRT by ~15–65%), and ultimately complete freezing in Feb 2000. These were warmer (still extremely cold), but much drier than usual winters, and even applying about 0.5 m of straw to the existing plant cover was insufficient insulation without normal snow accumulations to prevent freezing. A similar problem occurred at Grand Lake in late winter 2000 and has forced us to add a 15 cm insulating layer of reed-sedge peat for winter 2001. Performance improvements noted for CW1 in summer 1998 when the flow was reduced by 35% (to decrease the organic loading rate) were not evident in 1999 and we attribute this, at least in part, to poor plant growth this year and invasion by terrestrial weeds. Bulrush declines in cell-2 were particularly apparent and may have been associated with freezing damage the previous winter. We actively removed invasive species in summer 2000 and this, along with the addition of insulation will be important recommendations to homeowners and contractors in the future.

Grand Lake – TSS/BOD$_5$/nutrients/pathogens

Performance data are reported in Table 2 and Figure 2 for Years 2–4. This CW-seepage cell system has successfully solved a lakeshore problem although performance of the CW has again been poorer than expected. As for NERCC, wastewater BOD$_5$ and TN have been higher than design criteria, and the calculated first order rate constants (Table 2) were far below those used for the design. Concentration based performance (no outflow sensing) for the system has been generally acceptable with average (mean of annual means) summer effluent quality of 5 mg TSS/L (82% removal), 45 mg BOD$_5$/L (76% removal), and 443 cfu/100 mL (99.7% fecal removal). Nutrient treatment was relatively poor, with summer values of 48 mgTN/L (only 20% removal) 5.9 mgP/L (30% removal). Winter performance was reasonably good for TSS (6 mg/L;73%) and fecal coliforms (1265 cfu/100 mL; 98.9%) but was again greatly reduced for BOD$_5$ (86 mg BOD$_5$/L;49% removal) and poor for N (45 mgTN/L; only 21% removal) and P (6.6 mgP/L; only 15% removal). Despite somewhat lower strength STE, the Grand Lake CW has had generally poorer performance than at NERCC for several reasons including: (1) Poorer vegetative growth – cattails have not grown well since initial planting in fall 1995 and the CW has been replanted annually, including a variety of species, with only moderate success. Initial construction delays necessitated filling the cell with local bogwater that clearly led to nutrient deficiency for the cattails and then later problems as the denser wastewater stratified within the bed

Table 2 Pathogen removal by NERCC constructed wetlands for septic tank effluent (STE). Fecal coliform and coliphage data compiled from routine monitoring (Winter = Nov–Apr; Summer = May–Oct). Salmonella experiments based on seeding experiments where removal efficiency is based on the total seeded cells recovered following ~7 d of dosing (Pundsack et al., 2001). %-removal (R) was converted to Log factors by Log Removal = – log [1-(%R/100)]. Standard deviations are for seasonal variation (for FC's), for duplicate wetlands (Salmonella experiment), or for trench lysimeters (FC and coliphage). n = number of measurements. [ ][a] = geometric mean of effluent concentrations for both CW's, in cfu/100 mL

	NERCC CW Pathogen Removal (log reductions relative to STE influent)					
	SUMMER (May–Oct: 15.7°C)			WINTER (Nov–Apr: 2.6°C)		
Site	Fecal coliforms (1996–00)	Salmonella choleraesuis (seeded 1998)	somatic coliphage (1998–00)	Fecal coliforms (1996–00)	Salmonella choleraesuis (seeded 1998)	somatic coliphage (1998–00)
CW1 & CW2 (14d & 9d)	2.8 ± 0.6 (n = 64) [491][a]	5.2 @ HRT 14d 2.4 @ HRT 9d	1.7 ± 0.2 (n = 20)	1.9 ± 0.2 (n = 88) [6211][a]	1.4@ HRT 22d 1.3@ HRT 17d	1.2 ± 1.0 (n = 19)
Trench – CW2 0.3 m deep	3.8 [58][a]	5.4 ± 0.4	2.0	3.7 [79][a]	3.3 ± 0.7	1.2
0.6 m deep	3.9 [43][a]	5.8 ± 0.4	2.3	3.8 [26][a]	4.0 ± 10	1.2
0.9 m deep	4.4 [13][a]	6.0 ± 0.1	2.6	4.3 [16][a]	5.9 ± 0.1	1.8

and flowed beneath the root zone (note – these problems were corrected after Years 1 and 3, respectively. The low aspect ratio of the cell (0.6) necessitated by the site likely exacerbated this short-circuiting. (2) Partial (1998/99) and complete (March 2000) winter freezing also reduced the HRT and perhaps affected subsequent plant growth and net oxygen translocation to the bed; and lastly, (3) the bed depth was 0.60 m, relative to NERCC's 0.45 m. The bed was deeper in order to minimize potential for freezing, but the trade-off is that less water is exposed to the root-zone. Our experiences at these and other Minnesota SSF CWs has been that typical rooting depths have only been ~0.25–0.40 m (Axler et al., 1999, unpubl.).

The entire system, CW + seepage cell, has however, consistently met design criteria for TSS, BOD and fecals in summer, and for TSS and fecals in winter (Axler et al., 1999; Henneck et al. 2001; Kadlec et al., 2001). Summer averages for the seepage cell for 1997–1999 were 4 mg TSS/L (system removal 88%), 29 mg BOD/L (84% removal) and 30 cfu/100 mL (99.98% removal) and winter averages were 8 mg TSS/L (69%), 63 mg BOD (4%) and 62 cfu/100 mL (99.94% removal), respectively. TN and TP means for summer were 41 mgN/L (32%) and 3.9 mgP/L (50%) and for winter were 34 mgN/L (43%) and 5.1 mgP/L (35%).

Conclusions

Overall, despite poorer performance and more operation and maintenance problems than originally anticipated, SSF constructed wetlands offer significant potential for effectively treating small-flow, domestic residential wastewater at sites with poor and/or shallow soils, or limited drainfield areas in the cold climate of northern Minnesota. Previous analyses have shown them also to be viable financially (McCarthy et al., 1997) and they also offer the potential for significant nitrogen reduction, unlike many other alternative technologies. The data to date suggest that literature summaries of first order plug flow reaction rate constants and Arrhenius temperature coefficients are too high for our systems although the reasons may relate to the relatively high strength septic tank effluent in our systems. Although

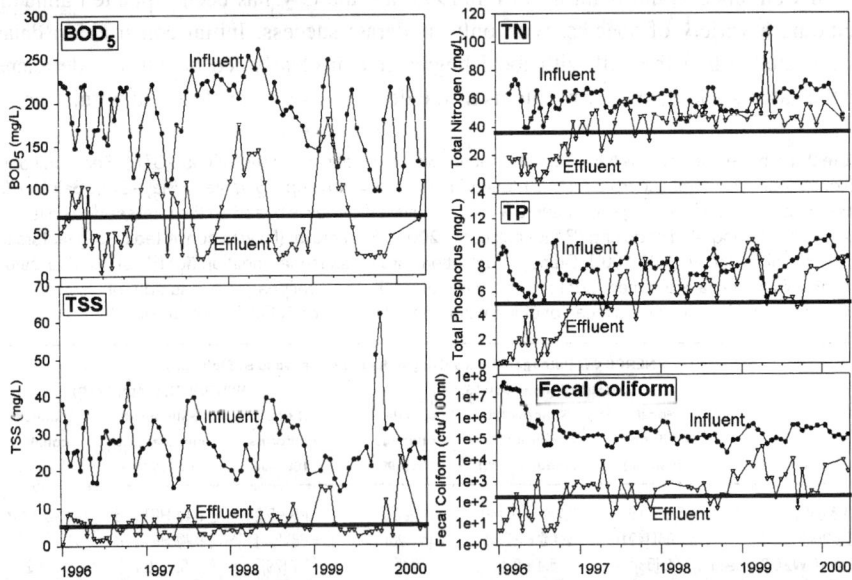

Figure 2 Grand Lake wetland influent and effluent values for 1996–2000. The solid lines are the mean effluent values (mg/L unless noted); BOD_5 = 69, TSS = 5, TN = 37, TP = 5, and fecal coliform = 323 cfu/100 ml

higher STE values than reported in the literature (e.g. Reed *et al.*, 1995; Kadlec and Knight, 1996; Crites and Tchobanoglous, 1998) one would expect higher values in areas constrained by site conditions due to voluntary water conservation.

Winter in northern Minnesota presents special challenges, particularly in light of the fact that our recent warm winters were much more of a problem than the first two cold winters because of the lack of snow cover (cf Kadlec, 2001). Maehlum and Jenssen (1998) offer some cold weather design considerations based on their experiences in Norway, but at odds are the need to provide longer retention times to meet performance expectations due to the colder temperatures and the potential for freezing due to the longer retention time and the potential for water to pass below the root zone if the beds are deepened. The shorter growing season also allows the plants less time to fill in densely in a climate where insulative growth is critical. After determining the flow characteristics (both rate and wastewater strength) and the performance level required, the wetland size must be balanced between being large enough for adequate treatment but not so large as to allow freezing problems to occur. Local building codes require that uninsulated water pipes be buried five feet (1.5 m) below the surface to prevent freezing so special care must be taken since pipes will hold water just below the surface of the wetland. The outlet structure is an area of special concern as the wastewater is most exposed to the environment and at its coldest temperature. We have now added an insulating layer (15 cm) of reed-sedge peat to be left in place throughout the year as per Wallace *et al.* (2001). Other construction techniques to minimize freezing include foam around and on top of the septic tank to help maintain the wastewater heat, grass over the piping to provide additional insulation, and limiting traffic over the pipes to minimize compaction of the earth and lessen frost depth. Plant growth is critical by providing insulating mulch and trapping snow. A layer of foam around the perimeter of the wetland beds may also be prudent.

Acknowledgements

We thank S. Bamford, P. Drevnick, S. Monson-Geerts, D. Nordman, J. Pundsack, S. Syria and the Western Lake Superior Sanitary District for lab and field assistance; the Legislative Commission for Minnesota Resources, Minnesota Technologies Inc., Iron Range Resources and Rehabilitation Board, St. Louis County Health Dept. and Minnesota Sea Grant for funding. Dr. R. Kadlec helped with design and data analysis from the project's inception and Dr. R. Hicks co-directed the *Salmonella* studies. This is Contribution # 199 from the Center for Water and the Environment at NRRI.

References

Ameel, J., Ruzycki, E. and Axler, R.P. (1998). *Analytical chemistry and quality assurance procedures for natural water samples.* 6th edition. Central Analytical Laboratory, NRRI Tech. Rep. NRRI/TR-98/03.

Anderson, J. (1998). Another alternative: Lake Washington cluster drainfield, *Focus 10,000- Minnesota's Lakeside Magazine* (July 1998) 7, 16–19.

APHA (1995). *Standard Methods for the Examination of Water and Wastewater*, 19th ed, Amer. Public Health Assoc./Amer. Water Works Assoc./Water Environment Federation, Washington, D.C., USA.

Axler, R., Henneck, J., Nordman, D., McCarthy, B. and Monson Geerts, S. (1999). Operation and Maintenance Experiences with Constructed Wetlands in Minnesota. Pages 219–223. In: *Proc. National On-Site Wastewater Recycling Assoc.* (NOWRA) 8th Annual Conference, Nov. 3–6, 1999 Jekyll Island, Georgia, USA.

Brix, H. (1998). Denmark. In: *Constructed wetlands for wastewater treatment in Europe*, J. Vymazal, H. Brix, P.F. Cooper, M.B. Green and R. Haberl (eds.), Backhuys Publ., Netherlands, pp.123–152.

Cooper, P.F. and Green, M.B. (1998). UK. In: *Constructed Wetlands for Wastewater Treatment in Europe*, J. Vymazal, H. Brix, P.F. Cooper, M.B. Green and R. Haberl (eds.), Backhuys Publ., Netherlands, pp.315–335.

Crites, R. and Tchobanoglous, G. (1998). *Small and Decentralized Wastewater Management Systems.* McGraw-Hill, Companies, Inc., 1084 p.

Crosby, J., McCarthy, B., Gilbertson, C. and Axler, R. (1998). A regulatory perspective on impediments and solutions to the use of performance standards for on-site wastewater treatment. Pages 259–267, In: *Onsite Wastewater treatment*, Proc. 8th Internat. Symp. on Individual and Small Community Sewage Systems, Orlando, FL, March 1998, Amer.Soc. of Agricultural Engineers, St. Joseph, Missouri. 49085-9659 USA.

Henneck, J., Axler, R., McCarthy, B., Monson Geerts, S., Heger Christopherson, S., Anderson, J. and Crosby, J. (2001). Onsite treatment of septic tank effluent in Minnesota using SSF constructed wetlands: Performance, costs and maintenance. *Proc. of 9th Nat. Symp. on Individual and Small Community Sewage Systems*, Ft Worth, TX, 2001, March 11–14, 2001. Amer. Soc. of Agric. Engineers., St. Joseph, MI. 49085-9659 USA.

International Organization for Standardization (IOS). (1993). Water Quality – Detection and Enumeration of Bacteriophages, Method 10705-1 Amended 1997.

Kadlec, R. (2000a). The inadequacy of first-order treatment wetland models. *Ecol. Engineer.* **15**, 105–119.

Kadlec, R.H. (2001). Thermal environments of subsurface treatment wetlands.*Wat. Sci. Tech.*, **44**(11–12), 251–258 (this issue).

Kadlec, R. and Knight, R. (1996). *Treatment Wetlands.* Lewis Publ., Boca Raton, FL, 893 p.

Kadlec, R.H., Axler, R., McCarthy, B. and Henneck, J. (2001). Subsurface treatment wetlands in the cold climate of Minnesota. In: *Advances in Ecological Sciences*, U. Mander and P. Jenssen, Editors, Constructed Wetlands for Wastewater Treatment in Cold Climate Areas. WIT Press, Southampton, U.K.

Maehlum, T. and Jenssen, P. (1998). Norway. In: *Constructed Wetlands for Wastewater Treatment in Europe*, J. Vymazal, H. Brix, P.F. Cooper, M.B. Green and R. Haberl (eds.), Backhuys Publ., Netherlands, pp. 206–217.

McCarthy, B., Axler, R., Monson Geerts, S., Henneck, J., Crosby, J., Nordman, D., Weidman, P. and Hagen, T. (1997). Development of alternative on-site treatment systems for wastewater treatment: A demonstration project for Northern Minnesota. Tech.Rep. NRRI/TR-97/10, U. of Minnesota, Duluth, MN 55811, USA.

McCarthy, B., Axler, R., Monson Geerts, S., Henneck, J., Nordman, D., Crosby, J. and Weidman, P. (1998). Performance of alternative treatment systems in Northern Minnesota. Pages 446–457, In: *Onsite Wastewater treatment*, Proc. 8th Internat. Symp. on Individual and Small Community Sewage Systems, Orlando, FL, March 1998, Amer. Society of Agricultural Engineers, St. Joseph, Missouri. 49085-9659 USA.

McCarthy, B.J., Axler, R., Monson-Geerts, S., Henneck, J., Crosby, J. and Weidman, P. (1999). Cold weather operation and performance of alternative treatment systems in Northern Minnesota. p. 37–44 In: *Proceedings for the National On-Site Wastewater Recycling Association (NOWRA) 8th Annual Conference*, November 3–6, 1999 Jekyll Island, Georgia, USA.

Pundsack, J., Axler, R., Hicks, R., Henneck, J., Nordman, D. and McCarthy, B. (2001). Seasonal Pathogen Removal by Alternative On-site Wastewater Treatment Systems. *Water Environ. Research* **73**(2), 204–212.

Reed, S., Crites, R. and Middlebrook, E. (1995). *Natural Systems for Waste Management and Treatment.* McGraw-Hill Companies, Inc., 433 p.

USEPA (1994). Wetlands treatment database. Version 1.0. Risk Reduction Engineering Laboratory, Cincinnati, OH. (June 1994).

Wallace, S., Parkin, G. and Cross, C. (2001). Cold climate wetlands: design and performance. *Wat. Sci. Tech.*, **44**(11–12), 259–265 (this issue).

On-site domestic wastewater treatment by reed bed in the moist subtropics

L. Davison, T. Headley and M. Edmonds

School of Environmental Science and Management, Southern Cross University, PO Box 157, Lismore, 2480, Australia. (E-mail: *ldavison@scu.edu.au*)

Abstract This paper summarises the results of studies on four subsurface flow wetlands (reed beds) located in the moist sub-tropical north eastern corner of the Australian state of New South Wales. The reed beds, which are subjected to a variety of effluent types, all have a gravel substrate planted with *Phragmites australis*. All four units were found to maintain satisfactory treatment performance year round. Mean removal efficiencies ranged from 56% to 90% (SS), 70% to 93% (BOD), 38% to 66% (TN), 87% to 99.8% (Faecal coliforms), and 42% to 70% (TP – with one seasonal result of 0% for the eight year old unit) for the four reed beds. After eight years in operation the oldest reed bed was showing signs of phosphorus saturation with outlet TP concentrations exceeding inlet concentrations on some occasions. The youngest reed bed studied appeared to be operating efficiently after five months. A summer water balance on one of the reed beds revealed an average crop factor of 1.6 and a moisture loss to atmosphere of 40% of influent flow. Treatment performance (particularly for TN and SS) was found to be negatively correlated with rainfall during one study. The paper discusses the implications of the above results for on-site system designers and regulators and identifies areas for further investigation.

Keywords Moist subtropics; on-site wastewater treatment; reed bed; subsurface flow constructed wetland; treatment performance

Introduction

Some twelve per cent of Australia's population (about two million people) use on-site systems for the treatment and disposal of wastewater (Geary and Gardner, 1996). The traditional approach to on-site domestic wastewater management in Australia has been to use a septic tank to remove some suspended solids and BOD prior to infiltrative disposal by subsurface land application, usually via a relatively small gravity flow leach field (absorption trench). Recent studies have indicated a high level of failure in these systems due to either poor design, inappropriate location or lack of maintenance (Geary, 1992). A major problem has been the leaching of nitrate into ground and surface waters. An additional issue has been the clogging of absorption trench surfaces by an impermeable film or "biomat" leading to surface ponding of potentially pathogenic effluents. Because of these problems and the associated threats to environmental and human health most Australian states have begun to revise their on-site system regulations. In the state of New South Wales (NSW) the release of a new set of Environment and Health Protection Guidelines (1998) has given impetus to an already existing trend towards innovation in the design of on-site systems. The new guidelines encourage the use of source control approaches such as waterless toilets. They also attempt to overcome the problem of clogged absorption trenches and groundwater pollution by encouraging evaporative disposal of wastewater to the atmosphere via irrigation fields which are relatively large compared to the traditional absorption trench. The new state guidelines delegate considerable regulatory discretion to local government bodies (councils) which are each required to prepare their own on-site wastewater management strategy.

Lismore City and Byron Shire are two adjoining local government areas in north eastern

New South Wales which have been a centre for innovation in relation to on-site system design for some years. The region experiences a moist subtropical climate (mean annual precipitation 1,700 mm, mean summer max. temp. 29°C). Local on-site system designers, faced with the challenge of developing cost effective systems which reduce the hydraulic and pollutant loads discharged into the environment, have recently begun to experiment with the inclusion of horizontal flow constructed wetlands (reed beds) as elements in both black and grey water treatment systems. Both Byron Shire and Lismore City Councils make provision for the use of reed beds in their respective draft on-site wastewater management strategies. This paper:

- summarises the results of studies on four reed beds, using *Phragmites australis* planted into gravel, employed in various on-site wastewater treatment applications in the Lismore/Byron area;
- discusses the relevance of the findings of these studies to local regulatory instruments;
- discusses the question of the applicability of reed beds in moist subtropical on-site systems; and
- identifies areas for further investigation.

System descriptions and methods

A summary of the main features of the four systems studied is provided in Table 1. Systems 1 and 2 were monitored from July to November 1997 while Systems 3 and 4 were studied in from March to August 1999.

Reed Beds 1 and 2 (RB1 and RB2) each treat the wastewater from schools of 250 students. Septic tanks provide primary treatment for both systems. RB1 (see Figure 1) consists of two separate wings which are loaded on alternate weeks. It was completed just prior to the study conducted in 1997 while RB2 (see Figure 2), the first reed bed built in the region, was eight years old at that time. The ten cells in RB2 are loaded in pairs from a flow diverter which is shifted each weekday morning. One of the objectives of this study was to compare the performance of a mature (RB2) and a new (RB1) system. A second objective was to assess the effect, if any, of season on treatment performance, so sampling was conducted during winter (July), spring (September) and summer (late November). Both reed beds discharge into open storage ponds prior to disposal so a third objective was to investigate the degree of treatment provided by the ponds. Five effluent samples were collected during each of the three seasonal sampling periods from the inlet and outlet of

Table 1 Summary of system features for the four reed beds

System Feature	System 1	System 2	System 3	System 4
Date commissioned	April 1997	Feb 1989	Jan 1997	June 1998
Study period	Jul–Sept 1997	Jul–Sept 1997	Feb–Aug 1999	Feb–Aug 1999
Social situation and type of load	school / mainly toilets	school / mainly toilets	shower, kitchen, dairy	home / greywater
People served	250 students	250 students	12 part time	1 full time
Primary treatment	septic tanks	septic tanks	grease trap	gravel filter
Reed bed surface area	80 m^2	80 m^2	8.8 m^2	3.6 m^2
Hydraulic loading mm/d	22	28	34	28
Reed bed depth (metres) and no. of samples taken *	0.5 m N = 2 × 5 = 10 excludes winter	0.5 m N = 3 × 5 = 15	0.4 m N = 11	0.43 m, N = 8 0.34 m, N = 10 0.5m , N = 5
Mean nominal HRT	9.1 days	7.1 days	4.5 days	5.7 days
Pre-disposal storage	pond	pond	tank	tank
Disposal	absorption trench	irrigation	absorption and irrigation	irrigation

* Water level was varied in System 4

each reed bed. Composite samples were collected from each pond. A peak loading on System 1 was also monitored. A summer water balance was conducted on RB1 with a view to determining the crop factor of the *Phragmites australis* reeds. Flows into RB1 were estimated by timing the duration of dosing pump operation. Flows out of RB1 were measured using tipping bucket flow meters. Piezometers were used to measure water level drawdown as a cross check to the water balance. Rainfall data was collected on-site and Class A pan evaporation collected at a weather station 30 km from each site. A more detailed description of the studies on RB1 and RB2 is given in Headley (1998).

Reed Bed 3 (RB3, see Figure 3) treats greywater from the communal house with compost toilet on a rural community. Some twelve people use the house on a part time basis. On a typical day three or four showers are taken there and an evening meal for between twelve and twenty people is prepared. Dairy products (cheese, butter etc.) are also produced there so a relatively greasy, high BOD, low TN effluent is produced. Primary treatment is by a 360 L grease/sediment trap. System 3 was sampled at the grease trap outlet, at a point one metre from the reed bed inlet, and at the reed bed outlet. The fourth reed bed studied (RB4, see Figure 4) treats greywater from a single person dwelling. Primary treatment is by vertical flow aerobic gravel filter. After passing through the reed bed the greywater is stored in a tank from which it is pumped through a UV disinfector and a set of flowforms prior to disposal by irrigation. System 4 was sampled in the reed bed inlet and outlet chambers. Samples (11 for RB3 and 23 for RB4) were collected on a weekly basis during the autumn-winter period of 1999. Rainfall data was collected on-site in each case and Class A Pan evaporation data was obtained from a weather station about 30 km from each site. For RB3 weekly outflow volumes were estimated from the level change in the effluent storage tank

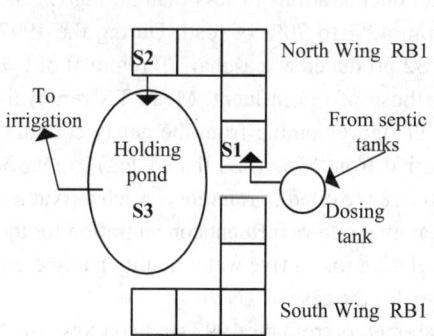

Figure 1 System 1 schematic with North Wing of RB1 operational

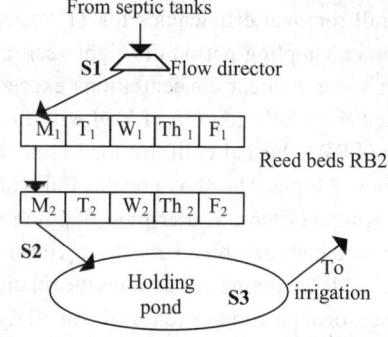

Figure 2 System 2 schematic with Monday cells of RB2 operational

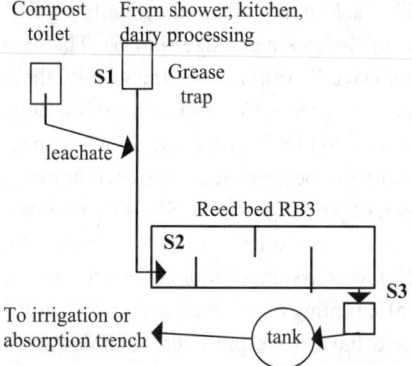

Figure 3 System 3 schematic

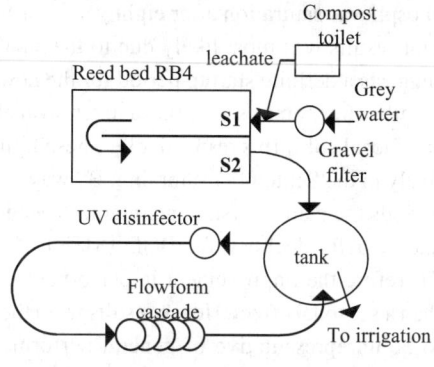

Figure 4 System 4 schematic

and an assumed crop factor of 1.3 applied to estimate inflowing volume. For RB4, inflowing volumes were obtained using a water meter on the house supply line and outflows estimated using the assumed crop factor of 1.3. This assumption was based on the results of an earlier study (Headley, 1998). The main objectives of the study were to assess the treatment performance of the reed beds with respect to nutrients, SS, BOD and disinfection. Six of the System 3 samples were also tested for oils and grease. The water level in RB4 was varied to assess the effect of lowered levels on treatment performance. As the rainfall during the sampling period turned out to be much higher than average, the opportunity was taken to investigate the effect of precipitation and resultant decreased detention time on treatment performance. A more detailed description of the studies on RB3 and RB4 is given in Edmonds (1999).

Water quality samples from all four reed beds were analysed at Southern Cross University's Environmental Analysis Laboratory for suspended solids (SS), BOD_5, faecal coliforms, total nitrogen (TN) and total phosphorus (TP) in accordance with procedures described in Standard Methods (1992).

Results and discussion
Reed bed studies

Results for the four systems are summarised in Table 2. Of note is the fact that the two greywater systems (RB3 and RB4) had an influent TN concentration an order of magnitude lower than the school systems. In relation to treatment performance, three of the reed beds removed between 85% and 90% of the SS load. In all four cases effluent SS concentrations of 15 mg/L or less were achieved. BOD removal efficiencies varied from 70% to 93% of load with three of the units achieving effluent concentrations of less than 20 mg/L. The overall removal efficiencies for TP varied from 42% to 70% of load. During the 1997 summer sampling period the eight year old RB2 produced a seasonal TP removal of 0% with some effluent concentrations exceeding those of the influent. Mean TN removals varied from 38% to 66% of load with the lower figure coming from the newly commissioned RB1. Faecal coliform load removal varied from 87% (less than 1 log) to 99.8% (almost 3 logs). The above results fall within the range cited in references such as Kadlec and Knight (1996). Scatter-plots of pollutant removal rate vs influent concentration for the four reed beds exhibit a very strong linear correlation for all five water quality parameters (see Table 2). Figure 5 illustrates the relationship for the case of TN.

Seasonal percentage removals for all five parameters from the 1997 study on Systems 1 and 2 were subjected to a two-way ANOVA to investigate differences in performance on the basis of system maturity and season. The only significant result ($p<0.05$) was for TP removal between the old and new reed bed. It was apparent that RB2 was approaching phosphorus saturation after eight years service. The lack of any other statistically significant results was most likely due to the smallness of the treatment size ($n = 5$). The data suggests a definite startup period for the new system (RB1). Initial sampling in July, three months after commissioning, gave removal efficiencies of 56%, 35% and 70% for SS, BOD and faecal coliforms respectively. These figures increased to 85%, 85% and 99.9% respectively in the September sampling. By way of comparison: between the same two sampling periods the mature system maintained a steady level of performance for SS (93% to 95%), and actually declined for BOD (81% to 61%) and faecal coliforms (97.9% to 67.1%). Therefore the improvement in performance in RB1 was assumed to be a maturity rather than a seasonal effect. Hence results from the initial sampling of this reed bed were deemed to be unrepresentative of general performance and have therefore been excluded from the overall analysis given in Table 2. The overall impression was that the new system had reached operating efficiency within five months of commissioning and that the expected winter decline in performance had not occurred. The study revealed that the ponds were very effective at removing nutrients from the waste stream with TN removal in the System 1 pond averaging 77% and TP removal averaging 66%. Overall TN removal from the combined reed bed/pond system exceeded 85%. On the other hand the pond was found to produce an algal induced resurgence in SS and BOD concentrations which was particularly marked in summer.

Seven months after commissioning, System 1 was subjected to a daily hydraulic loading rate some 2.4 times normal during an open day at the school. Samples were taken every second day for 10 days before and after the peak event. Reed bed influent showed concentration spikes considerably in excess of pre-event levels with SS jumping from 40 mg/L to 125 mg/L, TN from 95 mg/L to 145 mg/L and faecal coliforms from 10^6 cfu/100 ml to 10^7 cfu/100 ml. In the days after the event reed bed effluent showed no noticeable change in SS concentration, TN jumped from 55 to a peak of 95 mg/L five days after the event before dropping back to 70 mg/L a day later, while faecal coliforms jumped from a pre-event average of 1,100 to 100,000 cfu/100 mL. In the pond the only parameter to change after the event was faecal coliforms which jumped from its pre-event mean of 880 to 14,800 cfu/100 ml four days later, before dropping to more typical levels six days after the event. The combination of reed bed and pond therefore appears to be quite robust under peak loading. The water balances were conducted on RB1 during December 1997, some eight months after commissioning. Daily water balances over a one week period gave crop factors of 0.9 for sparsely vegetated and 1.6 for densely vegetated wetland cells. During this week the average daily maximum temperature was 29.5°C and some 40% of the hydraulic load was evapotranspired.

In the 1999 study, RB3 was found to remove 97% of its oil and grease load. It was found that 89% of this removal actually occurred in the first metre of the reed bed. The corresponding figures for the other parameters were SS 68%, BOD 61%, faecal coliforms 89%, TN 71% and TP 63% suggesting that the majority of treatment was being achieved by physical removal in the entrance zone. After two years of use the top 50 mm of gravel in the entrance zone of RB3 was showing signs of clogging and the rich surface sediment around the inlet had been densely colonised by several species of earthworm. Although no overland

Table 2 Summary of mean removal rate and mean per cent load removal for the measured water quality parameters. Correlation coefficients for influent concentration vs. removal rate are shown in column 1

Parameter	variable	R. Bed 1 n = 10	R. Bed 2 n = 15	R. Bed 3 n = 11	R. Bed 4 n = 23
SS	influent conc. mg/L	45.5	125	102	40
$r = 0.99$	removal rate g/m^2/d	1.3	3.5	2.9	0.6
	% load removal	88	90	85	56
BOD	influent conc. mg/L	116	117	264	54
$r = 0.998$	removal rate g/m^2/d	3.2	2.9	8.3	1.1
	% load removal	88	80	93	70
TP	influent conc. mg/L	15	22	5.1	5.2
$r = 0.998$	removal rate g/m^2/d	0.23	0.29	0.12	0.12
	% load removal	49	42	70	50
TN	influent conc. mg/L	137	187	19	14
$r = 0.999$	removal rate g/m^2/d	2.2	3.0	0.43	0.22
	% load removal	38	51	66	55
Faecal col.	Influent conc. mg/L	4.8×10^5	160×10^3	2.1×10^5	1.4×10^6
$r = 0.91$	removal rate cfu/m^2/d	15×10^6	44×10^6	6.6×10^6	1.4×10^8
	% load removal	99.8	87	99.8	96
Oil and grease	Influent conc. mg/L			72.5	
	Removal rate g/m^2/d			2.2	
	% load removal			97	

flow or short circuiting has yet been observed, measures have since been implemented to reduce the oil and grease load by augmenting the grease/sediment trap at the front end of the system. A significant correlation ($r = 0.67$, $p<0.05$) was found to exist between the reductions in TN and SS concentrations in the first metre of this reed bed suggesting that much of the TN load in this influent was probably associated with solids. Future monitoring will attempt to determine whether the worm activity in the entrance zone is returning this trapped nitrogen to the effluent or to the atmosphere.

One way to facilitate breakdown of upper level organic material in a reed bed is to lower its water level to create aerobic conditions. Water level lowering can also be used to encourage deeper penetration of roots. One of the objectives of the RB4 study was to assess the effect of water level lowering (and the resulting reduction in hydraulic residence time (HRT)) on treatment performance. Periods of unusually heavy rain throughout the study period also had an effect on residence time. The assumed hypotheses were that load removal efficiency during any one of the 23 week-long periods between sample collections would be positively correlated to the actual theoretical hydraulic residence time (taking into account additional rain induced loading) and negatively correlated to precipitation during that time. Table 3 lists the relevant correlation coefficients for the five parameters tested. As expected all parameters were negatively correlated with precipitation.

Implications for system designers and regulators

The vast majority of on-site wastewater management systems in Australia dispose of effluent by land application in order to take advantage of the treatment capability of the local soil. Where there is insufficient area of suitable soil available to provide adequate treatment it may be economically and environmentally desirable to include reed bed treatment in the system design. In NSW an additional consideration is the fact that the new guidelines discourage disposal by the downward flow of effluent to groundwater and encourage the maximisation of opportunities for evapotranspirative return of water to the atmosphere. The question of disposal field design and size is dealt with differently by each local government body. In the case of Lismore City Council's Draft On-site Sewage and

Table 3 Correlation coefficients for weekly load removal efficiency vs. (a) weekly precipitation and (b) hydraulic residence time for RB4

pollutant	(a) Precip.	(b) HRT
BOD	−0.30	0.49*
SS	−0.79**	0.37
F. Col.	−0.09	0.5**
TN	−0.52*	0.02
TP	−0.24	−0.03

*$p<0.05$, **$p<0.01$

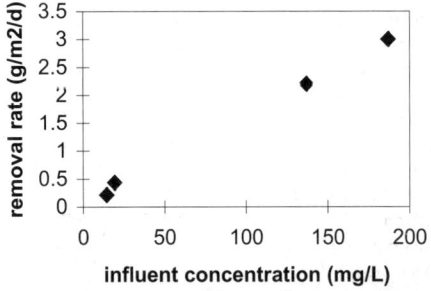

Figure 5 TN removal rate vs. TN influent concentration for the 4 reed beds

Wastewater Management Strategy (1999) the size of any given disposal field is determined separately for assumed hydraulic, nitrogen and phosphorus loadings using mass balance models. The largest of the three areas determined by the models is the size chosen. On the basis of the results outlined above, Lismore City Council has agreed to halve the design TN loading to the disposal area if a reed bed with a seven day HRT is included in the design. In systems with flush toilets TN loading is usually the factor which determines disposal area so the inclusion of a reed bed may therefore be indicated in such systems where suitable disposal space is limited.

The treatment performance of reed beds with respect to faecal coliforms and TP points to a role for them in locations where soil is unlikely to provide adequate disinfection and phosphorus removal. Examples would be sites with high water tables and/or bedrock, areas with highly permeable soils and disposal areas close to waterways. A major failure mode for trench based systems is the clogging of soil pores by a biomat which forms under saturated conditions at the soil/trench interface. A study by Converse and Tyler (1994) found that twelve of fifteen clogged disposal trenches recovered when a BOD removing element (in this case an aerated system) was inserted between the septic tank and trench. This suggests that the BOD and SS removing capacity of reed beds may be used to advantage in renovating clogged absorption trenches and in preventing failure of new trench systems.

The high crop factor found in the RB1 study indicates that reed beds may be capable of reducing hydraulic loadings and therefore disposal area size where hydraulic loading is the determining factor (e.g. most systems with dry toilets under Lismore City Council's Draft Strategy). A problem in wet climates is the fact that reed beds also take in most of the incident precipitation leading to reduced treatment performance as found in the study on RB4. The question that arises here is: "how can designers take advantage of the excellent evaporative performance of reed beds while avoiding ingress of precipitation into the aqueous waste stream?". One approach might be to build reed beds under wide overhanging eaves on the sun facing side of a house.

Conclusions and further work

The reed beds studied exhibited treatment performances in line with those found in the literature and showed little variation in seasonal performance under the subtropical climatic conditions. Daily removal rates for all five parameters tested were found to be strongly correlated with influent concentration. The new system RB1 took about five months to settle in, while the oldest reed bed RB2 was showing signs of phosphorus saturation after eight years. Of particular importance in relation to sizing of disposal fields under NSW On-site Guidelines is the nitrogen removing capacity of reed beds which was found to be in excess of 50% of TN load in the three mature systems. This has led to Lismore City Council allowing a halving of the design TN load applied to any disposal field which is preceded by a reed bed with seven day HRT, resulting in the need for a smaller disposal field when TN is the determining factor. The combination of reed bed and pond in System 1 was found to remove in excess of 85% of TN. This combination of treatment elements was also found to be particularly robust under peak loading. Entrance zone clogging in RB3 after only two years service illustrates the importance of efficient primary treatment and the need for the development of "declogging" management strategies. The water balance on RB1 found that densely growing *Phragmites australis* had a crop factor of 1.6 leading to the conclusion that evaporative disposal of wastewater through reed beds may be economically feasible in certain situations. In wet climates both treatment and hydraulic performance should improve if design strategies that minimise ingress of rainfall can be developed.

Results such as those reported in this paper have been sufficiently encouraging for the NSW Department of Local Government to allocate funds under its "Septic Safe" program for further research into reed bed performance in the moist subtropics. The research will have two directions. The first will involve further monitoring of domestic on-site reed beds with a view to obtaining locally relevant data on loadings as well as on reed bed performance. The second direction involves the construction of a well instrumented test facility at the South Lismore Sewage Treatment Plant where a number of reed bed designs will be monitored on a long term basis under relatively controlled conditions. The test facility will also serve as a demonstration site for innovative and alternative on-site system technologies.

References

Converse, J.C. and Tyler, E.J. (1994). Renovating failing septic tank-soil absorption systems using aerated pretreated effluent. In *Proceedings of 7th International Symposium on Individual and Small Community Sewage Systems*, 11–13 Dec. Atlanta, Georgia, USA.

Draft On-site Sewage and Wastewater Management Strategy (1999). Lismore City Council, Lismore, NSW, Australia.

Edmonds, M. (1999). *The Performance of Two Reed Bed Greywater Treatment Systems in North Eastern New South Wales*. Honours thesis, School of Resource Science and Management, Southern Cross University, Lismore, Australia.

Environment and Health Protection Guidelines – On-site Sewage Management for Single Households (1998). NSW Dept. of Local Government, NSW EPA, NSW Dept. of Health, NSW Dept of Land and Water Conservation and NSW Dept. of Urban Affairs and Planning, Sydney, Australia.

Geary, P.M. (1992). Diffuse pollution from wastewater disposal in small unsewered communities. *Australian J. Soil and Water Conservation*, **5**(1) 28–33.

Geary, P.M. and Gardner, E.A. (1996). On-site Disposal of Effluent. *In Innovative Approaches to the On-site Management of Waste and Water*. Conference 26 November – Southern Cross University, Lismore, Australia.

Headley, T. (1998). *The Effect of Season, System Maturity and Peak Loading on Treatment of School Wastewater in On-site Reed Beds*. Honours Thesis, School of Resource Science and Management, Southern Cross University, Lismore, Australia.

Kadlec, R.H. and Knight, R.L. (1996). *Treatment Wetlands*. Lewis Publishers, Boca Raton, Florida, USA.

Standard Methods for the Examination of Water and Wastewater (1992). 18th edn, American Public Health Association, American Waterworks Association/Water Environment Federation, Washington, DC, USA.

Removal of hydrogen sulphide BOD from brackish water using vertical flow wetlands in a Caribbean environment

E. Giraldo* and E. Zárate**

CIIA, Centro de Investigaciones en Ingeniería Ambiental. Departamento de Ingeniería Civil y Ambiental. Universidad de los Andes. A.A. 4976. Bogotá, Colombia

Abstract Wastewater from a 550-inhabitant community had been treated and discarded using an anaerobic filter. Due to seawater intrusion in the aquifer that supplies the water, high concentrations of hydrogen sulphide were detected in the effluent. A vertical flow wetland was designed in 1998 for treating this effluent. Four parallel reed beds with a total area of 556 m² were constructed. During the first months of operation, a mean BOD_5 removal efficiency of 91% was obtained, with loads to the wetland system up to 4g/m²/d of grease and oils (G&O). In 1999, problems of soil clogging were found due to high G&O content in the wastewater, with loads up to 15g/m²/d of G&O, which highly influenced the hydraulic conductivity of the beds, generating the clogging problems. The low hydraulic conductivity and the high effluent G&O content, caused low BOD_5 and COD removal efficiencies. As G&O accumulated in the soil, the removal efficiencies decreased. Despite the clogging problems, there has been a high sulphide removal throughout the system operation. The wetlands removed sulphides successfully, under loads up to 20 $gS^=/m^2/d$. Four native species of macrophytes were planted: *Paspalum penisetum*, *Typha* sp, *Conocarpres erectus* and *Scirpus lacustris*. All of them but *Typha* sp. were established in the system.

Keywords Brackish water; Caribbean coast; grease and oil content; hydraulic conductivity; soil clogging; sulphide BOD; vertical flow constructed wetlands

Introduction

San Andrés is a Colombian archipelago of small islands in the Caribbean Sea, which lies about 770 km northwest of the Colombian mainland. Fresh water for consumption is obtained from the aquifers existing in the Main Island. In 1997, a 550-inhabitant community was compelled to improve its wastewater treatment, which consisted of grease traps and an anaerobic filter.

Due to seawater infiltration in the aquifer, high levels of sulphates are observed in the water used for toilet flushing. Consequently, high concentrations of hydrogen sulphide (H_2S) were detected in the effluent of the anaerobic filter. A subsurface vertical flow wetland system was then implemented to treat this effluent, taking into account the aerobic characteristics needed to oxidise the H_2S present.

The goals of the system were:
- To improve the effluent quality, in terms of reduction of Biological Oxygen Demand (BOD_5), Chemical Oxygen Demand (COD) and Total Suspended Solids (TSS)
- To control offensive odour emissions (such as H_2S) to the atmosphere
- To compare the performance and affinity of the macrophytes planted in the beds
- To maintain the aesthetics of the island

Wastewater treatment using wetlands is a fairly new technology in Colombia, and it has a large potential domain of application. After more than two years of operation, experience has been gained and a large amount of data has been collected, throughout different inflow conditions, which enhance the knowledge for design and implementation of wetlands in Colombia. This paper describes the treatment performance results along different conditions of operation.

Design, construction and initial performance

System description

Four parallel vertical flow reed beds were constructed, following the grease traps and the anaerobic filter. These beds work intermittently, one each day, having a three-day resting period. The beds are 60 cm deep, the first being 20 cm sand, and the last 40 cm gravel. Each bed was planted with a different native plant: *Paspalum penisetum* (known as king grass), *Typha* sp., *Conocarpres erectus* (known as Mangrove: Mangle Botón) and *Scirpus lacustris*. Figure 1 shows a diagram of the system. The effluent coming out from the beds is treated with chlorine before being discharged to the environment.

The system was built in 1998, and the operation parameters are shown in Table 1. The initial design calculations included a 1,100 m² wetland system, but due to land limitations, it was only possible to build 556 m². This generated higher organic loads, over the recommended loading values found in the literature (Platzer and Mauch, 1997). A pump,

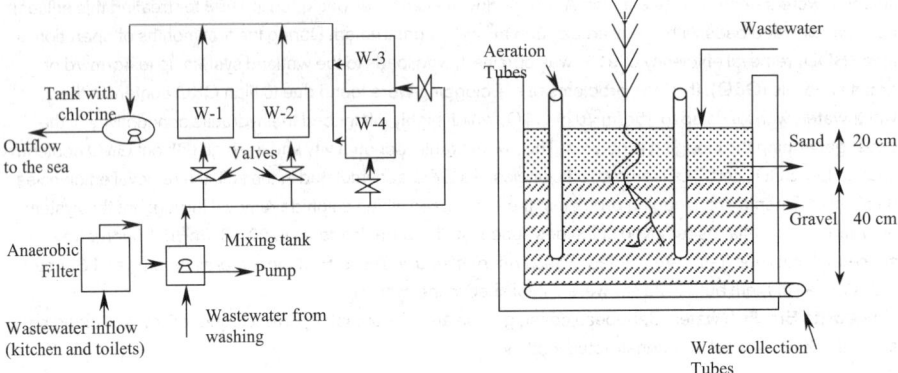

Figure 1 General diagram of the wetland system in San Andrés, Colombia

Table 1 Operating parameters for the wetland system

Parameter	Units	Value	Recommended values**
Wastewater average BOD$_5$	mg/l O$_2$	290*	
Wastewater average COD	mg/l O$_2$	502*	
Wastewater average TDS	mg/l	9894*	
Wastewater average sulphide content	mg/l	93.1*	
Wastewater average salinity	º/$_{oo}$	5.7	
Total area	m²	556	
Length/width ratio	M/m	1.72	
Design BOD$_5$ loading	g BOD$_5$/m²/d	21.4	
Operating BOD$_5$ loading	g BOD$_5$/m²/d	42.3	
Design COD loading	g COD/m²/d	37.0	25.0$^{(1)}$
Operating COD loading	g COD/m²/d	73.1	
H$_2$S loading	g H$_2$S/m²/d	13.6	
Initial G&O loading	g G&O/m²/d	1.31	
Final G&O loading	g G&O/m²-d	14.6	
Hydraulic load	l/m²/h	6.1	
Design flow	l/s	0.5	
Operating flow	l/s	0.63–0.94	
Specific area	m²/person	1	1–5
Resting period	days	3	
Mean water temperature	°C	30	
Mean air temperature	°C	30	

* Average BOD$_5$, COD and TDS out of the anaerobic filter, entering the beds. ** Cooper and Green (1995), Platzer and Mauch (1997). (1) For subtropical climates

which is located inside the mixing tank, discharges automatically to the wetland in use as soon as the tank gets full. The mixing tank (4.5 m × 4.5 m × 1 m = 20.25 m^3) discharges 3 to 5 times throughout the day to the bed. The design was made for 550 inhabitants, consuming 200 l/(person-d) approximately, which represents 1.5 times the maximum average flow rate per day measured.

Table 1 also shows a few parameters of the wastewater characterisation. It is worth mentioning that the results for the Salinity and Total dissolved solids (TDS) tests, measured in the laboratory, verify the sea water infiltration in the aquifers used as potable water sources. Additionally, the value presented for sulphide content approximately contributes with 75% of the BOD$_5$ entering the wetland system.

Performance

Table 2 shows the treatment efficiency for different samples taken since the construction of the system. The values presented as "Inflow" are those entering the wetland system, which means that they had already passed through the anaerobic filter. This filter presents an average BOD$_5$ removal efficiency of 50%. The maximum and minimum values for BOD$_5$ entering the whole treatment system were 630 and 361 mg/l (with standard deviation of 114.3); while the maximum and minimum BOD$_5$ values leaving the filter were 328 and 175 mg/l respectively (with a standard deviation of 64.1).

As it is shown in Table 2, during 1998 and the first half of 1999, the system had high removal efficiencies. Nevertheless, during the second half of 1999 and the first half of 2000, the results show a decrease in the overall percentage of removal. Despite this decrease, the lowest overall percentage of removal of the system remained 76%. Clogging problems were found in two of the beds, which explains the decrease. As is seen in Table 2, there was a drastic increase in the effluent BOD$_5$.

Methods

In order to study the clogging problems, a historical analysis of the Grease and Oil (G&O) content of the wastewater was made. Those results were compared with the BOD$_5$ and TSS data. On the other hand, unaltered samples of the soil column were taken and sliced into several portions, in order to determine their G&O and Total Volatile Solids (TVS) content, as well as their hydraulic conductivity. The methods used for the results presented in this paper were carried out employing the *Standard Methods* Procedures (AWWA-APHA-WPCF, 1992).

Table 2 Performance data of the wetland treatment system

Date	BOD$_5$-I mg/l	BOD$_5$-E mg/l	R (%)	COD-I mg/l	COD-E mg/l	R (%)	TSS-I mg/l	TSS-E mg/l	R (%)	G&O-I mg/l	G&O-E mg/l	R (%)	O %
Oct-98	200	55	72.5	–	–	–	30	22	26.7	25	20	20.0	87.2
Nov-98	–	33	–	–	–	–	–	10	–	–	–	–	90.9
Feb-99	175	22	87.4	–	–	–	45	22	51.1	–	–	–	–
Dic-99	328	96	70.7	638	184	71	112	25.5	77.2	43	10	76.1	–
Feb-00	277	83	70.2	351	138	61	92	40	56.5	100	50	49.6	86.9
Abr-00	290	117	59.6	502	214	57	148.8	25.25	83.0	42	36	15	76.0
Av.	254	68	–	497	179	–	86	24	–	52	50	–	–

BOD-I: Influent BOD$_5$ (out of the anaerobic filter), BOD-E: Effluent BOD$_5$, COD-I: Influent COD (out of the anaerobic filter), COD-E: Effluent COD, TSS-I: TSS in the influent (out of the anaerobic filter), TSS-E: TSS in the effluent, G&O-I: Grease and Oil in the influent (out of the anaerobic filter), G&O-E: Grease and Oil in the effluent, O: Overall BOD$_5$ removal efficiency, Av: Average, R: Removal percentage for the wetland system

Results and discussion
Grease content in wastewater

Table 3 shows the results obtained for the G&O content of the influent to the whole system. This content has been increasing continuously, contributing to the increase in COD and TSS. It is worth mentioning that the average content of G&O in the composed samples of wastewater entering the whole system, was 150 mg/l, with a maximum value that never exceeded 400 mg/l, during 1998.

Soil analysis

Table 4 shows the results for the G&O content and the TVS of the soil samples, taken along the soil column, for each wetland. Figures 2 and 3 show these results. It is seen how the grease content increases with depth, meanwhile, as a general tendency, the total volatile solids increase with the grease content.

Hydraulic conductivity

After having collected the G&O and TVS data, hydraulic conductivity analysis were performed, in order to identify the effect of grease content over clogging. Each portion of the soil column (for each one of the beds) was analysed. Figures 4 and 5 show the results found.

Table 3 Characterisation of the influent to the whole system

Date	G&O (influent to the system), mg/l	COD (influent to the system), mg/l	BOD$_5$ (influent to the system), mg/l	TSS (influent to the system), mg/l
Sep-97	136	1005	482	198
Oct-98	400	–	430	430
Nov-98	570	–	361	1252
Feb-00	966	1059	630	650
Apr-00	–	979	490	124

Table 4 Soil analysis for the wetlands

Wetland 1			Wetland 2			Wetland 3			Wetland 4		
Depth, cm	G&O, mg/kg	% TVS	Depth, cm	G&O, mg/kg	% TVS	Depth, cm	G&O, mg/kg	% TVS	Depth, cm	G&O, mg/kg	% TVS
10		1.77	15	2765	2.91	20	550	2.63	15	177	2.16
20	1934	3.23	30	1911	3.93	40	1098	2.96	30	623	2.84
30		2.93	45	2519	3.76	60	7608	3.05	45		2.56
40	5699	2.21	60	5264	4.72				60	659	3.25
50	6794	2.89									
60	6914	4.22									

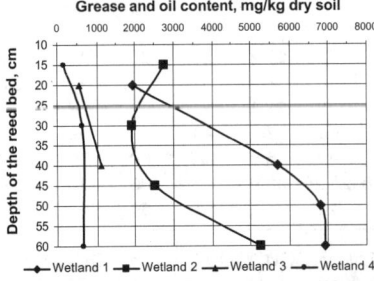

Figure 2 Grease content varying for each wetland with depth

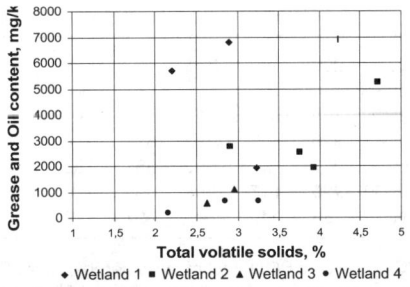

Figure 3 Total volatile solids varying with grease content for each wetland

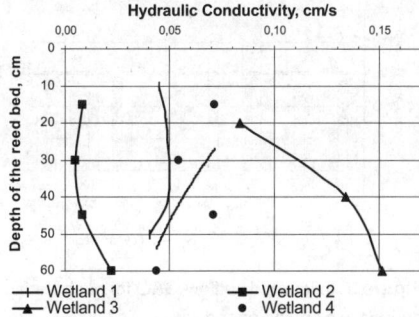

Figure 4 Variation of hydraulic conductivity of the sand with depth for each wetland

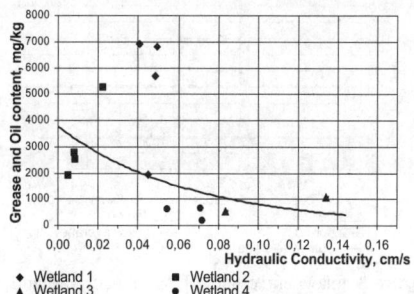

Figure 5 Variation of hydraulic conductivity of the sand with grease content for each

As a general behaviour, it was found that the beds 1 and 2 present the highest values of grease content, which agrees with the low values for hydraulic conductivity found for the same two beds. It was expected to find higher values for hydraulic conductivity in deeper zones, because of the increase of particle size with depth. The results show that the grease is being accumulated in the deeper parts of the wetland, which affects partially the hydraulic conductivity. Furthermore, there exists a clear decrease in the hydraulic conductivity of wetlands 1 and 2, which agrees with the results shown in Table 4.

Comparison of the bed performances

Figures 6, 7, 8 and 9 show the influent and effluent BOD_5, COD, TSS and G&O contents for each bed, during the last few months of operation.

As we can see from these figures, as well as in Table 5, the lowest efficiencies occur for wetlands 1 and 2, especially for the last month. For wetland 3, a good performance is found, in spite of the high loads to it. Therefore, it is clear that the grease content has affected both the hydraulic conductivity and the overall efficiency of the system.

Sulphide removal

The sulphate present in the wastewater is then converted to sulphide inside the anaerobic filter. 75% of the BOD_5 entering the wetland corresponds to sulphide BOD. Additionally, at the moment of discharge, it generates bad odour problems. As the time has passed and the plants have grown, a canopy has been formed, which helps with bad odour control.

Since vertical flow wetlands work as aerobic systems, due to their high oxygen transfer capabilities (Cooper, 1999), the sulphide is then oxidised to sulphate within the bed. Figure 10 and Table 6 show the removal results of sulphide for each wetland.

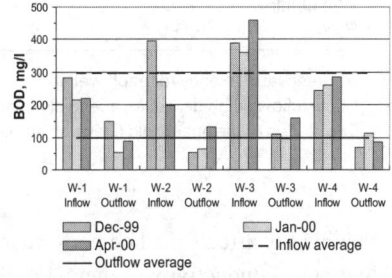

Figure 6 Inflow and outflow BOD_5 for each wetland during the clogging period

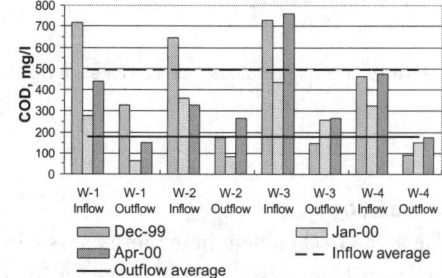

Figure 7 Inflow and outflow COD for each wetland during the clogging period

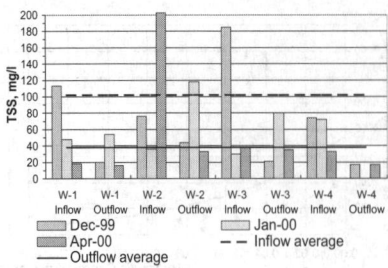

Figure 8 Inflow and outflow TSS for each wetland during the clogging period

Figure 9 Inflow and outflow G&O for each wetland during the clogging period

Table 5 Removal efficiencies for each wetland

	W-1	W-2	W-3	W-4
	Removal percentage for BDO_5, mg/l			
Dec-99	46.8	86.4	71.8	71.2
Jan-00	75.0	75.6	73.3	56.2
Apr-00	59.1	33.3	65.3	69.1
Average	**60.3**	**65.1**	**70.1**	**65.5**
	Removal percentage for COD, mg/l			
Dec-99	54.5	73.0	79.9	80.7
Jan-00	77.9	76.8	41.0	54.3
Apr-00	65.9	19.7	65.2	63.2
Average	**66.1**	**56.5**	**62.0**	**66.0**
	Removal percentage for G&O, mg/l			
Dec-99	83.8	29.8	88.8	78.8
Jan-00	16.2	63.7	52.6	22.4
Average	**50.0**	**46.7**	**70.7**	**50.6**

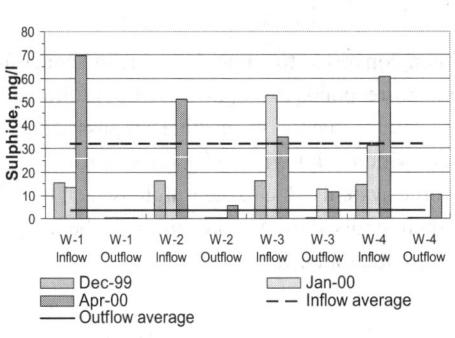

Figure 10 Inflow and outflow sulphide for each wetland

Table 6 Sulphide removal in each bed

	Sulphide content, mg/l	
	Jan-00*	Apr-00
W-1 Inflow	13.6	69.8
W-1 Outflow	<0.5	<0.4
R, %	>95	>95
W-2 Inflow	10	51
W-2 Outflow	<0.5	5.6
R, %	>95	89
W-3 Inflow	52.8	35
W-3 Outflow	12.8	11.6
R, %	76	67
W-4 Inflow	31.2	60.6
W-4 Outflow	<0.5	10.4
R, %	>95	83

R: Removal efficiency for each wetland
* Low inflow sulphide values due to cleaning: The sludge of the anaerobic filter was emptied

Discussion

The high G&O content in the influent is a factor that clearly affects the bed performance and contributes to soil clogging, by affecting the hydraulic conductivity of them. For this reason, it is very important to control the influent G&O to the beds. Additionally, the results show that the percentage of G&O removal is low once the beds experiment clogging, which means that they mostly go out with the effluent. Subsequent investigations (data not shown) have indicated that the amount of G&O retained affects the aerobic microbial population, because clogging generates restrictions in oxygen transfer. The high content of G&O in the effluent contributes to increasing the effluent BOD and COD, decreasing the overall removal efficiency.

In spite of the clogging problems and the decrease in the BOD_5 removal efficiency, the sulphide removal remained high, minimising the odour problems in the outlet to the sea, and the final effluent contained very low sulphide amounts. On the other hand, the odour problems presented at the beginning, when discharging to the wetlands, have been handled throughout the operation, performing smooth flushes within the canopy formed by the plants.

The planted macrophytes proved to be an effective mechanism for the control of undesirable odours, which result from the anaerobic treatment of brackish water. The macrophytes planted in this case showed different behaviours throughout the system operation. Three of the four species planted, presented full growth and development. *Typha* sp. did not adapt to the system conditions and did not establish in it. *Scirpus Lacustris*, adapted very well.

Conclusions

According to the data presented, we conclude that:
- During 1998, the system showed very good performance, with average BOD_5 overall removal efficiencies of 91%, in spite of the high organic loads to it (73.1 g COD/m^2-d).
- During the first year of operation (1998), the wetland system received up to 4 g/ m^2-d of grease and oils, still showing adequate treating efficiencies and no clogging problems. When the clogging problems were experienced, G&O loads to the wetland system were 15 g/m^2-d approximately.
- G&O highly affected the hydraulic conductivity of the beds, which generated the clogging problems. At the same time, the low hydraulic conductivity and the high G&O effluent content, generated low BOD_5, and COD removal efficiencies. As G&O accumulate in the soil, the removal efficiencies decrease.
- There has been a good sulphide removal throughout the operation of the system, regardless of the clogging problems. The constructed vertical flow wetlands removed sulphides successfully, with loads up to 20 g $S^=/m_2$-d.
- *Conocarpres erectus* (mangrove: "mangle botón") and *Scirpus lacustris,* two of the macrophytes planted, showed very good adaptation to the given conditions of salinity and high sulphide and TDS content. *Paspalum penisetum* also presented a good performance. However, *Typha* sp. did not resist these conditions and did not established in the system.

Acknowledgements

Special thanks to the *Armada Nacional de Colombia* for the financial and operational support given to this project.

References

AWWA-APHA-WPCF (1992). Standard Methods for the examination of water and wastewater.

Cooper, P. (1999). A review of the design and performance of vertical-flow and hybrid reed bed treatment systems. *Water Science and Technology* **40**(3), 1–9.

Cooper, P. and Green, B. (1995). Reed bed treatment systems for sewage treatment in the United Kingdom – the First ten years of experience. *Water Science and Technology* **32**(3), 317–327.

Platzer, C. and Mauch, K. (1997). Soil clogging in vertical-flow reed beds-mechanisms, parameters, consequences and . . . solutions? *Water Science and Technology* **35**(5), 175–181.

Constructed wetlands for wastewater treatment in the Czech Republic

J. Vymazal

Ecology and Use Wetlands, Říčanova 40, 169 00 Praha 6, Czech Republic. (E-mail: vymazal@yahoo.com)

Abstract The first constructed wetland (CW) for wastewater treatment was built in the Czech Republic in 1989. This recent survey shows that at the end of 1999 101 systems are in operation and several more are under construction. 95 CWs are designed with sub-surface horizontal flow, 6 systems are hybrid with a combination of vertical and horizontal flow beds. Most systems (56) were designed for the treatment of municipal and domestic sewage while 38 CWs were designed for the treatment of wastewater from combined sewer systems. The most commonly used size of vegetated beds is 1,001–2,500 m^2 (31 systems) followed by the area between 51–250 m^2 (19%). The area of vegetated bed of the largest system is 4,493 m^2. Size distribution is quite evenly spread from very small systems (PE = 3 or 4) up to 1,000 PE. However, most systems (44) were sized to treat wastewater from sources between 101 and 500 PE. The most commonly used macrophyte is Common reed (*Phragmites australis*) which is used in 34 systems as a monotypic stand and in 44 systems in combination with other macrophytes, most frequently with Reed canarygrass *(Phalaris arundinacea)* (31 systems) and cattails (*Typha* spp.) (8 systems).
Keywords BOD$_5$; Czech Republic; horizontal flow; nutrients; *Phragmites australis*

Introduction

Wetlands have been intensively studied in the Czech Republic for more than three decades. However, most studies were aimed primarily at wetland ecology, or ecophysiology of wetland plants, i.e., on primary productivity, biomass, nutrient cycling or evapotranspiration. Several experiments were carried out in wetland sites affected by sewage outfalls or in experimental hydroponic systems using a defined nutrient medium (Vymazal, in press). However, the first pilot-scale constructed wetland with emergent macrophytes which treated sewage was put in operation only in 1988 (Vymazal, 1990).

The first full scale constructed wetland (CW) for wastewater treatment was put in operation in the Czech Republic in May 1989. During the two following years only three constructed wetlands were built (Vymazal, 1993, 1998). However, after 1991 a steady increase in the number of CWs occurred with the peak in 1995, when 22 CWs were put in operation. At present, about 10 CWs are built every year (Figure 1).

Major design parameters

Type of wastewater

Most constructed wetlands in the Czech Republic are designed to treat municipal or domestic sewage (56 systems) or sewage combined with stormwater runoff (39 systems). Other types of wastewater include those from dairies, bakeries, abattoir facilities and a goat farm. One CW is a part of the system designed for the treatment of landfill leachate and one CW has been designed to treat stormwater runoff only. Ninety five CWs have been designed as a secondary step while 6 systems are tertiary steps.

Type of constructed wetlands

Ninety five CWs have been designed with horizontal sub-surface flow (HSF). Six CWs are hybrid systems with vertical flow-horizontal flow configuration. Five of those systems,

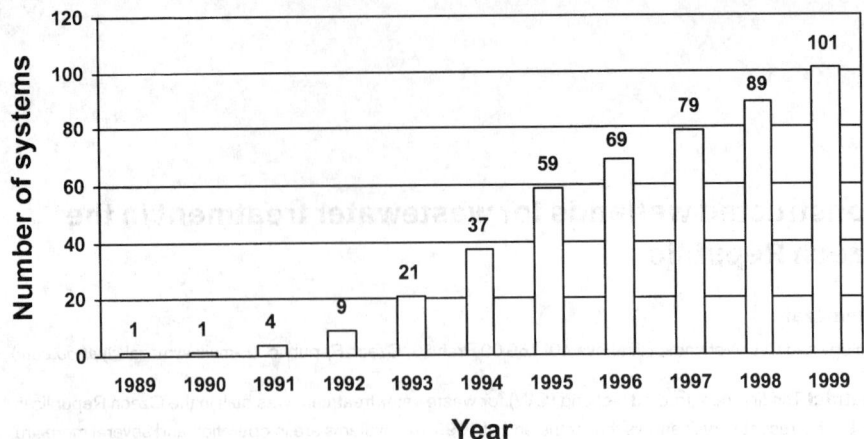

Figure 1 Cumulative numbers of full-scale constructed wetlands in the Czech Republic between 1989 and 1999

however, were designed as horizontal flow wetlands with pretreatment designed as a vertical flow reed bed. Therefore, those systems are classified as a hybrid in this survey.

Size of constructed wetlands

The size of constructed wetlands according to population equivalent (PE) ranges between 3 and 1,000. However, most systems (44) were designed to treat wastewater from sources of pollution between 101–500 PE (Figure 2). Recently, the number of on-site CWs for single households has increased and at present 17 systems for less than 10 PE are in operation. It seems that the number of these small systems could be underestimated because small on-site systems are sometimes not registered by water authorities.

Area of vegetated beds

Surface area of all secondary systems in the Czech Republic has been designed according to a simple formula (Cooper, 1990; Vymazal, 1998; Vymazal et al., 1998):

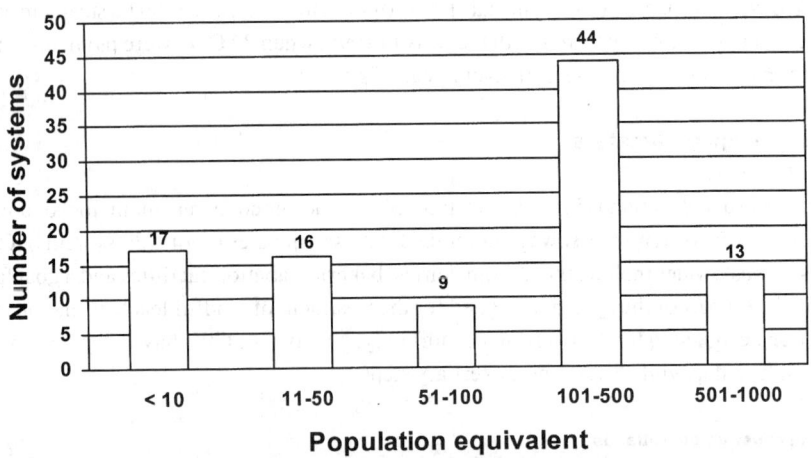

Figure 2 Distribution of the Czech constructed wetlands according to population equivalent. Landfill leachate and stormwater systems are not included

$$A_h = Q_d(\ln C_o - \ln C)/K_{BOD},$$

Where A_h = surface area of bed (m²)
  Q_d = average flow (m³ d⁻¹)
  C_o = influent BOD_5 (mg L⁻¹)
  C = effluent BOD_5 (mg L⁻¹)
  K_{BOD} = rate constant (m d⁻¹)

The value of K_{BOD} is usually set at 0.1 m d⁻¹ according to the recommendation of Cooper (1990). Despite some criticism this value has proved to be sufficient where the systems are designed to remove organics (BOD_5 and COD) and suspended solids (Vymazal, in press). For mechanically pretreated domestic and municipal sewage this generally means that the specific area is about 5 m² PE⁻¹. This survey revealed that the average specific area of the Czech constructed wetlands designed for secondary treatment of sewage (n = 94) is 5.06 m² PE⁻¹ (range: 2.1 – 10.0 m² PE⁻¹) The distribution is shown in Figure 3. Specific area of systems designed for tertiary treatment (n = 5) ranges between 0.27 and 1.5 m² PE⁻¹. However, it should be noted that instead of PE, the number of connected people better fits the reality. In the design, the number of connected people equals PE which is set at 60 g BOD_5 d⁻¹ person⁻¹ but at present, one person produces only 30–40 g BOD_5 per day.

The area of vegetated beds varies between 18 and 4,493 m² with 47 systems having bed area between 501 and 2,500 m². In Figure 4, distribution of the Czech CWs according to the area of vegetated beds is shown. Small and on-site systems usually use only one bed. Systems with the bed surface area larger than about 300 m² use more beds, mostly 2 to 4. These systems are often built with the possibility of both parallel or series configuration.

Aspect ratio (i.e., length:width) varies between 0.33 and 5.3 with the majority of systems built with aspect ratio < 2 although many of them have aspect ratio < 1 (Vymazal, 1998). The major reason for low aspect ratio is the need for wastewater distribution to as wide a profile as possible in order to avoid clogging of the inlet zone. Clogging of the inlet zone and subsequent surface flow may deteriorate the treatment effect of the process.

Filtration media
Filtration media in constructed wetlands with horizontal sub-surface flow should 1) facilitate macrophyte growth, 2) provide high and sustainable filtration effect, and 3) main-

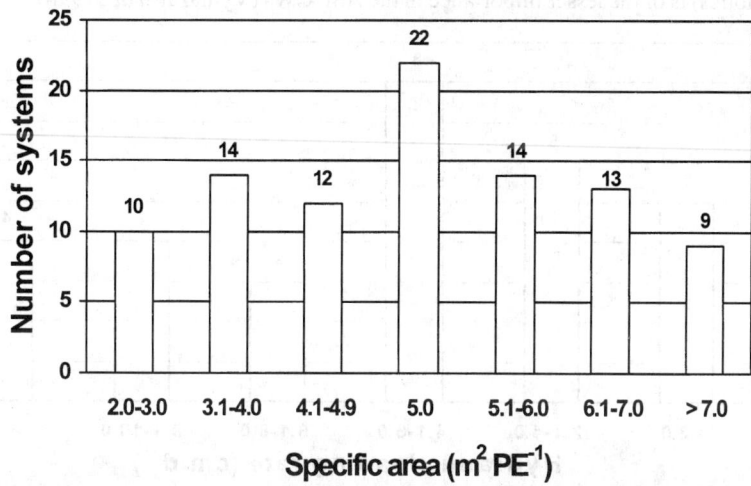

Figure 3 Specific area used for constructed wetlands in the Czech Republic

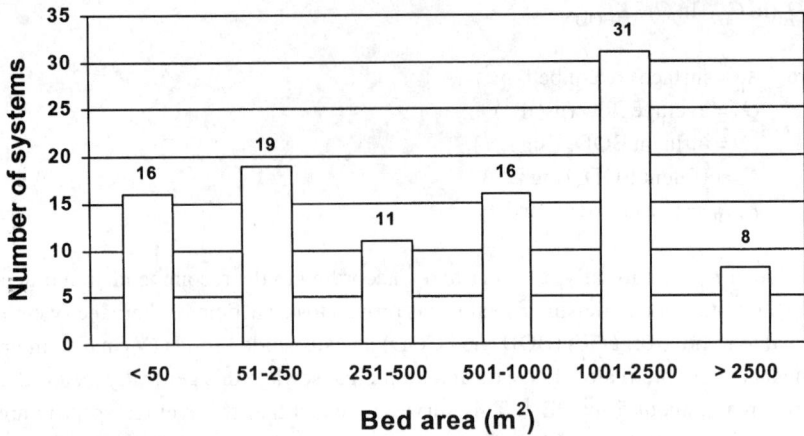

Figure 4 Surface area of vegetated beds of constructed wetlands in the Czech Republic

tain high hydraulic conductivity. Czech constructed wetlands use, with few exceptions, coarse media – gravel, crushed rock or gravel-sand. The early systems usually used fractions > 8 mm. At present, gravel or crushed rock fraction 8–16 mm is preferred because it has been shown that this fraction provides sufficient hydraulic conductivity while supporting a healthy macrophyte growth and good treatment efficiency.

Hydraulic loading

Constructed wetlands in the Czech Republic are designed to treat wastewater flows between 0.3 and 415 m^3 d^{-1} with only 11 systems designed to treat more than 100 m^3 d^{-1}. However, exact flow measurements are available from only 30 systems. The hydraulic loading rate varies between 0.6 and 19.8 cm d^{-1} for secondary treatment wetlands with an average value of 5.9 cm d^{-1} (Figure 5). For tertiary systems flow rates are not available.

Vegetation

The most important effects of macrophytes in relation to treatment process in sub-surface horizontal flow CWs are the physical effects of the plant tissue (i.e., erosion control, filtration effect, provision of surface area for attached microorganisms, insulation of the bed surface). The metabolism of the macrophytes (plant uptake, oxygen release and release of antibiotics) is of the lesser importance in the HSF CWs (Vymazal *et al.*, 1998).

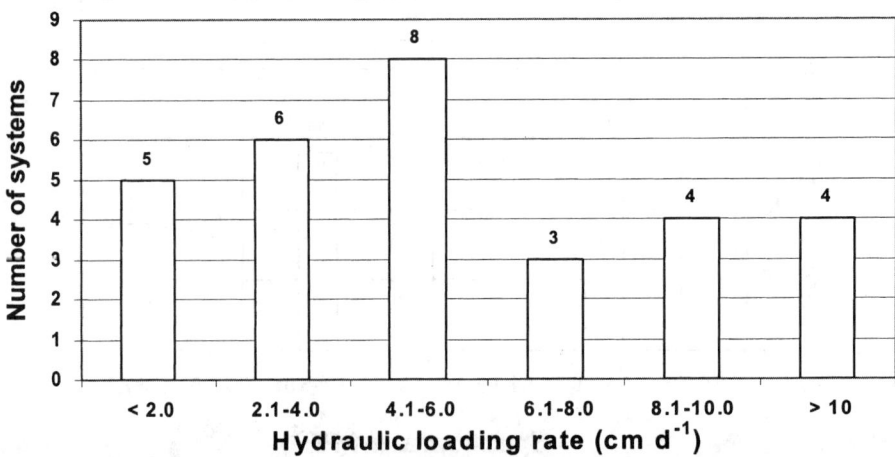

Figure 5 Hydraulic loading rate of the Czech constructed wetlands (data based on field measurement only)

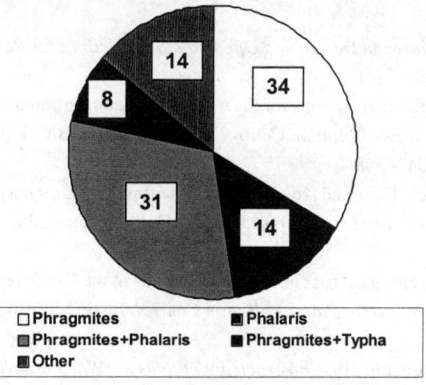

Figure 6 Plant species used in Czech constructed wetlands

The most frequently used plant in the Czech HSF CWs is *Phragmites australis* (Cav.) Trin. ex Steud. (common reed) which is used either singly or in combination with other macrophyte species (Vymazal 1996, 1998) (Figure 6). Other species used singly are *Phalaris arundinacea* L. (reed canarygrass), *Glyceria maxima* (Hart,) Holm. (mannagrass), and *Typha latifolia* L. (common cattail). The early systems mostly used rhizomes from natural localities. However, this method did not provide rapid vegetation coverage. It also introduced many weed species and in addition, the planting was limited due to climatic conditions to the period of May-August. At present, most systems are planted with seedlings grown in a nursery from seeds. The use of seedlings enables the planting of beds from early April to late October for *Phalaris* and from late April to late September for *Phragmites* (Vymazal, 1998). This method is very successful and provides fast coverage of the bed surface. The density 4–8 seedlings per 1 m^2 is sufficient. Weeds, if present, are restricted to vegetated bed margins. Vegetation is not harvested in most cases. The major reason for that is that litter provides an excellent insulation of the surface of vegetated beds during periods of cold weather (Vymazal, in press).

Treatment efficiency

Treatment effectivness of Czech constructed wetlands is shown in Table 1. It is important to note that most CWs have been designed for removal of BOD_5 and SS only and removal of these parameters is very high. Removal of nutrients is much lower but comparable with conventional technologies without a special regime for nutrient removal. It has been shown that removal of organics, suspended solids, and surprisingly, also of nutrients is affected by season only very little in the Czech Republic (Vymazal, 1999, 2000).

Table 1 Treatment efficiency (EFF in %) of the Czech constructed wetlands (each count (N) represents annual mean). RAW = raw wastewater, IN = inflow to vegetated beds after pretreatment, OUT = final effluent from beds

		Concentration (mg L^{-1})						Loading rate (kg ha^{-1} d^{-1})			
	N	RAW	OUT	EFF	N	IN	OUT	EFF	N	IN	OUT	EFF
BOD_5	97	160	15.4	90.4	59	92	10.9	88.2	53	35.2	4.75	86.5
COD	79	352	59	83.2	57	222	53	76.1	52	93.4	22.7	75.7
SS	90	165	14.5	91.2	55	76	10.4	86.0	46	42.8	4.6	89.0
TP	48	5.8	2.95	49.1	38	6.5	3.5	46.2	37	3.1	1.9	39.0
TN	31	54	27	50.0	35	48	28	42.0	31	25.4	15.6	39.0
NH_4-N	56	31	18.4	41.0	39	29	17.2	41.0	37	13.6	8.8	35.0

References

Cooper, P.F. (ed.) (1990). *European Design and Operational Guidelines for Reed Bed Treatment Systems*. WRc Report UI 17, Swindon, U.K.

Vymazal, J. (1990). Use of reed-bed systems for the treatment of concentrated wastes from agriculture. In: *Constructed Wetlands in Water Pollution Control*, P.F. Cooper and B.C. Findlater (eds.), Pergamon Press, Oxford, U.K., pp. 347–358.

Vymazal, J. (1993). Constructed wetlands for wastewater treatment in Czechoslovakia: State of the art. In: *Constructed Wetlands for Water Quality Improvement*, G.A. Moshiri (ed.), CRC Press, Boca Raton, FL, pp. 255–260.

Vymazal, J. (1996). Plant species used for constructed wetlands in the Czech republic. In: *Proc. 5th Internat. Conf. on Constructed Wetlands for Water Pollution Control*. Institut für Bodenkultur, Vienna, Austria, Chapter II/5.

Vymazal, J. (1998). Czech Republic. In: *Constructed Wetlands for Wastewater Treatment in Europe*, J. Vymazal, H. Brix, P.F. Cooper, M.B. Green and R. Haberl (eds.), Backhuys Publishers, Leiden, The Netherlands, pp. 90–117.

Vymazal, J. (ed.) (1999). *Nutrient Cycling and Retention in Natural and Constructed Wetlands*. Backhuys Publishers, Leiden, The Netherlands.

Vymazal, J. (in press). The use of constructed wetlands for wastewater treatment in the Czech Republic: 10 years experience. *Ecol. Eng.*

Vymazal, J., Brix, H., Cooper, P.F., Haberl, R., Perfler, R. and Laber, J. (1998). Removal mechanisms and types of constructed wetlands. In: *Constructed Wetlands for Wastewater Treatment in Europe*, J. Vymazal, H. Brix, P.F. Cooper, M.B. Green and R. Haberl (eds.), Backhuys Publishers, Leiden, The Netherlands, pp. 17–66.

Subsurface-flow constructed wetlands treatment in the plains: five years of experience

M.F. Dahab* and R.Y. Surampalli**

* University of Nebraska-Lincoln, Department of Civil Engineering, Lincoln, NE 68588-0531 USA
** US Environmental Protection Agency, Region VII, Kansas City, KS, USA

Abstract This paper documents the performance of a subsurface-flow constructed wetlands system during its initial five years of operation under variable loading and operating conditions associated with a northern midwestern US climate. The results indicate that effective and sufficient CW seasonal removals of TSS, VSS, $CBOD_5$, COD, and fecal coliform were achieved. Wastewater temperatures seemed to affect $CBOD_5$ and COD removal rates. Nitrogen and phosphorus reductions were not as effective and varied seasonally, as well as with wastewater temperature. The addition of a sand filter, to aid in further nitrification and disinfection following CW treatment, markedly improved the performance of the wetlands system. After a few years of operation, the remarkable performance of the CW system was dampened by apparent clogging and subsequent eruption of wastewater at the head-end of the treatment cells. While clogging was partially caused by biomass build-up in the wetlands substrate, visual observations suggest that excessive vegetation coupled with relaxed maintenance may also be responsible for clogging.
Keywords Clogging; horizontal flow; seasonal variation; subsurface

Introduction

Most research and use of constructed wetlands for the treatment of polluted waters have occurred within the past 25 years. Through the use of physical, chemical, and biological mechanisms, CWs offer higher effluent quality than that from typical pond systems. The common removal mechanisms associated with wetlands that are often cited include sedimentation, coagulation, adsorption, filtration, biological uptake, and microbial transformation (Kadlec and Knight, 1996). Constructed wetlands maintain and perform these treatment mechanisms by supporting plant and bacterial life that cycle excess nutrients through successive seasons of plant growth, death, and decay (Vanier and Dahab, 1997; Dahab and Vanier, 1998; Sumrall *et al.*, 1994).

Often, individuals and communities do not have adequate resources to construct and maintain conventional wastewater treatment facilities The use of CWs can offer a high degree of process control while allowing for the development of generally applicable design and cost criteria for a given and desired level of wastewater treatment (Gersberg *et al.*, 1984).

The use of CWs for domestic wastewater treatment has been increasing in the United States, including the northern and midwestern US. In the US, numerous small facilities have been reportedly built in South Dakota, North Dakota, Kentucky, Iowa, and Missouri. Because this treatment alternative has been proven to be an effective low-cost technology, its use is extremely advantageous to small communities with limited resources for wastewater treatment. It has been well documented that CWs provide consistent treatment in temperate climates. However, the effectiveness of this technology in areas with severe temperature extremes, such as Nebraska and other northern states, needed to be investigated because of the limiting factor of colder temperatures on physical and biological activity.

Objectives

The goal of this study is to evaluate and document the relative performance of a CW treatment system as a municipal wastewater treatment system for a small community in southeastern Nebraska. To that end, data collected at a wetlands facility in Nebraska are presented to illustrate the effectiveness of CWs, particularly for use at small and remote localities. The study objectives were to evaluate: 1) the effectiveness of the CW treatment system in treating domestic wastewater with respect to seasonal effects, and 2) the relative long-term viability and sustainability of this technology.

Methodology

The treatment system consists of four parallel gravel-based cells, each 50 m × 25 m, with horizontal subsurface flow through a depth of 0.6 m (Figure 1). The wetland cells are composed of a 0.45 m bed of 15 to 40 mm gravel layer topped by a 0.15 m layer of 10 mm gravel that functions as a support medium for wetland vegetation. Each cell was planted with three vegetation zones, equal in cover area, of common cattails (*Typha latifolia*) and woody cattails (*Typha domingensis*), alkali bulrush (*Scripus acutus*), and common reeds (*Phragmites communis*). The front portions of the cells were planted with cattail sprigs while the middle and end portions were planted with bulrush and reed sprigs, respectively. The influent to the wetland cells consists of the effluent from a series of two pretreatment tanks. Flow is distributed and collected equally across the width of each of the wetlands cells. The design flow rate for the treatment system was calculated at 225 m^3/d with a current typical flow rate of approximately 120 m^3/d. Wetland cell effluent flows into a two-cell (each 15 m × 18 m) sand filter with a depth of 1.1 m. The sand filter is comprised of a 0.76 m layer of 1 to 2 mm clean masonry sand topped by a 0.34 m layer of 6 mm washed pea-sized gravel. Sand filter effluent is discharged to Stevens Creek, a small tributary of the Platte River. Additional details on the wetlands system were outlined by Vanier (1997).

Sampling and analysis

To meet the objectives of this study, a monitoring program of the CW treatment system was developed to document the facility's progress from start-up into steady-state operation. The sampling and analysis program reported herein was carried out from June of 1996 through December 1999. This represents 3.5 years of operation of the CW system. Wastewater samples were collected regularly (twice per month) throughout each year of operation.

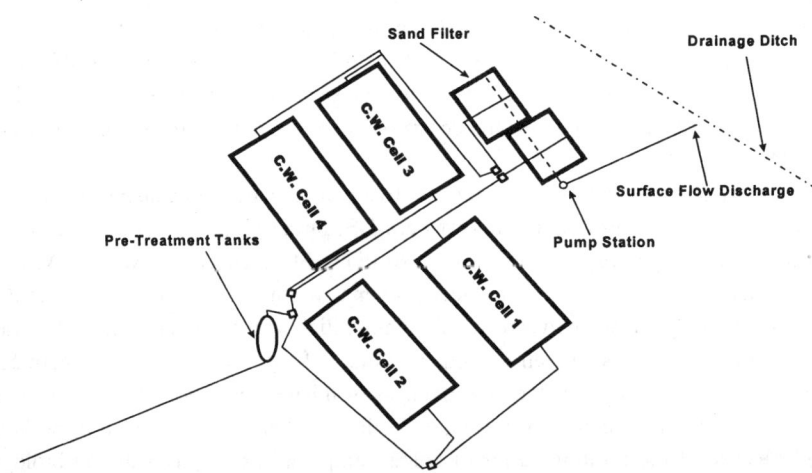

Figure 1 Constructed wetlands plan

The sampling points in the CW treatment system were selected at six locations along the treatment train and consisted of: 1) wetland cells influent (pretreatment tank and primary effluent); 2) fractional distances of 0.15, 0.40, and 0.70 through wetland Cell 3; 3) wetland cells effluent; and the treatment plant effluent (sand filter effluent). Cell 3 was initially randomly selected for the fractional-distance withing the wetlands cells sampling points. Each set of samples were analyzed for: 1) temperature, 2) pH, 3) DO, 4) TSS and VSS, 5) 5-day total and soluble $CBOD_5$, 6) total and soluble COD, 7) nitrogen (TKN, NH_3-N, and NO_3-N), 8) total phosphorus, and 9) fecal coliform (FC). All wastewater samples were grab samples representing a particular time and location in the treatment train. All analysis protocols followed *Standard Methods* (APHA, 1995) or EPA recommended methods when such methods were available.

Results

For comparison between extreme temperature conditions and seasonal variations, two separate periods were developed for data presentation and analysis: summer (June through October) and winter (November through March). Results within this presentation focuses on constructed wetland treatment, as opposed to total system treatment. The sand filter was added to the CW system to aid in ammonia removal, through nitrification, and for disinfection purposes. The additional effects of sand filter treatment, as well as removal efficiencies through the CW treatment profile, are detailed by Vanier (1997), Lionberger (1999), and Dahab and Surampalli (1999).

Constructed wetland treatment

The CW influent and effluent concentrations, as well as sand filter effluent concentrations, for the summer and winter periods are presented in Tables 1 and 2. The impact of the sand filter system on the CW performance is demonstrated by data summarized in Table 3.

Influent TSS concentrations ranged from a mean of 53 mg/L in the summer to 55 mg/L in the winter. TSS concentrations varied widely from about 18 to 152 mg/L during the summer period and 25 to 205 mg/L during the winter period. Effluent TSS concentrations averaged about 3 mg/L under both summer and winter conditions.

Influent total $CBOD_5$ concentrations ranged from a mean of 116 mg/L in the summer to 95 mg/L in the winter. During the seasonal periods, the influent concentrations ranged widely from 58 to 188 mg/L in the summer and from 18 to 200 mg/L during the winter period. Effluent total $CBOD_5$ had a mean of 19 mg/L under both summer and winter conditions. Mean summer and winter influent soluble $CBOD_5$ concentrations were 72 and 55 mg/L, respectively. Effluent soluble $CBOD_5$ concentrations had a mean of 23 mg/L in the summer and 16 mg/L in the winter.

Table 1 CW and sand filter effluent performance summary (summer)

Water Quality Parameter	CW Influent Mean	CW Influent Range	CW Effluent Mean	CW Effluent Range	Sand Filter Effluent Mean	Sand Filter Effluent Range
Temperature, °C	17.7	26.2–6.6	19.2	28.9–6.1	19.5	27.3–6.0
TSS, mg/L	53.1	152–17.8	3.3	8.6–1.4	2.9	5.8–1.2
Total $CBOD_5$, mg/L	116	188–58.5	18.7	48.0–2.3	2.8	7.4–ND
Sol. $CBOD_5$, mg/L	71.7	146–46.2	22.6	73.8–3.7	3.8	15.2–ND
Soluble COD, mg/L	167	311–111	64.1	171–18.5	27.2	81.1–5.4
NH_3-N, mg/L	19.7	34.1–7.0	13.7	23.2–2.1	3.9	9.3–0.1
NO_3-N, mg/L	3.8	5.7–2.2	2.2	4.0–0.6	6.3	14.5–0.7
Total P, mg/L	2.9	5.8–1.4	2.3	4.5–0.6	1.9	3.4–1.0
Fecal Coliform, # MPN, 100 mL	1,478,000	8,000,000–130,000	18,800	110,000–<200	5,500	90,000–<20

Table 2 CW and sand filter effluent performance summary (winter)

Water Quality Parameter	CW Influent Mean	CW Influent Range	CW Effluent Mean	CW Effluent Range	Sand Filter Effluent Mean	Sand Filter Effluent Range
Temperature, °C	10.1	15.3–2.8	4.8	10.3–0.8	4.6	10.8–0.0
TSS, mg/L	54.9	205–25.3	2.4	3.8–1.4	2.3	4.2–1.0
Total $CBOD_5$, mg/L	95.0	200–18.0	19.3	31.2–8.1	3.2	4.6–2.0
Sol. $CBOD_5$, mg/L	54.9	102–19.2	16.1	23.0–9.8	3.0	6.9–0.6
Soluble COD, mg/L	130	204–56.0	53.8	83.–26.8	17.8	38.3–6.2
NH_3-N, mg/L	17.8	24.–11.4	15.3	17.8–8.6	6.2	10.9–2.0
NO_3-N, mg/L	3.5	7.8–1.6	2.5	4.5–1.7	6.7	11.6–1.6
Total P, mg/L	2.5	4.2–0.8	2.2	3.0–0.9	1.8	3.2–0.7
Fecal Coliform, # MPN, 100 mL	811,000	2,700,000–170,000	13,600	50,000–1,700	1,600	5,000–200

Table 3 CW and additional sand filter removal efficiencies

Water Quality Parameter	CW (Summer and Winter)	CW (Summer)	CW (Winter)	CW & Sand Filter (Summer)	CW & Sand Filter (Winter)
TSS (%)	94.4	93.7	95.7	94.5	95.9
Total $CBOD_5$ (%)	82.6	83.9	79.7	97.6	96.7
Soluble $CBOD_5$ (%)	68.9	68.5	70.8	94.7	94.6
Soluble COD (%)	60.8	61.7	58.5	83.8	86.2
NH_3-N (%)	25.6	30.4	14.0	80.2	65.4
NO_3-N (%)	39.4	43.7	30.4	+65.3	+89.0
Total P (%)	18.7	21.4	12.6	34.8	28.6
Fecal Coliform (Log Reduction)	1.87	1.90	1.78	2.43	2.72

Influent NO_3-N concentrations averaged 3.8 mg/L in the summer and 3.5 mg/L in the winter. Concentrations ranged from 5.7 to 2.2 mg/L in the summer and from 7.8 to 1.6 mg/L in the winter. Constructed wetland treatment yielded a mean effluent NO_3-N concentration of 2.2 mg/L during summer conditions to 2.5 mg/L.

Influent NH_3-N concentrations had means of about 20 mg/L in the summer and 18 mg/L in the winter. Observed NH_3-N concentrations ranged from about 7 to 34 mg/L during the summer and from 11 to 24 mg/L during the winter. The effluent mean concentrations during the summer and winter periods were 13.7 and 15.3 mg/L, respectively. Therefore, the examination of effluent concentrations would suggest that CW NH_3-N concentrations were comparable during both periods. However, NH_3-N loadings were higher during the summer period and it would appear that greater removal occurred during the warmer period.

Influent total P concentrations were somewhat higher under summer conditions, with a mean of 2.9 mg/L, when compared to winter conditions. Winter loadings yielded a mean concentration of 2.5 mg/L. Constructed wetland treatment yielded mean effluent total P concentrations of 2.3 mg/L during the summer period and 2.2 mg/L in the winter. Therefore, it would appear little phosphorus removal actually occurred throughout the CW system.

Fecal coliform density determinations were only made at the CW influent, effluent, and sand filter/treatment plant effluent locations. The CW influent densities varied from 0.13 million to 8.0 million MPN/100 mL during the summer period and from 0.17 to 2.7 million MPN/100 mL during the winter period. Mean densities of 1.5 and 0.81 million MPN/100 mL were experienced during the summer and winter seasons, respectively. Mean effluent densities for the summer and winter periods were 18,800 and 13,600 MPN/100 mL, respectively. The typical regulatory requirement for fecal coliform concentration in the final effluent, (i.e. 200 MPN/100 mL, when disinfection is required) was not satisfied.

The impacts of the addition of the sand filter system to the CW are summarized in Table 3. Generally, the sand filter system resulted in improved removals of suspended solids BOD, fecal coliform and ammonia. The nitrification potential of the sand filter system is clearly evident year around, but with much improved performance particularly during the summer period. On average, ammonia removal was increased from about 14% in the CW alone to about 65% with sand filter during the winter and from about 26% to about 80% during the summer.

Long-term operation

As indicated previously, after a few years of operation, the remarkable performance of the CW system was dampened by apparent clogging and subsequent eruption and pooling of wastewater at the head-end of the treatment cells. These "eruptions" first appeared after two years of operation and the frequency of occurrence increased with time. Eruption and pooling of wastewater tended to impart odor and thus create undesirable conditions. Clogging appeared to be the main cause of these eruptions. Clogging was probably caused in part, by a combination of influent TSS loading and biomass accumulation, and in part by the presence of some fine sand within the wetlands cells gravel beds (Jillson, 2000). Furthermore, visual observations suggested that excessive vegetation coupled with relaxed maintenance may also be responsible for clogging. Ultimately, the head-end of some of the wetlands cells had to be rebuilt.

Discussion and conclusions

This paper documents the performance of a horizontal flow subsurface constructed wetland system treating domestic wastewater in eastern Nebraska. The system was monitored starting from June of 1996 up to the present. The results from this study have helped to further increase the knowledge of the treatment abilities for this type of system under variable climate conditions with a wide range of temperature extremes.

The CW treatment system demonstrated effective removal of soluble $CBOD_5$, ammonia nitrogen, total phosphorus, and fecal coliform. Average study removal efficiencies for the wetland cell were reported as 69% for soluble $CBOD_5$, 26% NH_3-N, 39% for NO_3-N, 18% for total P, and 98% (near 2 log reductions) in fecal coliform. The sand filter added an additional amount of treatment for all of the analyzed water quality parameters. For the system as a whole, the removals were found to be nearly 95% for soluble $CBOD_5$, 75% for NH_3-N, 72% for NO_3-N, 33% for total P, and 99.5% (2.5 log reduction) for fecal coliform. The average summer wastewater temperature was calculated to be 18.1°C with an average dissolved oxygen concentration of 0.4 mg/l within the wetland cell. The average winter wastewater temperature was found to be 4.1°C with an average dissolved oxygen concentration of 0.9 mg/l.

The wetlands treatment system satisfied the preliminary permit requirements of $CBOD_5$<30 mg/l and NH_3-N<22 mg/l. It was noted that the constructed wetland itself was not able to meet the $CBOD_5$ limit only a very few times during the sampling period. Addition of the sand filter provided additional removal of all water quality parameters except NO_3-N.

Removal rates in the wetland system were higher in the summer periods for all of the parameters with very few exceptions for suspended solids and soluble $CBOD_5$. Total $CBOD_5$ removal was slightly higher during summer, 84%, than during the winter, 79%. NH_3-N, NO_3-N, and total P removal rates exhibited more variability between summer and winter. Summer removals were registered as about 30% for NH_3-N, 43% for NO_3-N, and about 21% for total P. Winter rates were determined to be 14% for NH_3-N, 30% for NO_3-N, and a meager 13% for total P. The wetland system exhibited more effective removal of nitrogen and phosphorus during the warmer temperatures experienced during the summer periods than during the winter. The initial average summer removal for fecal coliform was found to be 98.7% while the winter period provided was 98.3% removal.

Acknowledgements

Credits: The research work presented herein was funded, in part, by the US EPA and, in part, by the University of Nebraska-Lincoln (UNL) Departments of Civil Engineering and Biological Systems Engineering, the UNL Water Center through the Nebraska Mandates Management Initiatives Program. M.F. Dahab is Professor and Chair, and R.Y Surampalli is Adjunct Professor, Department of Civil Engineering, University of Nebraska, Lincoln, NE 68588-0531.

Disclaimer: Views or opinions expressed in this article are strictly those of the authors and should not be construed as views or opinions of the US EPA.

References

American Public Health Association (APHA) (1995). *Standard Methods for Examination of Water and Wastewater*, 19th Edition, Washington D.C.

Dahab, M.F. and Surampalli, R.Y. (1999). Predicting subsurface-flow constructed wetlands performance a comparison of common design models, *Proceedings of the Annual Conference of the Water Environment Federation*, New Orleans, LA, Oct 9–13.

Dahab, M.F. and Vanier, S.M. (1998). Temperature effects on subsurface flow constructed wetlands performance in the Midwest, *Proceedings of the ASCE National Conference on Wetlands and River Systems*, Denver, CO, March 23–25.

Gersberg, R.M., Elkins, B.V. and Goldman, C.R. (1984). Use of artificial wetlands to remove nitrogen from wastewater, *Journal Water Pollution Control Federation*, **56**, 152–156.

Kadlec, R.H. and Knight, R.L. (1996). *Treatment Wetlands*, Lewis Publishers, Boca Raton, FL.

Lionberger, H.S. (1999). Performance evaluation and temperature effects on a subsurface-flow constructed wetlands in eastern Nebraska, Master of Science Thesis, University of Nebraska-Lincoln Libraries, Lincoln, NE.

Jillson, S. (2000). Pathogen indicator organism removal in subsurface-flow constructed wetlands, Master of Science Thesis, University of Nebraska-Lincoln Libraries, Lincoln, NE.

Sumrall, L.G., Surampalli, R.Y. and Sievers, D.M. (1994). Performance evaluation of constructed wetland treatment systems, *Journal of Cold Regions Engineering*, **8**(2), 35–47.

Vanier, S.M. (1997). Evaluation of a subsurface-flow constructed wetland system for small-community wastewater treatment in southeastern Nebraska, Master of Science Thesis, University of Nebraska-Lincoln Libraries, Lincoln, NE.

Vanier, S.M. and Dahab, M.F. (1997). Evaluation of subsurface flow constructed wetlands for small community wastewater treatment in the plains, *Proceedings of the 70th Annual Conference of the Water Environment Federation*, Chicago, IL, October 18–22.

Application of constructed wetlands for wastewater treatment in Nepal

R.R. Shrestha*, R. Haberl**, J. Laber **, R. Manandhar* and J. Mader**

* Environment and Public Health Organization (ENPHO), P.O.Box – 4102, Kathmandu, Nepal
** Institute for Water Provision, University of Agricultural Sciences, Muthgasse 18, 1190 Vienna, Austria

Abstract Surface water pollution is one of the serious environmental problems in urban centers in Nepal due to the discharge of untreated wastewater into the river-system, turning them into open sewers. Wastewater treatment plants are almost non-existent in the country except for a few in the Kathmandu Valley and even these are not functioning well. Successful implementation of a few constructed wetland systems within the past three years has attracted attention to this promising technology. A two-staged subsurface flow constructed wetland for hospital wastewater treatment and constructed wetlands for treatment of greywater and septage is now becoming a demonstration site of constructed wetland systems in Nepal. Beside these systems, five constructed wetlands have already been designed and some are under construction for the treatment of leachate and septage in Pokhara municipality, wastewater in Kathmandu University, two hospitals and a school. This paper discusses the present condition and treatment performance of constructed wetlands that are now in operation. Furthermore, the concept of the treatment wetlands under construction is also described here. With the present experience, several recommendations are pointed out for the promotion of this technology in the developing countries.
Keywords Constructed wetlands; developing countries; technology transfer; wastewater treatment

Introduction

Discharge of untreated wastewater into nearby streams, lakes or any other water body is a common practice in most of the developing countries, including Nepal. Although most deaths in these countries are due to water-borne diseases, treatment of wastewater is rarely practiced until conditions became unmanageable. In Nepal, many of the urban rivers have already been converted into open sewer canals due to the discharge of untreated wastewater from households and in some cases toxic industrial waste. The concept of wastewater treatment or recycling is often regarded as unaffordable technology. Over twenty years ago, four sewerage treatment plants were constructed around the Valley, but are no longer functioning. Due to the failure of these large treatment plants, small and decentralized treatment plants are in high demand. If such plants could be installed in communities and maintained by such communities, water quality improvements to receiving water could be expected.

Introduction of constructed wetlands in Nepal

One of the authors with a group of Nepali professionals from the Environment and Public Health Organization (ENPHO) initiated the introduction of Constructed Wetlands (CW) in Nepal in 1995. With the technical collaboration of the Institute for Water Provision, University of Agricultural Sciences, Vienna, Austria, a pilot scale treatment plant in Dhulikhel Hospital was designed and constructed in 1997 as a first CW in Nepal (Laber *et al.*, 1999, Shrestha, 1999, and Shrestha, *et al.* 2001). Currently, there are three CWs operating successfully in three different locations in and around the valley, which have done for the past three years. This paper highlights the efficiency of CW in operation and also describes briefly other systems under construction.

CW for Dhulikhel Hospital
Vertical flow bed just after completion   Vertical flow bed after reed growth

Design flow rate   : 20 m³/day
Three chambered settlement tank   : 18 m³
Horizontal flow bed   : 140 m² (7 × 20 m²), filled with 60 cm of crushed broken gravel
                                      (pore volume 39%)
Vertical flow bed   : 121 m² (11 × 11 m²), filled with 90 cm of clean sand $K_f = 10^{-3}$ m/s
Vegetation   : Both beds planted with *Phargmites karka* (local reeds)

A two-staged subsurface flow CW (horizontal flow followed by vertical flow bed) was built as the first full scale CW system in Nepal to treat Dhulikhel Hospital (45 beds) wastewater.

The system operates without electricity, as water flows through the system due to gravity and the intermittent feeding of wastewater is done by a simple hydro-mechanical system. It has been in operation since July 1997, the total costs of the system is only 16 400 USD.

Treatment efficiency. The system has shown high treatment efficiency since its operation (Table 1). More than 95% of major pollutants such as suspended solids, organic pollutants (BOD), ammonia-nitrogen are removed by this system. Removal of *E. coli* is even higher i.e. 99.9999% (Shrestha and Manandhar, 1998, Laber *et al.*, 1999 and Shrestha 1999).

Table 1 Summary statistics of inlet and outlet concentrations and mean reduction efficiencies of Dhulikhel Hospital constructed wetland system (1997 to 2000)

Month	Q m³/d	TSS IN mg/L	TSS OUT mg/L	NH_4-N IN mg/L	NH_4-N OUT mg/L	PO_4-P IN mg/L	PO_4-P OUT mg/L	BOD_5 IN mg/L	BOD_5 OUT mg/L	COD IN mg/L	COD OUT mg/L	E.coli IN col/ml	E.coli OUT col/ml
nos. of reading	13	12	12	12	11	12	12	13	13	13	11	11	11
min	7	25.67	0.3	17	0.04	2.177	0.637	31.1	0	62.55	4.34	39000	3
Max	40	230	6.7	52.32	5.4	26	17.5	210	10	1048	40	8E+08	987
Avg.	19.6	82.92	2.283	33.296	1.604	7.908	4.223	109.9	3.287	324.5	20.2	1E+08	148
Median	10.7	41	1.8	19.167	0.04	2.177	0.67	40.93	4.167	79.2	18.2	1E+05	38
Std. Deviation	11.1	58.19	1.946	12.206	2.18	7.483	5.757	62.28	2.968	272.8	14.2	2E+08	307
Reduction Efficiency (%)			97.25		95.183		46.6		97.01		93.8		99.9999

IN – Wastewater water after primary treatment and inlet to CW; OUT – Final effluent of CW

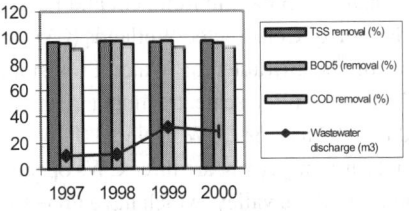 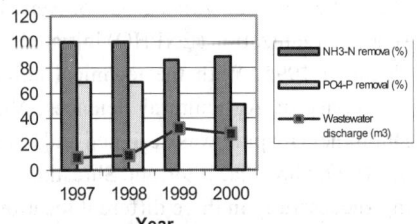

Figure 1 Concentration reduction of Dhulikhel Hospital constructed wetland system at different time interval

Though this system was initially designed for 20 m³/day of wastewater, it now treats 35–40 m³/day. Increase in hydraulic loading does not affect the removal efficiency for TSS, BOD$_5$ and COD (Figure 1). However, removal efficiency of ammonia, phosphorus and pathogens is reduced with the increase in hydraulic load and time interval. Loss of drainage capacity at the upper layer of the vertical flow bed and decrease in hydraulic loading interval reduces the oxygen flow into the vertical flow bed thus reducing the ammonia removal efficiency (Haberl et al., 1995; Felde vand Kunst, 1997 and Platzer, 1999). There is also a strong relationship between hydraulic retention time and removal of pathogens in the CW system (Green et al., 1997). Decrease in elimination efficiency of phosphorus with age is common in almost all kinds of CW systems which can also be seen in Dhulikhel Hospital CW system (Brix, 1987; Perfler and Haberl, 1993).

Demonstration of greywater recycling through constructed wetland in Kathmandu

Shortage of drinking water is a serious problem in Kathmandu. Present water demand in the valley is about 150 million litres per day (MLD), whereas production is only 90 MLD. In spite of this major problem, drinking water is being used for several non-drinking purposes such as flushing, washing, gardening, vehicle cleaning, etc.

Realizing this problem, one of the authors has installed a greywater treatment system through CW at his own house to demonstrate water-recycling technology. The system was built in April 1998 soon after successful implementation of CW at Dhulikhel Hospital (Shrestha, 1999; Shrestha, 2000).

Description of greywater treatment plant. A household with seven members produces about 500 litres/day of greywater. It consists of wastewater from bathroom, shower, washing machine and wastewater from kitchen. For the separation of greywater from blackwater, separate plumbing has been installed. Thus separate greywater is collected into a settling tank for further treatment. The system consists of a two-chambered settling tank (500 L) as a pretreatment, a feed tank (200 L) and a vertical flow bed (1 × 6 m) filled with coarse sand and planted with common reed (*Phragmites karka*) and *Canna* sp. as the main treatment unit.

The system does not need any electrical devices. Water is flushed hydro-mechanically into the bed 3 to 4 times a day. Finally the treated water is collected in an underground tank. The collected water is used for flushing, gardening and cleaning. Thus, this house saves nearly 400 to 500 litres of water everyday. The system costs around US $ 430 for construction. The operating cost is negligible.

Treatment efficiency. The system has been operating since April 1998 giving excellent results. Removal efficiencies of nutrients and organics like BOD are constant even after 2 years of operation (Table 2).

Table 2 Summary statistics of in-and outlet concentrations and mean elimination rates of greywater treatment system (April, 1998 to May, 2000)

	TSS IN mg/L	TSS OUT mg/L	NH$_4$-N in mg/L	NH$_4$-N out mg/L	PO$_4$-P in mg/L	PO$_4$-P out mg/L	BOD$_5$ in mg/L	BO$_5$ out mg/L	COD in mg/L	COD out mg/L
nos. of reading	8	8	9	9	7	7	9	7	9	9
Minimum	52	0.5	3.66	0.02	0.79	0.78	100	0	177	6.8
Maximum	188	6	25.7	1.983	4.9	3.62	400	12	687	72
Average	97.9	2.6	13.3	0.5	3.1	2.0	200.1	5.2	411.4	29.1
Median	78.0	2.3	11.7	0.3	3.2	2.3	187.8	2.5	362.0	24.0
Std. Deviation	53.4	2.0	8.0	0.6	1.4	1.2	93.6	4.6	174.0	19.9
Elimination rate (%)		97		96		35		97		93

Constructed wetland for septage treatment in Kathmandu City

There are about 40,000 septic tanks in Kathmandu. Several private companies are providing septic tank cleaning services in the valley. They discharge the collected septage into nearby streams or rivers without any treatment. When the Mayor from KMC observed the Dhulikhel hospital constructed wetland system, he decided to build a CW system for septage treatment as a demonstration plant for Kathamndu City.

Designing and efficiency of the system. Septage has not yet been widely treated in constructed wetlands. Therefore design and operating criteria do not exist as yet. However, there is some experience on septage treatment by using CW (Cooper *et al.*, 1996; Summerfelt *et al.*,1996). These experiences were considered for designing the CW for KMC. The system consists of a three-chambered septic tank (200 m^3), three units of sand and gravel filter beds (each unit 75 m^2) in parallel as a pretreatment, and a vertical flow bed (362 m^2) as a main treatment unit.

The system has been operating since January 1999. Regular monitoring and maintenance is lacking in this wetland. As a result, problems like breakage of water distribution pipes, break down of electric submersible pump, untimely removal of sludge from settlement tank, etc. are often encountered. However, the treatment efficiency is still found to be good enough to discharge the effluent into the river according to Nepalese Standards. The few analytical results show high removal of TSS, BOD and COD (Table 3).

New systems under construction

The operating CW systems are now becoming demonstration sites for researchers, professionals, students, local leaders and others. Encouraged by this success, a school has recently installed a two-staged CW system (for 250 pe) and another one is under construction at Kathmandu University (for 700 pe). Similarly, CWs have recently been designed for two hospitals. New systems have been designed on the basis of lessons learned from the existing systems.

Apart from above mentioned system, a large CW system has been designed for treatment of septage and leachate for Pokhara sub-metropolis. The system has been designed to treat 100 m^3/day of septage and 40 m^3/day of landfill leachate. The design consists of a settlement tank (200 m^3), a sand drying bed for primary treatment of septage and a two staged CW system (horizontal flow bed – 1,105 m^2 followed by a vertical flow bed – 2,203 m^2). The design was based on an outlet BOD$_5$ concentration less than 100 mg/L. (East Consult/CEAD/ENPHO, 1998).

Problems encountered in implementation, operation and maintenance of CW
- Due to the lack of awareness about the system, it took a lot of effort to start the first CW system.
- Wastewater treatment is still not on the priority list for the community and industries due to lack of strong legislation on effluent criteria.

Table 3 Some analytical results of constructed wetland for treating septage

Month	A TSS mg/L	B TSS mg/L	Removal rate (%)	A NH$_4$-N mg/L	B NH$_4$-N mg/L	Removal rate (%)	A BOD$_5$ mg/L	B BOD$_5$ mg/L	Removal rate (%)	A COD mg/L	B COD mg/L	Removal rate (%)
February, 99	430	20	95.3	660.0	122.0	81.5	650	70	89.2	2950	350	88.1
March, 99	1820	110	94.0	721.0	242.0	66.4	525	34	93.5	2500	380	84.8
Sept, 99	870	44	94.9				1092	63	94.2	5800	330	94.3

A – Septage after settlement tank; B – Final effluent

- Capital investment is high although the system is natural and materials are locally available.
- A specified type of filter media like sand and gravel is not easily available.
- Sealing material for the beds is still a question.
- There is also a security problem especially regarding distribution pipes of the vertical flow bed which are exposed above ground.
- Carelessness in operation and maintenance.
- Unavailability of funds for further research to improve the system.

Conclusion

Although the need for appropriate wastewater management technologies is undoubtedly a realized issue by the government, planners and policy makers, so far there has been very little action taken towards wastewater treatment in Nepal. CW has recently been introduced in the country as an appropriate technology because of its simple and natural system. The present experience suggests that the system is appropriate for a country like Nepal where cities are growing without any demographic, municipal and regional planning. After five years of research and development, this technology has proved to be useful in Nepal and is now ready for wide scale application.

In this way this system has been able to gain popularity in the country within a very short time. This however, is only the beginning as there are possibilities to install these systems throughout the country and even expand their application in the South Asian region.

Recommendation

- For the exploration of CW technology in developing countries, a separate design guideline is necessary. Developed countries are now concentrating more on the removal of nitrogen, phosphorus and other chemical constituents whereas the present goal of developing countries is limited to removal of organic load.
- CW should also be used for industrial effluents like tanneries, dairies, breweries and other industries that are feasible for CW.
- Research should also be focused on minimization of construction cost for developing countries.
- International experts should lobby for CW technology to International Development Aid Organizations who are investing substantial amounts of money in the water and sanitation sector.
- Professionals and organizations who are working in this field should be encouraged by
  - providing research grants and training
  - sending research students or researchers to perform a joint research program
  - supplying reading materials, especially international proceedings on the topic
  - providing opportunities for disseminating their findings

References

Brix, H. (1987). The Applicability of the Wastewater Treatment Plant in Othfresen as Scientific Documentation of the Root Zone Method. *Wat. Sci. Tech.* **19**(10), 19–24.

Cooper, P.F., Job, G.D., Green, M.B. and Shutes, R.B.E. (1996). *Reed Beds and Constructed Wetlands for Wastewater Treatment.* WRc Publications, UK.

East Consult/CEAD/ENPHO (1998). Pokhara Sub-Metropolis Second Tourism Infrastructure Development Project. Pokhara Environment Project. Septage and Solid Waste Management and Access Road Imrovement Component. Draft Report. Vol. 1. Executive Summary. East Consult (P) Ltd., Kathmandu in association with CEAD Consultant and Environment and Public Health Organization.

Felde v, K. and Kunst, S. (1997). N-and COD- Removal in Vertical Flow Systems. *Wat. Sci. Tech.*, **35**(5), 79–85.

Green, M.B., Griffin, P., Seabridge, J.K. and Dhobie, D. (1997). Removal of bacteria in subsurface flow wetlands. *Wat. Sci. Tech.* **35**(5), 109-116.

Haberl, R., Perfler, R. and Mayer, H. (1995). Constructed Wetland in Europe. *Wat. Sci. Tech.*, **32**(3), 305–315.

Laber, J., Haberl, R. and Shrestha, R.R. (1999). Two Stage Constructed Wetland for Treating Hospital Wastewater in Nepal. *Wat. Sci. Tech.*, **40**(3), 317–324.

Perfler, R. and Haberl, R., (1993). Actual Experiences with the use of Reed Bed Systems for Wastewater Treatment in Single households. *Water Sci. Tech.* **28**(10), 141–148.

Platzer, C. (1999). Design recommendation for subsurface flow constructed wetlands for nitrification and denitrification. *Wat. Sci. Tech.*, **40**(3), 257–263.

Shrestha, R.R. (1999). Application of Constructed Wetlands for Wastewater Treatment in Nepal. A dissertation for the fulfillment of degree of the Doctor of Applied Natural Sciences. University of Agricultural Sciences, Vienna, Austria.

Shrestha, R.R (2000). Greywater Treatment : An Option for Water Recycling. Environment, A Journal of the Environment Ministry of Population and Environment, *HMG/NEPAL.* Vol 5 No 6 pp 97–105.

Shrestha, R.R. and Manandhar, R. (1998). Constructed Reed Bed – Technology Introduced fir the First Time in Nepal. A paper presented on the national workshop on *9th National Plan and Implementing Strategy in the WSS sub-sector*. 22–23 March 1998, Kathmandu.

Shrestha, R.R., Haberl, R. and Laber, J. (2001). Constructed Wetland Technology Transfer to Nepal. *Wat. Sci. Tech.*, **43**(11), 345–350.

Summerfelt, S.T., Adler, P.R., Glenn, D.M. and Kretschmann (1996). Aquaculture Sludge Removal and Stabilization within Created Wetlands. In: *Proc. 5th Internat. Conf. Wetland Systems for Water Pollution control*. Universität für Bodenkultur Wien, Austria. Chapter XIII/2.

Experimental results on constructed wetland pilot system

J. M. González, G. Ansola* and E. Luis

Area de Ecología, Universidad de León, Campus de Vegazana, E-24071, León, Spain.
(E-mail: deggag@unileon.es)

Abstract Research into a constructed wetland for wastewater treatment using M.H.E.A. (Hierarchical Mosaic of Artificial Ecosystems) pilot system was carried out over a vegetative period in 8 different flow and vegetable composition series. The system consisted of a free water pond as a first step working as primary treatment followed by a zone with *Typha* sp. and surface flow and finally a woody zone with a subsurface flow and planted with ligneous species (*Salix* sp., *Populus* sp., *Fraxinus* sp. and *Alnus* sp.).
Removal efficiency in the study reflects an optimal result: 80–99% total suspended matter removal, 82–98% organic matter removal, 70–98% nutrients removal and up to 99.9% faecal bacterial disinfecting. Effluent characteristics were in accordance with European Union legislation criteria for wastewater treatment systems.
Keywords Constructed wetlands; experimental sewage treatment; pilot system; removal efficiency; wastewater

Introduction

Wastewater treatment plant viability is a pressing problem, especially for small communities in regions characterised as "sensible" for eutrophication. It is therefore necessary to develop new low installation and operation cost technologies; these must efficiently remove organic and nutrient loads, allowing for water reuse and biomass utilisation in order to reduce operating investments. In most cases, the economic costs should cover the proposed objectives, such as the inclusion of tertiary treatment for sensitive designated zones, is economically unapproachable for small villages in rural areas. In Spain wastewater is reused for irrigation in many areas.

The European Union Directive 91/271/EEC (ECOD, 1991a) established a series of requirements for wastewater treatment plant effluents and removal efficiency, which must be applied before the end of December 2005 for populations between 2,000 to 15,000 equivalent inhabitants. Nutrients must therefore be removed in rural areas in the north-west of Spain where wastewater is very diluted (BOD_5 between 20 and 50 mg/l in 70% of the villages), but with moderate to high concentration of Nitrogen (N) and Phosphorus (P) due to use of fertilisers (González and Cabo, 1998).

The province of León localities with less than 10,000 inhabitants come to about 97%, and those with less than 2,000 inhabitants represent 82.5% of the total population of the province. Most of these small localities are located in disperse zones or in mountain and riverside zones, many of them being catalogued as eutrophication sensitive and consequently will require tertiary treatment (nutrients elimination). In another case, these localities are on the Spanish meseta, but very tied to agricultural or cattle-raising activities and they are equally conditioned to treat their wastewater to tertiary level according to European legislation EU 91/676 (ECOD, 1991b).

It has been widely demonstrated in Spain that conventional (activated sludge) treatment systems applied to small rural municipalities have many operational and management problems, which is at present greatly inactive or abandoned. As an alternative to the conventional treatment systems, low-cost and natural systems are increasingly applied in central Spain, where stabilisation ponds are one of the most adequate and applied systems. Constructed wetlands have not up to now been considered as a treatment option by the administration but mainly due to the absence of previous experiences on pilot plants. The aim of this work was to study the applicability of this system to the wastewater and climatic conditions of the region and to know the parameters for the design of a full-scale process.

Materials and methods

An M.H.E.A. (*Hierarchical Mosaic of Artificial Ecosystems*) model developed by Michel Radoux at Fondation Universitaire Luxembourgeoise, (Radoux and Kemp, 1982; Radoux et al., 1995) in Viville (Belgium), was followed up by the experimental pilot plant design situated in the Northwest of Spain.

Eight series of three glass fibre basins of $0.85 \times 1.3 \times 0.55$ m, were installed with a capacity of 0.6 m^3 and a surface of 1.1 m^2 each basin. They were preceded by a pond like primary treatment (WSP) with a detection time of 24 hours. First basins on series were filled to 25 cm high with 6–8 mm diameter gravel bed. Water is over 20 cm of gravel. Effective volume was reduced to 0.29 m^3. These basins had a superficial flow and they were planted with 30 units of *Typha latifolia* collected from their natural system (Ansola et al., 1995a, 1995b).

Two following steps in the series were totally filled with inert gravel, they had a subsuperficial flow and were planted with 15 units/m^2 of *Salix atrocinerea*, *Alnus glutinosa*, *Populus alba* or *Fraxinus excelsior*, each being one year old. The different series, each composed of a surface of 4.4 m^2, were as follows (Table 1).

Raw wastewater was taken and pumped directly from a municipal sewer to supply the system and distributed to the series basins by peristaltic pumps in continuous action. The hydraulic load differed between series. For the series 1, 5 and 7 the operative water-flow was 160 l/d for a surface of 4,4 m^2. The series 2, 6 and 8 had a water-flow of 250 l/d with the same surface. There were two special series, 3 and 4 that represented the experimental operations designed flows as a reply to the minimum and maximum flow of the full-scale pilot plant constructed in Bustillo de Cea (León, Spain) (Ansola et al., 1998).

Physico-chemical and microbiological analysis in the influent, pond effluent and final effluent were carried out for a vegetative period (March to October) in the pilot scale experimental system. The analysis was based on *Standard Methods* (1989).

In Table 2, we show the European Union established elimination removal levels and limited effluent values of different parameters proceeding of wastewater treatment plants. There is no legislation which refers to microbiological parameters or the treatment removal efficiency with regard to this topic.

Results and discussion

Influent characteristics

The average electrical conductivity detected was 436 μS/cm, which corresponds to a typically domestic wastewater (Metcalf and Eddy, 1991) and low mineral composition (Henze et al., 1997), with a maximum of 498 μS/cm and a minimum of 412 μS/cm. Concentration of total suspended solids (TSS) varied between 287 mg/l and 124 mg/l, the

Table 1 Configuration of experimental series and their designed water-flow

Series 1	Pond-*Typha-Salix-Salix*	P-T-S-S	Water-flow = 160 l/d
Series 2	Pond-*Typha-Salix-Salix*	P-T-S-S	Water-flow = 250 l/d
Series 3	Pond-*Typha-Salix-Salix*	P-T-S-S	Water-flow = 80 l/d
Series 4	Pond-*Typha-Salix-Salix*	P-T-S-S	Water-flow = 200 l/d
Series 5	Pond-*Typha-Alnus-Populus*	P-T-A-P	Water-flow = 160 l/d
Series 6	Pond-*Typha-Alnus-Populus*	P-T-A-P	Water-flow = 250 l/d
Series 7	Pond-*Typha-Fraxinus-Fraxinus*	P-T-F-F	Water-flow = 160 l/d
Series 8	Pond-*Typha-Fraxinus-Fraxinus*	P-T-F-F	Water-flow = 250 l/d

Table 2 European Council Directive, May, 21, 1991

DIRECTIVE (91/271/EEC)	Units	Maximum daily concentration	Removal (%)
Biochemical Oxygen Demand (BOD$_5$ at 20°C)	mg O$_2$/l	25	70-90
Chemical Oxygen Demand (COD)	mg O$_2$/l	125	75
Total Suspended Solids (TSS)	mg/l	35	90
Total Phosphorus	mg/l P	2*	80
Total Nitrogen	mg/l N	10 or 15**	70-80

* populations higher than 100,000 eq. inhab. The value will be 1 mg/l
** 10 ppm for populations of higher than 100,000 eq. inhab. and 15 ppm for those between 10,000–100,000

average concentration was 164 mg/l, all these values less them of 350 mg/l, which allowed for the quantification of this wastewater as weak concentration (Metcalf and Eddy, 1991) or diluted wastewater type (Henze et al., 1997). The organic load also expressed as the total Chemical Oxygen Demand (COD) varied between 271 and 112 mgO$_2$/l, with average concentration of 156 mgO$_2$/l, which also confirms its classification as weak concentration wastewater since the average concentration for this parameter in the typical domestic wastewater (Henze et al., 1997) is located between 250 and 500 mgO$_2$/l for very diluted types. Biochemical Oxygen Demand (BOD$_5$) varied between a maximum of 93 mgO$_2$/l and a minimum of 72 mgO$_2$/l, with an average concentration for the study time of 83 mgO$_2$/l and values below 110 mgO$_2$/l for very diluted wastewater types (Henze et al., 1997).

Nutrients concentrations showed the same dilution effect of the wastewater used as an influent. Nitrate presented a range of variation, with a maximum of 3.9 mg/l and a minimum of 2.4 mg/l, below the typical domestic wastewater (15–29 mg N/l; Metcalf and Eddy, 1991). The total nitrogen concentration in the influent, measured as Total Kjeldahl nitrogen (TKN), had a maximum of 19.6 mg/l and a minimum of 7.8 mg/l, below the 20 mg/l described (Henze et al., 1997) for very diluted wastewater types. With regard to phosphorus concentrations, (TP) was presented with a minimum of 4.2 mg/l and a maximum of 5.0 mg/l, close to 4 mg/l for very diluted types (Henze et al., 1997). Average concentration and the standard deviation of the different parameters measured are shown in Table 3.

This very diluted influent is a common pattern in the north of Spain owing to the infiltration of ground water into the sewage (González and Cabo, 1998).

Pond effluent

The effect of pond treatment as a first step into the series, operating as a sedimentation tank with only one day of hydraulic detention time, reduced the average concentration levels of solids up to 31 mg/l for TSS and 26 mg/l for TVS. The organic load was reduced to 47 mg/l

Table 3 Average influent concentration and standard deviation for the measured parameters

Parameters (units)	Average concentration	Standard deviation
Temperature (°C)	7.5	0.1
pH	16.6	1.8
Conductivity (µS/cm)	463	22.9
DO (mg O$_2$/l)	2.1	0.4
TSS (mg/l)	164	40.1
TVS (mg/l)	145	19.2
COD (mg O$_2$/l)	156	39.1
BOD$_5$ (mg O$_2$/l)	83	7.0
Nitrate (mg NO$_3^-$/l)	3.4	0.4
Ammonia (mg NH$_4^+$/l)	5.1	0.9
TKN (mg/l)	14.6	2.6
TP (mg P/l)	4.7	0.3
Faecal coliform (col./100 ml)	600285	257602

average concentration for COD, but it was not sufficient in BOD_5 with an average concentration of 38 mg/l.

The average concentration levels of nutrients after pond treatment were insufficient for the requirements according to European legislation. Thus, we obtained average concentrations of 4 mg/l for ammonia, 2 mg/l for nitrate, and with regard to total nutrients forms, 11 mg/l for total nitrogen and 3 mg/l for total phosphorus were obtained.

Disinfecting efficiency, owing to the short detention time applied, was insignificant depending on waste stabilisation pond possibilities, though an average concentration of 169,044 col/100 ml was obtained.

Final effluent

Values of concentration for the most important parameters in the final effluent are given in Table 4, showing good results obtained for the final effluent quality and they agree with the requirements of the European Council legislation.

Removal efficiency

Average removal efficiencies obtained in the study are shown in Figure 1, these results show a high reduction level in all the series for the different analysed parameters.

The suspended solid removal (Figure 1a) was high in the first pond stage with an average removal of 81% in TSS. An average removal between 97% of series 4 (PTSS-200) and 99% of series 8 (PTFF-250) was obtained in the final effluent of the series. This last series, which was operating with the higher water-flow, is significantly different to others applying a statistical "t" test for independence variables (significant level of 95%), except with series 6 (PTPA-250), for the solids removal efficiency. Suspended (TSS) and volatile (TVS) solids concentration in the influent and the effluent have positive correlation, therefore the degree of removal found for each one of these parameters does not vary.

Organic matter removal efficiency (Figure 1b) was represented for the obtained elimination in COD and BOD_5. In the pond stage an average removal of 69% to COD and 54% to BOD_5 was obtained. These results are not sufficient. Treatment in the following stages showed a high removal efficiency and was more efficient with higher assayed hydraulic detection time. Thus removal levels varied between 91% in series 6 (PTPA-250) to 95% in series 1 (PTSS-160) for the COD average elimination and between 98% in series 7 (PTFF-160) to 99% in case of BOD_5. According with the "t" test assayed for the variables independence of results obtained in COD reduction. Series 1 (PTSS-160) is significantly independent of the rest also occurs with the series 3 (PTSS-80).

Table 4 Average effluent concentration for the main parameters in the different series tested

Parameters (units)	Effluent							
	Series 1 PTSS 160 l/d	Series 2 PTSS 250 l/d	Series 3 PTSS 80 l/d	Series 4 PTSS 200 l/d	Series 5 PTPA 160 l/d	Series 6 PTPA 250 l/d	Series 7 PTFF 160 l/d	Series 8 PTFF 250 l/d
pH	7.38	7.45	7.36	7.30	7.39	7.44	7.39	7.32
DO (mg O_2/l)	3.0	3.5	5.8	3.4	3.8	3.9	3.7	3.7
Conductivity (µS/cm)	366	355	364	368	359	360	358	353
TSS (mg/l)	3.4	3.6	4.9	5.0	3.7	2.5	3.3	1.9
COD (mg O_2/l)	7.8	10.9	9.4	11.0	12.2	13.2	9.9	12.0
BOD_5 (mg O_2/l)	1.1	1.2	0.9	1.3	1.1	0.9	0.7	1.1
Nitrate (mg NO_3^-/l)	0.3	0.3	0.1	0.1	0.3	0.5	0.2	0.5
Ammonia (mg NH_4^+/l)	0.4	0.7	0.1	2.0	0.2	1.0	0.1	0.2
TKN (mg N/l)	0.4	1.4	0.3	2.2	0.5	0.9	0.2	0.9
TP (mg P/l)	0.2	0.2	0.1	0.4	0.2	0.3	0.04	0.3
FC (col./100 ml)	167	245	14	73	71	96	113	116

With regard to the distribution of woody species in the third stage of the series for COD elimination, the use of *Salix atrocinerea* was more adequate than other species. However, with regard to the DBO$_5$ removal, the use of *Fraxinus excelsior* had a greater efficiency, although significant differences were not detected between the data of this parameter.

Nutrients (N and P) removal (Figure 1c and 1d) is very low in the pond stage, obtaining 25% for the TP and 23% for TN; with regard to the nitrogen forms studied, an average removal of ammonia of 29% and 30% in the case of nitrate was obtained in the first stage.

The nitrate assimilation in the series was produced with greater hydraulic detention times and fundamentally in the case of series that used *Salix* sp. as woody species, a maximum average removal in series 3 (PTSS-80) of 98% was obtained. All the results obtained for this parameter were found to be 85% of average removal (minimum in series 6 PTPA-250 and series 8 PTFF-250).

The removal of ammonia reached maximum average levels in larger used surfaces (99% in the case of series 3 PTSS-80) and it was minimal in greater applied water flow (81% in the case of series 6 PTPA-250). Significant differences for this parameter were not found in the different flows tested with the series of *Fraxinus* sp. (series 7 and 8) or between those which used the minor flow of *Salix* sp. series (series 3).

In total nutrients forms (TN and TP), the use of *Fraxinus* sp. in series with the lowest water flow applied (series 7 PTPP-160) obtained the best average results (98% for TN and 99% for TP), these results are significantly different to the other series studied, except for TN with series 3 (PTSS-80) and series 1 (PTSS-160) which obtained 98% in TN removal and 97% in TP removal for series 3, and 97% for TN and TP removal in series 1.

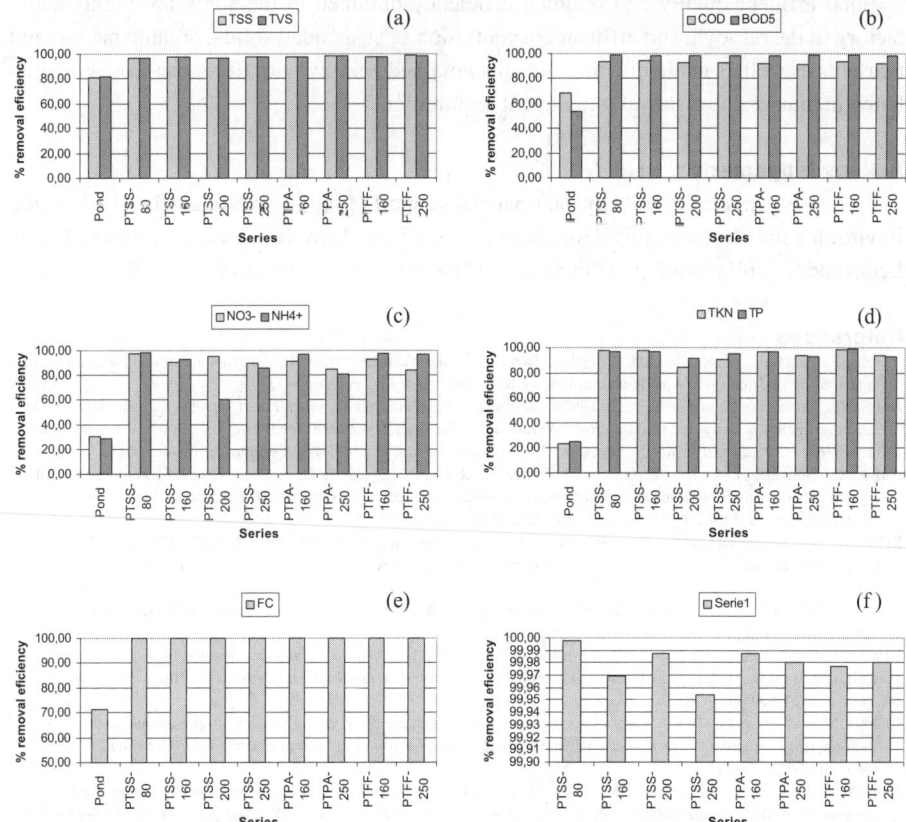

Figure 1 Removal efficiency in series for different parameters studied. a) Total solids; b) Organic matter; c) Nitrogen species; d) Total nutrients; e) Faecal coliform disinfecting; f) Detail of faecal coliform disinfecting in assayed series

Disinfecting (Figure 1e) in the pond stage obtained average results of 71%. The use of macrophytes and woody species notably improves the results, with average disinfecting from faecal coliform bacteria (Figure 1f) in all the series studied up to 99.9% being obtained. The maximum results for disinfecting were obtained with the lowest flows tested, with 99.99% maximum being observed in the series 3 PTSS-80 (99.99%), however there did not seem to exist a preference for the disinfecting in the use of one or another kind of woody species.

Conclusions

Wastewater characteristics used as an influent in the study reflected a weak concentration. However this kind of water is common in rural areas of north west Spain.

Removal of solids from the diluted wastewater was satisfactorily obtained with any of the following operational strategies and selected species in the study. If the objective for the application of this model at full scale is solids removal, we could reduce the surface to system establishment. Organic matter removal was obtained with high efficiency levels and is procured in a significantly superior way to more commonly used surfaces and with the use of *Salix* sp. a woody species at the end of the system.

The removal efficiency of nutrients was accomplished with statistical significance at higher hydraulic detection times. Statistical differences observed with the use of *Fraxinus* sp. as a third system stage made it possible to reduce the surface requirements as opposed to the use of *Salix* sp., with equal results. The hydraulic detention time is very significant, with disinfecting being higher at greater times.

Final effluent quality and removal efficiency obtained in the study are highly satisfactory in the removal and effluent concentration of suspended solids, organic matter, and nutrients as well as in disinfecting. Results obtained comply with European Council legislation in municipal wastewater treatment requirements.

Acknowledgement

This work was made possible by the financial support of the Junta de Castilla y León to the Environmental Biotechnology Research group of the Area de Ecología, Universidad de León, and the collaboration of "Fundation Universitaire Luxembourgeoise" (Belgium).

References

Ansola, G., Fernández, C. and de Luis, E. (1995a). Removal of organic matter and nutrients from urban wastewater by using an experimental emergent aquatic macrophyte system. *Ecological Engineering* **5**, 13–19.

Ansola, G., Fernández, C. and de Luis, E. (1995b). Removal of nutrients and organic matter from urban wastewater by using an experimental constructed wetland. *Proc. Natural and Constructed Wetlands for Wastewater Treatment and Reuse. Experiences, Goals and Limits*. R. Ramadori, L. Cingolani and L. Cameroni (ed.) Perugia. pp. 39–46.

Ansola, G., González, J.M., Soto, F., García, M., López, G., Radoux, M., Cadelli, D. and Luis, E. (1998). Natural integrated systems using constructed wetlands for treating wastewater in Northwest Spain. *Procc. 6th International Conference on Wetlands systems for water Pollution Control*. Brasil. pp. 680–687.

ECOD (1991a). European Council Directive about urban wastewater treatment. *DOCE n° L 135, (91/271/CEE)*. 40–52.

ECOD (1991b) European Council Directive of contaminated water protection by nitrates used in agriculture *DOCE (91/676/CEE)*.

González, J.M. and Cabo, A. (1998). Libro Blanco sobre calidad de agua en el Bierzo (León, Spain). *Universidad de León*. León (Spain). 222 pp.

Henze, M., Harremoës, P., La Cuor Jansen, J. and Arvin, E. (1997) *Wastewater treatment*. Springer, Berlin.

Metcalf, L. and Eddy, H.P. (1991). *Wastewater Engineering, Treatment, Disposal and Reuse*. 3rd edition. McGraw-Hill Inc. New York. 1,334 pp.

Radoux, M. and Kemp, D. (1982). Aproche écologique et experimentale des potentialités épuratrices de quelques hélophytes: *Phragmites australis* (cav.) Trin. Ex Steud. *Typha latifolia* L. et *Carex acuta* L. Trib. *Cebedeau*. 465–466(35), 325–340.

Radoux, M., Cadelli, D. and Nemcova, M. (1995). A mosaic of artificial ecosystems as a wastewater treatment plant: Evaluation of the pilot plant of Lallaing (France). *Proc. Natural and Constructed Wetlands for Wastewater Treatment and Reuse. Experiences, Goals and Limits*. R. Ramadori, L. Cingolani and L. Cameroni (ed), Perugia. pp: 275–284.

Standard Methods for the Examination of Water and Wastewater (1989). 17th edn. American Public Health Association – American Water Works Association – Water Pollution Control Federation. Ed. L.S. Clesceri, A.E. Greenberg and R.R. Trussell.. Baltimore. Marylan, USA.

Reed beds: constructed wetlands for municipal wastewater treatment plant sludge dewatering

J.S. Begg, R.L. Lavigne, P.L.M. Veneman

Department of Plant and Soil Sciences, University of Massachusetts, Amherst, Massachusetts, 01003, USA

Abstract Reed beds are an alternative technology wastewater treatment system that mimic the biogeochemical processes inherent in natural wetlands. The purpose of this project was to determine the effectiveness of a reed bed sludge treatment system (RBSTS) in southern New England after a six-year period of operation by examining the concentrations of selected metals in the reed bed sludge biomass and by determining the fate of solids and selected nutrients. Parameters assessed in both the reed bed influent and effluent: total suspended solids, biochemical oxygen demand, nitrate-nitrogen and total phosphorus. In addition, the following metals were studied in the reed bed influent, effluent and *Phragmites* plant tissue and the sludge core biomass: boron, cadmium, chromium, copper, iron, lead, manganese, molybdenum, nickel, and zinc. The removal efficiencies for sludge dewatering, total suspended solids and biochemical oxygen demand were all over 90%. Nitrate and total phosphorus removal rates were 90% and 80% respectively. Overall metals removal efficient was 87%. Copper was the only metal in the sludge biomass that exceeded the standards set by the Massachusetts Department of Environmental Protection for land disposal of sludge. The highest metal concentrations, for the most part, tended to be in the lower tier of the sludge profile. The exception was boron, which was more concentrated in the middle tier of the sludge profile. The data and results presented in this paper support the notion that reed bed sludge treatment systems and the use of reed beds provide an efficient and cost effective alternative for municipal sludge treatment.
Keywords Constructed wetlands; reed beds; sludge dewatering

Introduction

Constructed wetlands for wastewater treatment are comper biological systems that mimic self cleansing processes inherent in natural wetlands. The Reed Bed Sludge Treatment System (RBSTS) is one type of constructed wetland. The RBSTS is a vegetated submerged bed that can be used to dewater. The RBSTS is planted with wetland plants such as *Phragmites australis* Cav. (common reed). Drying and mineralization reduce the liquid sludge volume by as much as 90%, and physical filtration removes almost 100% of the total suspended solids from the RBSTS effluent (Seidel, 1976). The final product is a well decomposed, stabilized compost suitable for land application (Krueger, 1991).

The general hypothesis of this research was that RBSTSs serve as an effective sludge dewatering system producing underdrain effluent quality that meets National Point Discharge Environmental Standards (NPDES) for direct discharge into waterways. The project examines the fate of metals in the sludge biomass and the possible accumulation of metals in *Phragmites* plants. Concentrations of metals in the sludge are a major concern for final disposal, affecting issues such as town economics, public safety, and state and federal compliance.

The reed bed at the Shelburne/Buckland wastewater treatment plant located in Buckland, MA served as the project site. The data series began in the fall of 1992 and continues until the spring of 1999. The specific objectives of this project were: i) quantitatively show that RBSTSs effectively dewater sludge; ii) measure total suspended solids, biochemical oxygen demand, nitrate-nitrogen, and total phosphorus in the underdrain effluent, and compare, where relevant, to DEP discharge standards; and iii) to determine the concentrations of boron, cadmium, chromium, copper, iron, manganese, molybdenum, nickel, lead and zinc in RBSTS sludge biomass and in *Phragmites* plant tissue to determine the fate of these metals.

Materials and methods
Background
The Shelburne/Buckland wastewater treatment plant services the business districts and densely populated residential areas of two towns. During dry weather, flows to the plant typically average about 570 m^3/d, but during storm events flow rates can exceed 2000 m^3/d. After initial screening and settling processes, an extended aeration system treats the raw wastewater followed by aerobic digestion of the sludge in an aerated 200 m^3 above-ground tank with a residence time of 25 to 30 days. During that time the volatile solids are reduced to between 60% and 70%. The results reported herein only pertain to the first reed bed planted in 1992.

Bed Design. The reed bed is 16.5 m wide by 30 m long with an overall depth of 2.4 m. Each reed bed is lined with an impervious material and has a 20–30 cm layer of pea stone (3–6 mm) or gravel (>2 mm) around the 100 mm perforated underdrain pipes. On top of the pea stone is 30 cm of coarse sand (0.5–1 mm). The pea stone and sand are the initial growing medium for the *Phragmites* plants and the primary physical filtration medium for the sludge. The rhizomes and root system of *Phragmites* penetrate the growing medium and help channel wastewater flow. The effective depth of sludge storage is 1.2 m and the expected length of time between reed bed cleanings, limited by the effective depth of sludge storage, is 8–10 years. The underdrain pipes surrounded by pea stone and underlain by the impervious liner direct the effluent to a concrete cistern where a sump pump forces the effluent from the cistern back to the headworks for chlorinating prior to discharge.

Analyses
Dewatering efficiency and total suspended solids. Measurements of biomass volume were taken in March of 1995 and 1999 from ten random locations. Biomass volume was compared to sludge volume data. A mass balance determination relates the applied liquid sludge volume to residential biomass for a six-year period. Fifteen samples were collected to determine the bulk density of the sludge. Samples were dried at 103–105°C until they reached a constant weight and the bulk density was calculated. Solids content was determined by gravimetric procedures in the laboratory for the sludge influent and the underdrain effluent following Standard Methods 2540 A-G (*Standard Methods for the Examination of Water and Wastewater*, 1995).

Biochemical Oxygen Demand (BOD$_5$). For each loading of the reed bed BOD$_5$ samples were collected. The samples were analyzed following Standard methods 5210 B (*Standard Methods for the Examination of Water and Wastewater*, 1995).

Nitrate-nitrogen and total phosphorus. Nitrate and total phosphorus concentrations were determined with a modified colorimetric analysis utilizing a spectrophotometer (Hach model DR-2000).Nitrate analysis followed cadmium reduction (Standard Method 4500-NO$_3$-E). Samples for phosphorus analysis were digested with persulfate (Standard Method 4500 P B 5) prior to colorimetric analysis (*Standard Methods for the Examination of Water and Wastewater*, 1995).

Metals. Sludge biomass samples (1998) were taken at five points in the reed bed. *Phragmites* plants were harvested at random in December of 1994, 1995, and 1998 to be analyzed for metal content. The plants were severed at the sludge surface. Thirty plants formed a composite sample. Both sludge and tissue samples were dried at 100°C until constant weight was reached. The dried plant material was milled and washed in a muffle furnace at 500°C. Both dry sludge and washed tissue samples were then digested in nitric acid (Standard Method 3030 and 3030 E, respectively), and analyzed by ICP plasma emission spectrometry following Standard Method 3120 B (*Standard Methods for the Examination of Water and Wastewater*, 1995) to determine the concentrations of selected elements, including boron, cadmium, chromium, copper, iron, manganese, molybdenum, nickel, lead and zinc.

Results and discussion

Sludge dewatering efficiency

The sludge volume applied from 1992–1995 was 2637 m^3 (Lavigne, 1996), the volume of sludge applied from 1996–1999 was 2704 m^3 for a total of 5341 m^3. In an earlier study of the Shelburne/Buckland reedbed system during the period of October 1992 to October 1995 dewatering efficiency was 96.9% (Lavigne, 1996). After a total of six years of operation the dewatering efficiency is 92.6% (Table 1).

In addition to the influent applied to the reed bed, precipitation, averaging 1.2 m/y over the last six years, added approximately 3569 m^3 of water to the reed bed. The total (8909 m^3) was either evapotranspired through the *Phragmites* plants or pumped back to the headworks. While sludge application and dewatering are year-round processes there is approximately 189 m^3 more treated effluent per winter month returned to the headworks via the underdrain than during the growing season months. This amount likely represents the evapotranspiration taking place during the summer (see Figure 1). This indicates an average evapotranspiration rate during the summer months of 39 cm^3 per month.

A water balance for the reed bed (Table 2) shows that 85% of the water is accounted for. If the volume unaccounted for (1375 m^3 over a six-year period) is averaged for the seven months of winter usage per year it would equal 32.7 m^3/month (4.7 cm/month). During those winter months evaporation still takes place; it is reasonable to conclude that the 4.7 cm/month remainder are evaporated during the winter months. The measured water loss at the Shelburne/Buckland reed bed was 39 cm/month during the growing season and 4.7 cm/month during the winter.

Monthly underdrain records were compared to inflow data (Figure 1) showing that 80%

Table 1 Dewatering efficiency for the Shelburne/Buckland reed bed from the start up in the fall of 1992 to the spring of 1999.

	Total sludge applied	Existing depth of dewatered sludge	Efficiency
Fall 1992 to fall 1995	2637 m^3 = 4.8 m	0.15 m	96.9%
Fall 1992 to spring 1999	5341 m^3 = 10.2 m	0.75 m	92.6%

Table 2 Water balance fall 1992 to spring 1999

Source	Quantity (m^3)	Yearly average (m^3)
Influent	5341	890
Precipitation	3569	595
Effluent	1486	248
Evapotranspiration	5677	946
Sludge water content	372	62
Unaccounted volume	1375	229

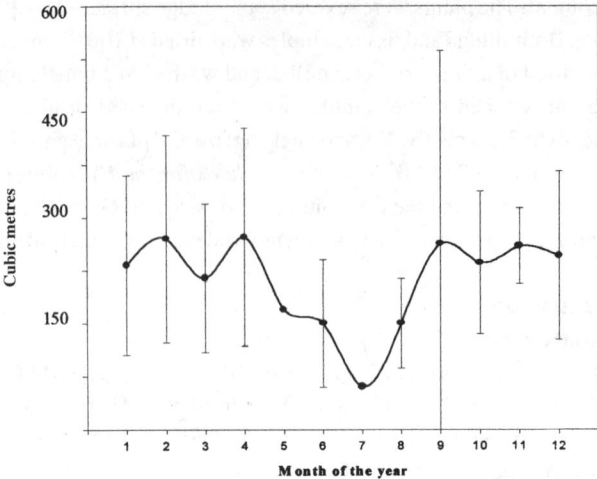

Figure 1 Average seasonal variation in the underdrain return flow based on data from fall 1995 through the spring 1999

of the total volume of effluent returned to the headworks was returned during the non-growing season when little transpiration is happening.

The bed was designed with a 1.2 m effective biomass storage depth anticipating a storage life of 8–10 years. The application of 1,411,100 gallons of sludge over a 6-year period resulted in 0.75 m of reed bed residual biomass, 62% of its storage capacity.

Total suspended solids

The Shelburne/Buckland wastewater treatment plant has a DEP discharge limit for total suspended solids of 30 mg/l, a limit they must meet daily in order to avid fines and revocation of their permit. The average reed bed influent TSS concentration is 12,352 mg/l with the average effluent TSS concentration 14.2 mg/l, which is well below discharge standards. Removal rates for total suspended solids average over 98%.

Biochemical oxygen demand (BOD_5)

The state DEP discharge permit for the SBWWTP sets the BOD_5 limit at 30 mg/l. Results from the Shelburne/Buckland BOD_5 data set, covering a 3-year period with 43 samplings of the reed bed's influent and effluent, show that the reed bed effluent BOD_5 quality consistently passed regulatory standards. The average reed bed influent BOD_5 concentration was 1,342 mg/l and the average reed bed effluent BOD_5 concentration was 6 mg/l, resulting in a BOD_5 treatment efficiency of 99%.

Nitrate-nitrogen and total phosphorus

RBSTSs show positive removal capacities for nitrogen and phosphorus with the removal mechanisms ascribed to plant adsorption, plant uptake, or chemical transformation and precipitation (Revitt et al., 1997). Removal efficiencies for nutrients vary considerable and have been reported ranging from 10 to 93% for total nitrogen and from 9 to 94% for total phosphorus (Reed, 1990; Yang et al., 1995). At the Shelburne/Buckland reed bed nitrate and total phosphorus were not monitored on a regular basis. For purposes of this study, 11 samples were taken over a six-month period (March through October 1998) from the reed bed influent and effluent. The results show a 90% reduction in nitrate-nitrogen and an 80% reduction in total phosphorus (data not included).

Metals

It is reasonable to assume that metals associated with a wastewater/sludge stream may concentrate in the biosolids. This may be specially true for reed beds because of the reduction in sludge over time. Sludge core analyses of the RBSTS were used to assess to what extent this happens. Pollutants concentrated in the reduced biosolids mass could complicate final disposal because the material may exceed regulatory standards.

The results from the Shelburne/Buckland RBSTS show an average metal concentration removal rate of 87% (influent to effluent) (Table 3). By analyzing influent and effluent in addition to analyzing sludge and plant tissue an attempt was made to determine the fate of metals in the reed bed. The concentrations of metals in the sludge biomass and in the *Phragmites* tissue are indicative of the wastewater quality of the treatment plant influent. In this case the metals concentrations are relatively innocuous as there is little industry in either Shelburne or Buckland. Of the ten elements studied iron consistently had the highest concentration, followed by copper, zinc, nickel, manganese, lead, boron, chromium, cadmium and molybdenum.

At the Shelburne/Buckland treatment plant the roots and rhizomes were not harvested nor analyzed. During regular operation only the above ground portion of the *Phragmites* plants are harvested, thus the contribution of the root biomass to metal attenuation is ignored. Metal attenuation in the above ground portion of the plants can also be ignored as the standard operating procedure at the Shelburne/Buckland treatment plant for disposal of the upper portion of the *Phragmites* plants is to burn them right in the reed bed with the residual ash incorporated into the reed bed sludge.

Conclusions

The dewatering efficiency of the Shelburne/Buckland reed bed system after a 6-year period is 93%. Of the 7535 m^3 of sludge and rain received over a six-year period, 5678 m^3 were lost due to evapotranspiration during the growing season, 1486 m^3 were returned to the headworks, approximately 372 m^3 of water was held in the sludge biomass. While the *Phragmites* plants do not transpire significant amounts of water during the winter months there is still water loss from the reed bed due to evaporation. To account for the remaining 1375 m^3 we estimate that winter evaporation would have to equal approximately 32.7 m^3/month (6 cm/month).

TSS and BOD$_5$ effluent concentrations are well under the regulatory standards, showing that reed beds can produce an effluent of high quality that needs no further treatment and

Table 3 Concentrations and removal efficiencies of selected elements in sludge, plant tissue, and reed bed influent and effluent from the Shelburne/Buckland reed bed. Values represent averages.

Parameter	Sludge	*Phragmites*	Method detection limit	Influent	Effluent	Average removal efficiency
	mg/kg	mg/kg	mg/l	mg/l	mg/l	%
Boron	60	10	0.05	0.41	0.11	73
Cadmium	3	2	0.,02	BDL	BDL	100
Chromium	42	10	0.02	0.140	BDL	97
Copper	1906	155	0.02	11	0.17	98
Iron	11592	3263	0.02	19	0.07	98
Lead	154	32	0.02	0.5	0.29	40
Manganese	457	1443	0.025	2	0.29	86
Molybdenum	4	1	0.0015	0.09	BDL	100
Nickel	23	9	0.02	0.82	0.17	80
Zinc	684	453	0.02	3	0.14	95
Mean removal rate of all elements combined:						86.7

could be discharged directly to our waterways. The Shelburne/Buckland reed bed average BOD_5 effluent concentration was 6 mg/l, well below the regulatory discharge limit of 30 mg/l. With less transpiration taking place in the winter months there was a greater amount of effluent returned to the headworks, however there was no significant difference in TSS and BOD_5 treatment results for winter or summer months. This clearly shows that, even when located in northern climates, reed bed technology is a viable alternative to conventional sludge treatment.

Nitrate-nitrogen reduction was 90% and total phosphorus reduction was 80%, these results are encouraging and seem to follow reports in the literature.

The overall average metal removal efficiency was 87%. Metals tend to concentrate in the reed bed sludge and not in the *Phragmites* plant tissue. All metal concentrations in the sludge except for copper meet DEP standards for Type 1 sludge, suitable for land application. The metal concentrations in the *Phragmites* plants are below the disposal standards set by the DEP and do not pose a problem if disposal of separately from the sludge.

The data presented in this paper confirm the reported success of RBSTS in cold climates and show that while RBSTS are not a panacea for all parameters of wastewater treatment, they are certainly a viable alternative to current technology.

References

Kim, B.J., and Smith, E.D. (1997). Evaluation of Sludge Dewatering Reed Beds: A Niche for Small Systems. *Water Science and Technology*, **35**(6), 21–28.

Krueger, A. (1991). Reed Beds: A Low-Cost Sludge Treatment System. Engineers Proposal for RBSTS at Shelburne/ Wastewater Treatment Plant, pp. 1–10.

Lavigne, R. (1996). Shelburne/ Reed Bed Sludge Treatment Facility, Three Year Operation Report. Howard Laboratories, Inc., Amherst, MA, pp. 1–11.

Reed, S.C. (1990). *Natural Systems for Wastewater Treatment*. Water Pollution Control Federation, Alexandria, VA.

Revitt, D.M., Shutes, R.B., Llewellyn, N.R., and Worrall, P. (1997). Experimental Reed Bed Systems for the Treatment of Airport Runoff. *Water Science and Technology*, **36**(8–9), 385–390.

Seidel, K. (1976). Macrophytes and Water Purification. In: *Biological Control of Water Pollution*, Torubier, J. and Pierson, R.W. (eds.), University of Pennsylvania Press, Philadelphia, PA, pp. 109–121.

Standard Methods for the Examination of Water and Wastewater (1995). 19th edn, American Public Health Association/American Water works Association/Water Environment Federation, Washington, DC, USA.

Yang, Y., Zhensheng, X., Kangping, H., Junsan, W., and Guizhi, W. (1995). Removal Efficiency of the Constructed Wetland Wastewater Treatment System at Bainikeng, Shenzhen. *Water Science and Technology*, **32**(3), 31–40.

Reciprocating constructed wetlands for treating industrial, municipal and agricultural wastewater

L. Behrends*, L. Houke*, E. Bailey*, P. Jansen* and D. Brown**

* Tennessee Valley Authority, P.O Box 1010, Muscle Shoals, AL 35662-1010, USA
** Environmental Protection Agency, Cincinnati, Ohio, USA

Abstract Scientists at the Tennessee Valley Authority (TVA), and in collaboration with the U.S. Environmental Protection Agency (EPA), are continuing to develop and refine an innovative wastewater treatment system referred to as *reciprocating* subsurface-flow constructed wetlands. Reciprocation relates to patented improvements in the design and operation of paired subsurface-flow constructed wetlands, such that contiguous cells are filled and drained on a frequent and recurrent basis. This operating technique turns the entire wetland system into a fixed-film biological reactor, in which it is possible to control redox potential in alternating aerobic and anaerobic zones. Reciprocating systems enable manipulation of wastewater treatment functions by controlling such parameters as hydraulic retention time, frequency of reciprocation, reciprocation cycle time, depth of reciprocation, and size and composition of substrate. These improved wetland technologies have been used for treating municipal/domestic wastewater, high strength animal wastewater, and mixed wastewater streams containing acids, recalcitrant compounds, solvents, antifreeze compounds, heavy metals, explosives, and fertilizer nutrients. Results from selected treatability studies and field demonstrations will be summarized with respect to conceptual design and treatment efficacy.

Keywords Biochemical oxygen demand; denitrification; nitrification; reciprocating wetlands; total ammonia nitrogen; total phosphorus

Introduction

Scientists at the Tennessee Valley Authority (TVA), and in collaboration with the U.S Environmental Protection Agency (EPA), are continuing to develop and refine an innovative wastewater treatment system referred to as *reciprocating* subsurface-flow constructed wetlands. Reciprocation relates to patented improvements in the design and operation of paired subsurface-flow constructed wetlands such that paired wetland cells are filled and drained on a frequent and recurrent basis (U.S. Patent 5,863,433; Behrends, 1999a). This novel vertical-flow wetlands process provides limited aeration to the bulk water (2–4 g O_2/m^2/day), but provides rapid and frequent oxygenation of the root-zone and substrate biofilms (Behrends *et al.*, 1993, 1996a). Fill and drain cycles in conventional vertical-flow wetlands are generally dependent on the influent flow rate (Cooper *et al.*, 1997). In contrast, fill and drain cycles for recurrent reciprocating systems are independent of influent flow rate and allow frequent exposure of microbial biofilms to aerobic, anoxic and anaerobic environments. Sequential aerobic-anaerobic environments are required for removal of mixed metals, nitrification/denitrification, biological phosphorus removal, and breakdown of many recalcitrant compounds (Zitomer and Speece, 1993, Reddy and D'Angelo, 1997).

During the reciprocation drain cycle, thin water films surrounding plant roots and substrate biofilms are exposed to atmospheric oxygen. Diffusion of oxygen through such thin films is rapid, with saturation occurring within a matter of seconds. In contrast, it takes days to weeks to saturate shallow deoxygenated water columns, such as those associated with conventional surface- and subsurface-flow wetlands (Behrends *et al.*, 1993). The drain cycle also allows supersaturated metabolic gases, such as carbon dioxide (CO_2), to pass from biofilms into the atmosphere. Off-gassing of CO_2 helps maintain near neutral pH, and promotes abiotic precipitation of calcite and phosphorus compounds of low solubility.

During the subsequent fill cycle microbial biofilms are bathed in anoxic and/or anaerobic water, where reducing conditions can be near optimum for microbial-induced reduction of metals, nutrients and other recalcitrant compounds (Reddy and D'Angelo, 1997). Because the contiguous treatment environments are alternated as often as six to twelve times daily, the system is in perpetual flux and no particular microbial consortia can predominate. This promotes development of a stable, high-diversity microbial biomass, which is key to treatment of complex mixed wastewater. This paper will summarize design and operational features of reciprocating wetlands, and review results from a variety of treatability studies and demonstration-scale field trials.

Methods

Reciprocating systems are usually designed as paired rectangular symmetrical cells, but can be designed as free-form asymmetrical cells that conform to the landscape. Multiple systems can be operated in parallel or serial arrangements. Depth of individual cells can range from < 1 m to > 3 m, and possibly deeper, where land area is limited. Treatment cells are lined with either 30–40 mil synthetic liners or low-permeability clays to eliminate percolation to groundwater. Various graded substrates, including limestone, river gravel and pea-gravel are used to back-fill cells, and at least 30 cm of freeboard is incorporated into the design to accommodate precipitation. Large substrates on the bottom of the cell in conjunction with underdrain pipes facilitate rapid movement of water to pump stations, while smaller overlying rock substrates significantly increase substrate surface area for microbial biofilm development and gas diffusion. Selected aquatic and terrestrial species of vegetation can be planted to enhance removal of nutrients, metals, and recalcitrant compounds such as explosives and pesticides.

Reciprocating systems can be designed and operated as either a stand-alone technology, as a retrofit to existing conventional treatment systems, or as a component of complex integrated wastewater treatment systems. Variations of reciprocating systems have been under development since 1993 and have demonstrated potential for treatment of a diversity of waste streams. Various design and operational modes have been used for treating municipal, industrial, and agricultural wastewater (Table 1), and have led to significant improvements in removal of water borne pathogens, total suspended solids (Behrends, 1999a); biochemical oxygen demand (BOD_5), chemical oxygen demand (COD), and nutrients (Behrends *et al.*, 1993, 1996a, 1999a); and metals, solvents, and recalcitrant compounds (Behrends, 1999a; Behrends *et al.*, 1993, 1996b and Sikora *et al.*, 1996, 1997, 1998).

Initial system designs and treatability studies used u-tube airlifts to move water between contiguous cells. However, low- head high-volume centrifugal pumps have been substituted for air-lifts, and have proven to be more practical for commercial systems. Pump selection is based on flow rates required to accommodate a specified draw-down depth over a specified time period. Dual-channel programmable timers and double float pump switches are used to operate pumps with respect to pump on/off cycles and duration of pumping.

Flow-through and batch loaded treatability studies have been conducted in replicated microcosms to provide estimates of pollutant removal rates as a function of hydraulic retention time (HRT), and elapsed time (Behrends *et al.*, 1996b and Sikora *et al.*, 1997). HRTs required for optimum treatment are influenced by strength of wastewater, type of contaminant(s), and desired levels of treatment, and have ranged from <1 to >15 days.

Results and discussion

The following section provides annotated information on treatment designs and efficacy for various wastewater streams, and is intended to illustrate the general utility and scaleability of reciprocating systems for wastewater treatment.

Coupled anaerobic/aerobic systems for treating explosives contaminated groundwater

Explosives-contaminated groundwater from the Milan Army Ammunition Plant (MAAP), Milan, Tennessee, was continuously loaded into 38 L glass microcosms that were backfilled with river gravel to a depth of 22 cm (pore volume = 40%). Inlet concentrations of the two primary explosives contaminants, TNT and RDX, were 1,540 ppb and 3,000 ppb, respectively. All microcosms were planted with canary grass (*Phalaris arundinacea*), at 600 g wet wt. Hydraulic loading rate was continuous at 1.4 ml/min (16 L/m^2/day), which provided a six day HRT. The microcosms were partitioned into 4 cells with the first two cells being operated as anaerobic subsurface horizontal-flow wetlands (each 416 cm^2), while the second two cells (each 215 cm^2) were operated as aerobic vertical-flow cells. Anaerobic conditions were maintained in the first two cells by amending the contaminated water with either 350 or 700 mg/L powdered milk, and amending on an as-needed-basis to maintain low redox conditions. Aerobic conditions in the final two cells were maintained via recurrent reciprocation by sequentially air-lifting water between contiguous cells every two hours. Reciprocation was full depth, such that all gravel substrate in each aerobic cell was exposed to the atmosphere every two hours for a duration of 105 minutes.

Table 1 Reciprocating Wetlands R & D: Municipal, Industrial and Agricultural Applications. Tennessee Valley Authority, 2000

Municipal Domestic	Application	Design flow (m^3/day)	Comments
Benton, TN	Municipal wastewater treatment	190	Retrofit[1]
Lawrence Co, AL	Day use facility	6.8	On site treatment
Huntsville, AL. Botanical Garden	Day use facility	1.9	On-site treatment
Oahu, HI	Domestic wastewater	380	AEC demonstration[2]
Morgan Co. AL.	50 house subdivision, on-site wastewater treatment	85	National On-site demonstration[3]

Industrial	Application	Design flow (m^3/day)	Comments
Milan, TN	Explosives contaminated groundwater	27	AEC Demonstration[2]
Savannah, TN	Food processing wastewater	80	Commercial demonstration
Wilmington, OH	Airport deicing water	10–50	Pilot-scale test
Port Said, Egypt	Mixed wastewater: municipal, industrial, agricultural	2000	UNDP demonstration[4]
Muscle Shoals, AL	Acid mine drainage	<0.1	Microcosm tests[5]

Agricultural	Application	Design flow (m^3/day)	Comments
Muscle Shoals, AL	High density aquaculture wastewater	0.6	Pilot-scale demonstration
Aliceville, AL	Confined swine wastewater	190	Field demonstration

1. Reciprocating system retrofit to existing 4 cell subsurface flow wetland.
2. U.S. Army Environmental Center, Aberdeen, Maryland.
3. Co-funded National On-site Decentralized Wastewater Treatment Demonstration, Joe Wheeler Electric Cooperative, Tennessee Valley Authority, Electric Power Research Institute.
4. United Nations Development Program, Cairo, Egypt.
5. Simulated wastewater containing Mn, Fe, Pb, Ni, and Zn.

The coupled anaerobic-aerobic treatment system provided excellent treatment with respect to removal of explosives and daughter compounds (Table 2). RDX treatment efficacy was highly correlated with oxidation-reduction potential in the anaerobic cells ($r = 0.89$), with best removal rates achieved at redox levels < –250. The reciprocating aerobic cells enhanced removal of nutrients, BOD_5 and daughter compounds of explosives. BOD_5 was reduced from 600 to less than 10 mg/L within three days. Results from this study and an earlier batch loaded study (Sikora *et al.*, 1997), were used to design, construct and operate a field-scale demonstration to treat the equivalent of 570 m³/ha/day of explosives contaminated groundwater at the MAAP facility. The 27 month field-scale demonstration provided excellent removal of primary explosives, their respective daughter compounds, BOD_5, manganese, nitrate and other fertilizer nutrients (Sikora *et al.*, 1998; Behrends *et al.*, 1999a.)

Reciprocating wetlands for municipal and domestic wastewater treatment
A 0.5 ha. gravel-based subsurface-flow constructed wetland was constructed in 1992 to provide secondary treatment for up to 190 m³ d⁻¹ of primary treated municipal wastewater. The facility, located in Benton, Tennessee USA, consists of a 380 m³ septic tank followed by two pairs of constructed wetlands, with each pair operated in series. Each of the four cells measured 15.2 × 61 m and contained approximately 0.1 ha surface area. Depth of monotypic limestone gravel in the primary cells ranged from 0.3–0.6 m; depth of gravel in the secondary cells ranged from 0.3 to 0.5 m. Within six months of start up, the subsurface-flow wetland system was chronically out of compliance with respect to BOD_5, dissolved oxygen, total ammonia nitrogen and fecal coliform bacteria. Ammonia nitrogen levels in the effluent were also elevated, ranging from 15 to 30 mg/L.

In November of 1995, a series of centrifugal pumps, under-drains and timers were installed to affect recurrent reciprocation between contiguous cells. Pumps were operated under various experimental repeating cycles accordingly: on for 1 to 2 hours: off for 1–2 hour. Within six months of installation, water quality criteria improved significantly. Figures 1a and b. illustrate improvements in effluent values for BOD_5, non-compliance violations (NCVs), and total ammonia nitrogen (TAN).

Beginning in 1996, reciprocating wetland designs were changed to reflect deeper layered substrates (1.2–1.8 m deep). This design change allowed increased fill and drain rates, improved treatment and a significant reduction in land area requirements.

Reciprocating wetlands for treating high strength aquaculture wastewater
A novel two-stage reciprocating wetland was designed and operated over a period of six months to treat wastewater from a pilot-scale intensive tilapia aquaculture operation (Figure 2, Table 3). Aquaculture system water volume was approximately 17,100 L. Fish manure, particulate matter and wasted feed were continuously removed from the recirculating water by a 60 μm screen filter and stored as a liquid slurry. The manure slurry (10–14% solids, v:v) was batch loaded via pump every four hours to either an intermittent

Table 2 Removal of TNT and RDX in coupled anaerobic / aerobic treatment wetlands as a function of organic fertilization rate. (After Behrends et al., 1996b)

Explosive	Fertility rate (mg/L)[1]	Influent conc. (ppb)	Effluent conc. (ppb)	Reduction (%)
TNT	350	1,540	4.6	99.7
TNT	700	1,540	4.0	99.7
RDX	350	3,000	561	81.3
RDX	700	3,000	314	89.5

[1] Powdered milk replacement starter, dry matter basis

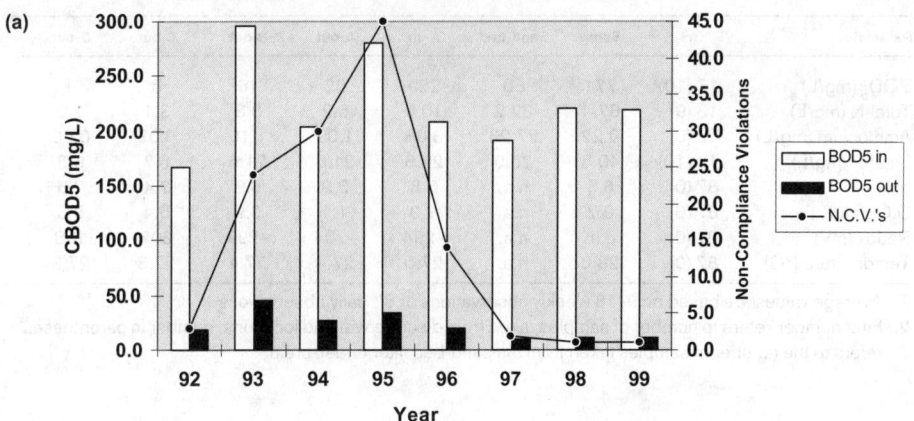

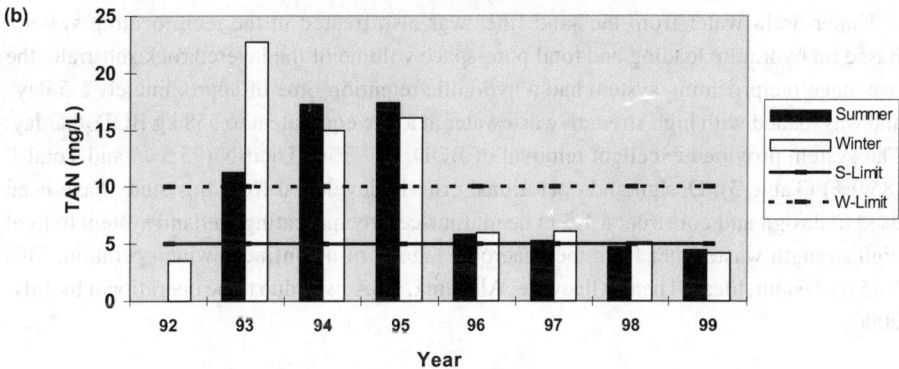

Figures 1a and **b** illustrate historical effluent values for $CBOD_5$ and TAN respectively. Notice the significant reduction in values for the years 1996 to 1999, which correspond to the time period in which recurrent reciprocation has been in use. In Figure 1b, S-Limit and W-Limit refer to summer limit and winter limit, respectively (after Behrends, 1999b).

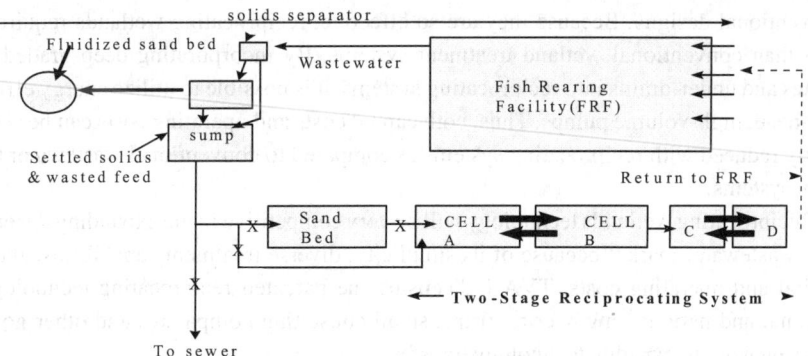

Figure 2 Plan view of aquaculture production and wastewater treatment facility (not to scale). Bold arrows indicate recurrent reciprocation, X's indicate gate valve locations, lined arrows refer to direction of water flow. Note: treated wastewater was returned to the aquaculture production system. Pilot system was housed in an environmentally controlled greenhouse (After Behrends et al., 1999b)

Table 3 Average parameter values[1] for a pilot-scale two-stage reciprocating system for treating high strength aquaculture wastewater. (After Behrends et al., 1999b)

Parameter	n[2]	Sump	Sand-bed	A-in	A-out	B-out	C-out	D-out
BOD_5 (mg/L)	16 (10)	771	80	229	62	16	5	4
Total-N (mg/L)	13 (9)	67.4	32.2	10.0	6.2	3.3	3.1	3.0
Ammonium (mg/L)	9 (9)	0.22	27.26	3.08	1.03	1.10	0.07	0.03
Total-P (mg/L)	13 (10)	40.5	25.0	22.8	21.6	21.5	6.7	6.3
pH	87 (0)	6.5	n.a.	6.8	6.9	6.8	7.4	7.5
D.O. (mg/L)	87 (0)	0.7	n.a.	1.0	1.1	0.9	5.4	5.5
Redox (mV)	87 (0)	n.a.	n.a.	−294	−204	−299	307	158
Temperature (°C)	87 (0)	28.6	n.a.	27.5	27.4	27.4	27.8	27.8

1. Average values are based on 9–18 weekly observations or 87 daily observations.
2. First number refers to number of samples taken from discrete wetland locations; number in parentheses refers to the number of samples taken from the sand-bed filter under-drain.

sand filter, or to the influent manifold of the experimental two-stage reciprocating subsurface-flow wetland system.

Under-drain water from the sand filter was also treated in the reciprocating system. Based on hydraulic loading and total pore-space volume of the layered rock substrate, the two-stage reciprocating system had a hydraulic retention time of approximately 5.5 days and was loaded with high strength wastewater at a rate equivalent to 558 kg BOD_5/ha/day. The system provided excellent removal of BOD_5 (99.5%), Total-N (95.5%) and Total-P (85.9%) (Table 3). Design and operational criteria developed from this study have been used to design and construct a 1.5 m deep, four-cell reciprocating wetland system to treat high strength wastewater from the anaerobic lagoon of a confined swine operation. This 0.45 ha system, located near Aliceville, Alabama, USA, was due to be operational by July, 2000.

Conclusions

Reciprocating wetland systems have proven to be scaleable and highly efficient at removing wastewater contaminants including BOD_5, TSS, metals, solvents, explosives, nutrients, and pathogens. The exposure of diverse biofilms to anaerobic and aerobic conditions on a frequent and recurrent basis provides novel treatment environments that are not available in conventional designs. Because they are so effective, reciprocating wetlands require less land than conventional wetland treatment systems. By incorporating deep graded substrates and under-drains into reciprocating systems, it is possible to utilize energy efficient low-head, high-volume pumps. Thus, both capital costs and operating costs can be significantly reduced with reciprocating systems as compared to conventional wastewater treatment systems.

Reciprocating wetlands technology will be very competitive in the expanding decentralized wastewater market because of its simplicity, diverse treatment capabilities, and low capital and operating costs. TVA is licensing the patented reciprocating technology to regional and national engineering firms, small consulting companies, and other government agencies to expedite technology transfer.

References

Behrends, L.L., Coonrod II, H.S., Bailey, E. and Bulls, M.J. (1993). Oxygen diffusion rates in reciprocating rock biofilters: potential applications for subsurface flow constructed wetlands. 12 pp. *Proceedings Subsurface Flow Constructed Wetlands Conference*, University of Texas at El Paso.

Behrends, L.L., Sikora, F.J., Coonrod, H.S., Bailey, E. and Bulls, M.J. (1996a). Reciprocating Subsurface-flow Wetlands for Removing Ammonia, Nitrate, and Chemical Oxygen Demand: Potential for Treating Domestic, Industrial and Agricultural Wastewater. Vol. 5: 251-263. *Proceedings Water Environment Federation*, 69th Annual Conference and Exposition. Dallas, Texas, October 5–9, 1996.

Behrends, L.L., Sikora, F.J., Phillips, W.D., Bailey, E., McDonald, C. and Coonrod, H.S. (1996b). Phytoremediation of explosives-contaminated groundwater in constructed wetlands: II. Flow through study. U.S. Army Environmental Center report no. SFIM-AEC-ET-CR-96167.

Behrends, L.L. (1999a). Reciprocating subsurface-flow constructed wetlands for improving wastewater treatment. U.S. Patent 5,863,433. January 1999.

Behrends, L.L. (1999b). Reciprocating subsurface-flow wetlands for municipal and on-site wastewater treatment. In: *Wetlands and Remediation: An International Conference*, J.L. Means and R.E. Hinchee (eds.), Battelle Press, Columbus, Ohio, pp.179-186.

Behrends, L.L., Almond, R.A., Sikora, F.J. and Bader, D.F. (1999a). Phytoremediation of explosives-contaminated groundwater using constructed wetlands. In: *Wetlands and Remediation: An International Conference*, J.L. Means and R.E. Hinchee (eds.), Battelle Press, Columbus, Ohio, pp. 375–383.

Behrends, L.L., Houke, L., Bailey, E and Brown, D. (1999b). Reciprocating subsurface-flow wetlands for treating high-strength aquaculture wastewater. In: *Wetlands and Remediation: An International Conference*, J.L. Means and R.E. Hinchee (eds.), Battelle Press, Columbus, Ohio, pp.317–325.

Cooper, P., Smith, M. and Maynard, H. (1997). The design and performance of a nitrifying vertical-flow reed bed treatment system. *Wat. Sci. Tech.*, **35**(5), 215-223.

Reddy, K.R. and D'Angelo, E.M. (1997). Biogeochemical indicators to evaluate pollutant removal efficiency in constructed wetlands. *Wat. Sci. Tech.*, 35(5), 1–10.

Sikora, F.J., Behrends, L.L., Brodie, G.A. and Bulls, M.J. (1996). Manganese and trace metal removal in successive anaerobic and aerobic wetlands. *Proceedings of American Society for Surface Mining and Reclamation 1996 Annual Meeting*, May 18–23, 1996, Knoxville, TN.

Sikora, F.J., Behrends, L.L., Phillips, W.D., Coonrod, H.S., Bailey, E. and Bader, D.F. (1997). A microcosm study on remediation of explosives-contaminated groundwater using subsurface-flow constructed wetlands. *Ann. New York Acad. Sci.* **829**, 202–218.

Sikora, F.J., Behrends, L.L., Almond, R., Phillips, W.D., Coonrod, H.S., Bailey, E. and Bader, D.F. (1998). Demonstration of results of phytoremediation of explosives contaminated groundwater using constructed wetlands at the Milan Army Ammunition Plant, Milan, Tennessee. U.S. Army Environmental Center Report Number SFIM-AEC-ET-CR-95090.

Zitomer, D.H. and Speece, R.E. (1993). Sequential environments for enhanced biotransformation of aqueous contaminants. *Environ. Sci. Tech.*, **27**, 227–244.

Zero-discharge of nutrients and water in a willow dominated constructed wetland

P. Gregersen* and H. Brix**

* Centre of Recycling, Forsomho Skolevej 5, DK-6870 Ølgod, Denmark
** Department of Plant Ecology, University of Aarhus, Nordlandsvej 68, 8240 Risskov, Denmark

Abstract A novel constructed wetland system has been developed to treat sewage, evaporate water and recycle nutrients from single households at sites where effluent standards are stringent and soil infiltration is not possible. Main attributes of the willow wastewater cleaning facilities are that the systems have zero discharge, the willows evapotranspire the water, and nutrients can be recycled via the willow biomass produced in the system. The willow wastewater cleaning facilities generally consist of c. 1.5 m deep high-density polyethylene-lined basins filled with soil and planted with clones of willow (*Salix viminalis* L.). The surface area of the systems depends on the amount and quality of the sewage to be treated and the local annual rainfall. For a single household the area needed typically is between 200–300 m^2. Settled sewage is dispersed underground into the bed under pressure. When correctly dimensioned, the willow will – on an annual basis – evapotranspire all water from the sewage and rain falling onto the system, and take up all nutrients and heavy metals from the sewage. The stems of the willows are harvested on a regular basis to remove nutrients and heavy metals and to stimulate the growth of the willows. Initial experiences from full-scale systems in Denmark show promising results.

Keywords Constructed wetland; evaporation; heavy metal; nutrient; recycling; Salix; water treatment; willow; zero-discharge

Introduction

Discharge of domestic sewage from single households to streams and lakes in the countryside is resulting in poor freshwater quality in many areas of Denmark. Therefore new legislation requires adequate treatment of sewage from single households. Soil infiltration is the preferred solution (EPA, 1999a) but at many locations this is not possible because of clayish soil conditions or high ground water tables. Other treatment solutions, like sand-filters and reed beds (EPA 1999b, c), may not provide the required treatment, and consequently other solutions are needed. This paper describes a willow-based constructed wetland system with no outflow, which has been developed to treat sewage and recycle nutrients from single households at sites where effluent standards are stringent and soil infiltration is not possible.

Willow plantations have been successfully used as recipients for municipal wastewater, sewage sludge and landfill leachate (e.g. Rosenqvist *et al.*, 1997; Hasselgren, 1998, 1999; Venturi *et al.*, 1999). By these techniques the resources in the wastewater, namely water and nutrients, are used for biomass production, but excess nutrients and water are discharged to receiving water bodies. The treatment concept in the willow wastewater cleaning facility described in this paper is that all the nutrients contained in the sewage are used to produce plant biomass, and all the water is evapotranspired to the atmosphere by the willows. Hence there will be no outflow from the systems (Figure 1).

A willow wastewater cleaning facility as constructed in Denmark generally consist of approximately 1.5 m deep high-density polyethylene-lined basins filled with soil and planted with clones of willow (*Salix viminalis* L.) (Figure 2). The surface area of the systems depends on the amount and quality of the sewage to be treated and the local annual rainfall, and range typically between 200–300 m² for a single household. Settled sewage is dispersed underground into the bed under pressure. When correctly dimensioned, the willow should – on an annual basis – evapotranspire all water from the sewage and rain falling onto the system, and take up all nutrients and heavy metals from the sewage. The stems and leaves of the willows are harvested every third year (one third of the bed area is harvested every year) to keep a healthy vegetation and a high production of bark, which is known to contain high concentrations of phosphorus and heavy metals (Sander and Ericsson, 1998). Consequently, a bed should always contain one and two year old willow plants that can evapotranspire water (Figure 3).

Materials and methods

In 1997 six facilities receiving sewage from single households were constructed in Denmark to investigate the performance and operation of the systems. The surface areas of the systems varied between 150 and 500 m² depending on number of inhabitants connected, their water consumption and the local precipitation. Three different clones of *Salix viminalis* ("Björn", "Tora" and "Jorr") were planted in May as 20-cm cuttings with 5 cm above the soil surface. Wastewater discharges into the systems were monitored by flow meters in the pump lines and were also estimated based on water consumption in the households. The water levels within the willow beds were monitored with inspection wells placed vertically in the soil of the beds. Precipitation was measured at two sites and compared with data from the nearest meteorological station. For all sites precipitation data was available from meteorological stations. Water content in the saturated and non-saturated soils was measured by drying soil samples. Biomass production was measured by weighing the harvested biomass (stems after defoliation) and by weighing representative stems with leaves before defoliation. Stems and leaves were analysed for content of dry matter, total nitrogen, phosphorus, potassium, and the heavy metals cadmium, lead, mercury, zinc, nickel, copper and chromium. Soils were sampled and preserved for future analysis. Conductivity of the water in the beds was analysed to elucidate if accumulation of salts occur. At visits every month the willow health as indicated by the colour of leaves was evaluated. Furthermore, the number of plants, the number of stems per plant before and after the first, second and third harvests was counted, and the time of foliation and defoliation was recorded.

Results

Water balance

One of the most important aspects of the willow wastewater cleaning facilities is their ability to evapotranspire all the sewage discharged into the systems as well as the rain

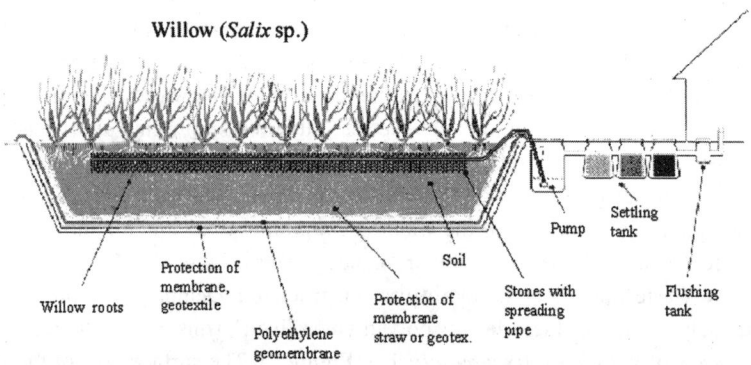

Figure 1 Cross-section through a willow wastewater cleaning facility

Figure 2 Excavation for establishment of willow wastewater cleaning facility showing placement of geotextile and high-density polyethylene membrane

falling onto the systems. Table 1 presents data on the estimated evaporation from the six systems in the two initial years of operation. The wastewater loading into the systems was 450 to 600 mm per year. During the second year the precipitation was approximately 400 mm higher than the "normal" 30-year average (1,150 mm). Facilities No. 1 and 5 had relatively poor growth of willow because of vigorous growth of weeds in the beds. Facility No. 6 had some surface water flowing into the system because of construction problems. The high rate of precipitation the second year resulted in completely saturated conditions (water on the bed surface) in some of the systems, and hence the systems were hydraulically overloaded.

Biomass production and contents of nutrients and heavy metals

Data on biomass production and the contents of nutrients and heavy metals in the stem and leaves of one-year and two-year old shoots was collected in facility No. 4. Here the plantation consists of 3 rows of the clone "Jorr", 2 rows of the clone "Bjørn", and 2 rows of the clone "Tora" (Table 2). Unfortunately we have no accurate measurement of the nutrient and heavy metal discharged into the system. Using "normal" contents in "normal" household wastewater, i.e. 30 mg l^{-1} total-N, 10 mg l^{-1} total-P (Henze, 1982), and 30 mg l^{-1} K, it can be seen that the amount of N, P and K in the harvestable biomass almost exactly balances the amount discharged into the system with the sewage. Only for P the amount

Table 1 Estimated evaporation rates (mm per year) for six willow wastewater cleaning facilities in Denmark during the first (April 1997 to March 1998) and the second year (April 1998 to March 1999) of operation

Facility	Year 1	Year 2
1	980	1,470
2	1,270	2,090
3	1,140	1,650
4	1,130	1,690
5	980	1,660
6	1,020	1,880

discharged into the system was approx. 30% higher than the amount in the harvestable biomass. The balance for P will however depend on the use of phosphate-containing detergents in the specific household. For heavy metals, it is not possible, based on the available data, to evaluate the mass balance. But usually sewage from single households contains low levels of heavy metals. "Normal" levels of heavy metals in domestic sewage have been reported to be Cd: 2 μg l^{-1}; Pb: 40 μg l^{-1}; Zn: 130 μg l^{-1}; Cu: 40 μg l^{-1}; Ni and Cr: 15 μg l^{-1}; and Hg: 1 μg l^{-1} (Henze, 1982). If these levels are used to make up the mass balance, it can be calculated that some accumulation of heavy metals may occur in the system over time. However, it is know that the uptake of heavy metals by willows depends on the levels in the soil as well as on the clone (Greger and Landberg, 1995; Landberg and Greger, 1996), and therefore removal by harvesting may be higher than indicated by the present data. A worst case scenario, based on the present removal data and the concentration levels cited above, shows that after 25 years of operation the heavy metal levels in the soil will not exceed the present legislative standards for use of soil for agricultural purposes (Cd: 0.5 mg kg^{-1} dry matter; Pb: 40 mg kg^{-1} dry matter; Zn: 100 mg kg^{-1} dry matter; Cu: 40 mg kg^{-1} dry matter; Ni: 15 mg kg^{-1} dry matter; and Cr: 30 mg kg^{-1} dry matter).

Accumulation of salts

Conductivity of the water within the beds was analyzed and the data do not as yet show an increase in conductivity. However, it is very likely that the contents of salts in the system will increase over time, but the rate of increase is unknown and will depend on the amount of salts in the sewage and hence the habits of the sewage producers. If the contents of salt in the system increase to unacceptable levels it may be possible at some later stage to discharge the salt-containing water from the system.

Discussion

The first experiences indicate that willow wastewater cleaning facilities have the ability to evapotranspire all water, and recycle nutrients and heavy metals from household wastewater by uptake in willow stems and leaves. Composting biomass from willow wastewater cleaning facilities and using it as a soil amendment could bring nutrients back to the food chain. Using harvested biomass for energy purposes could reduce emission of carbon dioxide to the atmosphere because it replaces fossil fuels. By extracting heavy metal from the ashes it may be possible to make a source of fertilizer with a high content of potassium. Potassium is a limiting resource in organic farming.

Table 2 Biomass of three clones of willow (*Salix viminalis*) at a willow wastewater cleaning facility (facility No. 4) September 1998, after two growing seasons. Some growth occurred after harvesting

	"Bjørn"			"Jorr"			"Tora"		
	Stem	Leaves	Total	Stem	Leaves	Total	Stem	Leaves	Total
Biomass (tonnes DM ha^{-1})	9.4	1.4	**10.8**	10.0	2.1	**12.1**	10.1	1.5	**11.6**
Nutrients (kg ha^{-1})									
N	120	50	**170**	102	68	**170**	89	48	**137**
P	26	7	**33**	27	11	**38**	26	4	**30**
K	85	62	**147**	121	92	**213**	123	75	**198**
Heavy metals (g ha^{-1})									
Cd	1.7	0.2	**1.9**	3.6	0.3	**3.9**	3.4	0.3	**3.7**
Pb	0.4	0.3	**0.7**	0.5	0.7	**1.2**	–	0.4	–
Zn	201	25	**226**	206	49	**255**	253	32	**285**
Cu	15	3	**18**	23	6	**29**	15.6	4.4	**20**
Ni	1.9	0.1	**2.0**	1.6	0.3	**1.9**	1.3	0.1	**1.4**
Cr	7.7	0.6	**8.3**	19.6	1.3	**20.9**	9.3	0.6	**9.9**
Hg	0.2	0.1	**0.3**	0.2	0.2	**0.4**	0.2	0.1	**0.3**

Figure 3 Three-year-old willow wastewater cleaning facility showing one year old stems (at the right) and two year old stems (at the left) at three year old root stocks

The initial experiences from the Danish systems show that it is important to keep a new-established bed free from weeds the first year after planting. Vigorous growth of weeds will significantly reduce the production of willow stems the first year. Usually the willow stems are cut the first year to increase the number of stems per plant, but if the willows have had a low number of stems the first year they will also have a low number in the second and following years. Hence biomass production will be lower and evapotranspiration and nutrient uptake will be affected. It is therefore urgent to keep the facilities free of weeds the first year. The second year the willows will outcompete the weeds if kept clean the first year.

The parameters of importance when designing a willow wastewater cleaning facility include: (1) the exact amount of wastewater during the first year of operation; (2) the amount of rainfall at the site of construction, and (3) the ability of the selected willow clones to evapotranspire water and accumulate nutrients and heavy metals in the aboveground harvestable biomass. To exemplify: in an area where the annual mean precipitation is 700 mm per year, it is assumed that the willow can evapotranspire 1,200 mm per year. The difference between precipitation (700 mm) and evapotranspiration (1,200 mm), i.e. 500 mm or 500 l m^{-2}, is equal to the amount of sewage that can be loaded into the system on an annual basis. Assuming a water discharge rate of 100 l per person per day or 36,500 l per person per year, it can be calculated that the surface area needed to evapotranspire the sewage equals 36,500 l year^{-1} divided by 500 l m^{-2} year^{-1} = 73 m^2 per person. The seasonal variation in precipitation and evapotranspiration must also be considered as the system should have volume (depth) enough to be able to store the sewage and rain during winter. In addition, the amount of nutrients discharged into the system should balance the amount that can be removed by harvesting aboveground biomass.

Our data show that when optimal growth of willow is achieved during the first year of operation the evapotranspiration in the system may increase by at least 300 mm under Danish conditions the following year, i.e. from 1,200 mm to 1,500 mm per year. Therefore, willow wastewater cleaning facilities designed for 2–3 persons may be able to receive higher amounts of sewage than designed for the following years. However, there is still some uncertainty about the long-term performance of the systems, particularly the potential accumulation of salts and the sustained health of the willows. Research is presently being carried out to evaluate these aspects and to further optimize the systems.

Acknowledgement

We thank Uffe Jørgensen, Danish Institute of Agricultural Sciences, for elemental analysis of biomass.

References

EPA (1999a). Soil infiltration systems up to 30 person equivalents (in Danish). Environmental Guidelines 2, pp. 1–44, Miljø- og Energistyrelsen, Copenhagen.

EPA (1999b). Root zone treatment systems up to 30 person equivalents (in Danish). Environmental Guidelines 1, pp. 1–46, Miljø- og Energistyrelsen, Copenhagen.

EPA (1999c). Biological sandfilters up to 30 person equivalents (in Danish). Environmental Guidelines 3, pp. 1–44, Miljø- og Energistyrelsen, Copenhagen.

Greger, M. and Landberg, T. (1995). Use of willow clones with high Cd accumulating properties in phytoremediation of agricultural soils with elevated Cd levels. *Proc. 3rd Int. Conf. on the Biogeochemistry of Trace Elements*, Contaminated Soils, Paris, 1995, pp. 505–511.

Hasselgren, K. (1998). Use of municipal waste products in energy forestry: highlights from 15 yesrs of experience. *Biomass and Bioenergy* **15**, 71–74.

Hasselgren, K. (1999). Utilization of sewage sludge in short-rotation energy forestry: a pilot study. *Waste Management and Research* **17**, 251–262.

Henze, M. (1982). Husspildevands sammensætning. Stads- og Havneingeniøren **12**, 386–387.

Landberg, T. and Greger, M. (1996). Differences in uptake and tolerance to heavy metals in *Salix* from unpolluted and polluted areas. *Applied Geochemistry* **11**, 175–180.

Rosenqvist, H., Aronsson, P., Hasselgren, K. and Perttu, K. (1997). Economics of using municipal wastewater irrigation of willow coppice crops. *Biomass and Bioenergy* **12**, 1–8.

Sander, M.-L. and Ericsson, T. (1998). Vertical distributions of plant nutrients and heavy metals in *Salix viminalis* stems and their implications for sampling. *Biomass and Bioenergy* **14**, 57-66.

Venturi, P., Gigler, J.K. and Huisman, W. (1999). Economical and technical comparison between herbaceous (*Miscanthus x giganteus*) and woody energy crops (*Salix viminalis*). *Renewable Energy* **16**, 1023–1026.

Long-term performance summary for the Boot Wetland Treatment System

J.R. Martin*, C.H. Keller**, R.A. Clarke, Jr.* and R.L. Knight***

* CH2M HILL, 3011 SW Williston Road, Gainesville, Florida 32614, U.S.A.
** CH2M HILL, 4350 W Cypress Street, Suite 600, Tampa, Florida 33607, U.S.A.
*** Environmental Scientist, 2809 NW 161 Court, Gainesville, Florida 32609, U.S.A.

Abstract The Boot WTS is a 46.5-ha, hydrologically altered cypress-gum wetland in Polk County, Florida. Poinciana Wastewater Treatment Plant No. 3 has discharged advanced secondary treated effluent to the Boot WTS since August 1984. Comprehensive operational monitoring has been ongoing since 1990. The Boot WTS has provided consistent removal of nitrogen and phosphorus. Influent total nitrogen (TN) and total phosphorus (TP) concentrations averaged approximately 10.0 mg/L and 2.5 mg/L at an average hydraulic loading rate (HLR) of 0.2 cm/d. Wetland effluent concentrations for TN and TP averaged 1.8 mg/L and 1.2 mg/L. Available flow and water quality data were used to develop estimates of the first-order removal rate, k, for TN (14 m/y) and TP (1.8 m/y). These removal rates are within the range of values for other forested treatment wetlands. Biochemical oxygen demand (2.2 mg/L) and total suspended solids (4.9 mg/L) in the influent are near background levels for forested wetlands and are not significantly reduced with passage through the system.

Keywords Cypress dome; natural treatment wetland; nitrogen; phosphorus; wastewater

Introduction

The Boot Wetland Treatment System (WTS) is a 46.5-hectare, natural cypress dome that has been receiving secondarily treated municipal wastewater since 1984. Because the wetland had been completely bermed to support the retention and treatment of wastewater, the entire wetland has been continuously inundated during this period with water depths typically ranging from 75 to 90 cm. Before 1985, forestry, drainage canals, and surrounding land development had adversely affected this wetland. In the early 1980s, peat was rapidly oxidizing, cypress trees were leaning and dying, and groundcover vegetation had changed to invasive upland plants. The Boot WTS now supports a healthy and growing stand of trees, diverse wetland groundcover species, and a stable population of fish and macroinvertebrates (Martin *et al.*, 2000).

The Boot WTS is located in Poinciana, Polk County, Florida. It lies within the Marion Creek/London Creek watershed, which is a subbasin of the Reedy Creek and Kissimmee River basin. Severn Trent-Avatar Utility Services, LLC. owns and operates the Poinciana Wastewater Treatment Plant No. 3 (WWTP No. 3) which discharges secondarily treated municipal wastewater via a submerged discharge pipe into the north end of the Boot WTS. The treated wastewater discharge to the Boot WTS has been nearly continuous since March 1985. The wetland seasonally overflows from its south end to the M-7 drainage canal.

The Florida Department of Environmental Protection operating permit allows for discharge of up to 1,325 cubic metres per day (m^3/d) from WWTP No. 3 to the Boot WTS. A special permit condition requires operational monitoring of the wetland in accordance with the Wetlands Application Rule (Chapter 62–611, F.A.C.) with a requirement for annual reporting. Operational monitoring of the Boot WTS under the guidelines of the permit was initiated in October 1990 and continues at present.

This paper summarizes the water quality data and treatment performance of the Boot WTS for the period from October 1990 through June 1998.

Methods
Site description
The Boot wetland is a 46.5 hectare altered cypress-gum wetland located in Polk County approximately 1.6 km south and 0.8 km west of the Polk-Osceola County line. Overflow from the Boot WTS discharges into M-7 canal which runs along the northern and western edge of the wetland. The M-7 Canal continues its course southward approximately 3.2 km and abruptly ends at the edge of a forested wetland on the east side of the Marigold Avenue bridge. There is no surface water channel through the wetland at this junction. Surface water from the M-7 canal flows intermittently through the wetland, during wet weather only, and exhibits no identifiable connecting channel to London Creek, a natural channel approximately 2.4 km from the M-7 Canal. Water from London Creek then discharges into Lake Hatchineha approximately 8 km south of the Boot WTS outflow point.

The Boot WTS is a natural cypress dome that has been historically altered by drainage and logging. Dominant canopy tree species within the Boot WTS are pond cypress (*Taxodium ascendens*) and black gum (*Nyssa biflora*). A well developed shrub layer exists in the wetland and is primarily composed of buttonbush (*Cephalanthus occidentalis*), primrose willow (*Ludwigia* sp.), fetterbush (*Lyonia lucida*), and wax myrtle (*Myrica cerifera*). Emergent and floating aquatic herbaceous species are common throughout the Boot WTS. Leaning and fallen cypress trees in the Boot WTS reflect drainage impacts prior to 1985 and resulting peat subsidence. Although partially drained in the past, the Boot is now impounded by a perimeter berm that maintains a nearly constant water depth within the wetland of approximately 60 cm. Increasing canopy dominance has been observed in the Boot Wetland since monitoring commenced in 1990.

Sampling stations are located adjacent to three transects in the Boot WTS. Transect 1 is adjacent to the submerged influent distribution pipe, beginning near the northeast corner of the Boot WTS. Transect 2 is located in the approximate center of the Boot WTS, beginning on the west side. Transect 3 is located near the outlet of the Boot WTS, at the southwest corner of the wetland.

Hydrology
Daily inflow to the Boot WTS is measured with a flow totalizer at the WWTP No. 3 effluent pump station. Discharge from the Boot WTS into M-7 Canal is through a 30-cm rectangular weir. Wetland outflow measurements were estimated using stage readings once per month by CH2M HILL.

Water quality
Analytical samples for surface water quality were collected monthly from the influent to the Boot WTS and discharge from the wetland. Influent samples were collected from Poinciana WWTP No. 3 at the pump station (Station 1) which delivers treated effluent directly to the Boot WTS. Discharge from the Boot WTS was sampled upstream from the overflow weir (Station 3). Samples of discharge water were collected even when overflow was not occurring on the sampling date.

At the two water quality-sampling stations (Stations 1 and 3), field measurements were made for total depth, dissolved oxygen (DO), pH, temperature, and conductivity. Monthly surface water samples were collected for color, five-day carbonaceous biochemical oxygen demand ($cBOD_5$), total suspended solids (TSS), total phosphorus (TP), orthophosphorus (OP), total Kjeldahl nitrogen (TKN), total ammonia nitrogen (NH_3), nitrate plus nitrite nitrogen (NO_3+NO_2), and fecal coliforms. Quarterly, additional samples were collected for sulfate (SO_4) and chlorophyll *a*. Semiannually, surface water samples were also collected for total metals including mercury (Hg), lead (Pb), cadmium (Cd), copper (Cu), zinc (Zn), iron (Fe), nickel (Ni), and silver (Ag). Annually, samples were collected for non-metallic priority pollutants including volatiles, pesticides, PCBs, phenol, and cyanide.

Dawn-to-dusk DO and temperature measurements were collected monthly at the same two stations. At each station a permanent wooden platform has been constructed to facilitate monitoring, and a small area of emergent and floating vegetation is kept cleared to minimize sampling interference. Dissolved oxygen and temperature were measured at 5 to 7 cm below the water surface and within 5 to 7 cm of the bottom. Measurements were conducted at a maximum of four hour intervals, dawn to dusk, during a 48 hour period.

Results and discussion

This section reports the results of the operational monitoring completed in the Boot WTS from October 1990 through December 1998. Annual averages for water quality parameters presented below only include data corresponding to periods of measurable discharge from the wetland.

Hydrology

Table 1 summarizes annual average surface water inflows and outflows for the Boot WTS from October 1990 to June 1998. Annual average inflows ranged from 402 to 1,185 m³/d. Annual average outflows ranged from 275 to 1,523 m³/d. The long-term average inflow hydraulic loading rate [HLR] of 0.20 cm/d is the lowest HLR reported for natural treatment wetlands operating in the southeastern United States (Knight et. al., 1995). Annual average rainfall at the Lake Alfred weather station, was about 1.46 m/y (0.40 cm/d). Annual average evapotranspiration was about 1.22 m/y (0.33 cm/d).

Water quality

Table 2 presents annual averages for physical parameters. Water temperature, dissolved oxygen, pH, and conductivity are all reduced between the inlet and outlet. Temperature reduction is attributable to shading effects from the cypress/gum canopy and nearly complete surface water cover provided by floating species such as frog's bit (*Limnobium spongia*), mosquito fern (*Azolla caroliniana*), and duckweed (*Lemna* spp.). Final effluent DO levels (0.5 mg/L) are consistent with literature values for natural cypress wetlands in the southeast (Kadlec and Knight, 1996). Figure 1 shows the typical diurnal variation in DO concentration at Station 1 with highest concentrations measured in the afternoon and lowest concentrations measured near sunrise.

Table 1 Summary of inflows and outflows at the Boot WTS

Year	Boot Inflow (m³/d)	Boot Outflow (m³/d)	Inflow HLR (cm/d)	Outflow HLR (cm/d)	Rainfall (cm/d)	ET (cm/d)
1990	932	275	0.20	0.06	0.16	0.14
1991	1083	742	0.23	0.16	0.47	0.28
1992	402	550	0.09	0.12	0.41	0.34
1993	921	508	0.20	0.11	0.34	0.33
1994	799	1065	0.17	0.23	0.47	0.32
1995	921	952	0.20	0.20	0.40	0.35
1996	1109	903	0.24	0.19	0.39	0.34
1997	1009	1107	0.22	0.24	0.44	0.32
1998	1185	1523	0.25	0.33	0.37	0.41
Grand Mean	**929**	**847**	**0.20**	**0.18**	**0.40**	**0.33**

Table 2 Summary of annual averages for physical parameters at the Boot WTS

Year	Inflow Temp (°C)	Outflow Temp (°C)	Inflow DO (mg/L)	Outflow DO (mg/L)	Inflow pH (s.u.)	Outflow pH (s.u.)	Inflow Cond. (μmhos/cm)	Outflow Cond. (μmhos/cm)
1990	24.6	18.1	3.6	0.3	6.8	5.4	453	202
1991	24.9	21.1	2.1	0.3	6.8	5.6	432	224
1992	22.9	19.0	1.6	0.8	6.7	5.5	416	178
1993	24.4	18.4	3.8	0.4	6.5	5.8	418	193
1994	25.6	19.7	5.7	0.6	5.6	5.6	425	184
1995	25.6	19.9	7.2	0.5	6.5	5.7	430	175
1996	25.7	20.7	5.5	0.4	6.6	6.0	401	205
1997	25.6	20.0	5.2	0.8	6.6	6.1	405	217
1998	23.8	16.7	5.1	0.6	6.3	5.9	368	168
Grand Mean	**24.8**	**19.3**	**4.4**	**0.5**	**6.5**	**5.7**	**416**	**194**

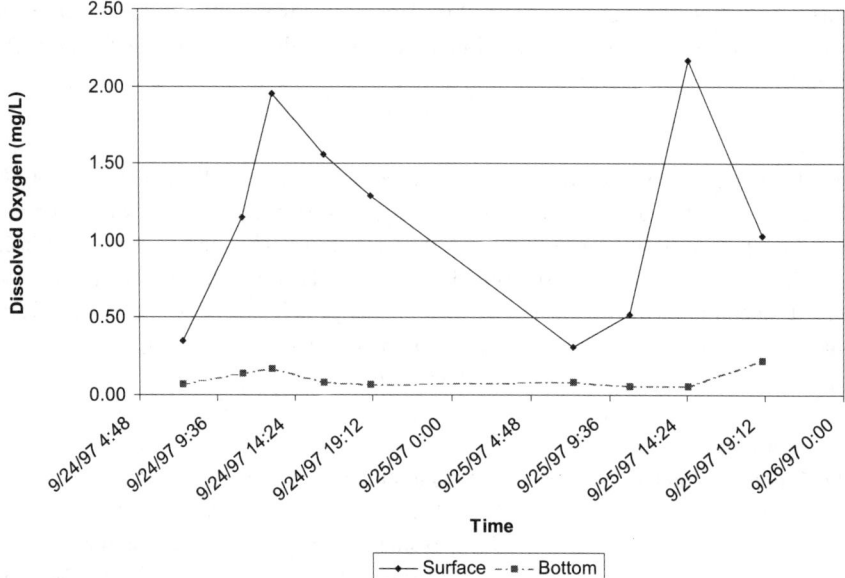

Figure 1 Diurnal variation in DO concentration at the Boot WTS

Average values for pH and conductivity also decrease across the wetland due to dilution with rainfall, assimilation of SO_4^-, and the formation of humic and fulvic acids. Typical values for pH and conductivity in natural cypress domes are 4.5 and <100 μmhos/cm, respectively (Kadlec and Knight, 1996).

Average color increased from 29 to 360 color units and the increase is due to the presence of organic acids that are released as plant material decomposes. Chlorophyll a concentrations decrease from 7.2 to 0.8 mg/m^3 indicating that planktonic algae from the treated effluent holding pond die-off primarily in response to shading of the water column by canopy and herbaceous species. Sulfate ion concentration decreased from 27.4 to 7.5 mg/L as SO_4^- is converted to hydrogen sulfide by sulfate reducing bacteria or reduced under anaerobic conditions and binds with metal ions.

Of the 8 metals sampled, only copper, iron, and zinc were routinely detected. Influent total copper averaged 0.014 mg/L. Copper was detected at the outflow only once at a concentration of 0.013 mg/L and was below the detection level of about 0.01 mg/L for all other samples. Influent and effluent total iron concentrations averaged 0.07 and 0.40 mg/L, respectively. Average total zinc was reduced from 0.029 mg/L to 0.011 mg/L. Annual priority pollutant analyses show periodic occurrences of chloroform, bromodichloromethane, and dibromochloromethane, which are believed to be byproducts of wastewater chlorination at the WWTP. These compounds were not detected in the outflow from the wetland. Methylene chloride was detected once at Station 3 at a concentration of 10 μg/L. The pesticides Aldrin and Lindane were each detected once at the inflow station, but not at the outflow station. Phenol and cyanide were detected at both stations on several occasions, however measured concentrations were typically very near the detection limit of 0.005 mg/L.

Table 3 presents annual average inflow and outflow concentrations, mass loadings, and mass removal percentages for the most commonly regulated wastewater parameters BOD, TSS, TN, and TP. Mass balance calculations for TN and TP include the lumped effects of wetfall (i.e. rainfall) and dryfall loads. Mass loads to and from the wetland were estimated as follows:

$$\text{Mass In} = Q_{IN}C_1 + Q_R C_R$$
$$\text{Mass Out} = Q_{OUT}C_2$$

where
- Q_{IN} = input wastewater flow rate, m^3/d
- Q_{OUT} = measured flow at outlet weir, m^3/d
- C_1 = concentration measured at inflow (Station 1), mg/L
- C_R = rainfall concentration, mg/L
- C_2 = concentration measured at outflow (Station 3), mg/L

The C_R values used in the analysis were 2.15 mg/L for TN (Zarbock et al., 1994), and 0.03 mg/L for TP (Kadlec and Knight, 1996).

Inflow concentrations of BOD (2.2 mg/L) and TSS (4.9 mg/L) were low and reflect the high quality of the effluent produced by the WWTP. Outflow BOD (1.7 mg/L) and TSS (3.7 mg/L) were also low suggesting that this system operates near the natural wetland background level (C^*) for both BOD and TSS. The average mass removal efficiencies for BOD and TSS were 32 and 38 per cent, respectively.

Total nitrogen discharged to the Boot WTS averaged 10.03 mg/L, of which 7.68 mg/L was in the form of NO$_{2/3}$, 1.00 mg/L was NH$_3$, and the remaining 1.35 mg/L was organic nitrogen. Since most of the nitrogen load was in the more easily reducible NO$_{2/3}$ form, the mass removal efficiency for TN was consistently high, averaging 90 per cent. This TN radically changed form between the wetland inlet and outlet with about 80 per cent of the TN in the organic form at the outlet. Figure 2 shows that outlet TN was inversely proportional to WWTP inflow, indicating an internal loading effect related to detention time.

Approximately 2.19 mg/L of the 2.46 mg/L TP was in the form of orthophosphate. The long-term mass removal of TP averaged 49 per cent. Slightly higher mass removal percentages were observed during the first several years of operational monitoring, but the system appears to have stabilized with respect to phosphorus removal. Outlet TP was predominantly in the organic form.

These data indicate that there may be substantial dilution occurring in the Boot WTS due to direct rainfall inputs (approximately 50% dilution as estimated by average changes in specific conductance). Dilution may partially explain the observed net change in TP concentration and mass (average 49%). The relatively low TP removal rates during recent years are surprising considering the apparently long detention time (estimated as about 300 days based on observed hydraulic loading and average water depth). This low net TP mass reduction rate may be due to the effect of declining inlet TP concentrations from the WWTP through time, allowing the release of exchangeable P from sediments that were loaded at a higher rate in the past (Kadlec and Knight, 1996).

Table 3 Summary of wastewater constituent removal at the Boot WTS

Year	BOD In (mg/L)	BOD Out (mg/L)	Mass BOD In (g/m²/y)	Mass BOD Out (g/m²/y)	Mass Removed (g/m²/y)	% Mass Removed
1990	1.1	1.5	0.79	0.33	0.45	58
1991	1.3	1.3	1.18	0.73	0.45	38
1992	4.9	3.5	1.50	0.80	0.70	47
1993	2.7	1.4	2.04	0.76	1.28	63
1994	1.5	1.1	1.00	1.25	−0.25	−26
1995	1.2	1.4	0.88	0.92	−0.05	−5
1996	2.3	3.0	2.05	1.53	0.51	25
1997	1.6	1.0	1.39	0.95	0.44	32
1998	2.8	1.0	2.52	1.79	0.72	29
Grand Mean	**2.2**	**1.7**	**1.48**	**1.01**	**0.47**	**32**

Year	TSS In (mg/L)	TSS Out (mg/L)	Mass TSS In (g/m²/y)	Mass TSS Out (g/m²/y)	Mass Removed (g/m²/y)	% Mass Removed
1990	8.4	2.3	6.19	0.51	5.68	92
1991	3.6	2.1	3.23	0.98	2.25	70
1992	7.0	11.6	2.09	2.13	−0.04	−2
1993	2.3	3.1	1.67	0.95	0.71	43
1994	5.9	2.1	3.47	2.43	1.04	30
1995	4.3	3.9	2.92	2.25	0.67	23
1996	3.1	2.5	2.68	1.50	1.18	44
1997	3.8	3.0	3.27	4.28	−1.01	−31
1998	5.3	2.5	4.79	3.82	0.97	20
Grand Mean	**4.9**	**3.7**	**3.37**	**2.09**	**1.27**	**38**

Year	TN In (mg/L)	TN Out (mg/L)	Mass TN In (g/m²/y)	Mass TN Out (g/m²/y)	Mass Removed (g/m²/y)	% Mass Removed
1990	11.60	3.34	9.83	0.72	9.11	93
1991	6.86	1.89	8.99	1.10	7.89	88
1992	3.21	3.08	4.49	1.33	3.16	70
1993	8.94	1.53	8.97	0.61	8.36	93
1994	13.70	1.27	12.04	1.06	10.98	91
1995	15.80	1.63	14.48	1.22	13.26	92
1996	8.60	1.46	10.50	1.03	9.47	90
1997	11.13	1.39	12.11	1.21	10.90	90
1998	10.44	0.73	12.61	0.87	11.73	93
Grand Mean	**10.03**	**1.81**	**10.60**	**1.02**	**9.58**	**90**

Year	TP In (mg/L)	TP Out (mg/L)	Mass TP In (g/m²/y)	Mass TP Out (g/m²/y)	Mass Removed (g/m²/y)	% Mass Removed
1990	3.39	1.16	2.50	0.25	2.24	90
1991	3.51	1.62	2.93	0.96	1.97	67
1992	2.34	1.58	0.79	1.09	−0.30	−38
1993	2.83	1.52	2.06	0.81	1.25	61
1994	2.15	1.04	1.38	1.18	0.20	14
1995	2.20	1.07	1.62	0.81	0.81	50
1996	1.71	0.95	1.53	0.70	0.83	54
1997	2.18	1.17	1.76	1.05	0.71	40
1998	1.80	0.88	1.69	1.53	0.16	9
Grand Mean	**2.46**	**1.22**	**1.81**	**0.93**	**0.88**	**48**

In spite of the apparent dilution, concentrations of some elements decline by more than 50% (sulfate, $NO_{2/3}$, NH_3, and metals), indicating that there are active depuration and transformation processes at work in this wetland. On the other hand, relatively conservative parameter concentrations such as iron, color, and organic N increased due to apparent internal loadings. These increases were inversely proportional to hydraulic loading from the

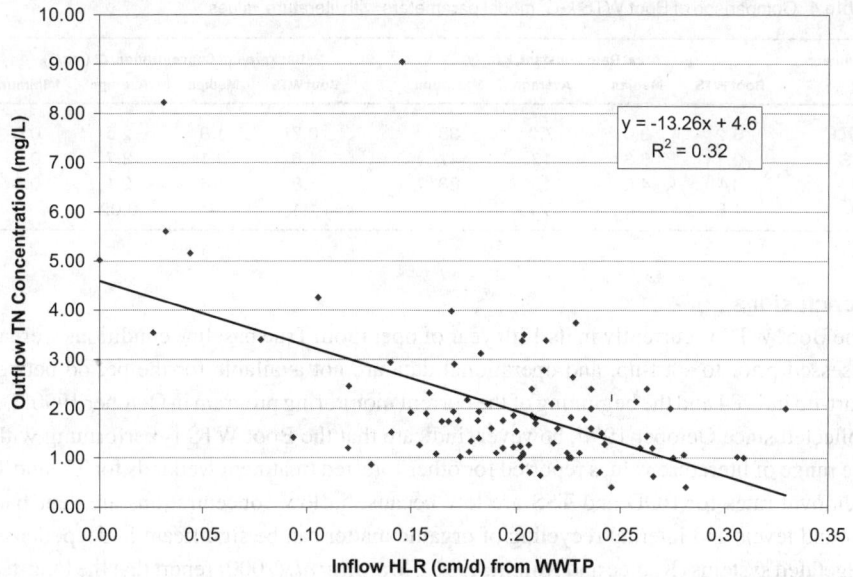

Figure 2 Relationship between inflow HLR and outflow concentration for TN

WWTP and must be a response to very long detention times during low inflow periods. The very low or negative removal rates observed in 1992 are probably a response to the very low hydraulic loading during that year.

k-C* model calibration

The performance of the Boot WTS can be compared to other natural treatment wetlands using the area-based first-order k-C^* model (Kadlec and Knight, 1996). The plug-flow form of the model is:

$$\frac{-k}{HLR} = \ln \frac{(C_2 - C^*)}{(C_1 - C^*)}$$

where k = area-based, first-order removal rate, m/y
HLR = hydraulic loading rate, m/y
C_1 = inflow concentration, mg/L
C_2 = outflow concentration, mg/L
C^* = irreducible background concentration, mg/L.

Knight et al. (1995) summarized areal removal rate constants (k) and background concentrations (C^*) for 8 southeastern forested natural treatment wetlands for BOD, TSS, TKN, and TN. The data set included one year (1993) of operational data from the Boot WTS. Updated k-C^* values for the Boot WTS were estimated for BOD, TSS, TN, and TP based upon the annual average concentrations presented in Table 3. Table 4 presents a comparison of the estimated Boot WTS model parameters with the reported results for BOD, TSS, TN (Knight et al., 1995) and TP (Kadlec and Knight, 1996).

Values for k and C^* for TN and TP fall within the range of values shown in Table 4 and reported by Kadlec and Knight (1996) for other southeastern forested systems. Values for C^* for BOD and TSS are also within the range of values reported in the literature, but the k values are very low and reflect the low influent concentrations and an assumed high rate of internal recycling of organic material.

Table 4 Comparison of Boot WTS k-C^* model parameters with literature values

Parameter	Areal Rate Constant, k (m/y)				Background Concentration, C* (mg/L)			
	Boot WTS	Median	Average	Maximum	Boot WTS	Median	Average	Minimum
BOD	0.2	3.0	7.2	38	0.7	1.8	2.5	0.5
TSS	0.1	6.3	13	47	1.6	2.1	3.7	0.6
TN	14	4.0	15	83	1.8	1.6	2.1	0.8
TP	1.8	–	3.1	–	1.1	–	0.02	–

Conclusions

The Boot WTS is currently in its 15th year of operation. True baseline conditions were not assessed prior to start-up, and operational data are not available for the period between start-up in 1984 and the beginning of the current monitoring program in October 1990. Data collected since October 1990, however, indicate that the Boot WTS is performing within the range of literature values reported for other forested treatment wetlands for TN and TP. Removal rates for BOD and TSS are low because inflow concentrations are near background levels and internal recycling of organic matter can be significant in deep, densely vegetated systems (Kadlec and Knight, 1996). Martin *et al*. (2000) report that the long-term discharge of wastewater to the Boot WTS has stimulated a diverse and productive mix of wetland flora and fauna.

References

Kadlec, R.H., and Knight, R.L. (1996). *Treatment Wetlands*. Lewis Publishers. Boca Raton, FL. 893 pp.

Knight, R.L., Bays, J.S., Baughman, D. and Dunn, W.J. (1995). "Natural Wetlands for Wastewater Management in the Southeastern Coastal Plain." Presented at Innovative Wastewater Strategies for the Future, Water and Pollution Control Association of South Carolina. Hilton Head Island, SC. October 1995.

Martin, J.R., Clarke, R.A. and Knight, R.L. (2000). "Ecological Characteristics of a Natural Wetland Receiving Secondary Effluent." *Wat. Sci. Tech.* **44**(11–12) 317–324 (this issue).

Zarbock, H., Janicki, A., Wade, D., Heimbuch, D. and Wilson, H. (1994). Estimates of Total Nitrogen, Total Phosphorus, and Total Suspended Solids Loadings to Tampa Bay. Prepared for the Tampa Bay National Estuary Program.

Diel changes in iron concentrations in surface-flow constructed wetlands

R.R. Goulet* and F.R. Pick**

* INRS-Eau, 2700 Einstein,, P.O.Box 7500, Ste-Foy, Québec, G1V 4C7, Canada,
(E-mail: richard_goulet@inrs-eau.uquebec.ca)
** Ottawa-Carleton Institute of Biology, University of Ottawa, Ottawa, Ontario, K1N 6N5, Canada,
(E-mail: frpick@science.uottawa.ca)

Abstract Diel changes in Fe concentrations were examined from spring to late fall at two surface-flow wetlands. The highest concentrations of ferrous, dissolved and total Fe were measured at night in the littoral zone, when oxygen and pH were low. The lowest Fe concentrations were measured during the day when oxygen and pH were highest. The amplitude of change over the day-night cycle was greatest in July and lowest in May and October. These diel changes were also observed at the outlet of both wetlands. Overall, O_2 and pH explained 60% of the observed seasonal and diel variation in water Fe ($R^2 = 0.60$, $p = 0.004$). The treatment performance of wetlands can be overestimated when based on samples collected during the day.
Keywords Diel variations; iron; photosynthesis; respiration; surface-flow wetland

Introduction

The retention of metals by wetlands is generally estimated by the periodic collection of inlet and outlet water samples (Kadlec and Knight, 1996). A wetland is effectively retaining metals when the mass of metals at the outlet is lower than at the inlet (Wieder, 1993). However, this conclusion is frequently based on monthly or at best, weekly collections of water samples. Rarely is retention calculated with more frequent sampling points because it is assumed that a single water sample represents the water conditions of a week or more (Wieder, 1994). However, such an assumption may not be valid for many elements and particularly for iron.

Significant diel changes in Fe concentrations can occur in aquatic ecosystems (McKnight *et al.*, 1988; McKnight and Bencala, 1988; Fuller and Davis, 1989; Bourg and Bertin, 1996; Brick and Moore, 1996). Increases in metal concentrations during the day can be the result of the photoreduction of Fe^{3+} into water soluble Fe^{2+} (e.g. Colienne, 1983; Madsen *et al.*, 1986; McKnight *et al.*, 1988; McKnight and Bencala, 1988). However, increases in metal concentrations at night are also observed and cannot be the result of photoreduction (e.g. Bourg and Bertin, 1996; Brick and Moore, 1996). In four wetlands, Wieder (1994) observed an increase in Fe concentration at night and suggested that oxygenation of surface water by photosynthetic algae and cyanobacteria created conditions under which Fe^{2+}, produced from Fe^{3+} reduction or FeS_2 oxidation, was readily oxidized back to Fe^{3+}.

The solubility of Fe in water is closely linked to pH and redox potentials. At a redox of about 300 mV, Fe^{2+} is oxidized to Fe^{3+} and precipitates at pH values higher than 5.5 (Faulkner and Richardson, 1989). Photosynthesis affects the precipitation/dissolution of Fe by inducing changes in pH and redox potentials (Wetzel, 1983). During the day, photosynthesis increases pH and redox potentials. At night, biological respiration is intense because aquatic plants consume oxygen along with heterotrophic organisms resulting in rapid pH and O_2 depletion in the water column.

In this study, we hypothesized that diel changes in O_2 and pH resulting from photosynthesis and respiration would lead to diel variations in metal concentrations in surface-flow wetlands. We predicted an increase in iron concentrations at night and a decrease in iron concentrations during the day. Secondly, we examined the seasonal importance of diel variations in metal concentrations and examined how these variations affected iron retention in these constructed wetlands.

Method
Study sites

The experiments were conducted in two different surface-flow wetlands. The Monahan wetland, within the city of Kanata, Ontario, Canada (45°16' N, 75°51' W), was built in 1995 and completed in 1996 as a stormwater management facility (Table 1). The wetland has three consecutive cells, which were lined initially with a thick layer of clay to prevent infiltration. No organic substrate was added to the clay liner and a significant amount of sediment has accumulated since construction. The Falconbridge wetland is located in the city of Falconbridge, Ontario, Canada (46°57' N, 80°80' W). Its construction in 1981 resulted from the flooding of old mine tailings on which cattails quickly established and now represent 70% of the surface area. It was built to treat acid mine drainage originating from a 1,200 hectares drainage basin (Table 1).

Experimental approach

At the Monahan wetland, we conducted diel experiments on a monthly basis (May 19, June 10, July 13, August 16, September 27) during the summer of 1999. At the Falconbridge wetland, we conducted diel experiments on July 22 and October 14. We examined the influence of photosynthesis by comparing diel changes in metal concentration when photochemical reactions were blocked and when surface sediments were partly exposed to sunlight (because of macrophyte shading). The experiments were conducted in shallow (~0.5 metre) littoral areas. The dark treatment consisted of a black, opaque and bottom-less chamber, which was deposited onto surface sediments in each wetland the night before the startup of the experiment. Near the dark treatment, a light treatment was established as a reference site within a cover of cattails (*Typha latifolia*) and floating duckweed (*Lemna minor*) where photosynthesis was not inhibited. At each site or within the dark chamber duplicate water samples were collected every 4 hours over a 24 hours period using high-density polyethylene tubing fixed 5 cm over the sediment-water interface. At the same time, we measured oxygen, temperature, pH and redox using a data-logging hydrosonde (Hydrolab, Austin, TX, USA). We also collected duplicate water samples at the inlet and outlet in the same temporal manner. At the Falconbridge wetland, inlet waters were not collected and we collected water samples in a deep (~1.5 metres) littoral zone on July 22 while on October 14, we collected water samples in a shallow (~0.5 metre) littoral zone.

Table 1 Description of the Monahan and Falconbridge wetlands

		Monahan		Falconbridge
Age (years)		1		18
Maximum depth (m)		2.0		2.5
Mean depth (m)		1.0		1.5
Water storage (m^3)		53 000		830 000
Drainage area (ha)		637		1200
Yearly average total suspended solids (mg L^{-1})	In	50.5	In	–
	Out	20.3	Out	–
Yearly average Fe (µg L^{-1})	In	360	In	–
	Out	270	Out	530
Yearly average pH	In	7.82	In	–
	Out	8.04	Out	6.89

Fe analyses

Unfiltered water samples were digested by microwave with ultra pure HNO_3 (10% of the sample volume) and then analyzed for the total iron content using an ICP-AES (Thermo Instrument Atomscan 25) at the Department of Earth Sciences of the University of Ottawa. Filtered water samples (0.45 μm Nylon membrane, Chromatographic Specialties) were analyzed by ICP-MS (model VG2000) at the Geological Survey of Canada, Ottawa. An aliquot of the filtrate of each water sample was immediately analyzed on site for Fe^{2+} concentration using the ferrozine method (Stookey, 1970).

Results and discussion

Diel changes in Fe concentration

The effect of sunlight on diel changes of Fe is depicted in Figure 1. In the dark treatment, the concentration of dissolved and total iron increased steadily when oxygen and pH decreased because photosynthesis was blocked. Much of the increase in dissolved iron is ferrous iron (Table 2) and is likely the result of developing anoxic conditions at the sediment-water interface (e.g. Stumm, 1992), which triggers microbially mediated Fe oxide reduction into Fe^{2+} (Lovley et al., 1987).

The concentration of ferrous, dissolved and total iron showed strong diel changes in the light (Figure 1). The oxygen concentration in the littoral zone of the Monahan wetland ranged from a high of 11 mg L^{-1} at noon to a low of 2 mg L^{-1} at 4h00 in July. The pH of the water also followed a similar trend. In response to these variations in oxygen and pH, the concentrations of ferrous, dissolved and total iron reached a maximum at 4h00 and a minimum at 20h00 at the end of the day. The night time release of Fe can be explained by the microbial reduction of Fe^{3+} oxides because of low oxygen concentration and pH at the water-sediment interface (Lovley et al., 1987). During the day, there was a decrease in Fe because photosynthesis produces oxygen and increases pH, which will oxidize Fe^{2+} into Fe^{3+} oxides that precipitate at the water-sediment interface (Stumm, 1992).

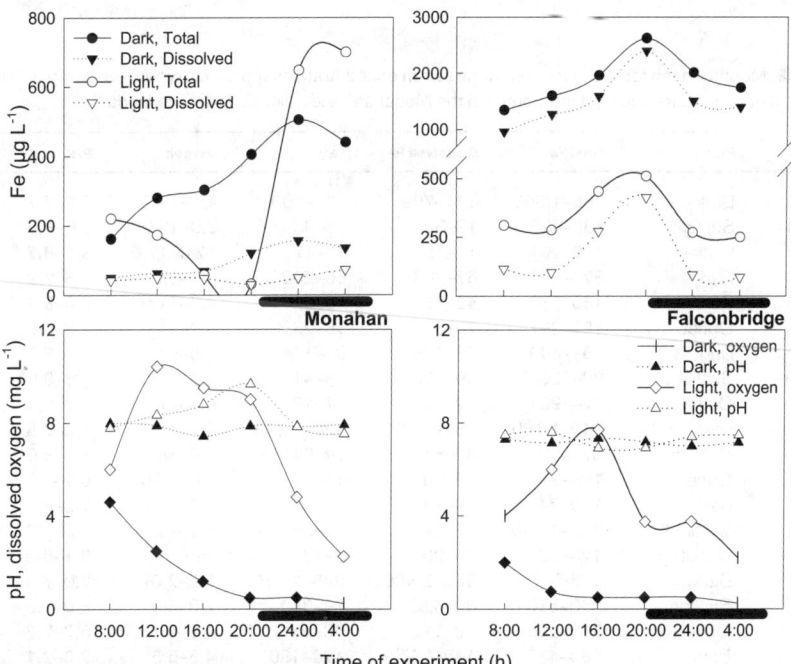

Figure 1 Diel changes in Fe concentrations (top), dissolved oxygen and pH (bottom) in dark treatments and in the light treatment at the Monahan and Falconbridge surface-flow wetlands in July 1999

In contrast, at the Falconbridge wetland, there was an increase in dissolved and total iron in the light despite an increase in oxygen levels (Figure 1). A small decrease in pH during the day suggests some photoreduction of Fe oxides at the sediment-water interface. The increase in Fe occurred only from 16h00 to 20h00 perhaps because duckweed shaded the sediments prior to 16h00. The Monahan littoral sediments were always shaded by duckweed.

The amplitude of the diel changes in iron varied from spring to late fall (Table 2). In May, the amplitude was low but increased in June to peak in July and then decreased throughout August and September and disappeared in October. These changes in Fe follow the diel changes in oxygen and pH induced by photosynthesis. Water temperature and solar radiation drive photosynthesis, which both peak in July (Table 2). The low water temperature combined with low solar radiation will limit photosynthesis, which translates into less difference in O_2 concentration between day and night in spring and fall. In addition, lower water temperatures lead to a homogenized O_2 profile throughout the water column because of rapid vertical mixing and because the solubility of O_2 in cold water is high (Wetzel, 1983).

At the outlet of both wetlands, the concentrations of dissolved and total Fe increased to reach a maximum to reach a maximum at 8h00 when oxygen and pH were low (Figure 2). During the day, the concentrations of Fe species decreased to reach a minimum at midnight when oxygen and pH were high (Figure 2). The outlet of both wetlands showed analogous diel changes of Fe concentrations as those occurring at the shallow littoral sites. However, the amplitude of change was reduced probably because the outlet water results from the mixing of surface pelagic water (where diel variations in metal concentration are minimal) and littoral water (where diel variations are significant). In the summer, the deep water is not mixing with surface water because some thermal stratification occurs in these wetlands. As further evidence of this, the increase in Fe in the deep water (1.5 m) at Falconbridge was not measured at the outlet. Therefore, the shallow littoral zones probably contribute significantly to the treatment efficiency of wetlands.

Table 2 Monthly mean temperature and range in iron concentrations (µg L^{-1}), oxygen levels (mg L^{-1}) and pH at the outlet and in dark and light treatments at the Monahan[1] and Falconbridge[2] wetlands

	Place	Total Fe	Dissolved Fe	Fe^{2+}	Oxygen	PH
May[1]	Dark	785–1,300	141–709	135–700	4.0–10.2	7.2–7.7
(20°C)	Sunlight	184–525	19–36	16–33	9.0–18.0	7.6–8.5
	Outlet	160–201	14–18	13–17	12.0–17.0	8.1–8.7
June[1]	Dark	324–658	67–266	60–208	0.4–6.2	7.6–7.9
(22°C)	Sunlight	186–283	42–73	40–62	4.0–14.0	7.9–8.4
	Outlet	122–267	20–35	16–26	5.0–14.0	7.8–8.6
July[1]	Dark	161–606	49–156	31–134	0.5–5.0	7.4–7.9
(24°C)	Sunlight	32–1260	59–105	23–41	2.0–11.0	7.5–9.6
	Outlet	77–496	12–19	11–18	8.5–11.0	8.4–9.0
Aug.[1]	Dark	340–1,160	132–1,331	120–1,118	0.5–2.0	6.9–7.6
(20°C)	Sunlight	87–420	32–85	24–64	2.0–14.0	7.4–8.0
	Outlet	180–370	12–28	10–15	8.0–15.0	8.2–8.5
Sept.[1]	Dark	160–240	20–33	19–28	3.0–6.0	7.2–7.3
(16°C)	Sunlight	320–1,840	16–44	14–36	8.0–11.0	7.5–7.7
	Outlet	120–230	13–20	8–13	12.0–13.0	8.1–8.3
July[2]	Dark	1,350–2,630	960–2,400	396–1,110	0.5–2.0	7.0–7.3
(24°C)	Sunlight	250–510	80–420	78–340	2.0–8.0	6.9–7.6
	Outlet	255–295	70–110	63–93	4.0–8.1	6.7–7.7
Oct.[2]	Dark	380–430	145–205	120–180	4.5–5.5	7.0–7.1
(7°C)	Sunlight	160–200	82–97	76–89	6.6–7.6	7.0–7.1
	Outlet	130–170	59–89	53–89	6.4–7.9	7.0–7.1

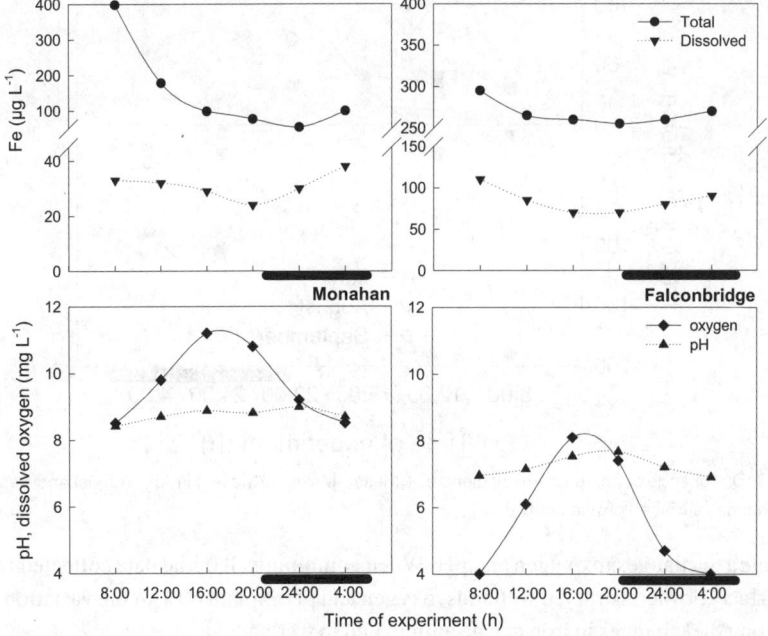

Figure 2 Diel changes in Fe concentrations (top), dissolved oxygen and pH (bottom) at the outlet of the Monahan (left) and the Falconbridge (right) surface-flow wetlands in July 1999

Other metals

Like Fe, total and dissolved Mn concentrations showed diel trends at the Monahan wetland (Data not shown) perhaps because the redox behavior of Mn in aquatic systems is similar to Fe. Other dissolved trace elements such as Cu, Zn, Pb, and As also varied on a diel basis but the amplitude of change in concentration between day and night was less important than for Fe and Mn (Goulet, 2001). Fe and Mn oxides are important binding phases for these trace elements in natural oxic sediments (Benjamin and Leckie, 1981; Tessier *et al.*, 1996). During the microbially mediated reductive dissolution of Fe and Mn oxides, adsorbed trace elements were possibly released to the water column during night time at the Monahan wetland. Diel changes in dissolved Zn, Cu and As have previously been observed in rivers (Fuller and Davis, 1989; Bourg and Bertin, 1996; Brick and Moore, 1996).

Implications for iron retention

The diel changes in Fe concentrations at the outlet of wetlands will affect the overall retention. In July, the treatment efficiency of Fe by the Monahan wetland ranged from –190% at 8h00 to 75% at 20h00 at night (Figure 3). Towards the end of summer, the difference in treatment efficiency between night and day decreased but the Monahan was still a source of metals overnight. Similar trends in Mn retention were also observed at the Monahan wetland (Goulet, 2001). This result emphasizes the need to integrate water samples over a 24 hour period in order to get a more accurate measure of retention. Otherwise, retention is either significantly overestimated or underestimated depending on when the samples are collected.

Conclusion

Constructed wetlands are believed to be efficient sinks for Fe. This affirmation stems from mass balance calculations that generally do not consider diel variations in Fe concentrations. However, this study showed that significant diel variations occur in wetlands, which

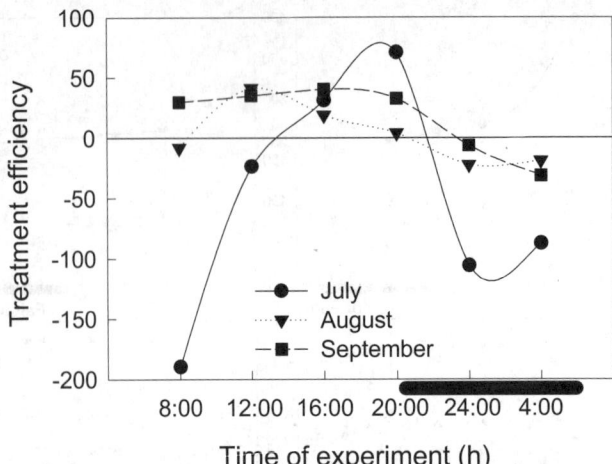

Figure 3 Diel change of Fe treatment efficiencies ([Mass$_{in}$-Mass$_{out}$]/Mass$_{in}$) in July, August and September at the Monahan surface-flow wetland

are related to changes in oxygen and pH. When combining all of the data collected from the two surface-flow constructed wetlands, oxygen and pH explain 60% of the variation in total Fe. Strong diel changes in iron can be anticipated in wetlands:

1. if intense photosynthesis results in significant diel changes in oxygen and pH
2. during warm summer months and not during spring, late fall or winter
3. in the absence of a macrophyte cover in littoral areas, iron increases can be observed in the light that are likely due to photoreduction of Fe^{3+} to Fe^{2+} at the sediment-water interface.

To determine if wetlands are sinks or sources for metals, the calculation of retention should be based on integrated samples taken over a 24 hours interval.

References

Benjamin, M.M. and Leckie, J.O. (1981). Multiple-site adsorption of Cd, Cu, Zn, and Pb on amorphous iron oxyhydroxides. *J. Colloid. Interf. Sci.*, **79**(1), 209–221.
Bourg, A.C.M. and Bertin, C. (1996). Diurnal variations in the water chemistry of a river contaminated by heavy metals: natural biological cycling and anthropic influence. *Water Air Soil Pollut.*, **86**, 101–116.
Brick, C.M. and Moore, J.N. (1996). Diel variation of trace metals in the Upper Clark Fork river, Montana. *Environ. Sci. Technol.*, **30**(6), 1953–1960.
Collienne, R.H. (1983). Photoreduction of Fe in the epilimnion of acidic lakes. *Limnol. Oceanogr.*, **28**(1), 83–100.
Faulkner, S.P. and Richardson, C.J. (1989). Physical and chemical characteristics of freshwater wetland soils. In: *Constructed Wetlands for Wastewater Treatments: Municipal, Industrial, and Agricultural*, D.A. Hammer (ed.), Lewis Publisher, Chelsea, MI, pp. 41–74.
Fuller, C.C. and Davis, J.A. (1989). Influence of coupling of sorption and photosynthetic processes on trace element cycles in natural waters. *Nature*, **340**(6), 52–54.
Goulet, R.R. (2001). *Mechanisms affecting metal retention by surface-flow wetlands in cold temperate climates*. Ph.D. thesis, Ottawa-Carleton Institute of Biology, University of Ottawa.
Lovley, D.R., Stolz, J.F., Nord, G.L. and Phillips, E.J.P. (1987). Anaerobic production of magnetite by a dissimilatory iron-reducing microorganism. *Nature*, **330**, 252–254.
Madsen, E.L., Morgan, M.D. and Good, R.E. (1986). Simultaneous photoreduction and microbial oxidation of Fe in a stream in the New Jersey Pinelands. *Limnol. Oceanogr.*, **31**(4), 832–838.
McKnight, D. and Bencala, K.E. (1988). Diel variations in iron chemistry in an acidic stream in the Colorado rocky mountains, USA. *Arctic Alpine Res.*, **20**(4), 492–500.
McKnight, D.M., Kimball, B.A. and Bencala, K.E. (1988). Iron photoreduction and oxidation in an acidic mountain stream. *Science*, **240**, 637–639.
Tessier, A., Fortin, D., Belzile, N., DeVitre, R.R. and Leppard, G.G. (1996). Metal sorption to diagenetic iron and manganese oxyhydroxides and associated organic matter: Narrowing the gap between field and laboratory measurements. *Geochim. Cosmochim. Acta*, **60**(3), 387–404.
Stookey, L.L. (1970). Ferrozine-A new spectrophotometric reagent for iron. *Anal. Chem.*, **42**(7), 779–781.
Stumm, W. (1992). *Chemistry of the solid-water interface: Processes at the mineral-water and particle water interface in natural systems*. John Wiley and Sons, New York.
Wetzel, R.G. (1983). *Limnology*. 2nd ed., CBS College Publishing, Madison, NY.
Wieder, R.K. (1993). Ion input/output budgets for five wetlands constructed for acid coal mine drainage treatment. *Water Air Soil Pollut.*, **71**, 231–270.
Wieder, R.K. (1994). Diel changes in iron (III)/iron(II) in effluent from constructed acid mine drainage treatment wetlands. *J. Env. Qual.*, **23**, 730–738.

Wastewater treatment by algal turf scrubbing

R.J. Craggs

National Institute of Water and Atmospheric Research, P O Box 11 115, Hamilton, New Zealand
(E-mail: *r.craggs.niwa.co.nz*)

Abstract Algal turf scrubbing (ATS) is a novel wetland technology that has been designed and engineered to promote natural wastewater treatment processes. Algal turf scrubbing improves water quality by passing a shallow stream of wastewater over the surface of a gently sloped floway. The floway is colonised by a natural heterogeneous assemblage of periphyton consisting of cyanobacteria, filamentous algae and epiphytic diatoms together with aerobic bacteria and fungi. Algal photosynthesis provides oxygen for aerobic breakdown of wastewater by heterotrophic bacteria. Pollutants are extracted from the wastewater by several processes including assimilation, adsorption, filtration and precipitation. The algal turf is harvested periodically to remove the accumulated periphyton biomass and associated pollutants from the system. This paper will present results from a demonstration ATS facility in Patterson, California which was used to polish secondarily treated wastewater. The design and operational factors that influence the treatment performance of ATS systems is discussed. Results indicate the potential of the ATS for nutrient removal from secondarily treated wastewater and agricultural drainage waters.
Keywords Algal turf scrubber; nutrient removal; periphyton; wastewater treatment

Introduction

Algal turf scrubbing (ATS) was developed by Adey and associates at the Smithsonian Institution, Washington, D. C. as a more natural means of maintaining water quality in model ecosystems (Adey, 1983; Adey and Hackney, 1989; Adey and Loveland, 1998). ATS has since been used successfully to maintain water quality in several ecosystem models (including the 3,400 m^3 mesocosm of the Space Biosphere 2 project in Arizona, USA) and large aquarium systems such as the 2,500 m^3 Great Barrier Reef Aquarium in Townsville, Australia (Adey and Loveland, 1998). Initial small-scale laboratory experiments on wastewater treatment indicated the potential of ATS systems to remove both refractory organic compounds and heavy metals from polluted water (Adey *et al.*, 1996). The capability of ATS to remove nutrients from agricultural run-off and aquaculture effluent has also been demonstrated at pilot scale (Adey *et al.*, 1993; Adey and Hackney, 1989; Adey and Loveland, 1998). This paper summarizes results from the first large-scale ATS wastewater treatment system and discusses the design and operational factors that influence the treatment performance.

Methods

Patterson wastewater treatment system

The city of Patterson (latitude 37°, 30', 21" and longitude 121°, 04', 58") is situated in the Central Valley of California, approximately 70 miles south east of San Francisco. The ATS system at Patterson was used to polish secondary treated effluent that is discharged to evaporation and infiltration ponds. The ATS system consisted of a single floway (152.4 m long, 6.5 m wide and 1012 m^2 surface area) that was formed from graded (0.25% to 0.5% slope), compacted earth covered with an impervious liner and an overlying biomass retention screen between two raised side walls (Figure 1). Flow rates varied from 108.9 m^3 d^{-1} to 1336 m^3 d^{-1}.

Sampling protocol and analytical methods. Evaluation of the ATS system was conducted over a three-year period between August 1993 and March 1996 which enabled performance to be related to seasonal variations. Influent and effluent water quality samples were taken twice weekly and analysed by standard methods (APHA, 1992) for BOD_5, TSS, TP, SRP, TKN, NO_3-N, NH_4-N, faecal coliforms, alkalinity and hardness. Physical parameters of the wastewater including DO, temperature, pH and conductivity were also measured. Accumulated solids (algal and bacterial biomass, trapped particulates and precipitated compounds) were sampled from five random 0.093 m^2 sites on the ATS floway prior to harvesting. The samples were dried in an oven at 105°C and used to calculate the mean for the floway.

Results and discussion

Several design and operational variables were tested during the course of the study and the influence of each variable on the treatment performance of the ATS system is discussed below.

Pulsed inflow

Pulsed inflow using counter weighted dump trough was found to be important in maintaining the treatment efficiency of marine ATS systems (Adey and Hackney, 1989; Adey and Loveland, 1998). However, research at Patterson, showed that it may not be necessary for freshwater wastewater treatment systems, since freshwater algae are probably more adapted to the continuous turbulent flow of streams and rivers.

Residence time

The hydraulic residence time, the time that the wastewater is in contact with the algal turf, greatly influences the treatment efficiency of ATS systems and can be controlled by either changing the hydraulic loading velocity (HLV) or the length of the floway. Reducing the HLV of the Patterson ATS system from 1.36 m d^{-1} to 0.44 m d^{-1} improved removal of nutrients, total suspended solids (TSS) and biochemical oxygen demand (BOD). For a particular HLV, a minimum length of floway will be required for efficient treatment to be achieved. Some parameters (for example, DO, temperature and concentrations of all forms of nitrogen) have a linear correlation with floway length. Other parameters (particularly, pH, soluble reactive phosphorus concentration, hardness and conductivity) all change more rapidly at or after a particular floway length.

Figure 1 The large-scale ATS system at Patterson, California, USA

Algal species

Algal turf species can be divided into three ecological types: mat forming cyanobacteria, canopy forming green filamentous periphyton and epiphytic diatoms (Figure 2). The dominant algal species is strongly influenced by season. For much of the year the predominant algal species are cyanobacteria (particularly, *Oscillatoria* sp.) and diatoms (including *Navicula* sp., *Nitzschia* sp. and *Cyclotella* sp.). It is only in the summer that the green filamentous periphyton *Ulothrix sp.* and *Stigeoclonium* sp. are dominant with *Cladophora sp.*, *Spyrogyra* sp., *Tribonema* sp., *Oedogonium* sp. and *Rhizoclonium* sp. present to a lesser extent. The population density and algal species diversity of algal turfs in wastewater treatment systems are much lower than those in systems treating agricultural drainage water and seawater (Adey and Hackney, 1989; Adey and Loveland, 1998; Adey *et al.*, 1993; Craggs *et al.*, 1996a). This is due to the absence of algal spores in the secondary treated wastewater to seed the system compared to agricultural drainage water (Adey *et al.*, 1993) and the lack of recirculation, which provided a source of algal spores in the model ecosystems (Adey and Loveland, 1998). Algal turf development and species diversity are much improved by seeding the flow.ay with periphyton from near-by streams and rivers. This was achieved most successfully by collecting algal turf, breaking it up and evenly distributing it over the floway surface, which can bring a new floway into full productivity after only a few weeks (Craggs *et al.*, 1996b). Species diversity could probably be maintained by recirculating a portion of the algal turf scrubber effluent to introduce algal spores with the influent.

Invertebrates

Several invertebrate species are found in wastewater algal turfs. The most abundant are amphipods and chironomid midge larvae. Chironomid infestation can be a significant problem since the larvae, which live in cocoons on the floway surface, dislodge the surrounding algal turf causing a reduction in algal turf standing crop. Chironomid problems have been observed in several other periphyton growth systems (Lock *et al.*, 1984; Cook *et al.*, 1986; Davis *et al.*, 1990). A simple method of controlling chironomid populations on algal turf scrubber floways is by drying out the floway following harvest of the turf, especially as the productivity of the turf can be quickly restored by seeding. However, further research is needed on less severe methods of Chironomid control for ATS systems.

Algal standing crop

Sufficient standing crop of periphyton in exponential growth phase is required to maintain effective treatment. Several operational parameters influence the standing crop of ATS systems and these are harvest technique and interval, and growth surface. Harvesting the floway has several roles. It physically removes all the pollutants accumulated by the algal turf from the system before they naturally slough off. It simulates heavy grazing of the plant

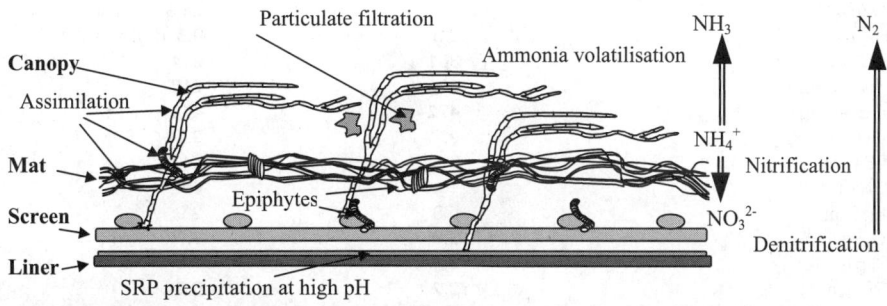

Figure 2 Schematic drawing of nutrient removal processes that occur on an algal turf scrubber

community, which has been shown to stimulate algal growth and nutrient removal (Sladeckova et al., 1983; Davis et al., 1990; Adey and Loveland, 1998). It maintains the exponential growth rate by preventing self-shading by accumulated turf biomass. Harvesting also controls populations of most invertebrate grazers so that pollutants are not reintroduced to the water. Harvest is achieved by initially stopping the flow of wastewater and allowing the water to drain from the floway. Several harvest methods have been employed successfully including manual scraping or vacuum suction of the screen surface on the floway, although vacuum harvest was found to reduce the population density of filamentous algae. Since the growth of algal turf varies with season, treatment performance can be improved by varying the harvest interval over the year to maintain sufficient biomass on the ATS floway. A harvest interval of less than one week was used when the highest productivities were achieved during the summer, while intervals of up to a month were required during the winter. The texture of the floway liner seems to be of particular importance for the maintenance of algal species diversity and population density. Particularly, the addition of a biomass retention screen to the floway surface improves treatment performance by increasing the surface area available for turf colonisation and retaining a larger standing crop of filamentous algae beneath the screen following harvest (Adey and Hackney, 1989).

Algal turf productivity

There is a gradual decline in the productivity of the algal turf down the length of the floway, which probably results from a reduction in nutrient availability and temperature inhibition of algal growth. Algal turf productivity (total solids dry weight) averaged for the whole floway can attain values of up to 61 g m^{-2} d^{-1}, although the annual mean productivity (24 g m^{-2} d^{-1}) is similar to that previously reported for periphyton water treatment systems (Davis et al., 1990; Adey et al., 1993; Craggs et al., 1996b). Sedimentation, filtration and precipitation of particulates on the floway surface all contribute to the accumulation of algal turf solids, such that nearly 50% of the turf dry weight is non volatile containing high concentrations of calcium and magnesium.

Wastewater treatment

Algal turf scrubbing improves wastewater quality in many ways (Table 1). Dissolved oxygen and pH are increased during the day, while alkalinity, conductivity and hardness are all reduced. BOD and TSS of the wastewater can be removed by a combination of particulate filtration by filamentous algae and bacterial oxidation. Bacterial oxidation of BOD varies

Table 1 Mean annual influent and effluent concentrations of water quality variables measured over the three-year study (n = 144)

Water quality variable	Influent	Effluent
DO (g m^{-3})	4.9	24.3 (daytime)
pH	8.3	9.6 (daytime)
Alkalinity (g m^{-3})	301	214
Conductivity (µS)	223	212
Hardness (g m^{-3})	472	444
BOD$_5$ (g m^{-3})	9.7	9.7
TSS (g m^{-3})	35.1	24.5
TON (g m^{-3})	10.0	6.2
NH$_4$ (g m^{-3})	5.8	2.9
NO$_3$ (g m^{-3})	2.9	1.7
TP (g m^{-3})	3.1	1.5
SRP (g m^{-3})	2.2	0.9
Faecal coliform MPN (100 ml)$^{-1}$	5.6 × 10^4	2.0 × 10^4

diurnally, correlating with dissolved oxygen concentration. The mean annual BOD_5 concentration was minimally affected and is probably close to background level for ATS systems. Removal of BOD and COD by ATS systems was shown by Adey *et al.* (1996), although the operational parameters of the recirculating mesocosm system used in that study are very different to those of a single pass wastewater treatment floway. The capability of ATS floways to significantly remove BOD remains to be fully evaluated.

Disinfection. Because of the very short residence time (typically 20–40 minutes) of wastewater on the ATS floway there is little reduction of faecal coliform bacteria concentrations by this system.

Nutrient removal. Both nitrogen and phosphorus are removed effectively from secondary treated wastewater with greater removal achieved at higher influent concentrations. Nutrient removal is a result of several microbiological and physical processes (Figure 2) which are affected by a number of parameters including hydraulic loading, seasonal variation in turf growth and species, and attainment of a pH greater than 9.0 in the floway effluent. Reduction of nutrient concentrations to mg g^{-3} levels was probably limited by insufficient organic carbon to enable the nutrients to be assimilated into algal turf biomass. Therefore, addition of a carbon source to the wastewater on the floway when carbon is limiting may enhance nutrient removal by the system.

Nitrogen removal. All forms of nitrogen can be removed by ATS wastewater treatment systems. Particulate nitrogen removal is related to the dominance of filamentous species on the floway. The simultaneous removal of both ammoniacal-N and nitrate is due to the mixed assemblage of algal species.

Phosphorus removal. Periphyton can remove phosphorus from wastewater by a combination of filtration, adsorption, assimilation (including luxury uptake) and precipitation (Swift and Nicholas, 1987). Most of the phosphorus removal in the ATS wastewater treatment systems was by assimilation into algal biomass or precipitation of soluble reactive phosphorus (SRP). Precipitation of SRP with cations such as Ca^{2+}, Mg^{2+}, and Al^{3+} is known to occur at pH between 8.9–9.5, depending upon the buffering capacity of the water (Belsare and Belsare, 1987). SRP removal is therefore dependent upon the wastewater pH reached while the wastewater is on the floway. The increase in the pH of the ATS effluent is a result of carbon limitation of the turf algae, and their subsequent use of bicarbonate as a carbon source for photosynthesis (Soeder and Hegewald, 1988). At Patterson a threshold pH of approximately 9.0 was required before SRP removal occurred to any great degree.

Reducing the HLV of the ATS system from 1.4 to 0.11 m d^{-1}, increased the residence time of the wastewater on the floway and the pH attained in the floway effluent, hence SRP removal by precipitation was improved. SRP precipitation on the algal turf scrubber floway is indicated by the reduction of alkalinity, hardness and conductivity that occurs as the pH of the effluent increases. Maintenance of the pH of the ATS system effluent above the level at which phosphate precipitates may provide an effective means of controlling SRP removal by ATS systems. The threshold pH could be maintained by controlling the length of time the wastewater is in contact with the algal turf. This may be done either by altering the hydraulic loading rate of the floway, or by passing the wastewater down different lengths of floway. However, other factors need to be considered in determining the residence time of the floway, such as the temperature attained in the ATS effluent, which may restrict algal growth.

Night-time decline of the ATS system effluent pH to below the threshold pH results in dissolution of some of this precipitate and release of SRP back to the ATS effluent. Discharge of this released SRP can be prevented by either recirculating the ATS effluent at night, or greatly reducing the flow at night such that little or no water is discharged until the following morning when the effluent pH has risen above the threshold again. ATS systems could be designed to enhance phosphorus removal from wastewater by a two-stage process. The first floways would be only operated during the day and have low hydraulic loading to promote pH mediated precipitation with cations. This would be followed by the second floways with high hydraulic loading and perhaps recirculation to promote rapid algal growth and assimilation of phosphorus into algal biomass. The efficiency of treatment under both of these modes of operation is affected by the amount of biomass on the floway. Frequent harvesting would maintain exponential growth of the biomass in the second floways, while a longer harvest interval would maintain a large standing crop to raise pH in the first floways.

Algal turf nutrient content. The mean annual nitrogen and phosphorus content of the accumulated solids of the ATS floway at Patterson were 3.96% and 1.83% respectively. The values for nitrogen are typical of algal biomass, however those for phosphorus are higher than that (< 1%) normally associated with periphyton biomass (Kesler, 1982; Davis *et al.*, 1990). Adey *et al.* (1993) found a phosphorus content of 0.4% in periphyton grown on agricultural run-off. The high phosphorus content of the algal turf of ATS wastewater treatment systems further indicates the role of pH mediated SRP precipitation in phosphorus removal. Based on the percentages of N and P in the accumulated solids, and the mean annual turf solids accumulation rate (24 g m^{-2} d^{-1}) the mean removal of nitrogen was 0.95 ± 0.48 g m^{-2} d^{-1} and the mean removal of phosphorus was 0.44 ± 0.22 g m^{-2} d^{-1}. The pilot ATS system treating agricultural run-off achieved phosphorus removal rates of 0.12 g m^{-2} d^{-1} (Adey *et al.*, 1993). Mass balance calculations of the wastewater treatment system show that more nitrogen is removed than is accumulated in the algal turf and indicate that denitrification does occur, perhaps in anoxic micro environments of the algal turf.

Algal turf biomass. The solids content of harvested turf biomass usually does not exceed 5% and the fibrous material may simply be dried using drying racks or sand beds. Finding a use for the turf biomass remains unresolved, although there are several potential applications, such as fertiliser (dried or composted), energy (by fermentation to alcohol or methane), or as a feed additive for ruminants, poultry and fish.

Conclusions
Algal turf scrubbers are capable of removing nitrogen and phosphorus from secondary wastewater without added chemicals and with a minimum of energy expenditure. When covered with filamentous algae, ATS systems are capable of filtering significant amounts of suspended solids and the BOD, TSS, TN and TP associated with these suspended solids. Inorganic nitrogen removal is due to assimilation and some denitrification while soluble reactive phosphorus removal is a result of assimilation and precipitation. Maintenance of the pH of the ATS system effluent above the threshold at which precipitation occurs could provide a simple means of controlling phosphorus removal by the ATS. This may be achieved by altering the length of time the wastewater is in contact with the algal turf, either by reducing the hydraulic loading velocity of the floway or passing the wastewater down a longer floway. Stopping overnight flow until the threshold pH is regained the following day also prevents resolution of precipitated phosphorus at night.

Application of ATS. ATS treatment systems have good potential for the removal of nutrients and other contaminants from wastewaters. The simplicity of ATS treatment systems and the ease with which configuration and operational parameters such as hydraulic loading, floway length and harvest period can be changed should enable process control and optimisation of the system for treatment of different wastewaters. Appropriate use of ATS technology may be for the remediation of eutrophic reservoirs and lakes, for polishing secondary effluents, and for algal biomass production for aquaculture.

Acknowledgements
This research was conducted under the direction of Professor W. Oswald at the University of California, at Berkeley. Funding was provided by Aquatic Bio Enhancement Systems Inc. Texas. Algal turf scrubbing is a patented environmental control system developed by W. Adey and held by the Smithsonian Institution. Aquatic Bio Enhancement Systems Inc. holds an exclusive license for the technology.

References
Adey, W.H. (1983). The microcosm: A new tool for reef research. *Coral Reefs*, **1**, 193–201.

Adey, W. and Hackney, J. (1989). The composition and production of tropical marine algal turf in laboratory and field experiments. In: *The Biology, Ecology, and Mariculture of Mithrax spinosissimus utilising cultural algal turfs*, Adey, W. (ed.), The Mariculture Institute, Los Angeles, California.

Adey, W. and Loveland, K. (1998). *Dynamic Aquaria: building living ecosystems.* 2nd edn. Academic Press, New York.

Adey, W., Luckett, C. and Jensen, K. (1993). Phosphorus removal from natural waters using controlled algal production. *Restoration Ecology*, March, 29–39.

Adey, W., Luckett, C. and Smith, M. (1996). Purification of industrially contaminated ground waters using controlled ecosystems. *Ecological Engineering Journal of Ecotechnology*, **7**, 191–212.

Belsare, D.K. and Belsare, S.D. (1987). High-rate oxidation pond for water and nutrient recovery from domestic sewage in urban areas in the tropics. *Arch. Hydrobiol. Beih. Ergebn. Limnol*, **28**, 123–128.

Cook, P.E., Hoffmann, J.P., Morris, J.W. and Stuart, L.S. (1986). *Feasibility of tertiary wastewater treatment by attached microbes (periphyton) in cold regions.* Completion report, project 84-06 VWRRC, University of Vermont.

Craggs, R.J., Adey, W.H., Jensen, K.R., St. John, M., Green, F.B. and Oswald, W.J. (1996a). Phosphorus removal from wastewater using an algal turf scrubber. *Water Science and Technology*, **33**(7), 191–198.

Craggs, R.J., Adey, W.H., Jessup, B.K. and Oswald, W.J. (1996b). A controlled stream mesocosm for tertiary treatment of sewage. *Ecological Engineering*, **6**, 149–169.

Davis, L.S., Hoffmann, J.P. and Cook, P.W. (1990). Production and nutrient accumulation by periphyton in a wastewater treatment facility. *J. Phycol.*, **26**, 617–623.

Kesler, D.H. (1982). Periphyton phosphorus concentrations in a small New England Lake. *J. Freshwater Ecol.*, **1**, 507–514.

Lock, M.A., Wallace, R.R., Costerton, J.W., Ventullo, R.M. and Charlton, S.E. (1984). River epilithon: toward a structural-functional model. *Oikos*, **42**, 10–22.

Sladeckova, A., Marvan, P. and Vymazal, J. (1983). The utilization of periphyton in waterworks pretreatment for nutrient removal from enriched influents. In: *Periphyton of Freshwater Ecosystems*, Wetzel, R.G. (eds), Dr W. Junk Publishers, The Hague, pp. 299–303.

Soeder, C.J. and Hegewald, E. (1988). Scenedesmus. In: *Micro-algal Biotechnology*, Borowitzka, M.A. and Borowitzka, L.J. (eds), C.U.P., Cambridge, pp 59–84.

Swift, D. and Nicholas, R. (1987). *Periphyton and water quality relationships in the Everglades Water Conservation Areas 1978–1982.* Technical Publication 87-2. South Florida Water Management District, West Palm Beach, Florida.

Rerating capacity of a constructed wetland treatment system

J.A. Jackson* and M. Sees**

* PBS&J, 482 South Keller Road, Orlando, FL 32810-6101, USA
** City of Orlando, 25155 Wheeler Road, Christmas, FL 32709, USA

Abstract The 482-hectare (ha) City of Orlando (Florida) Easterly Wetlands (OEW) was designed to reduce nutrient concentrations in 0.90 m³/s of wastewater from the Iron Bridge Regional Water Reclamation Facility. Design influent nutrient concentrations were 6 mg/L total nitrogen (TN) and 0.75 mg/L total phosphorus (TP).

Actual TN and TP concentrations have been less than design, averaging 2.6 mg/L and 0.29 mg/L, respectively from January 1988 through December 1999. If influent concentrations remain at these levels, the OEW may have the potential to treat significantly higher flows since less than 20% of the total area was utilized for nutrient reduction.

To test this theory, a capacity study was performed for approximately nine months in 1997 and 1998. Simulated flows of approximately 1.26 m³/s, 1.66 m³/s, and 1.93 m³/s were tested. It was found that approximately 15% of the area was utilized for nutrient reduction during the 1.26 m³/s simulation, 35% in the 1.66 m³/s, and 1.93 m³/s simulations. Based on these testing results, an application was submitted to the state in early 2000 to increase the permitted capacity to 1.57 m³/s.

Keywords Capacity; constructed wetland; efficiency; nitrogen removal; phosphorus removal; rerating; wetland

Introduction

The Orlando Easterly Wetlands (OEW) began receiving highly treated reclaimed water from the City of Orlando's Iron Bridge Regional Water Pollution Control Facility (WPCF) in July 1987. The OEW was permitted to provide additional treatment for an annual average daily flow of 0.90 m³/s (20 mgd). It is a constructed wetlands with three vegetative communities: a 166-ha (410-ac) deep marsh composed primarily of cattail (*Typha spp*) and bulrush (*Scirpus spp*) and designed to accomplish nutrient removal; a 154-ha (380-ac) mixed marsh composed of over 60 submergent and emergent herbaceous species designed to provide nutrient removal and wildlife habitat; and a 162-ha (400-ac) area originally intended to be a hardwood swamp with a herbaceous understory. This final area remains primarily composed of herbaceous plants with sparse areas of hardwoods. An approximately 36-ha (90-ac) lake is located within the hardwood swamp. The OEW is divided into 17 cells (Figure 1). There are three general flow paths or trains.

- Northern flow path – water enters at Cell 1, flows through Cells 3 and 4 to Cells 7 and 8, to Cell 13, then through Cell 17 to the final discharge point.
- Central flow path – water enters at Cell 2, flows through Cells 5 and 6, to Cells 9 and 10, to Cell 14, to Cell 16B, through the lake, and then through Cell 17 to the final discharge point.
- Southern flow path – water enters at Cells 11 and 12, flows through Cell 15, to Cell 16A, through the lake, and then through Cell 17 to the final discharge point.

The performance of the OEW has been better than expected. Effluent concentrations have been well below the permit limits of 2.31 mg/L TN and 0.20 mg/L TP. Nearly all of the nutrient removal has consistently occurred in the first 10% to 20% of the site. In the

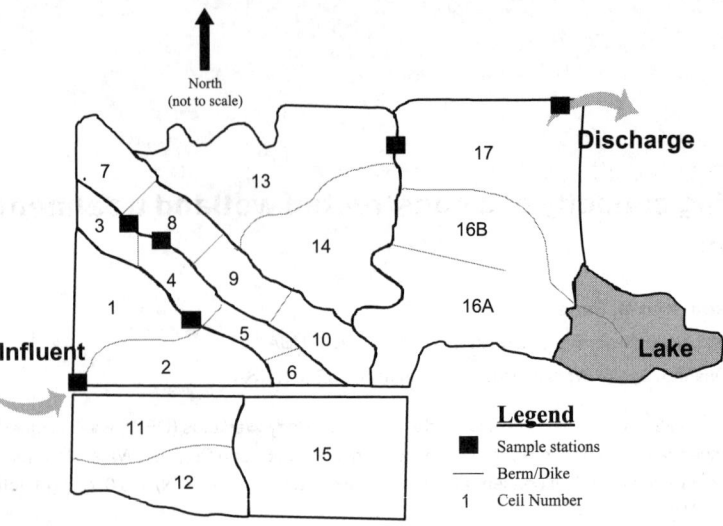

Figure 1 OEW site plan

mid-1990's, as the influent flow approached 0.90 m³/s (20 mgd), it appeared that the system could treat additional flow. In August 1997, a capacity testing study was initiated to assess the treatment capabilities of the OEW under various increased flow regimes.

Methods

Capacity testing of the OEW was accomplished by field simulation of higher flows through a portion of the system. The northern flow path was selected to receive the higher loading because it has the shortest and most direct path to the final discharge point. The northern flow path has been monitored since start up of the OEW, providing a good record of historical performance. The total area of the northern train is approximately 162 ha (400 ac), or about one-third of the total site.

To test the northern flow path at simulated higher flow conditions, the central and southern flow paths were removed from service completely by closing weirs at structures that separate each cell. Three simulated flow regimes were tested for a period of eight weeks each, as summarized in Table 1.

Four weeks prior to loading, the northern flow path was sampled on a weekly basis to establish pre-loading conditions (study background) to supplement the historical background data. Grab samples were collected at six sample stations at the discharge of the following treatment cells: 1, 3, 4, 7, 8, and 13. The OEW influent and discharge were monitored from continuous composite samplers used for permit monitoring. All samples collected were analyzed by the City's laboratory at the WPCF or in the field for the following parameters:

Table 1 Flows used in capacity testing

	Actual flow, m³/s	Simulated flow, m³/s
Pre-loading Background	0.86	N/A
Historical Average	0.56	N/A
Simulation 1	0.42	1.26
Simulation 2	0.56	1.66
Simulation 3	0.64	1.93

- PH (field
- Dissolved Oxygen (field)
- Conductivity (field)
- Temperature (field)
- Total Suspended Solids
- Total Dissolved Solids
- Turbidity
- Total and Ortho Phosphorus as P (TP)
- Total Nitrogen as N (TN)
- Ammonia Nitrogen as N
- Total Kjeldahl Nitrogen as N
- $NO_2 + NO_3$ Nitrogen as N

Once simulated loading of the northern flow path was initiated, weekly grab water samples were collected at the same stations and analyzed for the same parameters as the pre-loading sampling regime. Water quality sampling continued on a weekly basis until the capacity testing was finalized in March 1998.

Results and discussion
Simulation 1
Profiles of TN and TP concentrations during 1.26 m³/s (28-mgd) capacity testing are illustrated in the graphs on Figures 2 and 3, respectively. During capacity testing, TN concentrations were less than or equal to the pre-loading background and historical average. Nearly all the TN removal occurred within the first 450 m to 750 m (1,500 feet to 2,500 feet) of the influent structure. By approximately 450 m from the influent structure, most of the TP removal occurred. This is also consistent with the historical average, where most of the TP removal occurred within the first 600 m (2,000 ft) of the influent structure. In the remainder of the system, TP concentrations during the 1.26 m³/s capacity testing closely tracked concentrations during the pre-loading background.

Simulation 2
Profiles of TN and TP concentrations during the 1.66 m³/s (37-mgd) capacity testing are illustrated in Figures 4 and 5, respectively. TN concentrations did not stabilize in this second simulation until about 1,000 m (3,500 ft) from the influent, as compared to 450 m (1,500 ft) to 750 m (2,500 ft) under pre-loading background and historical average conditions. After 1,000 m, TN concentrations closely tracked concentrations during the pre-loading study. TP concentrations also did not stabilize until about 1,000 m from the influent

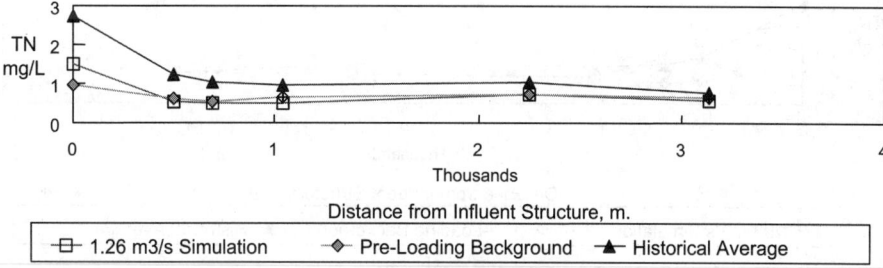

Figure 2 Total nitrogen profile through OEW during 1.26 m³/s flow simulation

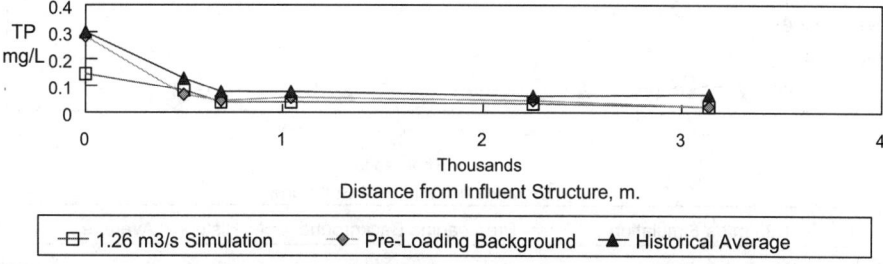

Figure 3 Total phosphorus profile through OEW during 1.26 m³/s flow simulation

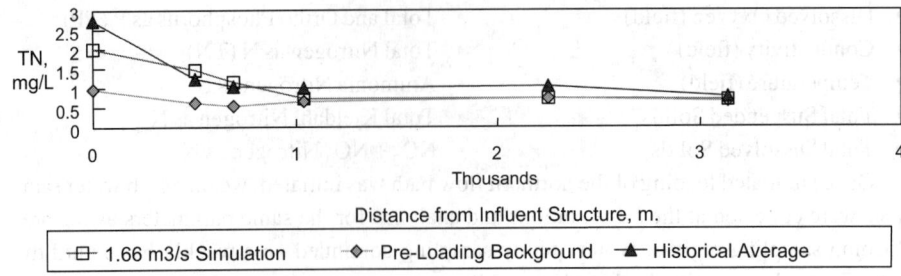

Figure 4 Total nitrogen profile through OEW during 1.66 m³/s flow simulation

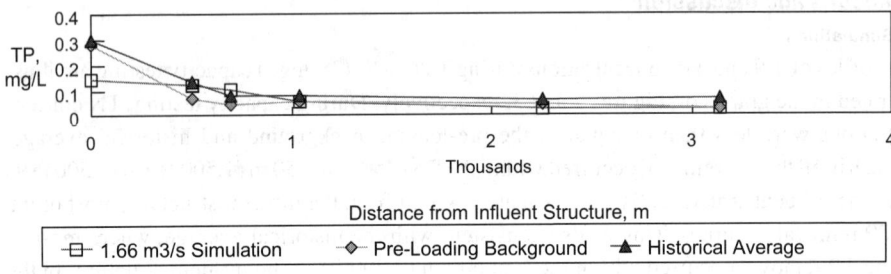

Figure 5 Total phosphorus profile through OEW during 1.66 m³/s flow simulation

structure, as compared to 600 m (2,000 ft) under pre-loading and historical average conditions. After 1,000 m, TP concentrations closely tracked concentrations during the pre-loading study.

Simulation 3

Profiles of TN and TP concentrations during the 1.93 m³/s (43-mgd) capacity testing are illustrated in Figures 6 and 7, respectively. The testing results are similar to the results of

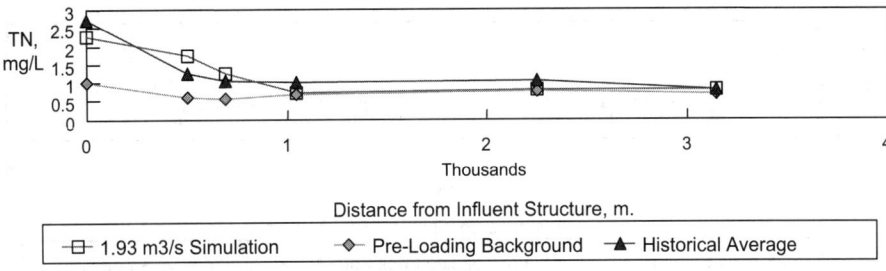

Figure 6 Total nitrogen profile through OEW during 1.93 m³/s flow simulation

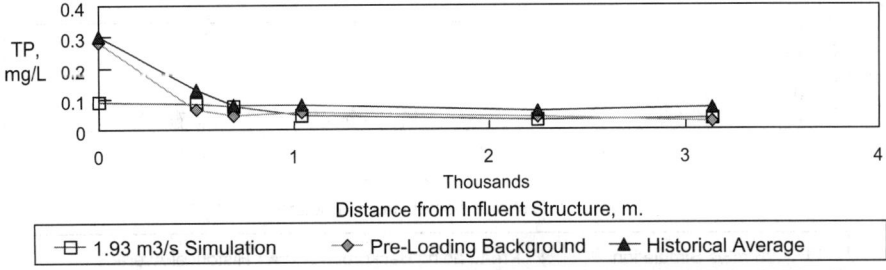

Figure 7 Total phosphorus profile through OEW during 1.93 m³/s flow simulation

the 1.66 m³/s testing. Nearly all of the TN and TP removal occurred by approximately 1,000 m (3,500 ft) of the influent structure. After 1,000 m, both TN and TP closely tracked the results of the pre-loading background testing.

Conclusions

At 1,000 m (3,500 ft) from the influent structure, reclaimed water has passed through approximately 36 per cent of the OEW northern flow path. Even under the highest flow simulation of 1.93 m³/s (43 mgd), nearly all of the TN and TP treatment appears to be occurring within this area. This correlates to a 68.4% reduction in TN and a 66.7% reduction in TP. Because so little of the system appears to be utilized, there is evidence that even higher flows or concentrations could be successfully treated. Under all flow simulations, final effluent TN and TP concentrations were less than historical average conditions and significantly less than permit limits.

A summary of nutrient uptake during the capacity testing compared to the original design, historical average, and pre-loading background conditions is presented in Table 2. For this comparison, the area utilized was calculated as the area upstream of the point where most of the nutrient removal occurred (point where the slope in the nutrient concentration profile changed). During the capacity testing, estimated TN uptake rates were greater than during pre-loading background and historical average conditions. TN removal efficiency appeared to improve at higher flows. TP uptake rates, however, were lower during the capacity testing than during preloading background and historical average conditions. This is thought to be due to the extremely low TP influent concentrations during capacity testing, which left very little TP available for uptake and removal.

If influent TN and TP concentrations remain consistent to that measured during the testing, it can be concluded that the OEW is capable of accepting and treating up to 1.93 m³/s (43 mgd). Based on these results, an intent to issue a permit to increase the capacity of the OEW to 1.57 m³/s (35 mgd) was received from state regulators in July 2001.

Table 2 Estimated TN and TP uptake during capacity testing compared to design, historical and background conditions

Simulated flow, m³/s	Influent mg/L	Effluent[1] mg/L	Uptake mg/L	Area[2] utilized, ha	Uptake[3], g/m²/yr
Total nitrogen					
Design	6.00	2.31	3.69	482	21
Historical average[4]	2.72	1.05	1.67	119	26
Background[5]	0.99	0.67	0.32	34	8
1.23	1.52	0.53	0.99	23	56
1.66	1.99	0.76	1.23	58	36
1.93	2.28	0.72	1.56	58	52
Total phosphorus					
Design	0.75	0.20	0.55	482	3
Historical average[4]	0.30	0.08	0.22	119	3
Background[5]	0.28	0.07	0.21	34	5
1.23	0.14	0.06	0.08	34	3
1.66	0.15	0.06	0.09	58	3
1.93	0.09	0.03	0.06	58	2

[1] Concentration at point in the OEW where most of the nutrient removal has occurred. Determined by location of change in slope in nutrient concentration profile graphs
[2] Area utilized is based upon the total area upstream of the point where most of the nutrient removal has occurred (se note (1))
[3] Uptake = (uptake concentration × actual flow)/area utilized
[4] 1987 through 1997
[5] Results obtained during background testing that preceded capacity study

References

PBS&J (1998). Compliance and Performance Review for the City of Orlando's Easterly Wetland Treatment System – '98 Monitoring Report. Prepared for the City of Orlando, Orlando, Florida.

Treatment of a molasses based distillery effluent in a constructed wetland in central India

S.K. Billore*, N. Singh*, H.K. Ram*, J.K. Sharma*, V.P. Singh*, R.M. Nelson** and P. Dass*

*Institute of Environmental Management and Plant Sciences, Vikram University, Ujjain 456010, MP, India
**Nelson Environmental Consultants, 211 Gregory Road, Wilmington, NC 28405, USA

Abstract A field-scale 4-celled, horizontal subsurface constructed wetland (CW) was installed to evaluate removal efficiencies of wastewater constituents in an industrial distillery effluent. Total and dissolved solids, NH_4-N, TKN, P and COD were measured. This CW design provides four serial cells with synthetic liners and a river gravel base. The first two unplanted cells provide preliminary treatment. Specific gravel depths and ensuing biofilm growth provides anaerobic treatment in Cell 1 and anaerobic treatment in Cell 2. Cell 3 was planted with *Typha latifolia* with an inserted layer of brick rubble (for phosphorus removal). Locally grown reed, *Phragmites karka* was planted in Cell 4. COD was reduced from 8420 mg/l 3000 from Cell 1 to the outlet of Cell 4. Likewise other parameters: total and dissolved solids, ammonium and total nitrogen, and total P, indicated declining trends at the 4-celled CW effluent. This study reveals how high strength distillery wastewater strongly impacts morphology, aeration anatomy in the chiseled plant tissues, reed growth; and composition of the biofilm in the specialized substratum. The reliability of a CW for organic and nutrients reduction, in association with a poorly performing conventional system is discussed. There is an immense potential for appropriately designed constructed wetlands to improve high strength wastewaters in India.
Keywords COD; constructed wetland; distillery effluent; methane; molasses; *Phragmites karka*; tidal flow system

Introduction

Use of constructed wetlands (CWs) now has been established as a method for treating domestic and low strength industrial wastewaters that are a low cost alternative to conventional systems. Over the past decade, there has been a growing appreciation of the multiple values and functions of constructed wetlands increasingly used for the treatment of a variety of wastewaters (Haberl *et al.*, 1995; Kadlec and Knight, 1996). In India, production of alcohol by fermentation from sugarcane molasses has been used for the disposal problem of molasses. The raw material for production of ethyl alcohol is 'molasses' which is the byproduct of the sugarcane industry. In nearly all distilleries, the bath fermentation mode is adopted with about 12–15 litres of spent wash generated per litre of alcohol produced. The untreated distillery effluent (UDE) is a high strength wastewater and can severely affect the environment if not properly treated. The UDE from the distilleries is characterized with a very high organic load with BOD and COD levels in the range of 35000–60000 mg/l and 60000–120000 mg/l, respectively. A routine conventional treatment system reduces the high organic load utilizing anaerobic biological digestion (biomethanation). Distilleries characteristically utilize biomethanation to generate methane (as fuel to compensate for energy needs). This is followed by conventional secondary treatment requiring continued aeration, which is an energy intensive process.

The performance of the conventional effluent treatment plant (ETP) is such that the outfalls of the secondary treatment system still contain high strength effluent quality. The general practice is to transfer the secondary treatment effluent (STE) into several open earthen lagoons for further natural treatment and/or to place on an open field for sun drying. Therefore, the present study was initiated for evaluating treatment performance of STE in a field scale horizontal subsurface flow (HSF) wetland to investigate removal efficiencies of organics in terms of COD & BOD and N-P. During this study special design considerations for the 'constructed wetland' are intended to provide: 1) an additional aeration system in the medium voids filled with influent and 2) insertion of a 'brick rubble band' in the gravel bed to enhance phosphorus removal from the distillery influent.

Molasses scenario

There are 285 distilleries in India producing 2.7 billion litres of alcohol and generating 40 billion litres of wastewater/vinasse, annually. The raw material for producing ethyl alcohol is the 'molasses', which is the byproduct of the sugarcane industry. During the production of sugar, about 0.40–0.45 tones of molasses per tonne of sugar is produced. Molasses production in India has increased from 2.46 million tonnes in 1984–85 to nearly 9.00 million tonnes in 2000 AD (Mall and Kumar, 1979). About 90% of molasses is used for the production of alcohol by fermentation with the remaining 10% used for cattle feed, foundries and manufacture of citric acid. The population equivalent of distillery wastewater based on BOD has been reported to be as high as 6.20 billion and clearly indicates that India's distillery waste contributes approximately 7 times more organic pollution than the entire Indian population. With little motivation to impose stringent standard on effluent quality, the untreated or partially treated effluents very often find access to surface and ground water courses. Marginal lands in the vicinity of distilleries are commonly sought to spread distillery effluents. The retention of effluents in several open earthen anaerobic lagoons and solar drying ditches are the primary sources for the odor one encounters while passing through a distillery (Joshi, 1999).

Methods

Site descriptions

The horizontal subsurface flow (HSF) CW was built for the distillery effluent emanating from a private distillery, Associated Alcohols and Breweries Ltd. The facility is located in Khodigram village in the outskirts of Baraha town in Central India. The distillery supports a conventional treatment system (CTS) with a wastewater capacity of 750 m^3/day. The distillery factory effluent is treated in two phases i.e. anaerobic digestion, which produces methane, followed by secondary treatment systems (the extended aeration system). The four-celled CW was constructed during June-July 1999 as a demonstration project to treat 10 m^3 distillery effluent per day. No earlier attempts have been made in India to apply CW technology for treatment of high strength distillery wastewater effluent. The distillery occupies more than 200 hectares of land. After secondary treatment, the effluent is retained in several shallow open lagoons for approximately two months to undergo natural treatment and air-drying. During this period, the waste is partially irrigated and re-irrigated after plowing on uncropped fields for sun drying. It is partly being used for crop irrigation after mixing with freshwater from the tube wells. The area climate is typically a dry tropical zone characterized with monsoons with annual rainfall amounts up to 800–900 mm. Rainfall occurs mostly (about 95%) in the rainy season i.e. June-September. The summer season's (March-mid June) day temperatures range between 40–45°C and between 56–62°C and between 56–62°C around the CW gravel beds.

CW design

Earlier successful experience with HSF for treating domestic wastewater (Billore et al., 1999) as compared to the free water surface (FWS) initiated the development of the HSF design for the distillery's high strength wastewater. In tropical climates, suitable year-round temperatures offer favorable conditions to support growth of emergent plants under an ever-availability of surface water in the gravel medium. A subsurface CW will function as a bioreactor combining the plants and the ensuing biofilm that actively grow in the gravel medium. Specific sizing of the CW as based on BOD (Cooper et al., 1996) was not followed in the present study for sizing formulation. For this design, sizing of the CW is based on a temperate climate (rather than tropical) in which the biological growth and CW treatment performance is inhibited due to the presence of year-round low temperatures. A four-celled rectangular earthen CW (surface area 364.21 m^2) with the daily effluent treatment capacity of 10 m^3 (approximately 3.64 hectares 1000 m^{-3}) was installed on the natural terrain near the distillery end-of-pipe final effluent discharge after secondary treatment (ST). The effluent after ST was regulated to enter into a pretreatment chamber (PC) before entering into the four-celled CW. Figure 1 indicates the design for two parallel CWs; only the upper 4-celled CW was operational in the first phase.

Pretreatment

Distillery effluent even after ST contains solids and floating debris. An open three-partitioned pretreatment chamber (size 5.5 × 3.4 × 2.5 m) was constructed to receive a regulated portion of the ST effluent. The bottom of the first compartment of the PC was filled with round gravel (8–12 cm) for half the depth for development of the biofilm and to capture suspended and dissolved solids. The second chamber had a baffle of a half partition wall to allow the transfer of only the supernatant to the third chamber in the PC. The settled sludge was manually removed routinely from the bottom of the first chamber.

Cell-1

This cell was designed to provide anaerobic treatment to the influent and to provide greater detention time for better settlement of organics. The cell had a volume of 148.6 m^3, medium porosity 45%, water retention of 66.87 m^3, and a residence period of 6.6 days (for a daily flow of 10 m^3 influent). To prevent clogging and to maintain proper hydraulic conductivity in the gravel medium of the cell for water flow, a 'desludging' facility was installed for frequent removal of the periodically settled bottom sludge by suction.

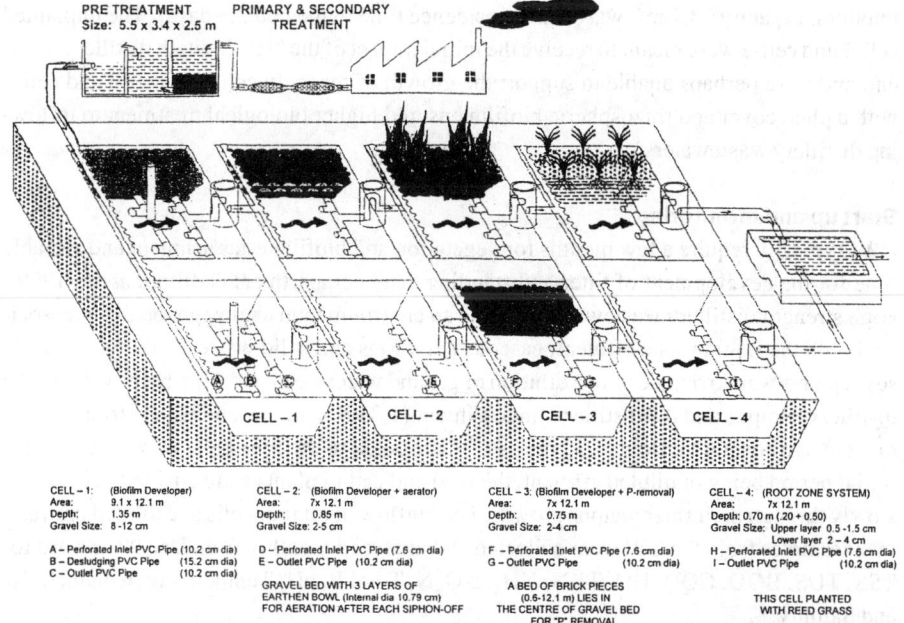

Figure 1 Design of 'constructed wetland' for Barwaha Distillery wastewater

Cell-2 as a Tidal Flow System (TFS)

This special system was installed to provide extra aeration (other than that routinely provided by voids in the gravel medium) to water in the voids and for the biofilm in the gravel medium. Besides round gravel (size 2–5 cm) the bed had 6200 inverted porous 'baked earthen bowls' (BEB) or 'earthen lamps' arranged horizontally in three tiers one above the other in the bottom among the travel upward to 15 cm (Figure 1). Each BEB had an internal diameter of 01.79 cm with 200 cc of air holding capacity. The porous earthen bowls (depth 6.35 cm) were manufactured manually by kneading local yellow clay, horse-dung and coal ash in the proportion of 4:1:0.5. The hand made bowls are air-dried for three days and baked overnight in a coal-fired furnace. The horse-dung particles after baking resulted in perforation of the earthen wall with fine micro-pores. The cell had a volume of 71.99 m^3, medium porosity 40%, water retention capacity 28.7 m^3, and a residence period of 3 days (for a daily influent flow of 10 m^3).

For TFS, the plumbing arrangement in the end pipe of the cell has been designed to siphon all the wastewater from cell-2 into cell-3, as soon as the water level reaches just below the gravel surface in the former. The completely drained water allows the gravel bed voids to fill with the air completely including air pockets in the 6200 BEB. The influent flow takes three days to completely fill cell-2 until again it siphons into cell-3 in a 12-hour period. This phenomenon has also been reported as 'tidal flow reed beds' (TFRB), a new system gaining attention in recent years (Sun *et al.*, 1999).

Cell-3 and Cell-4

Ninety rhizome pieces from *Typha latifolia* (cattail) plants were sown in cell-3 on 20 September 1999, at equidistance. These rhizomes were obtained from 6-year old floating tussocks of cattail in the wetland garden of the Botany department, Vikram University. The cell (gravel depth 0.75 m, volume 63.52 m^3, bed porosity 40%, water retention capacity 25.41 m^3, water residence time 2.5 days, influent flow 10 m^3 day^{-1}) had a wide strip of brick pieces across the width (size 0.6 m × 12.1 m) to adsorb phosphorus in the influent. Cell-4 supports *Phragmites karka* sown on 19 August 1999 (3 months old nursery raised 138 plants) in two-layered gravel medium (cell volume 59.27 m^3, porosity, *ca* 40%, water retention capacity 23.7 m^3, wastewater residence time in the bed 2.3 days). The unplanted cell-1 and cell-2 were meant to receive the initial impact of the high strength distillery influent, and were perhaps unable to support the growth of reeds. In contrast, cell-3 and cell-4 with a plant cover and rhizospheric biofilm ensured higher biological treatment to inflowing distillery wastewaters.

Start up and monitoring

CWs typically require a few months for vegetation and biofilm establishment and sizeable time for the development of litter and standing dead compartments (Billore *et al.*, 1999). High strength distillery wastewater presents an environmental challenge locally as associated with the liquid waste effluent, gaseous emissions and solid waste. Initially the newly sown plants were irrigated with a dilution of ground water extracted from bore-wells on the distillery campus and the distillery final effluent (1:3 effluent: ground water) from Dec 99 onward at a constant inflow of 10 m^3 day^{-1} distributed to the four-celled CW. During this initial period being of diluted effluent, the reed and cattail plants were enabled to establish slowly through their rhizomatous growth. The outflow of all four cells and diluted pretreatment samples from the CW were collected and analyzed monthly from Dec 99 onward for TSS, TDS, BOD, COD, DO, TKN, NH_4, NO_3N, Total P, pH, Temperature, Conductivity and Salinity.

Results and discussion

The CW system has been in operation and monitoring since Dec 99. Each cell contains inflow and out-flow structures required to provide even distribution of water, reduction in channelization, and specific treatment of pollutants, case by case. The bottom of each cell is sealed initially by yellow clay (Billore et al., 2000), overlaid with a LDP synthetic liner, one percent slope, and sufficient freeboard to contain stormwater during the rainy season, and to protect the tall reed and cattail plants from strong winds. The CW gravel media was from the nearby flowing perennial River Narmada. This gravel is round and smooth and of the finest quality for full development of a biofilm. Except for color removal, the treatment performance of the 4-celled CW has been showing promise, initially by treatment of the major wastewater pollutants. A total of 14.4 days was regulated as the retention period when pretreated effluent enters cell-1 until it leaves as effluent from cell-4 for renovation.

Plant growth: Reed and cattail

In fact, the distillery effluent (spent wash) has a highly polluting nature containing almost all the characteristics of original molasses except the fermentable sugar. Establishment of a successful 'root zone' system in the gravel bed was a crucial 'acid test' in the application of CW technology for treating high strength distillery effluents. Of highest importance is the survival of reed and cattail plants under the influence of the strong organic loading, solids, nitrogen, and soluble salts (sulphates, chlorides). Dilution of the distillery effluent was a prerequisite to ensure growth of the reed and cattail plantation in cell-3 and cell-4. The distillery effluent even after the conventional secondary treatment contained a BOD as high as 2539 and a COD Of 13866 mg/l^{-1}. The plant growth responses indicated 7 to 12 times increase in shoot and belowground (roots and rhizomes) biomass in both the planted species of CW when they achieved a growth period from 2 to 5 months. In contrast, the plant morphology of Reed grass initially exhibited the impact of a stressed environment in the habitat due to the influent loading. The length of tiller internodes and of leaves was 3.3 and 4.4 times smaller during the 5-month growth period, as compared to recorded observations at 8 months. Under the stressed environment, the leaves were plentiful in number (188 per tiller) on the grass, but appeared as small green spines (length 5.2 cm) as a rosette on shortened stems (internode length 2.8 cm). However, the stressed morphological indicators disappeared (recorded in the 8 month observations) when new growth in the Reed grass assumed healthy leaves (length 23.1 cm) and higher internode size (length 9.2 cm) in each tiller.

Solids, COD and BOD

The percentage removal of end parameters for solids (TS, TSS, TDS), COD, BOD, nitrogen species, phosphorus, conductivity and salinity parameters from the individual cell of the CW treatment system are presented in Figure 2. Among the solids, dissolved solids were 89% with suspended solids comprising the remaining portion. Decreases in solid concentrations to 6337 mg/l^{-1} (about 40%) were observed in the outflow from cell-4 with 80% of the removal occurring in the macrophyte planted cell-3 and cell-4 of the CW. Plants seem to constitute a substratum for the fixation of decomposing microorganisms that act as filters for the dissolved organic matter.

Traditionally, organic matter in the wastewater is characterized by COD, BOD and TOC and often divided into a particulate and dissolved fraction. The spent wash contains numerous amounts of dead microbial biomass, besides volatile organic acids due to fermentation (Poggi-Varaldo, 1992). In addition, the byproduct remaining after distillation during the production of alcohol is characterized by a high concentration of organic acids and polyphenols (Lalov et al., 2000). These are the reasons for the relatively high values of the

COD in the liquid distillery effluent. Cell-1 of the CW having a maximum depth is performing as an 'anaerobic digester' removing significantly, 44% of the COD for a total removal of 64%. Table 1 reveals the results of preliminary monitoring completed during January 2000 for methane fluxes as an indicator of anaerobic/aerobic in all four cells, clearly indicating highest methane flux, i.e. anaerobic stage in cell-1 (442.16) compared to the relatively aerobic stage in cell-2 (102.26); plant mediated aerobic cell-3 (56.64) and cell-4 (22.08 CH_4.mg m-day-1) respectively, in the CW.

Cell-1 and cell-2 have been left unplanted for full development of a biofilm on the round, smooth-surfaced river gravel. The measurement of the COD as the oxidisable matter in the biofilm is an indirect estimation of the fixed biomass of total biofilm amount (Brayers and Characklis, 1981). Wastewater biofilms are very complex systems consisting of microbial cells and colonies embedded in a polymer matrix whose structure and composition is a function of biofilm stage and environmental condition (Lazarova and Manem, 1995).

Nitrogen and phosphorus removals

The highest removal of organic and ammonium nitrogen which are present in significant amounts in the pretreatment effluent after the dilution, occurred in the unplanted anaerobic cell-1 and cell-2 in the CW. However, ammonium N continues to be removed in subsequent planted cell-3 and cell-4 in a slight lower magnitude (Figure 2). Nitrate nitrogen which is very low in concentration in the pretreatment effluent (11.9 mg/l^{-1}), and initially removed by 38% in the anaerobic cell-1, formed again in cell-2 having the 'tidal flow system' and subsequently removed considerably in cell-4 with the reed plantation. Differences in per cent removal in nitrogen species can be explained by the residence time in the cell, aerobic-

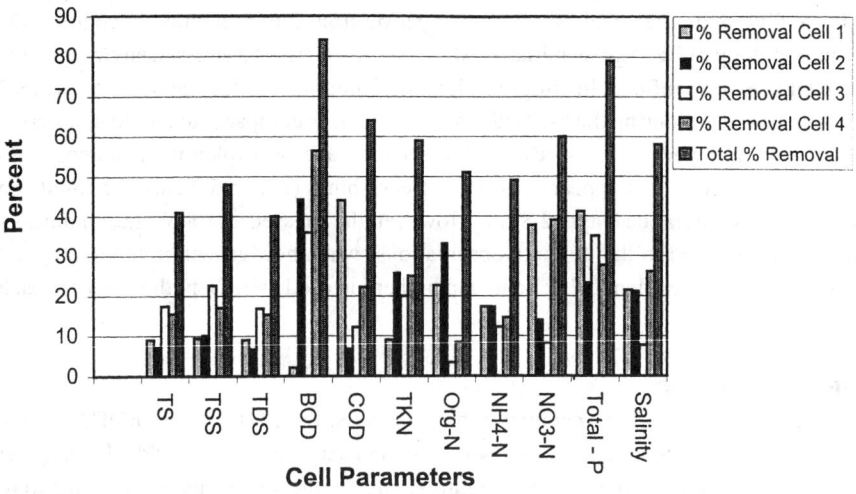

Figure 2 CW cell-wise % removal of distillery effluent parameters

Table 1 Preliminary study on methane emission from different cells in the constructed wetland for the treatment of distillery effluent (January 2000)*

Cell No.	METHANE FLUX			
	CH_4.mg/m^{-2} h^{-1}			CH_4.mg/m^{-2}day^{-1}
	Morning	Evening	Average	
1	11.62	23.55	17.59	442.16
2	3.72	4.79	4.26	102.26
3	1.66	3.06	2.36	56.64
4	0.82	1.01	0.96	22.08

*measured by close chamber, GC FID column

anaerobic controlled nitrification-denitrification pulse (Reddy et al., 1989), plant uptake, volatilization (Billore et al., 1994), or biofilm immobilization. The surface temperature of the gravel bed in each cell with temperatures reaching 56–62°C during the summer season (March to mid-June) significantly enhances ammonia removal through volatilization. The HSF constructed wetland in the present study was capable of removing 50–60% of the nitrogen species within a short span of time after its operation. This occurred before full development of a biofilm in the unplanted gravel-bed cells and in the reed-bed-rhizospheres of poor cattail and reed covers. For stepwise nitrate nitrogen reduction in each cell through denitrification, the endogenous carbon sources, such as ethanol, acetate and sugars were already in the distillery effluent to affect the removal process (Werner and Kayser, 1991). Methane may also present a possible sole carbon source for denitrification. Methane production during wastewater anaerobic treatment has been assumed to be 0.25 kg $CH_4.kg^{-1}$ BOD removed (Thalasso et al., 1997).

Cell-2 in the CW has been characterized by special facility of supplementing air into the gravel bed by a 'tidal flow system' (TFS), an eco-engineered process that was further improved in the present study by introducing a more porous and diffusive reed bed. Results clearly illustrate an increase in DO (233%) and nitrate concentration (14.20%) in the out flow of cell-2. In the TFS, the bed is alternately filled with water and then drained completely. During the water filling process, the air is expelled and the reed bed matrix is gradually submerged. Once the whole bed is saturated, the water begins to drain. Fresh air is drawn positively from the atmosphere into the bed during the draining process. Thus, the water behaves similarly to the action of the 'piston' in an air compressor (Sun et al., 1999). Therefore, the air drawn into the TFS is used to remove the pollutants, enable maximum media-water contact, and the problem of poor water distribution is avoided (Sun et al., 1999). Likewise, the cattail-planted cell-3 has been improved horizontally by inserting a wide strip of 'brick rubble' across the gravel medium to remove the phosphorus from the distillery influent by surface adsorption. Figure 2 displays a relatively higher removal of phosphorus (35%) from the 'brick rubble' banded cell, compared to cell 2 and 4. This supports the sorptive role of an adsorbent low cost inert substratum besides other removal mechanisms involved (Richardson, 1985; Mann, 1990), thus removing a total of 79% phosphorus from the distillery influent in the CW.

Conclusion

Conventional wastewater treatment technologies are generally based on highly optimized physical, chemical and microbiological processes. Conventional technologies may not offer affordable solutions to a developing country like India for achieving the ultimate "end-of-pipe" purification of wastewater effluents eventually discharged to India's rivers and streams. The molasses based distillery is an immensely water intensive process. The present study to treat distillery effluent from the conventional secondary treatment, has demonstrated the potential for application of 'constructed wetland' technology in association with a poorly performing conventional treatment process. The constructed wetlands also provide eco-friendly treatment alternatives, instead of large unlined lagoons that are built without proper engineering and environmental care for wastewater storage and subsequent groundwater contamination.

Acknowledgement

The authors are grateful to Mr. H.K. Bhandari, Director, Associated Alcohols & Breweries Ltd. in appreciation for finding this project and for his constant encouragement.

References

Billore, S.K., Singh, N., Sharma, J.K., Dass, P. and Nelson, R.M. (1999). Horizontal subsurface flow gravel bed constructed wetland with *Phragmites karka* in Central India. *Wat. Sci. Tech.* **40**(3), 163–171.

Billore, S.K., Singh, N., Sharma, J.K., Krishnamurthi, R., Kobayashi, T. and Yagi, R. (2000). Ujjain clay as low cost sealant and liner for artificial ponding and bentonite alternative. *Current Science* **78**(11), 1381–1383.

Cooper, P.F., Job, G.D., Green, M.B. and Shutes, R.B.E. (1996). *Red beds and constructed wetland for wastewater treatment*. WRc Swindon. UK, 184 p.

Haberl, R., Perfler, R. and Mayer, H. (1995). Constructed wetlands in Europe. *Wat. Sci. Tech.*, **32**(3), 305–315.

Joshi, H.C. (1999). Bio-energy Potential of Distillery effluents. *Bio Energy News* **3**(3), 1–8.

Kadlec, R.H. and Knight, R.L. (1996). *Treatment Wetlands*. CRC Press/Lewis Publishers, Boca Raton, Florida.

Lalov, I.G., Guerginov, I.I., Krusteva, M.A. and Fartsov, K. (2000). Treatment of wastewater from distilleries with chitosan. *Wat. Res.* **34**, 1503–1506.

Lazarova, V. and Manem, J. (1995). Biofilm characterization and analysis in water and wastewater treatment. *Wat. Res.* **29**, 2227–2245.

Mall, I.D. and Kumar, Vivel (1997). Removal of organic matter from distillery effluent using low cost adsorbent. *Chemical Engineering World*, **32**(7), 1–9.

Mann, R.A. (1990). Phosphorus removal by constructed wetlands: Substratum adsorption. In: *Constructed Wetlands in Water Pollution Control*. Cooper, P.F. and Findlater, B.C. (eds), Pergamon Press, Oxford, UK, pp. 97–105.

Poggi-Veraldo, H.M. (1992). A comparison of treatments for high strength distillery slops from the sugar cane industry. In: 47 *Purdue Industrial Water Conference Proceedings*, Lewis Publishers, Inc., Chelsea, Michigan, USA, Chapter 80: 789–799.

Reddy, K.R., Patric, W.H. and Lindau, C.W. (1989). Nitrification-denitrification at the plant root-sediment interface in wetlands. *Limnol. Oceanogr.*, **34**, 1004–1013.

Sun, G., Gray, K.R., Biddlestone, A.J. and Cooper, D.J. (199). Treatment of Agricultural wastewater in a combined tidal flow-down flow reed bed system. *Wat. Sci. Tech.* **40**(3), 139–146.

Thalasso, F., Vallecillo, A., Garcia-encia, P. and Fdz-polanco, F. (1997). The use of methane as a sole source for wastewater denitrification. *Wat. Res.*, **31**, 55–60.

Werner, M. and Kayser, R. (1991). Denitrification with biogas as external carbon source. *Wat. Sci. Tech.*, **23**(4–6), 701–708.

Use of constructed wetlands for acid mine drainage abatement and stream restoration

F.J. Brenner

Biology Department, Grove City College, Grove City, PA 16127, USA

Abstract Constructed wetlands have been used for over two decades for the treatment of acid mine drainage (AMD). Through a variety of physical, chemical, biological processes, these wetlands are effective in reducing acidity and removing up to 99% of iron and aluminium from AMD, but they only remove 20–30% of the manganese loading. The Slippery Rock Creek watershed in northwestern, Pennsylvania has been adversely impacted by acid mine drainage (AMD) for over 100 years with 74 mine discharges contributing a total of 1,228.8 kg, 282 kg and 69 kg/day of sulfuric acid, iron and aluminium, respectively to receiving streams. In the Slippery Rock Creek Watershed, aerobic and vertical flow wetlands, along with limestone drains and vertical flow limestone beds help to restore acid mine drainage impacted streams. Since 1995, the construction of seven passive treatment systems currently contribute 192.8 kg of alkalinity and remove 39% of the acid loading to a 4.83 km section of Slippery Rock Creek. When the eight passive treatment currently under construction are in operation, it is anticipated that there will be an additional 34.7% reduction in acidic loading to streams within the watershed. The cost of restoring all streams currently impacted by acid mine drainage within the Slippery Rock Creek watershed is currently estimated at $8,929,500.
Keywords Acid mine drainage; limestone drains; watershed restoration; wetlands

Introduction

Discharges from abandoned coal mines are major contributors to the degradation of surface and groundwater systems throughout the mining regions of the world. In Pennsylvania alone, over 4,050 km (2,400 miles) of streams are currently being degraded by acid mine drainage. In the absence of neutralizing compounds, these drainages are acidic, as well as containing high concentrations of iron, manganese, aluminium and sulfate. Over the last 20 years, several different passive treatment technologies, including constructed wetlands, limestone drains and vertical flow wetlands, have been employed to abate the impact of acidic discharges on freshwater streams. The constructed wetland concept originated from observations that natural *Sphagnum* wetlands receiving AMD often improved in water quality without any visible ecological effects on the wetland system (Huntsman *et al.*, 1978; Wieder and Lang, 1986). Although problems may occur with these wetland systems, current studies indicate that overall they are more cost effective and ecologically beneficial than conventional chemical treatments (Brenner *et al.*,1993).

For over four decades, the impacts of acidic discharges have been addressed on a site by site basis without consideration of their cumulated impacts on a watershed. Watershed restoration should involve the identification of the major problem areas and then the application of the best management practices to improve water quality and increase biological diversity in the impacted streams. In the current paper, aerobic and vertical flow wetlands will be examined for their effectiveness in the abatement of acidic discharges and their impact on receiving streams as they relate to watershed restoration.

Historical perspective and construction
Aerobic wetlands

Initially, most of the research on constructed wetlands focused on utilizing *Sphagnum* and peat as a medium for removing acidity and metals from mine drainages. Although the early laboratory results were promising (Kleinmann et al., 1983; Burris et al., 1984; Gerber et al., 1985), almost all constructed *Sphagnum* wetlands ceased to be effective in improving water quality after several months. According to Hedin (1989), these wetlands failed for a variety of reasons including the stress related to transplanting *Sphagnum*, abrupt changes in water chemistry, excessive or insufficient water depth, and iron toxicity. As the result of this early work, almost all acid mine drainage wetlands are planted with cattails (*Typha latifolia*) since they have been shown to be tolerant of a wide range of water conditions (Sencindiver and Bhumbla, 1988; Samuel et al., 1988; Brenner, et al., 1993, 1995). Although numerous wetlands were constructed during the decade between 1970 and 1980, little if any research was available to enable the designer to determine the size required to remove the acidity and metals. The size of the wetland was thought to be a function of flow volume without regard to acidity and metal loading into these systems. Wetlands were generally constructed based on the concept that between 5 to 15 m^2 of wetland were required per L/min flow to remove acidity and metals for mine discharges (Kleinmann et al., 1983). These sizes were also based on discharges with pH>4.0 and iron <50 mg/l and the more acidic discharges and/or those with high iron and manganese concentrated were chemically treated. Hedin (1989) indicated that for discharges with a pH <3.0 and iron concentrations exceeding 150 mg/L, the incorporation of metal loading is necessary in the design criteria. These wetlands were generally designed using the criteria of 0.1 to 2.0 m^2 of wetland per kg/Fe/yr, but most systems had ratios of less than 0.5 m^2 per kg Fe/yr (Hedin, 1989). In addition to being undersized, many of these wetlands became channelized; reducing the area of the wetland for treatment (Brenner et al., 1993, 1995). In 1991, Hedin proposed the size of the wetland required to treat acid mine drainage be based on the following criteria: Wetland Size = iron load/10 + manganese load/0.5. This formula is based on the assumption that the entire surface area is being used as a treatment system and does not address the problem of channelization. Brenner and Pruent (1999) indicated that a serpentine design reduced the problem of channelization and leaves the entire wetland surface available for treatment.

Limestone drains and channels

Within the last decade, anoxic limestone drains (Hedin and Wetzlaf, 1994) and limestone channels (Skousen, 1977; Brenner and Pruent, 1999) have been used in conjunction with aerobic wetlands to treat acidic mine discharges. In these systems, acidity is reduced resulting in increased pH and retention of ferric iron and aluminium, allowing the wetlands to operate more effectively in removing additional metals. The mass of limestone required for treatment may be calculated by the following equation: $M = Q p_b t_d / V_v + Q C T/x$ where Q is the volume of water flow, p_b is the bulk density of the limestone, t_d is the desired retention time (14–23-hrs), V_v is the bulk void expressed as a decimal, C is the predicated concentration of alkalinity in the effluent, T is the designed life of the system and x is the $CaCO_3$ of the limestone expressed as a decimal (Hedin and Wetzlaf, 1994).

Vertical flow wetland systems

In 1994, Kepler and McCleary proposed a vertical flow wetland, termed Successive Alkalinity Producing System (SAPS) that theoretically was designed to be a continual source of alkalinity to treat acidic mine drainages. However, the design and effectiveness of these systems was based on limited data obtained over a period of one year or less. In the three systems designed by Kepler and McCleary (1994), the total area and volume varied from 150 m^2 to 1500 m^2 (10 m × 15m × 1.5 m) and 225 m^3 to 2250 m^3 (60 m × 25 m × 1.5 m) and averaged 534 m^2 and 801m^3 (26.7 m × 20 m × 1.5 m), respectively. The substrate in all three systems consisted of 45 cm of mushroom compost and between 45 and 60 cm of 1.3 to

1.9 cm diameter limestone with an average of 1.8 m (1.9–1.8 m) of free standing water over the substrate. In the first year of operation, all three systems were effective in removing acidity and iron from the discharges, but the authors did not report the effectiveness of these systems in manganese removal (Kepler and McCleary, 1994). A follow up study of these systems by Demchak (1998) indicated that all these systems began to fail after 18–24 months due to limestone depletion and iron accumulation in the compost. Demchak (1998) also reported that these systems do not remove manganese from acidic mine discharges and that their effectiveness in treating AMD varies seasonally, especially as the systems age. Studies using scale models of these systems also indicated that their effectiveness in removing acidity and metals from AMD also declined in the second year of operation due to limestone depletion and plugging of the compost by iron oxides. Several vertical flow systems are being designed with a PVC pipe distribution system that can be back flushed if plugging occurs and based on preliminary studies, these systems have to be flushed approximately at 6 to 8 month intervals (Brenner et al., 2000).

Wetland processes in acid mine drainage
Aerobic wetlands
Several different physical, chemical and biological processes are involved in the chemical changes in mine waters as they flow through constructed wetlands. Although some dilution occurs through the addition of surface and groundwater in the wetlands, this is probably a minor component in the reduction of acidity and heavy metals. Wieder and Lang (1986) reported that the substrates used in wetland construction may remove metals from mine discharges via a process of absorption, chelation and cation exchange. However, detailed studies on these mechanisms have not been completed But in most constructed wetlands, the capacity of substrate to remove heavy metals should be exhausted within several months of construction. A considerable amount of the original research on constructed wetlands focused on metal accumulation in plants (Burris et al., 1984; Gerber et al., 1985) and, although *Sphagnum* has been shown to accumulate iron, metal accumulation reaches toxic levels within a single growing season (Spratt and Wieder, 1988). The common cattail (*Typha latiflora*) has been shown to be more tolerant of mine drainage than *Sphagnum*, probably because it does not accumulate metals to toxic concentrations (Sencindiver and Bhumbla, 1988). Because of the lack of metal accumulation, the role of cattails as an iron sink is negligible (Hedin, 1989) usually less than one per cent of the annual iron loading (Sencindiver and Bhumbla, 1988; Brenner et al., 1993, 1995). It has been speculated that blooms of *Oscillatoria* (Kepler, 1986), *Microspora, Oedogonium* (Dionis and Stevens, 1985) and *Ulothrix* (Robbins et al., 1999) have been associated with decreased manganese concentrations. Attempts at periodic fertilization of open water wetlands to stimulate algal blooms for enhancing manganese removal are confounded by abiotic metal removal reactions that occur when phosphate is added to AMD (Hedin, 1989). Overall algal removal of manganese is probably a minor component in wetland function, except when manganese occurs in low concentrations. Based on the assumption that algal productivity is 2,500 g/m/yr and if manganese accumulation by algae approaches 50,000 μg/g, then 4.2 m^2 would be required to remove a kg Mn annually (Hedin, 1989).

The most accepted theory for the removal of metal by wetlands involves oxidation and hydrolysis reactions resulting in the precipitation of metals. Wieder and Lang (1986) reported that 93 per cent and 27 per cent of iron and manganese accumulation, respectively, was in oxidized forms. In aerobic wetlands, abiotic processes and bacterial action oxidize hydrolyzed ferrous iron forming iron oxyhydroxides, as well as the precipitation of aluminium oxides. In wetlands with a pH greater than 6, abiotic precipitation of iron and aluminium compounds occurs rapidly. Therefore, the use of either anoxic limestone drains or limestone channels to reduce acidity and increase pH in conjunction with aerobic wetlands will enhance iron and metal removal. In addition, the longer the retention time of the acidic discharge within the wetland, the greater will be the precipitation of iron and aluminium compounds. But in drainages with a pH less than 6, these abiotic reactions are reduced and the iron oxidizing bacteria become an important component of wetland function (Kleinmann and Crerar, 1979; Brenner et al., 1995). These chemoautotrophic and chemoheterotrophic bacteria have been reported to increase the oxidation of ferrous iron, thereby enhancing the formation of iron precipitates. Brenner et al. (1995) reported a greater amount of iron oxidizing bacteria occurred in association with cattail rhizosomes than elsewhere in the substrate suggesting increased iron oxidization activity. The increase in oxidation associated with the rhizosphere may be due to a combination of plant induced oxygenation and iron bacteria (Sencindiver and Bhumbla, 1988; Brenner et al., 1995). These authors further stated that higher bacteriological activity occurred at pHs between 5.5 and 6.5 than at pHs <5.0, again indicating the value of limestone drains or channels to increase the pH prior to discharging AMD into aerobic wetlands.

Although numerous studies have demonstrated the effectiveness of aerobic wetlands in removing in excess of 95 per cent of iron and aluminium from mine discharges (Brenner et al., 1993, 1995; Brenner and Pruent, 1999; Brenner et al., 2000), they are only effective in removing between 20–30 per cent of manganese from AMD (Brenner et al., 1993, 1995; Demchak, 1998; Brenner and Pruent, 1999) According to Owens (1963), manganese oxidation is limited in water with a pH of less than 9.5 and in chemical treatment systems, some removal of manganese occurs due to coprecipitation with iron oxyhydroxides (Watzlaf, 1988), but in constructed wetlands, manganese oxidizing bacteria and fungi appear to be the primary source of manganese removal from AMD (Brenner et al., 1995; Robbins et al., 1999). As with iron bacteria, manganese activity was greater in rhizosphere of cattails.

In addition to iron and manganese oxidizing bacteria, Brenner et al. (1995) isolated the anaerobic sulfur reducing bacteria from both water and substrate in constructed wetlands. In anoxic environments, these bacteria use sulfate to oxidize organic matter and release, as a waste product, carbonate resulting in additional alkalinity and hydrogen sulfate which ionizes to bisulfate, which combines with ferrous iron and perhaps manganese, forming insoluble sulfate compounds. Although numerous studies have demonstrated the importance of bacteria in the oxidation and reduction processes occurring in AMD treatment wetlands, little is known about the ecology, physiology and temporal and spatial relationships of these micro-organisms in these systems.

Vertical flow wetland systems

The processes occurring in vertical flow wetlands (VFWS) are similar to those described for aerobic systems with alkaline addition occurring by the dissolution of limestone and the sulfate reduction and precipitation of iron and aluminium in the substrate. Although VFWS will remove up to 99+ percent of the iron and aluminium (Brenner et al., 2000), they are not effective in removing manganese (Demchak, 1998). It is interesting to note that in their original one year study, Kepler and McCleary (1994) did not address the removal of manganese by their systems. In theory, these systems function anaerobically, but in a recent study by the author, iron, manganese and sulfate bacteria were isolated under both anaerobic and aerobic conditions suggesting the occurrence of both oxidative and reduction processes in VFWS (Brenner et al., 2000). Despite the promising first year results reported by Kepler and McCleary (1994), systems based on their design begin to fail after 18–24 months of operation (Demchak, 1998) and similar results occurred in scale models of these systems (Brenner et al., 2000). When these scale models were dissected, the upper third of the substrate was plugged with iron precipitates. The efficiency of these systems may be improved by modifying the design with limestone drains and/or an aerobic wetland to remove the iron and aluminium prior to the VFWS. This would reduce the accumulation of iron and aluminium thereby, allowing the VFWS to provide additional alkalinity to enhance manganese removal. Although VFWS have only been used in the treatment of AMD for a few years in combination with aerobic wetlands and limestone drains with some design modifications, they may be an effective method of treating acid mine drainages.

Watershed restoration

The ultimate goal of the installation of AMD passive treatment systems is restoration of watersheds impacted by acidic mine discharges. In the Slippery Rock watershed in northwestern Pennsylvania. a total of 74 mine drainages have been identified within the watershed and of these, 59 are contributing acid loading to receiving streams with approximately 90% of the acid loading occurring from 35 discharges. The total acid loading in the watershed is 1288.8 kg/day and iron and aluminium loading averages 282 and 69 kg/day, respectively. Since the initiation of the restoration efforts within the Slippery Rock Creek watershed in 1995, seven passive treatment systems have been installed which are currently removing 39% of the acid loading to a 4.83 km section of Slippery Rock Creek. In addition, during peak flows, these systems contribute approximately 192.8 kg/day of alkalinity to the stream systems. The passive treatment systems installed in this portion of the watershed include two anoxic limestone drain/wetland systems, two vertical flow wetlands (VFWs), one retention pond/wetland system, one ALD and one retention pond with a total cost of $272,000. In addition, reclamation efforts in the watershed have restored approximately 32.4 ha of an abandoned coal tipple, refuse piles and open pits through the application of fly ash and landowner reclamation programs.

An additional, eight passive treatment systems, including four anoxic limestone drain/wetland systems, and four VFWs, are proposed for the watershed at a total cost of $712,500. When these projects are completed, there will be an anticipated acidity reduction of 34.7% to Slippery Rock Creek. The severity of the iron and aluminium loading in the watershed reveals that passive treatment alone may not be sufficient for long term metal reduction. It is anticipated that an additional $5,029,500 would be required for continued alkaline addition, along with passive treatments to remediate the 50 point source discharges and improve water quality in the affected 50.6 km of streams in the watershed. In addition, an estimated $2,943,000 and $954,000 would be required for abandoned surface mine and refuse pile reclamation, respectively. Based on these figures, an estimated cost of $8,926,500 would be required to implement the entire reclamation/remediation plan for the Slippery Rock Watershed.

Conclusions and recommendations

Constructed wetlands are an effective means of addressing stream restoration on a watershed basis. But, prior to the installation of any pollution abatement programs, the effect of these measures on water quality and habitat restoration must be considered for the entire watershed and not on a site by site basis. Not only must the construction, installation and effectiveness of these pollution abatements be considered, but the long term maintenance as well.

References

Brenner, E.K., Brenner, F.J. and Bovard, S. (1995). Comparison of bacterial activity in two constructed acid mine drainage wetland systems in western Pennsylvania. *J. Penn. Acad. Sci.* **69**, 88–92.

Brenner, E.K., Brenner, F.J., Bovard, S. and Schwartz, T.S. (1993). Analysis of wetland treatment systems for acid mine drainage. *J. Penn. Acad. Sci.* **67**, 85–93.

Brenner, F.J., Kosick, K.D., Gardner, C.A., and Tippie, C. (2000). Metal removal and the role of bacteria in a model vertical flow and aerobic wetland systems. Unpub. Report. Urban Wetlands Institute. 22 pp.

Brenner, F.J. and Pruent, P. (1999). Evaluation of a limestone channel and wetland system for treating acid mine drainage. In: *Proceedings of Mining and Reclamation for the Next Millennium*. American Society for Surface Mining and Reclamation. **2**, 584–591.

Burris, J.E., Gerber, J.W. and McHerron, L.E. (1984). Removal of iron and manganese from water by Sphagnum moss. pp.1–13. In: *Treatment of Mine Drainage by Wetlands*. J.E. Burris (ed) Contribution 264. Department of Biology, Pennsylvania State University. University Park, Pa.

Demchak, J.L. (1998). Analysis of vertical flow wetlands in the treatment of acid mine drainage. Unpub. M.S. Thesis. Clarion University of Pennsylvania. 152pp.

Dionis, K. and Stevens, S.E. Jr. (1985). Removal of dissolved iron and manganese ions by a *Sphagnum* moss system. In: R. Brooks, D. Samuel and J. Hill (eds). *Proc. on Wetlands and Water Management on Mined Lands*. Pennsylvania State University, University Park, PA. pp 365–372.

Gerber, D.W., Burris, J.E. and Stone, R.W. (1985). Removal of dissolved iron and manganese ions by a Sphagnum moss system. pp. 365–372. In: *Wetlands and Water Management on Mined Lands*. R.P. Brooks, D.E. Samuel and J.B. Hill (eds). Pennsylvania State University, University Park, PA.

Hedin, R.S. (1989). Treatment of coal mine drainage with constructed wetlands. In: *Wetlands Ecology and Conservation: Emphasis in Pennsylvania*. S.K. Majumdar, R.S. Brooks, F.J. Brenner and R.W. Tiner, Jr. (eds). Pennsylvania Academy of Science. Easton, PA. pp 349–362.

Hedin, R.S. and Wetzlaf, G.R. (1994). The effects of anoxic limestone drains on mine water chemistry. *Proceeding International Land Reclamation and Mine Drainage Conference and Third International Conference on the Abatement of Acidic Drainage*. **1**, 185–194.

Huntsman, B.E., Solch, J.G. and Porter, M.D. (1978). Utilization of *Sphagnum* species dominated bog for coal acid mine drainage abatement. *Geol. Soc. Amer. Annual Meeting Abst.*, Toronto, Ontario. p 322.

Kepler, D.A. (1986). Manganese removal in drainage by artificial wetland construction. In: *Proc. of the 8th Annual National Abandoned Mine Lands Conference*. Billings, MT. pp 74–80.

Kepler, D.A. and McCleary, E.C. (1994). Successive alkalinity-producing systems (SAPS) for the treatment of acidic mine drainage. *Proceedings International Land Reclamation and Mine Drainage Conference and Third International Conference on the Abatement of Acid Drainage*. **1**, 195–204.

Kleinmann, R.L.P. and Crerar, D. (1979). *Thiobacillus ferrooxidans* and the formation of acidity in simulated coal mine environments. *Geomicrobiology* **1**, 373–388.

Kleinmann, T.L.P., Tierman, T.O., Solch, J.G. and Harris, R.L. (1983). A low-cost, low-maintenance treatment system for acid mine drainage using Sphagnum moss and limestone. In: *Proc. of the 1983 Symposium on Surface Mining, Hydrology, Sedimentology and Reclamation*. S.B. Carpenter and R.W. Devore (eds). University of Kentucky, Lexington, KY. pp 241–245.

Owens, L.V. (1963). Iron and manganese removal by split flow treatment. *Water and Sewage Works*. **10**, 76–87.

Robbins, E.I., Brant, D.L. and Ziemkiewicz, P.F. (1999). Microbial, algae, and fungal strategies for manganese oxidation at a shade township coal mine, Somerset County, Penna. In: *Proc. Mining and Reclamation for the Next Millennium. American Society for Surface Mining and Reclamation*. **2**, 634–640.

Samuel, D.E., Sencindiver, J.C. and Rauch, H.R. (1988). Water and soil parameters affecting growth of cattails: pilots studies in West Virginia mines. Mine Drainage and Surface Mine Reclamation, Vol. 1 Mine Water and Mine Waste., U.S. Bur. Mines IC 9183. pp.367–374.

Sencindiver, J.C. and Bhumbla, D.K. (1988). Effects of cattails (*Typha*) on metal removal from mine drainage. In: *Mine Drainage and Surface Mine Reclamation*. Vol.1: Water and Mine Waste. Bur. Mines IC 9183. pp 317–324.

Skousen, J.G. (1997). Overview of passive treatment systems for treating acid mine drainage. *Greenlands* **27**, 34–43.

Spratt, A.K. and Wieder, R.K. (1988). Growth responses and iron uptake in *Sphagnum* plants and their relations to acid mine drainage (AMD) treatment. In: *Mine Drainage and Surface Mine Reclamation*. Volume I: Mine Water and Mine Waste. BuMines IC 9183. pp 317–324.

Wieder, R.K. and Lang, G.E. (1986). Fe, Al, Mn, and chemistry of Sphagnum peat in four peatlands lands with different metal and sulfur input. *Water, Air and Soil Pollution*. **29**, 309–320.

Nutrient and heavy metal uptake and storage in constructed wetland systems in Arizona

M.M. Karpiscak*, L.R. Whiteaker**, J.F. Artiola*** and K.E. Foster*

* Office of Arid Lands Studies, University of Arizona, 1955 East 6th St., Tucson Arizona 85719 USA. (E-mail: *karpisca@ag.arizona.edu* and *kfoster@ag.arizona.edu*)

** University of California, Riverside, Geology 2217, Riverside, California 92521 USA. (E-mail: *laska@citrus.ucr.edu*)

*** Department of Soil, Water and Environmental Science, University of Arizona, Shantz Bldg. 429, Tucson Arizona 85721 USA. (E-mail: *jartiola@ag.arizona.edu*)

Abstract The Constructed Ecosystems Research Facility (CERF) was conceived in the early 1980s as a test facility to explore the potential for using plants to treat wastewater in the arid west of the USA. One of the major issues that has been identified in the use of constructed wetland technology is plant nutrient uptake and tissue storage of nutrients as well as heavy metals. Our approach to understanding plant uptake and storage has been to look at both controlled conditions in constructed systems and background concentrations in natural systems. Plant tissues have been collected and analyzed from natural systems and from controlled systems receiving either wastewater or municipal water. Plants studied included the herbaceous species *Anemopsis californica* (Yerba mansa), *Scirpus* spp. (bulrush) and *Typha domingensis* (cattail), and tree species *Fraxinus velutina* (ash), *Populus fremontii* (cottonwood) and *Salix* spp. (willow). Data indicate that uptake varies not only among plant species, but also among chemical species, depending upon water quality within the wetlands. Leaf tissues of *Fraxinus*, *Salix* and *Populus*, contained the lowest amounts of nutrients and heavy metals studied (Na, P, K, Cu, Pb and Zn), while the root tissues of the herbaceous plants generally had the highest concentrations.

Keywords Constructed wetlands; heavy metals; natural wetlands; nutrients; plant uptake; wastewater

Introduction

There are many reasons for investigating the uptake and storage of nutrients and heavy metals in wetland plant species. One of the most important of these is the potential bioconcentration of nutrients and heavy metals that may pose a health risk to indigenous animals as well as humans. Previous studies have shown that essential and non-essential elements are more concentrated in plant tissues than in the soils supporting the plants (Outridge and Noller, 1991; Albers and Camardese, 1993). Other research reports also have indicated that herbaceous species are very efficient not only in removing heavy metals from mine and industrial wastewater (Gupta *et al.*, 1994; Goodrich-Mahoney, 1996; Knight *et al.*, 1999), but also in reducing pathogens and other selected parameters indicative of good water quality (Reed *et al.*, 1988; Martin and Fernandez, 1992; Karpiscak *et al.*, 1996; Knight *et al.*, 1999).

By understanding differences in concentration of nutrients and heavy metals in plants growing in effluent and potable water as well as the background concentrations of heavy metals in natural systems, it may be possible to engineer improved and more efficient wetland ecosystems to treat wastewater. In constructing new wetlands, the plant species of choice usually are aquatic macrophytes, herbaceous plants whose root systems tend to be extensive and fairly close to the soil surface in the water, whereas the root systems of riparian tree species tend to be more concise and rooted deeper in the soil. The emergent aquatic macrophytes that dominate in engineered systems are *Scirpus* spp. (bulrush), *Typha* spp. (cattail) and *Phragmites* spp. (common reed) (Cooper and Green, 1995; Cole, 1998).

Floating aquatic species such as *Lemna* spp. and *Eichhornia crassipes* (water hyacinth) have been used because the plants grow rapidly and are easy to harvest (Karpiscak *et al.*, 1994; Sharma and Gaur, 1995), but they tend to be grown alone in open areas of water rather than in complex ecosystems and although hyacinth has been documented as being very efficient in heavy metal uptake (Wolverton and McDonald, 1978), it is considered a noxious weed in Arizona and in many other areas.

Physiological differences with respect to elemental uptake sites and uptake mechanisms have been reported in the literature (Outridge and Noller, 1991; Kabata-Pendias and Pendias, 1992). Outridge and Noller (1991) wrote that maximum concentrations of trace elements in freshwater macrophytes from non-polluted environments were one to two orders of magnitude lower than in freshwater macrophytes from polluted environments. They also reported that root tissues usually were found to contain higher concentrations of most elements compared to above-ground plant parts.

Methods
Research sites
The constructed ecosystems located at CERF are subsurface flow wetland systems where water moves through gravel beds. The raceways used in this study measure 61 m long by 8.3 m wide and 1.4 m deep. The constructed ecosystems are planted identically with woody trees and shrubs mixed with emergent aquatic macrophytes to mimic a riparian habitat. Parallel systems receive secondary, unchlorinated wastewater effluent from a nearby sewage treatment facility while additional systems receive potable municipal water. The flow rate is regulated so that volume and detention times are similar in all raceways. Previous research and additional site descriptions are available in Karpiscak *et al.* (1996). Atomic absorption spectrophotometric analysis of the secondary effluent flowing into the effluent-fed systems indicated that the concentration of copper and lead were below their respective method detection limits (0.03 and 0.2 mg/l), but zinc was detected at 0.067 mg/l (Karpiscak *et al.*, 1991).

Seven natural wetland systems consisting of cienegas, woody swamps, cattail marshes or a combination thereof located in and around southern Arizona were sampled. These natural systems include woody trees and emergent aquatic macrophytes. Flow rates into these systems vary considerably, but the quality of the water entering these "natural" systems is considered to be relatively uncontaminated. Of the 16 water samples taken and analyzed for each of the metals reported here, none exceeded the method detection limit (Rohovit, 1999).

Tissue sampling
Plant tissue samples including leaves, shoots and roots were taken from controlled and natural ecosystems, examined for total concentration of nutrients and selected metals, and the results compared between the kind of tissue and system. The nutrients and metals studied include sodium (Na), phosphorus (P), potassium (K), copper (Cu), lead (Pb) and zinc (Zn).

Tissues were studied from selected herbaceous and tree species found in both the controlled and natural systems. Tree species included *Fraxinus velutina* (ash), *Populus fremontii* (cottonwood) and *Salix* spp. (willow). The macrophytes or herbaceous species included *Anemopsis californica* (Yerba mansa), *Scirpus* spp. (bulrush) and *Typha domingensis* (cattail).

Controlled systems. Leaf (trees) and shoot (herbaceous plants) tissue was cut from the growing tips. Roots were dug out of the soil and cleaned thoroughly. One to five aliquots of plant tissue from each species were taken for a combined total sample weight of approximately 125 g. Plant material for each sample was then cut into small pieces (approximately 5- to 7.5-cm square), placed in a paper bag to air dry for 1 to 2 days, followed by drying for another 3 to 5 days at 60°C in a drying oven. Samples were taken over a 3 to 4 year period and the results averaged.

Natural systems. Three aliquots of leaf tissue were taken from several trees of the same species near the designated water sampling location within the natural wetland. Shoots (leaves) were cut off 2 cm above ground level. Roots were dug out of the soil and cleaned thoroughly. The tissues were transferred in paper bags from the field to the laboratory where they were dried for 24 hours at 60°C in a drying oven. Sampling occurred during the last two weeks of May 1998. Not all of the species were present in each of the wetlands.

Analytical methodology

After drying, each sample was ground in a Wiley-mill fitted with a 40-mesh screen. When there was more than one tissue sample aliquot, the ground samples were composited and homogenized to present a uniform mixture. Aliquots of this mixture were then taken for chemical analysis. Root tissue was sampled in the same manner as leaf tissue.

Plant tissue samples were acid digested in a closed-vessel microwave oven (modified EPA Method 3052). The digests were then analyzed for nutrients and metals using ICP emission spectrophotometry except for phosphate-phosphorus, which was determined using colorimetry. The analyses were conducted at The University of Arizona's Soil, Water and Plant Analysis Laboratory.

Results and discussion

Nutrient concentration

A summary of the analytical results for the herbaceous and tree species by chemical parameter and tissue type is presented in Table 1. Macro nutrient data were obtained only for tree species growing in the controlled potable- and effluent-fed ecosystems while the heavy metals data were determined for all species in all three types of ecosystems.

Table 1 Mean Concentration of Nutrients and Heavy Metals in Plant Tissues by Species

		Anemopsis		Typha		Scirpus		Fraxinus	Salix	Populus
		Roots	Leaves	Roots	Leaves	Roots	Leaves	Leaves	Leaves	Leaves
				NATURAL WETLANDS[a]						
(µg/g)	Cu	9.81	6.12	9.50	4.32	9.40	2.58	1.41	3.02	3.24
	Pb	3.36	1.08	0.811	2.95	10.1	0.249	0.863	1.57	1.29
	Zn	14.4	15.8	30.5	12.9	26.1	9.69	8.18	45.5	46.1
				POTABLE SYSTEM[b]						
(%)	Na	0.259	0.393	0.524	0.711	0.314	0.491	(		)
	P	0.114	0.122	0.207	0.163	0.108	0.117	(not determined)		
	K	1.76	3.72	1.21	2.22	1.16	2.03	(		)
(µg/g)	Cu	16.4	6.06	16.7	8.34	19.2	7.49	7.53	10.3	6.05
	Pb	11.6	<1.25	5.62	2.34	2.20	1.21	4.10	2.28	<1.25
	Zn	16.4	13.1	49.3	23.5	20.5	36.4	14.3	67.5	45.2
				EFFLUENT SYSTEM[b]						
(%)	Na	0.444	0.746	1.19	1.08	0.584	0.494	(		)
	P	0.359	0.360	0.514	0.486	0.297	0.182	(not determined)		
	K	3.04	5.44	2.51	4.76	1.38	1.96	(		)
(µg/g)	Cu	3.74	4.26	13.8	3.54	16.1	2.89	3.44	4.37	2.24
	Pb	4.73	0.975	3.42	1.44	5.94	1.10	4.42	1.21	0.340
	Zn	23.7	15.0	49.4	21.7	32.6	14.3	11.3	21.8	14.4

[a] Data from Rohovit (1999)
[b] Data from Karpiscak *et al.* (1997, 1998)

Mean sodium concentrations determined in the tissues of *Anemopsis*, *Scirpus* and *Typha* and grown in the effluent-fed system ranged from 0.444 to 1.19 per cent of dry weight analyzed. Sodium concentrations determined in the tissues of the herbaceous species in the potable-fed system varied from 0.259 to 0.711 per cent. Mean phosphorus concentrations in herbaceous macrophytes taken from the effluent-fed system ranged from 0.182 to 0.514 per cent. In the potable system, mean phosphorus concentrations varied from 0.108 to 0.207 per cent. Mean potassium concentrations determined in herbaceous species grown in the effluent-fed system were 1.38 to 5.44 per cent, while plants grown in the potable system had a mean concentration range of 1.21 to 3.72 per cent.

It is clear from inspection of Figure 1 and a comparison of the effluent and potable systems that in both leaves and roots, the concentration of the parameters, sodium, phosphorus or potassium, is larger in almost all tissues taken from herbaceous plants in the effluent-fed raceway. The exception is *Scirpus* where the concentration of potassium is slightly lower in the leaves of plants grown in the effluent raceway. Potassium is also the only chemical species that appears more concentrated in the leaves (in some cases, more than twice) than in the roots and is seen in both the potable and effluent systems.

Heavy metal concentration
Copper. The mean concentration of copper for *Anemopsis californica* growing in the naturally-occurring wetlands was 9.81 µg/g in its roots and 6.12 µg/g in shoot tissues (Figure 2). The emergent macrophyte species *Scirpus* and *Typha* both concentrated copper to a high degree in root tissues, with means of 9.40 µg/g and 9.50 µg/g, while the shoot tissues had mean copper concentrations of 2.58 µg/g and 4.32 µg/g, respectively. Leaves from trees in the natural wetlands were found to have a mean copper of 2.17 µg/g, with the highest concentrations in the species *Populus* and *Salix*.

Copper typically was found at higher concentrations in the root tissues than in shoots and leaves of the herbaceous species growing in all three water systems, except for *Anemopsis* grown in wastewater (Figure 2). *Anemopsis* had a slightly higher concentration in leaf biomass at 4.26 µg/g vs 3.74 µg/g in root tissue. The mean concentration of copper determined in naturally occurring wetland macrophytes in Arizona was 5.40 µg/g, lower than that reported for freshwater vascular plants, 13 µg/g (Outridge and Noller, 1991).

Copper concentrations in the leaf and root tissue of potable water supplied plants

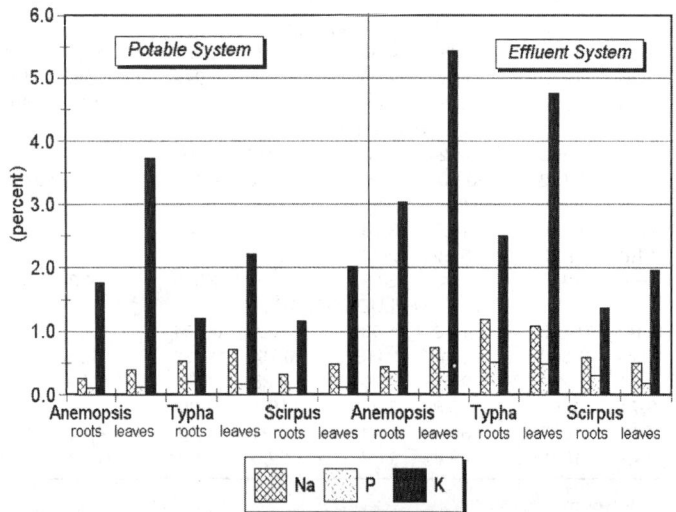

Figure 1 Concentration of selected nutrients in herbaceous plant tissues

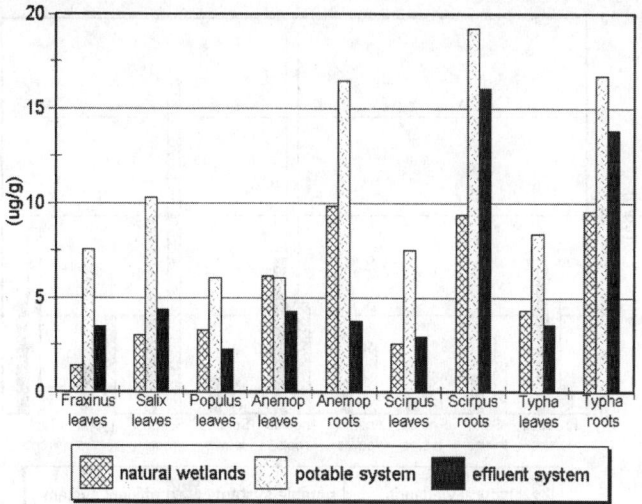

Figure 2 Copper concentration in plant tissues by wetland system

equaled or were typically greater than those found in the tissues of plants grown in natural wetlands and wastewater effluent. Effluent-grown plants did not uniformly have higher copper concentrations in either leaf or root tissue than those found in natural wetland grown plants.

Copper concentrations determined in naturally occurring wetland macrophytes were similar to those reported for the leaves of *Anemopsis* grown in a constructed raceway supplied with potable water and to *Populus* and *Typha* shoots grown in effluent-supplied raceways. *Fraxinus*, *Salix*, *Scirpus* roots and shoots and *Typha* roots sampled from naturally occurring wetlands were determined to have lower copper concentrations than those found for plants grown at CERF supplied with potable or effluent water.

As determined in this study, *Anemopsis*, *Scirpus* and *Typha* are reported to concentrate copper to a higher degree in root tissues than in shoots or leaves in constructed wetland systems.

Lead. In the natural wetlands, *Anemopsis californica* was determined to have a mean lead concentration of 3.36 μg/g in root tissues and 2.15 μg/g in leaves (Figure 3). The emergent species *Scirpus* concentrated lead forty times higher in root tissues than in shoots, with mean concentrations of 10.1 μg/g and 0.24 μg/g, respectively. *Typha* concentrated lead to a much higher degree in its shoot tissues in comparison to its root tissues with means of 2.95 μg/g in the shoots and 0.81 μg/g in the roots. Of the natural wetland trees sampled, a concentration range of <0.16 to 18.8 μg/g was determined for the tree species *Populus* with *Salix* having the highest concentrations (Rohovit, 1999).

All herbaceous species studied except *Scirpus* fell below the minimum normal lead range for terrestrial plants reported by Kabata-Pendias and Pendias (1992) of 10 μg/g, and all species analyzed fell within the reported range for freshwater vascular plants reported by Outridge and Noller (1991) of 0.3 to 35 μg/g. The high concentrations determined for *Scirpus* may indicate that it is a lead accumulator. The mean concentration of lead determined in naturally occurring wetland macrophytes was 2.57 μg/g, lower than that reported for freshwater vascular plants, 8.1 μg/g (Outridge and Noller, 1991).

Lead concentrations determined in naturally occurring samples of *Populus* and *Salix* were similar to those reported for the same species grown in CERF raceways supplied with potable or effluent water. *Anemopsis* leaves, *Scirpus* roots and *Typha* shoots were

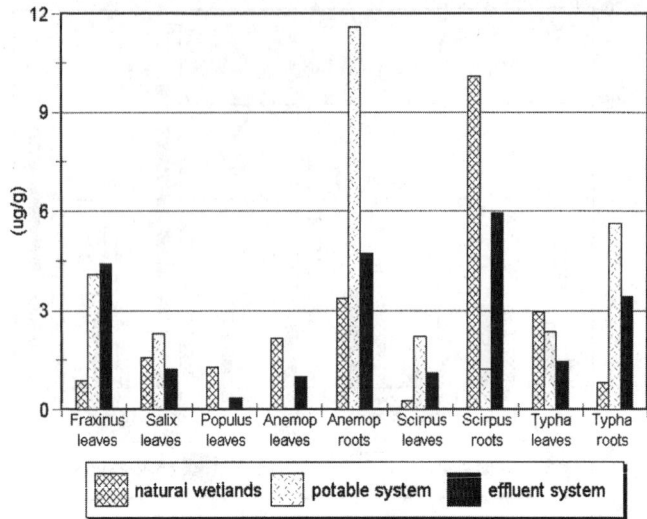

Figure 3 Lead concentration in plant tissues by wetland system

determined to have greater lead concentrations in naturally occurring samples than those reported for the same species grown at CERF. *Anemopsis* roots, *Fraxinus*, *Scirpus* shoots and *Typha* roots were determined to have lower lead concentrations in naturally occurring plants compared to data reported for those grown at CERF.

Zinc. In the naturally occurring wetlands, the tree species *Populus* and *Salix* were determined to have high levels of zinc in leaf tissues with a mean concentration of 46.1 µg/g and 45.5 µg/g, respectively (Figure 4). *Fraxinus* accumulated the least amount of zinc with a mean concentration of 8.18 µg/g. *Scirpus* and *Typhus* accumulated more zinc in roots than in shoots with mean concentrations for *Scirpus* of 26.1 µg/g and 9.69 µg/g and for *Typha* of 30.5 µg/g and 12.9 µg/g, respectively. The zinc concentration in *Anemopsis* leaf tissue was slightly higher than in roots with a mean of 15.8 µg/g and 14.4 µg/g, respectively.

The concentration of zinc determined in the tissues from *Anemopsis*, *Fraxinus* and *Scirpus* was below the maximum reported concentration of 300 µg/g for crop and terrestri-

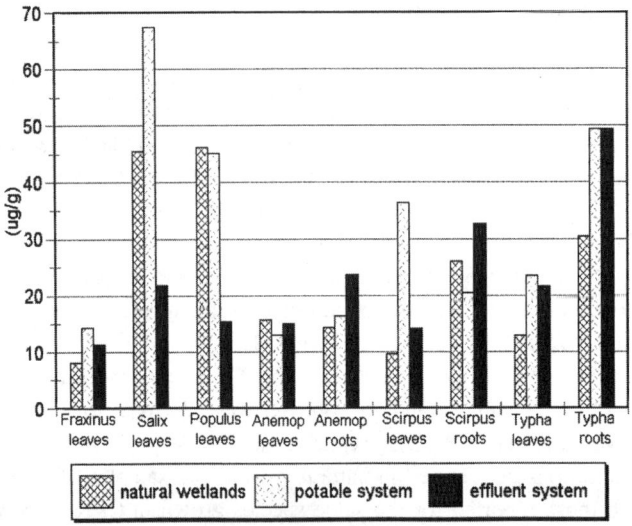

Figure 4 Zinc concentration in plant tissues by wetland system

al plants (Kabata-Pendias and Pendias, 1992). All species analyzed were within the reported range of 11 to 250 μg/g for zinc in freshwater vascular plants (Outridge and Noller, 1991), with the exception of *Fraxinus*, which had a higher concentration. The mean concentration of zinc determined in naturally occurring macrophytes was 26.1 μg/g, lower than the 52 μg/g reported for freshwater vascular plants.

Zinc concentrations determined in naturally occurring samples of the tree species *Fraxinus*, *Populus* and *Salix* were greater than those reported for the same species grown in the effluent- and potable-fed raceways at CERF (Table 1, Figure 4). *Anemopsis* leaves sampled from natural wetlands also had greater concentrations of zinc than those taken from the raceways, although *Anemopsis* roots had lower zinc concentrations compared to those from the constructed raceways. Significantly less zinc was found in the tissues of *Typha* sampled in the natural wetlands compared to that in *Typha* tissues from the constructed ecosystems. The zinc concentration in the roots of natural wetland-grown *Scirpus* was less than in roots from the effluent-fed raceways, and the concentration in *Scirpus* shoots was lower than in shoots collected from the potable-fed raceways when grown in wastewater containing low levels of these parameters.

Conclusions

In general, the leaf tissues of the tree species, *Fraxinus*, *Salix* and *Populus*, contained the lowest amounts of nutrients and heavy metals studied (Na, P, K, Cu, Pb and Zn), while the root tissues of the herbaceous plants generally had the highest concentrations. Some elevated concentrations of nutrients and heavy metals also were found in the shoots of *Anemopsis* and *Typha*.

For the chemical parameters studied, the concentrations determined in constructed wetlands were lower or equal to values determined in potable and natural systems, with the exception of lead in the leaves of *Fraxinus*, and zinc in *Anemopsis* and *Scirpus* roots. These values are not significantly greater, however, and do not seem to indicate that constructed wetland systems accumulate these elements to elevated concentrations.

Acknowledgements

The authors would like to thank the owners of the natural wetlands for their cooperation in allowing sampling at their facilities: Empire Cienega, Canelo Hills Cienega and Patagonia-Sonoita Creek Preserve owned by The Nature Conservancy; Cook's Lake and St. David Cienega owned by the U.S. Bureau of Reclamation; and Bingham Cienega and Agua Caliente Park owned by Pima County. Funding for research activities at CERF has been provided by Pima County Wastewater Management Department.

References

Albers, P.H. and Camardese, M.B. (1993). Effects of acidification on metal accumulation by aquatic plants and invertebrates. 1. Constructed wetlands. *Environmental Toxicology and Chemistry* **12**, 959–967.

Cole, S. (1998). The emergence of treatment wetlands. *Environmental Science and Technology* **32**(9), 218–223.

Cooper, P. and Green, B. (1995). Reed bed treatment systems for sewage treatment in the United Kingdom – the first 10 years' experience. *Water Science and Technology* **32**(3), 317–327.

Goodrich-Mahoney, J.W. (1996). Constructed wetland treatment systems applied research program at the Electric Power Research Institute. *Water, Air and Soil Pollution* **90**, 205–217.

Gupta, M., Sinha, S. and Chandra, P. (1994). Uptake and toxicity of metals in *Scirpus lacustris* L. and *Bacopa monniere* L. *Journal of Environmental Science and Health* **A29**(10), 2185–2202.

Kabata-Pendias, A. and Pendias, H. (1992). *Trace Elements in Soils and Plants*. CRC Press, Boca Raton, Florida.

Karpiscak, M.M., Hopf, S.B. and Foster, K.E. (1991). Pima County Constructed Ecosystems Research Facility Progress Report II, Pima County Wastewater Management Department.

Karpiscak, M.M., Foster, K.E., Hopf, S.B., Bancroft, J.M. and Warshall, P.J. (1994). Using water hyacinth to treat municipal wastewater in the desert southwest. *Water Resources Bulletin* **30**(2), 219–227.

Karpiscak, M.M., Gerba, C.P., Watt, P.M., Foster, K.E. and Falabi, J.A. (1996). Multi-species plant systems for wastewater quality improvements and habitat enhancement. *Water Science and Technology* **33**(10–11), 231–236.

Karpiscak, M.M., Hopf, S.B. and Foster, K.E. (1997). Pima County Constructed Ecosystems Research Facility Annual (1995) Progress Report VII, Pima County Wastewater Management Department.

Karpiscak, M.M., Hopf, S.B. and Foster, K.E. (1998). Pima County Constructed Ecosystems Research Facility Annual (1996) Progress Report VIII, Pima County Wastewater Management Department.

Knight, R.L., Kadlec, R.H. and Ohlendorf, H.M. (1999). The use of treatment wetlands for petroleum industry effluents. *Environmental Science and Technology* **33**(7), 973–980.

Martin, I. and Fernandez, J. (1992). Nutrient dynamics and growth of a cattail crop (*Typha latifolia* L.) developed in an effluent with high eutrophic potential – application to wastewater purification systems. *Bioresource Technology* **42**, 7–12.

Outridge, P.M. and Noller, B.N. (1991). Accumulation of toxic trace elements by freshwater vascular plants. *Reviews of Environmental Contamination and Toxicology* **121**, 1–63.

Reed, S.C., Middlebrooks, E.J. and Crites, R.W. (1988). *Natural Systems for Waste Management and Treatment*. McGraw-Hill, New York.

Rohovit, L. (1999). *Micronutrient and heavy metal concentrations observed in natural wetland macrophytes in Arizona*. Master of Science Thesis, Department of Soil, Water and Environmental Science, The University of Arizona, Tucson, Arizona.

Sharma, S. and Guar, J.P. (1995). Potential of *Lemna polyrrhiza* for removal of heavy metals. *Ecological Engineering* **4**, 37–43.

Vymazal, J. (1995). *Algae and Element Cycling in Wetlands*. Lewis Publishers, Boca Raton, Florida.

Wolverton, B.C. and McDonald, R.C. (1978). Bioaccumulation and detection of trace levels of cadmium in aquatic systems. *Environmental Health Perspectives* **27**, 161–164.

Retention of selected heavy metals: Cd, Cu, Pb in a hybrid wetland system

H. Obarska-Pempkowiak*

* Faculty Hydro and Envrionmental Engineering, Technical University of Gdańsk, ul. Narutowicza 11/12, 80-952 Gdańsk, Poland. (E-mail: hoba@pg.gda.pl)

Abstract The budget of heavy metals was investigated in a constructed wetland in a hybrid wetland system near Gdańsk. It is a pilot wastewater treatment plant (WWTP) designed for 150 PE (person equivalent). The system consists of two sections: a vegetated submerged bed (VSB) with horizontal flow of sewage and a cascade filter situated on a slope of a hill. Total area of the constructed wetland is about 870 m^2. Domestic sewage, after a conventional pretreatment (consisting of an Imhoff tank and a trickling filter) is pumped to the VSB filter located on the top of the hill and then flows through subsequent segments of the constructed wetland. In the period 1995–98 the measurements of several heavy metals (Cd, Cu, Pb) were carried out in sewage inflowing, outflowing and collected from the in between sections of the system. Moreover analysis of sediment collected in ditches of the cascade filter, VSB filter and reed were carried out. The content of heavy metals in suspended solids decreased along the course of treatment, starting from VSB filter, through the first ditch to the last ditch. Measurable concentrations of dissolved heavy metals were found in sewage collected from several subsequent ditches. Sorption was deemed the main mechanism of dissolved metals removal in subsequent ditches.
Keywords Accumulation; heavy metals; reed; retention; wetland

Introduction

The potential of constructed wetlands for removal of nutrients from domestic wastewater is well documented. Much less is known on their capacity for the removal of toxic heavy metals and persistant organic substances.

The agricultural regions of Poland are inhabited by approximately 30% of the country's population and these areas also contribute sigificantly to water pollution problems. It is estimated that about 1 km^3 of wastewater originating in these regions is discharged to surface and groundwaters, while only 0.07 km^3 is treated.

The removal of sewage from the households by sewerage systems ecompasses only about 8.5% of the rural population, while only 1.5% of them is treated biologically and 0.3% only is treated mechanically. Wastewater treatment plants (WWTP) with high removal of biogenic substances ecompass only 1.2% of the population of villages.

The local treatment of sewage usually takes place in septic tanks that provide only mechanical treatment. Often these tanks are leaking and the outflowing sewage, in many cases, is not sufficiently purified.

The quality standards for sewage discharged to surface waters are restrictive in Poland, also in the case of small volumes of sewage. The situation requires research and the application of new technologies and solutions. One of the new methods applied for sewage treatment is constructed wetland systems.

Constructed wetland systems have become a very popular technology for the removal of nutrients from domestic sewage. However, an application of these systems is their utilization for the removal of heavy metals. The potential of wetland systems for removal of heavy metals is the result of the lack of a biological barrier in the macrophyte plants, due to which the passive adsorption of heavy metal ions occurs. The retention of heavy metals in the wetland system may be also the result of binding by microorganisms according to one of two mechanisms: formation of complex extracellular compounds composed of heavy metals and substances released by the cells and biosorption of the metal ions on the cell surface (Forstner and Witmann, 1981; Vymazal *et al.*, 1998; De Maeseneer, 1997; Olivie-Lauquet *et al.*, 2001). According to Vymazal *et al.* (1998), heavy metals precipitate in the form of carbonates and sulphides, produced in the redox reactions.

This paper evaluates the removal efficiency of three heavy metal ions: Pb, Cu and Cd in a constructed wetland ecosystem planted with reed. Metals are supplied with domestic sewage from the Przywidz village near Gdańsk. The proportion of heavy metals retained in the reed biomass was estimated. Solid speciations of heavy metals in the filling material through which the sewage percolates, and in bottom sediments were also estimated.

Site description

A pilot-scale wetland system was constructed in Przywidz (the Pommeranian region) for treating domestic sewage. Prior to inflowing to the wetland system, the sewage is treated in a conventional wastewater treatment plant consisting of an Imhoff tank and trickling biofilter. The amount of inflowing sewage equals 22.5 m^3d^{-1}, which corresponds to 150 person equivalent (PE). The wetland system consists of two parts: a vegetated submerged

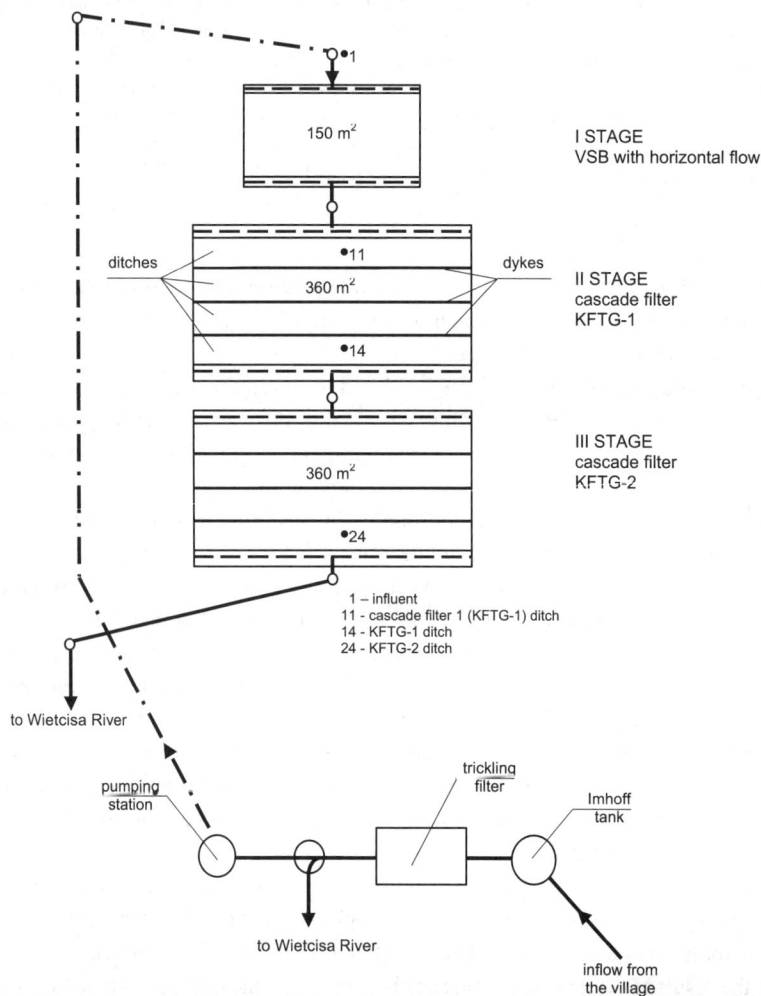

Figure 1 Schematic of constructed wetland system in Przywidz and localization of sampling points

bed (VSB) with horizontal flow (area 150 m²) and a cascade filter (KFTG), planted with reed, area 720 m². The cascade filter itself consists of two parts: upper cascade (KFTG-1) and lower cascade (KFTG-2). Each cascade is composed of three internal dams which are separated by four ditches. The length of each dam is 30 m. The bottom of each subsequent ditch is placed 25 cm below the bottom of the preceding one.

Gravitational flow of sewage is secured by the filter situated on a slope of a hill. After passing through the VSB filter, the sewage is directed to the first ditch of the upper cascade, where it is aerated. While the ditch is filling, the sewage percolates through the first dam to the lower ditch, percolates through the second dam and so on. Sewage from subsequent parts of the constructed wetland was drained with a drainage pipe and collected in an inspection chamber. The outflow was directed to the next part of the wetland.

Methods

In the period 1996–98 measurements of the contents of heavy metals (Cd, Pb and Cu) from sewage were performed at the following sampling points, shown in Figure 1:

1 – the inflow to the VSB (the inflow to the wetland system),
11 – the first ditch of the upper cascade KFTG-1 (effluent from the VSB),
14 – the last (fourth) ditch of the upper cascade KFTG-1 (effluent from the KFTG-1),
24 – the last ditch of the lower cascade KFTG-2 (the effluent from the wetland system).

The contents of heavy metal ions were measured both in the solution and in the suspended solids, according to the procedure described by Obarska-Pempkowiak and Klimkowska (1999).

In order to investigate the distribution of the analysed heavy metals in each of the component parts of the wetland system, the measurements of the contents of these elements in the bottom sediments stored in the ditches and in the reeds were performed.

Samples of the bottom sediments were taken at the following points:

11 – the first ditch of the upper cascade (KFTG-1),
14 – the last (fourth) ditch of the upper cascade (KFTG-1),
24 – the last (fourth) ditch of the lower cascade (KFTG-2).

The samples of sewage were collected in spring, summer and autumn. The plant's material (stalks, leaves, roots, rhizomes) was collected from the VSB, KFTG-1 and KFTG-2 filters during spring, summer and autumn. The samples of bottom sediments were collected at the beginning and at the end of the investigation period (1995–1998).

Sequential extraction of heavy metals was carried out according to a procedure described earlier (Forstner and Witmann, 1981; Obarska-Pempkowiak and Klimkowska, 1999). In short: 1.000 g subsamples of dried (105°C, 4 hours) sediment were shaken for 16 hours with 10 cm³ of 0.1 mol dm⁻³ CH_3COOH. The solution was separated by centrifugation. The extract was used for heavy metal determination (labile fraction of heavy metals), while the solid residue was shaken with 10 cm³ of 0.5 mol/dm³ $NH_2OH \cdot HCl$ (pH = 2.0), the solution was again separated by centrifugation. The extract was used for heavy metals determination (hydroxide fraction of heavy metals). This was followed by digestion of the solid residue with 8.8 mol dm⁻³ H_2O_2 and extraction with 1.0 mol dm⁻³ CH_3COONH_4, pH = 2.0 (organic fraction of heavy metals). The last stage of the procedure was digestion of solid residue with hydrofluoric acid, evaporation to dryness, dissolution in 0.1 mol dm⁻³ HNO_3 and determination of heavy metals in the obtained solution (residual fraction of heavy metals).

Results and discussion

The highest accumulation of heavy metals was found in the rhizomes. The highest amounts of Pb and Cu equal to 19.40 and 10.60 $\mu g\ g^{-1}$ (d.m.$^{-1}$), respectively, were present in summer. The highest amounts of Cd, 0.98 μg g^{-1} (d.m.$^{-1}$), were recorded in spring. The accumulation of these elements at the beginning of the wetland was higher than in the last ditch of the lower cascade (KFTG-24).

The daily loads of the analysed metals after the subsequent segments of the wetland system are presented in Figure 2. Concentrations of dissolved and suspended metals in sewage at subsequent stages of treatment are presented in Figure 2 and Figure 3. The content of heavy metals decreased in the subsequent compartments of the wetland system. The highest amounts of all three metals were retained in the first stage of the system (the VSB filter), while in the next stages (KFTG-1 and KFTG-2) the elimination of analysed elements was much smaller. The daily loads of Cd, Cu and Pb in sewage after subsequent stages of treatment are presented in the Figure 4. The results indicate a significant role of the first part of the wetland system – the VSB filter in removal of heavy metals (69.6% Cd, 69.2% Cu and 64.3% Pb). The average removal efficiency of the analysed elements in the whole wetland system was: 92.3% Cd, 89.4% Cu and 87.8% Pb.

The results allowed for counting a simplified balance for the analysed heavy metals: Cd, Cu and Pb in the wetland ecosystem. The results are given in Table 1. They confirm that a substantial part of the analysed elements is retained in the filtering materials of the dykes

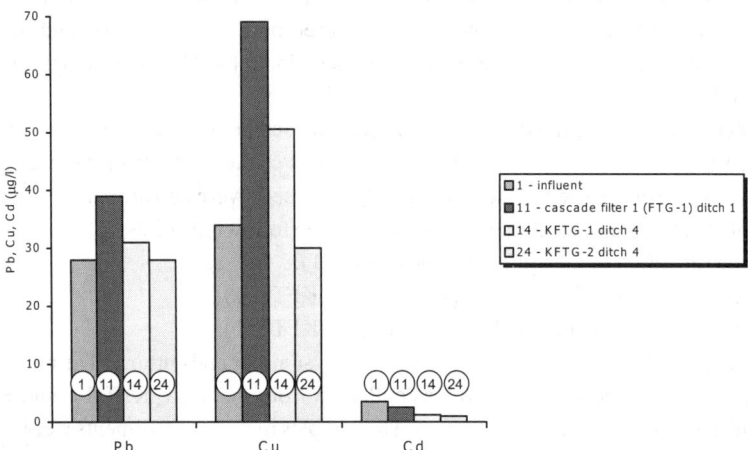

Figure 2 Changes of average concentrations of heavy metals in sewage after subsequent stages of treatment in Przywidz

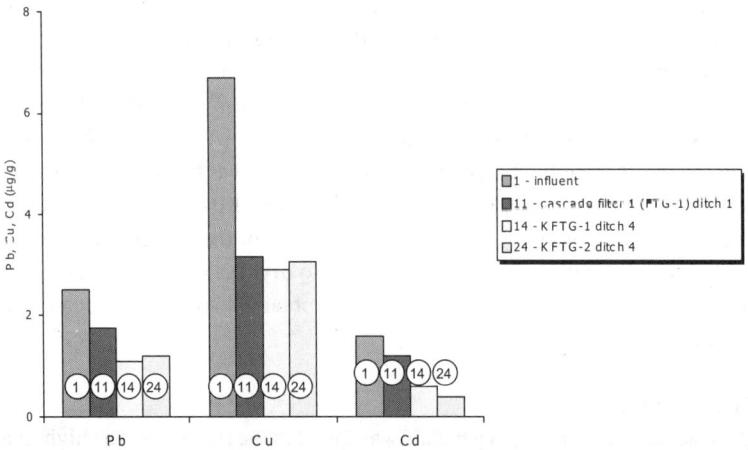

Figure 3 Average contents of metals in suspended solids

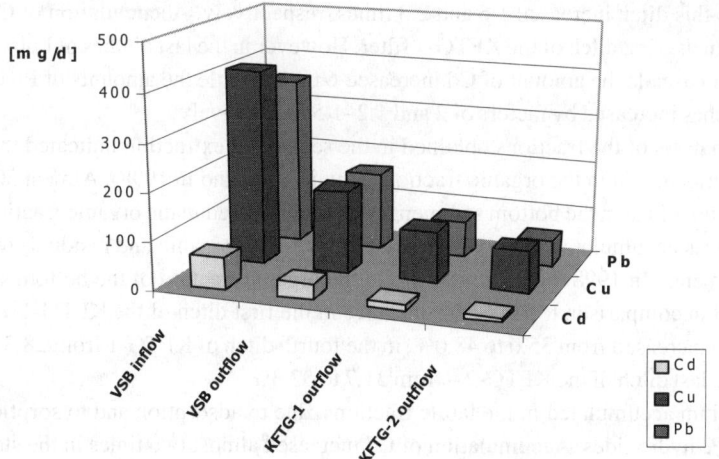

Figure 4 The daily loads of Cd, Cu and Pb in sewage after subsequent stages of treatment in Przywidz

Table 1 Balance of heavy metals in constructed wetland system in Przywidz, in g a^{-1}

Metals	Influent	Reed	Effluent
Cd	26.56	0.12	3.28
Cu	148.92	2.15	31.39
Pb	128.48	1.41	20.07

and in the bottom sediments of the ditches. Accumulation in the reed biomass was lower; 14.8% Cd, 22.1% Cu and 13.8% Pb.

The results of sequential extraction for the bottom sediments are given in Figure 5. Copper showed the highest variations, the content of cadmium – lower, while the content of lead changed slightly. Both the lowest amounts and the lowest variations of analysed elements were found in the filtering materials of the dykes. The highest variations were recorded in the bottom sediments.

The amount of elements retained in the ditches increased in 1998 in comparison to 1996.

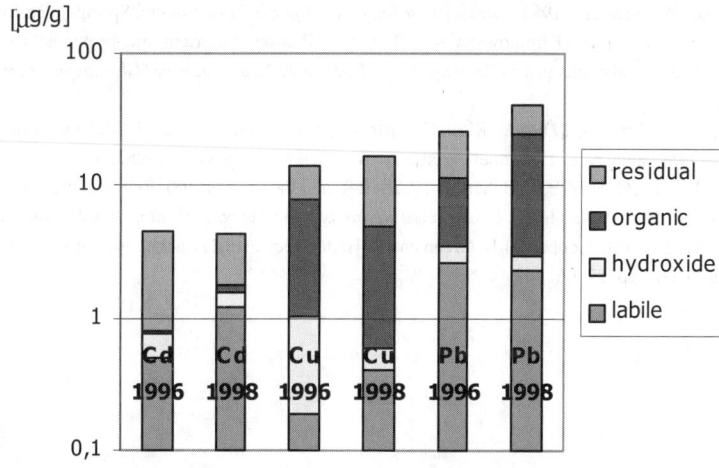

Figure 5 Results of sequence extraction of the bottom sediments of the ditches in the constructed wetland system in Przywidz

The highest retention took place in the first ditch of the KFTG-1 filter. Accumulation of Cu and Pb in this ditch increased 1.6 and 2.3 times, respectively. Accumulation of Cd did not increase in the first ditch of the KFTG-1 filter. However in the last ditches of both the upper and lower cascade the amount of Cd increased 6 times, while the amounts of Pb and Cu in these ditches increased by factors of 2 and 1.2–1.5, respectively.

The analysis of the fractions obtained in the sequential extraction indicated the highest accumulation of Cu in the organic fraction (both in 1996 and in 1998). Almost 50% of the total amount of Cu in the bottom sediments was accumulated in the organic fraction.

Lead was accumulated mainly in the stable fractions (organic and residual) of the bottom sediments. In 1998 the content of Pb in the organic fraction of the bottom sediments increased in comparison to 1996. For instance in the first ditch of the KFTG-1 accumulation of Pb increased from 35.0 to 48.0%, in the fourth ditch of KFTG-1 from 28.3 to 48.0% and in the last ditch of the KFTG-2 – from 31.7 to 52.3%.

Cadmium accumulated in the labile fractions, due to adsorption and to sorption on the Mn and Fe hydroxides. Accumulation of Cd increased almost two times in the stable fractions of the bottom sediments in the second year of the investigation.

Conclusions

1. The removal of suspended solids, organic matter and heavy metals took place in the VSB filter and in the subsequent first cascade.
2. Sorption was the main mechanism of heavy metal removal from sewage, while the accumulation in the reed biomass played a minor role (1.0% – Cd, 3.4% – Pb, 1.9% – Cu).
3. Solid speciation revealed that Pb and Cu were present in organic forms (37–57% and 41–71% respectively for Pb and Cu), while Cd was present in residual forms (60–88%). The proportion of labile forms, amounting to a few per cent in the case of Cd indicates a probable increase of bioavailability, and potential for migration.

Acknowledgements

Funding support from the Committee of Scientific Research of Poland for constructed wetland (3TO9C 040 19) is acknowledged.

References

De Meseneer, J.L. (1997). Constructed wetlands for sludge dewatering. *Wat. Sci. Tech.*, **35**(5), 279–285.
Forstner, V. and Witmann, L. (1981). *Metal Pollution in the Aquatic Environment.* Springer. Berlin. pp. 453.
Obarska-Pempkowiak, H. and Klimkowska, K. (1999). Distribution of nutrients and heavy metals between sediments, water and plants in a wetland system in Przywidz, Poland. *Journal Chemosphere*, **39**(2), 303–312.
Olivie-Lauquet, G., Gruau, G., Dia, A., Riou, C., Jaffrezic, A. and Henin, O. (2001). Release of trace elements in wetlands: role of seasonal variability. *Wat. Sci. Tech.* **35**(4), 943–952.
Vymazal, J., Brix, H., Cooper, P.F., Haberl, R., Perfler, R. and Laber, J. (1998). Removal mechanisms and types of constructed wetlands. In: *Constructed wetlands for wastewater treatment in Europe*, J. Vymazal, H. Brix, P.F. Cooper, M.B. Green and R. Haberl (ed.), vol.1, Backhuys Publishers, Leiden, The Netherlands, pp. 17–66.

The integration of constructed wetlands into a treatment system for airport runoff

D.M. Revitt*, P. Worrall** and D. Brewer***

* Urban Pollution Research Centre, Middlesex University, Bounds Green Road, London N11 2NQ, UK.
(E-mail: *m.revitt@mdx.ac.uk*)
** Penny Anderson Associates, 60 Park Road, Buxton, Derbyshire, SK17 6SN, UK.
(E-mail: *pw@paa-ecol.demon.co.uk*)
*** Heathrow Airport Ltd, Bath Road, Harlington, Middlesex, UK. (E-mail: *David_Brewer@baa.co.uk*)

Abstract A new surface runoff treatment system has been designed for London Heathrow Airport, which incorporates separate floating constructed wetlands or reedbeds and sub-surface flow constructed wetlands as major pollutant removal systems. The primary requirement of the newly developed treatment system is to control the concentrations of glycols following their use as de-icers and anti-icers within the airport. The ability of reedbeds to contribute to this treatment role was fully tested through pilot scale, on-site experiments over a 2 year period. The average reductions in runoff BOD concentrations achieved by pilot scale surface flow and sub-surface flow reedbeds were 30.9% and 32.9%, respectively. The corresponding average glycol removal efficiencies were 54.2% and 78.3%, following shock dosing inputs. These treatment performances are used to predict the required full scale constructed wetland surface areas needed to attain the desired effluent water quality. The treatment system also incorporates aeration, storage and, combined with reedbed technology, has been designed to reduce a mixed inlet BOD concentration of 240 mg/l to less than 40 mg/l for water temperatures varying between 6°C and 20°C.
Keywords Aeration; airport runoff; BOD removal; constructed wetlands; glycol

Introduction

London Heathrow Airport, which is situated 24 km to the west of Central London, UK, is one of the world's busiest international airports. The existing surface water drainage system transports the runoff to balancing ponds for storage and, in some cases, aeration prior to discharge to the receiving environment. Efficient treatment of the runoff is particularly important during the winter months when de-icers and anti-icers are applied to runway areas and aircraft, respectively to prevent ice formation and build-up. Ethylene and diethylene glycols are widely used as the active component in de-icers (e.g., under the trade name *Konsin*) whereas propylene glycol is a common constituent of anti-icers (e.g., under the trade name *Kilfrost*). The adhesion properties of anti-icers are increased by the addition of polymeric thickening agents but it has been reported that up to 80% of the applied liquid runs off the aircraft (O'Connor and Douglas, 1993). Relatively few airports have recovery systems for glycols (Sabeh and Narasiah, 1992) and their discharge can have serious problems for receiving water quality due to microbial metabolism processes exerting a high biochemical oxygen demand (Ellis *et al*., 1997). Koryak *et al*. (1998) have reported the presence of dense biological slimes on stream beds and the existence of severely stressed invertebrate and fishery communities downstream of an airport runoff discharge. Laboratory studies have demonstrated the high toxicity of anti-icers and de-icers with the former exhibiting an enhanced toxicity by two orders of magnitude (Hartwell *et al*., 1995). Fisher *et al*. (1995) have shown that airport runoff from winter storm events caused acute toxicity to both the fathead minnow (*Pimephales promelas*) and the daphnid (*Daphnia magna*) with LC50 values as low as 1.0% of the effluent.

The application rates of de-icing and anti-icing agents can result in the use of glycol loadings approaching 175,000 kg at major international airports during a severe winter. The subsequent concentrations in runoff can be highly variable with levels as high as 4,500 mg/l for BOD and 6,000 mg/l for total glycols (Ellis *et al.*, 1997). Constructed wetlands have been shown to be capable of efficiently removing BOD when applied at high loading rates. Vymazal (1999) has reported average BOD removal efficiencies of 86.6% by horizontal sub-surface flow systems at average BOD loading rates of 33.5 kg/ha/day. The main removal mechanism was reported to be by both aerobic and anaerobic microbial biodegradation in the rhizosphere region of the vegetated beds. Other studies have demonstrated BOD removal efficiencies of between 60 and 95% by constructed wetlands for a variety of different wastewaters including domestic sewage (Hiley, 1995), refinery effluent (Huddleston *et al.*, 2000), wildfowl effluent (Worrall *et al.*, 1997), paper mill waste (Knight *et al.*, 1994) and dairy farm waste (Tanner *et al.*, 1995; Geary and Moore, 1999; Newman *et al.*, 2000).

In order to identify the most appropriate technology for dealing with the glycol runoff problem, London Heathrow Airport have investigated a number of options including reverse osmosis, carbon filtration, modified cellulose filtration, UV catalytic oxidation and bioremediation techniques including constructed wetlands. In this paper the contribution of the experimental results from pilot trial reedbed systems at Heathrow towards developing the design and subsequent construction of the full-scale system is described. In this process the design team has applied sustainable principles to materials acquisition, energy usage and future management procedures. In addition, it has been important to be aware of the inherent problems of constructing wetlands close to airports due to the increased risk of bird strike.

Methodology

Full details of the pilot scale constructed reedbed systems, situated adjacent to the Eastern Balancing Reservoir at Heathrow Airport, have been given previously (Revitt *et al.*, 1997; Chong *et al.*, 1999). Water from the Eastern Balancing Reservoir was continuously pumped at known flow rates to two substrate based systems, a soil surface flow wetland and a gravel sub-surface reedbed, to support two different types of experimental strategy. The responses of the reedbeds to continuous but relatively low levels of pollution were monitored by assessing the BOD removal capacity from diluted Heathrow runoff through determination of the BOD concentrations in appropriately timed inlet and outlet water samples over a 2 year period. The typical loading rates to both beds were approximately 26 g/m^2/day. The reedbeds were also subjected to shock dosing experiments in which neat mixtures of *Konsin* and *Kilfrost* were introduced at the front ends of the beds in order to simulate high pollutant loadings, typical of those encountered during winter de-icing events. Outlet water samples were collected automatically and the glycol concentrations were determined using gas chromatography–mass spectrometry. Similar experiments were conducted on adjacent rafted reedbed tanks in order to identify the role reedbed rafts might play in glycol pollution attenuation.

The surface water drainage system at Heathrow divides the airport into four catchment areas of which by far the largest are the eastern and southern catchments. The southern catchment includes terminal, cargo and runway areas extending over an area of 290 ha (78% impermeable) compared to the eastern catchment which includes terminal, maintenance and runway areas covering 309 ha (80% impermeable). The runoff from these catchments currently discharges into two separate balancing reservoirs. The Southern Balancing Reservoir is a man-made water body containing a nominal volume of 962,000 m^3 which eventually discharges to the River Thames. The Eastern Balancing

Reservoir consists of three separate ponds and has an overall capacity of 227,000 m³. The runoff water quality from the eastern catchment is generally lower due to pollutant inputs from maintenance areas which are located therein (Consultants in Environmental Sciences Ltd, 1995).

Results and discussion
Pilot scale study

The average BOD removal efficiencies for the 2 different types of pilot scale reedbeds were 30.9% (for the surface flow bed) and 32.9% (for the sub-surface bed) based on the data collected over a 2 year monitoring period. The average BOD loading rate was 44.9 mg/s and comparison of this experimental loading rate with the monitored BOD loadings to the Eastern Balancing Reservoir provides an estimation of the reedbed surface areas required for different degrees of treatment. During a 6 day period in March 1995, 22,406 kg of BOD entered the Eastern Balancing Reservoir at a loading rate of 43.2 g/s and 2,472 kg of BOD left the Eastern Balancing Reservoir at a loading rate of 4.77 g/s (Consultants in Environmental Sciences Ltd, 1995). The estimated surface areas required for both types of reedbed are shown in Figure 1 according to whether they are to be used as initial front-end treatment systems or as a final treatment option following initial storage and aeration. The results from the BOD data (Figure 1) indicate that unacceptably large reedbed surface areas would be needed to provide efficient treatment of direct airport runoff under routine operating conditions but that they are suitable as a final polishing treatment facility for airport runoff following an initial breakdown of the biodegradable organics within, for example, an aeration lagoon. The derived reedbed surface areas (Figure 1) indicate that 90% BOD removal could be achieved using a 4.3 ha sub-surface flow reedbed or a 4.64 ha surface flow reedbed in this role. From a practical point of view, the use of sub-surface flow reedbed systems would be most suitable as a final polishing treatment facility after aerated lagoon treatment.

The average glycol removal efficiencies determined from 3 dosing experiments were 54.2% (for the surface flow bed) and 78.3% (for the sub-surface flow bed). The increased performance of the latter is related to the presence of higher treatment contact areas resulting from flow pathways directly through the gravel substrate. The total weight of glycols introduced into each reedbed was 6,627g and comparison of the subsequent loading rates with those arriving at the Eastern Balancing Reservoir predicts the surface treatment areas shown in Figure 2 for the purposes of full treatment of raw runoff or as a final polishing system. The glycol dosing experiments are representative of the situation occurring during heavy rainfall conditions following a heavy winter application of de-icing fluids within the airport. The calculated results shown in Figure 2 indicate that both sub-surface flow and surface flow reedbed systems have the capacity to efficiently remove high glycol loadings over short time periods and therefore can provide a "front-end" treatment facility with regard to airport runoff. The recommended reedbed surface areas are 1.04 ha for the sub-surface flow system and 1.30 ha for the surface flow system to achieve 90% glycol removal from highly polluted runoff. However, a concern would be the ability of a substrate-based constructed wetland to cope with the hydraulic conditions which might accompany this situation. The absence of flow restrictions in rafted systems would make them better able to deal with higher hydraulic loadings but the overall evidence from these experiments is that reedbeds should not be used for front end treatment of airport runoff. However, they represent an appropriate treatment facility for use in conjunction with other BOD removal processes such as storage with accompanying aeration. Pilot studies using rafted tanks (5 m by 3 m) produced BOD reductions of 34% during routine treatment of airport runoff over a 1 year period.

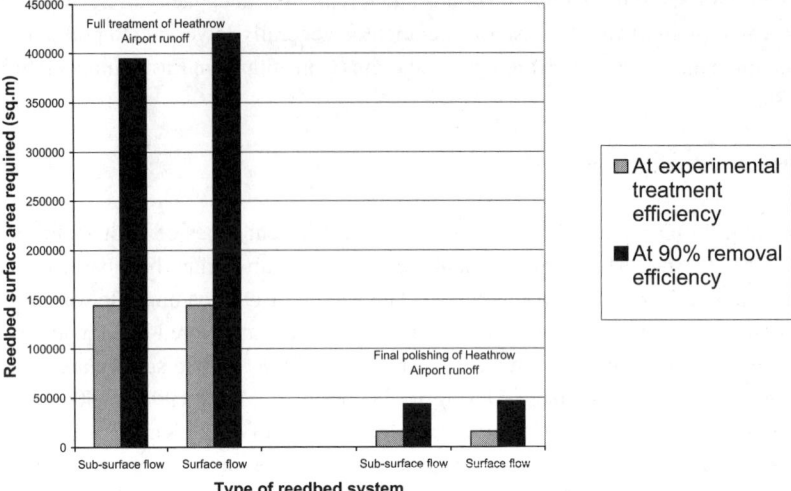

Figure 1 Predicted surface areas of different types of reedbeds for treatment of Heathrow Airport runoff based on BOD monitoring data

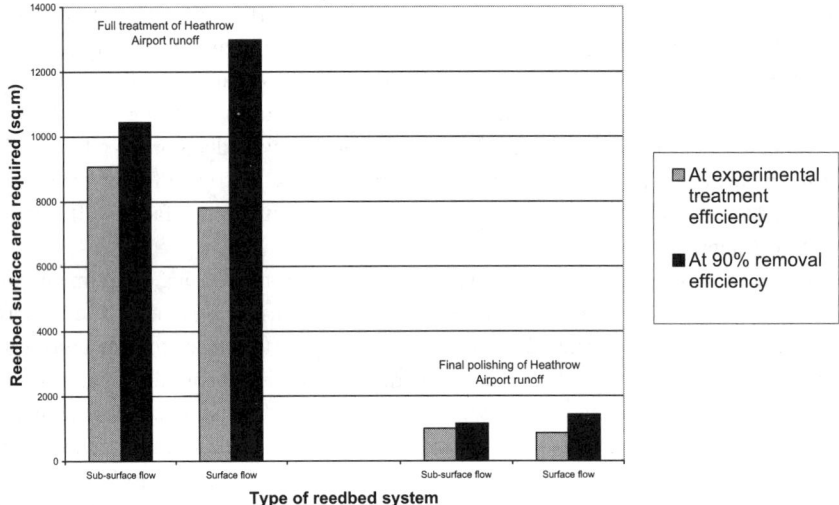

Figure 2 Predicted surface areas of different types of reedbeds for treatment of Heathrow Airport runoff based on glycol monitoring data

Design of new treatment system

A diagrammatic representation of the overall treatment system is shown in Figure 3 together with the planned flow rates between the individual treatment components and the BOD concentrations prior to transfer where these are relevant. The contaminated runoff from the eastern catchment will be intercepted in the top pond of the Eastern Balancing Reservoir and diverted into a separate area of the existing middle pond where initial treatment will be carried out by aeration. Polluted water from the southern catchment will be diverted to a new treatment facility (Mayfield Farm; located 2 km south of the airport) where it will initially be held in an aerated storage reservoir (Main Reservoir). The partly treated runoff from the eastern catchment will be transferred to the Mayfield Farm treatment facility and, after passing through the floating reedbed system, will combine with the partly treated runoff from the southern catchment. The combined runoff will be well mixed in the Balancing Pond before treatment in the sub-surface reedbed system prior to discharge to the Southern Balancing Reservoir and subsequently to the receiving river.

The runoff arriving at the top pond of the Eastern Balancing Pond will be continuously monitored for BOD concentration and during clean water conditions (defined by BOD <100 mg/l in winter and BOD <50 mg/l in summer) the runoff will be directed towards the "Clean" side of the middle pond where aeration will occur prior to discharge to the lower pond and then to the receiving river. The "Clean" side of the middle pond will be separated from the "Dirty" side by a flexible butyl curtain which enables the capacity of the "Dirty" side to be maximised for winter storage (up to 109,000 m^3). Under these conditions a maximum uniform BOD concentration of 240 mg/l could be achieved as determined by modelled determinations of glycol wash-off efficiencies (HR Wallingford, 1999).

A distinctive part of the overall treatment system is the existence of a 3 km transfer pipeline to transport the water from the "Dirty" side of the middle pond (Eastern Balancing Reservoir) to the Mayfield Farm treatment facility. This transfer, to a maximum volume of 42,700 m^3, will be initiated when the BOD concentration falls below a trigger level of 170 mg/l. An aeration system will be installed in the "Dirty" side to facilitate efficient degradation of the glycols during the storage phase. The rate of degradation will be dependent on the water temperature. The minimum and maximum water temperatures monitored during a 2 year survey were 6°C and 20°C, respectively. The relevant degradation rates for *Kilfrost* at these two temperatures are estimated to be 0.114 day^{-1} and 0.300 day^{-1}, respectively and based on these values the respective storage times to achieve a reduction in BOD concentrations in the "Dirty" side from 240 mg/l to 170 mg/l are 3.03 days and 1.15 days (see Table 1). The aeration system will operate for a fixed time period during every 24 hours. This will be determined by the water temperature, measured at a depth of 30 mm, at mid-day and related to the theoretical amount of oxygen which is required to enable efficient degradation of the glycols. A similar time dependent aeration process will also be employed in the two storage ponds located within the Mayfield Farm treatment facility.

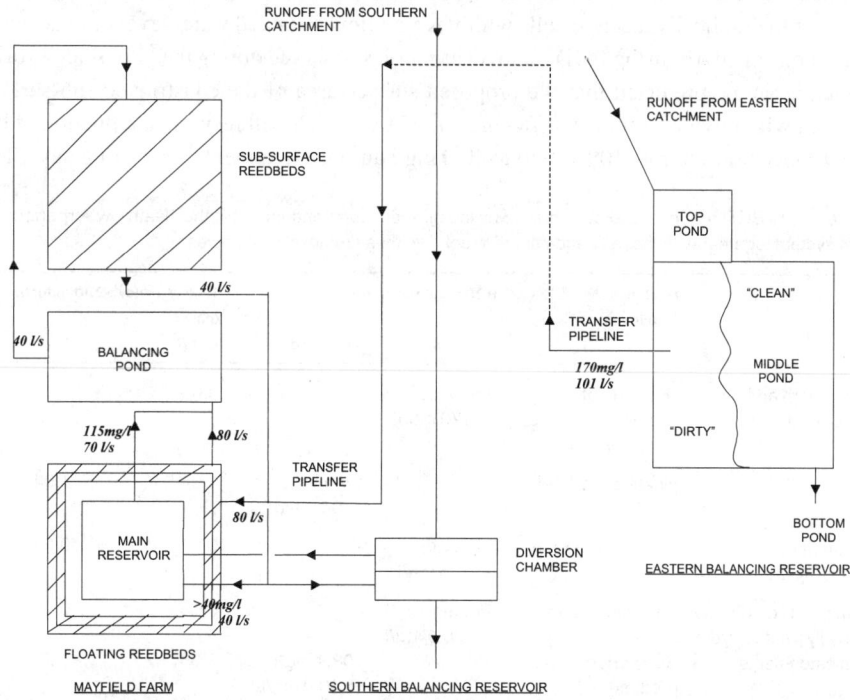

Figure 3 Diagrammatic representation of the new treatment system for runoff from Heathrow Airport

The water transferred to Mayfield Farm from the Eastern Balancing Reservoir flows directly into the floating reedbed system, which covers a surface area of 1.20 ha and consists of two parallel canals, 10 m wide and approximately 500 m in circumference. The flow directions around each of the reedbed canals can be alternated in order to vary the reeds exposed to the inlet flows. The residence time within these reedbeds at the design flow rate of 80 l/s will be 1.04 days. By comparison with the pilot scale studies of BOD reduction by floating reedbeds, it can be predicted that the surface area of the proposed treatment system would achieve a BOD removal efficiency of 43.4% and hence a reduction of BOD concentration as indicated in Table 1.

Runoff from the southern catchment will be intercepted in a new diversion chamber (see Figure 3) with a trigger level of above 40 mg/l determining transfer to the Main Reservoir at the Mayfield Farm treatment facility. Continuous monitoring of the BOD concentration of the effluent within the diversion chamber will enable runoff with a BOD less than 40 mg/l to continue directly to the Southern Balancing Pond. The dirty water, which is diverted to the Main Reservoir, will be subjected to aeration to reduce the glycol levels from an estimated maximum BOD equivalent value of 128 mg/l. The calculated storage times within this water body to reach the BOD trigger level of 115 mg/l are shown in Table 1 and when achieved will initiate a pumping operation to transfer water (45,000 m^3) to the Mayfield Farm Balancing Pond. The partially treated waters from the two catchments will come together in the Mayfield Farm Balancing Pond, which has a capacity of 19,000 m^3. The installed aeration system is required to reduce the initial maximum mixed BOD concentration of 110 mg/l to 108.1 mg/l. The times to achieve this, at the extreme water temperatures of 6°C and 20°C, are given in Table 1. The hydraulic design of the treatment system indicates that the low indicated retention times estimated for the balancing pond will be considerably exceeded (typically approaching 3 days) resulting in a final predicted effluent BOD concentration from the overall treatment system of 30 mg/l at 6°C and 18 mg/l at 20°C.

The final part of the Mayfield Farm treatment facility involves a sub-surface flow constructed wetland covering an area of 2.08 ha. This gravel filled reedbed is divided into a series of hydraulically discrete cells within which flow rates and water levels can be accurately controlled. From the BOD removal capacities observed during the pilot scale experiments, it can be predicted that the proposed surface area of the constructed sub-surface reedbed will provide a removal efficiency of 63.0% which will result in a reduction of the BOD concentration from 108.1 mg/l to 40.0 mg/l during a treatment time of 1.08 days. The

Table 1 The BOD reductions to be achieved by the different components within the Heathrow Airport treatment system together with the residence times to achieve these removal efficiencies

	Treatment Component	BOD reduction (mg/l)		Time to achieve BOD reduction (days)	
		Required	Calculated	At 6°C	At 20°C
Runoff derived from eastern catchment	Dirty side of Eastern Balancing Reservoir	240 mg/l → 170 mg/l	–	3.03	1.15
	Floating reedbeds	–	170 mg/l → 96.2 mg/l	1.04	1.04
Runoff derived from southern catchment	Main Reservoir	128 mg/l → 115 mg/l	–	0.94	0.36
Combined Runoff (partially treated) from both catchments	Balancing Pond	110 mg/l →108.1 mg/l	–	0.15	0.06
	Sub-surface reedbeds	–	108.1 mg/l → 40.0 mg/l	1.08	1.08

examples of the BOD concentrations shown in Table 1, together with the associated reductions, represent the maximum values which are expected to be encountered in the overall Heathrow Airport runoff treatment system. In the majority of pollution runoff events, the initial BOD level in the "Dirty" side of the Eastern Balancing Reservoir will be less than 240 mg/l, therefore the final BOD will be less than the design maximum value of 40 mg/l. In the event that the combined BOD removal capacities do not achieve a reduction to 40 mg/l or below, the treated effluent will be diverted back into the Main Reservoir (see Figure 3) for further treatment by aeration and sub-surface reedbed technology.

Conclusions

A new treatment system has been designed to treat the runoff from a major international airport using a combination of storage, aeration and reedbed technology. The treatment processes have been optimised to enable the highly polluted runoff, which can be produced within the eastern catchment (maximum mixed BOD concentration of 240 mg/l), to be discharged with a BOD of 40 mg/l or less following two storage/aeration procedures and passage through both floating and sub-surface reedbed systems. The total treatment time to achieve clean-up to an effluent BOD of 40 mg/l is dependent on the water temperature and is estimated to be 6.2 days at 6°C and 3.7 days at 20°C. The runoff deriving from the southern catchment is predicted to be less polluted (maximum mixed BOD concentration of 128 mg/l) and the corresponding treatment times are 2.2 days and 1.5 days following consecutive combined storage and aeration treatments and a final polishing within the sub-surface reedbed system.

References

Chong, S., Garelick, H., Revitt, D.M., Shutes, R.B.E., Worrall, P. and Brewer, D. (1999). The microbiology associated with glycol removal in constructed wetlands. *Wat. Sci. Tech.*, **40**(3), 99–107.

Consultants in Environmental Sciences. (1995). Heathrow Airport Southern and Eastern Balancing Reservoirs monitoring study. Report to Heathrow Airport Ltd, CES Ltd, Kent, UK.

Ellis, J.B., Revitt, D.M. and Llewellyn, N. (1997). Transport and the environment: Effects of organic pollutants on water quality. *J. Inst. Wat. & Envir. Mangt.*, **11**(3), 170–177.

Fisher, D.J., Knott, M.H., Turley, S.D., Yonkos, L.T. and Ziegler, G.P. (1995). The acute whole effluent toxicity of storm water from an international airport. *Envir. Toxicol. & Chem.*, **14**(6), 1103–1111.

Geary, P.M. and Moore, J.A. (1999). Suitability of a treatment wetland for dairy wastewaters. *Wat. Sci. Tech.*, **40**(3), 179–185.

Hartwell, S.I., Jordahl, D.M., Evans, J.E. and May, E.B. (1995). Toxicity of aircraft de-icer and anti-icer solutions to aquatic organisms. *Envir. Toxicol & Chem.*, **14**(8), 1375–1386.

Hiley, P.D. (1995). The reality of sewage treatment using wetlands. *Wat. Sci. Tech.*, **32**(3), 329–338.

HR Wallingford (1999). Modelling of glycol wash-off. Report EX 3978, HR Wallingford Ltd, Wallingford, UK.

Huddleston, G.M., Gillespie, W.M. and Rodgers, J.H. (2000). Using constructed wetlands to treat biochemical oxygen demand and ammonia associated with a refinery effluent. *Ecotoxicol & Envir. Safety*, **45**(2), 188–193.

Knight, R.L., Hileke, J. and Grayson, S. (1994). Design and performance of the Champion pilot – constructed wetland treatment system. *Tappi Journal*, **77**(5), 240–245.

Koryak, M., Stafford, L.J., Reilly, R.J., Hoskin, R.H and Haberman, M.H. (1998). The impact of airport deicing runoff on water quality and aquatic life in a Pennsylvania stream. *J. Freshwater Ecol.*, **13**(3), 287-298.

Newman, J.M., Clausen, J.C. and Neafsey, J.A. (2000). Seasonal performance of a wetland constructed to process dairy milkhouse wastewater in Connecticut. *Ecol. Eng.*, **14**(1–2), 181–198.

O'Connor, R. and Douglas, K. (1993). Cleaning up after the big chill. *New Scientist*, **137** (1856), 22–23.

Revitt, D.M., Shutes, R.B.E., Llewellyn, N.R. and Worrall, P. (1997). Experimental reedbed systems for the treatment of airport runoff. *Wat. Sci. Tech.*, **36**(8–9), 385–390.

Sabeh, Y. and Narasiah, K.S. (1992). Degradation rate of aircraft de-icing fluid in a sequential biological reactor. *Wat. Sci. Tech.*, **26**(9–11), 2061–2064.

Tanner, C.C., Clayton, J.S. and Upsdell, M.P. (1995). Effect of loading rate and planting on treatment of dairy farm wastewaters in constructed wetlands. 1. Removal of oxygen – demand, suspended solids and fecal coliforms. *Wat. Res.*, **29**(1), 17–26.

Vymazal, J. (1999). Removal of BOD_5 in constructed wetlands with horizontal sub-surface flow: Czech experience. *Wat. Sci. Tech.*, **40**(3), 133–138.

Worrall, P., Peberdy, K.J. and Millet, M.C. (1997). Constructed wetlands and nature conversation. *Wat. Sci. Tech.*, **35**(5), 205–213.

Investigation of copper adsorption to peat using the simple metal sorption model

S. Dierks

Limno-Tech, Inc., 501 Avis Drive, Ann Arbor, Michigan, 48108, USA. (E-mail: *sdierks@limno.com*)

Abstract The Simple Metal Sorption (SiMS) equilibrium model was used to simulate the proton/cation exchange behavior of peat with dissolved copper. The SiMS model represents proton binding and metal binding as cation exchange for heterogeneous sorbents as a function of pH, salt concentration, total metal concentration and total ligand concentration. The SiMS model uses fewer parameters than other cation exchange models for multidimensional datasets and can be executed on a standard spreadsheet. The cation exchange selectivity coefficient, $K_{Me,app}$, is represented as $K_{Me,app} = K_{Me}\{H^+\}^\alpha (L_T/Me_T)^\beta I^\varphi$. The model is similar to standard surface complexation approaches, with an intrinsic relationship described by mass action laws (K_{Me} = metal equilibrium constant) and variable terms that are expressed as simple power functions of proton concentration, ligand to metal ratio (L_T/Me_T), and ionic strength (I). The model successfully simulated the proton exchange behavior of acid-washed, *Sphagnum* peats over a range of 4 to 8 pH units with ionic strength differing by three orders of magnitude (I = 0.001 to 0.1). Simulation of copper binding on five peat data sets and the dried biomass of *Potamogeton lucens* was also successful ($0.94 < r^2 < 0.99$). However, there was no apparent relationship between model parameters and peat characteristics. Incorporation of the SiMS model into a framework for predicting metals removals in wetlands will require more work.

Keywords Cation exchange selectivity coefficient; ion-exchange; peat; surface complexation model

Introduction

Constructed wetland systems are finding increasing use for treatment of high volume, dilute, heavy metal wastewaters, such as landfill leachate and mine drainage. However, predicting the removal or conversion of metals in a wetland, without site-specific data, is still beyond the state of the science. Most metals treatment research has focused either on relating influent and effluent concentrations or partitioning metal removal into wetland storages such as sediment, shoots, roots and detritus. These studies have led to the conclusion that heavy metal removal is highly variable and difficult to predict (Kadlec and Knight, 1996). Even site-specific partitioning coefficients are not useful when field conditions, such as total metal and ligand concentration, ion composition, pH, and E_H, fall outside tested conditions.

A structured approach for understanding the underlying source of this variability is rarely attempted. One reason for this is that understanding the fate of metals in a wetland is fundamentally different than understanding the fate of compounds that are broken down or consumed, such as organic contaminants. Metal contaminants are ultimately conserved and their fate can generally be understood by determining which species is more thermodynamically stable under field conditions. As conditions warrant, a metal can be mobile as a dissolved or complexed species or adsorbed or incorporated into minerals, and plant and animal tissue.

Studies have shown that wetland sediments are the biggest sink for metals in wetlands (Kadlec and Knight, 1996). These sediments capture settling particulate metal and adsorb dissolved metals. Predicting the behavior of the inorganic fraction of these sediments and their interaction with heavy metals is supported by several chemical speciation equilibrium models, such as MINTEQA2 and MINEQL. Predicting the behavior of heavy metals with organic matter under a wide variety of conditions has not been achieved.

One of the components of sediments that acts as a significant metals trap is peat. For peat, the most abundant and effective surface-active components that contribute to metal adsorption are humic and fulvic acids (Kadlec and Keolian, 1986). The sites of the phenolic and carboxylic structural groups that create this acidity are occupied by protons or cations like Ca^{2+}, Mg^+, etc. (the earth metals). As pH rises, or cations such as bivalent heavy metals, with higher selectivity for these sites are introduced, the protons and cations with lower site affinity are replaced (exchanged). These protons and cations can also be displaced by dissolved complexes, such as $Cu-NO_3$. While electrical attractive forces, like those arising from a diffuse ion swarm can also influence metal adsorption (by enhancement or reduction), ion exchange and complex exchange are probably the main mechanisms for heavy metal adsorption onto peat.

Most of the models striving to describe ion exchange relationships between peat, pH, metals, and other ions can generally be classified as electrostatic, discrete ligand or continuous distribution models. Electrostatic approaches (e.g., Debye-Huckel or Gouy-Chapman theory) require detailed structural information. Discrete site models sum the individual binding isotherms for each particular site to approximate the "smeared" adsorption curves of heterogeneous sorbents. Alternatively, the effective sum of all the sites is approximated with a continuous affinity distribution function. However, the latter two approaches tend to require the optimization or estimation of a wide number of model parameters (Ganguly *et al.*, 1999).

The SiMS (Simplified Metal Sorption) Model is an attempt to develop a relatively simple model accurate enough for engineering applications. It is a semi-empirical surface complexation model that uses empirical correction terms for the surface equilibrium constants that are expressed as simple power functions of H+, ionic strength and metal to ligand ratio. These correction terms are an attempt to represent the macroscopic system response to electrostatic and/or heterogeneity effects (Ganguly, 1999) without recourse to prior knowledge of the peat's particular structure. Ultimately, the goal is to incorporate the SiMS model into a larger equilibrium speciation model framework, such as MINTEQA2 or MINEQL.

Background

The SiMS model simultaneously represents proton binding and metal binding as a function of pH, salt concentration, total metal concentration and total ligand concentration. The SiMS model assumes that reactions involving protons and metal ions in a heterogeneous mixture of binding ligands can be represented by lumped macroscopic reactions. In a manner similar to standard surface complexation models, the equilibrium constant for adsorption in SiMS consists of an intrinsic component described by mass action laws, and variable terms (analogous to electrostatic correction factors) that are expressed as power functions of H+, metal to ligand ratio and ionic strength.

The macroscopic proton-binding stoichiometry of the ligand, L (peat), is written as a single monoprotic acid:

$$HL = L^- + H^+ \tag{1}$$

where HL is understood to be the protonated sites on peat. The apparent equilibrium constant for (1) includes an intrinsic constant (K_a) and correction factor that is assumed to be a power function of H+ (Note: [ ] = molar concentrations; { } = ion activities):

$$K_{a,app} = K_a \{H^+\}^{\alpha_1} = \frac{\{L\}\{H^+\}}{\{HL\}} \qquad (2)$$

Binding of the divalent metal ion (Me^{2+}) to the peat ligand is written as an exchange reaction with H+:

$$HL + Me^{2+} = MeL^+ + H^+ \qquad (3)$$

The apparent metal equilibrium binding (cation exchange selectivity) constant for (3), $K_{Me,app}$ also requires a correction factor that is a function of H^+. The apparent metal equilibrium binding constant also accounts for the effects of varying metal/ligand ratios and ionic strength on metal binding:

$$K_{Me,app} = K_{Me}\{H^+\}^{\alpha_2}\left(\frac{Me_T}{L_T}\right)^{-\beta} I^{\varphi} = \frac{[MeL^+]\{H^+\}}{[HL]\{Me^{2+}\}} \qquad (4)$$

where MeL^+ = bound metal species; K_{Me}, α_2, and b = fitting parameters; φ = constant (φ = –0.5); I = solution ionic strength and Me_T and L_T = total metal and complexing ligand concentrations, respectively. In some work, the metal ion is visualized as binding with two moles of ligand. But as Sposito (1989) pointed out, as long as the equations balance, prior knowledge of the structure of the acid functional groups is not required to determine the total peat charge. Therefore, either representation is suitable.

Mass balances are written for total metal and ligand concentration:

$$Me_T = [Me^{2+}] + [MeL^+] \qquad (5)$$

$$L_T = [HL] + [L^-] + [MeL^+] \qquad (6)$$

Solution phase metal complexes, such as carbonates or hydroxides, are included in the model, with the corresponding association constants taken from the literature. Activity correction coefficients are applied to all solution species using the Davies equation (Stumm and Morgan, 1996).

Methods

Data for this application of the SiMS model were taken from the literature. Studies by Marinsky *et al.* (1980) and Crist *et al.* (1999) were used for the peat acid-base titration data; while studies by Gossett *et al.* (1986), Chen *et al.* (1990), Ho *et al.* (1994), and Schneider *et al.* (1999) were used for the copper exchange data.

Data

Acid-base titrations. The peat (*Sphagnum*) samples in the peat acid-base titrations were shredded, sieved and pre-washed with a strong acid, such as HCl to protonate all available acid sites. In the experiments 1–3 g of peat (measured as dry weight) was immersed in 50–100 ml de-ionized water suspensions. For Marinsky *et al.* (1980) titrations were run at three ionic strengths: I = 0.1M; I = 0.01M and I = 0.001M NaCl. Crist *et al.* (1999) ran one titration at 1M NaCl. The peat suspensions were treated with small amounts of a strong acid or base, such as HCl and NaOH to adjust pH and affect proton binding. pH changes not attributable to the acid or base addition are assumed to be due to the adsorption or release of protons from the peat. However, binding capacity is not measured in terms of pH changes, but rather as equivalents exchanged between the peat and the solution. This method is an evaluation of the potential cation exchange capacity (CEC).

Copper adsorption studies. Three types of peat were used in the copper adsorption studies: *Sphagnum*, eutrophic and dried *Potamageton lucens* tissue (Schneider et al., 1999). The *Sphagnum* peat used by Gosset et al. (1986) and Chen et al. (1990) was described as oligotrophic, i.e., more acidic and containing more organic material than the eutrophic peat. Chen et al. (1990), characterized the eutrophic peat as poor in cellulose but rich in humic substances.

For the copper adsorption studies, peat (1–5 g dry weight) was shredded, sieved and washed with distilled water. This pretreatment was meant to standardize the form of the peat for experimentation without losing its initial electrochemical conditions, such as pH and cation composition. In some experimental work (Marinsky et al., 1980), metal exchange studies are run at different ionic strengths in the same manner as the titrations. However, in Marinsky et al. (1980), the reported calculations used to transform the data were not complete enough to recreate the historical data set for use with the SiMS Model. The other studies used here did not adjust ionic strength as part of the experiment.

The peat samples were then spiked with known concentrations of dissolved copper (Gosset = 10 mM; Chen = 10 mM; Schneider = 0.1 mM; Chen = 0.3–3 mM) and allowed time to equilibrate. Typically, equilibrium is reached on a time scale of minutes to hours. The samples are then centrifuged and the supernatant sampled for dissolved copper with the difference assumed to be adsorbed to the peat.

Calculations

The SiMS model was implemented on a standard spreadsheet (MS EXCEL). For the acid-base titrations the proton balance equation was used as the basis for fitting the model to the experimental data. Taking HL as the reference species, the calculated proton balance equation is given as:

$$T_{Hcalc} = [H+] - [OH-] - [L-] \qquad (7)$$

The experimental value T_{Hexp} (expressed in terms of equivalents) is defined by the quantity of strong acid (Ca) or strong base (Cb) added to the solution:

$$T_{Hexp} = Ca - Cb + T_H^0 \qquad (8)$$

T_H^0 represents the equivalent concentration of strong acid or base initially present as equivalents on the peat. However, determination of this value is problematic; therefore, following Ganguly et al. (1999) T_H^0 is a fitting parameter. EXCEL's SOLVER function is used to iteratively select appropriate parameter values that minimize the sum of squares between the fitted and experimental data. For the acid-base titrations L_T, K_a, α_1 and T_H^0, were the parameters fit. Using the Marinsky et al. (1980) acid-base titration parameter fit, K_{Me}, α_2 and β were then fitted to the metal exchange data.

Results

Figure 1 below shows an example of a titration fit while Table 1 summarizes the calculated parameter values. Plots of $K_{a,app}$ (for all ionic strength solutions) versus pH yield approximately the same slope, –0.86 and –0.92 respectively, for Crist et al. (1999) and Marinsky et al. (1980) (see Figure 2). Since $K_{a,\,app}$ and L_T are the only acid-base model parameters needed for the copper adsorption model, the results of these studies were considered similar. The differences for K_a and T_H^0 are probably due to differences in the initial acid characteristics of the peat before the titration. However, the L_T values, essentially the concentration of all cation exchange sites, seem high. Sposito (1989) gives cation exchange

Table 1 Parameter fits for acid-base titration data

Study	K_a	α_1	L_T	T_H^0	r^2
Crist et al.	0.00162	0.841	12.5	0.08	0.995
Marinsky et al.	0.817	0.942	11.9	−7.13	0.966

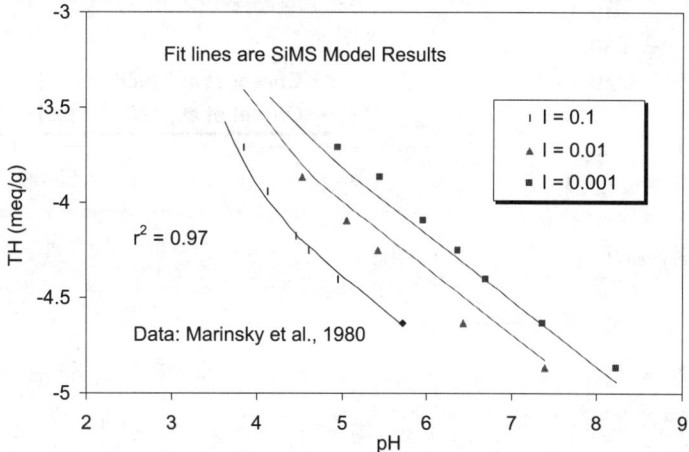

Figure 1 Example of acid-base titration model fit

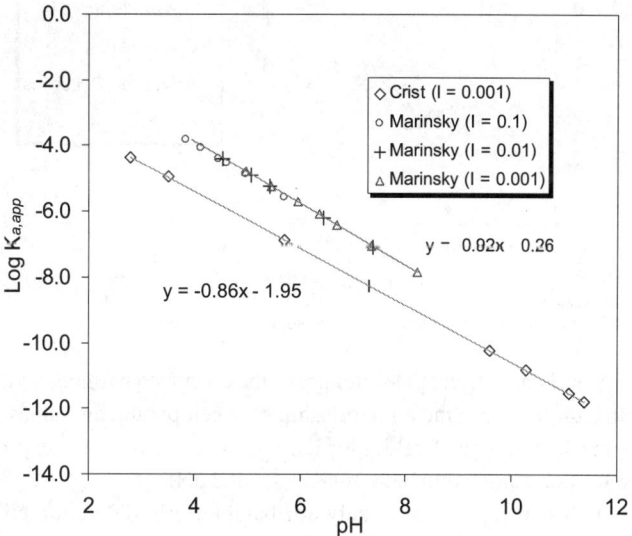

Figure 2 Variarion of the apparent proton exchange equilibrium constant with pH

capacities (CEC) for peat ranging from 1–4 meq/g. However, Sposito notes that CEC tests typically measure "readily exchangeable ions". There may be a quantitative difference between the site densities quoted by Sposito (as readily exchangeable ions) and the CEC estimated for completely protonated peat by Crist et al. (1999) and Marinsky et al. (1980).

Copper adsorption studies

Figure 3 below demonstrates two examples of the copper adsorption fits. As with the titration fits, SiMS does an adequate job of representing the range of behavior. However, as shown by the relationship between the calculated copper exchange selectivity coefficient

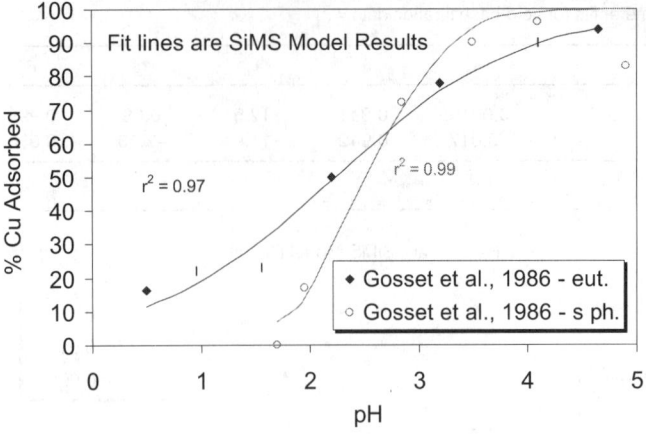

Figure 3 Examples of copper adsorption model fit

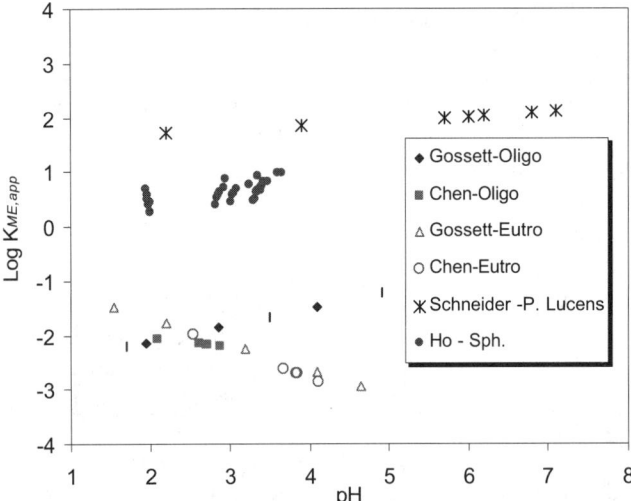

Figure 4 Comparison of calculated copper exchange selectivity coefficient

($K_{Me,app}$) and pH in Figure 4, and the summary of the estimated parameter values in Table 2 below, it is difficult to determine a relationship between parameter values and peat type. Not only is there a large range of values for $K_{Me,app}$ but there is also, inexplicably, a divergence of slope for the relationship between $K_{Me,app}$ and pH.

Differences between $K_{Me,app}$ are likely attributable to initial cation composition and copper ion interaction with other potential binding sites, such as dissolved organic acids or complexes. Chen et al. (1990) suggested that a Cu-NO_3 complex may have adsorbed to peat thereby increasing the apparent exchange of the copper ion form.

Table 2 Parameter fits for copper selectivity expressions

Study	K_{me}	α^2	β	r^2
Gosset, Sph.	0.255	−0.306	−0.463	0.973
Gosset, Eut.	0.288	0.474	0.118	0.985
Chen, Sph.	0.221	0.561	0.647	0.981
Chen, Eut.	0.523	0.163	−0.972	0.957
Ho, Sph.	1.55	−0.163	0.431	0.991
Schneider, P. lucens	5.88	−0.077	47.7	0.938

Conclusions

This investigation suggests that with additional work a model such as SiMS might offer a way to forecast metal removals with peat over a range of solution conditions. By incorporating SiMS into a chemical equilibrium model, a quantitative framework for advancing understanding of metals removals in wetlands may be achieved.

While this work also suggests that a preconceived idea of physical peat structure may not be necessary for understanding metals adsorption, some understanding of initial conditions, in terms of cation composition, is probably necessary. Wolf *et al.* (1977) used two different peat pretreatments to compare heavy metal ion selectivity for different ions, one an acid-washing and the other a calcium saturation treatment. There was a higher "selectivity" for the H^+ protons as compared to Ca^{2+} and a higher selectivity of heavy metal ions for Ca^{2+} sites than for H^+ sites. The Ca^{2+} ions are less tightly bound and are therefore more easily displaced by heavy metal cations.

The importance of representing the initial condition of cation exchange sites as occupied by protons or by cations requires more investigation. There is also an open question regarding model parameter uniqueness. It is unclear how much of the variability between fitted parameter values is due to true adsorption differences or to the fallibility of a technique that can generate a large set of reasonable fits. Additional modeling work will first focus on empirical adjustments to the one-site model to avoid adding sites and complexity. Also, structured peat laboratory studies are recommended that isolate the impacts of intial peat proton/cation composition, complexes and the competition between dissolved and solid organic material on metal ion fate.

Acknowledgement

The author is indebted to Chandra Ganguly for providing invaluable assistance with the SiMS Model. He would also like to thank Bob Martin and William J. Reimjsdek for providing data.

References

Chen, X., Gosset, T. and Thevenot, D.R. (1990). Batch Copper Ion Binding and Exchange Properties of Peat. *Wat. Res.*, **24**(12), 1463–1471.

Crist, R.H., Martin, R. and Crist D.R. (1999). Interaction of Metal Ions with Acid Site of Biosorbents Peat Moss and Vaucheria and Model Substances Alginic and Humic Acids. *Env. Sci. Tech.*, **33**, 2252–2256.

Ganguly, C., Huang, C., Rabideau, A.J., Van Benschoten, J.E. (1999). Simple Metal Sorption Model for Heterogeneous Sorbents: Applications to Humic Materials. *J. of Environmental Engineering*, **125**(8), 712–720.

Gosset, T., Trancart, J. and Thevenot, D.R. (1986). Batch Metal Removal by Peat. *Wat. Res.*, **20**(1), 21–26.

Ho, Y.S., Wase, D.A.J., and Forster, C.F. (1994). Adsorption of Divalent Copper Ions from Aqueous Solution by Sphagnum Moss Peat. *Trans. IChemE*, 72 (Part B): 185–194.

Kadlec, R.H. and Keolian, G.A. (1986). Metal Ion Exchange on Peat. In: *Peat and Water*, Fuchsman, C.H. (ed). Elsevier Applied Science Publishers, Ltd., 1986.

Kadlec, R.H. and Knight, R. (1996). *Treatment Wetlands*. CRC Press, Boca Raton, Florida.

Marinsky, J.A., Wolf, A. and Bunzl, K. (1980). The Binding of Trace Amounts of Lead(II), Copper(II), Cadmium(II), Zinc(II) and Calcium(II) to Soil Organic Matter. *Talanta*, **27**, 461–468.

Schneider, I.A.H. and Rubio, J. (1999). Sorption of Heavy Metals Ions by Nonliving Biomass of Freshwater Macrophytes. *Env. Sci. Tech.*, **33**, 2213–2217.

Sposito, G. (1989). *The Chemistry of Soils*. Oxford University Press, New York.

Stumm, W. and Morgan, J.J (1996). *Aquatic Chemistry: Chemical Equilibria and Rates in Natural Waters*, 3rd Ed., John Wiley and Sons, New York.

Wolf, A., Bunzl, K., Dierl, F. and Schmidt, W.F. (1977). The effect of Ca2+ ions on the adsorption of Pb2+, Cu2+ and Zn2+ by humic substances, *Chemosphere* **5**, 207–213.

Design and performance of experimental constructed wetlands treating coke plant effluents

N. Jardinier*, G. Blake*, A. Mauchamp** and G. Merlin*[1]

* TEPE, Ecole Supérieure d'Ingénieurs de Chambéry (ESIGEC), Université de Savoie, 73376 Le Bourget du Lac, France
** La tour du Valat, Le Sambuc, 13200 Arles, France
[1] To whom correspondance should be addressed

Abstract Reed beds were chosen to treat effluents from a coke plant in France (Usinor-Sollac, Fos/mer). The pilot is composed by a two-stage gravel bed with subsurface flow and *Phragmites australis* as plant. This experimental constructed wetland was monitored for one year at steady-state conditions. The composition of influent shows high concentrations of organic compounds. The hydraulic residence time was close to 10–12 days with a plug flow with longitudinal dispersion. Results show that global removal of nitrogen ranged from 54 to 94% of load removal efficiency, but corresponds easily to the regulation recommendations. Because of wintertime, the denitrification process was inhibited by aerobic conditions observed in the gravel bed with oxygen concentrations higher than 2–3 mg/L, and by small amounts of biodegradable carbon. The fate of mineral pollutants are linked to the complex ferric hydroxides balance and a lack of phosphorus was observed for reed plants, as this nutrient is dependent on iron compounds. Some necrosis was observed on plant tissues corresponding with anthocyanic pigments accumulation caused by phosphorus absorption deficiency due to its co-precipitation with iron.
Keywords Coke plant; constructed wetland; nitrogen; phosphorus; *Phragmites australis*

Introduction

Constructed wetlands are today widely used for wastewater management and water pollution control (Vymazal *et al.*, 1998; Haberl, 1999). Although the use of constructed wetlands is well established for wastewater such as domestic, stormwater, runoff or acid mine drainage waters, their use in treating specific industrial effluents is less developed. The Usinor Company (Sollac, Fos/mer, France) intended to develop this low cost and low risk technology to treat cokery effluents (gas washing liquor). These effluents have a high ammonium concentration, high levels of iron and traces of organic compounds. A pilot study was started in 1998 to evaluate the potentialities of constructed wetlands for pollutants removal and to evaluate the tolerance of reed. The main objective was to obtain with a low-cost biological treatment, a realistic and long-term possibility of decreasing high concentration of NH_3, which is toxic for the fauna of the sea, which receives the effluents.

This paper reports the results of a two years monitoring of a pilot-scale system for the treatment.

Methods

The pilot scale constructed wetland was established in spring 1998. The experimental system consisted in a two-stage subsurface flow, 2 m wide × 6 m long × 0.6 m deep each, for a total area of 24 m^2 and a total volume of 5 m^3 according to a mean porosity of 0.35. The bottom of each basin was designed with a 2% slope and all slide slopes of the wetland cells were constructed on a 2:1 ratio. Basins were sealed off by a synthetic rubber liner. On the first stage, at the inlet and the outlet of the basin there was free water and large stones zones, on the second stage another free water zone was added in the middle of the basin (Figure 1).

Between the large stones zones, basins were filled with coarse washed gravels (6–12 mm diameter) composed by 70% of SiO_2 and 30% of CaO. Common reeds (*Phragmites australis*) were planted in gravel zones, initial planting density was approximately 20 plants per m².

Water level was set about 5 cm below the bed surface. The influent flow rate was 0.75 m³/d pumped in a lagoon receiving coke plant effluent treated in a conventional activated sludge plant. Hydraulic residence time was determinated using fluoroscein at 5 g/L as tracer in a pulse test experiment.

Temperature, pH and Dissolved Oxygen were measured on site using portable instruments. Determination of TSS and COD was made according to *Standard Methods* for the examination of water and wastewater (APHA, 1992). Samples were collected weekly. Total Kjeldahl Nitrogen was determinated using French Standard procedures (AFNOR, 1995).

Phosphorus analysis

Plant samples were analysed for phosphorus by Laboratoire Wolff Environnement (Marseille, France) by microwave extraction according to procedure ISO 11885 03/98.

Metals analysis

Sediment and plant samples were analysed in duplicate by Laboratoire Wolf Environnement (Marseille, France). Samples were oven dried at 100°C for 24 hours and digested with concentrated nitric acid and with a concentrated nitric acid and perchloric acid mixture (9:1 v/v) respectively and analysed using inductively coupled plasma emission spectroscopy. For each pollutant under study, removal rates were calculated as follows: % difference = [(inlet – outlet)/inlet]

For COD and nutrients, removal efficiencies take account of retention time and values are the means of two weekly samples.

Results and discussion

Hydrological characteristics

Hydrology regulates all the characteristics of wetland structure and function; the vegetation and the microorganisms are adapted to the hydrologic regimes (Gopal, 1999). The mean hydraulic residence time (ts) calculated from the tracer response curves is 3.4 d for the first stage and 5.3 d for the second stage at a flow rate of 0.75 m³/d for the first stage and 0.39 m³/d for the second stage. An example of response curve for the first stage is given in Figure 2. The response curves at a different stage are intermediate between continuous still

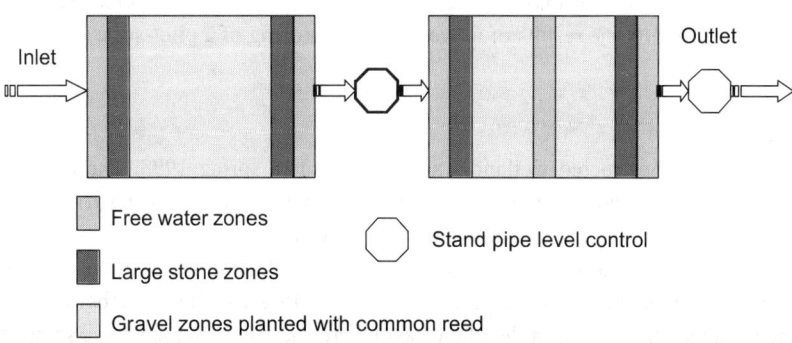

Figure 1 Schematic design of the pilot scale constructed wetland

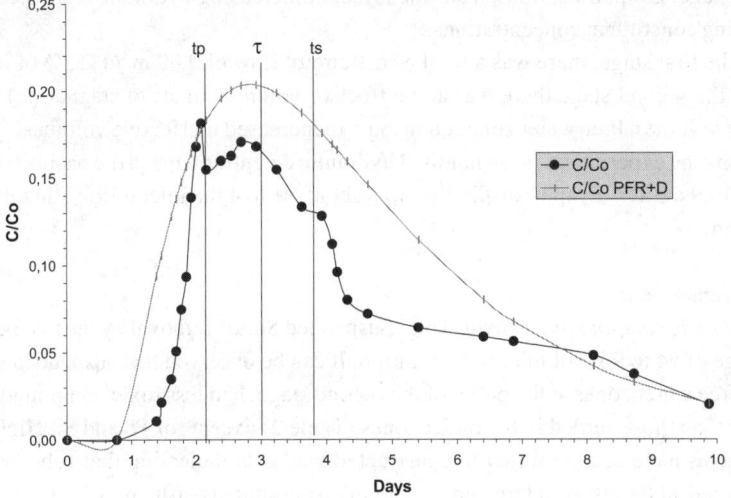

Figure 2 Tracer concentration time distribution in stage 1

tank reactor (CSTR) and plug flow reactor (PFR). A better modeling was obtained with a plug flow with longitudinal dispersion (Figure 2), using:

$$C/Co = 0.5 \, (Pe/\pi\tau t)^{0.5} \exp[-Pe(t-\tau)^2/4t\tau]$$

(Levenspiel, 1972; Gourlia, 1998) where $Pe = 1/D = u\,L/D$ and D = dispersion constant (m^2/d), u = velocity (m/d), L = length of the wetland (m); $\tau = V/Q$ (volume/volumetric flow rate) (d); t = time (d); $Co = n/V$ (tracer amount/volume); C = tracer concentration at time t.

A least squares fit of the model to experimental data results in a wetland dispersion number (D) of 0.25 reported as a typical value for Subsurface Flow constructed wetlands (Kadlec and Knight, 1996).

But this model doesn't describe adequately the tracer response because flow through the beds is certainly not uniform. The nominal hydraulic detention time (τ) is 3 d for the first stage and 7 d for the second at flow rates tested during tracer experiments. Difference in detention time between the two stages was due to evapotranspiration (ET). During tracer experiments, ET was estimated at 14 mm/d ± 10 for the first stage and 9 mm/d ± 5 for the

Table 1 The characteristics of coke plant effluent at the inlet of the reed-beds (mean values in 1998)

Parameters	Concentration (mg/L) or value	Loading (kg/d)
TSS	12.7	24
COD	140	248
BOD$_5$	6.3	10.6
TOC	25.8	45.6
Ntot	76.2	137
Ptot	0.21	0.112
Hydrocarbons	0.95	1.72
Cyanides	0.03	0.05
Fe	2.66	4.6
Al	0.5	0.9
PAH	<0.001	<0.002
Temperature (°C)	17.4	
pH	6.8	

second stage. Evapotranspiration has the effect of increasing hydraulic residence time and increasing constituent concentrations.

For the first stage, there was a weak short-circuit flow of 0.09 m³/d (12% of inlet flow) and for the second stage there was an ineffective volume (zones of stagnation) of 1.3 m³ (40% of volume). Free water zones contribute to increased ineffective volumes.

During the experiment, mean nominal hydraulic detention time in the basins (stage 1 and 2) was 10.9 d ± 1.7. Evapotranspiration was about 50% of the inlet inflow, thus 12.7 mm/d ± 2.8 mm.

Removal efficiencies

TSS and toxic compounds removal. Total Suspended Solids removal by the two beds was in the range of 94 to 96% of inlet concentration. It can be observed that suspended solids settled in free water zones at the outlet of the second stage, had less toxic compounds concentrations than those settled in the outlet zones (Table 2) except for Pb and Si. High removal efficiencies have been reported in constructed wetlands indicating that sub-surface flow constructed wetlands would treat metal and micro-pollutants effluents efficiently (Mungur et al., 1997).

Nitrogen removal. Total nitrogen concentration varied in the inlet between 60 and 110 mg/l. The variation in outlet was parallel and resulted in concentrations at the outlet ranging between 10 and 70 mg/l (Figure 3 and Table 3). In terms of loading, removal efficiency (taking into account the retention time) was up to 95% in October 98 and was never below 54%, and corresponds easily to the regulation recommendations (Figure 3).

Table 2 Metals and PAH concentrations in suspended solids settled down in free water zones at the inlet and the outlet (mean of two samples)

Toxic compounds	Inlet mg/kg	Outlet mg/kg
As	18	<5
Hg	<0,05	<0,05
Zn	179	148
Si	126	833
Pb	23	26
Al	30,393	21,655
Se	62	<5
PAH tot	13,6	7,5

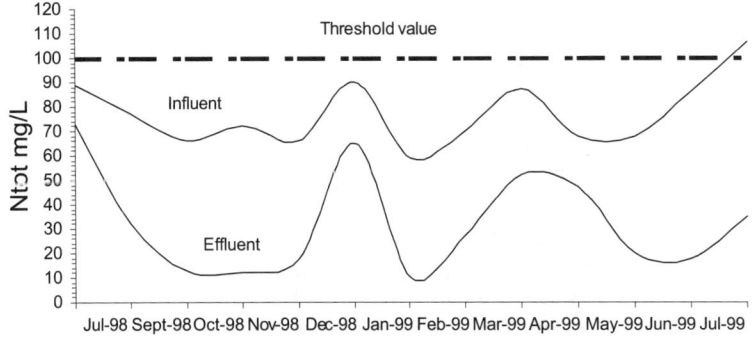

Figure 3 Influent and effluent concentrations of total nitrogen (values are mean of two weekly samples)

Seasonal changes can be observed (Figure 3), and variables may be represented by sinusoidal variation around an annual mean (Kadlec, 1999). Because of wintertime, the denitrification process was inhibited by aerobic conditions observed in the gravel bed with oxygen concentrations higher than 2–3 mg/L, and by small amounts of biodegradable carbon, carbon addition (about 10 kg of reed leaves in free water zones) in May 1999 increasing removal efficiency

Sedimentation of solids and mineralization were certainly the main processes contributing to the removal of organic nitrogen. Nitrification was indicated by the increase of N-NO2 concentration, and decrease of NTK concentrations underlined the loss of N over the whole process (Table 4).

COD removal. Removal of COD ranged from 5 to 86% of inlet loadings or from 35 to 52% of inlet concentrations (Table 5 and Figure 4). These relatively weak values and large variability can be explained by type of carbon pollutants (weak biodegradability). These kinds of organic pollutants, present at high concentration (Table 1) may be toxic to plants and micro-organisms or may form complex molecules with metals such as iron for example, in reducing biodegradation and removal efficiency. However, as for total nitrogen, outlet concentrations correspond easily to the regulation recommendations. These recommendations should be more severe in the future.

P removal and P contents in macrophytes. In the effluents, the P concentration was 0.21 mg/L on average for the year. This concentration represented a P load of about 6.6 mg/m^2.d and the mean year mass removal efficiency was 83.6%. The high values of iron in this kind of effluent can explain the lack of phosphorus in the macrophytes due to the precipitation of P on the iron oxide plaques of their roots. Iron can react with soil anions. In wetland situations, the rooting zone of emergent plants is oxidized when oxygen is present and reducing chemicals are in low concentrations. In our case, visual aspects of the roots showed both oxidized and reduced zones on roots (Crowder and St-Cyr, 1991).

Visual aspects of macrophytes have shown purple stains and spots on stems and leaves. These stains have been analysed and gave evidence of anthocyanic pigments. It has

Table 3 Highest and lowest removal efficiencies for nitrogen (in %)

Removal efficiency	Lowest (loading)	Lowest (concentration)	Highest (loading)	Highest (concentration)
Value (%)	54	23.5	95	83
Date	December 98	April 99	October 98	September 98

Table 4 Nitrogen outlet and inlet concentrations (mg/L) in summer 1999

| | Influent | | | | Effluent | | | | |
	N-NO2	N-NO3	NTK	N TOTAL	N-NO2	N-NO3	NTK	N TOTAL	N Removed
Jun. 99	0.9	4.9	84.1	89.9	5.1	2.7	33.6	41.4	48.5
Jul. 99	1.2	1.4	86.9	89.5	14.7	5.6	52.5	72.8	16.7
Aug. 99	1.0	1.8	74.0	76.8	6.0	2.5	23.0	31.5	45.5

Table 5 Highest and lowest removal efficiencies for COD (in %)

Removal efficiency	Lowest (loading)	Lowest (concentration)	Highest (loading)	Highest (concentration)
Value (%)	5	34.6	85.8	51.6
Date	July 99	July 99	January 99	September 98

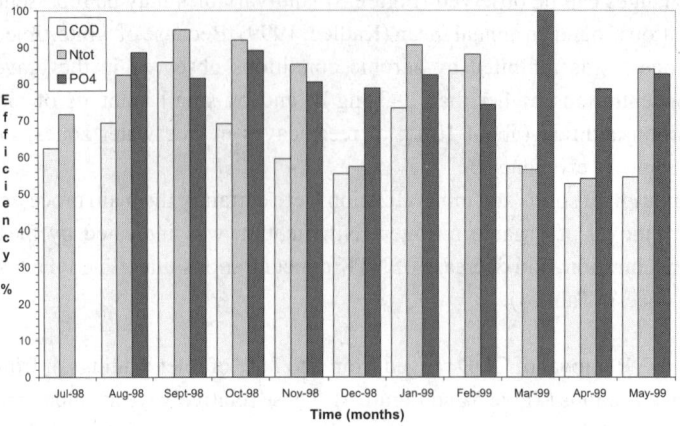

Figure 4 Loading efficiencies of elimination for COD, Ntot, and PO$_4$ (in % of mass removal)

been shown that these pigments are in high concentration when P is lacking (Moore *et al.*, 1998). The iron oxide plaques may also reduce Mn in plants, but no symptom of deficiency of this metal has been noticed. Iron oxide plaques reduce the availability of P for plants with synthesis of Fe(III) complex and these plaques settles itself complexation surfaces for metals and phosphorus compounds. These complexations are pH dependent and in our conditions, the pH values are suitable for this immobilisation of P (pH close to 7).

For the macrophytes of the pilot, all the conditions are present for decreasing capacities of P incorporation: a) low concentration of P in effluents, and b) high capacity of complexation on iron plaque (mainly in the upper parts of the substratum where drying conditions predominate).

As the question of the role of iron plaque to be a P resource or a P well may be asked, we analysed P content of *Phragmites australis*.

In our pilot scale experiments, the phragmites have shown deficiency in P stem and leaves contents versus lagoon or control situations (Table 6). Lagoon and control reeds are blank situations and were not exposed to normal concentration of iron. In this case and for this first experiment data, we can expect that the hypothesis made on the effects of the iron plaque is valid. Experiments carried on, in laboratory, with high concentrations of iron have reproduced similar stains on leaves, but with external conditions (unpublished results), more data for a long period of functioning is necessary to confirm and to assess this phenomenon.

Conclusion

In this pilot-scale experiment, we have shown that treating coke plant effluents by reedbeds may be a valid method in order to decrease nitrogen concentrations above the regular threshold and also to trap some toxic compounds as PAH and metals. The necrosis and

Table 6 P concentrations in different part of common reed

P contents (mg/kg)	Pilot	Lagoon	Control
Roots Feb-99	500	650	1,080
Apr-99	980	3,200	920
Rhizomes Feb-99	280	821	610
Apr-99	210	750	2,400
Leaves Feb-99	–	–	–
Apr-99	1,200	3,200	2,400

stains observed on leaves and stems of the plants may be linked to high concentrations of iron that are responsible for phosphorus deficiency due to co-precipitation on iron oxide plaques on roots or physiologic disruptions in nutrients assimilation. Despite those symptoms, reed plant grew well and may sustain the system in the long term. Beyond the removal of nitrogen, such a device traps other compounds and, even if removal of these chemicals are not yet required for regulatory demand at the moment, a large advantage of this system is to act on the whole effluent rather than on only one compound.

Aknowledgements

This work was supported by Sollac-Usinor, Fos/Mer (France).

References

AFNOR (1995). Essais écotoxicologiques. Recueil des normes Françaises. AFNOR Edition. Paris.

APHA (1992). *Standard Methods for the Examination of Water and Wastewater*. 18th edn, American Public Health Association/American Water Works Association/Water Environment Federation, Washington DC, USA.

Crowder, A. and St. Cyr, L. (1991). Iron oxide plaques on wetland roots. *Trends in Soil Science* **1**, 31–329.

Gopal, B. (1999). Natural and constructed wetlands for wastewater treatment: potentials and problems. *Wat. Sci. Tech.*, **40**(3), 27–35.

Gourlia, J.P. (1998). La modélisation en génie des procédés. In: *Techniques de l'Ingénieur*, vol 3, Paris, pp. 1–21.

Haberl, R. (1999). Constructed wetlands: a chance to solve wastewater problems in developing countries. *Wat. Sci. Tech.*, **40**(3), 11–17.

Kadlec, R.H. and Knight, R.L. (1996). *Treatment Wetlands*. Lewis Publishers, New York, USA.

Kadlec, R.H. (1999). Chemical, physical and biological cycles in treatment wetlands. *Wat. Sci. Tech.*, **40**(3), 37–44.

Levenspiel, O. (1972). Chemical Reaction Engineering, 2nd Ed., Wiley, New York.

Moore, R., Clarke, W.D. and Vodopich, D. (1998). *Botany*. McGraw-Hill, Boston., USA.

Mungur, A.S., Shutes, R.B.E., Revitt, D.M. and House, M.A. (1997). An assessment of metal removal by a laboratory scale wetland. *Wat. Sci. Tech.*, **35**(5), 125–133.

Vymazal, J., Brix, H., Cooper, P.F., Green, M.B., Haberl, R. (1998). *Constructed wetlands for wastewater treatment in Europe*. Backhuys Publishers, Leiden.

The effect of refinery effluent on the aquatic macrophytes *Scirpus californicus*, *Typha subulata* and *Zizaniopsis bonariensis*

A.R. Campagna* and D. da Motta Marques**

* Curso de Pós-Graduação em Ecologia, Universidade Federal do Rio Grande do Sul Av., Bento Gonçalves, 9500, CEP 91540-000, Porto Alegre-RS, Brazil

** Instituto de Pesquisas Hidráulicas, Universidade Federal do Rio Grande do Sul, Avenida Bento Gonçalves, 9500, Caixa Postal 15029, CEP: 91501-970, Porto Alegre, RS, Brazil
(E-mail: *dmm@iph.ufrgs.br*)

Abstract Experimental wetlands were built to follow the implementation and permanence of three species of aquatic macrophytes (*Scirpus californicus*, *Typha subulata* and *Zizaniopsis bonariensis*) under different treatments (water, water + nutrients and water+ nutrients + refinery effluent). Morphological variables (number of lateral shoots produced, height of the main lateral shoot and final density values) were used to check the influence of a petrochemical effluent. All the response variables showed significant differences (p = 5%) in their development, mainly between the water treatment and water + nutrient + effluent, followed by water with water + nutrients. In the Water treatment, the lowest variable values were found for the three species, possibly due to the lack of nutrients in the medium. Opposite results were found in the other treatments, indicating that the petrochemical effluent was not a limiting factor for the implementation of the species in the systems.

Keywords Constructed; growth; macrophytes; refinery effluent; wetlands

Introduction

Aquatic macrophytes are perennial plants, with a high biomass production capacity, and growing in diverse types of substrates and a large variety of wastewater (Tchobanoglous, 1987). Several aquatic plants have been proposed for the treatment systems of different types of wastewater due to their capacity for assimilation and long term storage of organic and inorganic components in the wastewater (Reddy and Debusk, 1987).

The importance of evaluating the deleterious effects on macrophytes and other organisms when exposed to pollutants is due to the fact that they are components of the food chain and may influence or control the physical conditions of the aquatic environment (Bowmer, 1986). This is especially important for the macrophytes used in constructed wetlands. Wetlands are increasingly used to treat refinery and petrochemical effluents (Litchfield, 1993; American Petroleum Institute (1998). However the information on how particular effluents or pollutants affect aquatic macrophytes growth is still limited.

This work looks at the effect of refinery wastewater on phenological aspects of three aquatic macrophytes to elucidate the capacity of different species to take part as components of constructed systems for the final treatment of refinery liquid effluents.

Methods

The experiment was performed at the Universidade Federal do Rio Grande do Sul, southern Brazil, under a subtropical climate.

The macrophytes chosen for this study were *Scirpus californicus* (C.A. Mey) Steud, *Typha subulata (*Crespo and Per, Mor) and *Zizaniopsis bonariensis (*Bal and Point) Speg. which are found in southern Brazilian wetlands, Estado do Rio Grande do Sul.

Fifteen experimental wetlands were used, with a useful area of 1.25 m² each and substrate consisting of coarse sand 30 cm deep. Propagules of the three species were collected by hand from natural stands to obtain the standardised dimensions of the root (+/− 7–10 cm), rhizome (+/− 10 cm) and aerial part, +/− 20 cm high. The propagules were transplanted into these systems at a density of 16 individuals/m² (Giovannini, 1997), and from then on considered as isolated genets.

The wastewater used in this study (Table 1) was the final effluent from the aerated lagoon of Alberto Pasqualini Refinery (REFAP, RS-Brazil). During the experiment, the effluent was stored in tanks covered by black plastic canvas and homogenised manually before each dosing. The load, added daily to the systems that were the components of the treatment with effluents, was calculated taking as its base an area that exists at REFAP, with a potential to constitute a constructed wetland, and the mean flow of effluent produced by the refinery.

The water and water + nutrients treatment systems received 126 litres of potable water and those of effluent treatment received 8.25 litres of effluent per day, diluted into 117.75 L of water. In the latter two treatments, nutrients (NPK) were added at intervals of approximately 20 days, at a constant proportion of 20 g, seeking to maintain a low concentration (Adriano et al., 1980). By the end of the experiment, each system had received 100 g of NPK. The experimental wetland water level control was maintained at an appropriate height to characterise the systems as being sub-surface flow.

The experimental design consisted of plots subdivided over time, and treatment and time were the two factors. The type of treatment determined the main plots at three levels (Water; Water + Nutrients; Water + Nutrients + Effluent). Time determined the sub-plots at 5 levels (5 measures).

For all the genets present in the different systems, the following variables were monitored using non-destructive methods: number of lateral shoots, height of the largest lateral shoot and final density of the species in the different treatments. Data were analysed by Analysis of Variance of Repeated Measures, and the LSD multicomparison test at $p = 5\%$. The production and growth rates of the lateral shoots were obtained by linear regression fitting.

Results and discussion

Production of lateral shoots

Generally, both in *Z. bonariensis* and in *T. subulata* an increased production of lateral shoots was observed, usually with significant differences ($p = 5\%$) in the Water + Nutrients and Water + Nutrients + Effluents treatments, as compared with the data obtained in the treatment in which only water was added (Figure 1). The productivity of *Typha*, in systems

Table 1 Petroleum refinery effluent characteristics*

Variable	Mean value	Standard deviation
pH	7.22	0.26
COD (mg.L^{-1})	157.14	83.06
BOD (mg.L^{-1})	42.03	17.44
Total Nitrogen (mg.L^{-1})	12.09	2.33
Total Phosphorus (mg.L^{-1})	0.427	0.207
Phenols (mg.L^{-1})	0.051	0.047
Suspended Solids (mg.L^{-1})	56.47	32.53
Sulphide (mg.L^{-1})	0.149	0.172
Mercury (µg.L^{-1})	29.83	24.27

* Mean values (August to December 1996)
Source: Campagna, 1998

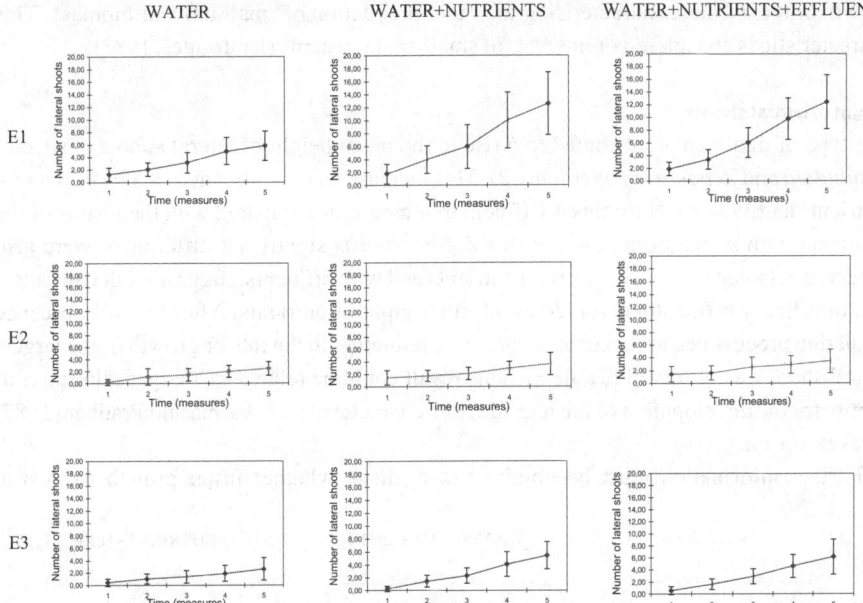

Figure 1 Variation in the mean number of lateral shoots produced by the three species in the different treatments as related to time (measures). (E1: *S. californicus*; E2: *T. subulata*; E3: *Z. bonariensis*)

receiving wastewater, can be higher than that of natural systems, which are, mostly limited by some type of nutrient (Cary and Weerts, 1984; Kadlec, 1989). The addition of effluents, on the other hand, did not cause any significant difference ($p = 5\%$), as compared to the treatment in which nutrients were added, showing that the effluent was neither a limiting nor a stimulating factor in the development of this variable in the three species. The observation time contributed to produce significant differences ($p = 5\%$) in the production of lateral shoots in the three species, and this was considered a normal result in accordance with what was expected for the development of this variable.

The production rates of lateral shoots, fitting regression equations, diminished (Table 2) for species *S. californicus* in water and water + nutrients treatments. The diminished production of lateral shoots could be associated with a displacement of resources for the formation of inflorescence (Esteves, 1988), in the case of both treatments, or in the treatment with water due also to a constraint on resources. This did not occur in the treatment with effluent for any of the species tested. The additional input of nutrients provided by the refinery effluent might have been the cause, which maintained the trend for vegetative

Table 2 Comparative data on the rates of production of lateral shoots, before and after the formation of inflorescence, for the three species when submitted to the different treatments

Treatments→	Water			Water + nut			Water + nut + EFF		
Specie↓	T_i	T_f	T_m	T_i	T_f	T_m	T_i	T_f	T_m
S. californicus	0.95	0.85	1.21	3.3	2.25	2.78	1.5	3.08	2.74
T. subulata	x	x	*	•	•	0.56	0.56	0.57	0.54
Z. bonariensis	x	x	*	1.2	2.35	1.61	1.05	2.04	1.58

T_i = initial rate of production (shoots/measure) observed before inflorescence formation;
T_f = rate of production (shoots/measure) observed after inflorescence formation;
T_m = mean rate of production (cm day^{-1});
* = no linear regression was possible;
x = they did not present inflorescence;
• = *T. subulata* in the water + nutrients treatment, inflorescence were only observed in the last measure.

growth of the species, characterising the rapid production of small units of biomass. This characteristic is considered of interest for small-scale systems (Krattinger, 1983).

Height of lateral shoots

The type of treatment contributed to a rise in the mean height of lateral shoots in species *T. subulata* and *Z. bonariensis* (Figure 2). This increase was significant ($p = 5\%$) in Water + Nutrients and Water + Nutrients + Effluent treatments, as compared with the results of the treatment with water alone. For species *Z. bonariensis* significant differences were also observed between the treatments with nutrients and with effluents, suggesting that the addition of refinery effluent has served as a further input of nutrients. After the inflorescence formation process began in the three species, a reduction in the rate of growth of the largest lateral shoot was observed (Table 3). This result could be related to a larger allocation of energy for the development of the reproductive characteristics (Sharma and Pradhan, 1983; Esteves, op. cit.).

Further information must be obtained to evaluate whether faster growth can cause

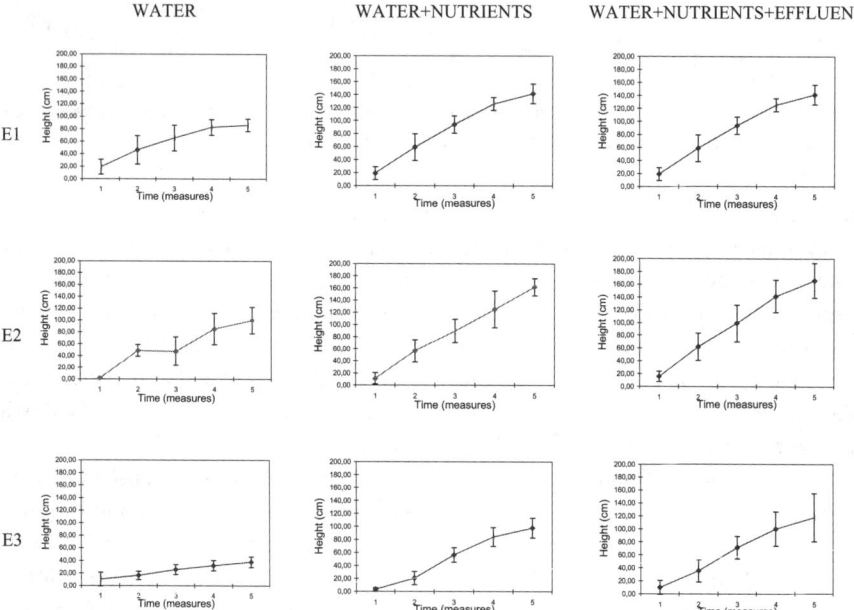

Figure 2 Variation of the mean height of lateral shoots in the three species of aquatic macrophytes, in the different treatments to which they were submitted (E1: *S. californicus;* E2: *T. subulata;* E3: *Z. bonariensis*)

Table 3 Comparative data on the growth rates of lateral shoots of the three species when submitted to the different treatments

Treatments→	Water			Water + nut			Water + nut + EFF		
Specie↓	T_i	T_f	T_m	T_i	T_f	T_m	T_i	T_f	T_m
S. californicus	0.93	0.17	*	1.59	0.86	1.21	1.87	0.60	1.15
T. subulata	x	x	0.93	•	•	1.44	2.21	1.21	1.44
Z. bonariensis	x	x	0.28	1.10	0.47	0.98	1.03	0.84	1.09

T_i = initial rate of growth (cm day^{-1}), observed before the formation of inflorescence;
T_f = rate of growth (cm day^{-1}), observed after the formation of inflorescence;
T_m = mean rate of growth (cm day^{-1});
* = it was not possible to perform linear regression;
x = no inflorescence were presented;
• = *T. subulata* in the water + nutrients treatment, inflorescence was only observed in the last measure.

changes in the different functions of constructed systems as compared with natural systems, and thus evaluate whether the oil refinery effluent is or is not a constraint for aquatic macrophytes in these systems.

Final density of individuals in the different treatments

In order to obtain final density data under the different treatments, propagules and lateral shoots were considered as individual units. *Scirpus californicus* was the species with the highest final density (101.6 individuals . m^{-2}) followed by *Zizaniopsis bonariensis* (56.8 individuals . m^{-2}) and by *Typha subulata* (39.2 individuals . m^{-2}) in treatment Water.

In the treatment Water + Nutrients *Scirpus californicus* was also the species with the highest final density (214.4 individuals. m^{-2}), followed by *Zizaniopsis bonariensis* with 100 ind/m^2 and by *Typha subulata* with 61.6 individuals. m^{-2}.

The same pattern was obtained in treatment Water + Nutrients + Effluent with *Scirpus californicus*, attaining a higher final density (202.4 individuals. m^{-2}), followed by *Z. bonariensis* with 108.0 individuals. m^{-2} and by *T. subulata* with 58.9 individuals. m^{-2}.

The lowest density values for the three species, found in the treatment with water (Figure 3), could be related to a degenerative phase, possibly caused by the exhaustion of nutrients available for growth (Krattinger, 1983). The presence of small clones could be an adaptive characteristic for the survival and permanence of species in homogeneously poor habitats (Wijesinghe and Hutchings, 1997).

Species *S. californicus* was the one which presented the best performance in the water treatment, since it developed all the phenological states evaluated (shoot, vegetative adult and adult with inflorescence) and also a greater productivity of lateral shoots. The final mean densities of the three species were significantly different ($p = 5\%$) between the treatments of water and Water + Nutrients, indicating the limitation of the first treatment. Species of *Typha* growing in nutritionally enriched environments, attained rapid growth in the density of the plants when compared with a control (Grace, 1988). A significant difference ($p = 5\%$) occurred in the density of *Scirpus californicus* between the treatment with nutrients and with effluent + nutrients, which could be explained by the higher mortality of propagules occurring during the last treatment.

Conclusions

The oil refinery effluent was not a limiting factor for the implementation, survival and permanence of macrophytes *Scirpus californicus*, *Typha subulata* and *Zizaniopsis bonariensis* in constructed wetlands.

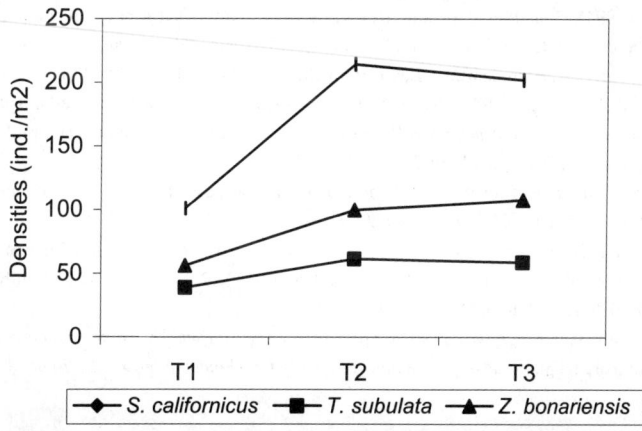

Figure 3 Mean final population densities of the three species found in the three treatments

The addition of effluents only had a significant influence on the increase in the mean height of the lateral shoots of *Zizaniopsis bonariensis* when compared with the results obtained in the treatment with nutrients. The refinery effluent, in this case, served as an additional input of nutrients.

In the different characteristics tested, the addition of effluents presented a nutrient value on the development of certain characteristics (variables), as compared with the results obtained in the treatment with water alone.

Independent of type of treatment used, the highest final density found in the different systems was attained by *Scirpus californicus* followed by *Zizaniopsis bonariensis* and *Typha subulata*.

Aknowledgements

To CNPq and CAPES for their financial support to authors research activities.

References

Adriano, D.C., Fulenweider, A., Shartz, R.R., Ciravolo, T.G. and Hoyt, G.D. (1980). Growth and mineral nutrition of cattail (Typha) as influenced by thermal alteration. *J. Environ. Qual.*, **9**(4), 649–653.

American Petroleum Institute. (1998). *The Use of Treatment Wetlands for Petroleum Industry Effluentes*. Prepared by CH2 HILL for the API Biomonitoring Task Force, Health and Environmental Sciences Department, Washington, D.C., 8 pp.

Bowmer, K.H. (1986). Rapid biological assay and limitations in macrophyte ecotoxicology: A review. *Aust. J. Mar. Freshw. Res.*, **37**, 297–308.

Campagna, A.A.R. (1998). *Efeito de Efluente de Refinaria de Petróleo sobre Macrófitas Aquáticas em Ecossistemas Construídos (Banhados)*. Dissertação de Mestrado. Curso de Pós-Graduação em Ecologia, Universidade Federal do Rio Grande do Sul.

Cary, P.R. and Weerts, P.G.J. (1984). Growth and nutrient composition of *Typha orientalis* as affected by water temperature and nitrogen and phosphorus supply. *Aquat. Bot.*, **19**, 105–118.

Esteves, F.A. (1988). Fundamentos de Limnologia. Rio de Janeiro, Interciência. FINEP, 575p.

Giovannini, S.G.T. (1997). Estabelecimento e Desenvolvimento das Macrófitas Aquáticas *Scirpus Californicus*, *Typha Subulata* e *Zizaniopsis Bonariensis* sob Condições Experimentais de Regimes Hídricos Diferenciados. Dissertação de Mestrado. Instituto de Pesquisas Hidráulicas, Universidade Federal do Rio Grande do Sul.

Grace, J.B. (1988). The effects of nutrient additions on mixtures of *T. latifolia* L. and *T. domingensis* PERS. along a water-depth gradient. *Aquat. Bot.*, **31**(1–2), 83–92.

Kadlec, R.H. (1989). Dynamics of inorganic and organic materials in wetlands ecosystems. In: *Constructed Wetlands for Wastewter Treatment: Municipal, Industrial, and Agricultural*, D.A. Hammer (ed.), Lewis Plubishers, Chelsea, MI, pp

Krattinger, K. (1983). Estimation of size and number of individual plants within populations of *Typha latifolia* L. using isoelectrofocusing (IEF). *Aquat. Bot.*, **15**, 241–247.

Litchfield, D.K. (1993). Constructed wetlands for wastewater treatment at Amoco oil company's Mandan, North Dakota refinery. In: In *Constructed Wetlands for Wastewter Treatment: Municipal, Industrial, and Agricultural*, D.A. Hammer (ed.), Lewis Plubishers, Chelsea, MI, pp 485–488.

Reddy, K.R. and Debusk, W.F. (1987). Nutrient storage capabilities of aquatic and wetland plants. In: *Aquatic Plants for Water Treatment and Resource Recovery*, K.R. Reddy and W.H. Smith (eds.), Magnolia Publishing, Orlando, FL. pp 337–357.

Sharma, K.P. and Pradhan, V.N. (1983). Study on growth and biomass of underground organs of *Typha angustata* Bory and Chaub. *Hydrobiologia*, **99**, 89–93.

Tchobanoglous, G. (1987). Aquatic plant systems for wastewater treatment: engineering considerations. In: *Aquatic Plants for Water Treatment and Resource Recovery*, K.R. Reddy and W.H. Smith (eds.), Magnolia Publishing, Orlando, FL. pp 27–48

Wijesinghe, D.K. and Hutchings, M.J. (1997). The effects of spatial scale of environmental heterogeneity on the growth of a clonal plant: an experimental study with *Glechoma hederacea*. *J. Ecol.*, **85**, 17–28.

Treatment of laboratory wastewater in a tropical constructed wetland comparing surface and subsurface flow

A.A. Meutia

Research and Development Center for Limnology, Indonesian Institute of Sciences, Jalan Raya Bogor Km 46, Cibinong 16911, Indonesia. (E-mail: *mizuno@idola.net.id*)

Abstract Wastewater treatment by constructed wetland is an appropriate technology for tropical developing countries like Indonesia because it is inexpensive, easily maintained, and has environmentally friendly and sustainable characteristics. The aim of the research is to examine the capability of constructed wetlands for treating laboratory wastewater at our Center, to investigate the suitable flow for treatment, namely vertical subsurface or horizontal surface flow, and to study the effect of the seasons. The constructed wetland is composed of three chambered unplanted sedimentation tanks followed by the first and second beds, containing gravel and sand, planted with *Typha* sp.; the third bed planted with floating plant *Lemna* sp.; and a clarifier with two chambers. The results showed that the subsurface flow in the dry season removed 95% organic carbon (COD) and total phosphorus (T-P) respectively, and 82% total nitrogen (T-N). In the transition period from the dry season to the rainy season, COD removal efficiency decreased to 73%, T-N increased to 89%, and T-P was almost the same as that in the dry season. In the rainy season COD and T-N removal efficiencies increased again to 95% respectively, while T-P remained unchanged. In the dry season, COD and T-P concentrations in the surface flow showed that the removal efficiencies were a bit lower than those in the subsurface flow. Moreover, T-N removal efficiency was only half as much as that in the subsurface flow. However, in the transition period, COD removal efficiency decreased to 29%, while T-N increased to 74% and T-P was still constant, around 93%. In the rainy season, COD and T-N removal efficiencies increased again to almost 95%. On the other hand, T-P decreased to 76%. The results show that the constructed wetland is capable of treating the laboratory wastewater. The subsurface flow is more suitable for treatment than the surface flow, and the seasonal changes have effects on the removal efficiency.
Keywords Constructed wetland; removal efficiency; subsurface flow; surface flow; tropical wetland; wastewater

Introduction

Although water pollution in Indonesia has been very serious, little effort has been made to treat wastewater. Both domestic and industrial wastewater is discharged almost directly into the water body (river/lake), and has caused significant environmental and public health damage. Many industrial establishments, especially the middle and small scales industries, have not installed wastewater treatment facilities, and many wastewater treatment plants have not been used at all, because the installment and operational costs are too expensive in Indonesia. The treatment plants used in developed countries are not fit for conditions in the developing countries, like Indonesia.

Constructed wetland is an alternative system. It is energetically sustainable because it uses only natural energy to reduce pollutants. It needs low construction and operational costs. The operation and maintenance is easy (Martin and Johnson, 1995). It is suitable for the developing countries where land and manpower are relatively abundant, and capital is scarce. Constructed wetland that uses microorganisms and plants maintains biodiversity and esthetic balance of wetlands. It improves water quality in the environmentally sustainable way. Constructed wetlands have good prospects in developing countries.

While knowledge on constructed wetlands in Europe, America and Australia has increased, there is little knowledge as well as research on constructed wetlands in tropical countries including Indonesia.

Constructed wetlands are categorized into two types: free water surface flow or subsurface flow (Crites, 1994). The surface flow system is popular in the United States, while the subsurface flow system is widely accepted by Europe countries, Australia and South Africa (Wood, 1995). Little information is available about appropriate flow conditions for constructed wetlands in tropical regions.

The aim of the research is to examine the capability of constructed wetlands in treating laboratory wastewater in tropical conditions, to investigate suitable flow, namely subsurface or surface flow, and to study the effects of seasons.

Methods
System design

Two series of small-scale constructed wetland systems were constructed at the premises of the Research and Development Center for Limnology, Cibinong, 30 km South of Jakarta, Indonesia in 1998. The systems were stabilized before the commencement of sampling in May 1999. One series is vertical subsurface flow and the other is horizontal surface flow (Figure 1). The constructed wetland is composed of three chambered unplanted sedimentation tanks followed by the first and second beds planted with *Typha* sp.; the third bed planted with floating plant *Lemna* sp.; and a clarifier with two chambers. Final treated water is kept in a container for other purposes. A sedimentation tank is intended primarily to reduce solids. The first and second beds are intended to reduce organic (COD), nitrogen and phosphorus compounds. The third bed is intended to further reduce the compounds that remain in low-level concentrations. Finally, the clarifier is intended to separate very fine suspended solid and treated water. The differences of flow types are involved only in the beds. In the subsurface flow system inflow is located 5 cm above the bottom of each bed and outflow is 20 cm from the top. In the surface flow system inflow and outflow are located around 20 cm from the top of each bed. Each bed has 3 m^2 surface area, filled with 2–3 cm gravel (15 cm in height) and sand of 10 cm in height. It has a water level of 30 cm above the sand. Wastewater was discharged from laboratories on the first and second floors, prototype laboratory and several sinks. The wastewater fluctuated widely depending on the laboratory activities. Most of the wastewater was discharged into the constructed wetlands in the daytime, five days a week. The wastewater flowed continually into the series of constructed wetlands by gravitation. Neither pump nor electricity was needed. The constructed wetland series worked as the main wastewater treatment without prior treatment. The hydraulic retention time in the wetland was 24 hours for a flow rate of 250 l/day.

In the dry season of the year 1999 monthly average of rainfall depth was 186.9 mm and it increased two times during the rainy season (365.0 mm). In the dry season air temperature ranged 29–34°C and 25–29°C in the rainy season. The transition period is indicated by unusually high winds beside the rain.

Sampling and analysis

Water samples were taken at the influent, outlets of the first, second and third beds, clarifier, and final effluent. Analysis of chemical oxygen demand (COD), NH_4-N, NO_3-N, NO_2-N, T-N, T-P, orthophosphate, soluble phosphate and dissolved oxygen were carried out according to *Standard Methods for the Examinations of Water and Wastewater* (APHA, 1995). Physical parameters of pH, temperature, turbidity and conductivity were measured at every sampling time with Water Quality Checker (Horiba). Flow rate measurement was also conducted at every sampling time. The data collected was then averaged. Each season was represented by average data from three sampling times.

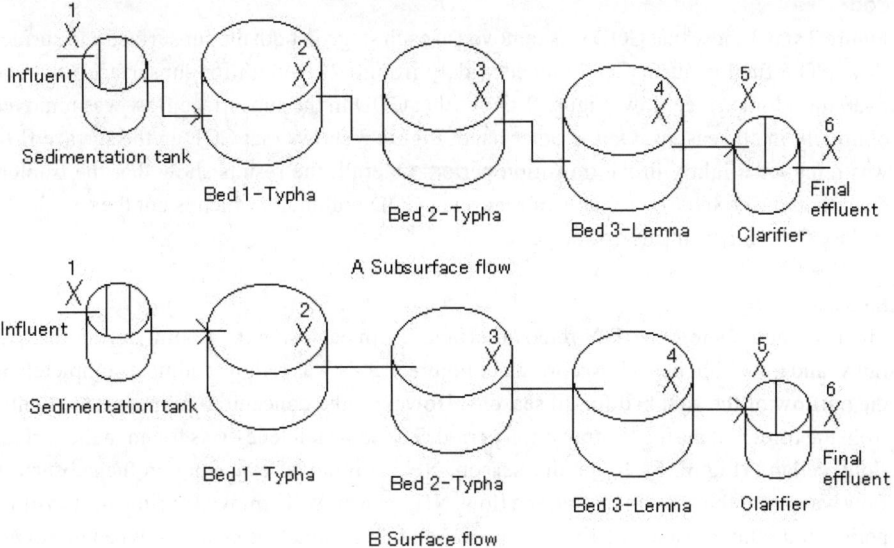

Figure 1 Schematic diagram of small-scale tropical constructed wetlands. A. vertical subsurface flow. B. horizontal surface flow. X sampling point

Results and discussion

Physical parameters

Concentration ranges of physical parameters are given in Table 1. Both subsurface and surface wetlands showed pH in the influent was around 2.0–4.2 in the dry season, 2.1–5.9 during the transition period from the dry season to the rainy season, and 5.7–6.5 in the rainy season. The pH varied widely depending on the chemical content of the laboratory wastewater. The pH tended to be higher in the rainy season. However pH at the final effluent was almost constant in these seasons, around 6.4. In the two series of tropical wetlands, temperature varied around 26–29°C at the influent and 28–34°C at the effluent (Table 1). Overall temperature decreased slightly in the rainy season. Conductivity was high at the influent in the dry season, and ranged 0.59–0.67 mS/cm at the beginning of the transition period. Conductivity remained at the lower level (0–21–0.25 mS/cm) in the transition period and rainy season. The conductivity at the effluent in the subsurface flow of the wetland was higher than in the surface flow for all seasons. Low turbidity at the influent decreased to zero level at the effluent during the dry season and transition period. It was shown that treated water was very clear. High turbidity decreased to about 1 NTU in the subsurface flow and 5 NTU in the surface flow in the rainy season.

During the transition period and the rainy season water level was almost constant. Water from rainfall automatically flowed to the outflow. However, in the dry season water level was affected by evapotranspiration as shown as decreasing water level about 5 cm below the normal level.

Table 1 Varied physical parameters of the subsurface and surface flows in tropical constructed wetlands

Parameter	Subsurface flow		Surface flow	
	Influent	Effluent	Influent	Effluent
pH [–]	2.1–6.5	6.3–6.6	2.0–6.0	6.2–6.6
Temperature [°C]	26–29	28–34	27–29	28–34
Conductivity [mS/cm]	0.17–0.65	0.15–0.40	0.13–0.67	0.13–0.24
Turbidity [NTU]	2–66	0–6	2–50	0–24

COD

Figure 2 and 3 show that COD was removed at each stage in both the subsurface and surface flows. The final effluent COD ranged widely from 0.2–18 mg/l for subsurface flow and 3–30 mg/l for surface flow. Figure 2 shows that COD in the subsurface flow was removed optimally in all seasons. On the other hand, Figure 3 shows that COD in the surface flow was removed slightly in the transition period. Overall, the results show that the tropical constructed wetlands are capable of removing COD pollutants which is not the case in the surface flow during the transition period.

Nitrogen

Figures 4 and 5 show the T-N removal efficiency in each sampling point during the dry, rainy, and transition period. According to Figure 4, T-N was removed almost completely at the outflow of the first bed for all seasons. However, the concentration increased slightly from the third bed during the transition period. The same tendency was found in the surface flow wetland (Figure 5). In the dry season, NH_4-N removal efficiency in the subsurface flow was lower than that in the surface flow. NH_4-N was well removed during the transition period and rainy season. Nitrifying bacteria seem to be much more active when dissolved oxygen concentration was sufficiently high for nitrification. Oxygen was realized by plant roots (Brix, 1993), while water and air mixtures influenced by rain could increase the DO concentration. As a consequence, significant amounts of nitrite and nitrate were often measured at the effluent of surface flow (Table 2). In the rainy season, almost all of the nitrogen compounds were treated. This proved that nitrification and denitrification were well conducted during this period.

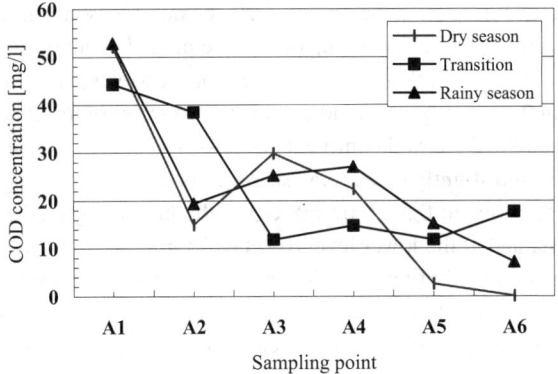

Figure 2 COD concentration in subsurface flow tropical constructed wetland

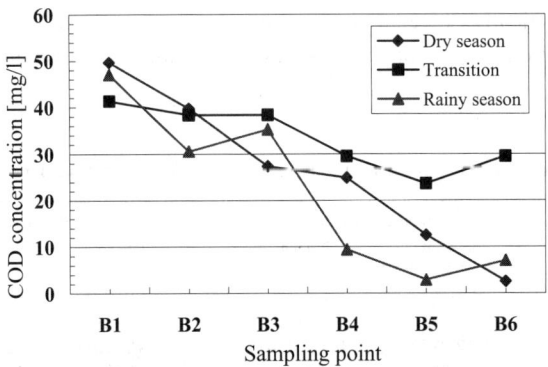

Figure 3 COD concentration in surface flow tropical constructed wetland

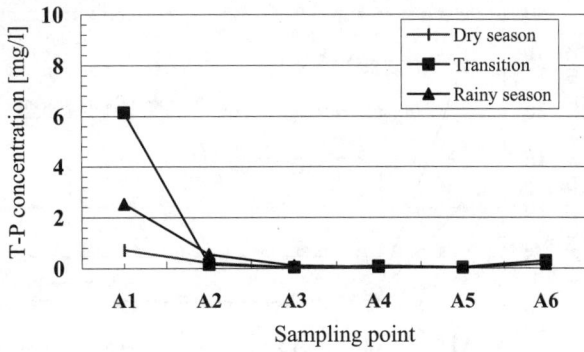

Figure 6 Total phosphorus concentration in subsurface flow tropical constructed wetland

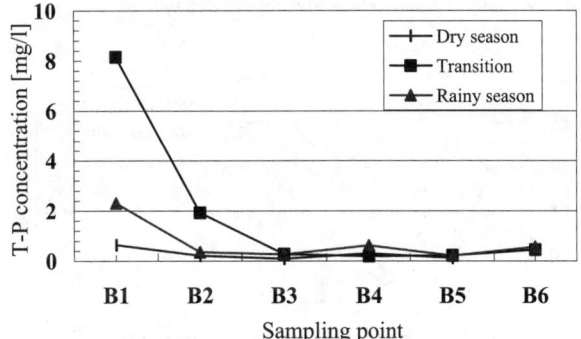

Figure 7 Total phosphorus concentration in surface flow tropical constructed wetland

Dissolved oxygen

Figures 8 and 9 show that dissolved oxygen concentrations increased after the treatment at the first bed. In the subsurface flow, the peak of the DO concentration was found in the second or third beds, while the peak in the surface flow was found at the first bed. The DO concentration then decreased because degradation processes such as nitrification started. Much of the dissolved oxygen concentrations brought about nitrification in the tropical constructed wetland. In addition, aerobic conditions kept phosphorus compounds settled in the sediment. Average effluent DO concentration was 5 mg/l for subsurface flow and 4 mg/l for surface flow.

The experiment results show that organic pollutant, nitrogen and phosphorus compounds contained in the laboratory wastewater could be treated in the tropical constructed wetland. However, the vertical subsurface flow was more suitable for the treatment than the horizontal surface flow. This is because the subsurface flow involved a filtration process besides the microbial degradation processes. The results show that treatment in the rainy seasons removed organic carbon, nitrogen, and phosphorus effectively. In the tropical conditions, seasonal changes have a significant effect on removal efficiencies regarding several parameters in both subsurface and surface flow tropical constructed wetlands. Especially treatment in the surface flow is much affected by the seasonal changes. This might be due to the rainfall having an effect on microorganisms and filtration processes.

Removal efficiency

Table 3 shows removal efficiencies of every parameter based on concentrations differences. The results showed that the subsurface flow in the dry season had the capacity of 95% organic carbon (COD) and total phosphorus (T-P) averaged removal respectively, and 82% total nitrogen (T-N) removal. During the transition period, COD averaged removal

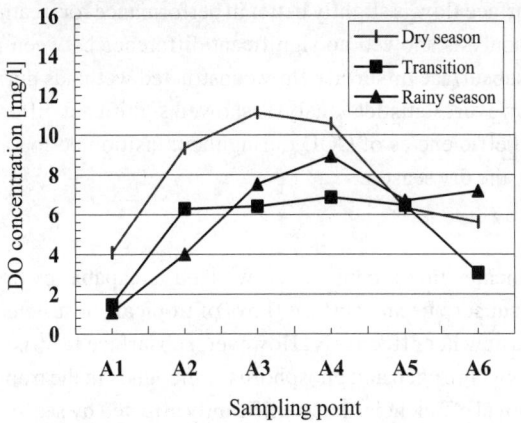

Figure 8 Dissolved oxygen concentration in subsurface flow tropical constructed wetland

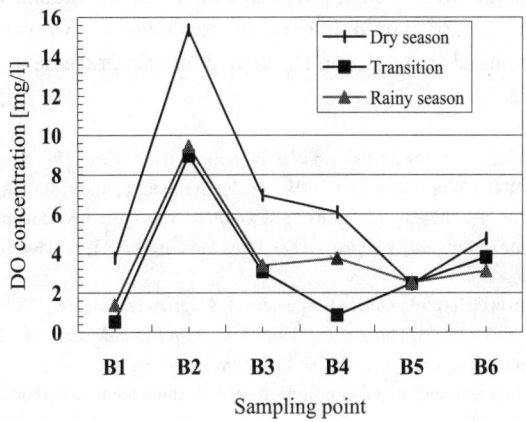

Figure 9 Dissolved oxygen concentration in surface flow tropical constructed wetland

efficiency decreased to 73%, on the other hand, T-N removal efficiency increased to 89%, and T-P removal efficiency was almost the same as that in the dry season. In the rainy season, COD and T-N removal efficiencies increased again to 95%, while T-P removal efficiency remained unchanged. In the dry season, the results of COD and T-P in the surface flow showed that the removal efficiencies were slightly lower than those in the subsurface flow. Moreover, T-N removal efficiency was only half as much as that in the subsurface flow. However, in the transition period, the COD removal efficiency decreased to 29%, while T-N removal efficiency increased to 74% and T-P removal efficiency was still constant, around 93%. In the rainy season, COD and T-N removal efficiencies increased again to almost 95%. On the other hand, T-P removal efficiency decreased to 76%.

Table 3 Removal efficiencies of the subsurface and surface flows in tropical constructed wetlands

Season	Removal efficiency of COD [%]		Removal efficiency of T-N (%)		Removal efficiency of T-P [%]	
	Subsurface	Surface	Subsurface	Surface	Subsurface	Surface
Dry	95	88	82	43	95	86
Transition period	73	29	89	74	94	93
Rainy	95	85	95	71	93	76

Although the subsurface flow is slightly better in performance for treatment than the surface flow, statistical analysis showed no significant difference between nutrient removal efficiencies in either subsurface or surface flow constructed wetlands during the transition and rainy seasons. However, statistical analysis showed significant difference at level 5% in the case of removal efficiencies of COD during the transition period and removal efficiencies of T-N during the dry season.

Conclusions

The present study indicates that a constructed wetland is capable of treating laboratory wastewater. Both the subsurface and surface flows of tropical constructed wetlands could treat the laboratory wastewater effectively. However, subsurface flow is more suitable for removing organic carbon, nitrogen and phosphorus compounds in the tropical wetland than the surface flow. Removal efficiencies are significantly affected by seasonal changes.

Acknowledgements

The small-scale wetlands were constructed and operated as Indonesian government research projects. The author would like to acknowledge E. Mulyana for technical assistance, Irma for chemical analysis, and K. Mizuno for encouragement.

References

APHA (1995). *Standard Methods for the Examination of Water and Wastewater* 19th edn, American Public Health Ass./American Water Works Association/Water Environment Fed., Washington DC, USA.

Brix, H. (1993). Macrophyte-mediated oxygen transfer in wetlands: transport mechanisms and rates. In: *Constructed wetlands for water quality improvement*, G.A. Moshiri (ed.), Lewis Publishers, Florida, pp. 391–398.

Crites, R.W. (1994). Design criteria and practice for constructed wetlands. *Wat. Sci. Tech.*, **29**(4), 1–6.

Martin, C.D. and Johnson, K.D. (1995). The use of extended aeration and in-series surface flow wetlands for landfill leachate treatment. *Wat. Sci. Tech.*, **32**(3), 119–128.

Wood, A. (1995). Constructed wetlands in water pollution control: fundamentals to their understanding. *Wat. Sci. Tech.*, **32**(3), 21–29.

Heavy metal contents and mobility of artificially inundated grasslands along River Weser, Germany

C. Erber* and P. Felix-Henningsen**

* Institute of Landscape Ecology, Westfälische Wilhelms-University, Robert-Koch Str. 26, D-48149Münster, Germany
** Institute of Institute of Soil Science and Soil Conservation, Justus-Liebig-University, Heinrich-Buff-Ring 26, D-35392 Giessen, Germany

Abstract At the beginning of the 20th century municipal wastewater was used to fertilize grassland in the freshwater marsh of the river Weser. In 1987, 150 ha of the marsh became part of a mitigation area with artificial inundation during winter and spring. Heavy metal input may be expected from former wastewater treatment and artificial flooding nowadays. In addition, inundation may increase the availability of heavy metals that were accumulated during municipal wastewater treatment. In order to get an idea of heavy metal content and availability, the content of Cd, Cr, Cu, Ni, Pb, and Zn of the soil, the vegetation, and the input due to inundation were determined. Metal enrichment in the epipedon is evident for Cr, Cu, Pb, and Zn. Total content of Pb and Zn exceed the precaution limit. Soils treated with wastewater seems to contain more heavy metals than the ones without. Inundation causes an input of metals, but it is very low and varies in a broad range. Metal input is higher by atmospheric deposition than the one due to inundation. Degree of enrichment can be arranged in the order: atmospheric deposition > municipal wastewater >> inundation. In shallow ditch soils heavy metals are becoming more available.

Keywords Artificial inundation; fractionated extraction; fresh-water marsh; heavy metals; mitigation; municipal water treatment

Introduction

The freshwater marsh of the Weser river near the Hanseatic City Bremen in Northwest Germany is built by sediments originating from the Harz Mountains. The sediments contain heavy metals. The Harz Mountains are known for their heavy metals incidence (Pb and Zn as well as Cu, Ag, Au, Cd, and Hg), which were mined since the early Middle Ages. Because of simultaneous intensive clearing for agriculture and logging for ore dressing in the Harz Mountains, erosion and sediment load of the Weser river increased. In the 12th century, diking of the freshwater marsh stopped regular floodings and sedimentation. For drainage of the area, a dense ditch system and a specific topography were created. The topography consists of ridges and depressions running in parallel. The level difference amounts to 15–30 cm, and the distance between ridges is 20–40 m. The depressions between the ridges are like shallow surface ditches with small or no connection to the main ditch system. In the following they are called shallow ditch. The ridges were mainly built up by the material removed from the ditches and their surfaces are cambered.

The whole area has mainly been used as grassland. During the first half of the 20th century, parts of the freshwater marsh near Bremen were fertilized with municipal water. Thereby the ridges were also flooded. In order to retain all nutrients the water was not purified. Documents about composition and quantity of the irrigation water are not available any more. Since the middle of the eighties of the last century about 150 ha grassland has been artificially inundated with surface water every year in order to provide mitigation for industrial settlement in the structure rich ditch-grassland area. The goal of mitigation was to create resting and breeding places for birds, and restore species-rich moist grassland by wetting, inundation and nutrient-impoverishment of the present grassland. The inundation water originates from the Ochtum, a tributary river of the Weser. About 90% of the grassland is inundated every year in the middle of November with a maximum water depth of 80 cm. In spring, water level is reduced in steps. At the end of May and sometimes even at the beginning of June, areas below 2.0 m elevation are still under water. These shallow water zones represent important food resources for breeding birds.

Since heavy metals cannot be decomposed they accumulate in soil feeding animals like earthworms (Yongcan et al., 1998). Since a higher total metal content in soil does not always mean higher mobility and plant availability, information on metal fraction is more important to organisms than total metal content in the soil (Canarutto et al., 1991).

This study is part of a 10 year monitoring-program designed for natural environmental studies on the influence of artificially inundation on soil, vegetation, and fauna. In the present study content and availability of some heavy metals (Cd, Cr, Cu, Ni, Pb, Zn) in soil and vegetation of the mitigation area were investigated.

Material and methods

Within the mitigation area five sites were selected. Three of them were treated with municipal water (site III, IV, and V) and four of them are inundated nowadays (site I, II, IV, and V). At each site one soil pit on a ridge and one in a shallow ditch were prepared for soil description and soil sampling. About 2 kg soil material were taken from each horizon and combined to one sample. The soils are eutric gleysols according to World Reference Base (Food and Agriculture Organization of the United Nations, 1998). They are slightly acidic (pH(H_2O) 5.8), clay content amounts of about 60% (40–75%), and content of organic carbon varies between 0.3 and 10.9%. Mean potential cation exchange capacity is high amounting to 31 $cmol_c$ kg^{-1}. The epipedon is characterized by high content of organic matter (8% C_{org}) and acid soil reaction (pH(H_2O) 4.7). For detailed soil characteristics see Erber (1998). Material accumulated during inundation at the bottom of shallow ditch sites was collected at two sites by six 1 m^2-polyethylene sheets for the entire inundation time in 1993/94 and 1994/95. In order to study aboveground biomass and its heavy metal content at six sites (three ridges and three shallow ditches) plant material was collected on three plots (1 m^2 each) at hay making time in June/July and October/November in 1994 and 1995.

Sample preparation

Soil samples and deposited material were air dried, sieved through a 2 mm mesh, and subsamples were powdered in an agate mortar. Air-dried plant material was powdered in a plate grinder. Because of contamination with Cr by powdering with the plate grinder, Cr content was not determined in plant material. Contamination with other analyzed heavy metals could not be found. Plant material was not washed before being analyzed.

Extraction

Soil samples were analyzed by three different extraction procedures. Plant and deposited material was only digested with aqua regia.
1. Aqua regia: 5 g powdered soil and plant material, respectively, was digested with a mixture of 18 ml HCl (36.5 M) and 6 ml HNO_3 (63 M) for 12 h at room temperature and then temperature was raised to 160°C for 2 h. After cooling the solution was filtered and filled up with aqua dest. to 100 ml (DIN ISO 11466, 1997). Even if digestion with aqua regia does not dissolve heavy metals present in the crystal lattice of highly resilistant silicates it has a strong recovering power for metals present in most sediments (VDLU-FA, 1991; Moalla et al., 1998), and contents of aqua regia extracts represent approximately the total content.

2. EDTA-extraction: 10 g sieved soil were extracted for 1.5 h by end-over-end shaking with 0.025 M Na$_2$-EDTA solution adjusted to pH 4.6 (ÖNORM, 1989). EDTA is a synthetic organic molecule, which forms more stable complexes with heavy metals than organic soil matter (Gardiner, 1976). The EDTA fraction includes also heavy metals bound at oxides (Andersson, 1975). Therefore, the EDTA fraction represents the potentially available fraction.
3. CaCl$_2$-extraction: 20 g sieved soil were extracted for 2 h by end-over-end shaking with 0.1 M CaCl$_2$ solution (Köster and Merkel, 1982). This fraction is generally considered readily bioavailable as the extracted elements were weakly bound and may equilibrate easily with the aqueous phase.

Metal concentration of all extracts was determined with a flame atomic absorption spectrometer (Perkin Elmer 1100). Reported metal concentrations are based on dry weight.

Results and discussion

Total heavy metal content of the soil

In the subsoil, the relative variation coefficient of the heavy metal contents ranges from 3 to 23% depending on the metal (Table 1). The observed variation is a result of soil heterogeneity (Qian *et al.*, 1996), especially of clay and organic matter contents due to different sedimentation conditions. In addition, one or two buried epipedons occur. Information about typical natural background levels of alluvial sediments is rare since heavy metal contents of diverse floodplains and marshes vary over a broad range as a result of different catchment areas. For all studied metals the subsoils have total concentrations which are within the contents of marsh soils given in Hindel and Fleige (1991). Therefore these heavy metal contents may be considered as geogenic.

With the exception of Ni, heavy metal contents in the epipedon are significantly higher

Table 1 Total content of heavy metals in the soil [mg kg^{-1}], level of significance of different metal concentration in epipedon and subsoil (*** p<0.001, ** p<0.01, * p<0.05), and organic carbon content [%] (r = ridge, s = shallow ditch, a: epipedon 0–10 cm soil depth, b: 40–60 cm soil depth, c: 90–110 cm soil depth, sites III–V were treated with municipal water)

		I r	I s	II r	II s	III r	III s	IV r	IV s	V r	V s	median of all sites
Cd	a		0.25	0.36	0.45	0.39	0.58	0.38	0.21	0.41	0.97	0.39*
	b		0.43	0.34	0.14	0.11			0.35		0.01	0.13
	c		0.13	0.11	0.12		0.19	0.09				
Cr	a	36	46	39	46	50	52	50	56	50	51	50**
	b	31/39+)	39	36/30	37	41/38	39/33	39	26	38	40	38
	c	29	30	33	36	40	40/36	46	36	38	40	
Cu	a	28	37	22	32	34	37	30	31	36	41	33**
	b	16/30	23	34/16	22	38/16	22/17	15/20	12	23	31	23
	c	13	19	24	29	27	29/41	24	26	22	30	
Ni	a	26	27	29	34	35	22	31	33	32	34	32***
	b	35/49	55/43	54/29	35	62/33	36/29	37/43	25/31	40	58/50	40
	c	31	36	34	42	36	37/60	56	43	40	62	
Pb	a	79	84	61	87	117	110	71	73	118	115	86***
	b	19/23	22	29/18	20	29/22	20	20	12	20	32/21	20
	c	15	14	21	22	18	22	19	18	20	23	
Zn	a	159	180	138	174	191	256	167	176	209	282	178***
	b	95/134	122	124/96	114	151/99	122/94	93/99	56/70	114	146/131	114
	c	90	86	112	114	105	114	115	132	108/118	147	
Pb/Zn	a	0.50	0.47	0.44	0.50	0.61	0.43	0.43	0.41	0.56	0.41	
	c	0.17	0.16	0.19	0.19	0.17	0.19	0.17	0.14	0.19	0.16	
C$_{org}$	a	5.2	6.2	5.5	7.1	9.9	10.4	7.7	6.4	10.9	10.8	

+) Two values are quoted if metal concentration of two horizons within the soil depth differs widely

Table 2 Median of heavy metal concentrations of the epipedon [mg kg^{-1}] separated to different treatments

	Cd	Cr	Cu	Ni	Pb	Zn
Treated with municipal water	0.40	51	35	33	103	200
Not treated with municipal water	0.31	42	30	28	82	167
Ridge	0.38	50	30	31	79	167
Shallow ditch	0.45	51	37	33	87	180

than in the subsoil (Table 1). Concentrations of Pb and Zn in the epipedon exceed even the precaution limits of Germany (BBodSch V, 1999). Significance level is high for Pb and Zn and low for Cd. The enrichment of the heavy metals is highest for Pb (fourfold concentration) and descends in the order Pb>Cd≫Zn>Cu>Cr. Since Pb/Zn ratios above 0.2 are typical for anthropogenic enrichment (Matschullat et al., 1991), and this is true only for the epipedon but not for the subsoil, anthropogenic influence is very likely. This enrichment may be the result of municipal wastewater treatment, inundation practice, or atmospheric deposition. The peculiarity of Ni content being lower in the epipedon than in the subsoil may result from low Ni concentration in domestic wastewaters and above ground water whereas Ni is a typical companion metal of the Harz Mountains rocks (Schimming, 1990).

Heavy metal enrichment as a result of municipal water treatment can be revealed by comparing median heavy metal content of differently treated epipedons (Table 2). For all studied metals the median content of the treated areas is 15–20% higher than of the non treated ones. Because of low database this cannot be evidenced on a statistically significant level.

For Cu, Cr, and Zn heavy metal concentration is always higher in the shallow ditch epipedon than in the ridge epipedon. For Cd and Ni there is one and for Pb there are two exceptions to this regularity. Enrichment in the shallow ditch epipedon may be a result of the ridge-shallow ditch topography. In comparison to the ridges, low lying parts are flooded longer and more often which promotes sedimentation. Additionally translocations from the ridges into the trenches are conceivable. This effect could be proved for nitrate (Erber, 1998).

Available heavy metal contents of the soil

The availability of metals, particularly the $CaCl_2$-extractable fraction, is mostly depending on soil acidity (Chuan et al., 1996). Soil pH ($CaCl_2$) of 4.7 is below the threshold for mobilisation of Cd, Ni, and Zn with pH 6, 5.5, and 5, respectively. For Cr and Pb pH is higher than the threshold with pH 4 and 3.5, respectively, and for Cu the threshold is nearly reached. Therefore Cr and Pb were not detectable and Cu contents are very low in the $CaCl_2$-extract, whereas Ni and Zn have noticeable percentages in the mobile fraction (Table 3). Furthermore, concentration of $CaCl_2$-fraction correlates with pH only for Ni and Zn (Table 4).

For Pb and Zn, EDTA-extractable contents and their percentage of the total content is higher at sites treated with municipal water than at the other sites. This holds true for the $CaCl_2$-extractable fraction of Zn. This statement cannot be proved for Pb, because the concentrations are too low. Besides Cd and Cr, which are not detectable, Pb and Zn are the most accumulated heavy metals in the epipedon. Therefore it can be concluded that heavy metals are not only enriched, they are even better available.

At sites treated with municipal water, content of organic matter has increased (Table 1). Similar effects were reported of other sewage fields (Salt, 1988). Since contents of clay and oxides, which are also important sorbents, are not different, this emphasizes the importance of organic matter for sorption and complexation of added metals (Leita et al., 1999), the

Table 3 Content of EDTA- and CaCl$_2$-extractable heavy metals [mg kg-l] and their percentage of total content (r = ridge, s = shallow ditch)

		I r	I s	II r	II s	III r	III s	IV r	IV s	V r	V s
Cu	EDTA	10.3	15.5	7.3	14.4	13.3	16.8	10.7	12.9	13.8	18.6
	CaCl$_2$	0.05	0.06			0.08	0.1	0.05	0.05	0.08	0.08
	% EDTA	37	42	33	46	40	45	36	42	38	45
	% CaCl$_2$	0.2	0.2			0.2	0.3	0.2	0.2	0.2	0.2
Ni	EDTA	3.0	3.0	3.5	5.0	5.0	4.5	3.5	4.0	4.0	4.0
	CaCl$_2$	1.00	0.75	0.75	0.88	1.13	0.88	0.25	0.25	1.0	0.50
	% EDTA	12	11	12	15	14	14	11	12	13	12
	% CaCl$_2$	3.8	2.8	2.6	2.6	3.3	2.8	0.8	0.8	3.1	1.5
Pb	EDTA	27.5	35	16.0	38.5	57.0	65.5	23.0	32.0	67.0	68.5
	% EDTA	35	42	26	44	50	60	32	44	57	60
Zn	EDTA	18.8	25.4	20.4	31.4	30	60.4	26.4	22.4	47.0	63.9
	CaCl$_2$	6.25	6.5	5.13	9.25	8.82	21.51	3.25	3.38	16.5	16.88
	% EDTA	12	14	15	18	16	24	16	13	22	23
	% CaCl$_2$	3.9	3.6	3.7	5.3	4.6	8.1	1.9	1.9	7.9	6.0

Table 4 Correlation coefficients of heavy metal fractions and some soil properties (level of significance: *** p<0.001, ** p<0.01, * p<0.05)

		pH	Corg	Clay	Fe-oxide	Mn-oxide	Al-oxide
Cu	EDTA	−0.42***	0.51***	**0.77*****	−0.47***	−0.22	0.25*
	EDTA%	−0.42***	0.17	0.53***	−0.36***	−0.28*	−0.10
	CaCl$_2$	−0.42*	0.05	−0.23	0.34	0.27	0.13
	CaCl$_2$%	−0.46*	−0.48*	−0.58**	0.43	0.33	−0.33
Ni	EDTA	−0.21	0.00	0.54***	−0.29**	0.16	−0.02
	EDTA%	−0.29**	0.15	0.45***	−0.38***	0.12	0.02
	CaCl$_2$	**−0.75***	−0.12	0.27*	−0.27*	0.07	−0.04
	CaCl$_2$%	**−0.81***	−0.07	0.01	−0.24	0.00	−0.03
Pb	EDTA	−0.34**	**0.96***	0.24*	−0.24*	−0.12	0.66***
	EDTA%	−0.39***	0.69***	0.30**	−0.30**	−0.14	0.42***
Zn	EDTA	−0.37***	**0.91***	0.29**	−0.29**	−0.07	0.59***
	EDTA%	−0.42***	0.88***	0.31**	0.32**	−0.05	0.51***
	CaCl$_2$	**−0.76***	0.85***	0.24**	−0.30**	−0.08	0.63***
	CaCl$_2$%	**−0.81***	0.78***	0.24*	−0.33**	−0.11	0.55***

more so as soil acidity enforces metal sorption at organic matter (Verloo and Cottenie, 1972).

Because inundation lasts much longer at the shallow ditch sites (up to 6 months) than at the ridge sites (about 2.5 months), comparison between metal concentrations of ridge and shallow ditch epipedon can indicate the effect of inundation. The EDTA-extractable fraction and its percentage of the total contents is higher in the shallow ditch soil than in the ridge soil. This is not reflected in the CaCl$_2$-extractable fraction. At anaerobic conditions, which prevail in soil during inundation (Erber *et al.*, subm.), enhanced availability of heavy metals by dissolution of manganic and ferric oxides is normal (Chuan *et al.*, 1996; Charlatchka and Cambier, 2000). This reasoning is supported by the fact that oxides are known to include more heavy metals than the surrounding soil (McKenzie, 1980; Brümmer *et al.*, 1983) and content of oxides is reduced in the epipedon of the shallow ditch sites (Erber, 1998). Under anaerobic conditions, small organic acids like citric acid are products of decomposition and form stable and soluble chelates with heavy metal ions (Leeper, 1978). This can explain why only the organic bound metal fraction increased. Nevertheless these metals are soluble and may be taken up by the vegetation and soil organisms like the mobile fraction.

Heavy metal content of the vegetation

The main plant species of the studied grassland are *Alopercurus pratensis*, *A. geniculata*, *Agrostis stolinifera*, *A. canina*, *Ranunculus repens*. Metal content of the vegetation varies broadly. The variation at one site is higher than between the sites. Therefore average and range of all samples are shown in Table 5. Normally, metal content of the second cut is higher than of the first cut. Heavy metal accumulation by plants depends on plant species and time in growing season (Jung and Thornton, 1996; Shallari *et al.*, 1998). At the study sites no increased uptake of heavy metals by the plants because of higher metal content in the soil can be stated. Heavy metal contents are within typical value for plants, especially for grassland species and therewith below critical values.

Heavy metal input by inundation

In addition to release of soil-borne heavy metals inundation causes a regular input of metals. Heavy metals listed in Table 6 originate from material deposited on the soil surface during inundation. Deposited material mainly consists of organic matter coming from algae, which are able to incorporate and accumulate heavy metals from water. Since algae easily decompose and disappear completely a few weeks after inundation, deposited metals are rapidly released. Algae growth depends on weather conditions which are different year by year. Consequently, metal input varies widely. Heavy metal input by inundation is very low compared to the total amounts (about 0.2 ‰ other than Zn), and it even is below the atmospheric deposition rate (Table 6). Output by hay making, however, is lower than input by inundation and atmosphere, whereas heavy metals continuously accumulate in the

Table 5 Heavy metal content of the vegetation [mg kg^{-1}]

	Cu	Ni	Pb	Zn
mean (range)	7.6 (2.2–14.4)	2.0 (0.2–8.0)	3.7 (1.0–23.4)	50.4 (25.1–86.1)
normal values[1]	5–15	0.1–5	2–7	20–100
critical values[2]	15–100	10–100	10–300	100–400
grassland[3]	7–13	3.1	1–38	23–99

[1]Bergmann (1992), Hock and Elstner (1995)
[2]Sauerbeck (1982)
[3]Salt (1988), Hornburg (1991)

Table 6 Input and output of heavy metals

	Mass [g m^{-2}]	Cu	Cr	Ni	Pb	Zn
Input by deposited material during inundation deposition [mg m^{-2}]						
Site I 1993/94	10.65–46.48	0.20–0.91	0.12–0.52	0.17–0.65	0.19–0.98	8.73–29.75
1994/95	25.21–75.14	0.32–1.29	0.69–1.55	0.33–1.29	0.63–2.03	3.95–12.14
Site II 1993/94	14.05–28.57	0.19–0.38	0.16–0.30	0.12–0.20	0.37–0.66	7.03–17.71
1994/95	52.90–153.18	0.67–2.79	0.79–5.94	0.58–4.17	1.20–7.66	6.51–24.37
Mean		0.8	0.9	0.7	1.4	22.6
Total content in epipedon [mg m^{-2}]		3168	4800	3072	8256	17088
Annual input by inundation [‰]		0.25	0.19	0.23	0.17	1.32
Annual atmospheric deposition[1] [mg m^{-2} a^{-1}]	9.9	1.1	1.8	3.9	34.5
Annual loss by hay making [mg m^{-2} a^{-1}]		5.4		1.3	1.7	37.5
balance [mg m^{-2} a^{-1}]	5.3		1.2	3.6	19.6

[1] Senator für Umweltschutz und Stadtentwicklung (1994)

epipedon. Leaching can be neglected because of low water conductivity of the soil (Erber, 1998), which is a result of high clay and high water content, which inhibits formation of cracks.

Conclusion

Heavy metal concentration has been enriched in the epipedon due to human activity. Above all, atmospheric deposition is a factor that should not to be underestimated as a source of metal input into the environment. Calculation of metal input by inundation for 10 years shows an increase of less than 2% for all accumulated elements. Metal input by atmospheric deposition for the last 100 years based on the data of 1991/92 (Senator für Umweltschutz und Stadtentwicklung, 1994) seems to be in the range of 20, 30 and 40% for Pb, Zn, and Cu, respectively. Cr would have been increased only by 3%. But if one takes the comparison of the epipedon and the subsoil of each site as the basis of calculation, metal enrichment at sites without municipal wastewater is 20 and 400% for Cr and Pb, respectively. For Zn and Cu, deposition rate seems to fit with the estimated 20 and 40%. The huge underestimation for Pb is a result of implementation of unleaded gas in the eighties wherefore Pb immission strongly reduced. Referring to the median comparison (Table 2), metal input by municipal wastewater tends to be 15–20%. For Cr and Zn, this enrichment seems to be correct for each site. But it cannot be proved for Cu and Pb because metal concentrations vary too much between the sites so that no effect of municipal water treatment can be estimated for these elements. Overall, degree of metal enrichment can be arranged in the following order: atmospheric deposition > irrigation with municipal water >> inundation. Mainly Pb and Zn were accumulated, which are also better available. Present artificial inundation for 3–6 months with surface water does obviously not increase exchangeable metals, but may enhance mobility by formation of soluble chelates since the EDTA-fraction rose at the inundated shallow ditch sites.

References

Andersson, A. (1975). Relative efficiency of nine different soil extractants. *Swed. J. agric. Res.*, **5**, 125–135.
BbodSch, V (1999). *Bundes-Bodenschutz- und Altlastenverordnung*. Bundesgesetzblatt Jahrgang 1999, Part 1 No.36.
Bergmann, W. (1992). *Nutritional disorders of plants. Development, visual and analytical diagnosis*. Spektrum Acad. Publ., Hdg.
Brümmer, G., Tiller, K.G., Herms, U. and Clayton, P.M. (1983). Adsorption-desorption and/or precipitation-dissolution processes of zinc in soils. *Geoderma*, **31**, 337–354.
Canarutto, S., Petruzzelli, G., Lubrano, L. and Vigna Guidi, G. (1991). How composting affects heavy metal content. *Biocycle*, 48–50.
Charlatchka, R. and Cambier, P. (2000). Influence of reducing conditions on solubility of trace metals in contaminated soils. *Water, Air, and Soil Pollution*, **118**, 143–167.
Chuan, M.C., Shu, G.Y .and Liu, J.C. (1996). Solubility of heavy metals in a contaminated soil: effects of redox potential and pH. *Water, Air and Soil Pollution*, **90**(3/4), 543–556.
DIN ISO 11466 (1997). *Soil quality – Extraction of trace elements soluble in aqua regia*.
Erber, C. (1998). *Bodeneigenschaften und Stoffhaushalt winterlich überstauter Flußmarschen im Niedervieland bei Bremen*. Boden und Landschaft, No 22.
Erber, C., Felix-Henningsen, P., Handke, K., Kundel, W. and Schreiber, K.-F. (subm.): Management of wet grassland in a fresh-water area near Bremen: effects on soil, vegetation and fauna. In: *Wetlands in Germany. Soil organisms, soil ecological processes, and trace gas emissions*, Broll, G., Merbach, W. and Pfeiffer, E.-M. (eds.), Springer.
Food and Agriculture Organization of the United Nations (1998). *World Reference Base for Soil Resources*. World Soil Resources Reports, No 84.
Gardiner, J. (1976). Complexation of trace metals by ehtylenediaminetetraacetic acid (EDTA) in natural waters. *Water Research*, **10**, 507–514.

Hindel, R. and Fleige, H. (1991). *Schwermetalle in Böden der Bundesrepublik Deutschland –geogene und anthropogene Anteile.* Texte 19/91, Umweltbundesamt, Hannover/Berlin.

Hock, B. and Elstner, E.F (1995). *Schadwirkungen auf Pflanzen. Lehrbuch der Pflanzentoxikologie.* 3rd edn, Spektrum Acad. Publ., Hdg.

Hornburg, V. (1991). *Untersuchungen zur Mobilität und Verfügbarkeit von Cadmium, Zink, Mangan, Blei und Kupfer in Böden.* Bonner Bodenkundliche Arbeiten. No 2.

Jung, M.C. and Thornton, I. (1996). Heavy metal contamination of soils and plants in the vicinity of a lead-zinc mine, Korea. *Applied Geochemistry*, **11**, 53–59.

Köster, W. and Merkel, D. (1982). Beziehungen zwischen den Gehalten an Zink, Cadmium, Blei und Kupfer in Böden und Pflanzen bei Anwendung unterschiedlicher Bodenuntersuchungsmethoden. *Landw. Forsch.*, Special Issue 39, 245–254.

Leeper, G.W. (1978). *Managing the heavy metals on the land. Pollution engineering and technology*, Vol 6, M. Dekker, New York.

Leita, L., De Nobili, M., Mondini, C., Muhlbachova, G., Marchiol, L., Bragato, G. and Contin, M. (1999). Influence of inorganic and organic fertilization on soil microbial biomass, metabolic quotient and heavy metal bioavailability. *Biol. Fertil. Soils*, **28**, 371–376.

Matschullat, J., Niehoff, N. and Pörtge, K.-H. (1991). Zur Element-Dispersion an Flusssedimenten der Oker (Niedersachsen); röntgenfluoreszenz-spektrometrische Untersuchungen. *Z. dt. geol. Ges.*, **142**, 339–349.

McKenzie, R.M. (1980). The adsorption of lead and other heavy metals on oxides of manganese and iron. *Austr. J. Soil Res.*, **18**, 61–73.

Moalla, S.M.N., Awadallah, R.M., Rashed, M.N. and Soltan, M.E. (1998). Distribution and chemical fraction of some heavy metals in bottom sediments of Lake Nasser. *Hydrobiologia* **364**, 31–40.

ÖNORM (1989). *Bestimmung von EDTA-extrahierbarem Eisen, Mangan, Kupfer und Zink. Österreichisches Normungsinstitut.* Wien.

Qian, J., Shan, X., Wang, Z. and Tu, W. (1996). Distribution and plant availibility of heavy metals in different particle-size fractions of soil. *The Science of the Total Environment*, **187**, 131–141.

Salt, C. (1988). *Schwermetalle in einem Rieselfeld-Ökosystem. Landschaftsentwicklung und Umweltforschung.* Schriftenreihe des Fachbereichs Landschaftsentwicklung der TU Berlin, No 53.

Sauerbeck, D. (1982). Welche Schwermetalle in Pflanzen dürfen nicht überschritten werden um Wachstumsbeeinträchtigungen zu vermeiden. *Landw. Forsch.*, Special Issue 39, 108–129.

Senator für Umweltschutz und Stadtentwicklung (1994). *Luftmessprogramm 1991/92 im Lande Bremen.* Schriftenreihe Luftreinhaltung. No 2.

Shallari, S., Schwartz, C., Hasko, A. and Morel, J.L. (1998). Heavy metal in soils and plants of serpentine and industrial sites of Albania. *The Science of the Total Environment*, **209**, 133–142.

Schimming, C.-G. (1990). Metalle. In: *Handbuch des Bodenschutzes*, H.-P. Blume (ed.), 2nd edn, ecomed, Landsberg/Lech, pp. 258–304.

VDLUFA (ed.) (1991). *Die Untersuchung von Böden. Methodenhandbuch*, Vol 1. 4th edn, Darmstadt.

Verloo, M. and Cottenie, A. (1972). Stability and behaviour of complexes of Cu, Zn, Fe, Mn and Pb with humic substances of soils. *Pedologie*, **22**, 174–184.

Yongcan, G., Zhenzhong, W., Youmei, Z. and Xiaoyang, M. (1998). Bioconcentration effects of heavy metal pollution in soil on the mucosa epithelia call ultrastructure injuring of the earthworm's gastrointestinal tract. *Bull. Environ. Contam. Toxicol.*, **60**, 280–284.

Tolerance towards explosives, and explosives removal from groundwater in treatment wetland mesocosms

E.P.H. Best, J.L. Miller and S.L. Larson

U. S. Army Engineer Research and Development Center, Environmental Laboratory, 3909 Halls Ferry Road, Vicksburg, MS 39180, USA

Abstract A short-term study was performed to determine the feasibility of using constructed wetlands to remove explosives from groundwater, and to assess accumulation of parent explosives compounds and their known degradation compounds in wetland plants. Tolerance towards explosives in submersed and emergent plants was screened over a range of 0 to 40 mg L^{-1}. Tolerance varied per compound, with TNT evoking the highest, 2NT the lowest, and 24DNT, 26DNT, and RDX an intermediate growth reducing effect. Submersed plants were more sensitive to TNT than emergent ones. A small-scale 4-month field study was carried out at the Volunteer Army Ammunition Plant, Chattanooga, TN. In this surface-flow, modular system, the influent contained high levels (>2.1 mg L^{-1}) of TNT, 2,4DNT, 2,6DNT, 2NT, 3NT, and 4NT, and the HRT was 7 days. The performance criteria of US EPA treatment goals for local discharge of 2,4DNT concentration <0.32 mg L^{-1}, and 26DNT concentration <0.55 mg L^{-1} were not met at the end of the experiment, although explosives levels were greatly reduced. Low levels of 2ADNT and 4ADNT were transiently observed in the plant biomass. Results of two other, older, constructed wetlands, however, indicated that in these systems treatment goals were met most of the time, residues of explosives parent compounds and known degradation compounds in plant tissues were low and/or transient, and in substrates were low.

Keywords Constructed wetlands; explosives; TNT; RDX; 24DNT; 26DNT; 2NT; groundwater

Introduction

2,4,6-Trinitrotoluene (TNT) and hexahydro-1,3,5-trinitro-1,3,5,-triazine (RDX) are important, widely used military explosives. Their manufacture and handling can lead to contamination of the environment by improper disposal. Concerns regarding the environmental fate of nitroaromatic compounds, such as TNT and its manufacturing compounds (mononitrotoluenes and dinitrotoluenes), and nitramine compounds, such as RDX, are arising because of their relative recalcitrance and potential mutagenicity (Marvin-Sikkema and De Bont, 1994; Gorontzy *et al.*, 1994). Efforts are under way to develop effective and cost-efficient remediation technologies to reduce human and ecological risk of these contaminants.

Most Department of Defense installations have sites with contaminated ground- and surface water. Traditionally, contaminated water has been remediated using industrial treatment technologies, including pump-and-treat systems. The feasibility of explosives removal from surface and groundwater using constructed wetlands is studied, since it is believed that this green technology may provide an environmentally-friendly and cost-effective alternative for traditional methods.

A prerequisite for plants to persist and function in an explosives remediation wetland is to tolerate explosives to such an extent that growth takes place. Several submersed and emergent plant species underwent short-term screening for survival in explosives-contaminated groundwater from Army Ammunition Plants (AAPs), notably from Milan AAP (MAAP) and Iowa AAP (IAAP; Best *et al.*, 1997a,b; TVA, 1995) before the treatment wetlands were implemented at these sites. Although these first screens allowed assessment of explosives disappearance from water, incubation periods were not long enough to evaluate plant growth, and, consequently, several plant species performed poorly in the wetlands. However, even if plants had been incubated long enough, the chemical characteristics of explosives-contaminated groundwater at AAPs vary considerably (Table 1), and, therefore, it is not likely that plants that tolerate explosives-contaminated groundwater from one particular site will also tolerate water from other sites. It is, consequently, important to screen several submersed and emergent plant species with known inorganic carbon and nutrient requirements for tolerance towards individual explosives and explosives mixtures for a long enough period to allow growth. Very few wetland plants have been properly screened so far. For this study, two submersed and two emergent plant species were screened for tolerance towards explosives.

At the Volunteer AAP (VAAP) in Chattanooga, TN, the groundwater was contaminated as a consequence of TNT manufacturing activities with a mixture of mono-, di- and trinitrotoluenes. 2,4DNT was the most widespread nitroaromatic compound detected, and generally occurred at higher concentrations (>1 mg L^{-1}) than 2,6DNT and TNT. Hexahydro-1,3,5-trinitro-1,3,5-triazine (RDX) and Octahydro-1,3,5,7-tetranitro-1,3,5,7-tetrazocine (HMX) were not found. The VAAP is one of the SERDP National Cleanup Technology Demonstration Sites used to investigate various options for the removal and remediation of explosives contamination within both soil and groundwater matrices. The use of constructed wetlands was designated as one of the options.

The objective of the current study was to quantify the ability of aquatic plants to phytoremediate local explosives-contaminated groundwater when planted in local sediment under flow-through conditions, at the VAAP. The results of this experiment were compared with two, longer, studies performed at other AAPs (Sikora *et al.*, 1998; Kiker *et al.*, 2000).

Material and methods

Aquatic plant tolerance towards explosives

Plant species were chosen based on their known short-term (10-day) survival in MAAP or IAAP groundwater. Submersed species included *Potamogeton pectinatus* (sago pondweed) and *Heteranthera dubia* (water stargrass) and emergent species included *Phragmites australis* (common reed) and *Phalaris arundinacea* (reed canary grass). Plants were exposed to aqueous explosives in concentration ranges believed to be relevant for both on-site plant survival and phytoremediation potential (0–40 mg L^{-1}). Three-week incubations were performed in a walk-in growth room with plants receiving approximately 500 to 600 μE m^{-2} s^{-1} in the photosynthetically active region, augmented with 2% UV-B, at their upper surface, at a photoperiod of 14 h of light and 10 h of darkness, and a temperature of 23°C. Explosives were administered 2× per week by complete refreshment of the medium. Plants were planted in N-amended silty sediment, and culture solution contained low concentrations of nutrients (Smart and Barko medium), and was aerated. All incubations were started with at least 5.7 g fresh weight per L or three whole plants. Plant response was determined by measuring initial and final plant mass, and expressed as relative growth rate. Relative growth rate was calculated from a differential equation commonly used to describe simple exponential growth. RGR = $1/t \times \ln(W_t/W_0)$; where RGR is relative growth rate (d^{-1}), W_i is plant weight at time t (g DW), and W_0 is plant weight at time 0 (g DW).

Explosives removal in outdoor wetland mesocosms

A short-term study was undertaken to assess explosives removal in wetland mesocosms under field conditions. This experiment involved a 115-day study (31 May–23 September, 1996) at the VAAP, in Chattanooga, TN (Best *et al.*, 2000). Groundwater from well MW78, located at the border of the TNT manufacturing valley, was used as influent. This water contained high 2,4DNT and 3NT levels, and was expected to contain sufficient levels of inorganic carbon and nutrients to support submersed plant growth (Table 1).

The effects of 3 treatments on aqueous explosives concentrations were compared, notably planted, non-planted, and UV-shielded. Shielding was done by placing plastic OP-2 sheets on frames 0.05 m above the top of the lagoons. OP-2 (Faro Industries, Stamford, ME) absorbs UV light between 200 and 400 nm and allows 98 per cent transmission of the visible light. Each treatment (lagoon) was replicated 3 ×. The wetland system, composed by 9 lagoons (Figure 1), was situated at a site 20 m removed from well MW78. Each lagoon consisted of a rectangular, polyethylene container (1.86 × 1.03 × 0.75 m, l × w × h; Bonar Inc., Newman, GA), with 72 1.9-L planted or non-planted glass jars on the bottom. Groundwater from the well was pumped into one, central, influent collection tank, equipped with high and low level safety switches to maintain a preset water level. Each lagoon received groundwater from the influent collection tank via self-priming, submersible, calibrated pumps. Effluent flowed by gravity into a sump pit. Water levels in the lagoons were adjusted to accomodate the plants, and included a final depth of 0.15 m above the jars (0.38 m above the bottom) for the planted and of 0.40 m above the jars (0.63 m above the bottom) for the non-planted treatments, giving total water volumes of 591.6 L and 1070.5 L, respectively. Explosives-contaminated groundwater was continuously pumped into the lagoons and a 7-day hydraulic retention time (HRT) was maintained. A granulated active carbon filter was used to polish lagoon effluents prior to discharge

Table 1 Chemical characteristics of the groundwater at three Army Ammunition Plants at the beginning of the experiments. At VAAP, TN, from well MW78. At MAAP, TN, from well MI146. At IAAP, IA, a 50/50 composite from well G19 and G20. Mean values and SD ($N = 3$)

Characteristic	VAAP	MAAP	IAAP
pH	7.1 ± 0	8.3 ± 0.1	7.5 ± 0.1
Macro-, micronutrients (mg L^{-1})			
Alkalinity	228 ± 1.7	0.015 ± 0.003	170 ± 1.4
Total Dissolved Solids	343 ± 7.1	NA	839 ± 58
Kjeldahl-N	<0.010	NA	1.6 ± 0.2
Nitrate-nitrogen	16.6 ± 0.4	5.8 ± 1.7	0.15 ± 0.09
Ammonium-nitrogen	0.36 ± 0.00	0.08 ± 0.08	0.17 ± 0.01
Total-phosphorus	0.003 ± 0.001	NA	<0.2
Phosphate-phosphorus	<0.001	0.179 ± 0.034	0.185 ± 0.020
Sulfate	15.5 ± 0.8	1.53 ± 0.16	56.35 ± 0.72
Calcium	72.8 ± 4.4	5.9 ± 1.3	121.16 ± 3.43
Manganese	2.1 ± 0	NA	0.06 ± 0
Explosives (mg L^{-1})			
Octahydro-1,3,5,7-tetranitro-1,3,5,7-tetrazocine (HMX)	NA	0.178[1]	NA
2, 4, 6-Trinitrotoluene (TNT)	2.727 ± 0.062	2.197 ± 0.068	0.681 ± 0.070
2-Amino-dinitrotoluene (2ADNT)	0.110 ± 0.003	0.043 ± 0	0.060 ± 0.004
4-Amino-, 2, 6-dinitrotoluene (4ADNT)	0.340 ± 0.017	0.036 ± 0	0.041 ± 0.002
2,4-Diamino-,6-nitrotoluene (2,4DANT)	<0.002	0.007 ± 0.002	0.023 ± 0.014
2,6-Diamino-,4-nitro-toluene (2,6DANT)	<0.002	0.074 ± 0.003	0.596 ± 0.036
2,4-Dinitrotoluene (2,4DNT)	16.663 ± 0.380	<0.002	<0.002
2,6-Dinitrotoluene (2,6DNT)	5.169 ± 0.174	<0.002	0.006 ± 0.007
2-Nitrotoluene (2NT)	42.635 ± 0.761	<0.002	<0.002
3-Nitrotoluene (3NT)	3.292 ± 0.095	<0.002	<0.002
4-Nitrotoluene (4NT)	30.456 ± 0.572	<0.002	<0.002
1,3,5-Trinitro-benzene (TNB)	1.422 ± 0.088	0.308 ± 0.017	1.422 ± 0.088
1,3-Dinitro-benzene (1,3DNB)	0.023 ± 0.003	0.029 ± 0.014	0.020 ± 0.004
1,4-Dinitro-benzene (1,4DNB)	<0.002	<0.002	<0.002
Nitrobenzene (NB)	0.017 ± 0.057	NA	0.015 ± 0.003
Hexahydro-1,3,5-trinitro-1,3,5-triazine (RDX)		3.002 ± 0.082	12.785 ± 1.744

NA, not analyzed; [1], Sikora et al., 1998

into two 285,000-L storage reservoirs to await transport to a disposal site. The plantings included *Typha angustifolia* (narrow-leaved cattail), with shoot length clipped to 0.20 m, 3 plants per jar. Sediment was excavated from a nearby, non-contaminated lake, mixed and transferred into jars prior to planting and placement into the lagoons.

Sampling and chemical analyses. In all systems in- and effluents were monitored for concentrations of explosives, pH, and various other water quality parameters. Aqueous explosives concentrations were measured using HPLC (USEPA, 1992). Other water quality parameters were determined following standard methods (APHA, 1992). Plant health and growth was recorded, and tissue residues of explosives parent compounds and known degradation compounds were determined in acetonitrile extracts using HPLC (USEPA, 1992).

Removal kinetics. Explosives removal kinetics were calculated using a simple mathematical model, based on the assumptions that the system has reached steady-state and plug-flow takes place. The following equation was used: $C_0 = C_i e^{-K_a/q}$; where C_0 and C_i are the constituent concentrations (mg L^{-1}) in the effluent and effluent, respectively, K_a is the areal removal constant (m d^{-1}), and q is the hydraulic loading rate (m d^{-1}; Kadlec and Knight, 1996a). In case of the VAAP study, for C_i of each target explosives parent compound, the median was calculated of all influent values. For C_0, the median was calculated of the effluent samples collected between 61 and 115 days of operation. q was 0.044 m d^{-1} in planted and 0.1 m d^{-1} in non-planted lagoons.

Results and discussion

Aquatic plant tolerance towards explosives

The toxicity of explosives for the aquatic plants varied per compound (Table 2). Most explosives reduced plant growth. Exceptions were: (1) relatively low concentrations of 2NT and RDX that stimulated growth in stargrass, and (2) 20 mg L^{-1} 24DNT and (3) TNT concentrations up to 10 mg L^{-1} TNT that stimulated growth in reed canary grass. TNT was usually toxic at lower concentrations for submersed plants than for emergent plants. Reed canary grass proved more sensitive to the highest concentrations of 24DNT and 2NT than the other species.

Explosives removal in outdoor wetland mesocosms

Rates of nitroaromatic removal from water varied with nitroaromatic chemical species, treatment lagoon, and time. Concentrations in the effluents of the following explosives

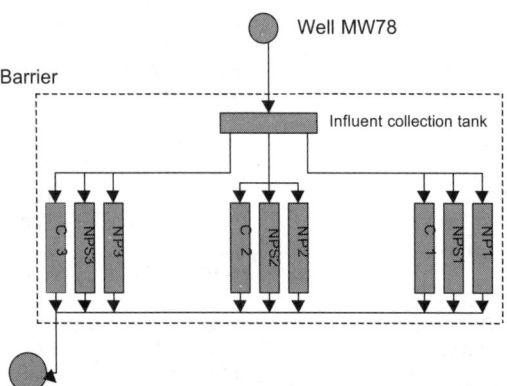

Figure 1 Diagram representing the constructed wetland system. C, cattail planted; NPS, non-planted shielded; NP, non-planted

Table 2 Growth responses of aquatic plant species exposed to aqueous explosives concentrations, as per cent of control. Growth response measured as relative growth rate (on DW basis). Average values and standard deviations ($N = 3$).

Explosives Concentration (mg L^{-1})	Plant species			
	Submersed		Emergent	
	Sago pondweed	Water stargrass	Common reed	Reed canary grass
TNT				
2.5	37 ± 151	1 ± 108	36 ± 0	160 ± 44
5	−468 ± 168	−46 ± 20	4 ± 0	116 ± 30
10	−605 ± 153	−85 ± 95	2 ± 2	1003 ± 16
24DNT				
5	−264 ± 169	66 ± 46	6 ± 9	81 ± 11
20	−787 ± 74	−114 ± 42	9 ± 4	131 ± 11
40	−341 ± 128	−66 ± 58	dead	dead
26DNT				
5	44 ± 108	90 ± 34	4 ± 2	60 ± 7
20	−254 ± 38	−21 ± 11	2 ± 5	51 ± 19
40	−137 ± 219	−117 ± 81	−3 ± 11	−1667 ± 19
2NT				
5	11 ± 106	167 ± 66	16 ± 20	32 ± 23
20	14 ± 81	15 ± 109	−2 ± 10	64 ± 100
40	9 ± 64	39 ± 30	−583 ± 579	dead
RDX				
5	−292 ± 147	136 ± 70	23 ± 0	6 ± 111
20	38 ± 118	123 ± 36	−2 ± 6	96 ± 63
40	58 ± 166	−6 ± 35	8 ± 12	60 ± 43
Control (d^{-1})	−0.007 ± 0.010	0.018 ± 0.005	0.020 ± 0	0.019 ± 0

decreased exponentially with time: TNT, 2,4DNT, 2,6DNT, 2NT, 3NT, 4NT, and TNB. Concentrations were far lower in the effluents of the planted than of the non-planted lagoons. The average total nitrobodies removal rates were 5.44 mg L^{-1} d^{-1} in the lagoons planted with cattails, 4.45 mg L^{-1} d^{-1} in the non-planted lagoons, and 4.01 mg L^{-1} d^{-1} in the UV-shielded non-planted lagoons (total nitrobodies in this case was defined to include the sum of TNT, 24DNT, 26DNT, 2NT, 3NT, and 4NT). The UV-part of sunlight contributed greatly to the removal rates in the non-planted treatments, for 30% of TNT, for 60% of 2,4DNT, and for 59% of 2,6DNT, but far less (10%) of 2NT, 3NT, and 4NT. The average levels of total nitroaromatics in the effluents from the lagoons were 14.0 in the planted, 28.1 in the non-planted, and 34.4 mg L^{-1} in the UV-shielded lagoons, while those in the influent were 94.6 mg L^{-1}. The areal explosives removal constants differed per explosives parent compound, and were highest for 2NT (Table 3). The thus-calculated areal removal constants may be considered as conservative estimates, because a sustainable wetland vegetation requires a period of 1–3 years to become established, wetland activity is usually lower during the first year than subsequently, and steady state was not yet reached.

The cattail biomass increased considerably and reached final weights of 0.3 kg fresh weight per jar, representing 21.6 kg per planted lagoon. Tissue residues in September were low; 2ADNT at 0.439 mg kg^{-1} fresh weight was found in shoots, and 4ADNT at 0.399 µg g^{-1} fresh weight in roots (Best et al., 2000). These compounds are both TNT degradation products. The lack of other explosives residues in cattails may be explained by either exclusion by, or rapid metabolism of these compounds inside these plants. Besides cattails, also three submersed plant species were tested in the lagoons, including *Elodea canadensis* (elodea), *Ceratophyllum demersum* (coontail) and *Potamogeton nodosus* (American pondweed). Submersed plants did not survive the test because of the high explosives levels. Explosives in sediments were not determined because their adsorption was expected to be low (Best et al., 1999).

Table 3 Explosives dynamics in constructed wetlands

Description	Age (months)	Concentration (mg L^{-1}) In	Concentration (mg L^{-1}) Out	Removal (%; avg. season)	Removal constant (m d^{-1})	Reference
Small-scale (1.9 m^2) wetlands mesocosms, SF- cattails; Volunteer AAP, TN	0–4					Best et al., 2000
Total nitrobodies		94.600				
TNT		2.727	0.300	79	0.086	
24DNT		16.663	6.000	58	0.043	
26DNT		5.169	1.400	61	0.046	
2NT		42.635	2.198	NC	0.123	
Intermediate-scale (352 m^2) wetlands, SSF-reed canary grass; Milan AAP, TN	12–24					Sikora et al., 1998
Total nitrobodies		9.200				
TNT		4.440	<0.002	99		
RDX		4.240	variab.	50–99		
TNB		0.330	<0.002	99		
HMX		0.091	variab.	50–99		
Large-scale (900 m^2) wetlands, SF-reed canary grass, coontail; Line 1 Iowa AAP, IA						Kiker et al., 2000
RDX	0–12	0.800	0.011	96[1]	NA	
RDX	12–24	0.800	<0.002	99[1]	NA	

Abbreviations: AAP, Army Ammunition Plant; NC, not calculated; NA, not applicable; [1], 1 June to 1 November

Wetland performance and outlook

At the VAAP explosives removal was considerable in the cattail-planted wetland lagoons, and explosives levels were far lower in the effluents from the planted than from the non-planted lagoons. However, several effluent explosives levels exceeded EPA treatment goals for local discharge, 2,4DNT concentration being <0.32 mg L^{-1}, and (2) 26DNT concentration <0.55 mg L^{-1}. Comparison of the current treatment system with both other systems is difficult because of the large differences between systems, being chemical composition of the influent (Table 1), surface area, age, type (surface-flow versus subsurface-flow), vegetation type (Table 3), and HRT (MAAP ~ 10 days, IAAP unknown). Moreover, performance criteria differed also: (1) for the MAAP, TNT concentration <0.002 mg L^{-1} (USEPA Health Advisory Level) and total nitrobodies concentration <0.050 mg L^{-1} (total nitrobodies being the sum of the concentrations of TNT, RDX, HMX, 24DNT, 26DNT and TNB); (2) for the IAAP, RDX concentration <0.002 mg L^{-1} (USEPA Health Advisory Level). In both MAAP and IAAP wetlands, explosives removal was usually higher and final aqueous concentrations lower. Better performance of the latter wetlands was attributed to (1) lower aqueous explosives concentrations in the influent, and (2) higher wetland age, allowing better plant acclimation and establishment (Kadlec and Knight, 1996b).

At the VAAP tissue residues of explosives parent compounds and known degradation products were low in the emergent plants and were not measured in submersed plants, because the latter did not survive. Poor performance of submersed plants in explosives-contaminated water, as found in the plant screens and outdoor lagoons of the current study, was also reported for the MAAP wetland. In the MAAP system plants generally accumulated TNT metabolites and RDX to a limited extent in winter, followed by less accumulation in summer. Final levels of explosives parent compounds and metabolites were <4 mg kg^{-1} dry weight. The quantity of total explosives (and of total nitrobodies) on the gravel was always < 1.3% of the mass of nitrobodies entering the gravel-based wetlands. The percentage of nitrobodies on the gravel decreased to <0.1% of influent nitrobodies in summer. In the IAAP system plant shoot tissue residues were all below detection, and sediment samples showed explosives levels at or below the remediation goals of 47.6 mg kg^{-1} TNT and 1.3 mg kg^{-1} RDX.

Based on the wetland performance reported so far, constructed wetlands vegetated by emergent plants can be considered as effective in removing explosives from water. It has to be cautioned, however, that the toxicities of effluents, and plant tissue and substrate residues are currently unknown, and toxicological testing remains to be done.

References

APHA (1992). *Standard methods for the examination of water and wastewater*. 18 ed., American Public Health Association, Washington, DC.

Best, E.P.H., Sprecher, S.L., Fredrickson, H.L., Zappi, M.E. and Larson, S.L. (1997a). Screening submersed plant species for phytoremediation of explosives-contaminated groundwater from the Milan Army Ammunition Plant, Milan, Tennessee. U. S. Army Engineer Waterways Experiment Station Technical Report EL-97-24.

Best, E.P.H., Zappi, M.E., Fredrickson, H.L., Larson, S.L., Sprecher, S.L. and Ochman, M.S. (1997b). Fate of TNT and RDX in aquatic and wetland plant-based systems during treatment of contaminated groundwater. *Ann. N. Y. Acad. Sci.* **829**, 179–194.

Best, E.P.H., Sprecher, S.L., Larson, S.L., Fredrickson, H.L. and Bader, D.F. (1999). Environmental behavior of explosives in groundwater from the Milan army Ammunition Plant in aquatic and wetland plant treatments. Uptake and fate of TNT and RDX in plants. *Chemosphere*, **39**(12), 2057–2072.

Best, E.P.H., Miller, J.L. and Larson, S.L. (2000). Explosives removal from groundwater at the Volunteer Army Ammunition Plant, TN, in small-scale wetland modules. In: J.L. Means and R.E. Hinchee (eds). *Wetlands and Remediation: an International Conference*, Salt Lake City, Utah, November 16–17, 1999, Battelle Press, Columbus, Richland, 365–373.

Gorontzy, T., Drzyzga, O., Kahl, M.W., Bruns-Nagel, D., Breitung, J., Von Loew, E., Blotevogel, K.H. (1994). Microbial degradation of explosives and related compounds. *Crit. Rev. Microbiol.*, **20**, 265–284.

Kadlec, R.H. and Knight, R.L. (1996). *Treatment Wetlands*. CRC Lewis Publishers, Boca Raton, New York; a, 194–263; b, 154, 690–91.

Kiker, J.H., Moses, D.D., Larson, S.L. and Sellers, R. (2000). Use of engineered wetlands to phytoremediate explosives contaminated surface water at the Iowa Army Ammunition Plant, Middletown, Iowa. National Defense Industrial Association Proceedings, *26th Symposium & Exhibition 'Sustaining DoD Readiness: Changes and Challenges in DoD Environmental Priorities*. March 27–30, 2000, Long Beach Convention Center, Long Beach, California.

Marvin-Sikkema, F.D. and De Bont, J.A.M. (1994). Degradation of nitroaromatic compounds by microorganisms. *Appl. Microbiol. Biotechnol.*, **42**, 499–507.

Sikora, F.J., Almond, R.A., Behrends, L.L., Hoagland, J.J., Kelly, D.A., Phillips, W.D., Rogers, W.J., Summers, R.K., Thornton, F.C. and Trimm, J.R. and Bader, D.F. (1998). Demonstration results of phytoremediation of explosives-contaminated groundwater using constructed wetlands at the Milan Army Ammunition Plant, Milan, Tennessee. Volume I, II, III, IV. Technical Report SFIM-AEC-ET-CR-97059; TVA Contract No. TV-99926V.

TVA (1995). Screening of wetland emergent species for remediation of explosives-contaminated groundwater. Tennessee Valley Report.

US Environmental Protection Agency (1992). Test Methods for Evaluating Solid Wastes, Proposed Update II, Method 8330. Rep. SW-846 (3rd Edn), Office of Solid Waste and Emergency Response, Washington, DC, U.S.A.

A constructed surface flow wetland for treating agricultural waste waters

M. Borin*, G. Bonaiti*, G. Santamaria* and L. Giardini*

* Dipartimento di Agronomia Ambientale e Produzioni Vegetali, Agripolis, Università di Padova, via Romea, 16, 35020 Legnaro (PD), Italy. (E-mail: *maurizio.borin@unipd.it*)

Abstract A study was conducted between December 1997 and December 1998 in NE Italy on a 3,200 m^2 surface flow vegetated wetland receiving agricultural drainage water from a cultivated field of about 6 ha and occasional applications of organic wastes. The study aimed at evaluating: 1) biomass and seasonal nitrogen dynamics in above- and below-ground biomass of *Phragmites australis* Cav. (Trin.) and *Typha latifolia* (L.) grown in separate zones; 2) the effectiveness of the wetland in removing nutrients and sediments coming from the fields; 3) the possibility that wetland could treat occasional applications of organic wastes and 4) to collect some general information on whether the wetland can receive heavy loads coming from storm water runoff.

Monthly observations showed that, in both species, aboveground biomass, nitrogen concentration and nitrogen content reached maximum values in summer and minimum values in winter. The contrary occurred in below-ground biomass. The total input of water in the wetland was 66,000 m^3 ha^{-1}, of which 7,700 were drained. Total nitrogen input was 526 kg ha^{-1}, of which 58 were discharged out of the wetland.
Keywords Agricultural pollution; constructed wetland; *Phragmites australis*; *Typha latifolia*

Introduction

To protect internal and sea waters from agricultural pollution, various strategies were initiated in the early 1990s in the Veneto Region (NE Italy), partly motivated by EU agro-environmental policies that foresaw economic benefits for farmers who agreed to adopt less intensive agricultural techniques or to implement buffer strips, wetlands and controlled drainage.

At present, the Regional Water Authorities have many projects under way on the implementation of wetlands to control diffuse pollution, increasing the need for local information. Most of the results from international research are encouraging, showing that many physical, chemical and biological processes take place in the wetland environment, effectively removing nutrients (e.g. Chescheir *et al.*, 1991; Olson and Marshall, 1992; Vellidis *et al.*, 1994; Lowrance *et al.*, 1995; Romero *et al.*, 1999) sediments (e.g. Chescheir *et al.*, 1991; Higgins *et al.*, 1993) and pesticides (Kao and Wu, 2000) from the water coming from agricultural fields. Wetlands have also been proved effective in controlling high organic loads coming from animal (e.g. Hunt *et al.*, 1995) or domestic wastewater (e.g. Gearheart, 1992). However, wetlands present a wide range of features such as environmental conditions, type of vegetation, hydrology etc. As wetland performance is affected by site conditions these make it difficult to extend the results to other situations (Kadlec and Knight, 1996).

Within this framework, an extensive research on wetland treatment of water from agricultural fields began in 1996 at the Department of Environmental Agronomy and Crop Production of Padova University. In this paper, the first results on the performance of a constructed wetland are presented and discussed, with the aim of evaluating: 1) seasonal dynamics of biomass and nitrogen in above- and below-ground organs of *Phragmites australis* Cav. (Trin.) (common reed) and *Typha latifolia* (L.) (cattail) grown in separate zones; 2) the effectiveness of the wetland in treating water coming from agricultural fields; 3) the possibility that wetland could treat occasional applications of organic wastes and 4) to collect some general information on whether the wetland can receive heavy loads coming from stormwater runoff.

The results refer to work done during 1998.

Methods
Experimental features

Site conditions. The experiment began in 1996 at the "L. Toniolo" Experimental Farm of Padova University at Legnaro (near Padova, 45° 21' N; 11° 58' E; 6 m a.s.l.) and is still ongoing. The climate of the site is sub-humid, with annual rainfall of about 850 mm fairly uniformly distributed throughout the year. The temperature increases from January (average minimum value: −1.5°C) to July (average maximum value: 27.2°C).

According to the FAO-UNESCO classification, the soil is fulvi-calcaric cambisol, with loamy texture in the upper 80 cm; the percentage of silt gradually increases with depth, reaching values of 68–75% at 2–2.4 m depth. A first layer of reduced permeability is located at 1.5–1.8 m and there is an impervious layer at about 3 m.

Wetland features. Six ha of land was divided into 12 plots, in which different drainage systems were installed to simulate the features of a typical agricultural basin on the Veneto low-lying plain (Borin *et al.*, 1998). The drainage water of the plots is diverted to a constructed surface flow wetland (SFW), covering 3200 m^2, for treatment before discharge into a stream. The wetland is almost square in shape and is fed water from a pump located at the SW corner. The outlet is located at the NW corner and a small bank forces the water to circulate in the basin (Figure 1). At the outlet, an upward curving pipe allows space for pipes of various lengths to be inserted to regulate the desired depth of water within the basin. To limit lateral water flow, a plastic film was installed vertically, to a depth of 1.5 m along the SFW perimeter.

The SFW was vegetated in spring 1997, dividing the area into 4 sectors: cattail was planted in sector I, common reed in II and III, and IV was left without vegetation. Cattail was planted as clumps with some developed plants, and common reed as rhizome cuttings

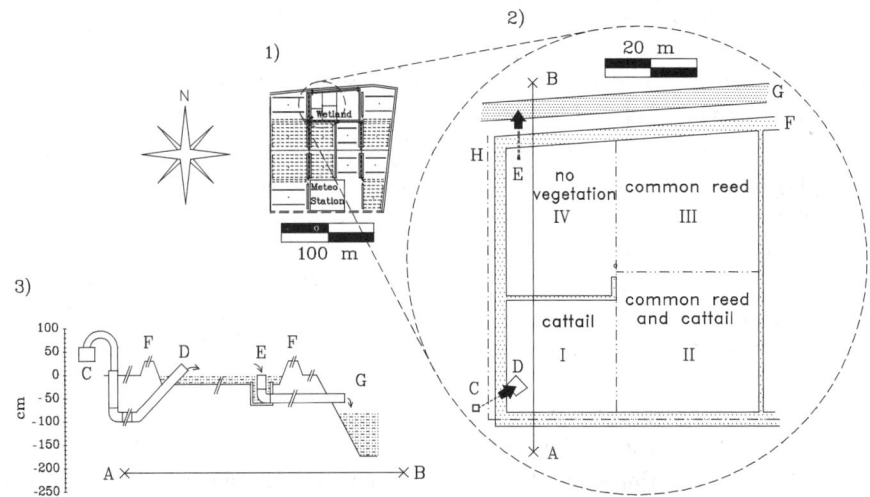

Figure 1 Layout of the experimental area. 1) map of the drainage basin; 2) map of the SFW; 3) longitudinal section of the SFW. A-B: longitudinal section; C: pump; D: wetland water inlet; E: wetland water outlet; F: bank; G: stream; H: plastic film. I, II, III, IV: sectors of the SFW

with 1–3 buds. During 1997, due to an exceptionally dry season, the hydrophytes were irrigated periodically, to allow them to establish. Soil moisture decreased from sectors I to IV on average during the year. Mechanical weed control was also carried out twice during the growing season. Due to favourable soil moisture and the availability of cattail seeds, sector II was planted with a mixture of cattail and common reed at the start of 1998.

Activity in 1998
Water balance and hydroperiod. All water inputs in the wetland (drainage from fields, rainfall, and irrigation) and discharged volumes were measured, together with the depth of flooding or water table depth. Mechanical flowmeters were installed at the pump and outlet, rainfall was measured at the local agrometeorological station very close to the wetland and the flooding depth was measured at the outlet. All measurements were taken daily.

Vegetation. Vegetation was sampled each month in the three vegetated sectors, taking 4 samples in each sector to determine dry matter (d.m.) and N contents of above- and belowground biomass (to a depth of 0.3 m). In addition, the height and number of shoots per plant were determined weekly on 16 plants of each species.

Water quality. To study the effects of the SFW on the drainage water coming from the fields, samples of input and output water were taken daily when water entered or was discharged from the wetland, to analyse NO_3-N concentration, pH and EC. The sampling was done by hand, taking water from the manhole of the suction lift of the pump and at the flowmeter installed at the discharge. The water samples were immediately frozen for storing until analysed.

In addition, 6 applications of an artificial sewage slurry were made with the aim of studying the performance of the SFW in treating occasional loads of organic wastes. The slurry was prepared by mixing a powdered commercial manure in water, in order to apply 50 kg ha^{-1} of N each time. Applications were followed by surface irrigation to allow movement of the slurry into the wetland, at the same time avoiding discharge of the water before the 15th day after distribution. After this time, the controlling pipe at the outlet was removed to allow any water discharge, which was then measured and sampled. The applications were done on the following dates: 12th June, 8th July, 30th July, 18th August, 2nd October and 2nd December. Every time, water samples were taken before the outlet by grab at 2, 4, 8 and 15 days after the slurry application to determine BOD, organic N, NH_4-N, NO_3-N, turbidity, electrical conductivity and pH. This paper presents and discusses the results regarding N.

N balance. Input and discharge of N from the SFW were calculated multiplying the daily measured water volumes by their relative concentrations.

Results and discussion
Water balance and hydroperiod
Total rainfall during the monitoring period was slightly lower than the median values of the area. The rainfall distribution was fairly irregular, with above-average values in December 1997, April, September and October 1998. Low rainfall was recorded in February, March, August, November and December 1998. The period of rainfall in October is worth noting, when more than 110 mm fell between the 4th and the 10th.

During the experiment, there was almost continuous water from agricultural drainage entering the wetland, but the volumes differed greatly over time. In particular, there were three peaks of intensive drainage, corresponding to the rainiest periods, and a period of very low drainage in August–September. In the latter, the water entering the wetland was almost completely due to the recycling of water lost from the wetland by lateral movement towards the neighbouring plots (Figure 2). This was also caused by the almost continuous irrigation during the period.

Taking all the water inputs together, the wetland received water on almost every day of the monitoring period, but the values and ratios among the different sources were very variable. In the first five months, almost all the water input was due to field drainage, with daily peak values of 1,500–1,600 m^3 ha^{-1} in December 1997-January 1998; during summer, the most important contribution to the input was from irrigation and, again, field drainage was of greater importance from October to December 1998. In this period, daily peak values entering the wetland were 1,600–1,900 m^3 ha^{-1} in the first few days of October. This was a very critical event, because the intensive drainage due to abundant rainfall occurred during a period of slurry distribution, and hence with the SFW flooded. In total, more than 66,000 m^3 ha^{-1} of water entered the SFW, 70% of it from field drainage, 18% from irrigation applied after slurry distributions and 12% from rain.

In the same period, only 7,700 m^3 ha^{-1} were discharged with a discontinuous time pattern. It was negligible from December 1997 to May 1998, then about half of the total discharge occurred gradually during the summer and the reminder was concentrated in the first few days of October, when the pipe controlling the water level in the SFW was removed to allow the water to be discharged. A maximum daily discharge of 2,900 m^3 ha^{-1} was recorded in this period.

It can be estimated that a significant part of the difference between input and output of water has been lost by evapotranspiration (ET), given that in another experiment seasonal ET of up to 2,400 mm for common reed grown in lysimeters without flooding was measured (Borin *et al.*, in press). The rest of the water may have been lost by lateral seepage, because the SFW was only partially isolated.

The hydroperiod was characterised by almost continuous flooding, except for the period February–March; water depth ranged between 5 and 20 cm, varying also in relation to the target level imposed.

Plant growth and subdivision of biomass and N

Cattail growth was very intensive from March to mid-June, after which the height remained constant. Plants produced new shoots during the growing season, but intensified this function after development stopped. Common reed stopped putting out shoots while it was still

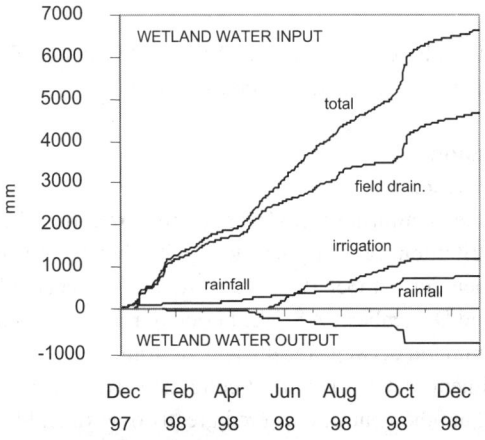

Figure 2 Cumulative water balance of the SFW

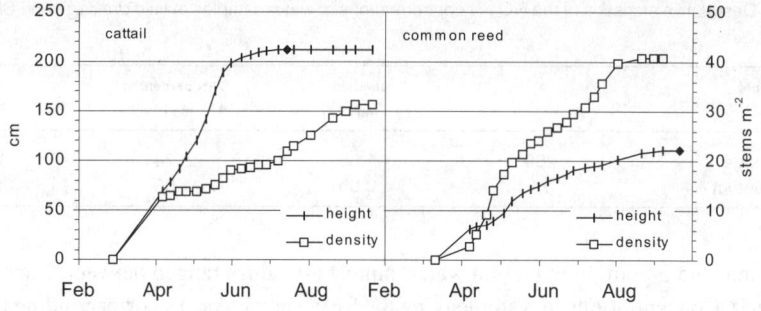

Figure 3 Seasonal patterns of height and density of cattail and common reed

increasing in height. It also had a longer growing season than cattail but it was smaller at the end of the growing season (Figure 3).

The percentage subdivision of biomass and N over time showed a prevalence of below-ground organs in cattail at the start and end of the growing season. Instead, common reed allocated both d.m. and N mostly in below-ground organs for longer (Figure 4). In both species, the amount of d.m. and N in below-ground organs may have been under-estimated, because other studies have shown a significant presence of roots and rhizomes even below a depth of 30 cm (Lawson, 1985; Borin and Bonaiti, 2000).

This different behaviour of the two species may be related to the fact that the establishment of cattail in 1997 was faster than that of common reed. As a consequence, at the end of the first season the plants were more mature and readier for re-growth in spring 1998. Taking the average of the two species, in September about 190 kg ha^{-1} of N were stored in the plant tissues, 120 of which in the aboveground biomass. At the end of the season, almost 300 kg ha^{-1} of N were uptaken by the two plants (on average), but less than 70 were allocated in the aboveground biomass (Bonaiti and Borin, 2000). Since the total N of the biomass of the two hydrophytes was less than 50 kg ha^{-1} at the start of the growing season, the net accumulation of N during the season was around 250 kg ha^{-1}.

Water quality and N balance

NO_3-N concentrations in the water from field drainage showed discontinuous time patterns during the experimental period, with peak values in December 1997–January 1998 (14–18 ppm) and spring 1998 (9–13 ppm). Between these periods and after the spring, concentrations were always low. In rainfall, NO_3-N concentrations fluctuated between less

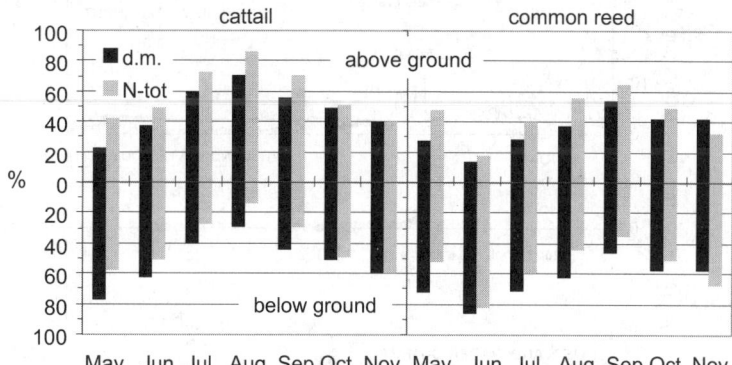

Figure 4 Percentage partitioning of d.m. and N in the above-ground (a.g.) and below-ground (b.g.) organs of cattail and common reed

Table 1 Descriptive statistics of the NO_3-N concentrations of water samples of field drainage and SFW output

Water sample	n.	median mg l⁻¹	75% percentile mg l⁻¹	max mg l⁻¹
field drainage	260	1.65	3.74	17.7
wetland output	71	0.05	1.00	9.0

than 1 ppm and 5 ppm. In irrigation water, almost all values ranged between 2 and 4 ppm (Figure 5). Concentrations in water leaving the SFW showed peaks corresponding to those in the inputs, with values being almost halved. The statistics reported in table 1 summarise the efficiency of the SFW in reducing NO_3-N concentrations.

After slurry applications, concentrations of nitric and ammoniac N measured at the SFW outlet were always lower than 4 ppm, while organic N showed wider oscillations. In general, the concentration of this form of N measured before the outlet increased rapidly after the slurry application and then declined over time. After all the applications, the imposed time lapse of 15 days meant that no form of N in water leaving the SFW exceeded the EU limit for discharge into surface water, i.e. 15 mg l⁻¹ for total N, 15 mg l⁻¹ for NH_4-N and 20 mg l⁻¹ for NO^3-N (D. Lgs 152/99) (Figure 5).

N inputs from field drainage were almost completely concentrated in the first half of the experimental period, with daily peaks of 26 and 15 kg ha⁻¹ in December 1997–January 1998. In this period, the SFW received 205 kg ha⁻¹ of N and discharged only 5 kg ha⁻¹, with nearly 100% removal. In the second part of the year, N loads in the SFW mainly came from the slurry applications (Figure 6). The lower overall reducing capacity (about 85%) in this period may also be related to the rainfall in early October, when the SFW was flooded by water from field drainage immediately after the slurry application. In total, the SFW received 526 kg ha⁻¹ of N, 50% from slurry and 40% from field drainage. It discharged 58 kg ha⁻¹, mostly with the rainfall in early October. The N accumulation in the plant biomass during the growing season can hence justify about half of the difference between N input and output of the SFW.

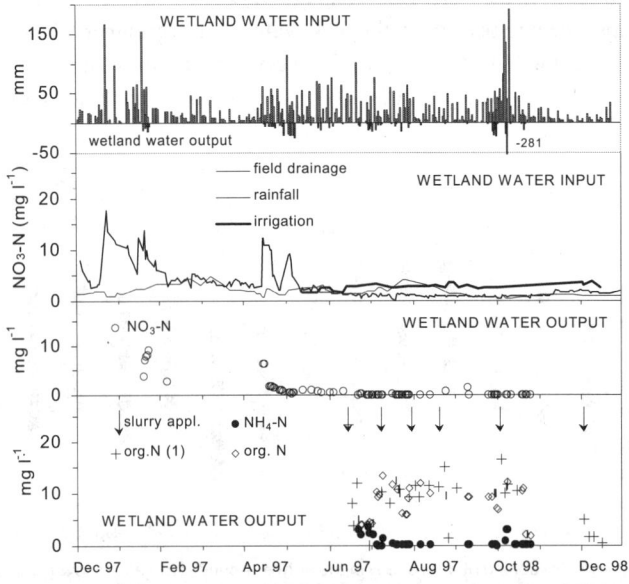

Figure 5 Seasonal pattern of amounts and quality of water entering and exiting the SFW

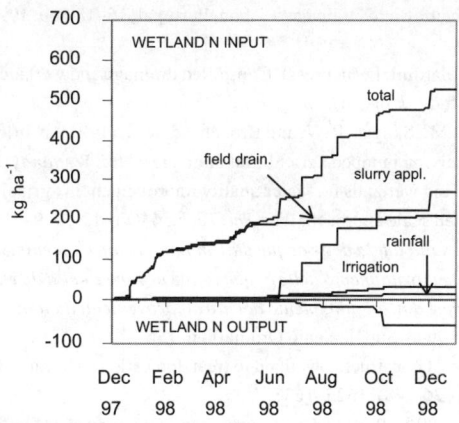

Figure 6 Cumulative N balance of the SFW

Conclusions

Observations of d.m. and N dynamics in above- and below-ground biomass of common reed and cattail showed typical seasonal trends (Ulrich and Burton, 1988; Hocking, 1989; Weisner *et al.*, 1994), with some differences between the species, probably due to their different establishment at the start of the season. Data on N uptake by the plants suggest that the maximum removal from the system can be achieved with early harvesting, at the end of summer, when about 120 kg ha^{-1} of N can be harvested with the biomass. Nevertheless, it has to be taken into account that, doing this, a reduction in the vigour of the hydrophytes can be expected (Hawke and Josè, 1996).

The SFW performed well in removing N from both agricultural fields and occasional organic loads, with an average reduction capacity of close to 90% for nitrates. These results are very encouraging if it is taken into account that the vegetation in question was still young and not yet fully established. In this case, the amount of N fixed in the plants during the monitoring period explains about half of the N removal obtained in the wetland, suggesting that, at least in the initial phase, plants have a quantitative importance in N disappearance with their relatively rapid uptake (Reddy and DeBusk, 1987). However, possible losses by deep seepage are not accounted for.

The small size of the SFW (only about 5% of the land surface served), together with its efficiency, suggests that small wetlands can be scattered over a farming area to offer treatment for diffuse pollution, provided that they are correctly sited to intercept significant amounts of agricultural drainage (Crumpton, 2001). In this way, they may also contribute towards creating a more complex agro-ecosystem, in terms of landscape, biodiversity and wildlife habitats. Given that the SFW can also accept occasional organic loads (arriving, for example, with storm waters), they could be a powerful tool for wastewater treatment and improved environmental quality.

Finalised Project PANDA, Subproject 2, Series 2, Paper No. 95.

References

Bonaiti, G. and Borin, M. (2000). Agronomic observations of *Phragmites australis* (Cav.) Trin. and *Typha latifolia* L.: biomass and nitrogen seasonal dynamics. *Macrophytes Newsletter*, **22**, 20–24.

Borin, M. and Bonaiti, G. (2000). Osservazioni sull'apparato ipogeo di Typha latifolia e Phragmites australis. *Inquinamento*, **17**, 48–52.

Borin, M., Bonaiti, G. and Giardini, L. (1998). First result on controlled drainage to reduce nitrate losses from agricultural sites. In: *Water and the environment: innovation issues in irrigation and drainage*, L.S. Pereira and J.W. Gowing (eds.), Selected papers of the 1st Inter-Reg. Conf. on Environment-Water:

Innovative Issues in Irrigation and Drainage, Lisbon, Portugal, 16–18 Sep. 1998. E & FN Spon, London, pp. 35–42.

Borin, M., Bonaiti, G. and Giardini, L. (in press). Controlled drainage and wetlands to reduce agricultural pollution: a lysimetric study. *J. of Environ. Qual.*

Chescheir, G.M., Gilliam, J.M., Skaggs, R.W. and Broadhead, R.G. (1991). Nutrient and sediment removal in forested wetlands receiving pumped agricultural drainage water. *Wetlands.* **11**, 87–103.

Crumpton, W.G. (2001). Using wetlands for water quality improvement in agricultural watersheds: the importance of a watershed scale approach. *Wat. Sci. Tech.*, **44**(11–12), 559–564 (this issue).

D.Lgs 152/99. *Disposizioni sulla tutela delle acque dall'inquinamento e recepimento della direttiva 91/271/CEE concernente il trattamento delle acque reflue urbane e della direttiva 91/676/CEE relativa alla protezione delle acque dall'inquinamento dei nitrati provenienti da fonti agricole*. Gazzetta Ufficiale n. 177, 30/07/1999 – Supplemento Ordinario n. 146.

Gearheart, R.A. (1992). Use of constructed wetland to treat domestic wastewater, City of Arcata, California. *Water Sci. and Technol.*, **26**(7–8), 1625–1637.

Hawke, C.J. and Josè, P.V. (1996). *Reed bed management for commercial and wildlife interest*. The Royal Society for the Protection of Birds, Sandy, Bedfordshire, UK.

Higgins, M.J., Rock, C.A., Bouchard, R. and Wengrezynek, B. (1993). Controlling agricultural runoff by use of constructed wetlands. In: *Constructed Wetlands for Water Quality Improvement*, G.A. Moshiri (ed.), Lewis Publishers, Boca Raton, pp. 359–367.

Hocking, P.J. (1989). Seasonal dynamics of production, and nutrient accumulation and cycling by *Phragmites australis* (Cav.) Trin. ex Stuedel in a nutrient-enriched swamp in inland Australia. *Australian Journal of Marine and Freshwater Research*, **40**, 421–444.

Hunt, P.G., Thom, W.M., Szogi, A.A. and Humelink, F.J. (1995). State of the art for animal wastewater treatment in constructed wetlands. In C.C. Ross (Ed.) *Proc. of 7th Intern Symp. on agricultural and food processing wastes*. ASAE St. Joseph, MI, pp 53–65.

Kadlec, H.K. and Knight, R.L. (1996). *Treatment wetlands*. Lewis Publishers, Boca Raton Fl.

Kao, C.M. and Wu, M.J. (2000). Evaluation of the non-point source pesticide removal by a natural wetland. *Preprints. 7th International Conference on Wetland systems for water pollution control*, Grosvenor Resort, Lake Buena Vista, Florida, 11–16 November. Vol. 3, 1437–1443.

Lawson, G.J. (1985). *Cultivating Reeds (Phragmites australis) for Root Zone Treatment of Sewage*, Report ITE Project 965 for Water Research Centre, Institute of Terrestrial Ecology, Grange-over-Sands, UK.

Lowrance, R., Vellidis, G. and Hubbard, R.K. (1995). Denitrification in a restored riparian forest wetland. *J. of Environ. Qual.*, **24**, 808–815.

Olson, R.K. and Marshall, K. (1992). The role of created and natural wetlands in controlling nonpoint source pollution. Spec. Issue. *Ecol. Eng.*, **1**, 1–171.

Reddy, K.R. and DeBusk, W.F. (1987). Plant nutrient storage capabilities. In: *Aquatic plants for water treatment and resource recovery*, K.R. Reddy and W.H. Smith (Eds.), Magnolia Publishing Inc., Orlando, Fla, 337–357.

Romero, J.A., Comin, F.A. and Garcia, C. (1999). Restored wetlands as filters to remove nitrogen. *Chemosphere*, **39**, 323–332.

Ulrich, K.E. and Burton, T.M. (1988). An experimental comparison of the dry matter and nutrient distribution patterns of *Typha latifolia* L., *Typha angustifolia* L., *Sparganium eurycarpum* Engelm. and *Phragmites australis* (Cav.) Trin. ex Steudel. *Aquatic Botany*, **32**, 129–139.

Vellidis, G., Lowrance, R. and Smith, M.C. (1994). A quantitative approach for measuring N and P concentration changes in surface runoff from a restored riparian forest wetland. *Wetlands*, **14**, 73–81.

Weisner, S.E.B., Eriksson, P.G., Graneli, W. and Leonardson, L. (1994). Influence of macrophytes on nitrate removal in wetlands. *Ambio*, **23**, 363–366.

Treatment of stormwater runoff from row crop farming in Ruskin, Florida

B.T. Rushton and B.M. Bahk**

*Southwest Florida Water Management District, 2379 Broad Street, Brooksville, FL 34609, USA
**1950 Treebark Dr., Charleston, SC 29414, USA

Abstract A wet-detention pond, constructed to treat agricultural runoff from winter vegetables, was studied to document constituent concentrations, measure hydrology and analyze conditions in the treatment system. The efficiency of the pond to remove pollutants was affected by the unseasonable amount of rainfall induced by the El Niño phenomenon and the succeeding dra La Niña year. During the two years of study (50 rain events), about 90 per cent of all the pollutant loads for potentially toxic metals entered the pond during five El Niño storms; and since higher pollutant loads are often more easily reduced these conditions contributed to the greater per cent reduction of metals during 1998. Another condition which may have enhanced constituent reduction was made possible by the newly evacuated sediments, since clean soils provide more attachment sites for constituent removal. Annual data show pollutant load reductions for 1998 were greater than 90 per cent for most metals, but for 1999 reduction was about 60 per cent. In contrast, inorganic nutrient removal was better in 1999 (> 80 per cent) than during 1998 (to 70 per cent). Organic nitrogen had the poorest removal for both years (20 to 40 per cent). Total phosphorus levels were measured at high median concentrations (1 mg/L at the inflow) compared to other studies for stormwater runoff.
Keywords Agricultural runoff; detention pond; sediments; stormwater; water quality

Introduction

Agriculture has been identified as a significant source of water pollution in the United States. The effects of agricultural pollution are numerous and include: sediment contamination and deposition, pesticide residues, eutrophication of surface waters and degradation of downstream water bodies. The Environmental Protection Agency (EPA) ranks agricultural activity as the greatest threat to water quality in streams and lakes (US/EPA, 1992). Given the problems associated with agriculture and the large amount of acreage in production in Florida, the Southwest Florida Water Management District (SWFWMD) initiated a study on the effectiveness of a wet-detention pond to treat stormwater runoff from an agricultural basin. This paper presents the results for the first two years of a four year study. The project was designed to characterize an agricultural stormwater treatment system: 1) by measuring the reduction (or increase) of pollutants in storm runoff treated by a wet-detention pond, 2) by comparing sediment samples for the two years, and 3) by analyzing conditions that affect constituent concentrations.

Methods

Site description

The stormwater project treats agricultural runoff in southern Hillsborough County near Ruskin, Florida and adjacent to Cockroach Bay. Located at longitude 82°30′3″ and latitude 27°41′15″, the 85 ha (210 ac) drainage basin consists entirely of active and fallow fields (Figure 1). The fields are irrigated using groundwater and most of the crops are winter vegetables such as tomatoes, mustard greens, lettuce, onions, and green peppers. No best management practices were used on the fields, but a pre-treatment ditch provided for some biological and chemical processes to take place as well as infiltration to groundwater before

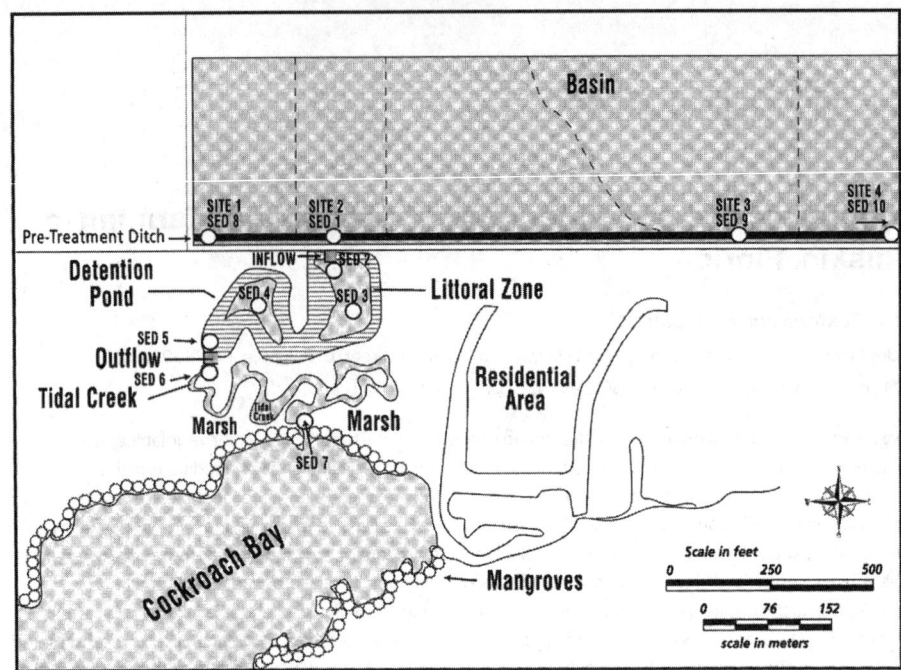

Figure 1 Site plan with sampling locations identified. The residential area is not part of the drainage basin. For the ditch, only sediment site 1 (SED 1) is included in this report.

runoff entered the pond. The detention pond is actually two ponds in series (cells) that are not connected until the pond level is above 0.61 metres (NGVD (2.0 feet) and only solidly connected when the pond begins to discharge at 0.76 metres NGVD (2.5 feet). The permanent pools were 25.5 cm (8.5 ft) deep when the water level in the ponds was at the control elevation of 0.76 metres NGVD. (NGVD = National Geodetic Vertical Datum and is essentially mean sea level).

Hydrology and stormwater quality samples

The hydrology of the basin was characterized by recording rainfall with a tipping bucket rain gauge and measuring pond levels with float and pulleys and bubbler type flow metres. Flow-weighed composite water quality samples were collected by converting water level to flow with weir formulas at the outflow and constructing a water budget using water levels for the inflow. Sensors were connected to data loggers that stored the data and averaged the measurements at fifteen minute intervals. Storm samples were collected and stored in automatic refrigerated units until retrieved, preserved with acid and transported on ice to the SWFWMD laboratory for analysis. Rainfall water quality was collected with a wet/dry precipitation collector, which sampled wetfall only, and stored it in a small refrigerator until collected as described above. Samples were analyzed using standard methods and following SWFWMD's approved quality assurance plan (SWFWMD 1998). Approved pre-washed bottles were used for storing nutrient and metal samples for lab analysis. Duplicate samples and blanks taken through the automatic equipment were periodically collected for quality assurance.

Sediment samples

Sediment samples were collected at seven locations in the ditch, pond and marsh in July 1997 and again in July 1988. See Figure 1 for sediment sampling sites identified with the prefix SED. Samples were extracted intact from the sediments with a two inch diameter hand driven corer which sampled the top 12.7 cm (5 in) of soil. For each sample, four to six replicate cores were taken in close proximity and composited. Samples were placed in EPA approved glass jars supplied by the Florida Department of Environmental Protection (FDEP) laboratory, covered with ice in insulated coolers and transported to Tallahassee for analysis using the FDEP approved quality assurance protocol (FDEP, 1996).

Calculations

Load efficiency was calculated by adding the individual storm events together for a given time period (in this case yearly) using the following formula:

Load efficiency (%) = ((SOL in – SOL out)/SOL in) × 100

where: SOL in = the sum of loads at the inflow for a given time period
SOL out = the sum of loads measured at the outfall for the same time period.

Loads are calculated by using Event Mean Concentrations (mg/L or μg/L) × flow volumes (m^3 or ft^3) × appropriate conversion factors. These were calculated for each storm event and added together for SOL.

Runoff Coefficient (RC)

RC = (inflow volume / total rain × basin area) × conversion factors

Results and discussion

A detention pond built to treat runoff from agricultural fields was studied to evaluate its ability to remove pollutants and to better understand the dynamics of the system. Most of the data in this preliminary report cover the first two years (January 1998 to January 2000) of a four year study. Some data before that time are also included for comparison purposes.

Hydrology

Continuous monitoring sensors recorded rainfall amounts and water levels and showed a close interaction between surface and groundwater.

Rainfall. El Niño produced above average rainfall from September 1997 through March 1998 and La Niña was implicated in the below average rainfall that followed (Figure 2). Southwest Florida normally has wet and dry seasons with about 60 per cent of the rainfall occurring during four summer months. In contrast, the El Niño storms began in September 1997 and ended in March 1998 (the normal dry season), and produced over twice as much rainfall (139 cm, 55 in) as the long term average for those months (55 cm, 24 in). As a comparison, rainfall measured in the same winter months during the following year was 43 cm (17 in).

Water budget. An analysis of the water budget lends some insight into the hydrology of the system. Calculations were made for individual storm events and terminated when the pond stopped discharging or when another storm began. For this time period, usually less than five days, evapotranspiration was minimal. Seepage into the pond occurred when the fields were irrigated and seepage out of the pond was caused by leaks in the weir (fixed in October 1998) and by a seepage area that was active when the pond was full. Although the seepage term is not exactly correct since it measured net seepage and also contained the error term, the results seem reasonable when the entire system is analyzed. Based on the estimated water budget for the inflow, about 72 per cent enters through the inflow structure, 27 per cent as rainfall directly on the pond and 1 per cent as seepage from irrigation water. For the outflow, about 71 per cent was discharged over the outflow weir, 10 per cent was lost by evapotranspiration and 19 per cent by seepage through the weir or the dike. When only the flows from the structures are calculated 24 per cent less water left the pond than was measured entering the pond. When rainfall directly on the pond is included as an input there was a 45 per cent reduction (Table 1).

Runoff coefficient. The runoff coefficient accounts for the integrated effect of rainfall interception, infiltration and depression storage. This sandy site with a pre-treatment ditch and flat terrain had low runoff coefficients. During periods when ample moisture was available, as seen when storms occur close together, the runoff coefficient showed between 20 to 30 per cent of the rainfall ran off the fields and into the pond. During dry periods only 1 to 7 per cent ran off. During extreme events such as El Niño storms, hurricanes and frequent irrigation 35 to 38 per cent of rainfall ran off.

Table 1 Estimated water budget during 50 rain events for the wet-detention pond

Yearly	Total rain amount cm	Total rain amount cu metre	Total inflow cu metre	Residuals & seepage cu metre	Total outflow cu metre	Total ET cu metre	Residuals & seepage cu metre	Total storage measured cu metre	Average runoff coeffic.
Amount			Inflow			Outflow			
1998	241	1,441	4,100	47	(3,147)	(367)	(957)	937	0.14
1999	183	1,041	2,573	40	(1,945)	(278)	(452)	978	0.10
Percent									
1998	25.79%	73.37%	0.84%	70.38%	8.22%	20.95%			
1989		28.49%	70.42%	1.09%	72.72%	10.38%	16.90%		

*ET = evapotranspiration, Seepage = net underground flow into or out of the pond, Residuals = error term (see text for more details)

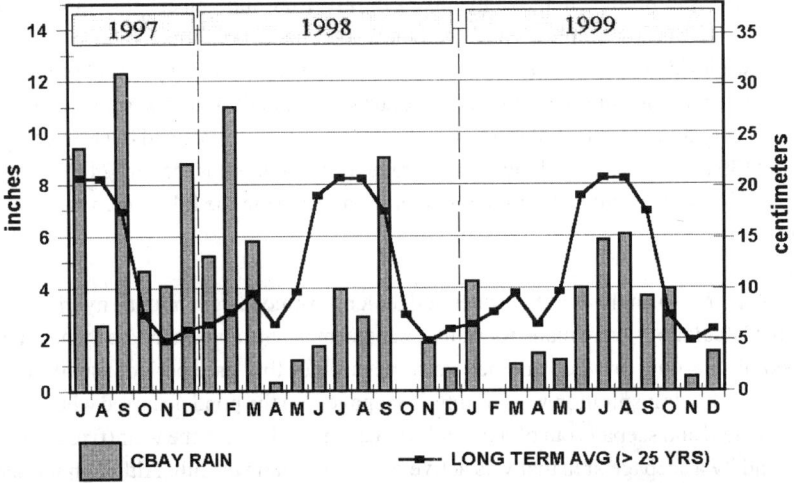

Figure 2 Rainfall amounts measured during the study compared to the long term average in Ruskin, FL

Water quality

Flow-weighed or timed composite water quality samples were collected for about 50 storm events during the two years of data anlyzed in this report. Much more water flowed out of the pond in 1999 than in 1998, but flow out of the pond never occurred until well into the rainy season during both years. This accounts for the fewer number of outflow samples (Table 2).

Concentrations. Five large intense storms created by El Niño produced elevated concentrations of metals in 1998 (Table 2). Many of these pollutants moved rapidly through the ponds accounting for high outflow value which are almost as high or sometimes even higher than the inflow concentrations for the following year. Rainfall often had some of the highest concentrations, such as inorganic nitrogen, zinc and possibly cadmium. Although metals exceeded state marine standards at the inflow only copper and iron were a problem at the outflow.

Pollutant loads. The efficiency of the system is characterized by the removal of pollutant loads from the inflow to the outflow (Figure 3). One of the goals of the State of Florida Water Policy (Chp 62–40 FAC) is an 90 per cent removal of pollutant loads by stormwater systems. The goal of 80 per cent reduction was met for metals and total suspended solids in 1998 and for nutrients, except total organic nitrogen, in 1999. Neither year shows consistent results and the wide variation in pollution removal demonstrates the differences that occur between years. As an example, TSS for 1999 could not be easily compared in Figure 3 because of its 45 per cent increase from the inflow to the outflow which is in stark contrast to the 98 per cent reduction in 1998. Poor removal of suspended solids in 1999 was probably caused by the low levels measured at the inflow since it is difficult to reduce concentrations below 10 mg/L by settling. Significant metal reductions in 1998 can be attributed to the high loads and the clean bottom sediments in this newly constructed pond. Improved nutrient reduction in 1999 may be the result of the wetland vegetation which became established in the littoral zone.

The impact of extreme events was explored by comparing pollutant loads contributed by five El Niño storms in 1998 to all fifty storms for the two years (Table 3). For metals of

Table 2 Average concentrations are calculated for each year. Numbers with asterisk exceed Marine Water Quality Standards (Chp 62–302 F.S.). Concentrations are compared to marine standards since the site discharges to salt water. (Abbreviations: bd = average values below the laboratory detection limit, na = not applicable, std = marine water quality standard).

Parameter	std	Rainfall		Inflow		Outflow	
		1998	1999	1998	1999	1998	1999
Number of samples		25	25	25	25	9	17
Ammonia (mg/L)	na	0.366	0.159	0.100	0.053	0.066	0.038
Nitrate+nitrite (mg/L)	na	0.406	0.189	0.427	0.131	0.144	0.006
Organic-N (mg/L)	na	0.657	0.152	1.110	0.902	0.828	1.132
Ortho-phosphorus (mg/L)	na	0.060	0.014	0.945	0.796	0.623	0.223
Total phosphorus (mg/L)	na	0.094	0.023	1.842	0.957	0.807	0.367
Aluminium (mg/L)	1.5	0.168	0.141	3.47*	0.322	1.30	1.36
Chromium (µg/L)	50	bd	bd	55.07*	2.42	4.70	6.49
Cadmium (µg/L)	9.3	0.35	0.19	1.340	0.180	0.157	0.150
Copper (µg/L)	2.9	4.0*	1.51	117.0*	8.24*	17.0*	3.6*
Iron (mg/L)	0.3	0.034	0.037	2.225*	0.246	0.472*	0.568*
Lead (µg/L)	5.6	bd	bd	16.6*	1.07	1.60	1.67
Manganese (mg/L)	0.1	0.003	0.002	0.255*	0.175*	0.054	0.088
Zinc (µg/L)	86	69.4	46.0	98.6*	13.0	16.9	16.0
Suspended solids (mg/L)	na	na	na	155	11	20	35

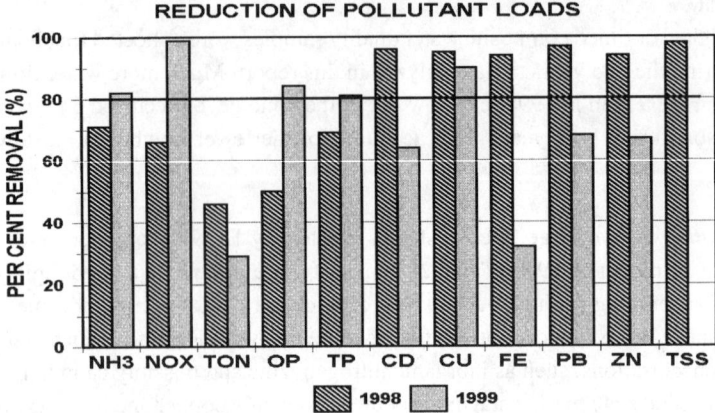

Figure 3 Reduction of pollutant loads from the inflow to the outflow of the wet-detention pond. TSS had a 45 per cent increase in n1999 which is not shown. (Key left to right: ammonia, nitrate+nitrite, total organic nitrogen, ortho-phosphorus, total phosphorus, cadmium, copper, iron, lead, zinc and total suspended solids)

Table 3 Per cent of pollution loads contributed by five El Niño storms compared to all fifty storms

Constituent	Inflow	Outflow	Constituent	Inflow	Outflow
Cadmium	83%	44%	Suspended solids	96%	35%
Copper	89%	69%	Ammonia	22%	0.31%
Iron	91%	46%	Nitrate+nitrite	30%	0.06%
Lead	93%	57%	Organic nitrogen	14%	1.05%
Manganese	71%	24%	Ortho-phosphorus	5%	0.29%
Zinc	90%	49%	Total phosphorus	3%	0.47%

concern, about 90 per cent of all the pollutant loads entered the system during these five events and over 25 per cent of loads left the pond during that same period. This effect was not seen for nutrients. These large events can flush out the system and contribute the majority of metal pollution measured for several years. The same pattern was also seen for pesticides, but not enough pesticide data was collected to make a quantitative comparison.

Sediment samples

Sediment samples provided insight into processes taking place in the stormwater system. Samples were collected twice, in July 1997 several months after the pond was constructed and a year later in July 1998. Sampling sites included not only the inflow, outflow and one ditch station, but also stations in the middle of the pond cells and two stations in the constructed marsh (see Figure 1). For ortho-phosphorus and total phosphorus (not shown) the greatest increase was in the center of the two ponds, probably caused by migration of the finest particles to the deepest part of the pond (Figure 4). For metals, both cadmium and zinc exhibited a pattern of rapid accumulation in the sediments similar to copper. The outlier for sediment site 2 (sed 2) in 1997 was also seen for zinc, but not for cadmium and it is not readily explained. Quite a few metals (mercury, zinc, lead, manganese, chromium and iron) showed rapid accumulation with elevated concentrations in the marsh, where freshwater from the pond first mixes with brackish water, as seen for the copper example in Figure 4.

The processes responsible for the increase in metals as water leaves the freshwater wet-detention pond and enters the brackish marsh are the responses to estaurine mixing where freshwater meets saltwater. This is similar to the saltwater wedge that occurs in rivers where large particle loads are trapped and re-suspended forming a turbidity maximum

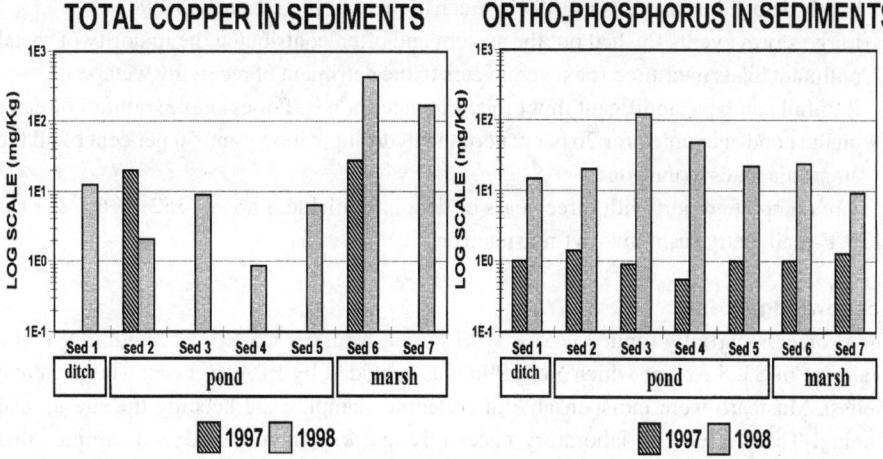

Figure 4 Copper and ortho-phosphorus are used for examples demonstrating differences between sediment sampling stations. Most sites had a dramatic increase in concentration accumulations from the first to the second year. (Note: concentrations are shown on a log scale).

zone. The dissolved and particulate matter may interact and respond in one of three ways that can affect the concentration of trace elements in solution as reported by Burton (1976).
1. The precipitation of dissolved material to give new solid phases.
2. The uptake of dissolved material onto solid phases already present.
3. The release of material into solution from particulate phases by dissolution, desorption and biological processes.

Trace metal (Al, Cu, Cd, Fn, Mn and Zn) removal by burial in the sediments varies among estuaries, but it can be substantial. Scavenging reactions involving hydrous oxides, such as hydrous manganese and iron oxides, have been linked to enrichment of trace metals in suspended matter and sediments of estuaries (Kennish, 1992). As salinity increases, the concentration of dissolved trace metals decreases as a result of gradual dilution and biological processing downstream. This is especially true for the constituents of most concern (Zn, Ni, Cu, Cd, Ag, Pb and Hg) (Church and Scudlark, 1998). In our study, the lower concentration of metals downstream, where the tidal creek discharges into Cockroach Bay, compared to concentrations where the pond discharges into the marsh demonstrates this dilution effect.

Summary of results

1. Considerable differences in average pollution levels were noted between years with extremely high concentrations flushed through the system during El Niño storms.
2. Annual average after quality concentrations for copper and iron exceeded Class II marine water standards at the outflow during both years, but since this is a mixing zone, standards do not apply.
3. Phosphorus concentrations were over twice as high as concentrations measured in other urban and agricultural studies conducted in the region.
4. Total suspended solids, total organic nitrogen, and total iron concentrations increased from the inflow to the outflow of the pond during 1999.
5. High concentrations of metals, with probably toxic concentrations for copper, were measured in the sediments where freshwater was discharged into the brackish marsh.
6. Concentrations in the sediments showed a dramatic increase from 1997 to 1998 demonstrating that the sediment pollutants in this newly constructed pond increased from

barely detectable amounts in 1997 to much higher concentrations in 1998.
7. Large storm events flushed out the system and often contributed the majority of metal pollutant loads measured for several years to the detriment of receiving waters.
8. Rainfall can be a significant input to stormwater ponds. For example, rainfall directly on the pond accounted for 26 per cent of the hydrologic inputs and 50 per cent of all the ammonia loads to the pond.

A more complete report with three years of data is available. Phone: (352) 796-7211 ext 4276. E-mail: betty.rushton@swfmd.state.fl.us

Acknowledgements

The Project is partially funded with a 319(h) grant from the US EPA through FDEP at a total cost of $328,327 of which $196,996 was provided by EPA. Rebecca Hastings and Melissa Musicaro were indispensable in collecting samples and keeping the site up and running. The SWFWMD laboratory cheerfully and expediently analyzed samples that could not be collected by any pre-determined schedule.

References

Burton, J.D. (1996). Basic Properties and Processes in Estuarine Chemistry. In *Estuarine Chemistry*, Burton, J.D. and Liss, P.S. (eds). Academic Press, London.

Church, T.M. and Scudlark, J.R. (1998). Trace Metals in estuaries: A Delaware Bay Synthesis. In *Metals in Surface Waters*, H.E. Allen, A.W. Garrison, and G.W. Luther, III (eds). Ann Arbor Press, Chelsea, MI 48118.

FDEP (1996). Comprehensive quality assurance plan, Florida Department of Environmental Protection State of Florida, Central Chemistry laboratory, Tallahassee, FL. Florida Department of Environmental Protection.

Kennish, M.J. (1992). *Ecology of Estuaries: Anthropogenic Effects*. CRC Press, Inc. Boca Raton, FL.

SWFWMD (1998). Comprehensive quality assurance plan. Southwest Florida Water Management District. 2379 Broad Street, Brooksville, FL 34602.

US EPA (1992). National Water Quality Inventory, 1990 report to congress. U.S. Environmental Protection Agency, Government Printing Office, Washington, DC EPA 503/9-92/006.

Evaluation of atrazine removal processes in a wetland

C.M. Kao*, J.Y. Wang** and M.J. Wu***

* Institute of Environmental Engineering, National Sun Yat-Sen University, Kaohsiung, Chinese, Taiwan
** School of Civil and Structural Engineering, Nanyang Technological University, Singapore
*** Institute of Marine and Agricultural Research, Inc., Raleigh, NC, U.S.A.

Abstract Agricultural non-point source (NPS) pollution is considered to be the largest single category resulting in water quality deterioration. Pesticide is one of the main detrimental agricultural NPS constituents causing the impairment of water bodies. In this study, a mountainous wetland located in McDowell County, North Carolina, was selected to demonstrate the effects of the natural wetland system on the removal of NPS pesticide (atrazine) pollution to maintain the surface water quality. The selected wetland receives water from two unnamed creeks, which drain primarily agricultural lands. The hydraulic retention time (HRT) for the wetland was approximately 10.5 days based on the results from a dye release study. Water quality monitoring of the wetland was conducted from March to October 1998. One major storm and baseline water quality samples were analyzed. Analytical results indicate that the wetland completely removed NPS atrazine flushed from the upgradient agricultural lands. Laboratory microcosm experiments were conducted to evaluate the feasibility of using the wetland sediments as the microbial sources to enhance the atrazine biodegradation. Microcosm results suggest that atrazine can be degraded under anaerobic or reductive dechlorinating conditions when sucrose was provided as the primary substrate. Atrazine can also serve as the nitrogen source for the growth of microorganisms under anaerobic conditions. Results from this study can provide us with further knowledge on evaluating the role of wetlands in controlling pesticide pollutants from stormwater runoff.

Keywords Atrazine; non-point source pollutants; pesticide; stormwater runoff; wetland

Introduction

Wetlands are an essential part of nature's stormwater management system. Important wetland functions include natural restorative processes, which improve water quality while being cost-effective; conveyance and storage of stormwater, which dampens effects of flooding; reduction of flood flows and velocity of stormwater, which reduces erosion and increases sedimentation; and modification of pollutants typically carried in stormwater. Accordingly, there is a great amount of interest in the incorporation of natural wetlands or constructed wetlands into stormwater management systems (Steven, 1998). This concept provides an opportunity to use, not abuse, one of nature's systems to mitigate effects of runoff associated with urbanization.

Non-point source (NPS) pollution is believed to be one of the major causes of impairment of water bodies. The US EPA estimated that NPS pollution contributes over 65% of the total pollution load to US inland surface waters including 332,000 km of rivers, 215,000 ha of lakes, and 1.5×10^6 ha of estuaries (Mitchell *et al.*, 1995). NPS mainly associated with stormwater runoff from urban, agricultural, or mining land uses can be quite diffuse and difficult to treat. Agricultural NPS pollution is considered to be the largest single category resulting in these environmental problems. Nutrients, pesticides, and sediments are the main detrimental agricultural NPS constituents.

Atrazine (2-chloro-4-ethylamino-6-isopropylamino-s-triazine) currently is one of the most widely used agricultural pesticides in the US. Atrazine is used for control of broadleaf and grassy weeds in corn, sorghum, asparagus, rangeland, pineapples, sugarcane, turf grass sod, forestry, grasslands, and grass crops. The maximum contaminant level (MCL) for atrazine is 3 µg/L. Atrazine may have the potential to cause cancer from a lifetime exposure at levels above the MCL (Detenbeck *et al.*, 1996; Daniel *et al.*, 1998; Shapir *et al.*, 1998).

In this study, atrazine was used as the target compound to evaluate (1) the effectiveness of a mountainous wetland on NPS pesticide removal, and (2) the bioavailability of atrazine using wetland soil slurry cultures. Results from this evaluation could be applied to demonstrate the ability of constructed wetlands to retain NPS pollutants (e.g., pesticides) from storm runoff. Pollutant removal efficiencies were to be calculated from the wetland flow and concentration data. Atrazine removal efficiencies, combined with results from the biodegradability study, would be useful in the design of wetland system for NPS pollutants removal.

Site description

The Catawba River Basin, along with the nearby Broad River basin in North Carolina, USA forms the headwaters of the Santee-Cooper River system, which flows through South Carolina to the Atlantic Ocean. The Catawba River is the eighth largest river basin in North Carolina, covering 852,106 ha in the southwestern region of the state. The headwaters of the Catawba River are formed by swift-flowing, cold water streams originating in the steep terrain of the mountains in Avery, Burke, Caldwell, and McDowell Counties. Many of these streams exhibit good to excellent water quality and are classified as trout waters. The studied wetland field test site, which occupies 7,822 m^2, is located in the McDowell County borders in the upper reaches of the Catawba River. Catawba River is a major freshwater resource that provides the potable water to the communities in both western North Carolina and South Carolina. Figure 1 presents the studied mountainous wetland site and the sampling locations.

The wetland receives water mainly from two unnamed creeks (Creek 1 and Creek 2), that drain primarily agricultural lands. Therefore, pesticide and nutrient discharges from the farmland become the major concerns effecting the wetland water quality. Herbicides containing atrazine, which are used for weed control, are generally applied in the Spring. The highest atrazine concentrations in surface water occur between April and June as a result of runoff from the short and intense Spring storms of the regions.

Results from previous study show that the flows into the wetland from two creeks were approximately 335 and 410 m^3/day for Creek 1 and Creek 2 during the test periods, respectively (Kao and Wu, 1999). The observed HRT was approximately 10.5 days based on the dye study. The depth of the mountainous wetland varies from 0.1 m to 1.5 m. Table 1

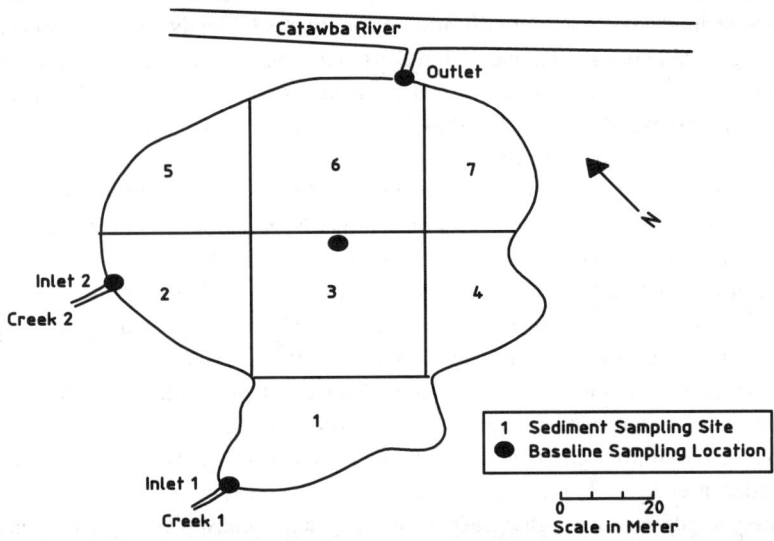

Figure 1 Site map showing the studied natural wetland

presents the baseline water quality monitoring data from the previous study. The baseline results were the averages of four sampling events. The diversity of physical and chemical niches and interspecific interactions present in this wetland results in a continuum of life forms from the smallest viruses to the largest trees. The tree species include cypress, red maple, willow, and spruce. The emergent herbaceous species include cattail and bulrush.

Materials and methods

Three water level/flow and water quality sampling stations (Inlet 1, Inlet 2, and Outlet) were established in the wetland. Sampling locations are shown in the site map (Figure 1). Baseline water samples were collected and analyzed over a two-month interval. An ISCO 3700 automatic water sampler was installed at each sampling station for stormwater sampling. Wetland sediments were collected from seven locations evenly distributed in the wetlands to evaluate atrazine accumulation on sediments (Figure 1). Sediments samples were collected and analyzed twice after the studied storm events. The first and second sampling events were performed at 12 and 30 days after the beginning of the storm event, respectively. Duplicate grab sediment samples were collected from the seven locations, which were marked on the site map (Figure 1). Collected sediments were analyzed for the mass of sediments, per cent of organic materials, and atrazine concentrations. The mass of sediments was determined to the nearest 0.01 g after drying in an oven at 105°C. Organic content was determined by burning the samples at 550°C and calculating the organic content as the preburn weight minus the postburn weight. All samples were analyzed by the Institute of Marine and Agricultural Research, Inc., and Environmental Quality Institute, North Carolina. Collected samples were analyzed for atrazine concentrations.

Sediment samples were extracted with dichloromethane. The extracts were filtered and evaporated under vacuum to dryness, and the residues were taken up in methanol. Atrazine was analyzed using a Waters HPLC, consisting of a Model 510 pump, an injector and a model 481 variable wavelength ultraviolet absorbance detector. The reverse phase column used was a 3.9 × 150 mm μBondapak C18 (10 mm) column. The mobile phase used was a 1:1 $CH_3CN:H_2O$, and the pump flow rate was 2 ml/min. The wavelength of the UV absorbency detector used was 254 nm and the injected sample was 40 μL.

Eight groups of microcosm experiments were conducted to examine the feasibility of atrazine biodegradation under aerobic (Groups A, B, C, and D) and anaerobic (Groups E, F, G, and H) conditions. Table 2 presents the characteristics of each microcosm. The inocula used in this microcosm study were sediments collected from the bottom of the wetland at the wetland center point. Groups A, D, E, and H microcosms were constructed with 35 ml nutrient medium, 15 g of sediments, and 5 ml atrazine solution in 70 ml serum bottles sealed with Teflon-lined rubber septa.

Table 1 Averaged baseline water quality monitoring data[1]

Parameter	Creek 1	Creek 2	Wetland center point	Outlet
TSS (mg/L)	52	69	35	25
NO_3^- (mg/L)	6	9	5	4
TP (mg/L)	3	4	4	3
NH_4^+ (mg/L)	6	11	6	5
NO_2^- (mg/L)	BDL	BDL	2	0
COD (mg/L)	25	29	21	18
pH	7.2	7.7	7.1	7.2
DO (mg/L) – 0.2 m bws[2]	3.6	3.1	2.5	2.2
DO (mg/L) – 1 m bws	–	–	1.4	–
DO (mg/L) – bottom	2.9	2.8	0.1	0.5

[1] results were averages of four sampling events
[2] below water surface

The medium contained the following components at the specified concentrations (units are in mg per litre of water): H_2PO_4, 326.4; Na_2HPO_4, 1263.8; $Mg_2SO_4 \cdot 7H_2O$, 98.6; $CaCl_2 \cdot 2H_2O$, 44.1; NH_4Cl, 10.7; plus 3.35 mg of trace elements which include $FeSO_4 \cdot 7H_2O$, 1; $MnSO_4 \cdot 4H_2O$, 1; $(NH_4)_6Mo_7O_{24} \cdot 4H_2O$, 0.25; $Na_2B_4O_7 \cdot 10H_2O$, 0.25; $CoCl_2 \cdot 6H_2O$, 0.25; $CuCl_2 \cdot 2H_2O$, 0.25; $ZnCl_2$, 0.25; NH_4VO_3, 0.1. However, in Groups D and H, nitrogen sources were removed from the medium to evaluate the feasibility of using atrazine as the nitrogen source by microorganisms. The pH of this buffer solution was 7.5. The medium solution was autoclaved before use.

Groups B, C, F, and G microcosms were constructed with 35 ml nutrient medium (containing 100 mg/L of sucrose), 15 g of sediments, and 5 ml atrazine solution in 70-ml bottles. In Groups C and G, nitrogen components were also removed from the nutrient medium. Thus, atrazine was the only nitrogen source in Groups C, D, G, and H microcosms. The final atrazine concentration in each microcosm was 1 mg/L.

Anaerobic microcosms were prepared in an anaerobic glovebox to preclude intrusion of oxygen. A redox indicator (0.0002% resazurin) and reducing agent (1 mM sodium sulfide) were added to each bottle. Sodium sulfide was chosen because it would not serve as a carbon source and it had a redox potential (−571 mV) low enough to completely reduce resazurin. In this microcosm study, both live and kill control microcosms were constructed for each group. Each kill control group contained $HgCl_2$ and NaN_3. Duplicate microcosms were sacrificed at each time point and analyzed for atrazine concentrations.

Results and discussions

One storm event during the wetland monitoring period was sampled. For this storm event, atrazine levels rose in response to runoff, which contained high levels of atrazine. Table 3 presents the summarized atrazine monitoring results for the storm event. Figure 2 shows the atrazine concentrations in the collected samples from two inlets and one outlet during the studied storm event. Peak inflow atrazine levels for Inlet 1 and Inlet 2 were around 85 and 131 μg/L, respectively for the studied storm event. Mass loadings were calculated by multiplying inflows and outflows by the measured pollutant concentrations. The calculated total inflow and outflow mass loadings during the storm event were 541 and 0 mg, respectively. Atrazine concentrations in all outflow samples were below detection limit (BDL) (1 μg/L). This indicates that the wetland has significant effects on NPS atrazine removal.

The sediment was collected from seven locations and measured for organic content and atrazine concentrations to evaluate possibility of atrazine accumulation on sediments. Sediments were high in organic content, averaging 16%. In general, the wetland has good vegetation growth including woody, herbaceous, emergent, and submerged species. The vegetation is probably a source of some of the organic matter. Analytical results for

Table 2 Characteristics of microcosms

Microcosm & control	Condition	Treatment	Primary substrate	Atrazine removal rate
A & AC[1]	Aerobic	Atrazine + N[2]	Atrazine	No
B & BC	Aerobic	Atrazine + Sucrose + N	Sucrose	medium
C & CC	Aerobic	Atrazine + Sucrose	Sucrose	medium
D & DC	Aerobic	Atrazine	Atrazine	No
E & EC	Anaerobic	Atrazine + N	Atrazine	Low
F & FC	Anaerobic	Atrazine + Sucrose + N	Sucrose	High
G & GC	Anaerobic	Atrazine + Sucrose	Sucrose	High
H & HC	Anaerobic	Atrazine	Atrazine	Low

[1] C: control microcosm (250 mg/L $HgCl_2$ + 300 mg/L NaN_3 were added in control microcosm)
[2] N: nitrogen source (contained in the medium) in the microcosm

Table 3 Summarized atrazine results for the studied storm event

Parameter	Creek 1 (Inlet 1)		Creek 2 (Inlet 2)		Outflow (Outlet 1)		Removal %
	Peak concentration µg/L	Total loading mg	Peak concentration µg/L	Total loading mg	Peak concentration µg/L	Total loading mg	
atrazine	85	213	131	328	BDL	0	100

[1] below detection limit

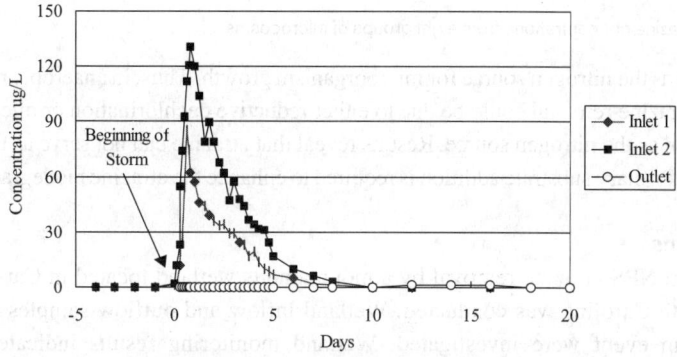

Figure 2 Atrazine concentrations from two inlets and one outlet for the studied storm

atrazine show that sediment samples yielded negative findings. Atrazine concentrations in all seven sediment samples were BDL (0.5 µg/kg of sediment). This indicates that atrazine did not accumulate in the bottom sediments of the wetland even though high atrazine levels exist in stormwater flows.

Aerobic microcosms (Groups A, B, C, and D) using inocula from bottom sediments were incubated for 30 days. The average concentrations of the duplicate analysis of atrazine versus time are presented in Figure 3. Results show that no significant atrazine degradation was observed in aerobic microcosms A and D without extra carbon (sucrose) addition. However, approximately 51% and 64% of the atrazine was removed in aerobic microcosms B and C, respectively with a measurable lag period (approximately 3 days). This suggests that sucrose addition in aerobic microcosms B and C significantly enhanced the atrazine removal, possibly due to the cometabolic processes. Because no nitrogen compounds were added in the nutrient buffer in aerobic microcosm C, atrazine could be used as the nitrogen source by the microcosms and caused higher atrazine removal rate in this group. In the control microcosms, no significant atrazine decrease (less than 9% decrease in all control bottles) was observed throughout this experiment (data not shown).

Anaerobic microcosms (Groups E, F, G, and H) with sediments as the inocula were also incubated for 30 days. Figure 3 presents the analytical results for this study. Approximately 18 and 24% of atrazine removal were detected in anaerobic microcosms E and H at the early incubation period (3 to 12 days). However, no further atrazine decrease was achieved during the remaining operating period from day 12 to 30. This suggests that the atrazine removal in anaerobic microcosms E and H might be due to the occurrence of the reductive dechlorination process. Existing organic contents in sediments may serve as the primary carbon sources and cause the degradation of atrazine. After the complete removal of bioavailable organic materials in the sediments, degradation of atrazine ceased.

Approximately 99% of atrazine was removed in anaerobic microcosms F and G within 15 days indicating that sucrose addition enhanced the atrazine biodegradation under anaerobic conditions. Sucrose might be used as the primary substrate and caused the atrazine removal via reductive dechlorination. Without nitrogen addition (anaerobic microcosms G and H), atrazine

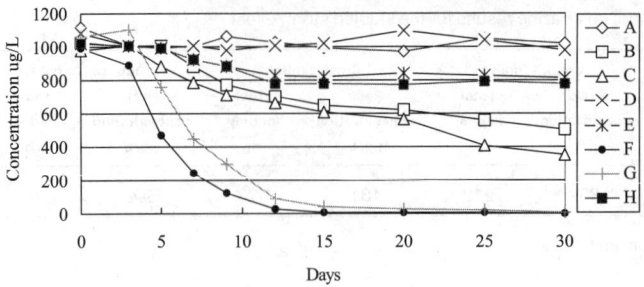

Figure 3 Atrazine concentrations from eight groups of microcosms

might serve as the nitrogen source for microorganism growth. Thus, in anaerobic microcosms G and H, atrazine removal could be due to either reductive dechlorination or microorganism consumption as the nitrogen source. Results reveal that atrazine can not serve as the sole carbon source. Primary substrate addition is required to enhance the atrazine biodegradation.

Conclusions

Research on NPS atrazine removal by a mountainous wetland located in Catawba River Basin, North Carolina was conducted. Wetland inflow and outflow samples during one major storm event were investigated. Wetland monitoring results indicate that high atrazine loadings in response to the storm event were observed. This might be attributed to the agricultural activities in the watershed. Results show that the wetland was able to completely remove NPS atrazine caused by the surface runoff.

Atrazine bioavailability was evaluated in microcosms under aerobic/anaerobic conditions. Wetland sediments were used as the inocula. Results show that atrazine biodegradation was most effective under anaerobic processes (possibly reductive dechlorination). Results also indicate that primary substrate addition is required to enhance the biodegradation of atrazine. Atrazine may serve as the nitrogen source for bacterial growth. Thus, the decrease in atrazine concentrations in this study could be due to either anaerobic biodegradation or microorganism consumption as the nitrogen source. No significant atrazine removal was observed in the microcosms prepared under aerobic conditions without primary substrate addition.

Atrazine accumulation on sediments was not observed based on the sediment analysis. This indicates that microbial activities and plant uptakes might be the two possible causes of atrazine removal during the studied storm event. The natural or influent carbon sources may serve as the primary substrates and enhance the atrazine removal in the wetland. The observed anaerobic conditions in the bottom of the wetland also created an optimal oxidation-reduction environment for atrazine biodegradation to occur. Results from this study show that the wetland system has a great capability to control the NPS atrazine and protect the downstream water bodies.

References

Daniel, R.S., Sadeghi, A.M. and Isensee, A.R. (1998). Effect of tillage on atrazine bioavailability. *Soil Science*, 163, pp. 891–896.

Detenbeck, N.E., Hermanutz, R., Allen, K. and Swift, M.C. (1996). Fate and effects of the herbicide atrazine in flow-through wetland mesocosms. *Environmental Toxicology and Chemistry*, 15, pp. 937–946.

Kao, C.M. and Wu, M.J. (1999). Sediments and non-point source pollutants removal by a mountainous wetland, Preprints 7th IAWQ Asia-Pacific Regional Conference, Taipei, Taiwan, October 18–20, pp. 1465–1470.

Mitchell, D.S., Chick, A.J. and Raisin, G.W. (1995). The use of wetlands for water pollution control in Australia: an ecological perspective. *Wat. Sci. Tech.* **32**, 365–373.

Shapir, N., Mandelbaum, R.T. and Jacobsen, C.S. (1998). Rapid atrazine mineralization under denitrifying conditions by Pseudomonas sp. Strain ADP in aquifer sediments. *Environ. Sci. Technol.*, 32, pp. 3789–3792.

Steven, C. (1998). The emergence of treatment wetlands. *Environ. Sci. and Technol. News*, May, pp. 218–223.

Treatment of swine wastewater in marsh-pond-marsh constructed wetlands

G.B. Reddy*, P.G. Hunt**, R. Phillips*, K. Stone** and A. Grubbs*

* Department of Natural Resources & Environmental Design, North Carolina, A&T State University, Greensboro, NC, USA
** USDA-ARS, Coastal Plains, Soil, Water and Plant Research Center, Florence, SC, USA

Abstract Swine waste is commonly treated in the USA by flushing into an anaerobic lagoon and subsequently applying to land. This natural system type of application has been part of agricultural practice for many years. However, it is currently under scrutiny by regulators. An alternate natural system technology to treat swine wastewater may be constructed wetland. For this study we used four wetland cells (11 m width × 40 m length) with a marsh-pond-marsh design. The marsh sections were planted to cattail (*Typha latifolia*, L.) and bulrushes (*Scirpus americanus*). Two cells were loaded with 16 kg N ha^{-1} day^{-1} with a detention of 21 days. They removed 51% of the added N. Two additional cells were loaded with 32 kg ha^{-1} day^{-1} with 10.5 days detention. These cells removed only 37% of the added N. However, treatment operations included cold months in which treatment was much less efficient. Removal of N was moderately correlated with the temperature. During the warmer periods removal efficiencies were more consistent with the high removal rates reported for continuous marsh systems – often > than 70%. Phosphorus removal ranged from 30 to 45%. Aquatic macrophytes (plants and floating) assimilated about 320 and 35 kg ha^{-1}, respectively of N and P.

Keywords Constructed wetlands; marsh; pond; swine wastewater

Introduction

North Carolina ranks second in the US in swine production and annually generates 42 billion pounds of manure per year. Many swine production facilities have flushing systems with fresh or recycled water to flush the liquid manure into an anaerobic lagoon. Excessive wastewater in lagoons is applied on the land with irrigation or spreading equipment, and this requires a large land area. Swine waste management is under scrutiny because of frequent lagoon spills and floods, which have caused surface water contamination of rivers. Therefore, alternate low-cost technologies are required to treat wastewater. Constructed wetlands with vegetation have received considerable interest as a method of wastewater treatment (Hammer, 1989). Constructed wetlands have been used to treat municipal wastewater (Kadlec and Knight, 1996), mine drainage (Kleimann and Girts, 1987), industrial effluents (Polprasert *et al.*, 1996), and animal wastewater (Payne *et al.*, 1992: Hunt *et al.*, 1994). Constructed wetlands are relatively low cost, operation/maintenance, and energy input technology; thus, they have good potential as a technology for treatment of swine wastewater.

Most of the N in the wastewater is in the form of NH_4 and its nitrification is dependent on the O_2 supply. Constructed wetlands with vegetation have aerobic microsites adjacent to the roots and at the water surface where NH_4 can be nitrified to NO_3 and subsequently denitrified in anoxic conditions (Brix, 1994; Hammer and Knight, 1994). On the other hand, P removal in constructed wetlands occurs through chemical precipitation, substrate adsorption, plant and algal uptake, and immobilization into organic matter (Swindell and Jackson, 1990; Moshiri, 1993; Reddy and Reddy, 1993).

Several authors have shown the efficiency of constructed wetlands in removing N, P,

COD, and TSS (Reddy and Debusk, 1985; Payne *et al.*, 1992; Hunt *et al.*, 1995). Animal wastewater contains high concentrations of nutrients and ammonia. High ammonia levels can kill aquatic plants in wetlands. Therefore, studies are required to determine the maximum or acceptable wastewater loading rates in constructed wetlands. Also, there is limited data for animal wastewater treatment in constructed wetlands. The purpose of this study was to 1) evaluate the high N loading rates for nutrient removal, 2) compare the performance of marsh-pond-marsh with the continuous marsh system, and 3) determine the temperature effect on N removal.

Methods

Site and wetland cells description

Six wetland cells (11 m wide × 40 m long) were constructed at the swine unit (130–250 sows) on the North Carolina A&T State University farm in 1995. Each cell has a 20 m middle pond section and 10 m section with marshes at the influent and effluent ends. Shallow sections with marshes at the influent and effluent ends and the deep section in the pond had an operating depth of 15 cm and 75 cm, respectively. The marsh sections were planted with cattails (*Typha latifolia*, L.) and bulrushes (*Scirpus americanus*) in March 1996. However, cattails have become dominant in the marsh areas. Plants grew from March through September.

Wastewater flow system

The waste from the swine house was flushed with recycled water into a two stage anaerobic lagoon [primary lagoon (L1) and secondary lagoon (L2)]. Flow from L2 was pumped by a submersible pump to an 8000 L storage tank. The wastewater from the storage tank was discharged by gravity into the wetland cells. Only four wetland cells were used for this study. The effluents from the cells were discharged into a holding pond for recycling into the swine house or application on land. The cells were operated in cold weather, and insulated covers were installed to prevent freezing of the inflow tipping buckets and the associated piping.

Nutrient loading

The wetland cells were operated with 5–6 kg N ha^{-1} day^{-1} in 1997 to establish the wetland ecosystem. From May 15, 1999 two cells received 16 kg N ha^{-1} day^{-1} with 6.06 m^3/day hydraulic load and 21 days theoretical retention and others received 32 kg N ha^{-1} day^{-1} with 12.13 m^3/day hydraulic load and 10.5 days theoretical retention. The experiment was conducted from June, 1999 to January, 2000.

Monitoring equipment

One tipping bucket wired to an electronic totalizer (volume counter) was installed at the inflow and outflow points of each wetland cell. Four ISCO 2700 (ISCO, Lincoln, NE) auto samplers were installed. The water sampler combined daily samples into weekly composites. Two mL concentrated HCl was added to each sampling bottle to lower the pH below 2.5. The samples were transfered to the laboratory and stored at 4°C.

Wastewater analysis

Weekly samples were carried to the laboratory and used for the following analysis. Ammonium (NH_4-N), nitrate-N (NO_3-N), total Kjeldahl-N (TKN), Ortho phosphate (o-PO_4), total phosphorus (TP), total solids (TS) were analyzed in accordance with the USEPA methodology by using TRAACS 800 Auto Analyzer (Kopp and McKee, 1983). Electrical conductivity, temperature, Eh and pH readings were recorded by electronic methods. Chemical oxygen demand (COD) was analyzed by use of the Hatch method (Gibbs, 1979).

Plant tissue analysis

Cattails/bulrushes and duckweed were sampled in the second week of September in marshes (0.25 m² area) and in the pond area (0.04 m²), respectively. The samples were dried at 60°C for 24 hours, and dry weights (biomass) were recorded. Total N was determined by using C-N-H-S analyzers (Perkin-Elmer model 2400), and total-P was analyzed with the perchloric acid digestion method and measured by using TRAACS 800 Auto Analyzer.

Results and discussion

Mean pH values ranged from 6.8 to 7.9. No seasonal changes of pH were observed (Table 1). The average temperature in the water ranged from 6.45 in winter to 21.9°C in the summer. The inflow water temperature was 1°C less than the outflow water temperature in late fall and winter months. The concentration of total Kjeldahl N (TKN), ammonium (NH_4), and total phosphorus (TP) were higher in the winter than in either the fall or summer for both the inflow and outflow wastewater. Redox ranges in inflow marsh, pond, and outflow marsh were −114 to −326, −15 to +150, and −194 to −320 −Eh (mV), respectively. The total rainfall was 683 mm and more than 25% of the total rainfall occurred during the three storms in September.

Treatment efficiency

The annual mean mass reductions of TKN, NH_4, TP, PO_4, COD, TSS during the experimental period are shown in Table 2.

Nitrogen removal. The TKN removal rate was 51 and 37% at 16 and 32 kg ha^{-1} day^{-1} loading rates, respectively. The TKN mass inflow and outflow ratio was approximately 2.1 for 16 kg and 3.9 for 32 kg N loading rate. When N loading rate was reduced from 32 to 16 kg ha^{-1} day^{-1}, the wastewater detention time was increased from 10.5 to 21 days. The decrease in loading rate with an increase in detention time improved the treatment efficiency of the wetland cell. Others (Nichols, 1983; Knight *et al.*, 1985; Hammer and Knight, 1994) have reported that an average mass reduction of TKN was in the range of 46–72%. Similar results were also found in our previous study (Phillips *et al.*, 1999).

Ammonium removal. Mass reduction of NH_4 was 60 and 43% at the low and high N loading rates, respectively. Most of the N removal occurred in warmer months. The reasonably high NH_4 in the outflow suggests that nitrification is limited in this system. This is consistent with the report of Hunt *et al.*, (1999) that the continuous marsh wetlands were oxygen limited for nitrification. Additionally, the high loading rates without dilution of the wastewater

Table 1 Wastewater characteristics

Average Inflow Concentration (mg L^{-1})			−Eh (mV) range	pH range
TKN	144.55	marsh inflow	−114 to −326	6.8 to 7.9
NH_4	80.53	pond	−15 to 150	
TP	74.81	marsh outflow	−194 to −302	
PO_4	61.87			
TSS	521.80			
COD	820.52			
NO_3	1.59			

Table 2 Performance of marsh-pond-marsh constructed wetlands

Loading rates (kg ha⁻¹ day⁻¹)		TKN	NH₄	NO₃ (g m⁻² day⁻¹)	TP	PO₄	TSS	COD
16	Inflow	1.52	0.67	0.01	0.66	0.53	4.03	6.74
	Outflow	0.74	0.27	0.01	0.37	0.33	1.27	3.14
	% Reduction	51.26	59.53		44.37	39.00	68.55	53.38
32	Inflow	3.10	1.47	0.02	1.35	1.21	10.72	15.47
	Outflow	1.97	0.84	0.03	0.94	0.78	3.60	8.88
	% Reduction	36.53	43.05		30.56	35.30	66.41	42.61

may have caused the high oxygen demand and slowed down the nitrification process. The N removal in the marsh-pond-marsh system coincided moderately well with the continuous marsh system operation in Duplin County, North Carolina, when cells were loaded with 15 kg N ha⁻¹ day⁻¹, particularly during the warmer months. However, the results do not indicate that a significant oxidative benefit was obtained from the pond section even though it had higher Eh values than the marsh sections.

Temperature effect on TKN removal. Since the significant seasonal differences were observed in N removal, the correlation was made between the weekly mean temperature and N removal. A linear relationship was observed between N reduction and water temperature for both N loading rates (Figure 1). The relation between the temperature and N removal was highly significant ($R^2 = 0.65$ for 16 kg and $R^2 = 0.46$ for 32 kg). The increased N removal in warmer months is directly related to the biological processes involving N-transformations. Similar results were also reported by Hunt *et al.* (1999).

Phosphorus removal. The total P and soluble-P (PO₄) removals were 44 and 39% at 16 kg N loading rate and 31 and 35% at 32 kg N loading rate, respectively. Phosphorus in wastewater is removed in wetlands by several physical, chemical, and biological mechanisms such as sorption, precipitation, and biological immobilization. These mechanisms may occur simultaneously in the wetland cells. In our study the P removal may be due to the sorption process and assimilation by the macrophytes.

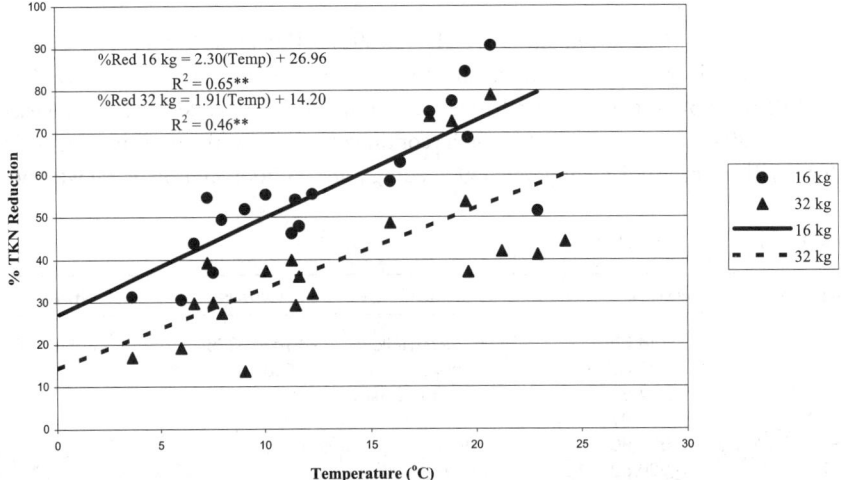

Figure 1 Temperature influence on TKN removal

Table 3 Aquatic macrophyte biomass, N and P uptake

Macrophytes	Biomass		N-uptake		P-uptake	
	\multicolumn{6}{c}{N-Loading Rates (kg N ha^{-1} day^{-1})}					
	16	32	16	32	16	32
	kg (dry wt.) m^{-2}		g m^{-2}			
Cattails and Bulrushes						
Inflow Marsh	1.62	2.51	30.00	33.00	3.30	3.70
Outflow Marsh	2.34	2.30	34.10	33.80	3.30	3.20
Duckweed (Pond Area)	0.19	0.08	9.45	3.70	0.60	0.20

Nitrogen and P removal by macrophytes. The nutrients (N and P) accumulation in macrophytes in autumn is shown in Table 3. Approximately 32 g N m^{-2} was assimilated by the plants grown in the marshes (Table 3). These results are in agreement with Szogi *et al.* (1999). Whereas, 9.3 g N m^{-2} and 3.3 g N m^{-2} was assimilated by duckweed in the pond area of the wetland cells loaded with 16 kg N and 32 kg N ha^{-1} day^{-1}. The plant biomass was 415 g m^{-2} higher in 32 kg N ha^{-1} day^{-1} than in the low N loading rate. The amount of N assimilated by duckweed in the wetland cell was directly related to the biomass. High N loading rate had a negative influence on duckweed growth. Aquatic macrophytes assimilated 3.3 to 3.9 g P m^{-2} in marsh areas. Duckweed assimilated 0.59 g P m^{-2} and 0.19 g P m^{-2} in pond area at 16 and 32 kg N ha^{-1} day^{-1} loading rates, respectively. The differences in P assimilation were related to the biomass.

Total suspended solids (TSS) removal. Total suspended solids were reduced by over 66 to 69%. However, no difference was observed in TSS removal based on N loading rate. Mass removal rates for TSS did not show a relationship to mass loading. The removal of TSS is primarily due to the physical processes such as sedimentation and filtration. Gearheart (1992) observed that approximately 75% of the TSS removal occurs in the first day of the retention.

Chemical oxygen demand (COD) removal. The removal of COD ranged from 43 to 53% across the N loading rates. The COD reduction followed the same pattern as TSS. This trend can be expected due to the association of organics with the solids.

Conclusion

The results indicate that constructed wetlands have a potential for nutrient removal and treatment of animal wastewater. The annual mean of TKN, NH$_4$, TP, and PO$_4$ at 16 and 32 kg N ha^{-1} day^{-1} loading rates were 51, 60, 44, and 39% and 37, 43, 31, and 35%, respectively. Similar results were shown in the N-data from the 17 constructed surface flow (SF) wetlands in the North American Database (Knight *et al.*, 1992). The average N efficiency of the SF wetlands was 49.9% for TKN and 33.9% for NH$_4$. Also Knight *et al.* (1985) reported a decline of total-N mass removal efficiency at higher loading rates (30 kg/ha/day). These low removal rates are due to the combined data of warmer and cooler months. However, high removal rates were observed in warmer months (June to August). Hydraulic loading was twice as high in 32 kg N with half of the retention days (10.5 days) as compared to 16 kg N loading rate. The high concentrations and mass of NH$_4$ in the outflow revealed that nitrification was a limiting factor. A significant reduction of TSS and COD occurred in wetlands. Temperature had a significant influence on TKN reduction. The outflow wastewater does not meet the discharging permit standards, however the large reduction in mass means that the wastewater can be applied on a smaller land area.

References

Brix, M. (1994). Function of Macrophytes in Constructed Wetlands. *Water Science and Technology* **29**(4), 71–78.

Gearheart, R.A. (1992). Use of Constructed Wetlands to Treat Domestic Wastewater, City of Arcata, California. *Water Science and Technology* **26**(7–8), 1625–1637.

Gibbs, C.R. (1979). Introduction to Chemical Oxygen Demand. Technical Information Series. Booklet No. **18**. pp. 1–16. Hatch Company, Loveland, Colorado.

Hammer, D.A. (ed.), (1989). *Constructed Wetlands for Wastewater Treatment – Municipal, Industrial, and Agricultural*. Lewis Publishers, Chelsea, MI. p. 831.

Hammer, D.A. and Knight, R.L. (1994). Designing Constructed Wetlands for Nitrogen Removal. *Water Science and Technology* **29**(4), 15–27.

Hunt, P.G., Humenik, F.J., Szogi, Rice, J.M. et.al. (1994). Swine Wastewater Treatment In Constructed Wetlands. Proceedings of ASAE, pp. 268–275.

Hunt, P.G., Thom, W.M., Szogi, A.A. and Humenik, F.J. (1995). State of the Art for Animal Wastewater Treatment in Constructed Wetlands. pp. 53–65. In: *Proceedings of the Seventh International Symposium on Agricultural and Food Processing Wastes*. C.C. Ross (ed.), ASAE. St. Joseph, Michigan.

Hunt, P.G., Stone, K.C., Humenik, F.J., Matheny, T.A. et al. (1999). In-Stream Wetland Mitigation of Nitrogen Contamination in a USA Coastal Plain Stream. *Journal of Environmental Quality* **28**, 249–256.

Kadlec, R.H. and Knight, R.L. (1996). *Treatment Wetlands*. Lewis Publishers, Boca Raton, CRC Press. pp. 111–112.

Kleimann, R.L.P. and Girts, M.A. (1987). Acid Mine Water Treatment in Wetlands: An Overview of an Emergent Technology. P. 255 In: *Aquatic Plants for Water Treatment and Resource Recovery*. K.R. Reddy and W.H. Smith (eds.), Magnolia Publishing Inc.

Knight, R.L., Winchester, B.H. and Higman, J.C. (1985). Carolina Bays-Feasibility for Effluent Advanced Treatment and Disposal. *Wetlands* **4**, 177–203.

Knight, R.L., Kadlec, R.H. and Reed, S. (1992). *Wetlands Treatment Data Base*. Presented at the Water Environment Federation 65th Annual Conference Exposition, New Orleans, LA. September 1992. AC92-009-003, pp.25–35.

Kopp, J.F. and McKee, G.D. (1983). Methods for Chemical Analysis of Water and Wastes. USEPA Report. No. EPA-600/4-79020. P. 521. Environmental Monitoring Supporting Lab., Office of Research and Development, U.S. EPA, Cincinnati, Ohio.

Moshiri, G.A. (eds). (1993). *Constructed Wetlands for Water Quality Improvement*. Lewis Publishers Inc., Boca Raton, Florida.

Nichols, D.S. (1983). Capacity of Natural Wetlands to Remove Nutrients from Waste-Water. *Water Pollution Control Federation Journal* **55**, 495–505.

Payne, V.W.E., McCasky, T.A. and Eason, J.T. (1992). Constructed Wetlands for Treating Swine Lagoon Effluent. Proceedings of ASAE, p. 32.

Phillips, R.L., Reddy, M.R. and Reddy, G.B. (1999). Development of a Marsh-Pond-Marsh Wetland. pp.73–74. In: *Proceedings of the Third National Workshop on Constructed Wetlands/BMP's for Nutrient Reduction and Coastal Water Protection*. Waste Management Program, North Carolina State University, Raleigh, NC.

Polprasert, C., Dan, N.P. and Thayalakumaran, N. (1996). Application of Constructed Wetlands to Treat Some Toxic Wastewaters Under Tropical Conditions. *Water Science and Technology* **34**(11), 165–171.

Reddy, K.R. and DeBusk, W.F. (1985). Nutrient Removal Potential of Selected Aquatic Microphytes. *Journal of Environmental Quality* **14**, 445–462.

Reddy, G.B. and Reddy, K.R. (1993). Phosphorus Removal by Ponds Receiving Polluted Water from Non-Point Sources. *Wetlands Ecology and Management* **2**, 171–176.

Swindell, C.E. and Jackson, J.A. (1990). Constructed Wetlands Design and Operation to Maximize Nutrient Removal Capabilities. pp. 107–114. In: *Constructed Wetlands in Water Pollution Control*. (*Advances in Water Pollution Control*. No. 11.) P.F. Cooper and B.C. Findlater (eds.), Pergamon Press, Oxford.

Szogi, A.A., Rice, J.M., Humenik, F.J., Hunt, P.G. et al. (1999). Constructed Wetlands for Confined Swine Wastewater Treatment. pp. 379–383. In: *Proceedings of Animal Waste Management Symposium*. Gerald B. Havenstein (eds.), North Carolina State University, Raleigh, NC.

Reed bed dewatering of agricultural sludges and slurries

J.K. Edwards*, K.R. Gray*, D.J. Cooper**, A.J. Biddlestone* and N. Willoughby**

* School of Chemical Engineering, University of Birmingham, Birmingham B15 2TT, UK
** A.R.M. Ltd., Rydal House, Colton Road, Rugeley, Staffordshire WS15 3HF, UK

Abstract In trials at Rugeley, UK, reed beds were used for dewatering agricultural sludges and slurries. Three beds, each of 3.5m^2, were employed, two planted with *Phragmites australis*, the third left unplanted as control. The sludge contained partly oxidised solids from a Biological Aerated Filter (BAF) treating weak pig slurry. It was supplemented with untreated settled pig slurry. Following reed establishment planted Bed A was fed at a constant similar rate to the unplanted Control Bed C. The second planted Bed B was fed at higher rates alternating with rest periods. On this bed the aeration pipes were blocked off. The trials were run for 16 months, which included two summer periods. The results showed that the planted Bed A had definitely better dewatering ability than the unplanted one fed at a similar rate. During the summer months Bed B could be fed at over twice the rate used for the constant input beds. The percolate from the control bed was more highly oxidised than from the planted beds, probably due to a longer holdup time in the absence of reeds. On Bed B the reed quality deteriorated during the second year, after use of untreated slurry as feed.

Keywords Percolate quality; primary sludge treatment; reed beds; sludge dewatering; sludge drying; sludge loadings

Introduction

Although far less numerous than reed beds for wastewater treatment, vertical flow reed beds have been employed for dewatering sludges for over 30 years. Such beds, in which Common Reed, *Phragmites australis*, is planted in soil, sand or gravel, appear to offer both economic and environmental advantages over alternative methods of sludge dewatering. They are far more effective than sand bed drying which is now almost extinct; they do not require chemical flocculants and centrifuges or belt presses. For small rural sewage treatment works (STW), particularly those employing only intermittent supervision, such reed beds can obviate the need for road tankers to convey the sludge to larger works for treatment.

In such dewatering operations sludge is fed onto the vertical flow reed bed periodically. It becomes dewatered by percolation through the preceding sludge layers and the matrix, by evapotranspiration through the reed roots, stems and leaves, and by evaporation from the sludge surface. Movement of the stems in the wind keeps open the drainage passages down through the sludge layer. With slow percolation of oxygen into the sludge layer, both via the reed plants and their root zone, and by diffusion through the air-sludge interface, the sludge gradually becomes oxidised/mineralised with a reduction in volatile solids and an increase in % ash or fixed solids. With gradual loss of moisture the % Total Solids (TS) of the dewatering sludge can build up towards 50% (De Maesener, 1996).

Most of the full-scale experience on reed bed dewatering comes from Denmark, much arising from the work of Nielsen (1990,1994). In Denmark about 25 such units have been installed on STW with another 35 under consideration. They serve communities from 5,000 to over 100,000 inhabitants. Reed bed dewatering appears to be economically competitive to the use of flocculants/centrifuges and requires far less supervision and maintenance. In Danish practice there is usually no primary settlement of sludge; all the sewage goes into an oxidation ditch. Thus the sludge fed to the multiple reed beds is well oxidised, usually contains less than 1% solids, spreads easily and permits some 10 years operation before bed emptying.

In France, Lienard and co-workers at CEMAGREF have done extensive work at pilot plant scale (Lienard et al., 1990, 1994). However, there does not appear to have been any major adoption of reed bed dewatering on a large scale in that country. In the USA, Kim (1993, 1997) has reported on 24 of the 50 small municipal STW in which reed beds have been used for sludge dewatering.

In the UK there now appear to be major opportunities for use of the technique (Cooper and Willoughby, 1999). Since 1998 the disposal at sea of sewage sludge has been banned. There is now a Landfill Tax of £11 per tonne for the disposal of waste to landfills; this tax is being increased by £1 per tonne annually for the next few years as the UK Government intends to greatly restrict the use of landfills for waste disposal.

So far in the UK few full size sludge drying reed beds are in operation. West of Scotland Water built a unit at Ballygrant on the Isle of Islay in 1995 and a second at Kiells in 1998; these take sludge from oxidation ditches. Both units were designed by Cooper at WRc (Cooper et al., 1996). Severn Trent Water has been carrying out sludge dewatering trials for several years at small STW using both primary and mixed sludges. In 1999 the Wildfowl and Wetlands Trust commissioned a unit at their bird sanctuary at Slimbridge. On a smaller scale, Burka installed a bed for a community at Oaklands Park in Gloucestershire in 1989 while Moodie constructed one for the National Trust at Dudmaston, Shropshire in 1992 (Nuttall et al., 1997).

For several years the University of Birmingham and A.R.M. Ltd. have studied the reed bed dewatering of sludges and slurries from agricultural sources (Edwards, 2000). The trials are the subject of this paper.

Methods

A pilot scale reed bed system was built at A.R.M. Ltd., Rugeley, Staffordshire, UK in December 1995 to dewater the settled humus sludge produced by a Biological Aerated Filter (BAF) unit when treating the wastewater from pig pens. The humus is the organic slime material which sloughs off from the plastic packings of the BAF reactors when operating. The reed bed system consists of a sludge holding tank and three identical vertical flow beds (Edwards, 2000). Two of the beds, A and B, were planted with *Phragmites australis*; the third bed, C, was left unplanted as a control.

Bed construction

The system was built above ground in blockwork. Each of the three beds is 2.6 m long by 1.27 m wide and 1.0 m deep comprising 500 mm depth of matrix and 500 mm of freeboard.

The matrix is of graded gravel as follows:

Top layer	100 mm deep	1–2 mm rounded sea gravel
Then	150 mm deep	5–10 mm pea gravel
Next	100 mm deep	20–30 mm gravel
Bottom layer	150 mm deep	30–60 mm stones

For the top layer rounded sea gravel was employed in preference to sharp sand which in the past has led to major blockage problems with reed beds at Rugeley treating wastewater.

The sludge in the holding tank was stirred by hand with a paddle and then pumped into a graduated vessel to ensure the required volume of the load; it was then hosed under hand control onto the bed spreading it evenly. With the first loads onto the beds it was difficult to form a sludge layer; it would have been better to blind the sea gravel surface with a thin layer of compost to form the initial filtration layer of solids.

After passing down through the bed matrix, in which some degree of treatment takes place, the percolate is discharged via a swivelling elbow to a 110 mm diameter pipe. This allows the water level in the bed to be controlled, being raised during the start-up period to aid reed establishment and during very hot weather to prevent the reeds becoming stressed due to lack of water in the root zone. During normal operation the bed is allowed to drain freely to help air diffusion back into the matrix. On discharge the percolate can be pumped to a nearby wastewater reed bed system for further treatment. Two aeration pipes were built into the matrix across the bed width; lower down a single aeration pipe runs the length of the bed. These are in 100 mm perforated plastic pipe. On Bed B the vertical risers from these pipes were capped in order to examine the effect of restricting matrix aeration on bed performance.

Two of the beds, A and B, were planted with *Phragmites australis* in January 1996 at a density of 4 plants/m^2. After planting, the beds were flooded to the surface with a weak solution of pig manure slurry diluted with tap water. This provided the reeds with nutrients for growth and adequate water to prevent stress in hot weather. During the summer months of 1996 further applications of solution were made. Within the first year the reeds spread to provide virtually a complete surface coverage. Bed C was left unplanted.

Characterisation of the sludges

Soon after operations started on the dewatering beds it became clear that the volume of humus sludge produced by the BAF system was much lower than expected and was insufficient to maintain an adequate loading regime on the beds. It became necessary to augment the volume of sludge available for dewatering.

The feed for the BAF unit was prepared from pig manure slurry sedimented in a storage tank, the supernatant being drawn off and diluted with tap water to the required level of Biochemical Oxygen Demand (BOD). The raw settled solids from the bottom of this tank were therefore blended with the treated humus sludge from the BAF unit to provide enough feed material for dewatering; the combination of these two sludges started on 30/7/97. Then in October 1997 the BAF unit was shut down; thereafter only the raw manure slurry solids were available for the dewatering trials. The various sludge feeds are shown in Table 1.

Over the trials period a series of samples were taken of these sludges and analysed for a range of characteristics, shown in Table 2. The analytical methods used are given in Edwards (2000).

Dewatering trials

Commissioning phase. Although Nielsen (1994) recommends a commissioning phase of 3 years to allow the reeds to grow and the roots and rhizomes to become well established in the bed matrix, this was clearly impossible with a PhD project. In the UK, Cooper *et al.* (1996) suggest a 1 year reed establishment phase; this was more appropriate to the present trials. During this commissioning year the two planted beds, A and B, were lightly loaded with two applications of BAF sludge to assess the potential dewatering capacity of the

Table 1 Sludges and slurries used for dewatering trials

Dates	Feeds
5/9/96 to 29/7/97	BAF humus sludge only
30/7/97 to 20/10/97	Combined BAF sludge and raw slurry solids
21/10/97 to 17/9/98	Raw slurry solids only

system and to help set up a sludge loading plan for the project. On 5/9/96 a volume of 87.5l of sludge was measured out and applied to Bed A, providing a layer 25 mm thick. Then on 26/9/96 a sample of the partially dewatered sludge layer was taken from Bed A and analysed for TS content. A further 25mm layer of sludge feed was added to Bed A and a 50 mm layer applied to Bed B. On the 10/10/96 samples of both dewatered layers were taken and analysed for TS. In addition to obtaining operating experience with the dewatering beds, these sludge applications greatly benefitted the reed growth. However, the total sludge loading used in this period was only about 3 kg dry solids (ds)/m^2.y, much less than the 18.3 kg ds /m^2.y recommended by Lienard *et al.* (1995).

Main trials. The loading of sludge for the main trials started in May 1997 and finished in September 1998, giving 70 weeks of investigations which included two summer periods. As shown in Table 1, during the initial part of this period only BAF Humus Sludge was used, then a combination of BAF Humus Sludge and Raw Slurry Solids; finally only Raw Slurry Solids were employed.

During the trial period Bed A and Control Bed C were loaded, as far as possible, with similar quantities of feed to compare the capabilities of a planted and an unplanted bed. Over this time of 70 weeks a total of 2,200 mm of sludge were added in 34 applications to Bed A, an average of 31 mm per week which, with the sludge averaging 4.07 w/w% TS, gave a loading of 67.7 kg ds /m^2.y. This figure is close to the loading of 60 kg ds/m^2.y suggested by Cooper *et al.* (1996) for an established bed. In practice, however, virtually no sludge was applied during the winter period from late October 1997 to mid March 1998. As a result the 60mm layers added during most weeks of the active season corresponded to a loading nearer 120 kg ds/m^2.y, right at the top of the range suggested.

Bed B was tested under different loading rates. Although the total sludge added over the period was only slightly more than on the other two beds, the 26 applications included layers of 120, 180 or 240 mm depth; these heavy applications were then followed by several weeks of rest. Although Bed B was constructed with aeration pipes through the matrix, these were blocked off in the trials.

Quality of percolates. An investigation was made of the quality of the percolates leaving the three beds between 5/3/98 and 18/8/98 to assess the effect of the presence of reeds on Beds A and B, the blockage of the aeration pipes on Bed B, and the heavier but less frequent sludge applications to Bed B. The samples were analysed for COD, BOD, SS, DS, pH, NH_4-N and o-PO_4.

Results and discussion
Characterisation of the sludges
As shown in Table 1, three different blends of settled humus sludge from the BAF unit and raw pig slurry solids were used to make up the feed to the dewatering beds over the 70 weeks of the investigation. Samples of the feed were taken frequently and analysed for a range of characteristics. The detailed results are given in Edwards (2000) but the mean results are shown in Table 2.

Using wastewater/sludge flows from a commercial farm operation it was only to be expected that the TS contents of the three feed sludges were highly variable, ranging between 1.18 and 9.20% w/w with a mean of 4.07. This led to the high standard deviations shown in Table 2 for many characteristics.

Commissioning phase
On 5/9/96, eight months after the reeds were planted, a 25 mm deep layer of sludge was

Table 2 Average characteristics of the 3 sludge feeds used for dewatering, see Table 1

Characteristic	Mean result	Standard deviation	Number of samples
Total solids (TS) w/w%	4.07	2.12	27
Suspended solids (SS) mg/l	31,364	21,062	20
Dissolved solids (DS) mg/l	4,645	1,682	16
Fixed solids (FS) as % of TS	25.8	2.8	13
Volatile solids (VS) as % of TS	74.2	2.8	13
pH	7.81	0.25	21
Specific gravity	1.020	0.009	4
Chemical Oxygen Demand (COD) mg/l	22,600	9,400	13
Biochemical Oxygen Demand (BOD) mg/l	7,000	3,440	13
Ammoniacal Nitrogen (NH_4-N) mg/l	659	125	10
Ortho Phosphate (o-PO_4) mg/l	5,650	2,956	9

added to Bed A. Then on 26/9/96 a sample of the sludge layer was taken and the TS content of the dewatered sludge found to be 38.9%, representing a high degree of water removal. The same day a further 25 mm layer was added to Bed A and 50 mm to Bed B. Then on 10/10/96 samples of the sludge layers were taken from both beds. That from Bed A had a TS of 21.4%, that from Bed B had 21.9%.

Dewatering trials

Sludge loadings and final residues. Following the preliminary loadings, the main sludge loading of the beds commenced on 16/5/97 and concluded with the final addition on 17/9/98, a total of 70 weeks. Over this period Bed A and Control Bed C were loaded at virtually the same rates in increments of 60 mm layers in most weeks, apart from the winter period from late October 1997 to mid March 1998 when few additions were made. Bed B was used to test the effect of much larger increments with some 120, 180 and 240 mm layers. Following the final sludge loadings the beds were left until 25/2/99 when the final residual sludge heights were recorded. The overall sludge loadings and final residue heights are given in Table 3.

The heights of the final residues indicate that planted Bed A achieved a dewatering ability of 84.3%, whereas that of unplanted Bed C was 81.1%. Planted Bed B, fed with larger and less frequent dosings, achieved a reduction of 86.3%. The final residue height of about 350 mm for the planted beds resulted from 70 weeks of sludge loading, an annual increase of 260 mm. This is much greater than the value of 150 mm/y stated by Nielsen (1994) for beds after 10 years of operation. A far more detailed analysis of the present sludge loadings and residue heights is given by Edwards (2000); in this the experimental period is divided into five seasons – Summer, Autumn and Winter 1997, Spring and Summer 1998. This analysis shows the major reduction in loading necessary during the winter months and gives a good comparison between the results in the two Summer periods.

A major finding was that during summer 1998 the reed growth was not as good as in summer 1997, with less bed coverage and lower stem heights. This was probably because in the second season of active growth some newly emerging shoots were overwhelmed by the raw inlet sludge when the loadings were increased in the spring following low loadings over winter. This contrasts with Danish experience when handling aerobic sludges from oxidation ditches and anaerobic sludges following digestion. Obviously the challenge for the UK is to find suitable sludge loadings and operating techniques for dewatering raw primary sludges, or their mixtures with secondary ones, without affecting reed growth.

Table 3 Overall sludge loadings and final residue heights, period May 1997 to February 1999

	Bed A	Bed B	Bed C
Sludge added, m^3	7.26	8.46	6.87
Height of added sludge, mm	2200	2561	2081
Final residue on bed, m^3	1.14	1.16	1.30
Height of final residue, mm	345	352	393
Reduction in height, %	84.3	86.3	81.1

Analyses of residual sludge layers. The final sludge loads were applied to the beds on 17/9/98. Then six weeks later, on 28/10/98, the height of the residual sludge layer was measured and samples of the dewatered sludge taken and analysed for TS and VS. The height measurements and analyses were repeated every 6–8 weeks for a period of 5 months until 25/2/99. The final heights are given in Table 3.

The sludge samples taken were core samples of the total depth of residue on each bed. The cores were then divided into 100mm deep segments to allow an estimation of the sludge characteristics at different depths in the residue and a mean overall value of the TS to be calculated. Table 4 shows the results from the final sludge cores taken on 25/2/99, 161 days after the final sludge was applied.

Table 4 shows that the mean values of TS were higher on the planted beds, A and B, than on the unplanted Control Bed C, indicating that the reeds aided the dewatering process. The TS value on Bed B, where the aeration pipes were blocked off, was slightly higher than on Bed A, suggesting that matrix aeration does not help dewatering. Analyses of the four horizontal slices of each core showed that generally the TS were highest at the top and bottom of the core, lowest in the middle section.

The VS were generally higher at the top of each core but the mean overall value showed little difference between the three beds. The mean values of VS, approximately 52.5% for the three beds, is a major decrease from the VS value in the fresh sludge, 74.2% in Table 2, showing that significant oxidation/mineralisation had occurred.

Examination of the cores of the final residual sludge layers showed that on both planted Beds A and B two distinct zones had formed – an anaerobic black upper layer, 300–350 mm deep, and a more oxidised lower brown layer, 50–100 mm deep. The two layers were also present on the unplanted Control Bed C but the oxidised lower layer was only 5–10 mm deep. These results indicate that, among other factors, the oxygen provided by the reeds aids the stabilisation of the sludge.

Quality of percolates. In Table 5 are summarised the mean results of the analyses of the percolates leaving the three dewatering beds between 5/3/98 and 18/8/98.

Lienard *et al.* (1995) and Hofmann (1990), using weak sewage sludges, have shown that planted beds give better percolates, with lower values of COD, BOD, SS and NH_4-N. In the present trials, using much stronger agricultural sludges, the planted Bed A has a lower

Table 4 Analysis of final sludge residue cores taken on 25/2/99

	Bed A		Bed B		Bed C	
Column depth mm	TS %	VS %	TS %	VS %	TS %	VS %
0–100	20.77	54.38	23.46	52.11	19.03	54.55
100–200	20.82	53.41	19.37	53.59	17.47	51.96
200–300	18.24	51.21	21.09	51.79	17.25	52.15
300–400	22.09	50.99	22.11	51.90	20.85	51.46
Mean value	20.48	52.50	21.51	52.35	18.65	52.53

Table 5 Characteristics of percolates from sludge dewatering beds, 5/3/98 to 18/8/98

Characteristic	Bed A		Bed B		Bed C	
	Mean	Std. Dev.	Mean	Std. Dev.	Mean	Std. Dev.
COD mg/l	1822	607	2220	1085	1962	1018
BOD mg/l	1123	588	1468	746	935	539
SS mg/l	809	419	917	455	770	429
DS mg/l	3674	624	4055	655	3290	601
pH	8.02	0.22	7.65	0.17	8.17	0.24
NH_4-N	471	226	589	221	474	177
o-PO_4	569	154	590	57	583	145

value for COD than the similarly loaded but unplanted Control Bed C but higher values for BOD, SS and TDS and comparable levels of NH_4-N and o-PO_4. Bed B with its higher sludge loadings had a definitely stronger percolate than did the other two beds. A possible cause of the better performance by the unplanted Bed C is that in the absence of reeds to keep open the passageways through the sludge layer, the hydraulic conductivity through the latter was much lower; this led to a longer holdup time in both the sludge and the gravel matrix, giving longer for percolate treatment.

Conclusions

1. The planted reed beds showed a definite improvement over the unplanted one in terms of higher % Total Solids and greater % Reduction in the height of the residual layer of sludge. They also showed a much deeper oxidised lower sludge layer. However, there was little difference in % Volatile Solids.
2. High loadings of raw sludges/slurries in the Spring can seriously damage the newly emerging shoots of reed plants.
3. In the Summer it is possible to dose dewatering reed beds with much higher loadings than 60 kg dry solids/m^2.y, often regarded as the maximum year round average in the UK.
4. In terms of the BOD, Suspended Solids and Dissolved Solids of the percolates, the planted beds performed worse than the unplanted one. This was possibly due to their higher hydraulic conductivity, giving less time for treatment in the sludge layer and the matrix.

Acknowledgements

The authors gratefully acknowledge the financial support and help given to the project by the BOC Foundation for the Environment, the Engineering and Physical Sciences Research Council and A.R.M. Ltd.

References

Cooper, D.J. and Willoughby, N. (1999). *Dewatering of sludge at small sewage treatment works using reed beds*. Paper presented to 4th European Biosolids and Organic Residuals Conference, Wakefield, Yorkshire 15–17 November 1999, Aqua-Enviro and C.I.W.E.M.

Cooper, P.F., Job, G.D., Green, M.B. and Shutes, R.B.E. (1996). *Reed Beds and Constructed Wetlands for Wastewater Treatment*. WRc Publications, Medmenham, UK.

De Maesener, J.L. (1997). Constructed wetlands for sludge dewatering. *Wat. Sci. Tech.*, **35**(5), 279–285.

Edwards, J.K. (2000). *Reed Bed Systems for the Treatment of Wastewaters and for Sludge Dewatering*. PhD thesis, School of Chemical Engineering, University of Birmingham, UK.

Kim, B.J. (1993). *Performance evaluation of existing reed beds for sludge dewatering*. Paper presented to Water Environment Federation 66th Annual Conference and Exposition, Anaheim, USA. 1993.

Kim, B.J. (1997). Evaluation of sludge dewatering reed beds: a niche for small systems. *Wat. Sci. Tech.*, **35**(6), 21–28.

Lienard, A., Esser, D., Deguin, A. and Virloget, F. (1990). Sludge dewatering and drying in reed beds: an interesting solution? In: *Constructed Wetlands in Water Pollution Control*, P.F. Cooper and B.C. Findlater (ed.), Pergamon Press, Oxford, pp. 257–267.

Lienard, A., Duchene, Ph. and Gorini, D. (1994). A study of activated sludge dewatering in experimental reed-planted or unplanted sludge drying beds. *Wat. Sci. Tech.*, **32**(3), 251–261.

Nielsen, S.M. (1990). Sludge dewatering and mineralisation in reed bed systems. In: *Constructed Wetlands in Water Pollution Control*, P.F. Cooper and B.C. Findlater (ed.), Pergamon Press, Oxford, pp.245–255.

Nielsen, S.M. (1994). *Biological sludge drying in reed bed systems – six years of operating experiences*. Paper presented to 4th International Conference on Wetland Systems for Water Pollution Control, Guangzhou, China, 6–10 November, 1994, 447–457.

Nuttall, P.M., Boon, A.G. and Rowell, M.R. (1997). *Review of the Design and Management of Constructed Wetlands*. Report 180. Construction Industry Research and Information Association, London.

Using wetlands for water quality improvement in agricultural watersheds; the importance of a watershed scale approach

W.G. Crumpton

Department of Botany, Iowa State University, Ames, IA 50011, USA

Abstract Agricultural applications of fertilizers and pesticides have increased dramatically since the middle 1960s, and agrochemical contamination of surface and groundwater has become a serious environmental concern. Since the mid-1980s, a variety of state and federal programs have been used to promote wetland restoration, and these continuing efforts provide a unique opportunity for water quality improvement in agricultural watersheds. However, wetland restorations have been motivated primarily by concern over waterfowl habitat loss, and model simulations suggest that commonly used site selection criteria for wetland restorations may be inadequate for water quality purposes. This does not lessen the promise of wetlands for water quality improvement in agricultural watersheds, but rather emphasizes the need for watershed scale approaches to wetland siting and design. Water quality is best viewed from a watershed perspective, and watershed scale endpoints should be explicitly considered in site selection for wetland restoration.

Keywords Nitrate; water quality; wetland restoration

Introduction

Agricultural applications of fertilizers and pesticides have increased dramatically since the mid-1960s, and agrochemical contamination of surface and groundwater has become a pressing environmental problem. Nitrogen (N) and pesticides are of foremost concern because of their potential impacts on both public health and ecosystem function, and because of the widespread use of N and pesticides in modern agriculture. The total amount of N applied in fertilizers far exceeds that of any other nutrient, and annual application of fertilizer-N in the U.S. has grown from a negligible amount prior to World War II to over ten million metric tons of N per year (Terry and Kirby, 1997). As much as 50% of the fertilizer-N applied to cultivated crops may be lost in agricultural drainage water, primarily in the form of nitrate (Neely and Baker, 1989). The environmental impacts of agriculture are a special concern in the US corn belt. Nonpoint loads of nutrients to surface waters in this region are among the highest in the country, and it is unlikely that these problems can be solved by chemical management alone (Baker *et al.*, 1997).

Wetland restoration is one of the most promising strategies for reducing surface water contamination in the corn belt (van der Valk and Jolly, 1992). Much of this region was historically rich in wetlands, and in some areas, farming was made possible only as a result of extensive wetland drainage. We now recognize the important functions that wetlands had served in this landscape, and there is considerable interest in recovering these lost functions. Since the mid-1980s, a variety of state and federal programs have been used to promote wetland restoration in the region, and over the next few years, thousands of wetland acres will be restored through a combination of federal, state, and private initiatives. These continuing efforts provide a unique opportunity for control of agricultural non-point source pollution. However, wetland restorations have been motivated primarily by concern over waterfowl habitat loss, and site selection criteria for wetland restorations have not primarily considered water quality functions. Of more than 500 wetland restorations in the southern prairie pothole region surveyed by Galatowitsch (1993), most drain very small areas and may intercept insufficient contaminant loads to significantly affect water quality at the watershed scale. This does not lessen the promise of wetlands for water quality improvement in agricultural watersheds but rather underscores the need for explicitly considering watershed scale endpoints when planning wetland restorations.

From a water quality perspective, siting and design criteria for wetland restoration require watershed scale approaches. However, there is little quantitative information on the effectiveness of wetland restoration in agricultural landscapes as a means for water quality improvement at the watershed scale. The principal objective of the research described here was to predict the effect of alternative wetland restoration scenarios on the concentration and mass load of nitrate-nitrogen in water exiting Walnut Creek Watershed, a typical agricultural watershed in the western corn belt ecoregion of the US Walnut Creek Watershed is characterized by gently undulating topography with moderately to poorly drained soils (Hatfield *et al.*, 1999). The upper part of the watershed is nearly level with poorly defined surface drainage channels and numerous depressions. Historically, the landscape was rich in wetlands, but these were drained for agriculture and most of the land is now planted in continuous corn and corn-soybean rotations (Hatfield *et al.*, 1999). However, as evidenced by the extent and distribution of hydric soils (Figure 1), there are opportunities for wetland restoration throughout the watershed.

Model framework and simulations

Our understanding of nutrient losses in treatment wetlands has progressed to a point that practical design models are now emerging for nitrogen and phosphorus removal (Kadlec and Knight, 1996; Kadlec *et al.*, 1997). In the case of wetlands receiving significant nitrate loads, loss rates can be described by a temperature dependent first-order model (Crumpton *et al.*, 1997; Crumpton and Goldsborough, 1998; Crumpton *et al.*, in preparation; Kadlec and Knight, 1996):

$$J = k_{20} * C * \theta^{(T-20)} \tag{1}$$

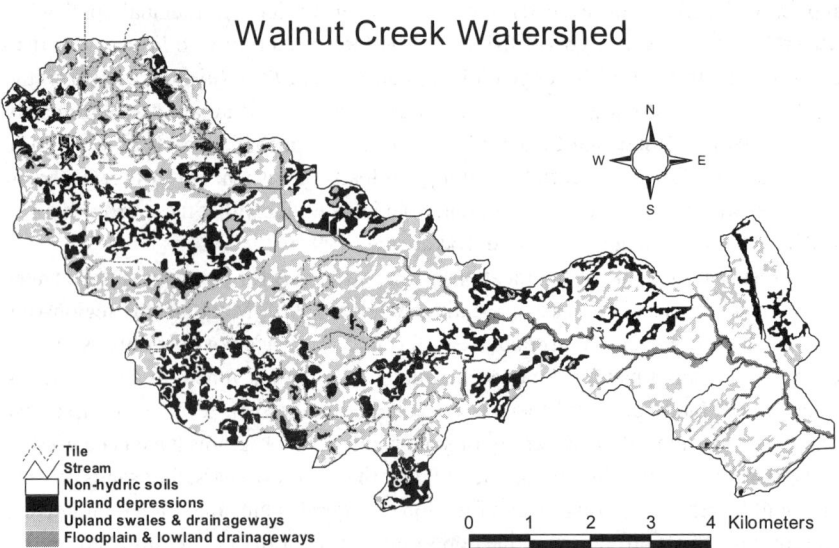

Figure 1 Map of Walnut Creek Watershed showing stream channels, drainage tiles, non-hydric soils, and 3 hydric soil groups representing (1) upland depressions, (2) upland swales and drainageways, and (3) floodplain and lowland drainageways

where,

J = the area-based nitrate-nitrogen loss rate, g N m^{-2} day^{-1}
k_{20} = the area-based first order loss rate coefficient for nitrate-N at 20°C, m/day
C = the concentration of nitrate-nitrogen, g N m^{-3}
θ = the temperature coefficient for nitrate nitrogen loss
T = the water temperature, °C

Wetlands receiving nonpoint source loads are subjected to widely varying hydraulic and contaminant loading rates, in addition to cyclic variations in other forcing functions such as temperature (Kadlec, 1999). Mass balance models must adequately represent these variations in load and in hydraulic residence time distributions in wetlands. For the current study, the reaction rate expression above was incorporated into a tanks-in-series (TIS) mass balance model of nitrate loading and loss for wetlands under different restoration scenarios in Walnut Creek Watershed. TIS models have been shown to realistically represent the residence time distributions of wetlands having a wide range of morphometries and aspect ratios (Kadlec and Knight, 1996). Model coefficients were estimated as k_{20} = 0.15 and θ = 1.09 based on mass balance measurements from experimental wetlands and wetland mesocosms (Crumpton et al., 1993; Crumpton et al., 1997; Crumpton and Kadlec, in preparation; Kadlec and Knight, 1996).

Forcing functions included inflow rates, inflow concentrations, and temperature, all of which were estimated from measured values in Walnut Creek. Stream flow rates and nitrate concentrations were measured as part of the Iowa Management Systems Evaluation Area Project (Hatfield et al., 1999; Jaynes et al., 1999). Discharge was measured continuously at three locations within the stream as well as from three county drains and two field tiles within the watershed (Jaynes et al., 1999). Weekly water samples were collected at each site where discharge was measured. Additionally, samples were collected using flow activated ISCO autosamplers at each of the three stream sites (Jaynes et al., 1999). Detailed sampling and analytical procedures are described in Hatfield et al. (1999). Temperature was measured using continuous water quality monitors (AQUA 2000 prototypes, Biodevices Inc., Ames, Iowa, USA) in Walnut Creek and in Goose Lake Marsh, a natural prairie pothole wetland located 40 km north of Walnut Creek (Rose and Crumpton, 1996).

Model simulations were run to examine nitrate loading and loss for two separate wetland restoration scenarios in a 2550 ha subwatershed of Walnut Creek Watershed. The scenarios represented hypothetical situations with restored wetlands occupying 25.5 ha. These scenarios differed primarily in the location of restorations within the subwatershed, which resulted in different drainage area:wetland area ratios. In the first scenario, referred to as the conventional approach, it was assumed that wetland restorations would be located in the same manner as has been typical since the mid-1980s. In general, wetland restorations in the region have not targeted watershed scale endpoints. Most frequently, small wetland basins at the terminus of lateral drainage networks are targeted, and as a result restored wetlands tend to drain small areas relative to the size of the wetland. An average drainage area:wetland area ratio of 4:1 was calculated from data for 538 restorations surveyed by Galatowitsch (1993). In the conventional approach scenario, wetlands sited to achieve this area ratio intercepted 4% of the total drainage prior to discharge to the stream. In the second scenario, referred to as the watershed approach, it was assumed that sites for wetland restorations would be selected using a watershed perspective that explicitly recognizes the importance of landscape position in the water quality functions of wetlands. In this scenario, wetlands occupied the same total area as in the first scenario, but were placed so as to intercept 50% of the total drainage prior to discharge to the stream.

Results and discussion

This study focused primarily on two watershed scale endpoints, the concentration and mass load of nitrate-nitrogen in water exiting the subwatershed. During 1995, nitrate-nitrogen concentrations in Walnut Creek displayed a pronounced seasonal pattern. The flow weighted average concentration of 11.3 mg L^{-1} was just above the USEPA drinking water standard of 10 mg L^{-1} and peak concentrations were 80% higher than the standard (Figure 2). Concentrations generally exceeded the standard from about mid-April through July, and with the combination of high nitrate concentrations and high flow rates, this same period accounted for the bulk of the annual nitrate exported from the subwatershed. The period between April 10 and July 31 accounted for 80% of the annual nitrate exported and 65% of the annual discharge (Figure 2). This general pattern is typical for agricultural watersheds in the region. Nitrate concentrations and mass loads are highest during high flow periods in spring and early summer, and in most years decline with declining flow in late summer and early fall. These patterns illustrate the variability in hydraulic and contaminant loading that can be expected for wetlands receiving non-point source loading from agricultural areas.

Although wetlands have significant nitrate removal capacity, their effect on watershed scale endpoints is largely determined by the fraction of the watershed's total nitrate load that the wetlands intercept. If not sited so as to intercept a significant fraction of the watershed load, restored wetlands would have very little effect on either nitrate concentrations or exported nitrate loads. This is the principal difference in the conventional approach and the watershed approach to restoration contrasted here. Wetlands sited so as to simulate a conventional approach would have very little effect on nitrate concentrations (Figure 3). The flow weighted average concentration of 10.9 mg L^{-1} would still exceed the drinking water standard, and concentrations would follow nearly the same pattern as under existing conditions, exceeding the standard for 111 days from mid-April through July. In addition, the wetlands would remove less than 4% of the annual nitrate load exported from the watershed (Figure 4). In contrast, wetlands sited to simulate a watershed approach would result in substantially reduced concentrations of nitrate in water exiting the watershed (Figure 3). Nitrate-nitrogen concentrations would average 7.4 mg L^{-1}, well below the drinking water standard, and peak concentrations would exceed the drinking water standard on only about 16 days. In addition, the wetlands would remove approximately 35% of the annual nitrate load exported from the watershed and would significantly reduce nitrate exports even under high flow conditions (Figure 4).

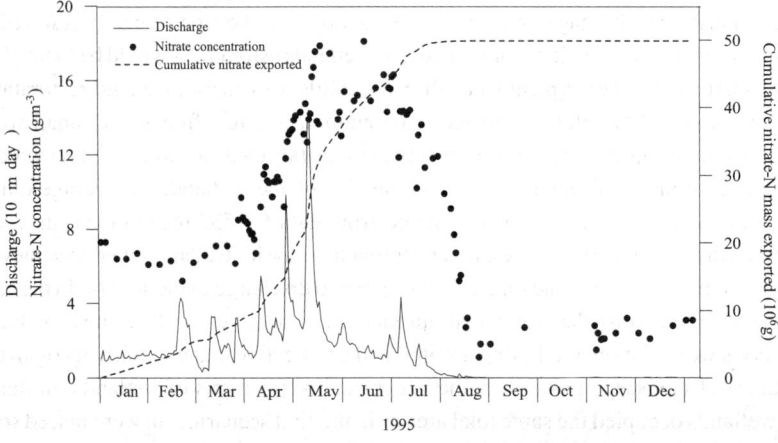

Figure 2 Nitrate-N concentrations, stream discharge, and cumulative nitrate export for the subwatershed

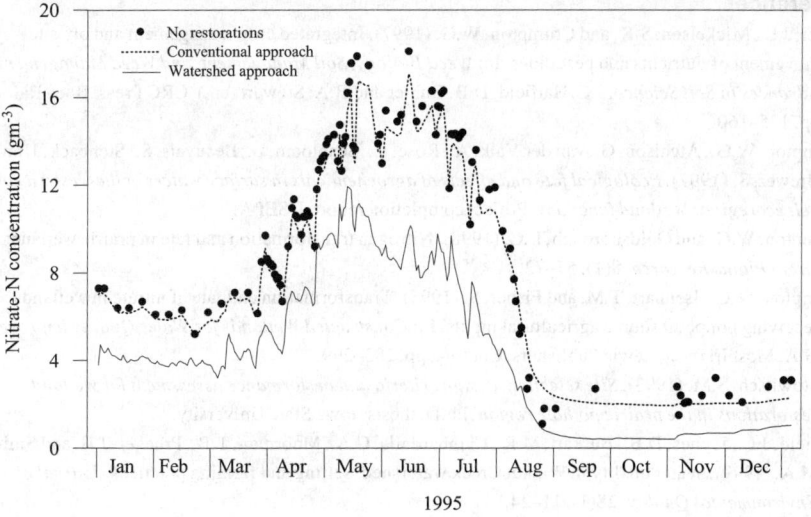

Figure 3 Nitrate-N concentrations at the subwatershed outlet under alternative restoration scenarios

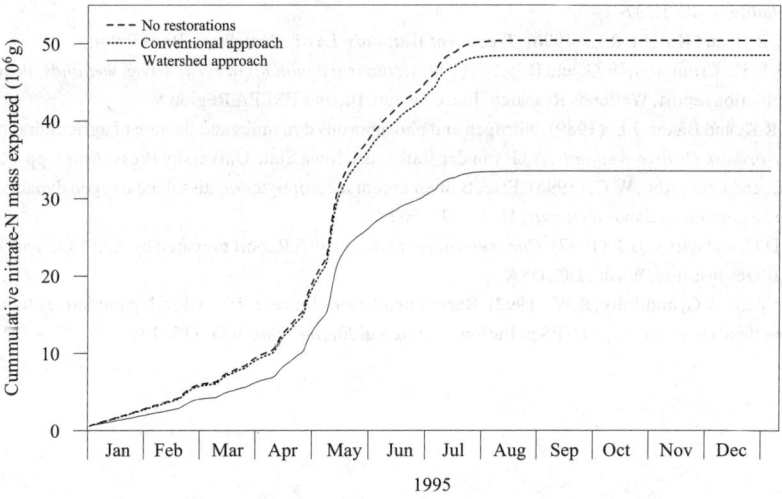

Figure 4 Cumulative nitrate-N export from the subwatershed under alternative restoration scenarios

Conclusions

These results suggest that commonly used site selection criteria for wetland restorations may be inadequate for water quality purposes. Water quality is best viewed from a watershed perspective, and watershed scale endpoints should be explicitly considered in site selection for wetland restoration.

Acknowledgements

This work was supported in part by USEPA, USGS, and USDA (MSEA) funds. I would like to thank Jerry Hatfield and staff at the National Soil Tilth Laboratory for technical assistance and use of data from the Iowa MSEA project and Jana Stenback and Shawn Richmond for help in preparing this manuscript.

References

Baker, J.L., Mickelson, S.K. and Crumpton, W.G. (1997). Integrated crop management and off-site movement of nutrients and pesticides. In: *Weed Biology, Soil Management, and Weed Management; Advances in Soil Science*, J.C. Hatfield, D.B. Buhler and B.A. Stewart (ed.), CRC Press, Boca Raton, pp. 135–160.

Crumpton, W.G., Atchison, G., van der Valk, A., Rose, C., Seabloom, E., Beauvais, S., Stenback, J. and Brewer, S. (1997). *Ecological fate and effects of agrochemicals in surface waters of the western corn belt ecoregion: Wetland functions*. Project completion report, USEPA.

Crumpton, W.G. and Goldsborough, L.G. (1998). Nitrogen transformation and fate in prairie wetlands. *Great Plains Research*, **8**(1), 57–72.

Crumpton, W.G., Isenhart, T.M. and Fisher, S. (1993). Transformation and fate of nitrate in wetlands receiving nonpoint source agricultural inputs. In: *Constructed Wetlands for Water Quality Improvement*, G.A. Moshiri (ed.), Lewis Publishers, Chelsea, pp. 283–299.

Galatowitsch, S.M. (1993). *Site selection, design criteria and performance assessment for wetland restorations in the prairie pothole region*. Ph.D. thesis, Iowa State University.

Hatfield, J.L., Jaynes, D.B., Burkart, M.R., Cambardella, C.A., Moorman, T.B., Prueger, J.H. and Smith, M.A. (1999). Water quality in Walnut Creek watershed: setting and farming practices. *Journal of Environmental Quality*, **28**(1), 11–24.

Jaynes, D.B., Hatfield, J.L. and Meek, D.W. (1999). Water quality in Walnut Creek watershed: herbicides and nitrate in surface waters. *Journal of Environmental Quality* **28**(1), 45–59.

Kadlec, R.H. (1999). Chemical, physical and biological cycles in treatment wetlands. *Water Science and Technology*. **40**(3), 37–44.

Kadlec, R.H. and Knight, R.L. (1996). *Treatment Wetlands*. Lewis Publishers, Boca Raton.

Kadlec, R.H., Crumpton, W.G. and Rose, C. (1997). *Nutrient dynamics in event driven wetlands*. Project completion report, Wetlands Research, Inc. Chicago, IL. and USEPA Region V.

Neely, R.K. and Baker, J.L. (1989). Nitrogen and phosphorous dynamics and the fate of agricultural runoff. In: *Northern Prairie Wetlands*, A.G. van der Valk (ed.), Iowa State University Press, Ames, pp. 92–131.

Rose, C. and Crumpton, W.G. (1996). Effects of emergent macrophytes on dissolved oxygen dynamics in a prairie pothole wetland. *Wetlands*, **16**(4), 495–502.

Terry, D.C. and Kirby, B.J. (1997). *Commercial Fertilizers 1997*. Report prepared by AAPFCO and The Fertilizer Institute, Wash., DC, USA.

van der Valk, A.G. and Jolly, R.W. (1992). Recommendations for research to develop guidelines for the use of wetlands to control rural NPS pollution. *Ecological Engineering* **1**(1), 115–134.

Stormwater treatment: do constructed wetlands yield improved pollutant management performance over a detention pond system?

H.J. Bavor, C.M. Davies and K. Sakadevan

Water Research Laboratory, Centre for Water and Environmental Technology, University of Western Sydney – Hawkesbury, Bourke Street, Richmond, NSW 2753, Australia

Abstract Constructed wetland systems have been proposed as representing an improved ecotechnological option over detention basins, in terms of their abilities to reduce stormwater bacterial and nutrient loads to receiving waters. Concentrations of microbial and pollutants were determined in inflow and outflow samples collected from each type of system. Removal efficiencies for the wetland although higher than for the pond, were lower than some previously reported values for the treatment of municipal wastewater by constructed wetlands. Performance of a number of constructed wetland systems for stormwater treatment is evaluated considering the functional components of the systems.
Keywords Constructed wetland; detention pond; faecal coliforms; phosphorus; stormwater

Introduction

Constructed wetlands and detention ponds are increasingly being used to reduce stormwater pollutant concentrations in urban catchments (Bavor and Mitchell, 1994). The two systems provide different treatment processes by virtue of their different macrophyte cover and density, depth, and water level fluctuation. Sedimentation is the main physical mechanism of pollutant removal in both wetlands and ponds and a significant proportion of pollutants present in stormwater are bound to particulate material ranging in size from gross solids to colloids. A large proportion of stormwater sediment-bound pollutants such as metals, nutrients, and microorganisms are associated with the fine particle fraction, i.e., less than 2 μm in size (Pitt and Amy, 1973; Mann and Hammerschmid, 1989; Davies and Bavor, 2000).

The long detention period of detention ponds promotes the sedimentation of solids in the coarse and medium size fractions, but fine particles are less effectively settled. In contrast, the presence of extensive vegetation in wetlands enhances sedimentation of the fine particle fraction as well as removing coarse and medium sized particles (Wong *et al*., 1999). This suggests that constructed wetlands may be more effective in removing particulate-bound pollutants than detention ponds. The aims of the present study were to compare the efficiencies for the removal of faecal coliforms from stormwater and to investigate the association of phosphorus and faecal coliforms with different particle size fractions in constructed wetland and detention pond sediments from systems treating urban stormwater.

Methods

Study sites

The two stormwater treatment systems studied are located in close proximity to each other and in similar catchments, thus providing a unique opportunity for comparison of two systems. Plumpton Park and Woodcroft Estate are two recently established residential developments approximately 40 km northwest of Sydney, New South Wales, Australia

(Figure 1). During storm events the catchments discharge large volumes of stormwater with high suspended solids and nutrient concentrations (Hunter and Claus, 1995).

The 0.45 ha constructed wetland system at Plumpton Park is located within a 75 ha residential catchment. It consists of a gross pollutant, a trashrack, and a linear horizontal flow wetland planted extensively with *Phragmites australis* (Figure 2a). The wetland consists of five cells, each approximately 40 m long separated by loose rock weirs 400 mm high. The minimum and maximum water depths are 200 mm and 600 mm, respectively. Stormwater enters the system via two inlets (PI1 and PI2) and there is a single outlet (PO). Sampling locations for inflow and outflow samples, and for sediment samples are indicated in Figure 2a.

The 1.5 ha detention pond system at Woodcroft Estate (Figure 2b) is located within a 53 ha residential catchment. The storage volume of the pond ranges from a minimum of 23 ML to 39 ML and consists of three cells of approximately 2.5 m in depth with an intervening depth of 1 m. Emergent *Typha* sp. are present around the periphery of the pond. The pond has a single inlet (WI), and a single outlet (WO). The soil landscape for each of the systems is typified by hard setting clays that are slightly saline and acidic with occurrences of soil which has a high potential for erosion along the watercourses (Hunter and Constandopoulos, 1997).

Sampling

Discrete inflow and outflow water samples were collected weekly in sterile containers from Plumpton Park wetland and Woodcroft pond. A box dredge sampler was used to collect sediment samples from each system (PP1, PP2, PP3, WC1, WC2, WC3 in Figure 2).

Pollutant analysis

Concentrations of presumptive faecal coliforms in the water and sediment samples, and total phosphorus and nitrogen in the sediment samples were determined according to standard methods (*Standard Methods for the Examination of Water and Wastewater*, 1998)

Sediment particle sizing and settlement

The particle size distribution of three sediment samples for each system was determined in duplicate using the pipette method. Settling times for particles <2, 2–5, 5–10, 10–20, 20–62, and >62 μm were 0 m, 26 s, 4 m 10 s, 16 m 40 s, 68 m 40 s, 416 m 40 s, respectively.

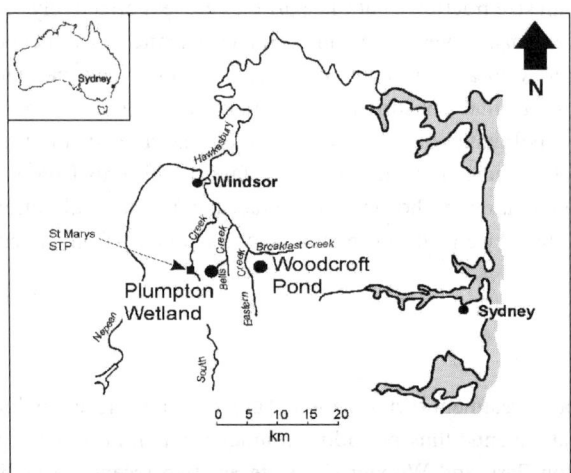

Figure 1 Location of stormwater treatment systems studied

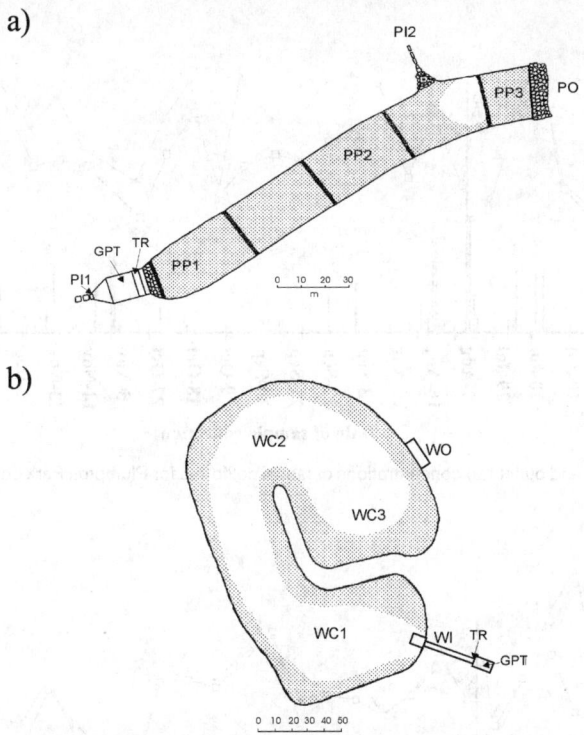

Figure 2 Schematic plan of a) Plumpton Park wetland and b) Woodcroft Estate detention pond systems with sampling locations and vegetation (shaded areas). TR trashrack, GPT gross pollutant trap, PI1 wetland main inlet, PI2 wetland secondary inlet, PO wetland outlet, WI pond inlet, WO pond outlet

The sediments (100 g) were mixed with distilled water and the suspensions allowed to settle in 1-l cylinders. At the determined settling times, 25 ml of sediment suspension was removed from a depth 10 cm below the surface and dried at 105°C for 24 h in a preweighed crucible. Simultaneously, the concentrations of faecal remaining suspended in the top 10 cm were determined from an additional subsample at each of the sampling times. The dried fractions were analysed for total phosphorus and nitrogen.

Data analysis
Analysis of variance was performed using Minitab Release 7.1 Data Analysis Software (Minitab Inc., State College, Pennsylvania, USA).

Results and discussion
Figures 3 and 4 show inflow and outflow faecal coliform concentrations for each of the two systems over a six-month period. The mean removal efficiency for faecal coliforms was 79% for the constructed wetland. The wetland was more effective at removing faecal coliforms than the pond. Indeed the outflow concentrations of faecal coliforms for the pond frequently exceeded inflow concentrations. The mean removal efficiency for faecal coliforms by the pond was −2.5%.

Figure 5 shows the particle size distributions for sediments collected at three different points in each system. The pond sediments contained significantly higher proportions of particles that were <2 μm and 2–5 μm in size than the wetland sediments ($P < 0.05$), whereas the wetland sediments contained significantly higher proportions of particles that were >62 μm in size ($P < 0.05$).

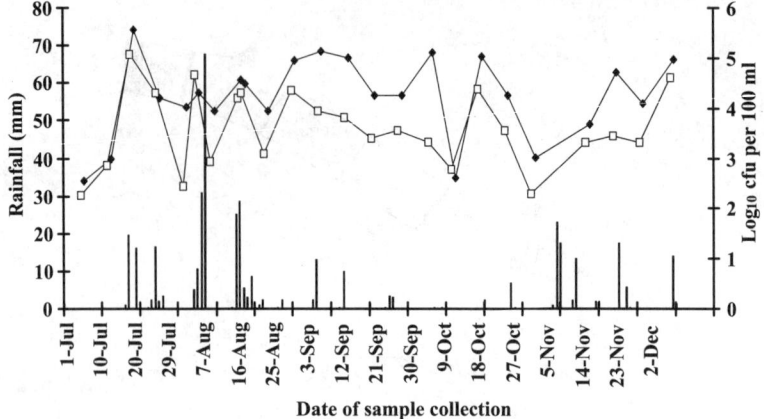

Figure 3 Inlet (♦) and outlet (□) concentrations of faecal coliforms for Plumpton Park constructed wetland, and rainfall (ι)

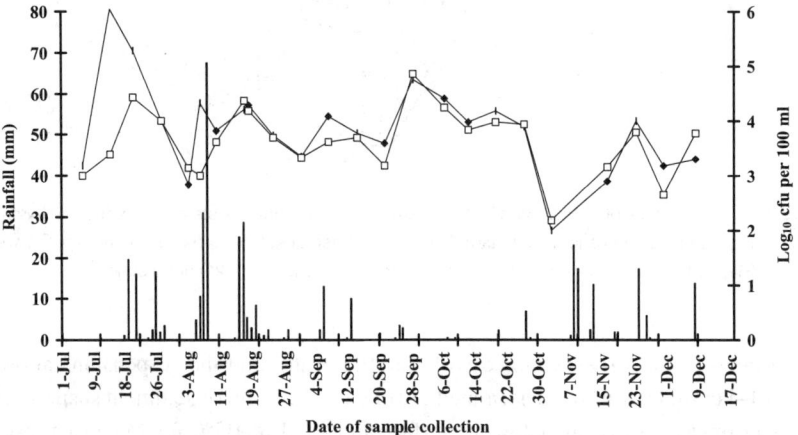

Figure 4 Inlet (♦) and outlet (□) concentrations of faecal coliforms for Woodcroft Estate detention pond, and rainfall (ι)

The difference in sediment particle size distribution in the two systems is most likely due to the different particle size inputs. Residential development within the wetland catchment has been established for several years and the soil has been stabilized to some extent. In contrast, construction work in the catchment of the detention pond was still in progress at the time of the study and consequently there were large areas of disturbed and exposed clay,

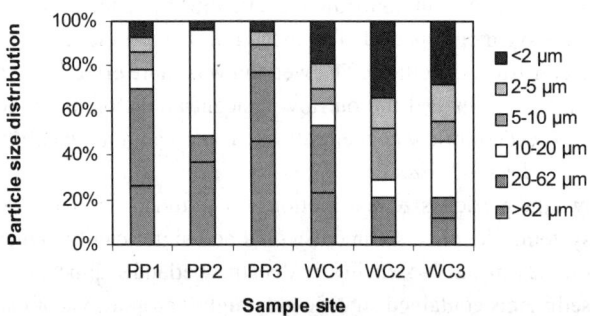

Figure 5 Particle size distribution for wetland and pond sediments

which may be easily mobilised and transported in stormwater. However, particle size determinations on the stormwater inputs to each system are required in order to determine whether the input of clay particles to the pond system was greater than for the wetland system.

Figure 6 shows that the concentrations of faecal coliforms present in the top 10 cm of six sediment suspensions remained relatively constant with time. This suggests that the bacteria were almost exclusively associated with the smaller particles (<2 μm) that remained suspended throughout the duration of the settling experiment, and not attached to the larger particles that settled out within the duration.

Figure 7 shows that that the majority of both phosphorus and nitrogen in the pond and wetlands sediments are associated with the <2 μm size fraction

Wetland design in Australia is currently commonly based on relationships derived by Lawrence (1986) using data for a range of lakes and ponds. Figure 8 shows the retention of total phosphorus by Plumpton Park constructed wetland (from Hunter and Constandopoulos, 1997) and the retention of total phosphorus by other wetland and pond systems (Lawrence, 1986). The curves suggest that there is greater retention of phosphorus by constructed wetlands in catchments where soils are sandy compared to catchments such as Plumpton Park where soil is more dispersible. However, both of the wetland systems shown performed better than the pond system.

The study presented here is ongoing with event sampling combined with flow measurement planned. In addition, the two systems are being studied for removal of viruses from stormwater using bacteriophages.

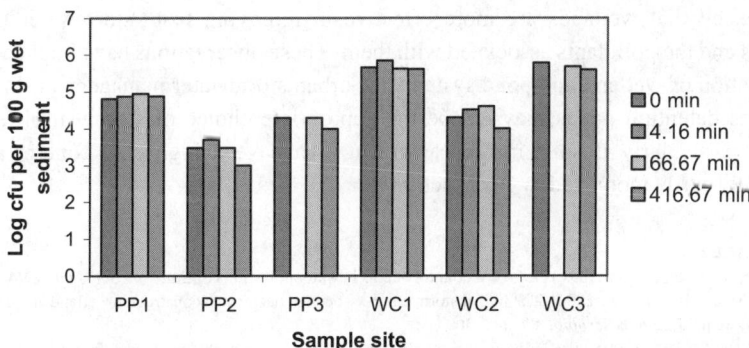

Figure 6 Concentrations of faecal coliforms suspended over time in the top 10 cm of sediment suspensions

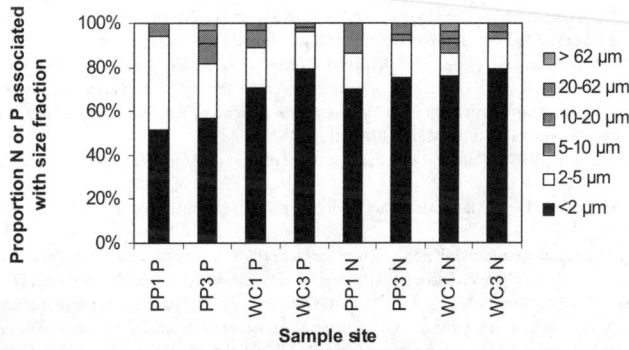

Figure 7 Concentrations of phosphorus and nitrogen associated with the different particle size fractions

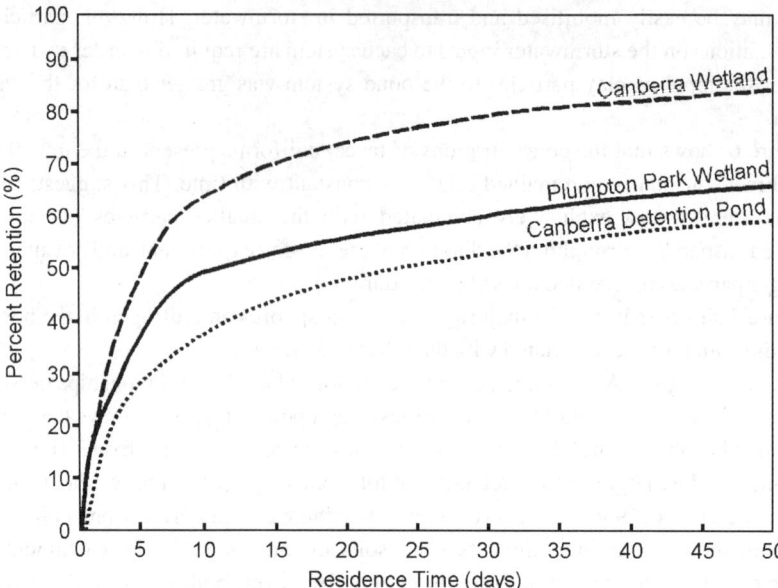

Figure 8 Retention of total phosphorus by Plumpton Park wetland (from Hunter and Constandopoulos, 1997) and relationship derived by Lawrence (1986)

Conclusions

Faecal coliforms, nitrogen and phosphorus in stormwater were largely associated with the fine particle size fraction ($< 2 \mu$m), which settles less easily than larger sized particles. It is suggested that wetlands are more effective in removing from stormwater the fine particles and the pollutants associated with them. These observations have implications in the selection of wetland and pond systems for urban stormwater management. In certain situations detention ponds may not be the appropriate choice of stormwater treatment system, particularly if soils in the catchment area have a high clay content and are potentially easily mobilised by storm activity.

References

Bavor, H.J. and Mitchell, D.S. (eds.) (1994). Wetland systems in water pollution control. *Wat. Sci. Tech.* **29**(4).
Davies, C.M. and Bavor, H.J. (2000). The fate of stormwater-associated bacteria in constructed wetland and detention pond systems. *J. Appl. Microbiol.* **89**, 349–360.
Hunter, G. and Claus, E. (1995). Preliminary water quality results from a constructed wetland at Plumpton Park Blacktown NSW. *Proceedings of the National Conference on Wetlands for Water Quality Control*. 25–29 September 1995. James Cook University, Townsville, QLD, Australia, pp. 265–274.
Hunter, G. and Constandopoulos, J. (1997). Development of pollutant load retention curves for a stormwater treatment wetland at Plumpton Park, Blacktown, Sydney. *Proceedings of the Geographical Society of New South Wales Conference* – Science and Technology in the Environmental Management of the Hawkesbury-Nepean Catchment. 10–11 July 1997. University of Western Sydney – Nepean, NSW, Australia, pp. 79–87.
Lawrence, A.I. (1986). Source and fate of urban runoff constituents and their Management. *Water Research Foundation of Australia, 12th Symposium on Stormwater Quality in Urban Areas*, Wollongong, Australia. 11 July 1986.
Mann, R. and Hammerschmid, K. (1989). Physical and chemical characteristics of urban runoff from two catchments in the Hawkesbury/Nepean basin. *Proceedings of the Australian Water and Wastewater Association 13th Federal Convention*, Canberra, Australia. 6–10 March 1989, pp. 121–125.
Palmer, R.G. and Troeh, F.R. (1995). *Introductory Soil Science Laboratory Manual*. 3rd edn, Oxford University Press, New York.
Pitt, R.E. and Amy, G. (1973). Toxic Materials Analysis of Street Contaminants, Report EPA-R2-73-283, USEPA, Washington DC, USA.
Standard Methods for the Examination of Water and Wastewater (1998). 20th edn, American Public Health Association/American Water Works Association/Water Environment Federation, Washington DC, USA.
Wong, T.F.H., Breen, P.F. and Somes, N.L.G. (1999). Ponds vs wetlands – performance considerations in stormwater quality management. *Proceedings of the Comprehensive Stormwater and Aquatic Ecosystem Management First South Pacific Conference*, vol 2. Auckland, New Zealand. 22–26 February 1999, pp. 223–231.

An experimental constructed wetland system for the treatment of highway runoff in the UK

R.B.E. Shutes*, D.M. Revitt*, L.N.L. Scholes*, M. Forshaw** and B. Winter**

* Urban Pollution Research Centre, Middlesex University, Bounds Green Road, London N11 2NQ, UK
** Environment Agency, Howbery Park, Wallingford, OX10 8BD, UK

Abstract This paper compares the performance of an experimental highway runoff treatment system, incorporating a subsurface flow constructed wetland, with that of a vegetated balancing pond. Both systems are located on the same major road in the UK which opened in November 1998. Copper, chromium and nickel total aqueous metal concentrations, although low, have been consistently removed (maximum efficiencies of 67.3, 69.8 and 87.0% respectively in the constructed wetland), particularly in the summer. Zinc showed the highest aqueous metal concentrations and the generally positive removal by the constructed wetland system (maximum efficiency of 60.6%) correlates with the expected metal uptake by *Typha latifolia* and *Phragmites australis*. Sediment levels for these metals and for lead progressively decreased over the one year monitoring period. For the two storm events monitored in the constructed wetland system, all metals showed evidence of removal (removal efficiencies of 24.2 to 99.4%) except for copper. Lead and cadmium demonstrated the highest removal efficiencies followed by nickel and zinc. For both storms, the wetland acted as a source of copper (removal efficiencies of –88.4 to –97.1%), which may be explained by the die-back of aquatic plants and consequent release of organically bound copper.
Keywords Constructed wetland; highway runoff; metals; *Phragmites australis*; *Typha latifolia*

Introduction

The use of constructed wetlands for the treatment of highway runoff is a relatively new technology in the UK although it has been established in the US for several years (Kadlec and Knight, 1996). The focus on fulfilling the objectives of Agenda 21 in Europe and the UK has encouraged the adoption of wetlands as a method of sustainable environmental management (UNCED, 1992). Initial recommendations for the incorporation of constructed wetlands into highway runoff treatment systems advocate the use of pre-treatment structures including oil separators, silt traps and settlement basins (CIRIA, 1994).

The Environment Agency for England and Wales (EA) is responsible for assessing the quality of rivers and streams. Schemes or mechanisms which have been introduced for the reduction of poor water quality will be monitored by the EA. The Agency is assessing methods for improving surface water management in terms of best management practices and plans to work in partnership with the UK Highways Agency to achieve this aim. Among the treatment options currently being considered are constructed wetlands and the development of criteria for their selection and design (Shutes *et al.*, 1999).

Although the performance of constructed wetland systems for the treatment of urban runoff has been studied (Scholes *et al.*, 1998, 1999; Revitt *et al.*, 1999), there have been no specific studies of highway runoff in the UK. A year long monitoring programme of a purpose-built constructed wetland and a more limited monitoring programme of a vegetated balancing pond was carried out between December 1998 and December 1999. A comparison of the two would indicate the degree of additional treatment provided by the subsurface flow system. The results of the study will be used to improve the design and operation of new and retrofitted constructed wetlands within conventional highway runoff treatment systems. A guidance manual will also be published for EA staff and environmental consultants to assist the selection and design of these systems.

Methods
Site description
A horizontal sub-surface flow system (known as Pond F/G) adjacent to the A34 Newbury Bypass (opened in November 1998) was retrofitted into a vegetated balancing pond preceded by an oil separator, silt trap, grass filter and settling pond. Thus, although Pond F/G is a sub-surface flow system, the size and layout of the pond means that storm flows will over-top the substrate and therefore turn the system from sub-surface flow to surface flow part way through an intense storm event. The constructed wetland within Pond F/G was planted with an initial section of *Phragmites australis* followed by *Typha latifolia*. A vegetated balancing pond (Pond B), which also contains an oil separator and silt trap but has no preliminary settling basin, was adopted as a control pond. The slope and depth of Pond B was graded and it was planted with a mixture of floating, emergent and submergent species.

Sampling programme
Water samples were collected from two sites on 20 occasions within the constructed wetland treatment system (Pond F/G). The sites were located at the wetland inlet in the silt trap, at the wetland outlet, at the base and at a water height of 30cm in two aeration pipes in the *Typha* section of the subsurface flow bed. Samples were also collected from the inlet and outlet of the vegetated balancing pond (Pond B) on 12 occasions over a one-year period. Two storm events were sampled at Pond F/G. In each case, water samples were collected manually at the inlet over intervals of 15–30 minutes during the extent of the storm event whereas outlet samples were collected automatically over a period of up to 54 hours after the commencement of the rainfall. The water samples were treated with nitric acid and duplicate samples were analysed for lead, copper, cadmium, chromium and nickel by Graphite Furnace Atomic Absorption Spectroscopy and for zinc by Inductively Coupled Plasma Atomic Emission Spectroscopy.

Plant and sediment samples were collected on 4 occasions, representing annual seasonal trends, from Pond F/G only. The plant sampling involved the collection of a single plant of *Typha latifolia* and *Phragmites australis* from the boundary area between the two species in the middle of the constructed wetland bed. Plant samples were washed thoroughly with tap water and separated into two parts (roots and rhizomes, leaves) and dried at 100°C for 24 hours. After grinding they were digested using concentrated nitric acid. Duplicate samples were analysed for lead, zinc, copper, cadmium, chromium, nickel, vanadium, and molybdenum by Inductively Coupled Plasma Atomic Emission Spectroscopy. Surface sediment samples were obtained from areas adjacent to the collected plants and were dried in an oven at 100°C for 24 hours, sieved to the fraction <250 μm and digested with aqua regia. Duplicate samples were analysed for lead, zinc, copper, cadmium, chromium, nickel, vanadium, molybdenum, platinum and palladium by Inductively Coupled Plasma Atomic Emission Spectroscopy.

Results and discussion
Water metal concentrations
The measured total aqueous metal concentrations were all lower than those that would be expected in the runoff from a road with the traffic density associated with the Newbury Bypass. This is probably caused by the relative newness of the porous asphalt road surface which still retains the ability to preferentially adsorb metal ions rather than allowing them

to be transported within drainage waters. Copper has a high affinity for organic components and therefore it might be expected that the concentrations of this metal would be highest in the autumn/winter samples due to the presence of decomposing plant tissue and that the removal performances within the ponds would be most efficient in spring/summer due to uptake by growing plants. There is evidence for the latter but not the former. Thus in the spring, Pond F/G exhibited a maximum copper removal efficiency of 67.3%. In contrast, Pond B copper concentrations increased at the outlet in all corresponding samples to a maximum of 163%. Therefore, it would appear that for Pond B, the increase in concentrations within the pond as a consequence of late release of copper ions from decomposing plant tissue was exceeding the rate of uptake of copper by new plant growth. In the summer months, both ponds demonstrated the ability to remove copper, but it was noticeable that Pond F/G functioned most efficiently (62% removal) when copper inlet concentrations were higher (11.2 μg/l).

Cadmium represents an exception compared to the other metals in that the monitored concentrations were equivalent to those expected in road runoff. However, on several occasions late in the monitoring period the metal was not detected. The low concentrations made the reliability of removal efficiencies difficult to authenticate but there was evidence of removal of cadmium in both ponds in January with 83% removal in Pond F/G and 67% removal in Pond B. There was a tendency for cadmium concentrations to decrease at the inlets to both ponds throughout the monitoring period and consequently positive removal efficiencies became more difficult to assess. The measured total aqueous chromium concentrations were lower than expected. No consistent removal trends were discernible at Pond B. In contrast, chromium was consistently removed by Pond F/G from spring to autumn with removal rates of above 60% being measured. The nickel concentrations in the runoff to both ponds were of similar magnitudes and covered the range of 2.8–14.3 μg/l. Consistent removal performances were observed in both ponds with the exception of three occasions. The highest monitored nickel removal efficiencies were in the summer (87% in Pond F/G; 80% in Pond B).

Lead was below the detection limit in many of the collected samples and when detected, concentrations were consistently low. Up to 90% of the lead in road runoff can be associated with particulate matter and the low levels observed may indicate the retention of lead containing particles within the new road structure material. The highest concentrations of lead were found in the sub-surface water samples collected from Pond F/G where suspended solids concentrations were also elevated. A similar trend was occasionally demonstrated by other monitored metals (copper, chromium and zinc) and tended to be most distinctive during low flow conditions when sediment particle metal concentrations might be expected to be at their highest as a result of established equilibrium conditions between the solid and aqueous phases. Because of the large number of non-detected lead concentrations, calculations of removal efficiencies could only be performed for a small number of data sets. Lead removal was shown to occur in Pond F/G on 3 occasions and in Pond B on 2 occasions. The maximum values were 73% in Pond F/G and 7% in Pond B.

The zinc concentrations were the highest of all the determined metal concentrations throughout the monitoring programme and this is consistent with previous studies of the behaviour of this metal in highway runoff. Pond B showed consistent removal of zinc between inlet and outlet on all sampling visits except three. An increase of 86% between inlet and outlet collection points in June coincided with the construction of galvanised fencing near to Pond B. The behaviour of Pond F/G towards zinc removal was initially intermittent but during the summer months there was clear evidence that positive removal had commenced. The monitored removal efficiencies on individual sampling dates reached a maximum value of 60.6%. The generally positive removal correlates with the expected metal uptake by plants during the summer months although there are still deviations from the expected overall trends in both ponds.

Metals in plant tissue

Zinc demonstrated the highest overall ability to be accumulated in all tissues of both *Phragmites* and *Typha* and this is consistent with its higher aqueous concentrations. Zinc, nickel and copper root/rhizome concentrations in *Phragmites* were higher in the autumn than the summer by a factor of two although this trend was not observed in *Typha*. The data suggest that the metals are accumulating in the root/rhizome area of *Phragmites* as the leaf ceases to grow and to provide a transport site for these metals from the roots and rhizome. Furthermore, the root biomass will be reduced in the autumn and this will tend to increase the metal concentrations. A similar process of metal concentration in the root/rhizome may occur at a later date in *Typha*. Lead concentrations remained low in both plant root/rhizome and leaf tissues reflecting the low concentration of bioavailable lead ions in the rhizosphere. The metals which continued to be consistently found at very low concentrations in plant tissues throughout the monitoring period were molybdenum, chromium, platinum, cadmium and palladium. The surprising element was vanadium which was consistently found at significant concentrations in both *Phragmites* and *Typha* tissue. This metal demonstrated a preference for bioaccumulation in *Typha* tissue (particularly the roots/rhizome) where it exceeded the levels found for lead which also showed a slight preference for the below ground tissue in this species.

Phragmites australis showed good growth in both ponds during the first year at Pond F/G. *Phragmites* grows well in brackish waters and therefore is suited to being located at the front end of Pond F/G where chloride concentrations from de-icing salts are high. The *Typha* bed at Pond F/G showed a less dense growth than *Phragmites* and evidence of recolonisation of this area by grasses and other weeds, including thistles, towards the end of the routine monitoring period. Yellowing of the leaves and fruit bearing stems in *Typha* was observed in the summer. A possible cause of this die-back is iron deficiency due to a limited source of this trace element in the wetland substrate, which is dominated by gravel. *Phragmites*, which is the initial plant, may be removing the iron. The recommended treatment for *Typha* would be to spray the leaves with a 2.5% ferrous sulphate solution. Hot and dry conditions may also be the cause of the observed die-back. The deeper root system of *Phragmites* enables it to extract sufficient water from the lower levels in the bed whereas *Typha* may not be able to do this so efficiently. However subsequent wet weather conditions did not rectify the condition of the *Typha* bed and it is possible that the combination of the iron deficiency resulted in stress levels from which *Typha* was unable to recover. Because of the weaker condition of the *Typha* bed it was more prone to invasion by grasses and other weeds.

Metals in sediment

It was not possible to detect either palladium or molybdenum in any of the sediment samples. The general temporal trend in sedimentary metal (chromium, copper, lead, nickel, zinc) concentrations was a decrease from the commencement of the sampling programme in winter through to the following autumn. The apparent progressive transfer of these metals from the sediment to plant tissue during the sampling year is encouraging as it indicates the ability of increasing plant biomass (particularly during the spring and summer seasons) to remove metals from the sediment. Platinum and cadmium concentrations showed a small increase in the autumn compared to the summer samples although the overall yearly trend still indicated a decrease. Although there was no evidence for significant uptake of platinum in either of the plant tissues, its presence in the sediment indicates the need for concern with regard to future environmental contamination by this metal. Vanadium was not detected in either the summer or autumn samples having been present at concentrations of 16.4 µg/g and 9.5 µg/g in the sediment samples collected in winter and spring, respectively. Sediment lead concentrations remained remarkably constant throughout the monitoring period which can probably be explained by the high affinity of lead for the particulate phase in comparison to the more bioavailable aqueous phase. The major contributing factors to the overall reduction in sediment metal levels in Pond F/G, during the monitoring period, with the exception of lead, were metal uptake by the two plant species and pre-treatment in the silt trap and grass filter zone.

Storm event monitoring

The first monitored storm event (Storm 1) was initiated by a rainfall volume of 2.8 mm occurring over a short period of approximately 30 minutes. Prior to the storm event, the flow at the outlet was 1.2 l/s and immediately after the storm this had increased to 8.8 l/s. Storm runoff commenced almost immediately after rainfall started and the peak runoff flow rate was monitored after 30 minutes. Figure 1 shows the outlet chemograph for zinc and it can be seen that there was an elevation of zinc loadings over the 24 to 54 hour period following the commencement of the storm event. This would be the expected pattern as the pollutant, in this case zinc, passed through the wetland system. Similar trends were also observed for suspended solids, copper and chromium. The outlet nickel loading pattern was less well defined and in the case of lead and cadmium there was little evidence for removal of pollutant loads over the identified time period although there is evidence for removal from the wetland immediately after the storm event. This is not considered to be related to the metal loads arriving with rainfall.

The calculated inlet and outlet loadings for each of the monitored pollutants are listed in Table 1 together with the corresponding removal efficiencies. The highest removal efficiencies are for lead and cadmium, both of which exhibited loading curves which were not compatible with the expected flow through pattern. In contrast, zinc and suspended solids both showed outlet loading patterns consistent with the expected hydrological behaviour of Pond F/G and therefore the monitored removal efficiencies of 66.2 and 75.0%, respectively are considered to be perfectly valid. The same is true for copper although in the case of this metal there was an increase of 97.1% in the loading leaving the

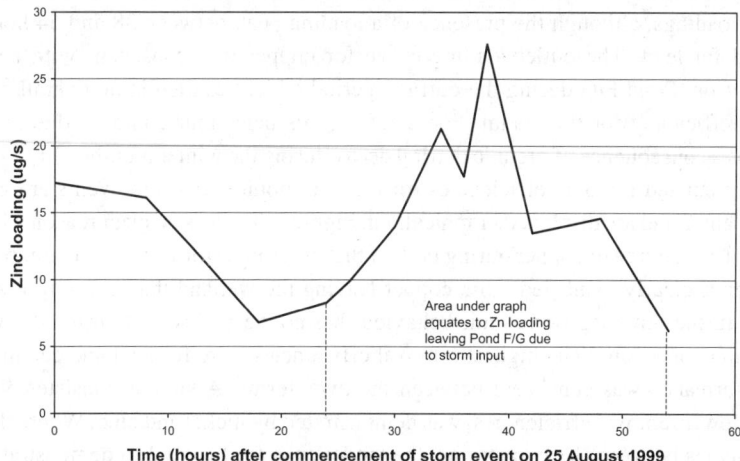

Figure 1 Outlet chemograph for zinc for Storm 1 at Pond F/G

Table 1 Pollutant removal efficiencies in Pond F/G following Storm 1

Parameter	Inlet load	Outlet load	Percentage removal efficiency
Suspended solids	11.3 kg	2.83 kg	75.0
Cadmium	0.114 g	0.014 g	90.3
Chromium	0.338 g	0.174 g	48.5
Copper	1.40 g	2.76 g	−97.1
Nickel	0.796 g	0.179 g	77.5
Lead	1.07 g	0.022 g	97.9
Zinc	4.79 g	1.62 g	66.2

wetland. The reason why Pond F/G acted as a source of copper during this storm event is not certain as copper is normally quite strongly bound to the organic content of soil particles.

In comparison to Storm 1, which was initiated by a short and intense summer rainfall event, Storm 2 occurred following a more prolonged rainfall volume (2.2 mm over a 3 hour period) during December. The distribution of this rainfall was 0.8 mm during the first hour followed by hourly rainfall depths of 0.2 mm and 1.2 mm.

Comparison of the inlet pollutant loadings for the two monitored storms shows reductions in metal loadings for Storm 2 for copper, chromium, nickel and cadmium which except for cadmium, are less exaggerated than those for suspended solids (see Tables 1 and 2). In contrast, zinc, and particularly lead loadings were higher for Storm 2. One explanation for this behaviour may be related to the large increase in chloride loadings for Storm 2 (94.2 kg) compared to Storm 1 (17.4 kg) due to the occurrence of winter road salting practices during December. The presence of chloride in runoff waters is known to have a mobilising effect on metals in terms of their solubilisation and it may be that this is particularly relevant for lead which is normally strongly associated with the particulate phase. This solubilisation effect would be less important for zinc due to its characteristically higher distribution in the dissolved as opposed to particulate phases when compared to lead.

The patterns of outlet metal loadings in Storm 2 are less consistent than that for suspended solids which demonstrated the expected behaviour over the 24 to 54 hour period. However, zinc, nickel and chromium show similar trends with fairly constant levels followed by a sharp elevation towards the end of the 30 hour period. The removal efficiencies of these three metals were 59.7%, 84.8% and 24.2%, respectively. Cadmium and lead both demonstrated very high removal efficiencies (99.4% and 97.6%) due to negligible outflow loadings, although the presence of a loading peak between 38 and 54 hours was observed for lead. The outlet loading curve for copper was consistent with enhanced removal from Pond F/G during the outflow period which resulted in an overall negative removal efficiency for this metal. For Storm 2, this behaviour could be due to copper release as a consequence of organic detrital decay during the winter months.

The calculated removal efficiencies for the two monitored storm events are given in Tables 1 and 2 and are displayed graphically in Figure 2. There is no clear reason why Pond F/G acted as a source of copper during both of the monitored storm events. The experimental evidence clearly identified more copper leaving the wetland than entering it and this requires further investigation as this behaviour has not been observed during dry weather monitoring conditions. The highest removal efficiencies were for lead and cadmium and this performance was consistent between the two storms. A similar consistency, but at slightly lower removal efficiencies, was demonstrated by nickel and zinc. Where there are discrepancies between the two storms, the tendency is for Storm 2 to demonstrate lower removal efficiencies than Storm 1. This can be explained by the less well defined character-

Table 2 Pollutant removal efficiencies demonstrated by Pond F/G following storm 2

Parameter	Inlet load	Outlet load	Percentage removal efficiency
Suspended solids	5.06 kg	3.02 kg	40.3
Cadmium	1.74 mg	0.10 mg	99.4
Chromium	0.120 g	0.091 g	24.2
Copper	0.888 g	1.67 g	−88.4
Nickel	0.656 g	0.100 g	84.8
Lead	4.12 g	0.097 g	97.6
Zinc	5.49 g	2.21 g	59.7

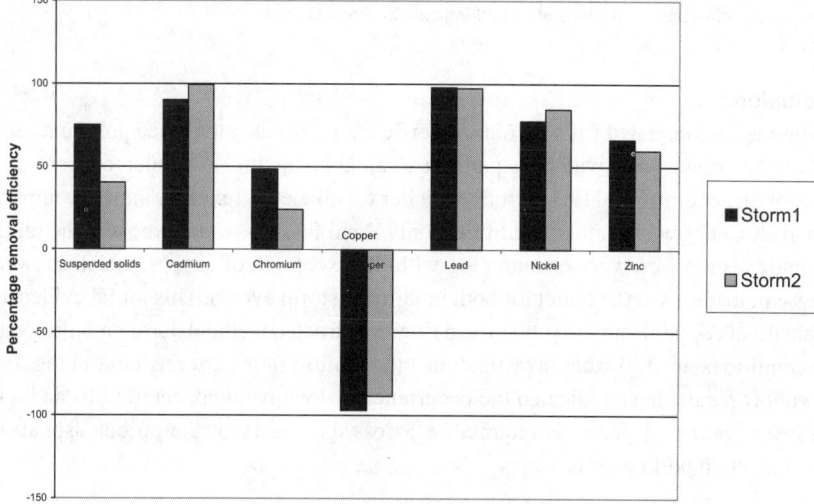

Figure 2 Comparison of the pollutant removal efficiencies for Storms 1 and 2

istics of Storm 2 and the lower flow rates and concentrations which lead to less confidence in the determined loadings and hence removal efficiencies.

Average dry weather pollutant removal efficiencies

Dry weather conditions for both ponds have been defined as those during which negligible rainfall occurred during the 48 hours prior to monitoring. Both chromium and nickel demonstrated consistent positive removal efficiencies by Pond F/G during these conditions (Table 3). Cadmium showed more evidence of removal within Pond F/G compared to Pond B but the behaviour of lead proved to be unpredictable. It is important to note the differing behaviours of copper within Pond F/G depending on whether dry or wet conditions were prevalent. The emphatic contribution of the pond to outflow copper loadings which was observed during the two monitored storm events was not reinforced by the behaviour of copper during dry weather conditions. The metal removal efficiencies for Pond B showed slightly improved removal for zinc compared to Pond F/G but a decreased performance for other metals, except lead, during dry weather conditions. Outlet metal loadings for Pond B were particularly enhanced for chromium and to a lesser extent for copper. Thus the behaviour of copper in Pond B during dry conditions was similar to that in Pond F/G during a storm event. In general, metal removal efficiencies were higher in storm events than in dry weather conditions which supports other studies reported in the literature (Kadlec and Knight, 1996; Mungur et al., 1999).

Table 3 Average pollutant removal efficiencies in Ponds F/G and B during dry weather conditions

Parameter	Percentage removal/enhancement efficiencies	
	Pond F/G[1]	Pond B[2]
Suspended solids	−75.8%	−6.4%
Cadmium	23.8%	8.2%
Chromium	49.8%	−337.5%
Copper	−15.9%	−75.2%
Lead	−9.0%	−0.1%
Nickel	66.4%	51.7%
Zinc	12.1%	29.2%

[1] Average values are based on 7 sets of monitoring data for Pond F/G
[2] Average values are based on 5 sets of monitoring data for Pond B

Conclusions

Both ponds demonstrated variable removal efficiencies for the monitored pollutants during dry weather conditions which is partly explainable by the low inlet concentrations. However, in general, Pond F/G performed better for all metals (except zinc), in comparison to Pond B. Under wet weather conditions, only Pond F/G was monitored and the removal efficiency results were very encouraging with the exception of copper, which showed an increase in loadings at the outlet for both monitored storm events. This initial evidence for the release of copper from the sub-surface flow constructed wetland during monitored high flow conditions needs further investigation and monitoring is currently continuing. Initial road runoff results have indicated the occurrence of low metal concentrations and a long term investigation (5 years) is required to assess the ability of the porous asphalt road surface to retain pollutants as it ages.

Acknowledgements

Funding to support this work was provided by Thames Region of the Environment Agency (EA) with site access facilitated by the Highways Agency. The project was managed by Dr Maxine Forshaw (EA) and Barry Winter (Consultant to the EA) and coordinated by the Halcrow Group Ltd.

References

CIRIA (1994). *Control of pollution from highway drainage discharges.* M. Luker and K. Montague (eds). CIRIA Report No. 142, London.

Kadlec, R.H. and Knight, R.L. (1996). *Treatment Wetlands.* Lewis, Boca Raton, USA.

Mungur, A.S., Shutes, R.B.E., Revitt, D.M., House, M.A. and Fallon, C.M. (1999). A constructed wetland for the treatment of urban runoff. In: *Remediation and Management of Degraded Lands*, M.H. Wong, J.W.C. Wong, and A.J.M. Baker, (eds.), Chapter 32, Ann Arbor Press, Michigan, USA, pp 329–341.

Revitt, M., Shutes, B. and Scholes, L. (1999). The use of constructed wetlands for reducing the impacts of urban surface runoff on receiving water quality. In: *Impacts of Urban Growth on Surface Water and Groundwater Quality*, J.B. Ellis (ed.), Publication No. 259, IAHS Press, Wallingford, UK, pp 349–356.

Scholes, L.N.L., Shutes, R.B.E., Revitt, D.M., Forshaw, M. and Purchase, D. (1998). The treatment of metals in urban runoff by constructed wetlands. *Sci. Tot. Env.*, **214**, 211–219.

Scholes, L.N.L., Shutes, R.B.E., Revitt, D.M., Purchase, D. and Forshaw, M. (1999). The removal of urban pollutants by constructed wetlands during wet weather. *Wat. Sci. Tech.*, **40**(3), 333–340.

Shutes, R.B.E., Revitt, D.M., Lagerberg, I.M. and Barraud, V.C.E. (1999). The design of vegetative constructed wetlands for the treatment of highway runoff. *Sci. Tot. Env.*, **235**, 189–197.

United Nations Committee on the Environment and Development (UNCED). (1992). *Agenda 21.* Conches, Switzerland.

Performance of constructed wetland system for public water supply

J.M. Elias*, E. Salati Filho*, E. Salati**

* Centro de Estudos Ambientais – Universidade do Estado de São Paulo, Av. 24-A, 1515 – Bela Vista, 13506-900 – Rio Claro, SP, Brazil
** Escola de Engenharia de São Carlos – Centro de Recursos Hídricos e Ecologia Aplicada – Universidade de São Paulo, Av. Carlos Botelho, 1465, CEP: 13560-970, São Carlos, São Paulo, Brazil

Abstract The project is being conducted in the town of Analândia, São Paulo, Brazil. The constructed wetlands system for water supply consists of a channel with floating aquatic macrophytes, HDS system (Water Decontamination with Soil – Patent PI 850.3030), chlorinating system, filtering system and distribution. The project objectives include investigating the process variables to further optimize design and operation factors, evaluating the relation of nutrients and plants development, biomass production, shoot development, nutrient cycling and total and fecal coliforms removal, comparing the treatment efficiency among the seasons of the year; and moreover to compare the average values obtained between February and June 1998 (Salati et al., 1998) with the average obtained for the same parameters between March and June 2000. Studies have been developed in order to verify during one year the drinking quality of the water for the following parameters: turbidity, color, pH, dissolved oxygen, total of dissolved solids, COD, chloride, among others, according to the Ministry of Health's Regulation 36. This system of water supply projected to treat 15 L s^{-1} has been in continuous operation for 2 years, it was implemented with support of the National Environment Fund (FNMA), administered by the Center of Environmental Studies (CEA-UNESP), while the technical supervision and design were performed by the Institute of Applied Ecology. The actual research project is being supported by FAPESP.

Keywords Aquatic macrophytes; constructed wetlands; filtering soil; public water supply

Introduction

The municipality of Analândia is located in the central region of São Paulo State, in Brazil. The population is estimated at 3,500 inhabitants, although as a tourist city the total population (fixed and floating) is approximately 6,500 inhabitants. Unfortunately less than 37% of the cities in Brazil have some type of water treatment for public supply (IBGE, 1997).

Most water resources for public supply in Analândia come from underground water, through artesian wells with poor production and are disinfected using the chlorinating method. During the dry seasons or even on days when there is a large number of inhabitants, the water supply is jeopardized.

Since the full installation of the Constructed Wetlands System for Water Supply, in early December 1997, the water production supplies 20% of the drinking water consumption in Analândia.

The constructed wetlands system for water supply consists of a channel with floating aquatic macrophytes, HDS system (Water Decontamination with Soil – Patent PI 850.3030), chlorinating system, filtering system and distribution. The decontaminating action of filtering soils function through their role as a mechanical filter, physico-chemical filter and biological filter (Salati, E. Filho et al., 1994). This system is projected to treat 15 L/s collected from a river (Retiro river) with unpolluted water. The catchment basin used is 7.77 km^2 in area, with average flow of 150 L/s. It has been in continuous operation for 30 months with previous assessment of the efficiency of the system (Salati et al., 1998). The project was implemented with support of the National Environment Fund (FNMA), administered by the Center of Environmental Studies (CEA-UNESP), while the technical supervision and designed were performed by the Institute of Applied Ecology.

Methods

The Analândia water treatment plant is designed to treat 15 L/s of water, collected from the Retiro river with unpolluted water. A dam was built on the river in order to collect the water to be treated. The water is pumped to the plant by two pumps with work capacity of 05 and 10 L/s, respectively. The first stage of the treatment is a channel of aquatic floating macrophytes (*Eichhornia crassipes, Eichhornia azurea and Salvinia auriculata*) with 1,590 m^2 of built area and average depth of 0.80 m. It has an average residence time of 3 days (72 hours). After the water passes through in the channel with floating macrophytes, it flows through the HDS system with filtering soils. The HDS system consists of 2 modules of filtering soils with 1,008 m^2 each, which operate alternately in a down flow vertical direction. The filtering soil consists of a drainage system (crushed stone, gravel and perforated pipes), a layer of gravel and a layer of soil on the top, cultivated with rice (*Oryza sativa* – Var. IAC 44/40). The soil was a mixture of sand-clay soil (red-yellow podzol – 30.5% of clay, 62.5% of sand and 7% of silt), 5% of the mixture in volume of sugar cane fiber and 5% of the mixture in volume of vermiculite.

After passing through the DHS system, the water is disinfected using the conventional chlorinating method with hypochlorite. The treated water is stored in a 100 m^3 tank and at the end of the treatment the water is filtered through a mixed filter of layers of sand. The constructed wetlands system for water supply in Analândia has been operating since December 1997. Figure 1 shows the plant of the system.

Water sample collection

Water samples were collected during the end of summer and autumn seasons (from 13 to 17 March and from 12 to 16 June 2000). Samples were collected at four different sites of the system: Entering flow – P01, Channel out flow – P02, Filtered out flow (after pass through downflow filtering soil) – P03 and Treated water after it passes through the sand filter – P04.

The same analyses of the water parameters, at the same sampling sites had been made by Salati during the Summer and Autumn seasons in the year of 1998.

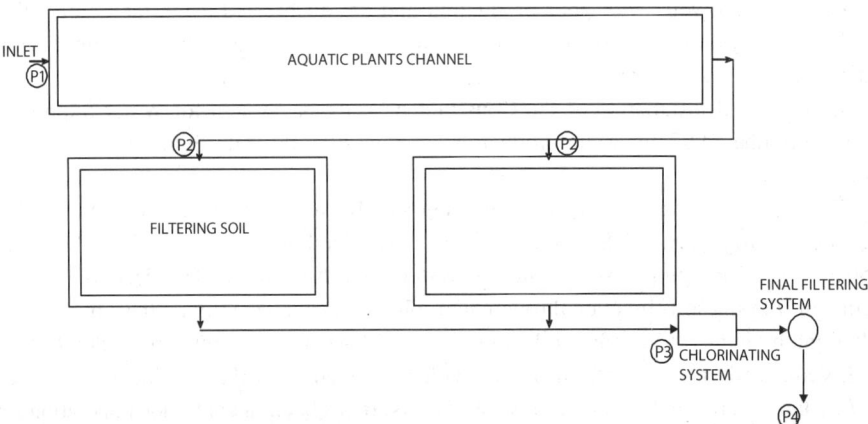

Figure 1 Plant of the constructed wetlands system for water public supply and sampling positions at several stages of treatment – Analândia – SP

Parameters analyzed
- pH
- Biochemical oxygen demand (BOD)
- Chemical oxygen demand (COD)
- Total and fecal coliforms
- Color
- Turbidity
- Nitrogen compounds: nitrite, nitrate, ammonia and total nitrogen
- Total phosphorus
- Inorganic Phosphorus
- Dissolved oxygen
- Chlorine

Macrophyte growth

The aquatic macrophytes present in the system were analyzed from September 1999 to June 2000. The rice growth and development were analyzed during one crop season (150 days), by measuring biometrics parameters such as number of leaves per plant, number of tillers per plant, height of plant, width of plant, dry weight of shoot, dry weight of root, number of panicle per plant, growth rate and productivity.

The experiments with aquatic floating macrophytes were conducted in a floating square (1 × 1 m) with three replications. Growth parameters were measured after 28 days. For each species of plant fresh weight was determined after draining for 5 min, and the dry weight after oven drying at 72°C for 48 h. Total number of daughter plants, number of leaves per plant, root growth and growth rate were recorded. Every week the plant coverage in the channel was measured and plant samples were collected to determine the plant density, the dry weight and the growth rate of the plants.

Results and discussion

Water analysis

Since its installation in mid 1997, the Constructed Wetlands System for Public Water Supply in Analândia has been monitored and several water analyses have been made by Center of Environment Studies (CEA-UNESP), Analândia municipal council, FAPESP and Institute of Applied Ecology and all the obtained results compared with all parameters in the GM 36 Ministry of Health Regulation referring to potable water.

The present work shows the comparison between the results obtained by Salati (1998) at the end of Summer and Autumn of 1998 and the results obtained for the same seasons in 2000 in a project supported by FAPESP.

Table 1 shows the average results obtained for the parameters under study analyzed during the same period in 1998 and 2000.

Table 2 shows the average efficiencies obtained for the different stages of the treatment during the experimental periods in the different years.

The obtained results show that the desirable quality of water has not been changed during 2 years of operation and monitoring. The obtained efficiencies for most parameters under study, at different stages of treatment didn't vary significantly.

Table 3 shows the average efficiencies obtained for the several parameters under study, at different stages of treatment, comparing the average efficiencies obtained in the Summer and in the Autumn of 2000.

Fecal coli present in the water has average removal efficiency over 90% after it passes through the HDS system in both years of studies. After passing through the HDS system, the water is disinfected in the chlorinating system with hypochlorite and filtered at the end

Table 1 Average results obtained in the Summer and Autumn of 1998 compared with the average results obtained at the same seasons in 2000, at the constructed wetlands system for water supply – Analândia – SP

Sampling sites Parameters	P01	P02	P03	P04	P01	P02	P03	P04	Units
	Summer and Autumn of 1998				Summer and Autumn of 2000				
Total dissolved solids	8.50	7.25	13.5	18.8	5.2	5.4	7.3	15.8	mg/L
Turbidity	35.0	33.2	1.96	1.55	5.14	5.65	1.05	0.81	UNT
Color	178	124	139	97	49	54	16	20	Pt/Co
Dissolved oxygen	11.7	10.2	7.00	8.39	6.12	6.08	3.34	4.21	mg/L
BOD_5	7.75	8.15	7.50	6.30	4.17	3.93	3.47	1.43	mg/L
COD	20.3	4.75	8.75	26.3	2.90	2.25	2.83	1.63	mg/L
Nitrate	0.18	0.16	0.49	0.16	0.31	0.21	0.08	0.26	mg/L
Nitrite	0.011	0.018	0.004	0.006	0.003	0.004	0.001	0.004	mg/L
Ammonia	0.310	0.330	0.040	0.040	0.097	0.106	0.035	0.017	mg/L
Total-N	0.21	0.18	0.18	0.14	1.60	1.40	1.00	0.90	mg/L
Total-P	0.044	0.040	0.032	0.020	0.019	0.023	0.027	0.022	mg/L
Inorganic – phosphorus	0.007	0.010	0.019	0.022	0.010	0.007	0.012	0.014	mg/L
Fecal coli	929	153	12	00	605	85	13	00	MNP/100 ml
Total coli	2685	2416	984	00	7140	4006	438	00	MNP/100 ml

P01 – Entering flow
P02 – Channel out flow
P03 – Filtered out flow (after pass through downflow filtering soil)
P04 – Treated water after it passes through the sand filter

Table 2 Average efficiency obtained in the Summer and Autumn of 1998 compared with the average efficiency obtained at the same seasons in 2000 of the constructed wetlands system for water supply – Analândia – SP

Sampling Sites Parameters	Efficiency (%) Summer and Autumn of 1998				Efficiency (%) Summer and Autumn of 2000			
	Partial P01-P02	Partial P02-P03	Partial P03-P04	Total P01-P04	Partial P01-P02	Partial P02-P03	Partial P03-P04	Total P01-P04
Total dissolved solids	14.7	−86.2	−38.9	**−121**	−3.85	−35.2	−116	**−204**
Turbidity	5.2	94.1	21.0	**95.6**	−10.0	81.4	22.5	**84.2**
Color	30.5	−12.4	30.2	**45.6**	−10.1	70.6	−26.2	**59.1**
BOD-5	−5.2	8.0	16.0	**18.7**	5.8	11.7	58.8	**65.7**
Nitrate	10.4	−201	67.2	**11.4**	32.3	61.9	−225	**16.1**
Nitrite	−74.3	80.9	−57.1	**47.6**	−27.6	73.0	−250	**−20.7**
Ammonia	−6.5	−21.2	00	**87.0**	−9.3	67.0	51.4	**82.5**
Total-N	14.3	00	22.2	**33.3**	12.5	28.6	10.0	**43.8**
Total-P	9.6	19.6	37.8	**54.8**	−22.6	−14.6	19.1	**−13.7**
Inorganic phosphorus	−55.4	−90.1	−15.6	**−242**	27.7	−65.8	−14.1	**−36.6**
Fecal coli	83.6	92.3	100	**100**	86.0	85.12	100	**100**
Total coli	10.0	59.3	100	**100**	43.9	89.0	100	**100**

P01 – Entering flow
P02 – Channel out flow
P03 – Filtered out flow (after pass through downflow filtering soil)
P04 – Treated water after it passes through the sand filter

of the system in a sand filter under pressure, becoming 100% potable according to the Ministry of Health's Regulation 36.

Macrophytes growth rate

The results were obtained during the Summer and Autumn of 2000 at the Constructed Wetlands System for Water Supply in Analândia. The experiments have shown that water

Table 3 Comparison of the average efficiencies obtained, at different stages of treatment in the Summer and Autumn of 2000 of the constructed wetlands system for water supply – Analândia – SP

Sampling sites Parameters	Efficiency (%) Summer of 2000				Efficiency (%) Autumn of 2000			
	Partial P01-P02	Partial P02-P03	Partial P03-P04	Total P01-P04	Partial P01-P02	Partial P02-P03	Partial P03-P04	Total P01-P04
Total dissolved solids	00	−116	−104	**−340**	−7.4	−13.8	−45.5	**−77.8**
Turbidity	−11.3	87.7	38.0	**91.5**	−7.3	69.2	10.26	**70.3**
Color	−8.9	68.1	−34.2	**53.4**	−12.8	75.6	−5.8	**70.9**
Nitrate	29.7	76.9	−333	**29.7**	40	33.3	−150	**00**
Ammonia	−16.4	66.9	42.5	**77.9**	−1.1	67.0	63.3	**87.8**
Total-N	00	38.9	−13.6	**30.6**	24.2	13.0	33.3	**56.0**
Total-P	−45.7	−25.5	39.3	**−11.0**	−5.1	−2.2	−8.2	**−16.3**
Inorganic phosphorus	36.6	−75.4	−17	**−30**	20.72	−61.4	−11.3	**−42.3**
Fecal coli	82.4	99.2	100	**100**	91.16	45.34	100	**100**
Total coli	12.61	99.0	100	**100**	71.01	63.22	100	**100**

P01 – Entering flow
P02 – Channel out flow
P03 – Filtered out flow (after pass through downflow filtering soil)
P04 – Treated water after it passes through the sand filter

hyacinth plants are not developing in the channel, because the collected water has a low concentration of nutrients (N and P). Otherwise, the floating aquatic macrophyte *Salvinia auriculata* has shown an adaptation to the water in the channel and a reasonable growth rate ranged from 9.5 to 12.5 g dry weight/m^2/day. For the HDS system the rice crop receives a soft fertilization to add a minimum nutrients supply to the development of the crop. The obtained rice productivity of the HDS system was 3,000 kg/ha.

Conclusions

The Constructed Wetlands System for Public Water Supply installed in the town of Analândia in early December 1997 has been operating since then achieving the desirable quality of water for the public supply, according with the GM 36 Ministry of Health Regulation referring to potable water. As it is a biological treatment system, the efficiency varies during the year.

The results obtained show that the efficiencies, at different stages of treatment in the Summer and Autumn of 2000 can vary according with the level of development of the plants present in the system and climatic conditions, as like the collected water can change its characteristics along the seasons of the year. The actual research project will be concluded in the Summer of 2001, comparing the obtained efficiencies during all the seasons of the year.

The water quality at Retiro river has maintained appropriate to be treated for public water supply, according with CONAMA/20, as a Class 1 river.

This water purification process may be used for public water supply and also to produce water for industrial purposes, specially when the quality of the untreated water is reasonable, as it in a Class 1 or 2 river (CONAMA/20).

References

Salati, E. (1987). Edaphic-phytodepuration: a new approach to wastewater treatment. In: K.R. Reddy and W.H. Smith, *Aquatic plants for water treatment and resource recovery*. Orlando, Florida, 199–208.

Salati, E., Salati, E., Tornisielo, S.M.T. and Salati, E. (1998). Public Water Supply Using Constructed

Wetlands Systems In: *6th International Conference on Wetland Systems for Water Pollution Control* – Águas de São Pedro, SP Brazil, 1998. Proccedings, p432–436.

IBGE – Recursos naturais e meio ambiente: uma visão do Brasil/IBGE, Departamento de Recursos Naturais e Estudos Ambientais. 2° ed. Rio de Janeiro: IBGE, 1997.

Application of a constructed wetland for non-point source pollution control

C.M. Kao*, J.Y. Wang**, H.Y. Lee* and C.K. Wen***

* Institute of Environmental Engineering, National Sun Yat-Sen University, Kaohsiung, Chinese Taiwan
** School of Civil and Structural Engineering, Nanyang Technological University, Singapore
*** Department of Environmental Engineering, National Chen-Kung University, Tainan, Chinese Taiwan

Abstract In Taiwan, non-point source (NPS) pollution is one of the major causes of impairment of surface waters. The main objective of this study was to evaluate the efficacy of using constructed wetlands on NPS pollutant removal and water quality improvements. A field-scale constructed wetland system was built inside the campus of National Sun Yat-Sen University (located in southern Taiwan) to remove (1) NPS pollutants due to the stormwater runoff, and (2) part of the untreated wastewater from school drains. The constructed wetland was 40 m (L) × 30 m (W) × 1 m (D), which received approximately 85 m^3 per day of untreated wastewater from school drainage pipes. The plants grown on the wetland included floating (*Pistia stratiotes* L.) and emergent (*Phragmites communis* L.) species. One major storm event and baseline water quality samples were analyzed during the monitoring period. Analytical results indicate that the constructed wetland removed a significant amount of NPS pollutants and wastewater constituents. More than 88% of nitrogen, 81% of chemical oxygen demand (COD), 85% of heavy metals, and 60% of the total suspended solids (TSS) caused by the storm runoff were removed by the wetland system before discharging. Results from this study may be applied to the design of constructed wetlands for NPS pollution control and water quality improvement.
Keywords Constructed wetland; non-point source pollutants; wastewater treatment; wetland

Introduction

Wetlands are an essential part of stormwater management systems. Important wetland functions include conveyance and storage of stormwater, which dampens effects of flooding; reduction of flood flows and velocity of stormwater; which reduces erosion and increases sedimentation; and modification of pollutants typically carried in stormwater (Steven, 1998; Kao and Wu, 1999). Accordingly, there is a great amount of interest in the incorporation of constructed wetlands into stormwater management systems. This concept provides an opportunity to use the wetland systems to mitigate effects of runoff associated with urbanization.

The US Environmental Protection Agency estimated that non-point source (NPS) pollution contributes over 65% of the total pollution load to US inland surface waters (US EPA, 1990). In Taiwan, the Taiwan Environmental Protection Administration reported that NPS pollution contributes more than 20% of the overall pollution load to Taiwan surface waters. NPS mainly associated with stormwater runoff from urban and agricultural land uses can be quite diffuse and difficult to treat. Nutrients, pesticides, and sediments are the main detrimental NPS constituents.

National Sun Yat-Sen University is located in Kaohsiung, the largest city in southern Taiwan. The school's secondary wastewater treatment plant has had a satisfactory record of compliance with environmental regulations since 1998. However, a small part of the untreated school wastewater is directly discharged from storm drains into the Hsitzu Bay (located in the west side of the campus) due to the unfinished wastewater collection system. Moreover, the campus stormwater runoff also drains into the Hsitzu Bay from storm drains. Thus, the untreated wastewater and NPS pollutants from storm runoff cause the contamination of the natural water body. The objectives of this study were to (1) monitor the baseline water quality and NPS pollution from storm drains during storm events, (2) evaluate the effectiveness of a pilot-scale constructed wetland system on NPS pollutants removal, and (3) design and operate a field-scale constructed wetland system for NPS pollutant removal and water quality improvements based on the results from the pilot-scale study. Pollutant removal efficiencies were calculated from the flow and concentration data. The removal efficiencies, combined with information on hydraulic retention time (HRT), may be useful in the design of constructed wetlands for field application.

Materials and methods

Wastewater and stormwater monitoring

Flow rate, water quality during storm events, and water quality from a school drainage pipe were monitored during the research period (Figure 1). Multiple-day sampling routines for two major storm events during the monitoring period were performed. Storm samples were collected from the outfall of the storm drain for up to 7 days for each studied storm event. All samples were placed on ice until transferred to the appropriate analysis bottles and were kept refrigerated until analyzed. Samples were analyzed for total suspended solids (TSS), nitrate (NO_3^-), total phosphate (TP), pH, nitrite (NO_2^-), ammonia (NO_4^+), chemical oxygen demand (COD), redox potential (Eh), and dissolved oxygen (DO). Instruments used for sample analyses included an Ion Chromatograph (Dionex) for nutrient and anion analyses, an Orion Ross pH meter for pH/Eh measurements, and an Orion DO meter (Model 840) for DO measurements. TSS and COD were analyzed in accordance with the methods in Standard Methods (APHA, 1995).

Pilot-scale study

A two-stage free surface and subsurface flow pilot-scale wetland system was constructed to (1) evaluate the effectiveness of the wetland system on NPS pollutants removal due to the stormwater runoff to meet the discharge standards, and (2) collect operational data for the design of the field-scale constructed wetland system. This system was designed using two continuous-flow experimentation tanks measuring 2 m (L) × 1.5 m (W) × 1 m (D): a first tank (first stage) with floating (*Pistia stratiotes* L.) plant species followed by the second tank (second stage) with emergent (*Phragmites communis* L.) plant species. The plant coverage area in the first tank was approximately 92%. The second tank contained 6 plants, which were placed at equal intervals in the media. Figure 2 shows the schematic diagram of the pilot-scale two-stage wetland system. Another set of the two-tank system was used as a control.

The system was filled with coarse sands to a depth of 0.35 m in the first tank and 0.75 m

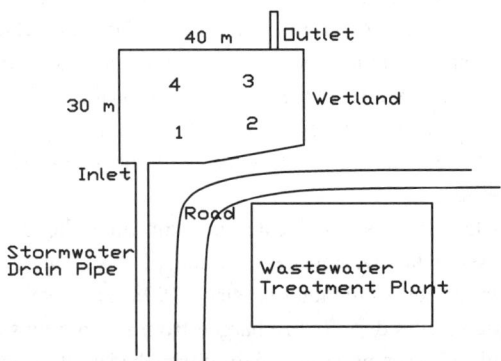

Figure 1 Site map showing the studied natural wetland

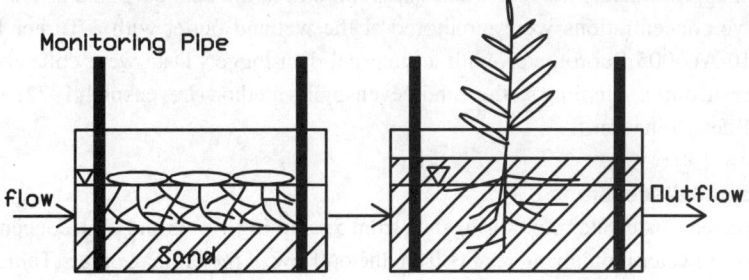

Figure 2 Schematic diagram showing the two-stage pilot-scale wetland system

in the second tank. The water levels were maintained at 0.65 m in both tanks. Both planted and unplanted (control) systems were connected to a reservoir containing synthetic sewage, via a peristaltic pump, to deliver the wastewater into the pilot-scale system. The system flow rate was approximately 0.1 m^3/d and the hydraulic retention time was approximately 7 days.

The synthetic sewage was prepared by adding nutrient and metals into the wastewater collected from the drainage pipe outlet. Thus, the final NO_3^-, TP, Pb, and Zn concentrations were approximately 45, 35, 20, and 20 mg/L. Four water quality monitoring piezometers (labeled as P1 to P4) were placed in the system to monitor the variations in DO, Eh, and pH. P1 and P3 were placed at 0.3 m from the center inlets of the first and second tanks, respectively. P2 and P4 were placed at 0.3 m from the center outlet of the first and second tanks, respectively. All piezometers had openings at 0.3 and 0.6 m in depth. Collected inlets and outlets samples were analyzed for TSS, NO_3^-, NO_2^-, NO_4^+, TP, pH, Pb, Zn, Eh, DO, and COD.

Field-scale study

A field-scale constructed wetland was built to evaluate the feasibility of using wetland on water quality improvement before discharging. A wetland which was 40 m (L) × 30 m (W) × 1 m (D) received approximately 85 m^3/d of untreated wastewater from the school drainage pipe. The plants grown on the front half and the rear half of the wetland were floating (*Pistia stratiotes* L.) and emergent (*Phragmites communis* L.) species. Water quality monitoring of the wetland was conducted from July 1998 to December 1999. Water quality data during one major storm event, baseline water quality, and sediment accretion were monitored in this wetland. Two water quality sampling locations (Inlet and Outlet) and four sediment accretion measurement locations were established in the wetland. Sampling and sediment accretion locations are shown in the site map (Figure 1). The wetland base was filled with coarse sands (with a porosity of 0.35) to a depth of 0.3–0.5 m from the bottom. The water depth was around 0.7 m from the bottom.

Sediment accumulation in the wetlands was measured using Plexiglas disks (Downer and Myers, 1996). The disks had a 100-cm^2 surface that had been abraded with sandpaper to provide a rough area to collect sediments. Four disks were laid evenly in this wetland (Figure 1). Disks were placed in standing water, approximately 0.1 m deep from the surface, in May of 1999. The disks were collected in November 1999 and analyzed for the depth of sediments, mass of sediments, and per cent of organic materials.

The depth of sediments was measured with a dial caliper to the nearest 0.1 mm. Three measurements were taken on the disks and then averaged to determine accretion. The mass of sediments was determined to the nearest 0.01 g after drying in an oven at 105°C. Organic content was determined by burning the samples at 550°C and calculating the organic content as the preburn weight minus the postburn weight. A dye study was conducted to estimate the HRT of the wetland. Three hundred millilitres of rhodamine WT 20% solution was mixed in approximately 90 L of water and poured in at the upstream side of the wetland inlet. Dye concentrations were monitored at the wetland outlet with a Turner Designs model 10-AU-005 fluorometer, with an internal data logger. Data were collected at the downstream outlet sampling station, and Levenspiel's method (Levenspiel, 1972) was used for HRT determination.

Results and discussions

The averaged wastewater monitoring data from 5 sampling events and peak concentrations of two stormwater monitoring results from the outflow of the drainage pipe (Table 1). For the monitored storm events, all parameter levels rose in response to the runoff. Monitoring results show that the untreated wastewater contained high levels of COD, NH_4^+, NO_3^-, and TP. The stormwater runoff contained high levels of TSS, Pb, and Zn. Thus, these constituents were used as the target compounds of this study.

The pilot-scale study was operated for 6 months by pumping the simulated wastewater into the two-stage system. The overall removal efficiencies for NH_4^+, NO_3^-, TP, and COD, were 96, 94, 54, and 83%, respectively. Results reveal that the wetland can effectively reduce nutrients, metals, and COD loadings. Moreover, results show that the second stage process (subsurface flow with emergent plants) removed more contaminants than the first stage process (free water surface with floating plants) (Table 2).

Results from the control system indicate that the removal efficiencies for NH_4^+, NO_3^-, TP, Pb, Zn, and COD, were less than 42, 36, 14, 32, 40, and 31%, respectively. This indicates that the microbial and physical/chemical processes also caused some levels of contaminant removal. Results from the pilot-scale study indicate that the two-stage system would be a feasible technique to remove the point and non-point sources contaminants and effectively control the water quality. The observed high ammonia/nitrate removal indicates that both nitrification/denitrification processes occurred simultaneously in the wetland.

The detected nitrite concentration (2 mg/L) in the wetland confirmed the occurrence of denitrification (Table 2). DO measurements along the vertical profile show that lower DO values were observed near the bottom of the wetland. Therefore, the decomposition/biodegradation of the organic materials in the sediments may cause the decrease in DO and enhance the denitrification process. The decrease in pH in the wetland and at the outlet also confirmed the decay of organic materials. Because the major routes for TP removal might be through plant uptakes and precipitation, thus, the removal for TP was only moderate compared to the nitrogen removal.

Flow into the wetland from the drainage pipe was determined by measuring the velocities and water levels. Mean flow for the drainage pipe during the test period was approximately 85 m^3/day. Results from the dye study show that the peak concentration of 23 μg/L occurred after 5 days of the dye release. One storm event during the wetland monitoring period was sampled. Water quality entering and leaving the wetland followed the same general pattern during the storm event. For the storm event, TSS, NH_4^+, NO_3^-, TP, Pb, Zn, and COD levels rose in response to runoff, which contained high levels of these constituents.

Peak inflow TSS level was around 76 mg/L for the studied storm event. The peak outflow was approximately 33 mg/L (Table 3). Mass loadings were calculated by multiplying inflows and outflows by the measured pollutant concentrations. The calculated total inflow and outflow mass loadings during the storm event were 15 and 6 kg, respectively. Therefore, the removal efficiency for TSS was approximately 60%. This indicates that the wetland has some effects on sediment removal. Maximum mass loading removal efficiencies for NH_4^+, NO_3^-, Pb, Zn, and COD were 91, 88, 85, 95, and 81%, respectively. Results indicate that the field-scale wetland can effectively remove contaminants from stormwater runoff.

Table 1 Averaged baseline water quality monitoring data and peak concentrations of stormwater monitoring results

Parameter	Wastewater	Stormwater 1	Stormwater 2
TSS (mg/L)	42	89	78
NO_3^- (mg/L)	19	11	7
TP (mg/L)	15	7	5
NH_4^+ (mg/L)	21	14	11
NO_2^- (mg/L)	0	0	0
COD (mg/L)	156	76	54
pH	6.2	7.7	7.3
Eh	81	103	89
DO (mg/L)	2	3.1	2.8
Zn (mg/L)	0.7	7.7	4.1
Pb (mg/L)	0.5	12.5	8.8

Table 2 Averaged water quality monitoring data from the two-stage system

Parameter	1st-stage Inlet	1st-stage Outlet	2nd-stage Outlet	P1	P2	P3	P4	1st [2] %	2nd [3] %	Overall %
TSS (mg/L)	52	45	33	–[4]	–	–	–	14	23	37
NO_3^- (mg/L)	17	11	1	–	–	–	–	35	59	94
TP (mg/L)	35	27	16	–	–	–	–	23	31	54
NH_4^+ (mg/L)	45	26	2	–	–	–	–	42	53	96
NO_2^- (mg/L)	BDL	BDL	2	–	–	–	–	–	–	–
COD (mg/L)	134	84	23	–	–	–	–	37	46	83
pH 0.3 m[1]	7.2	6.9	6.5	7.1	6.8	6.8	6.6	–	–	–
pH 0.6 m	–	–	–	7.2	7	6.6	6.4	–	–	–
Eh (mV) 0.3 m	117	22	8	52	29	17	11	–	–	–
Eh (mV) 0.6 m	–	–	–	43	21	16	–5	–	–	–
DO (mg/L) 0.3 m	2.5	1.9	1.1	1.7	1.3	1.0	0.5	–	–	–
DO (mg/L) 0.6 m	–	–	–	1.2	0.8	0.3	0.1	–	–	–

[1] below water surface;
[2] % of removal from the first stage [(1st-stage inlet – 1st-stage outlet)/1st-stage inlet];
[3] % of removal from the second stage [(1st-stage outlet – 2nd-stage outlet)/1st-stage inlet];
[4] not available

The sediment disks were collected and measured after a 6-month monitoring period (Table 4). Results of the sediment sampling are summarized in Table 3. The average disk sediment accretion was 8.7 mm and ranged from 5.7 to 11.5 mm. Results show that most sedimentation occurs near the inlet of the wetland. Sediments were high in organic content, averaging 10.5%. The vegetation is probably a source of some of the organic matter. Moreover, the inflow COD also caused the increase in organic material in the sediments.

Conclusions

Research on NPS pollutants removal by a two-stage pilot-scale wetland system was conducted. The system consisted of the first free surface stage with floating (*Pistia stratiotes* L.) plants, and the second subsurface flow stage with emergent (*Phragmites communis* L.) species. Results from this pilot-scale show that the overall removal efficiencies for NH_4^+, NO_3^-, TP, Pb, Zn, and COD, were 96, 94, 54, 95, 92, and 83%, respectively. Results reveal that the wetland can effectively reduce nutrients and COD loadings. Results from the pilot-scale study indicate that the two-stage system would be a feasible technique to remove the point and non-point sources contaminants and effectively control the water quality.

Based on the results from the pilot-scale study, a field-scale constructed wetland

Table 3 Summarized monitoring results for the studied storm event

Parameter	Inlet		Outlet		Removal %[2]
	Peak concentration mg/L	Total loading kg	Peak concentration mg/L	Total loading kg	
TSS	76	15	33	6	60
NO_3^-	9	1.6	1.3	0.2	88
NO_2^-	BDL[1]	0	BDL	0	–[3]
TP	3	0.5	1	0.15	70
NH_4^+	8	1.3	0.8	0.12	91
COD	60	14	17	2.6	81
Pb	14	2	1.2	0.3	85
Zn	9	1.7	0.5	0.08	95
pH (unitless)	7.5	–	6.6	–	–
DO	3.1	–	1.5	–	–
Eh (mv)	96	–	41	–	–

[1] below detection limit
[2] % of removal = [(inlet total loading – outlet total loading)/inlet total loading];
[3] not available

Table 4 Sediment measurement data

Disk	Average accretion mm	Sediment weight g / (0.01 m^2)	Sampling Site Area m^2	Area loading kg/m^2	Organic material %	Weight organicmatter g / (0.01 m^2)	Organic loading kg/m^2
1	11.5	35	300	3.5	13	4.6	0.46
2	5.7	71	300	7.1	9	6.4	0.64
3	9.6	59	300	5.9	8	4.7	0.47
4	7.9	49	300	4.9	12	5.9	0.59
Average	8.7	54	–	5.4	10.5	5.7	0.57

system was built inside the campus of National Sun Yat-Sen University. The constructed wetland was 40 m (L) × 30 m (W) × 1 m (D), which received approximately 85 m^3 per day of untreated water from school drainage pipes. The plants grown on the wetland also included floating (*Pistia stratiotes* L.) and emergent (*Phragmites communis* L.) species. Wetland inflow and outflow samples during a storm event as well as baseline water quality samples were investigated. The wetland was able to reduce 60% of TSS, more than 88% of N, 85% of heavy metals, and 81% of COD.

Sediment accretion monitoring in the wetland indicates that the accretion rate in the wetland was not high (8.7 mm/year). The high organic content of sediments indicates that the wetland built the characteristic organic layer on the bottom of the wetland. Such an organic layer is important in many wetland functions, including water quality improvements. Results from this wetland study can provide us with further knowledge on evaluating the role of wetlands in controlling NPS pollution.

References

APHA, American Public Health Association (1995). *Standard Methods for the Examination of Water and Waste Water*, 19th ed., APHA-AWWA-WEF, Washington, DC., U.S.A.

Downer, C.W. and Myers, T.E. (1996). Monitoring of Sediments and Nonpoint Source Pollution Removal at Spring Creek Wetland Project, Bowman-Hayley Lake. U.S. Army Corps of Engineers, Technical Report WRP-SM-18.

Kao, C.M. and Wu, M.J. (1999). Sediments and non-point source pollutants removal by a mountainous wetland, Preprint of *7th IAWQ Asia-Pacific Regional Conference*, Taipei, Taiwan, October 18–20, pp. 1465–1470.

Levenspiel, O. (1972). *Chemical Reaction Engineering*. Wiley Pub., New York.

Steven, C. (1998). The emergence of treatment wetlands. *Environ. Sci. and Tech./News*, May, pp. 218–223.

US EPA (1990). Managing Nonpoint Source Pollution: Final Report to Congress on Section 319 of the Clean Water Act. EPA/506/9-90. Washington, D.C., U.S.A.

Buffer zones promoting oligotrophication in golf course runoffs: fiddler crabs as estuarine health indicators

R.Y. George*[+], G. Bodnar*, S.L. Gerlach* and R.M. Nelson**

* Department of Biological Sciences and Center for Marine Sciences, The University of North Carolina at Wilmington, Wilmington, North Carolina 28403, USA

** Nelson Environmental Consultants, 211 Gregory Road, Wilmington, NC 28405, USA

[+] George Institute for Biodiversity and Sustainability (GIBS), 305 Yorkshire Lane, Wilmington, NC, 28409, USA

Abstract Nitrogen pollution above a threshold level induces a eutrophication process in coastal creek ecosystems and consequently impacts on the water quality. The remedy for this scenario is the introduction of methods to enhance oligotrophication by means of constructed wetlands and buffer zones. This paper discusses new data on nitrogen flux and population changes in the primary consumers in the Bradley Creek ecosystem, adjacent to the Duck Haven Golf Course in southeastern North Carolina. In 1998–99, over different seasons, density distribution of the field populations of the fiddler crab *Uca minax*, was monitored as an indicator of environmental health. A control site at Whiskey Creek, adjacent to the University Center for Marine Sciences, was monitored in the same period since this site is not influenced by any golf course nutrient flux. The results pointed out that threshold level for optimum population density in *Spartina grandiflora* salt marsh is 0.1 mg/L of nitrates. A dense crab population, adjacent to the golf course with a buffer zone, was indicative of restoration of the estuarine ecosystem. A model, involving the use of constructed wetlands for oligotrophication, is being prepared on the basis of studies conducted by the University of South Alabama for a stormwater wetland constructed adjacent to the university's golf course.

Keywords Constructed wetland; fiddler crab population density; golf course; oligotrophication

Introduction

In North Carolina, 60,000 hectares of estuaries along the rivers and coastal creeks are closed for shellfisheries because of eutrophication and high counts of coliform bacteria. In recent years, nitrogen pollution from hog lagoons has also caused blooms of toxic dinoflagellates and fish-kills in rivers and estuaries (Burkholder *et al.*, 1992). These estuaries are actually the nursery grounds for commercially important species such as blue crabs and panaeid shrimp. This paper deals with restoration ecology of the Bradley Creek ecosystem, located west of the Intracoastal Water Way and Wrightsville Beach in southeastern North Carolina.

Bradley Creek watershed (Figure 1) encompasses approximately 6,000 acres and is located in the middle of the city of Wilmington in New Hanover County with a population of 85,000. The creek is largely surrounded by residential areas (65%), a park (Arlie Gardens) (10%), commercial area (10%), Duck Haven golf course (10%) and a marina (5%). As early as 1947 the creek was closed for shell fishing because of fecal coliform contamination. The golf course is located down-stream at the very end of the creek where flux from fertilizers can influence the food chain. Nitrate pollution from golf course effluents has been investigated for Hewlett Creek where nitrate-nitrogen level, while passing through a golf course, increases from as low as 0.005 to as high as 1.462 mg/L (Mallin and Wheeler, 1998).

Spartina alterniflora marsh habitats are common, surrounding Bradley Creek. Flourishing in this marsh habitat, the red-jointed fiddler crab *Uca minax* lives between tides, always exposed during low tide and fully submerged during high tide. Burrows of a sympatric species *Uca pugnax* are less frequently seen at the high tide mark along the sand-mud shore of the creek. However, *Uca minax* is far more abundant and a good representative species as a primary consumer, grazing on benthic algae and diatoms during low tide. In most environmental health assessment studies thus far, coliform bacteria is used and very seldom this abundance of higher trophic levels is considered for evaluating the water quality and the health of the ecosystem. In this study, field population densities of the fiddler crabs are monitored in two different creek ecosystems throughout the year 1998 and in the spring 1999. Bradley Creek study sites are located in the vicinity of the Duck Haven golf course (Figure 1) whereas Whiskey Creek study sites are used as a control since these sites are not subjected to effluents from the golf course.

In recent years, an increase in the number of golf courses along the North Carolina coast has elevated environmental concern about how the region's golf courses are affecting natural ecosystems. Although golf courses do not contribute fecal coliform bacteria as sewage outfalls and floods as in hurricanes, nitrate-nitrogen flux (if there is a lack of any constructed wetlands or designated buffer zones) can induce eutrophication and trigger algal blooms. Paearl *et al.* (1995), in controlled experiments, reported that nitrate concentrations as low as 0.100 mg/L induced significant phytoplankton blooms in Neuse estuary waters. Land management practices, such as vegetated buffer zones between a golf course and a coastal creek, can regulate nutrient loading to encourage denitrification. In this study, we tested a new and challenging hypothesis that nutrient levels below a threshold can cause beneficial impact by increasing primary production to meet the demands of the primary

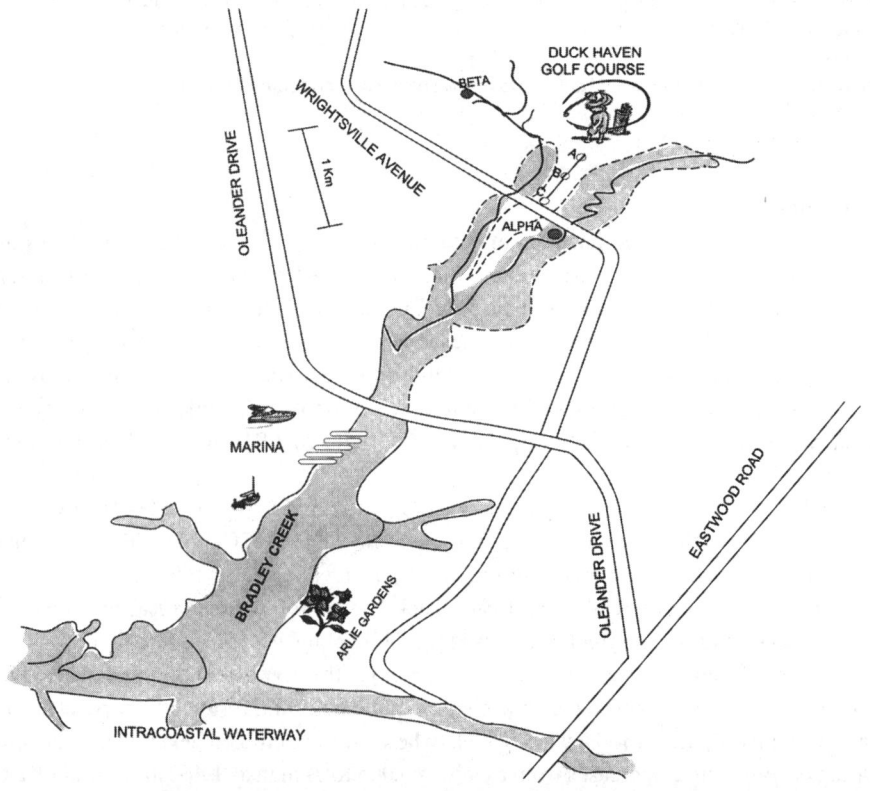

Figure 1 Bradley Creek Watershed

consumers such as the fiddler crab populations in the creek. This hypothesis is tested in the present study.

Materials and methods

The species of fiddler crab *Uca minax* has a range in the east coast of the United States from Massachusetts to Florida and along the northeastern coast of the Gulf of Mexico, extending to the Texas coast. These crabs inhabit the brackish waters that are typically the habitats of *Spartina alterniflora*, *Distchilis soicata* and *Spartina pattens* (Powers, 1977). In this region of tidal wetlands along the 300-mile coast of North Carolina, the red-jointed fiddler crab *Uca minax* is abundant. *Uca minax* is the largest of the local three species of the genus *Uca* and clearly distinguished from both *Uca pugnax* and *Uca pugilator* by the subtriangular carapace (small eye on thin elongated stalks) and by the morphology of the chelipeds. *Uca minax* has red patches at the joints of the articulation of the large claws.

The two coastal creeks, Bradley Creek and Whiskey Creek (Lat.34°52.37'N and Long.77°13.37'W), sampled for fiddler crab densities, are located in the southeastern part of North Carolina near the University of North Carolina at Wilmington. The lower reaches of Bradley Creek are 0.5 kilometres from the edge of the Duck Haven Golf Course. The watersheds from the golf course flow into the creek. Whiskey Creek, which is not influenced by any golf course, is located adjacent to the University's Center for Marine Sciences. Throughout 1998 and in the spring of 1999 the study sites in these two creeks were sampled twice a week during the low tide, only on sunny days.

Two sites were chosen in Bradley Creek (Figure 1) and each site was sampled in a triangular area at three different quadrates as illustrated in Figure 2. The same sampling procedure was conducted in Whiskey Creek. Each quadrate measured 25 × 25 cm. The numbers of crabs were counted in each quadrate; then the carapace width was determined for male and female crabs and the claw length was measured for male crabs. In spring 1999, two sample t-tests for mean density and sex ratio were performed on all triangles at each site using JMP IN. *In situ* measurements of temperature, salinity and dissolved oxygen content were made in the field. The interstitial water samples were taken in spring 1999 to determine nitrate-nitrogen and ammonia-nitrogen in a transect from the golf course to Bradley Creek.

Results and discussion

The results of the 1998 biweekly sampling of fiddler crabs/m^2 both in Bradley Creek (BC) and Whiskey Creek (WC) in the months of March, April, May, June, July, August and September are summarized in Table 1. It is evident from the data that the fiddler crab population density remained rather low, less than 14 per square metre in March. In the winter

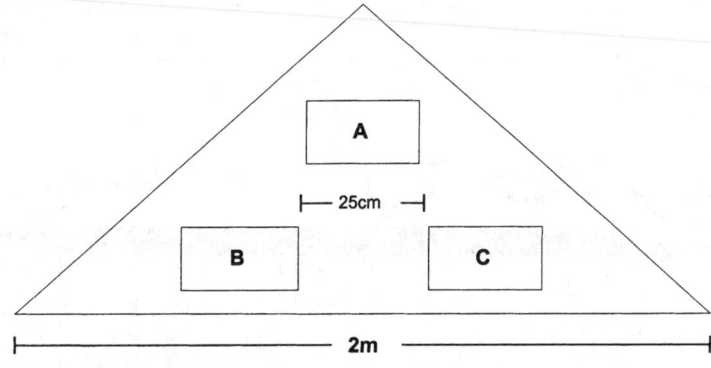

Figure 2 Diagram of sample quadrates

months of January and February, crabs are virtually absent. However, crab density reached a peak in both creeks in April, with as many as 35–52 crabs/m^2 (Table 1). This increase in density coincided with the spring diatom bloom and possibly linked to the breeding season in the temperate estuarine environment. In the late spring and summer, population density was rather moderate, less than 25 crabs/m^2 in both creeks (Figure 3). Nevertheless, crab density exhibited a pronounced increase only in Bradley Creek study sites in late summer and early fall in the months of July and August, 1998, which actually coincided with enhanced fertilizing activities at Duck Haven Golf Course during the peak golfing season. A maximum density of 45 crabs/m^2 was encountered in Bradley Creek adjacent to the golf course whereas density was below 22 crabs/m^2 in Whiskey Creek during the same summer months. The sexes are dimorphic with unusually large claws in the males. The results also pointed out that the sex ratio, when the crab populations of the two creeks are compared, shows a clear difference with greater abundance of females in the Bradley Creek (Figure 4). In the month of August, there is a substantial increase both in the number of females (40 per square metre) and males (72 per square metre) in Bradley Creek. It is difficult to conclude whether more energy available for the crab populations in Bradley Creek was channeled toward increase in population size and reproduction. Although this hypothesis is very interesting, it undoubtedly calls for further research. The carapace width data also supports the fact that Bradley Creek crabs were much larger than those studied in Whiskey Creek. The mean carapace width for Bradley Creek crabs was 0.829 cm. On the contrary, the mean carapace width for the Whiskey Creek crabs was significantly less, at 0.295 cm. The increase in the size of claws in the males progressively during the spring months is indicative of growth in the crabs coinciding with spring diatom blooms. Furthermore, the data on claw size suggests that Bradley Creek crabs have comparatively larger claws than Whiskey Creek crabs. The mean size of the claw of Bradley Creek crabs is 0.734 cm as opposed to 0.670 cm in Whiskey Creek crabs. However, the rapid shift in male claw growth, seen only in Whiskey Creek from August to September 1998 is hard to explain. One explanation is that the crabs in Whiskey Creek unlike Bradley Creek populations are expending more energy for claw growth since nutrient energy requirements are far below threshold levels for channeling energy toward reproduction and density increase as seen in Bradley Creek crab populations. The male claw size for Bradley Creek crabs did not exceed 10 mm but average claw size for Whiskey Creek crabs was as large as 16 mm. The claw size also correlates with carapace width data.

Spring 1999 investigations

In the spring 1999, the same study sites in Bradley Creek and Whiskey Creek were sampled quantitatively for determining the crab population size. The mean density of the fiddler

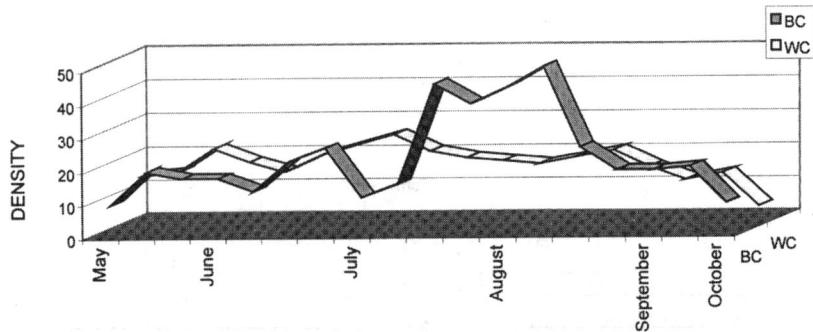

Figure 3 UCA crab species density per square metre

Table 1 Data on Fiddler crab population density* for Bradley Creek and Whiskey Creek

	BC X̄ SD	WC X̄ SD		BC X̄ SD	WC X̄ SD
March 1998	12 + 2	3 + 1	July	28 + 4	21 + 3
	14 + 2	5 + 3		17 + 3	18 + 2
April	30 + 5	12 + 3	August	48 + 4	19 + 3
	41 + 6	29 + 4		45 + 4	20 + 3
May	5 + 2	14 + 1	September	27 + 3	17 + 2
	18 + 3	13 + 2		24 + 3	15 + 2
June	21 + 3	15 + 2	October	21 + 4	16 + 1
	24 + 3	15 + 3		11 + 1	8 + 2

* Mean number per square metre with standard deviation ($N = 3$)

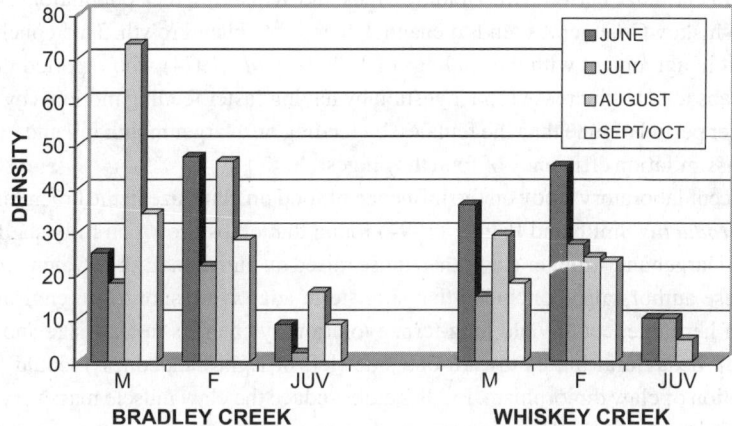

Figure 4 Sex ratios of crab populations compared in Bradley Creek and Whiskey Creek

crab population was far higher than in the previous year 1998 (Figure 5A and B). In Bradley Creek, the crabs/m² in February was 110; in March, it ranged from 50 to 140 and in April it ranged from 95 to 115. On the contrary, the crabs/m² was lower in Whiskey Creek with the mean of 18 crabs in March and 68 in April. The sex ratio was also different with more abundance of female crabs in the Bradley Creek and similar number of males and females in Whiskey Creek. The t-test on mean density between the two creeks clearly indicated that statistically the population density showed a higher level at Bradley Creek.

A transect was developed from Duck Haven Golf Course to the edge of Bradley Creek to sample both ammonia-nitrogen and nitrate nitrogen in four selected stations and the results revealed a distinct concentration gradient (Table 2). The nitrogen flux into the creek was significantly reduced from the golf course to the creek, with 10% of nitrate-nitrogen reaching the waters in the creek and above 25% of ammonia nitrogen reaching the waters in the creek. These data demonstrate that a buffer zone of 0.5 km efficiently filters nitrogen pollution and promotes oligotrophication in the creek. The optimum nitrate level promoted the diatom bloom and consequently was responsible for the increased population density of the fiddler crabs in the Bradley Creek estuarine ecosystem.

The red-jointed fiddler crab *Uca minax* inhabits waters of lower salinity than do the other two congeric species of fiddler crab *Uca pugnax* and *Uca pugilator* that are found commonly along the coast of North Carolina (Williams, 1965). The apparent difference between the two selected study sites in the present investigation, particularly with reference to fiddler crab densities and the increase in claw size in the Whiskey Creek populations

Table 2 Interstitial nitrate and ammonia levels in a transect from the Duck Haven Golf Course to the edge of Bradley Creek (See Figure 1 for transact location)

	Station A 0.5 km	Station B 0.25 km	Station C 0.1 km	Station D Creek	% Decrease
NH^+_4N (mg/L)	8.6 ± 1.4	14.2 ± 1.6	10.4 ± 0.5	4.6 ± 0.2	75%
NO^-_3N (mg/L)	1.6 ± 0.04	1.0 ± 0.06	0.6 ± 0.02	0.16 ± 0.01	90%

calls for some discussion. There is pronounced sexual dimorphism in the fiddler crabs. In the male crab, one claw is very similar in shape, size and function, as in the female crab. However, the other claw in the male is very much enlarged with strong muscles or zero manipulation during mating rituals, forming the architecture of the burrow and for defense against competing males in their territory. One possible inference from this study, showing disparity in claw size between the Bradley Creek and Whiskey Creek populations, is that in the fall Whiskey Creek crabs tend to channel energy for claw growth. This conclusion is somewhat in agreement with the findings of Valiela *et al.* (1974) who reported that male fiddler crabs tend to "increase food ingestion by having faster feeding motions, by feeding for a longer period of time than the females, by feeding on a larger mouth full and by having a higher assimilation efficiency of food they ingest."

In a recent laboratory study on the influence of food on claw size in the brachyuran crab *Cancer productus* Smith and Palmer (1994) found that crabs grown on fully shelled prey developed larger and stronger claws than those raised on nutritionally equivalent unshelled prey. These authors also concluded that short-term adaptive responses to environmental stimuli, if heritable, could yield long-term evolutionary changes in claw size and if combined with behavioral biases toward one side (left or right handedness), could promote the evolution of claw dimorphism. Fiddler crabs reduce the claw muscle mass very significantly prior to moulting. Experimental studies now in progress in the laboratory of the senior author provided some new data recently on metabolic demands of male fiddler crabs with enlarged chelae but the implications of this finding are beyond the scope of this discussion.

Previous studies correlated salinity of the water with abundance of fiddler crab *Uca minax*. Miller and Mauer (1973) found in Delaware rivers that *Uca pugnax* occurs in the high salinity zone of the estuaries, commonly found between 12 ppt and 29 ppt whereas *Uca minax* inhabits the zone of lower salinity from 16 ppt to as low as freshwater. However, in a salinity gradient there was a statistically significant positive association between salinity and density in *Uca pugnax* but not in *Uca minax*. The present study in both Bradley Creek and Whiskey Creek pointed out that the density of the fiddler crab *Uca minax* is linked closely to food availability and the selected study sites experienced wide fluctuations in salinity in any given day. These euryhaline crustaceans are adapted to withstand the rapid salinity changes.

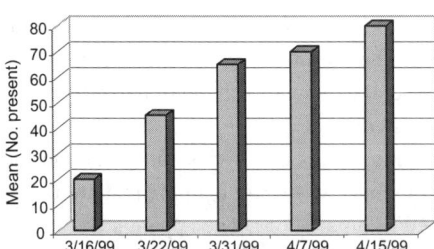
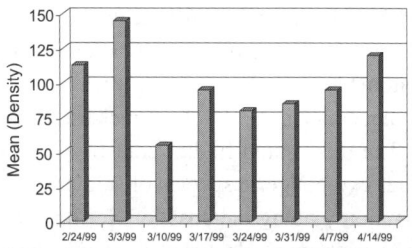

Figure 5A and 5B Measurement of crab population sizes in Bradley Creek and Whiskey Creek in 1999

The fundamental difference between Bradley Creek and Whiskey Creek is in reality the triggering influence of optimum nitrate flux from the golf course in the former and lack of it in the latter. Therefore, this study makes an important conclusion that in addition to a natural increase of fiddler crab density in the periods of spring diatom bloom there is evidently a boost to crab density in periods of increased golf course fertilizing activities in the summer. The presence of a significant buffer zone is responsible for the oligotrophication and the removal of excess nitrogen and is the cause for nitrate-nitrogen entering the ecosystem at optimum threshold levels, below the level for inducing eutrophication. Results from mesocosm experiments (Burkholder *et al.*, 1992) pointed out that nitrate concentration as low as 0.1 mg/L can cause sea-grass beds of *Zostera marina* to die. There is a genuine need to develop constructed wetlands as an effective method to induce oligotrophication. Constructed wetlands designed for stormwater treatment (Figure 6) are engineered ecosystems appropriate for removing pollutants from rainfall runoff including nitrogen flux from golf courses. The wetland removes pollutants by temporarily storing stormwater in shallow pools that create ideal growing conditions for wetland plants. The plants and associated microbiology act to filter sediments, uptake nutrients, and biodegrade carbonaceous materials (White, 1995). The hypothetical diagram illustrates a "Seasonal mesocosm-experiment-linked constructed wetland" (SMEL-CW) that is proposed on the basis of data from studies conducted by the University of South Alabama for a stormwater wetland constructed adjacent to the university's golf course (Figure 6).

Conclusion

This study concludes that the primary consumer population densities serve as an index of environmental health and water quality just as well as the coliform bacterial concentrations that are used currently for designating water bodies such as creeks, rivers and lakes as polluted ecosystems. In future restoration studies in ecosystems, that are closed for consumption of drinking water at low levels of coliform bacterial concentrations and closed for swimming or fishing activities at elevated levels of coliform bacterial concentrations, it is recommended that bioassessment of primary consumer population densities and also densities of animals of higher trophic levels (secondary and tertiary consumers) should be established to "resurrect" the polluted ecosystem to a status of normalcy.

Acknowledgements

The authors are grateful to the managers of the Duck Haven Golf Course and in particular, Mr. W. Trask, for their cooperation in our studies in Bradley Creek, adjacent to the golf course. We thank UNCW undergraduate marine biology students Andrea Stevens, Andrea Cowen and Lara Wise for initiating this study first in the spring and summer of 1996 and

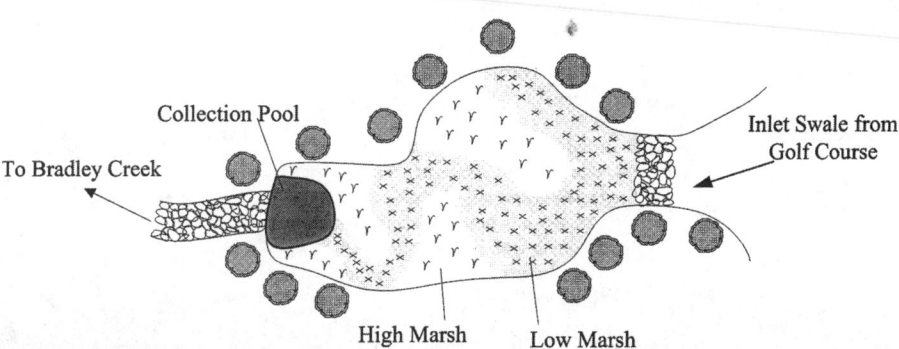

Figure 6 Hypothetical diagram to illustrate a "Seasonal" mesocosm-experiment-linked constructed wetland

1997. This program is a joint contribution in the cooperative program for promotion of constructed wetlands as buffer zones, between George Institute for Biodiversity and Sustainability (GIBS) and Nelson Environmental Consultants (NEC).

References

Burkholder, J.M., Mason, K.M. and Glasgow, H.B. Jr. (1992). Water-column nitrate enrichment promotes decline of eelgrass *Zostera marina:* Evidence from seasonal mesocosm experiments. *Marine Ecology Progress Series* **81**, 163–178.

Krebs, C.T., Valiela, I., Harvey, G.R. and Teal, J.M. (1974). Reduction of field population of fiddler crab by uptake of chlorinated hydrocarbon. *Mar. Poll. Bull.* **5**, 140–142.

Mallin, M.A. and Wheeler, T.L. (1998). Nutrient and fecal coliform discharge from Southeastern North Carolina Golf Courses. Symposium on "*Is golfing green?* – Impact of golf courses on the coastal environment", March 12, 1998, North Carolina Coastal federation and the University of North Carolina at Wilmington.

Mallin, M.A., Cahoon, L.B., Manock, J.J., Posey, M.H., Sizemore, R.K., Webster, W.D. and Alphin, T.D. (1998). A Four Year Environmental Analysis of New Hanover County Tidal Creeks, (1993–1997). CMSR Report No. 98-01, University of North Carolina at Wilmington.

Miller, K.G. and Mauer, D. (1973). Distribution of fiddler crabs, *Uca pugnax* and *Uca minax,* in relation to salinity in Delaware rivers. *Chesapeake Sci.* **14**, 229–231.

Paearl, H.W., Mallin, M.A., Donahue, C.A., Go, M. and Peierles, B.J. (1995). Nutrient loading sources and eutrophication of the Neuse River Estuary, North Carolina: Direct and Indirect roles of atmospheric deposition. Report No. 291, Water resources Institute of the University of North Carolina, Raleigh, N.C.

Powers, L.W. (1977). A catalogue and bibliography of the crabs (Brachyura) of the Gulf of Mexico. *Contributions Mar. Sci. Texas*, 48pp.

Smith, L.D. and Palmer, A.R. (1994). Effects of manipulated diet on size and performance of Brachyuran crab claws. *Science* **264**, 710–717.

Valiela, I., Babiec, D.F., Atherton, W., Seizinger, S. and Krebs, C. (1974). Some consequences of sexual dimorphism: Feeding in male and female fiddler crabs *Uca pugnax* (Smith). *Biol. Bull.* **147**, 652–660.

White, K.D. (1995). University of South Alabama Department of Civil Engineering Constructed Wetlands Page. *Stormwater Treatment*. Web page URL: http://www.eng.usouthal.edu/civil/stormw~1.htm.

Williams, A.B. (1965). Marine Decapod Crustacea of the Carolinas. *Fish. Bull.* **65**(1), 1–298.

River water quality improvement by natural and constructed wetland systems in the tropical semi-arid region of Northeastern Brazil

B.S.O. de Ceballos*, H. Oliveira**, C.M.B.S. Meira**, A. Konig*, A.O. Guimarães and J.T. de Souza**

*Department of Civil Engineering – Environmental and Sanitation Area, Federal University of Paraiba. Avenida Aprigio Veloso, 882; CEP 58109–753, Campina Grande, Paraiba, Brazil
**Technological Science Center, State University of Paraiba, Avenida des Baraúnas, 351; CEP 58109–753, Campina Grande, Paraiba, Brazil

Abstract The efficiencies of a natural *Typha* spp wetland (Wn) formed on a river bed and its effluent treatment in a constructed wetland (Wc, subsurface horizontal flow) were investigated in northeastern Brazil (Paraiba State). The Wc system (12 tanks with stone gravel, 4.13 m^2, 0.22 m^3, 20 *Typha* spp rhizomes. m^{-2} each, with 38, 29, and 19 mm.d^{-1} hydraulic loadings, and 5, 7, and 10 days HRT) was fed daily with effluent from a Wn. Wn removal presented the highest values after *Typha* spp were cut during the 5th week. Removal values were (1st and 2nd periods or before and after cutting): 75% and 81% BOD$_5$; 10–53% total phosphorus; 13%–55% ammonia; 89%–91% FC; 90–96% coliphages and bacteriophages. Wc removals increased with time with best results on 10d HRT. Removals were also higher in the 2nd period: 74%–78% BOD$_5$; 58%–82% ammonia; 90% FC; 94–98% FS; and 92%–96% coliphages and bacteriophages. Despite the high remaining values of FC (1.4×10^4 CFU/100 ml) and FX (4 × 10^3 CFU/100 ml), the removals were satisfactory and HRT dependent, suggesting a gradual optimization of the system with time. The Wc exhibited good efficiency for improving water quality from polluted river.

Keywords Bacteria and viruses removal; constructed wetland; natural wetland; river clean up

Introduction

Developing countries generally lack a policy for natural waters preservation, due probably to their successive financial crises leading to low levels of investments in sewage collection and proper treatment system, and, in some way, also due to their high human population density. A low cost and easily operated system would be necessary for removing nutrients, organic matter, and faecal contamination from polluted superficial water (Thomas *et al.*, 1995). Constructed wetlands play an important role in water pollution control and wastewater management using natural processes. Due to simplicity of their design, operation and maintenance they seem nowadays to be the most promising technology to be applied in developing countries. In addition, tropical regions have favourable climates for good wetland systems performance (Habrel, 1999).

Natural wetlands are temporary shallow waterbodies or large river floodplains acting as natural buffers and natural treatment systems for wastewater punctual and non-punctual pollution drainage. They are regulated by hydrological conditions such as frequency and duration of flooding depth and diurnal or seasonal water level changes. Constructed wetlands have a uniform water level and a different root substrate support where biofilm formation occurs, besides a restricted biodiversity (Gopal, 1999). Plants are important for a good wetlands performance because they absorb nutrients, their roots offer mechanical resistance to water flow, increase the hydraulic retention time (HRT), provide a large surface area for microbial growth, and transport oxygen to anaerobic layers. Plants biomass production could also be of economical importance.

The majority of the northeastern Brazilian urban rivers are highly contaminated and not useful for any practical purpose. They receive urban run-off polluted with sewage discharge, excreta of animal livestock and from agricultural uses. During the dry season those rivers are open channels for sewage transport with high turbidity, dark colour, nuisance odour and several health risks. Because of the low water availability the existing superficial waters are used for multiple purposes mainly irrigation. Most of the urban rivers run into lakes which are used for aquatic communities preservation, recreational activities, fishing, and industrial water supplies. Their fast eutrophication and faecal contamination are favoured by the influent polluted streams with high levels of BOD, nutrients and enteropathogenic microorganisms. Superficial waters for multiple purposes must preserve a good quality in order to provide safe and healthy conditions for bathing and fishing, and also to prevent human health risks, particularly in poor human population areas where diseases spread by faecal-oral route are endemic (OPS, 1995). Constructed wetlands are considered as an attractive alternative aiming at achieving a good river water quality with significant reduction of pathogens as well as low values of BOD and nutrients (phosphorus and nitrogen).

The present study aims to evaluate the performance of a natural and a constructed wetland which differ in hydraulic retention times (HRT) during polluted water treatment of an urban stream, as a rehabilitation method for that water body.

Methods

Both systems under investigation were situated in the Federal University of Paraiba (municipality of Campina Grande, 7°13'11" S, 35°52'1" W; 550 m above sea level), Paraiba State, northeastern Brazil. The natural wetland (Wn) system comprised a flood plain (7,666 m^2; 2.5 m average depth; 20 days HRT) on a river bed covered naturally with *Typha* spp (predominant plant), *Eicchornia crassipes*, and *Juncus* spp. The stream runs into Bodocongó lake which is currently used for recreational activities and fishing. A municipal fishery station to raise tilapia for repopulating northeastern lakes and dams and a research experimental station for irrigation are installed near by.

The constructed wetlands (Wc) comprised a subsurface horizontal-flow system with 12 tanks with a 19 mm thick stone gravel bed and hydraulic loadings of 38, 29, and 19 mm.d^{-1} generating retention times of 5, 7, and 10 days, respectively, with four repetitions each. All tanks were planted with *Typha* spp. (initial planting density of 20 rhizomes.m^{-2}). *Typha* spp were selected from the natural wetland and their leaves were cut off and their roots washed with clean water for removing any attached material. Tanks were fed daily with natural wetland effluent. Water samples from influents and effluents were collected weekly or fortnightly from the natural wetlands and from each tank during 12 weeks from 22nd December 1999 to 16th March 2000, and were analysed for temperature, pH, electrical conductivity, BOD$_5$ (biological oxygen demand), COD (chemical oxygen demand), SS (suspended solids), N (nitrogen), and P (phosphorus) levels, faecal coliforms (FC), faecal streptococci (FS) and coliphages (Cf) concentrations (APHA, 1998), and bacteriophages (Bf) (Debartolomeis and Cabelli, 1991).

Results and discussion

It rained during the analysed period (four months) which was a typical situation because it is usually dried from December to nearly the end of February. The differences obtained from the four replications for each HRT in the Wc were not statistically significant (one-way ANOVA test), so the results analyses were based on their average values.

Natural wetlands

The natural wetlands performance was evaluated with results obtained from P1 (influent) and P2 (effluent). The water quality at P1 was similar to raw diluted (1/10) sewage (Table 1, Figures 1 and 2). Both samples showed small temperature variations (from 24°C to 27°C) and pH variations (from 7.5 to 8.2) during the monitoring months and the highest values were registered at P1. The spatial and temporal results distribution showed decreasing concentration values for most parameters after the 6th week (2nd February 2000) clustering the data in two periods: a 1st period until the 6th week, with an erratic behaviour and higher values for all the parameters analysed, and a 2nd period after that date, presenting lower concentration values. The lower results were associated with dilution due to high rainfall during February (153.9 mm) and March (63.4 mm). Macrophytes assimilation may also have been stimulated after plants were cut off on the 5th week. The average electrical conductivity was high (Table 1), with highest values at P1 (2,462 μmhos.cm^{-1}) and lowest values at P2 (2,004 μmhos.cm^{-1}), which would indicate some ions absorption by the green biomass and/or some adsorption/desorption of particulate and dissolved material on rhizosphere and sediments.

The natural wetlands data showed efficient removals of BOD_5 (75% and 81%, average 78%); SS (53%, and 46%, average 50%), FC (89% and 91%, average 90%) and FS (94% and 98%, average 96%). Both Cf and Bf removals were also high: 94% (final concentrations of 4×10^4 and 8×10^4 PFU/100 mL, respectively). Although 1 log unit of FC and 2 log units of FS were removed, the remainder levels were still high which suggest human risks not attending to WHO guidelines for unrestricted irrigation (WHO, 1989). Low COD removals (8%–46%) were predominant. The large COD/BOD_5 quotient variations indicated great variables of water composition and of sewage influents into the natural wetland where the hardly biodegradable fraction ($COD/BOD_5 > 7$) was predominant.

Phosphorus and ammonia removals efficiency (Figure 1) varied within a very wide range in the 1st and 2nd periods as follows: total-P from 10% to 53% (average 24% and range 5.8–4.4 mg/L), ortho-P from 9% to 26% (average 14% and range 3.9–3.3 mg/L), and ammonia from 13% to 55% (average 27% and range 37.3–27.1 mg/L). N-compounds removal in wetlands is governed mainly by the microbial processes of nitrification and denitrification. The influent water ammonia concentration (average 37.3 mg/L) might have been partly oxidized to nitrate, partly absorbed by the plants and partly reduced to nitrogen gas in the denitrification process. The high ammonia removal during the 6th week could be associated with plant harvesting practised in the lake vegetation a few days before sampling, which probably stimulated plants growth and consequently, larger absorptions of nitrogen compounds. The Wn behaviour was strongly influenced by the climatic conditions.

Table 1 pH and electrical conductivity (EC) values and removal efficiencies of BOD_5 and COD in a natural wetland during 12 weeks monitoring (22nd December, 1999–16th March 2000), in Paraiba State, northeastern Brazil

Days	pH P1	pH P2	EC (μmho.cm^{-1}) P1	EC P2	BOD_5 (mg.l^{-1}) P1	BOD_5 P2	%P	COD (mg.l^{-1}) P1	COD P2	%P
15	8.0	7.9	2170	2310	88	29	67	1220	813	33
22	7.7	7.7	2110	1876	109	5	95	787	427	46
28	7.6	7.5	2930	2420	43	15	65	862	259	70
35	8.1	8.0	2260	2640	85	32	62	846	615	27
42	8.2	7.9	2580	2250	113	30	73	923	308	67
56	8.0	7.8	3000	1261	35	12	66	805	394	51
65	7.7	7.5	2940	1910	90	11	88	646	461	29
79	8.1	7.8	2390	1525	28	8	71	200	185	8
85	7.7	7.6	1781	1842	96	15	84	192	192	0

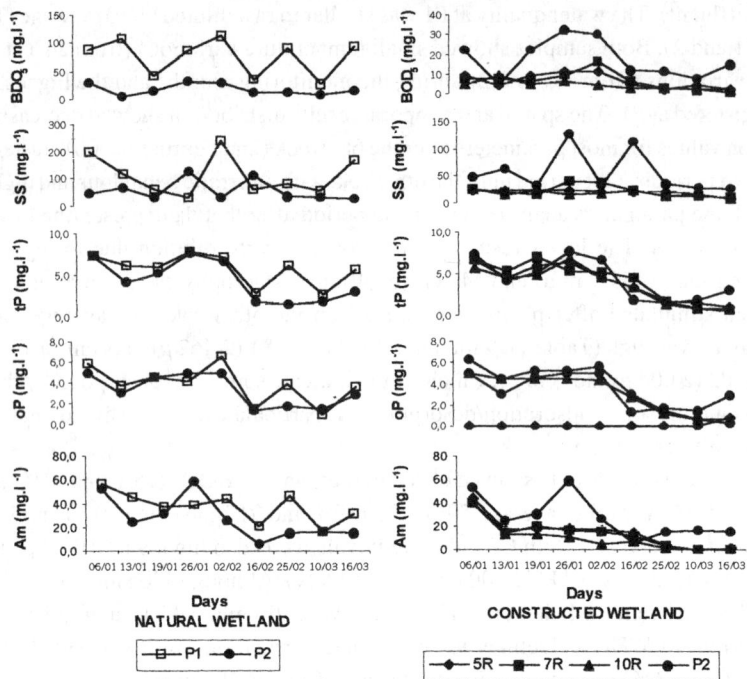

Figure 1 Average values distribution of some parameters in natural and constructed wetlands during 12 weeks monitoring (22nd December, 1999–16th March, 2000), in Paraiba State, northeastern Brazil (SS: suspended solids; tP: total phosphorus; oP: soluble orthophosphate; Am: ammonia; P1: influent; P2: effluent; R: HRT In days).

Constructed wetlands

On the third week, after *Typha* spp was transplanted to constructed wetlands (Wc), most plants showed to be well adapted, and few sites with dead specimens were re-planted. During the 6th monitoring week *Typha* spp reached an average 1.20 cm height and produced buds. Wc though with similar behaviour to Wn showed smaller fluctuations and presented also two removal periods (the 1st and the 2nd), with higher efficiencies on the 2nd during an intensive *Typha* growth. The 1st was dominated here as a maturation or stabilization period, considering the ecosystem adaptation when biofilm formation started on the gravel surface. Well developed and established plant roots, besides sedimentation of inlet organic matter, began to happen simultaneously to their degradation on the tank beds.

Temperature values of the effluents (average 24°C) were lower than of the influents for all the HRT determined; pH had a similar behaviour (7.3–7.9) and electrical conductivity was also reduced (Table 2). Higher retention time caused better removals of BOD_5 in both periods which were even higher during the 2nd period (60%–71% 5d HRT; 58%–74% 7d HRT; 74%–78% 10d HRT) and SS (64%–69% 5d HRT; 55%–71% 7d HRT; 69%–70% 10d HRT) with effluents values of $BOD_5 \leq 2$ mg/L and SS ≤ 10 mg/L for the three HRT considered (Figure 1). The BOD_5 final concentrations were lower than the recommended limit of the Brazilian regulatory standards established for discharges into the natural environment (Resolution 20/86 CONAMA, 1996), which may be to some extent due to low BOD_5 and SS loads fed by the influents. The SS volatile fraction was larger than the stable fraction, 9 mg/L (10d HRT) and 13 mg/L (7d HRT) with better removals during the second period (80% 10d HRT), due to earlier processes of biodegradation, adsorption, and sedimentation. The effluent BOD_5/COD rate, 12 weeks average, was different for each HRT

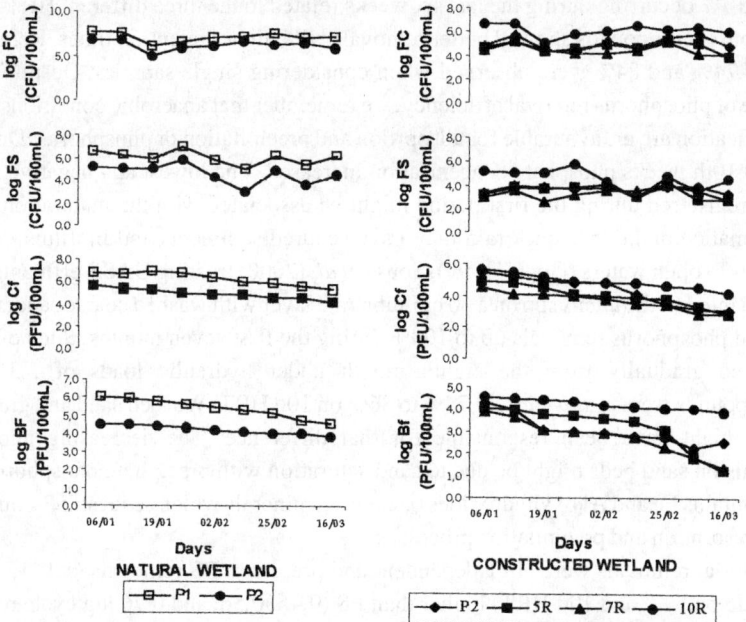

Figure 2 Average values distribution of microbiological parameters in natural and constructed wetlands during 12 weeks monitoring (22nd December, 1999–16th March, 2000), in Paraiba State, northeastern Brazil (FC: faecal coliforms; FS: faecal streptococci; Cf: coliphages; Bf: bacteriophages; P1: influent; P2: effluent; R: HRT in days).

(0.041 5d; 0.045 7d; 0.029 10d) showing a better biodegradation on 10d HRT, though they were similar on the last six weeks i.e. 0.021, 0.026, and 0.021 respectively for the mentioned three HRT. This suggests that after the system stabilization (biofilm formation and mature rhizosphere) the HRT would not affect strongly the removals of those parameters. In Table 2 the concentration and removal values of BOD_5 and COD are shown for the 12 monitoring weeks.

Ammonia concentrations decreased with increased residence time, particularly in the last six weeks (Figure 1). Removal of ammonia ranged average values from 47% (5d HRT) to 63% (10d HRT) and in some single samples removals were as high as 99% and 100% on 7 and 10d HRT, respectively. The lowest removals were observed with respect to total-P in both periods (9%–10%, 5%–10%, and 15%–24% on 5, 7, and 10d HRT, respectively). Negative results of ortho-P were obtained in the first period, with removal averages of 10%,

Table 2 pH, electrical conductivity (EC) values and removal efficiencies of BOD_5 and COD in constructed wetlands with different HRT during 12 weeks monitoring (22nd December, 1999–16th March, 2000), in Paraiba State, northeastern Brazil (R: HRT in days)

Days	pH		EC (µmho.cm^{-1})		$BOD_{5 (mg.l}$$^{-1})$				COD (mg.l^{-1})l			
	5R	10R	5R	10R	5R	%	10R	%	5R	%	10R	%
15	8.2	8.2	2325	2265	9	59	5	77	205	75	104	87
22	7.6	7.7	1614	1651	4	20	9	−80	156	63	176	59
28	7.7	7.5	2168	2280	9	59	5	77	140	46	143	45
35	8.0	7.8	2203	2130	11	66	6	81	83	87	94	85
42	8.0	8.0	2230	2233	11	63	4	87	358	−16	369	−20
56	7.6	7.7	1976	2328	5	58	3	75	79	80	161	59
65	7.1	7.3	1486	1361	2	82	3	73	183	60	79	83
79	7.3	7.5	1345	1347	4	50	2	75	177	4	137	26
85	7.2	7.4	1734	1791	2	87	2	81	29	85	89	24

6%, and 7% occurring during the last six weeks related to the three different HRT. Results from both P-compounds showed better removals with higher retention times. Removals as high as 74% and 84% were observed when considering single samples. Despite the low average of phosphorus removal efficiency, we remember that anaerobic conditions causing denitrification are unfavourable for adsoprtion and precipitation of phosphorus. During the 6th and 10th weeks phosphate concentration increased. The lowest and negative removal values registered during the first period might be associated with the maturation process and climatic conditions, since rain may have scoured sediments and biofilms, releasing nutrients to open waters (Gopal, 1999). Sousa *et al.* (2000) treating UASB effluents in wetlands planted with *Juncus* spp on a 60 cm substrate layer with washed coarse sand, reported elevated phosphorus removals up to 100% during the first seven months. Such efficiency decreased gradually from the eighth month under hydraulic loads of 2.3 cm.d^{-1}. N-compounds were removed from 74% to 86% on 10d HRT. Washed sand and gravel substrates might have been responsible for that difference. The decreasing phosphorus removals on sand beds might be due to sand saturation with precipitate phosphorus, since sand contains Fe and Al oxyhydroxides or calcium minerals which sequester P-compounds through sorption and precipitation processes.

Bacterial removals were HRT dependent and presented FC removals (1.1, 1.4 and 1.7 log cycle on 5, 7, and 10d HRT) higher than FS (0.46, 0.56, and 0.76 log cycle removals, respectively) with 3×10^4 CFU/100 ml and 2.1×10^3 CFU/100 ml final effluent concentrations on 10d HRT (Figure 2). The results suggest a stronger FS resistance to environmental conditions, as also reported in other aquatic environments like stabilization ponds and superficial dams with higher electrical conductivity. Removals in the first period were higher than or equal to removals in the second period. Bacteria removal mechanisms in wetlands are not well known and relevant mechanisms could be mechanical filtration, sedimentation, adsorption in biofilms, predation, natural die off, and antibiotics released by macrophytes roots (Rivera *et al.*, 1995). Wc reduced the numbers of enteric bacteria in polluted waters. However, such reductions were not sufficient to meet the bathing water quality standards or water reuse standards according to guidelines from WHO (1989), thus suggesting health risks. Coliphages and bacteriophages behaviour were similar to bacterial, whose removals increased with higher HRT. Coliphages presented removal values between FC and FS: 0.5, 0.8, and 1.0 log cycles (on 5, 7, and 10d HRT) while bacteriophage removals were shown to be more effective, i.e. 1.14, 1.0, and 1.47 log cycles, considering the three HRT, respectively. Bacteriophages have been proposed as tracer models for the fate of enteric viruses during sewage treatment in which the expected result would be a smaller or equal removal to pathogenic viruses removals (IAWPRC, 1991); thus bacteriophages seem to be interesting for Wc evaluation. However, there are few references about their removals in Wc and more data are required to compare results.

Conclusions

Removal efficiencies of constructed wetlands through quite variable and likely dependent upon different factors such as climatic conditions, HRT, macrophyte growth, root zone maturity and biofilm formation, suggest that this low cost system is a suitable method of treatment of polluted water in the tropical semi-arid region of Brazil, by reducing organic matter load, nutrients, and pathogens.

References

APHA (1998). *Standard Methods for the Examination of Water Wastewater*. 20th edn, American Public Health Association/American Water Works Association/Water Environment Federation, Washington, DC.

CONAMA-Conselho Nacional do Meio Ambiente (1996). Resolução no 20/1986. In: *Legislação de Conservação de Natureza*. 4th ed. FBCN/CESP, São Paulo.

Debartolomeis, J. and Cabelli, V. (1991). Evaluation of an *Escherichia coli* host strain for enumeration of F male-specific bacteriophages. *Appl. Environ. Microbiol.*, **57**(5), 1301–1305.

Gopal, B. (1999). Natural and constructed wetlands for wastewater treatment: potentials and problems. *Wat. Sci. Tech.*, **40**(3), 27–35.

Harbel, R. (1999). Constructed wetlands: a chance to solve wastewater problems in developing countries. *Wat. Sci. Tech.*, **40**(3), 1–17.

IAWPRC Study Group (1991). Bacteriophages as model viruses in water quality control. *Wat. Res.*, **25**(5), 529–545.

Rivera, F., Warren, A., Ramirez, E., Decamnp, O., Bonilla, P., Gallegos, E., Calderón, A. and Sanchez, J.T. (1995). Removal of pathogens from wastewaters by the root zone method (RZM). *Wat. Sci. Tech.*, **32**(3), 211–218.

OPS (1995). Organización Panamaericana de la Salud. La calidad del agua potable en America Latina. Craun F.C. and Castro R. (eds) OPS/OMS, I1SI Press, Washington DC.

Sousa, J.T. de, van Haandel, A.C., Cosentino, P.R. da S. and Guinarães A.V.A. (2000). Pós-tratamento de efluente de reator UASB utilizando sistemas "wetlands" construidos. *Rev. Eng. Agricola e Ambiental* **4**(1), 87–91.

Thomas, P.R., Glover, P. and Kalaroopan, T. (1995). An evaluation of pollutant removal from secondary treated sewage effluent using a constructed wetland system. *Wat. Sci. Tech.,* **32**(3), 87–83.

WHO-World Health Organization (1989). Health guidelines for the use of wastewater in agriculture and aquaculture. WHO Tech Report Series no 778, WHO, Geneva.

Metals in combined conventional and vegetated road runoff control systems

H. Pontier*, J.B. Williams* and E. May**

* Department of Civil Engineering, University of Portsmouth, Burnaby Road, Portsmouth, PO1 3QL, UK
** School of Biological Sciences, University of Portsmouth, Burnaby Road, Portsmouth, PO1 3QL, UK

Abstract Combined conventional and vegetated runoff treatment systems were installed on the rural Newbury Bypass in Southern England, which was estimated to carry an average daily traffic of 30,000 vehicles. The system components were arranged in series to give progressive pollutant removal, comprising an oil separator, sediment trap, grassy slope and constructed wetland. In the absence of specific design criteria for road runoff control, the system design was based on available guidance notes. Since the road opened, an 18-month monitoring programme has examined the performance of these systems. Early indications are that the systems, and especially the wetlands, promote removal of suspended solids (>1.2 µm) and effectively trap fine (<63 µm) sediments. The waters were generally oxidising with neutral to basic pH which may have favoured metal partitioning to the solid phase. The distribution pattern for Zn and Cu in fine sediments contrasted in two of the systems studied, indicating the complex behaviour of metals in wetlands. Further work will increase the data set and examine the adsorption behaviour of metals. These results will contribute to the development of future design criteria for the application of wetlands to road runoff treatment.

Keywords Constructed wetlands; metal partitioning; Newbury Bypass; road runoff control; sediments

Introduction

The environmental impacts of road runoff, especially from rural roads with traffic in excess of 30,000 vehicles per day, are of increasing concern. Construction of impermeable roads and efficient drainage systems increase the peak discharges from storms which causes stream bed erosion and increased flood risk. Traditionally this has been alleviated with balancing ponds and wet or dry detention basins (Luker and Montague, 1994). The quality of road runoff also has impacts on receiving waters. Pollutants accumulate on the road surface from vehicle wear, emissions, and accidental spillage and are transported in runoff. These pollutants include sediments, metals, salts, hydrocarbons and herbicides. Runoff quality is highly variable and depends on the amount of pollutants on the road and dilution by each particular storm event (Munger *et al.*, 1995; Nuttall *et al.*, 1997).

The physical presence of solids in runoff may damage river ecosystems due to increased turbidity and sediment blanketing. Solids also transport adsorbed pollutants, such as heavy metals, which can have acute or chronic ecotoxicological affects. The size of solids determines the ease of entrainment and transportation in the water column. The <63 µm grain size fraction is the most easily entrained and transported by runoff and has been associated with greater pollutant concentrations than large grain sizes (Luker and Montague, 1994).

Metals associated with runoff, such as zinc (Zn), copper (Cu) and iron (Fe), show partitioning between aqueous and solid phases which occurs by various physical, chemical and biological interactions (Bourg, 1989; Luker and Montague, 1994). Partitioning may occur by precipitation of inorganic metal compounds, complexation and adsorption to a variety of sorbent materials, such as organic matter, bacterial and algal cell walls, Fe/Mn oxyhydroxides and clay minerals. Partitioning depends on the concentration of ionic species and sorbents, the presence of competing cations or sorbents, redox potential, pH, temperature and contact time (Bourg, 1989; Jenne, 1998).

The successful use of constructed wetlands treating other wastewaters, has led to interest in their application to control of road runoff (Cooper et al., 1996; Nuttall et al., 1997). However, road runoff is an intermittent and highly variable feedstock and there are no established design criteria for constructed wetlands for this application.

The design and construction of the A34 Newbury Bypass included road runoff control systems to protect the high amenity, commercial (game fishing) and conservation value of the three rivers in the catchment. The systems were designed to provide flow balancing as well as pollution control and were based on the available guidance notes (Luker and Montague, 1994). The road runoff control systems at Newbury offered the opportunity to evaluate the behaviour of pollutants and system performance from the opening of the road in November 1998. The overall aim of the study was to improve design and operation codes for the use of constructed wetlands in treatment of road runoff.

Method
Site description
The 13.5 km long Newbury Bypass on the A34 in Berkshire, England is equipped with nine treatment systems, each with different catchment areas but with similar design features as shown in Figure 1. The road surface is porous asphalt to reduce traffic noise, but it may also provide pollution control at source by trapping soluble contaminants, particulates and water. The runoff is conveyed through the drainage to Klargester oil separators which are designed to retain oils and grease from the "first flush" of storm events. Rectangular reinforced concrete sediment traps allow sedimentation of larger particles (>80 μm). A grassed slope or swale is intended to reduce the flow velocity of water overflowing from the sediment trap and to prevent damage to the wetland plants as the flow enters the wetland.

The constructed wetland is effectively a balancing pond with wetland plants to enhance pollutant removal. It is nested within a storage basin designed to attenuate peak flows of up to a 1 in 100 years design storm. The basin is lined to protect the underlying aquifer. The design includes a permanent standing water body to sustain the plants, improve sedimentation and prevent resuspension of settled material. Different types of macrophytes are planted in ecological zones created on benches at different depths. The plants selected are pollutant tolerant, suited to the ecological zones, and contribute to the treatment in a variety of ways (Munger et al., 1995; Cooper et al., 1996).

Two systems, referred to as C and H, are the subject of this study and their design specifications are given in Table 1. Systems C and H differ mainly in the size of the catchment, and in the size, shape and vegetation of the wetland. Another System (F/G) was designed as a sub-surface flow wetland and is the subject of a complementary study by Middlesex University.

Table 1 Design specifications of systems C and H

Characteristic	System C	System H
Paved Catchment (Ha)	1.6	3.1
Discharge: 1:1y storm (l/s)	190	380
Total wetland area (m^2)	564	11,189
Pool area (1 m deep) (m^2)	7	2,905
Reed bed area (0.4–0.6 m deep) (m^2)	507	6,955
Substrate	chalky clay with flints	River gravels, fen peats
Main Plant Species	*Glyceria maxima*	*Phragmites australis*

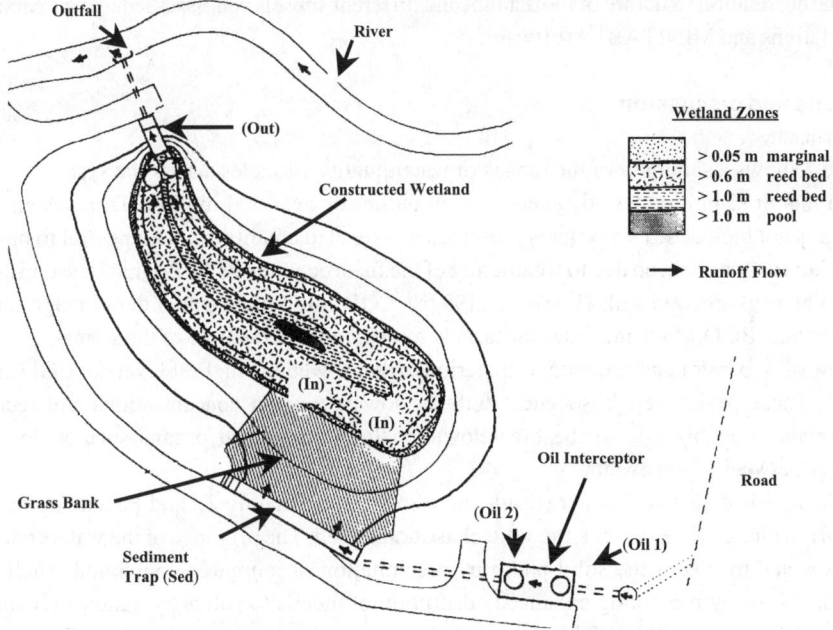

Figure 1 Basic design features of the treatment systems and sample locations

Sampling strategy and analysis

Systems C and H are being examined over an 18 month monitoring program. Results presented here are for sampling at 30–40 day intervals between December 1998 to July 1999 from duplicate fixed sampling locations in each part of the systems as indicated in Figure 1 (i.e. Oil 1, Oil 2, Sed, Wetland In and Out). Temperature, pH, dissolved oxygen and redox potential were recorded on site using hand held probes and portable meters. Samples were kept cool and dark and analysed or pre-treated within 24 hours of collection.

Grab samples of water were collected and 200–250 ml were immediately filtered in series through 1.2 μm pore glass fibre filters, then through a 0.45 μm nucleopore cellulose nitrate filter to separate the large suspended solids (TSS) and the ultra-fine residues, respectively. This filtrate was preserved with nitric acid and analysed for dissolved metals (< 0.45 μm). The mass of the large suspended solids was determined gravimetrically after drying at 80°C. The residues were digested in hot HNO_3 and analysed for metals using an acetylene flame Atomic Absorption Spectrophotometer (Pye Unicam SP9) according to Standard Methods (1992). BOD and COD were analysed using Standard Methods (1992) and the HACH^{TM} micro kit, respectively.

Solids settling through the water column were collected using catch traps consisting of plastic cups (50 mm deep, 95 mm inner diameter) which were placed just above the deposited sediment layers at fixed locations and left *in situ* between sampling occasions. The surface layer of deposited sediments was sampled, using a plastic funnel shaped scoop from randomly determined positions (but not from the same position twice) near the fixed sampling locations. Invertebrates and large solids were removed and the sediments were wet-sieved through a 63 μm mesh using sample point water. After a 24-hour settlement period, excess water was decanted off and the sediments were dried at 80°C. The mass of each fraction was recorded before metal analysis in the same manner as water residues. Blanks and standards including ground roofing slate (<63 μm) were taken through all analysis procedures.

The correlation structure of the data set and different subsets was assessed using Pearson correlations and MINITAB™ software.

Results and discussion

Water quality

Table 2 shows a summary of the ranges of water quality variables across the systems over the 6 month sampling period. There was an outlier of over 500 mg/l COD in system H which is not included. The systems presented oxygenated conditions, with neutral to basic pH which may have been due to weathering of the fresh concrete of the drainage system and the local chalk-derived soils (Lee et al., 1997b). COD was typically an order of magnitude greater than BOD which indicates the largely biologically inert nature of the waters.

The oil separator and sediment trap periodically showed the highest levels of BOD and COD. These peaks were associated with the lowest oxygen concentrations and redox potentials, probably due to the breakdown of the accumulated organic matter during storage between storm events.

The metals detected in water (results not shown) were mainly Zn and Fe and although dissolved phases were present, the solid phase dominated. The pH range of the waters could be expected to favour the solid phase by precipitation of inorganic compounds such as carbonates or by promoting enhanced adsorption of metals to solids by cation exchange (Bourg, 1989; Jenne, 1998). The greatest variation and highest concentrations of Zn and Fe in the water phases were seen in the oil interceptor and sediment trap.

The TSS concentration (> 1.2 μm) across the systems C and H are shown in Figure 3. The standard deviation is not shown and was generally <5%. There was no access to the oil separator in December and January.

The variability of solids loading entering the system and wetland component each month is demonstrated by the different concentrations in the two chambers of the oil separator and sediment trap. The smaller System C received greater suspended solids loads than System H. The outlet levels of TSS (and BOD and COD) were lower than the highest levels upstream on any sampling occasion. This indicates that the systems generally promote removal of TSS. Between the wetland inlet (in) and outlet (out) the greatest reductions in TSS was from 73 to 14 mg/l (81%) and from 96 to 30 mg/l (69%) in April and May for system C and H respectively.

The mass of fine (< 63 μm) settling solids caught by the catch traps and normalised for 30 day periods is shown in Figure 4. The variability in the 30 day catch mass of fines in the sediment trap and wetland inlet is partly due to the sediment loads received from the oil separator and partly due to a degree of resuspension and transport of settled, deposited material. Peak catches were made in the sediment trap in January and March in System C and in January, May and July in System H. System C showed the greatest catches in the early stages of the system. The catches at the outlet of System H from January-April were higher than at the inlet, partly due to some minor slippage of the banks of the balancing pond.

Peak levels of suspended solids in the oil separators and poor catches of fines were seen in the systems during April. This was probably due to a series of storms during March and

Table 2 Summary of ranges of water quality variables

System	Temp (°C)	DO (mg/l)	Redox (mV)	pH	COD (mg/l)	BOD (mg/l)
C	4–13	4.2–12.6	28–172	6.5–10.1	1–108	1.1–11.2
H	4–13	3.2–13.1	102–183	6.5–8.9	2–73	0.6–18.3

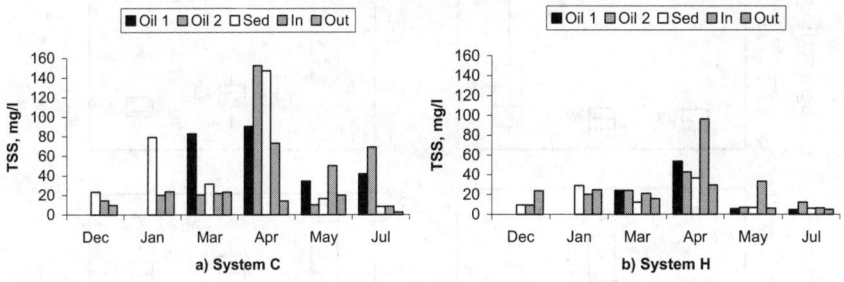

Figure 3 Mean concentration of suspended solids (> 1.2 µm) across systems C and H by month

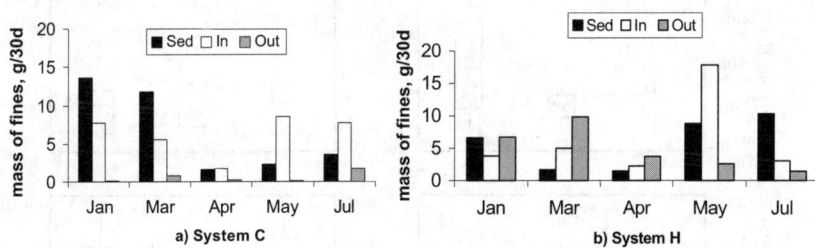

Figure 4 Mean 30 day catch mass of fine (< 63 µm) settling solids

April which may have flushed residual solids from the road, drainage system and the oil separator storage chamber (Oil 2).

Metals in sediments

The catch traps allowed settling fines to be sampled before they reached the deposited layers. This enabled the metal concentration of settling fines across the system to be compared with post depositional sediments. The concentrations of metals in settling solids showed considerable variation, but general patterns could be discerned from the data distributions. Cu and Zn were chosen as examples due to their contrasting distribution pattern and behaviour in each system. Figure 5 shows these distributions as box plots produced in MINITAB™. The boxes show the inter quartile ranges and the median as the central line. The extending vertical lines indicate upper and lower limits beyond the quartiles as ± 1.5 times the interquartile ranges and outliers are shown as asterisks. The mean values are printed besides the boxes.

Generally the metal concentrations in both systems, given in Figure 5, were comparable to other studies (Lee *et al.*, 1997a, b; Scholes *et al.*, 1998). System H was similar to these other studies in that the concentrations of Zn and Cu in settling fines and deposited sediments were greatest in the sediment trap, where Cu showed the greatest variability (Figure 5). However, in System C the concentration and variation of Zn and Cu in settling solids tended to increase towards the outlet of the wetland. These settling solids contained more Cu than the deposited sediments, whose mean concentrations of Cu and Zn were consistent across the system. These differences suggest that the settling material collected at the outlet may be different to that collected in the sediment trap. In the sediment trap the material is likely to be particles transported from the road surface, while at the outlet of the wetland the material could be from resuspended deposits, decaying plant material or transported solids from the road. Therefore the higher Cu concentrations of settling

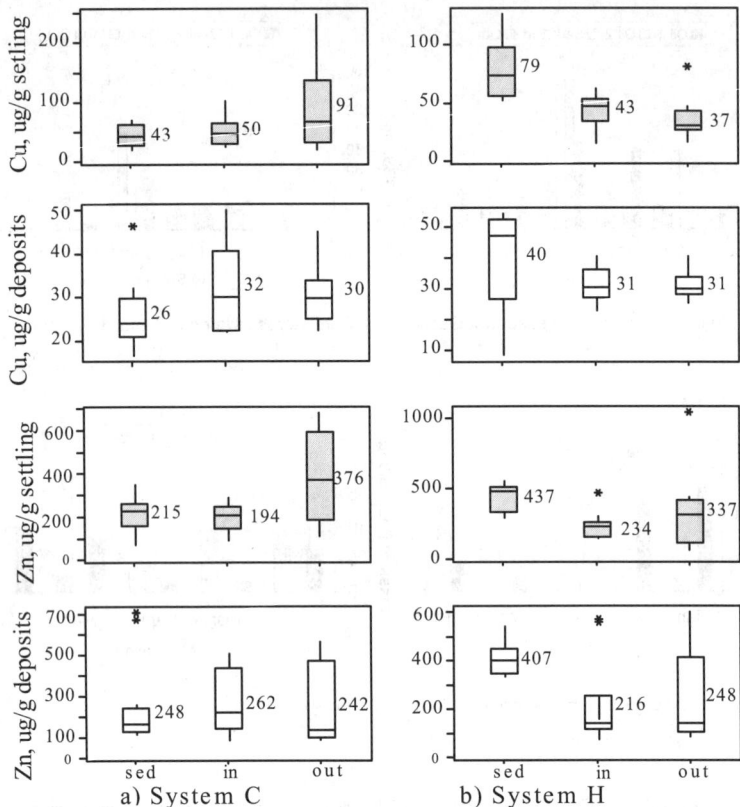

Figure 5 Data distribution of Cu and Zn in the fine (<63 μm) fraction of sediments (n = 10)

solids at the outlet of System C may indicate Cu enrichment during transport through the system.

There were significant associations (both with $r > 0.77$ and $p > 0.000$) between the Cu and Zn concentrations in fine settling solids in both ponds. The strength of these associations increased during the spring months. Associations between metal concentrations and water quality were examined to investigate the factors controlling the behaviour of metals in sediments. Although many factors, such as redox conditions and ionic strength, affect the retention of metals by sediments, pH is often considered to be the most important (Yousef et al., 1990). However there were no significant associations between Cu and Zn in sediments and pH in Systems C and H. This is probably due to the basic pH ranges encountered in the systems, which favour the solid carbonate phase and partitioning through adsorption to solids (Jenne, 1998; Bourg, 1989).

The correlation structure of the data set was examined to investigate the factors affecting the behaviour of metals. However the correlation structure differed between the ponds and in different parts of each pond. For example, in system C there is a positive association between redox in the water and Zn in fine deposits ($r = 0.88: p = 0.001$) whereas in system H there is a negative association between the two ($r = -0.62: p = 0.003$) which was strongest at the wetland outlet.

Zn and Cu are probably mainly associated with organic material, Fe oxyhydroxides and carbonates (Yousef et al., 1990; Lee et al., 1997a, b). The balance between these species may change across the systems. Organic metal complexes could have originated from organic debris on the road surface and in parts of the treatment system. The enrichment of settling solids across the smaller system C, could occur by diagenetic loss of organic material and scavenging by Fe/Mn oxyhydroxides as described by Lee *et al*. (1997a). The peak Zn and Cu levels in the sediment trap of System H with the larger catchment may indicate that settling solids arriving in the sediment trap may already be metal-enriched by diagenetic processes during the longer travel time. The redox sensitivity of Zn in the outlet of the larger system with longer in-pond residence times may indicate that Zn is more in the carbonate form and abiotic factors have a dominant influence on the solubility of inorganic material. The conditions influencing the enrichment and retention of metals by sediments are of interest in treatment system design and will be investigated further under laboratory conditions.

Conclusions

The systems appear to promote removal of suspended and settleable solids which represents an important improvement to the water quality. This not only provides mitigation of the potential physical effects of the sediment load but removes potentially toxic pollutants associated with the solids, such as metals.

The complex behaviour of pollutants was demonstrated by the different response of Zn and Cu in the two treatment systems on the same road. The wetlands of each system had different physical and chemical characteristics, which may have influenced the different behaviour patterns for metals. Some of the potential sediment metal enrichment and retention processes may involve adsorption to different sorbent materials which will be investigated under laboratory conditions.

The variability of road runoff quality and the complexity of the different processes involved in metal removal are major problems in standardising the design of treatment systems. Completion of the 18-month monitoring program will provide a sufficiently large data set to overcome seasonal and spatial variability and evaluate system performance. The research will provide a scientific basis for the improved design and operation of combined conventional and vegetated systems for the control and treatment of road runoff.

Acknowledgements

The research was funded by Mott MacDonald, the University of Portsmouth and the Highways Agency. Particular thanks goes to Peter Wilson (HA) and Colin Walker (MM) for their reliable support and discussions. Thanks to Dr. Cath Mant and Mike Woodhatch for technical support and help with sampling.

References

Bourg, A.C.M. (1989). Adsorption of trace inorganic and organic contaminants by solid particulate matter. In: *Aquatic Ecotoxicology: Fundamental Concepts and Methodologies*, A. Bidou and R. Ribeyre (eds.), vol 1, CRC Press. pp. 107–123.

Cooper, P.F., Job, G.D., Green, M.B. and Shutes, R.B.E. (1996). *Reed Beds and Constructed Wetlands for Waste Water Treatment*. Water Research Centre (WRc plc), Swindon, UK.

Jenne, E.A. (ed.) (1998). *Adsorption of Metals by Geomedia Variables, Mechanisms and Model Applications*. Academic Press.

Lee, P., Baillif, P. and Touray, J. (1997a). Geochemical behaviour and relative mobility of metals (Mn, Cd, Zn and Pb) in recent sediments of a retention pond along the A-71 motorway in Sologne, France. *Environ. Geol.* **32**(2), 142–153.

Lee, P., Touray, J. Baillif, P. and Ildefonse, J. (1997b). Heavy metal contamination of settling particles in a retention pond along the A-71 motorway in Sologne, France. *Sci. Total Env.* (201), 1–15.

Luker, M. and Montague, K. (1994). *Control of Pollution from Highway Discharges*. Report 142. CIRIA. London.

Munger, A.S., Shutes, R.B.E., Revitt, D.M. and House, M.A. (1995). An assessment of metal removal from highway runoff by a natural wetland. *Water Sci. Tech.* **32**(3), 169–175.

Nuttall, P.M., Boon, A.G. and Rowell, M.R. (1997). *Review of the Design and Management of Constructed Wetlands*. Report 180. CIRIA. London.

Scholes, L., Shutes, R.B.E., Revitt, D.M., Forshaw, M. and Purchase, D. (1998). The treatment of metals in urban runoff by constructed wetlands. *Sci. Total Env.* (214), 211–219.

Standard Methods for Examination of Water and Wastewater (1992). 18th edn, American Public Health Association/American Water Works Association/Water Environment Federation, Washington DC, USA.

Yousef, Y., Hvited-Jacobsen, T., Harper, H. and Lin, L. (1990). Heavy metal accumulation and transport through detention ponds receiving highway runoff. *Sci. Total Env.* (93), 433–440.

Nitrogen and phosphorus variation in shallow groundwater and assimilation in plants in complex riparian buffer zones

V. Kuusemets*, Ü. Mander*, K. Lõhmus* and M. Ivask**

* Institute of Geography, University of Tartu, 46 Vanemuise St, EE-51014, Tartu, Estonia
** Institute of Environmental Protection, Estonian University of Agricultural Sciences, 1 Akadeemia St, EE-51014, Tartu, Estonia

Abstract The study of purification efficiency and nutrient assimilation in plants was made in two riparian buffer zones with a complex of wet meadow and grey alder (*Alnus incana*) stand. In the less polluted Porijõgi test site, the 31 m wide buffer zone removed 40% of total nitrogen (total-N) and 78% of total phosphorus (total-P), while a heavily polluted 51 m wide buffer zone in Viiratsi retained 85% of total-N and 84% of total-P. The input of nutrients and purification efficiency displayed a significant relationship. The total-N removal in buffer zone was negative when the input value was less than 0.3 mg l^{-1} and the purification efficiency was always positive when the input value exceeded 5 mg l^{-1}. The purification efficiency of total-P was positive when the input value exceeded 0.15 mg l^{-1}. Grass vegetation plays an important role in nutrient retention in riparian buffer strips. The maximum phytomass production was measured in Porijõgi site where production of the *Filipendula ulmaria* community was up to 2,358 g m^{-2}, assimilation of N 32.1 and of P 4.9 g m^{-2}, respectively. This is much higher than the biomass production and N and P uptake of the grey alders (*Alnus incana*) at the same site – 1,730, 20.5 and 1.5 g m^{-2}, respectively.

Keywords *Alnus incana*; assimilation of nitrogen; assimilation of phosphorus; buffer zones

Introduction

Buffer zones have many functions that improve water quality, protect air and soil, increase biological and landscape diversity (Mander *et al.*, 1997b). One of the main functions of buffer zones and riparian wetlands, to purify water of contaminant substances, has been widely described in many regions (Peterjohn and Correll, 1984; Lowrance *et al.*, 1984; Uusi-Kämppä and Ylaranta, 1992; Haycock and Pinay, 1993; Vought *et al.*, 1994). However, our knowledge concerning the water quality buffering effects of riparian zones is far from adequate (Correll, 1997). Different studies indicate that buffer zones can retain 0.0043 to 13 g N m^{-2} day^{-1} and 0.000057 to 8.67 g P m^{-2} day^{-1} (Mander *et al.*, 1997b). Because retention and efficiency rates vary greatly under different climatic and physico-geographical conditions, few proposals have been presented with design criteria for buffer zones and their establishment and management (Dillaha and Inamdar, 1997; Mander *et al.*, 1997b). The optimal vegetation and the most effective buffer width is still unclear (Correll, 1997). Haycock and Pinay (1993) found that forested buffer zones are more effective than grass buffer zones, while other works show good purification in grass buffer zones (Peterjohn and Correll, 1984; Groffman *et al.*, 1991; Correll, 1997). Grass strips are considered sediment traps by which a large portion of nutrients, especially P is deposited from surface flow (Peterjohn and Correll, 1984; Young *et al.*, 1980). At the same time grass buffer strips can also remove dissolved nutrients (Peterjohn and Correll, 1984; Haycock and Burt, 1993; Vought *et al.*, 1994) which shows that grass strips can be an important part of a buffer zone. Although the assimilation of nutrients by vegetation in buffer zones has been described in only a few works, results indicate high nutrient assimilation ability. Van Oorschot (1994) measured above-ground N and P uptake in riparian communities up to

7.1 and 1.07 g m^{-2} year^{-1}, respectively. Prach and Rauch (1992) estimate the N and P removal by hay from a floodplain to be 15–30 and 2–4.5 g m^{-2}, respectively.

In the study described here we provide an overview on the purification efficiency of complex riparian buffer zones in Estonia consisting of native riparian plant communities with emphasis on nutrient assimilation and input versus retention calculations.

Materials and methods

Site description. To study buffer zone efficiency we established two transects in south Estonia with similar physico-geographical conditions. The transects are situated on slopes adjacent to streams and follow surface water flow across agricultural fields and different riparian plant communities. The Viiratsi transect is situated in the Sakala heights (Varep, 1964) consisting of moraine hills and undulated plains with a variety of glacial deposits. The transect is located on the moraine plain in the vicinity of the pig farm (about 30,000 pigs during the study). Almost all the slurry from the pig farm is spread on the neighbouring fields and whole area is heavily impacted by pig slurry. The transect crosses the following plant communities: a cultivated field on planosols and podzoluvisols where slurry was spread in autumn 1994; an 11 m wide strip of grassland (*Elytrigia repens-Urtica dioica*) and young alder (*Alnus incana*) trees on colluvial podzoluvisool; a 12 m wide wet grassland (*Filipendula ulmaria*) on gleysol and a 28 m wide grey alder (*Alnus incana*) forest on podzoluvic gleysol (Figure 1A).

The Porijõgi transect is situated in the plain of south-east Estonia (Varep, 1964), on the slope of a primeval valley where agricultural activities stopped in 1992, two years before we began our study. The Porijõgi transect crosses several plant communities: abandoned field (last cultivated in 1992) on planosols and podzoluvisols; abandoned cultivated grassland (last mowed in 1993) on colluvial podzoluvisol (dominated by *Dactylis glomerata* and *Alopecurus pratensis*); an 11 m wide wet grassland on gleysol (two parallel communities,

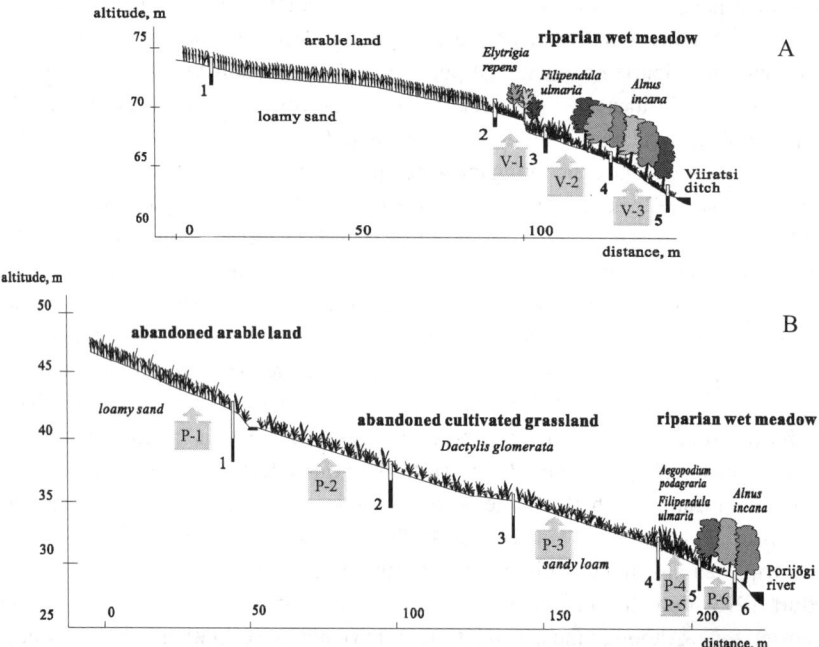

Figure 1 Study transects in complex riparian buffer zones in South Estonia. A – Viiratsi; B – Porijõgi. 3 – shallow groundwater sampling point; P-1 soil and phytomass sampling plots

one dominated by *Filipendula ulmaria,* another by *Aegopodium podagraria*); and a 20 m wide grey alder stand (*Alnus incana*) on gleysol (Figure 1B).

Water sampling and analysing. Shallow ground water samples from upper aquifer were collected once to twice a month from piezometers installed on the borders of plant communities. The piezometers on the borders of riparian communities were installed with 3 replicates in 20 (Pori) to 30 (Viiratsi) metres wide zones. The depth of groundwater varied 1–2 m in the field sampling points and 10–80 cm in riparian communities. Samples were taken from July 1994 to December 1995. Filtered water samples were analysed for NH_4^+-N, NO_2^--N, NO_3^--N, total Kjeldahl-N, PO_4^{3-}-P, total-P, SO_4^{2-}, Fe, Ca^{2+} in the laboratory of the Estonian Agricultural University following standard methods for examination of water and wastewater quality (*Standard Methods*, 1981).

Soil and phytomass sampling and analysing. Complex soil samples were taken in three replicates from two depths (0–10 cm and 10–20 cm) from all plant communities toposequent through the riparian buffer zones (Figure 1). Sampling was done twice a year: in spring (May) and autumn (October). A hand-held 4 cm diameter corer was used to collect samples. Soil pH value, organic matter (loss of ignition), and N and P content were analysed using standard methods (*Standard Methods*, 1981). The phytomass (i.e., standing crop) samples were collected from all riparian plant communities during the maximum flowering time of the dominant plant species (2nd and 3rd week in July; see Milner and Hughes, 1968). Sampling plots (six in Porijõgi and three in Viiratsi) were installed in typical areas of the community. The above ground phytomass was collected from three replicate quadrates (1 × 1 m) in each community. Below ground root phytomass was collected from soil cores taken by auger (diameter 158 mm) to a depth of 40–50 cm in three replicates from each location. Roots were washed of soil and from the dried roots and above ground phytomass, dry weight was measured and N and P content was analysed in the laboratory of the Estonian Agricultural University. In the grey alder community above ground phytomass and productivity of grey alders were estimated (without herb phytomass) using dimension-analysis techniques (Bormann and Gordon, 1984; Rytter, 1989). At both test sites (age 14 in Porijõgi transect and 40 years in Viiratsi transect) 17 and 5 sample trees per plot, respectively, were felled to collect data on the following tree components: stem (wood and bark), secondary branch growth (wood and bark), primary branch growth, leaves, generative organs (Lõhmus *et al.*, 1996). The relative increments of the wood and bark of an over-bark fraction were assumed to be equal. Root systems for 6 and 3 out of the sampled 17 and 5 trees, respectively, were excavated. The dried weight of all tree components was measured and N and P content in dried phytomass were analysed in the laboratory of Estonian Agricultural University.

Statistical analysis and calculations. The Kruskal-Wallis test was performed due to inhomogeneity of variances to analyse N and P concentration changes between measurement points and the test for Binary Sequences for load versus purification efficiency probability analyses using *Statgraphics Plus 7.1*. The regression analysis of relation between input load and appearance of negative removal of nutrients was performed by *Microsoft Excel 97*.

Results

The soil water nitrogen load in two transects was different. The total-N content in Viiratsi transect reached up to 76 mg l^{-1} in the cultivated field after slurry application. In Porijõgi transect the highest total-N concentration was 13.5 mg l^{-1}. The results show that there were considerable decreases in nitrogen content through the Viiratsi transect buffer zone (see Table 1). The average total-N decreased during the study period (1994 to 1995) from 19.1 mg l^{-1} in the field (transect point 1, Figure 1A) to 2.9 mg l^{-1} at the end of the buffer zone (transect point 5, Figure 1A). However, this decrease was not significant (Kruskal-Wallis test, $P>0.05$).

There was considerable decrease in nitrogen through the first 2 m of the buffer zone where the average total-N decreased from 19.1 to 11.8 mg l^{-1} in the grass community (transect point 2) and in the following 11 m wide wet grass community to 3.6 mg l^{-1} (in transect point 3). In the alder forest zone the nitrogen content decreased to 2.4 mg l^{-1}. This change from point 2 is highly significant ($P<0.01$) compared to the values at points 4 and 5. The change was highly significant also when comparing points 3 and 4, and 3 and 5.

Plant nutrient assimilation in the Viiratsi transect was higher in the wet meadow *Filipendula ulmaria* association (sampling plot V-2, Figure 1A), where an average of 21.1 g N m^{-2} yr^{-1} and 4.8 g P m^{-2} yr^{-1} was assimilated in grass (Table 1). This was higher than the nutrient uptake by alders (14.0 and 1.1 g m^{-2} yr^{-1}, respectively). In the Porijõgi

Table 1 Nutrient variation in shallow groundwater (average ± standard error), soil and the plant phytomass in the complex buffer zone in Viiratsi. Sampling points are given in brackets as shown in Figure 1A

	Field (1)	Grassland (2)(V-1)	Wet meadow (3)(V-2)	Alder I (4)(V-3)	Alder II (5)
Total-N in shallow groundwater (mg l^{-1})	19.1 ± 7.0	11.8 ± 3.2	3.6 ± 0.4	2.4 ± 0.5	2.9 ± 0.7
N assimilation by plants (g N m^{-2} yr^{-1})		17.5	21.1	14.0	
Topsoil (0–10 cm) N content (mg g^{-1})		2.16	6.76	9.87	
Soil (10–20 cm) N content (mg g^{-1})		1.66	5.37	5.61	
Total-P in shallow groundwater (mg l^{-1})	0.43 ± 0.28	0.17 ± 0.04	0.05 ± 0.005	0.05 ± 0.004	0.07 ± 0.01
P assimilation by plants (g P m^{-2} yr^{-1})		3.7	4.8	1.1	
Topsoil (0–10 cm) P content (mg g^{-1})		0.49	0.50	0.94	
Soil (10–20 cm) P content (mg g^{-1})		0.37	0.40	0.53	
Phytomass production (g m^{-2})		1320	1015	1060	

Table 2 Nutrient variation in shallow groundwater (average ± standard error, soil and the plant phytomass in the complex buffer zone in Porijõgi. Sampling points are given in brackets as shown in Figure 1B

	Field (1) (P-1)	Cultivated grassland I (2)(P-2)	Cultivated grassland II (3)(P-3)	Wet meadow I (4)(P-4)	Wet meadow II (5)(P-5)	Alder (6)(P-6)
Total-N in shallow groundwater (mg l^{-1})	1.2 ± 0.2	2.2 ± 0.7	2.0 ± 1.0	1.7 ± 0.3	2.0 ± 0.4	1.3 ± 0.2
N assimilation by plants (g N m^{-2} yr^{-1})	11.4	13.2	13.6	21.3*	18.8**	20.5
Topsoil (0–10 cm) N content (mg g^{-1})			2.02		2.82	10.74
Soil (10–20 cm) N content (mg g^{-1})			1.56		2.36	9.54
Total-P in shallow groundwater (mg l^{-1})	0.49 ± 0.3	0.29 ± 0.01	0.27 ± 0.15	0.06 ± 0.02	0.09 ± 0.03	0.06 ± 0.005
P assimilation by plants (g P m^{-2} yr^{-1})	2.7	2.1	2.6	3.0*	3.3**	1.5
Topsoil (0–10 cm) P content (mg g^{-1})			0.45		0.63	1.02
Soil (10–20 cm) P content (mg g^{-1})			0.35		0.53	0.9
Phytomass production (g m^{-2})	1113	1152	1493	1748	1977	1730

* *Aegopodium podagraria*
** *Filipendula ulmaria*

transect the assimilation of nutrients was also highest in the wet meadow where the *Filipendula ulmaria* association (sampling plot P-5, Figure 1B, Table 2) assimilated 21.3 and 3.0 g m^{-2} yr^{-1} of N and P, respectively. This was higher than annual N and P uptake by alders: 20.5 and 1.5 g m^{-2}, respectively. Maximum nutrient assimilation in 1994 reached 32.1 g N m^{-2} in the wet meadow of Porijõgi transect (sampling plot P-5) and 5.5 g P m^{-2} yr^{-1} in the wet meadow of Viiratsi transect (sampling plot V-2).

The N and P contents in soil increase in wet meadow riparian communities. The nitrogen content in the Viiratsi topsoil layer increased from 2.16 to 9.87 mg g^{-1} (sampling points V-1 and V-3, Figure 1) and the phosphorus content from 0.49 to 0.94 mg g^{-1}. In Porijõgi transect the increase of nitrogen and phosphorus was from 2.02 to 10.74 and from 0.45 to 1.02 mg g^{-1}, respectively (sampling points P-3 and P-6, Figure 1). The highest values were in the alder *Alnus incana* communities.

Discussion

Both transects show high purification efficiency. The average removal of total-N and total-P in Viiratsi was 85% and 84% and in Porijõgi 40% and 78%, respectively. The water quality had already improved within the first metres of the buffer zone. In Viiratsi transect the purification efficiency of total-N and total-P within first 2 m of the buffer grassland was 38% and 60%, respectively, whereas in Porijõgi transect the purification efficiency of total-N and total-P in the 2 m from the edge of the cultivated grassland was 18% and 78%, respectively.

To analyse the relation between purification efficiency and input concentration we calculated the purification efficiency separately for each sampling day for every buffer strip between two sampling points. Data were divided by input concentration into 8 classes with 11 to 36 measurements in each class. The Tests of Binary Sequences were performed to calculate the level of significance between occurrence of positive or negative removal efficiency. The probability of negative removal was calculated by dividing the number of negative removal cases with the total measurement number (%). Comparison of removal efficiency and input concentration shows that the removal of total-N was negative ($P<0.01$), when the input was less than 1.0 mg l^{-1} (Figure 2A). For input concentrations between 1.0 to 5.0 mg l^{-1}, removal of total-N showed no significant positive or negative tendency, although positive removal was more common – the probability of negative removal is 30.8 to 40.0% ($P>0.05$). For input concentrations greater than 5.0 mg l^{-1} the purification efficiency is significantly positive ($P<0.01$), for input concentrations greater than 42 mg l^{-1} purification efficiency is always positive ($P>0.001$). The relation between input concentration and appearance of negative removal of nitrogen (N_{neg}) is described by logarithmic regression:

$N_{neg} = 63.0 - 16.84 \ln(I_{Nmax})$
$R^2 = 0.78, P > 0.001$

where I_{Nmax} is maximum value of N input class.

For total-P there is no interval for negative removal (Figure 2B). The input concentration 0.01 to 0.15 mg l^{-1} yields positive or negative removal. Positive removal is prevalent for input from 0.05 to 0.15 mg P l^{-1} – the probability of negative removal is 18.8 to 38.9% ($P>0.05$). For input greater than 0.15 mg l^{-1}, the purification efficiency of total-P was significantly positive ($P<0.01$), for input concentrations greater than 2.1 mg l^{-1} purification efficiency is always positive ($P<0.01$). The relation between input concentration and appearance of negative removal of phosphorus (P_{neg}) is described by logarithmic regression:

$$P_{neg} = 7.9 - 10.20\ln(I_{P\max})$$
$$R^2 = 0.72, P<0.01$$

where $I_{P\max}$ is maximum value of P input class.

The load-retention relationship in various ecosystems has been discussed in earlier studies (Fleisher et al., 1991; Haycock and Pinay, 1993; Mander et al., 1997b). The results show strong positive correlation between nutrient load and removal. However, buffer zones have upper limits of purification and these regression formula can not been used in the planning in the case of high input values. This analysis provides limits for buffer strips as water purification systems. The input concentration range (1.0–5.0 and 0.01–0.15 mg l^{-1} of N and P, respectively) can be considered to represent natural conditions of studied buffer strips where water output quality also depends on natural processes taking place in the buffer and on increased nutrient content in soils. This can explain certain increases of total-N and total-P content in groundwater inside both studied buffers. However, the highest increase of N and P content in soil is taking place in *Alnus incana* community, while the content of nutrients in groundwater is decreasing. This indicates that the increased N and P contents in soil are not affecting the water quality directly. The increased nutrient content in soil in *Alnus incana* community can be explained by very high nutrient content in litter of alder (Mikola, 1958). To achieve good purification ability it is important to design a complex buffer zone consisting of different ecosystem strips. The average outflow values from both buffer zones were lower than 3.1 and 0.07 mg l^{-1} of total-N and total-P in Viiratsi, respectively and 1.5 and 0.06 mg l^{-1} in Porijõgi, respectively. Our estimation of denitrification intensity showed that this process does not play a substantial role in nitrogen removal (7.9–20.1 kg N ha^{-1} yr^{-1} in Porijõgi and 8.5–19.3 kg N ha^{-1} yr^{-1} in Viiratsi, Mander et al., 1997a).

The nutrient assimilation in plants indicates that grass communities play an important role for nutrient retention in buffer zones. The average N and P content in herbal shoots in the wet meadows were 11.6 and 1.6 g m^{-2}, respectively in the Porijõgi and 10.6 and 2.3 g m^{-2}, respectively in the Viiratsi transect. This gives good opportunity to remove a portion of nutrients by grass mowing and hay removing while felling of trees can be done with intervals of decades. Cutting should be done during the maximum flowering period of the dominant species when the nutrient content in the shoot phytomass is highest (Deinum, 1966). The mowed herbs should be removed after mowing to avoid rapid nutrient loss from hay (Schaffers et al., 1998).

The results show that complex buffer zones of grass and forest strips are very effective in N and P retention, which agrees with previous research (Schultz et al., 1995; Lowrance,

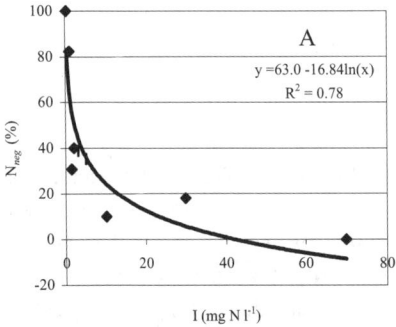
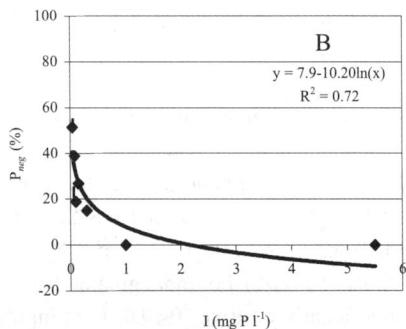

Figure 2 The relation between input concentration (I, mg l^{-1}) and probability of negative removal of nitrogen (A) and phosphorus (B) (N_{neg} and P_{neg}, %)

1991). This kind of complex can be recommended for buffer strip design where grass strips are considered as sediment traps but also as an important mechanism for dissolved N and P removal. Both features provide the opportunity to remove part of the nutrients from the system. In addition to efficient nutrient purification potential, forest buffer strips have many other environmentally important functions such as protection against soil erosion, filtering polluted air, canopy shading, and increasing biological and landscape diversity.

Conclusion

In general, buffer zones can be effective multifunctional tools to control nutrient losses from intensively used watersheds. The studied buffer strips showed good purification efficiency for N and P when input concentrations were high (>5 mg N l^{-1} and >0.15 mg P l^{-1}, respectively). At the same time the complex buffer zone consists of different buffer strips having high purification efficiency. For instance, a heavily loaded complex buffer zone consisting of grass and forest strips showed relatively low output concentrations for total-N and total-P, which are comparable with the output values from the unloaded transect. However, removal can be negative in case of lower input (<1 mg N l^{-1}). Some unpredictability of output concentration was observed when the initial input was intermediate. When planning and designing buffer zones, the role of grass buffer strips and wet meadows should be considered. Management of grasslands and forests can significantly decrease the load in complex buffer zones.

Acknowledgements

This study was supported by the International Science Foundation grants No LCU 100 (1994) and LLL100 (1995), and Estonian Science Foundation grants No 2471 and 3885.

References

Bormann, B.T. and Gordon, J.C. (1984). Stand density effects in young red alder plantations: Productivity, photosynthate partioning and nitrogen fixation. *Ecology*, **65**, 394–402.

Correll, D.L. (1997). Buffer zones and water quality protection: general principles. In: *Buffer Zones: Their Processes and Potential in Water Protection*, N.E. Haycock, T.P. Burt, K.W.T, Goulding and G. Pinay (eds.), Quest Environmental, Foundation for Water Research, Oxford, pp. 7–20.

Deinum, B. (1966). Climate, nitrogen and grass. Mededelingen Landbouwegeschool, Wageningen, **66**(11), 1–91.

Dillaha, T.A. and Inamdar, S.P. (1997). Buffer Zones as Sediment Traps or Sources. In: *Buffer Zones: Their Processes and Potential in Water Protection*, N.E. Haycock, T.P. Burt, K.W.T. Goulding and G. Pinay (eds.), Quest Environmental, Foundation for Water Research, Oxford, pp. 33–44.

Fleischer, S., Stibe, L. and Leonardson, L. (1991). Restoration of wetlands as a means of reducing nitrogen transport to coastal waters. *Ambio*, **20**, 271–272.

Groffman, P.M., Axelrod, E.A., Lemunyon, J.L. and Sullivan, W.M. (1991). Denitrification in grass and forest vegetated filter strips. *J. Environ. Qual.*, **20**, 671–674.

Haycock, N.E. and Burt, T.P. (1993). The sensivity of rivers to nitrate leaching. The effectiveness of near-stream land as a nutrient retention zone. In: *Landscape Sensitivity*, D.S.G. Thomas and R.J. Allison (eds.), Wiley, London, pp. 261–272.

Haycock, N.E. and Pinay, G. (1993). Groundwater nitrate dynamics in grass and poplar vegetated riparian buffer strips during the winter. *J. Environ. Qual.*, **22**, 273–278.

Jordan, T.E., Correll, D.L. and Weller, D.E. (1993). Nutrient interception by riparian forest receiving inputs from adjacent cropland. *J. Environ. Qual.*, **22**(3), 467–473.

Lowrance, R. (1991). Effects of buffer systems on the movement of to N and P from agriculture to streams. In: Npo-forkning fra Miljøstyrelsen. Conference Contributions, International Conference on N, P and Organic matter, Helsingør, Denmark.

Lowrance, R.R., Todd, R., Fail, J., Hendrickson, O., Leonard, R. and Asmussen, L. (1984). Riparian forest as nutrient filters in agricultural watersheds. *BioScience*, **34**, 374–377.

Lõhmus, K., Mander, Ü., Tullus, H. and Keedus, K. (1996). Productivity, buffering capacity and resources of gray alder forests in Estonia. In: *Short Rotation Willow Coppice for Renewable and Improved Environment*, K. Perttu and A. Koppel (eds.), Swedish University of Agricultural Sciences, Uppsala, pp. 95–105.

Mander, Ü., Lõhmus, K., Kuusemets, V. and Ivask, M. (1997a). The potential role of wet meadows and grey alder forests as buffer zones. In: *Buffer Zones: Their Processes and Potential in Water Protection*, N.E. Haycock, T.P. Burt, K.W.T. Goulding and G. Pinay (eds.), Quest Environmental, Foundation for Water Research, Oxford, pp. 35–46.

Mander, Ü., Kuusemets, V., Lõhmus, K. and Mauring, T. (1997b). Efficiency and dimensioning of riparian buffer zones in agricultural catchments. *Ecol. Eng.*, **8**, 299–324.

Milner, C. and Hughes, R.E. (1968). *Methods for the Measurement of the Primary Production of Grassland*. IBP Handbook No. 6. Blackwell Scientific Publications, Oxford, UK.

Mikola, P. (1958). Liberation of nitrogen from alder leaf litter. *Acta Forestry Fennica*, **67**, 1–10.

van Oorschot, M.M.P. (1994). Plant production, nutrient uptake and mineralization in river marginal wetlands: the impact of nutrient additions due to former land-use. In: *Global Wetlands: Old World and New*, W.J. Mitsch (ed.), Elsevier Science B.V., Amsterdam, pp. 133–150.

Peterjohn, W.T. and Correll, D.L. (1984). Nutrient dynamics in an agricultural watershed: observations on the role of a riparian forest. *Ecology*, **65**, 1466–1475.

Prach, K. and Rauch, O. (1992). On filter effects of ecotones. *Ekologia (CSFR)*, **11**(3), 293–298.

Rytter, L. (1982). Distribution of roots and root nodules and biomass allocation in young intensively managed grey alder stands on a peat bog. *Plant and Soil*, **128**, 363–365.

Schaffers, A.P., Vesseur, M.C. and Sykora, K.V. (1998). Effects of delayed hay removal on the nutrient balance of roadside plant communities. *J. of Appl. Ecol.*, **35**, 349–364.

Schultz, R.C., Colletti, J.P., Isenhart, T.M., Simpkins, W.W., Mize, C.W. and Thompson, M.L. (1995). Design and placement of a multi-species riparian buffer strip system. *Agroforestry Systems*, **29**, 201–226.

Standard Methods for the Examination of Water and Waste Water (1981). 15th edn, American Public Health Organization, Washington DC, USA.

Uusi-Kämppä, J. and Yläranta, J. (1992). Reduction of sediment, phosphorus and nitrogen transport on vegetated buffer strips. *Agric. Sci. Finl.*, **1**, 569–574.

Varep, E. (1964). The landscape regions of Estonia. Publications on Geography IV. *Acta et Commerstationes Universitatis Tartuensis*, **156**, 3–28.

Vought, L.B.-M., Dahl, J., Pedersen, C.L. and Lacoursiere, J.O. (1994). Nutrient Retention in Riparian Ecotones. *Ambio*, **6**, 342–348.

Young, R.A., Huntrods, T. and Anderson, W. (1980). Effectiveness of vegetative buffer strips in controlling pollution from feedlot runoff. *J. Environ. Qual.*, **9**, 483–487.

Keyword Index

2NT 515
2,4DNT 515
2,6DNT 515
accumulation 463
acid mine drainage 449
advanced treatment technology 101
aeration 9, 469
aerobic bacteria 207
agricultural pollution 523
agricultural runoff 531
airport runoff 469
algal turf scrubber 427
Alnus incana 615
alternative technologies 345
amended sediment 85
ammonia 157
anaerobic bacteria 207
aquatic macrophytes 39, 579
artificial inundation 507
assimilation 289
assimilation of nitrogen 615
assimilation of phosphorus 615
atrazine 539
Australia 303
avifauna 27

background levels 77
bacteria 211
bacteria and viruses removal 599
biochemical oxygen demand 399
biodiversity 289
biofilm 85
biomass harvesting 61
blast furnace slag 69
BMKO 331
BOD removal 469
BOD_5 211, 369
brackish water 361
buffer zones 615

C. perfringens 177
capacity 435
Caribbean coast 361
cation exchange selectivity coefficient 477
cattail 143
$CBOD_5$ 259

ciliates 191, 211
clogging 375
COD 441
coke plant 485
cold climate 237, 259, 345
coliforms 177
coliphages 177, 223
combined sewer overflow 171
combined system 137
conduction 251
constructed 311, 493
constructed wetlands 9, 27, 47, 55, 61, 69, 85, 93, 109, 117, 163, 171, 177, 183, 191, 215, 223, 231, 237, 251, 259, 267, 281, 295, 303, 325, 339, 345, 381, 387, 393, 407, 435, 441, 455, 469, 485, 499, 515, 523, 545, 565, 571, 579, 585, 591, 599, 607
contamination 325
convection 251
Cryptosporidium 191
cycles 251
Cyperus fabelliformis 137
Cyperus papyrus 199
cypress dome 413
Czech Republic 369

dairy 19
dairy farm wastewater 69
dairy wastewaters 183, 281
database 331
denitrification 19, 157, 163, 399
detention pond 531, 565
detritus 1
developing countries 381
die-off 223
diel variations 421
diffusion 157
dissolved organic carbon 267
distillery effluent 441

ecology 289
efficiency 435
energy 251
enterococci 177
environmental engineering 295
evaporation 407

evapotranspiration 267
Everglades 93, 101, 117, 123, 295
experimental sewage treatment 387
explosives 515

faecal coliforms 199, 565
fecal stroptococci 207
fen 143
fiddler crab population density 591
filter media 47
filtering soil 579
fish 317
fish farming 55
Florida 131
fractionated extraction 507
fresh-water marsh 507

glycol 469
golf course 591
grease and oil content 361
groundwater 515
growth 493

heavy metals 407, 455, 463, 507
highway runoff 571
hogs 157
horizontal flow 369, 375
horizontal subsurface flow 339
hydraulic conductivity 361
hydraulic residence time 77
hydrology 117
hydrophites 325

in situ gas measurement 273
indicator bacteria 183
indicator organisms 223
insulation 259
ion-exchange 477
iron 421
iron sand 149
iron-ore 69

kinetics 231

lake 131
limerock filter 101
limestone drains 449
loading 311

macroinvertebrates 317
macrophyte productivity 1
macrophytes 61, 303, 339, 493
marsh 545
mass retention 117

metabolism 273
metal partitioning 607
metals 571
methane 441
microbial metabolism 1
Miscanthidium violaceum 199
mitigation 507
modelling 231, 237, 273, 289
moist subtropics 353
molasses 441
mulch 259
municipal water treatment 507

Nakivubo wetland 199
natural systems 9
natural treatment wetland 413
natural wastewater treatment 331
natural wetlands 289, 317, 455, 599
nature study 27
Newbury Bypass 607
nitrate 559
nitrification 19, 399
nitrogen 77, 117, 137, 171, 259, 413, 485
nitrogen reduction 143
nitrogen removal 163, 435
nitrogen transformation 237
non-point source pollutants 539, 585
nursery runoff 77
nutrient 131, 407
nutrient balance 289
nutrient recycling 1
nutrient removal 9, 149, 303, 427
nutrients 245, 369, 455

oligotrophication 591
on-site wastewater treatment 353
oocyst removal 191
organic matter 1
organic matter composition 281
organic solids 281
oxygen consumption kinetics 273
oxygen transfer 259

P 131
pathogens 183, 215, 345
peat 477
percolate quality 551
performance 311
periphyton 1, 123, 427
pesticide 539
Phalaris arundinacea 207
phosphorus 19, 47, 77, 85, 93, 101, 109, 117, 131, 171, 413, 485, 565
phosphorus filter 149

phosphorus reduction 143
phosphorus removal 39, 55, 61, 123, 295, 435
phosphorus retention 69
phosphorus sorption 47
photosynthesis 421
Phragmites australis 207, 245, 369, 485, 523, 571
Phragmites karka 441
phytoremediation 9
pig farm wastewater 137
pilot system 387
plant biomass 303
plant uptake 137, 455
pond 331, 545
poplar plantation 331
pore clogging 281
potato processing wastewater 163
predation 191
primary sludge treatment 551
protozoa 191
public water supply 579

RDX 515
reciprocating wetlands 399
recirculation 137, 149
recreational use 27
recycling 407
redox 19
reed 245, 463
reed beds 47, 339, 353, 393, 551
refinery effluent 493
removal 131
removal efficiency 387, 499
rerating 435
respiration 421
respiration rates 273
respirometric test 273
restoration 93, 123, 131, 295
retention 463
rewetting 143
river clean up 599
road runoff control 607
roots 207

Salix 407
salmonella 215
saprobic index 211
seasonal variation 375
secondary effluent 267
secondary treatment 339
sediments 223, 531, 607
sludge dewatering 393, 551
sludge drying 551

sludge loadings 551
sludge treatment 55
soil 251
soil clogging 361
stormwater 85, 171, 223, 531, 565
stormwater runoff 93, 539
stormwater treatment areas 295
submerged aquatic vegetation 101
substrate 123
subsurface 375
subsurface flow 207, 211, 231, 499
subsurface flow constructed wetland 353
subsurface flow wetlands 149
subsurface horizontal-flow wetlands 77
subsurface wetlands 251
subtropical wetlands 109
sulphide BOD 361
sulphide production 273
surface complexation model 477
surface flow 499
surface flow wetland 157, 303, 421
surface waters 325
swine 19
swine waste 215
swine wastewater 545

technology transfer 381
temperature 251
terminal electron transport system (ETS) activity 245
tidal flow system 441
TNT 515
total ammonia nitrogen 399
total phosphorus 399
treatment performance 353
treatment wetlands 39, 123, 317
tropical wetland 499
trout 55
Typha latifolia 523, 571
Typha subulata 311

Uganda 199
urban runoff 171
USA 131

vegetation 317
vertical flow constructed wetlands 361
vertical flow wetlands 273
vertical-flow system 47, 171

wastewater 55, 157, 177, 223, 237, 345, 387, 413, 455, 499
wastewater effluent 317
wastewater reuse 339

wastewater treatment 27, 215, 231, 381, 427
wastewater treatment wetland 585
water quality 109, 531, 559
water table fluctuations 245
water treatment 47, 109, 171, 407
watershed restoration 449
wetland plants 9

wetland restoration 559
wetlands 1, 101, 131, 311, 331, 435, 449, 463, 493, 539
wildlife habitat 27
willow 407

zero-discharge 407
Zizaniopsis bonariensis 311

Scientific & Technical Reports Series
Up to 25% Discount for IWA Members

Sequencing Batch Reactor Technology
Scientific & Technical Report No. 10

Editors: Peter A. Wilderer, Robert L. Irvine, Mervyn C. Goronszy

This report provides an in-depth summary of and scientific background for periodic processes. Specifically, attention is given to what is commonly referred to as "Sequencing Batch Reactor (SBR) Technology". The report highlights various types of SBRs, design considerations and procedures, equipment required, and experiences gained from practical applications. This report will help both designers and operators of SBRs understand how to use this technology successfully.

The focus is on the application of fill-and-draw, variable volume, periodically operated, unsteady-state principles to activated sludge systems. The same concepts have been successfully applied to other systems including anaerobic reactors, biofilm reactors, biofiltration, solid phase and slurry reactors for the treatment of contaminated waters, wastewaters, gases, and soils. Summarized herein are results from bench, pilot and full scale SBRs. Also included is a description of trends for technological developments and a discussion of open questions regarding research, development, application, and operation that are in search of answers.

January 2001, c.100 pages
ISBN: 1-900222-21-3 Paperback

IWA MEMBERS PRICE	£37.50 / US$60.00
NON MEMBERS PRICE	£50.00 / US$80.00

ALSO AVAILABLE:

Respirometry in Control of the Activated Sludge Process: Principles

Scientific & Technical Report No 7
Editors: H. Spanjers, P.A. Vanrolleghem, G. Olsson, P.L. Dold
1998 48 pages, Paperback
ISBN: 1-900222-04-3

IWA MEMBERS PRICE	£24/$39.00
NON MEMBERS PRICE	£32/$52.00

The Activated Sludge Models (1,2,2d and 3)
Scientific and Technical Report No.9

By the IWA Task Group on Mathematical Modelling for Design and Operation of Biological Wastewater Treatment.

Editor: Mogens Henze

This book gives a total overview of the Activated Sludge Model (ASM) family, ASM1, ASM2, ASM2d and ASM3 together. In 1982 the International Association on Water Pollution Research established a Task Group on Mathematical Modelling for Design and Operation of Activated Sludge Processes. At that time modelling of activated sludge processes had been a discipline for about 15 years. The various models developed had only little use, partly due to lack of trust in them, partly due to the limitations in computer power, and partly due to the complicated way these models had to be presented in written form. To overcome this, Activated Sludge Model No.1 was published in 1987 and was received with interest.

The models have grown more complex over the years: from ASM1, which includes nitrogen removal processes, through the inclusion of biological phosphorus removal in ASM2, to ASM2d including denitrifying biological phosphorus organisms. In 1998 the Task Group decided to develop a new modelling platform, ASM3, to create a tool for use in the next generation of activated sludge models. The ASM3 in this book is based on recent developments in the understanding of activated sludge processes. Among these are the possibilities of following internal storage compounds that play an important role in the metabolism of the organisms.

June 2000 130 pages
ISBN: 1-900222-24-8 Paperback

IWA MEMBERS PRICE	£39.75 / US$63.00
NON MEMBERS PRICE	£53.00 / US$85.00

Secondary Settling Tanks: Theory, Modelling, Design and Operation

Scientific & Technical Report No 6
Editors: G.A. Ekama, J.L. Barnard, F W Günthert, P Krebs, J A McCorquodale, D S Parker, E J Wahlberg
1997 232 pages, Paperback
ISBN 1-900222-03-5

IWA MEMBERS PRICE	£37.50/$60
NON MEMBERS PRICE	£50.00/$80

Microbial Community Analysis — The Key to the Design of Biological Wastewater Treatment Systems

Scientific & Technical Report No 5
Editors: T.E. Cloete and N.Y.O. Muyima
1997 98 pages, Paperback
ISBN: 1-900222-02-7

IWA MEMBERS PRICE	£24/$39.00
NON MEMBERS PRICE	£32/$52.00

Constructed Wetlands for Pollution Control — Process, Performance, Design and Operation
Scientific and Technical Report No.8

By the IWA Specialist Group on Use of Macrophytes in Water Pollution Control.

Authors: Robert H. Kadlec, Robert L. Knight, Jan Vymazal, Hans Brix, Paul Cooper, Raimund Haberl

The book presents a comprehensive up-to-date survey of wetland design techniques and operational experience from treatment wetlands. It is the first and only global synthesis of information related to constructed treatment wetlands. Types of constructed wetlands, major design parameters, role of vegetation, hydraulic patterns, loadings, treatment efficiency, construction, operation and maintenance costs are discussed in depth. History of the use of constructed wetlands and case studies from various parts of the world are included as well.

April 2000 164 pages
ISBN: 1-900222-05-1 Paperback

IWA MEMBERS PRICE	£41.25 / US$66.00
NON MEMBERS PRICE	£55.00 / US$88.00

Real Time Control of Urban Drainage Systems: the State-of-the-Art.

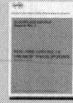

Scientific & Technical Report No 2
By the IAWPRC Task Group on Real-time Control of Urban Drainage Systems
Edited by W Schilling
1989 84 pages, Paperback
ISBN 0-08-040145-7

IWA MEMBERS PRICE	£25.00/$40
NON MEMBERS PRICE	£40.00/$64

To Order contact:
Portland Customer Services, Commerce Way, Colchester CO2 8HP, UK
Tel: +44 (0)1206 796351 Fax: +44 (0)1206 799331
Email: sales@portlandpress.com Web: www.portlandpress.com

To browse our publications visit:
www.iwapublishing.com

NEW PUBLICATION...NEW SERIES...

Decentralised Sanitation & Reuse
Concepts, Systems and Implementation

INTEGRATED ENVIRONMENTAL TECHNOLOGY SERIES

Editors
Dr. Ir. Piet Lens, *Wageningen University, The Netherlands*
Dr G Zeeman, *Wageningen University, The Netherlands*
Prof. Dr. Ir. Gatze Lettinga, *Wageningen University, The Netherlands*

Decentralised Sanitation and Reuse provides a definite and detailed discussion of the current-state-of-the-art of sanitation technologies. It further presents in detail how these technologies can be implemented to integrate domestic waste and wastewater treatment, in order to maximize resource recycling in domestic practise. Special attention is given to non process engineering aspects of sanitation, such as agricultural applications, health, regulations, economic impact, public participation, architecture and town planning.

The information density is unique owing to large amounts of figures, tables and case studies. A detailed subject index helps the reader to find their way through the multidisciplinary chapters, making it the perfect reference work for professionals and consultants dealing with sanitation and reuse of domestic waste (water).

Contents
The DESAR concept for environmental protection | *Waste and wastewater characteristics and its collection on the site* | *Technological aspects of DESAR* - Concepts of and technologies for DESAR | Anaerobic pre-treatment | Low strength wastewater (post) treatment | Water and mineral resource recovery | Agricultural reuse | **Environmental and public health aspects of DESAR** | *Sociological and economic aspects of DESAR* | *Architectural and urbanistic aspects of DESAR.*

March 2001, 650 pages, ISBN: 1-900222-47-7
IWA Member Price £63/US$99   Non Member Price £85/US$129.95

Order this publication from our distributors:
Portland Customer Services, Commerce Way, Colchester, CO2 8HP, U.K.
Tel: 44 (0)1206 796351   Fax: 44 (0) 1206 799331 sales@portlandpress.com
In North America
Technomic Publishing Co. Inc., 851 New Holland Avenue, Box 3535 Lancaster, 17604 Pennsylvania, USA.
Tel: (1) 717 291 5609   Fax: (1) 717 295 4538   Email: customer@techpub.com

IWA Publishing

www.iwapublishing.com

ContentsAlert

IWA Publishing

A FREE pre-publication alerting service, delivered via email.

QUICK

This valuable service will enable you to keep up-to-date with articles published in IWA Publishing journals.

- Journal of Hydroinformatics
- Journal of Water Supply: Research & Technology-AQUA

DIRECT

- Water Science & Technology
- Water Science & Technology: Water Supply

EASY

Delivers the contents pages of each journal issue approximately 2-4 weeks prior to publication.

Don't miss out on this fast and easy way to keep up-to-date with the latest developments in your field.

Register now at

www.iwapublishing.com

Journal of Hydroinformatics

Official Journal of the IAHR-IWA Joint Committee on Hydroinformatics

Editors:

Michael Abbott, IHE Delft, The Netherlands
Vladan Babovic, Danish Hydraulic Institute, Denmark
Roger Falconer, Cardiff University, UK
Peter Goodwin, University of Idaho, USA
Roland Price, IHE Delft, The Netherlands

Journal of Hydroinformatics is a peer-reviewed journal devoted to the application of information technology in the widest sense to problems of the aquatic environment. It promotes Hydroinformatics as a cross-disciplinary field of study, combining technological, human-sociological and more general environmental interests, including an ethical perspective.

Recent Papers include:

- **A hydroinformatician's approach to computational innovation and the design of genetic algorithms** - David E Goldberg
- **Forecasting high waters at Venice Lagoon using chaotic time series analysis and nonlinear neural networks** - J M Zaldívar, E Gutiérrez, I M Glaván, F Strozzi, A Tomasin
- **Subsymbolic methods for data mining in hydraulic engineering** - Anthony W Minns
- **Introducing hydroinformatics** - Michael B Abbott
- **Genetic programming as a model induction engine** - V Babovic, M Keijzer

To view a sample copy of **Journal of Hydroinformatics** visit **www.iwapublishing.com**

On-line Access

Journal of Hydroinformatics is available on-line in 2001. For details visit us at: **www.IWAPonline.com**

Subscription Information

Volume 3, 4 issues, 2001
ISSN: 1464-7141   **Institutional**
North America $US320.00
Rest of World   £200.00

Discounted Rates Available for IWA and IAHR members. Contact IWA or IAHR for details.

To subscribe contact:
Portland Customer Services, Commerce Way, Colchester, CO2 8HP, UK
Tel: +44 (0) 1206 796351 **Fax:** +44 (0) 1206 799331
Email: sales@portlandpress.com

To become a member of IWA contact:
IWA Alliance House, 12 Caxton Street, London SW1H 0QS
Tel: +44 (0) 207 654 5500 **Fax:** +44 (0) 207 654 5555
Email: members@iwahq.org.uk

Submit Your Paper!

For details on submitting a paper contact Emma Gulseven at IWA Publishing on
egulseven@iwap.co.uk Tel: +44 (0)20 7654 5500 Fax: +44 (0)20 7654 5555

www.iwapublishing.com